D1294650

INTRODUCTION TO BIOCHEMICAL TOXICOLOGY

THIRD EDITION

INTRODUCTION TO BIOCHEMICAL TOXICOLOGY

THIRD EDITION

Ernest Hodgson, PhD
William Neal Reynolds Professor of Environmental and
Molecular Toxicology
North Carolina State University
Raleigh, North Carolina

Robert C. Smart, PhD
Professor of Environmental and
Molecular Toxicology
North Carolina State University
Raleigh, North Carolina

A JOHN WILEY & SONS, INC., PUBLICATION

New York • Chichester • Weinheim • Brisbane • Singapore • Toronto

This book is printed on acid-free paper. ♾

Published simultaneously in Canada.

For ordering and customer service, call 1-800-CALL-WILEY.

Library of Congress Cataloging-in-Publication Data:

Introduction to biochemical toxicology / [edited by] Ernest Hodgson, Robert C. Smart—3rd ed.
p. cm.
Includes bibliographical references and index.
ISBN 0-471-33334-4 (cloth : alk. paper)
1. Biochemical toxicology. I. Hodgson, Ernest, 1932– II. Smart, Robert C., 1954–
RA1219.5 .I58 2000
615.9–dc21 00-043811

Printed in the United States of America.

10 9 8 7 6 5 4

CONTENTS

PREFACE

For the past 20 years we have taught a graduate course entitled "Biochemical Toxicology" to toxicology graduate students within the Department of Environmental and Molecular Toxicology at North Carolina State University. This volume, now in its third edition, has evolved from our experiences in this course and is designed as a teaching textbook. The book is intended as an advanced text for those students with a knowledge of the general principles of toxicology. Although the textbook is entitled *Introduction to Biochemical Toxicology*, it is much more than an introductory text and provides in-depth information describing the underlying biochemical, molecular, and cellular mechanisms through which toxicants produce their adverse responses. We have attempted to provide detailed mechanistic information in a straightforward uncomplicated manner to facilitate assimilation of knowledge and understanding rather than to provide a heavily referenced monograph.

Recent rapid advances in our understanding of how toxicants produce their adverse effects has been possible through the application of molecular and cellular biological techniques to toxicological problems. Accordingly, the third edition includes three new overview chapters on techniques entitled, "Molecular Techniques in Toxicology—Genes/Transgenes," "Immunochemical Techniques in Toxicology," and "Cellular Techniques in Toxicology." These chapters provide information that ranges from how transgenic and knockout mice are produced to the advantages and disadvantages of using primary cells and/or cell lines in toxicological studies to an entire array of RNA, DNA, and protein techniques. These chapters, taken together, are intended as a primer in methodology for the toxicology student and will provide the essentials to better understand the molecular and cellular aspects of the updated chapters of this text as well as the current literature in biochemical and molecular toxicology. These chapters will also provide a conceptual review for the more experienced reader.

In addition to including chapters on molecular and cellular techniques and updating the chapters from the second edition, we have also included five new additional chapters in the third edition on "Immunotoxicology," "Cutaneous Toxicology," "Reproductive and Developmental Toxicology," "Genetic Toxicology," and "Molecular Epidemiology."

We believe that both instructors and student researchers will find *Introduction to Biochemical Toxicology* to be a usable, integral component of their graduate curriculum as well as a valuable reference book. Many thanks to all of the contributing authors and to everyone at John Wiley and Sons who made this edition possible.

Ernest Hodgson
Robert C. Smart

Raleigh, North Carolina, USA

CONTRIBUTORS

Acosta, Daniel, College of Pharmacy, University of Cincinnati Medical Center, Cincinnati, Ohio 45267-0004.

Akunda, Jacqueline K., Department of Environmental and Molecular Toxicology, North Carolina State University, Raleigh, North Carolina 27695-7633.

Blake, Bonnie L., Neuroscience Center, Departments of Psychiatry and Pharmacology, University of North Carolina School of Medicine, Chapel Hill, North Carolina 27599-7250.

Bouldin, Thomas W., Department of Pathology, University of North Carolina School of Medicine, Chapel Hill, North Carolina 27599.

Branch, Stacy, Department of Environmental and Molecular Toxicology, North Carolina State University, Raleigh, North Carolina 27695-7633.

Combs, Alan B., Department of Pharmacology and Toxicology, University of Texas, Austin, Texas 78712.

Cunny, Helen C., Rhone-Poulenc Ag, Research Triangle Park, North Carolina 27709.

Dauterman, Walter C., (deceased), Department of Environmental and Molecular Toxicology, North Carolina State University, Raleigh, North Carolina 27695-7633.

Donaldson, William E., Department of Poultry Science, North Carolina State University, Raleigh, North Carolina 27695.

Germolec, Dori R., National Institute of Environmental Health Sciences, Research Triangle Park, North Carolina 27709.

Goldstein, Joyce A., National Institute of Environmental Health Sciences, Research Triangle Park, North Carolina 27709.

Goodrum, Jeffry F., Department of Pathology, University of North Carolina School of Medicine, Chapel Hill, North Carolina 27599.

Henderson, Rogene F., Lovelace Respiratory Research Institute, Albuquerque, New Mexico 87185-5890.

Hodgson, Ernest, Department of Enviromental and Molecular Toxicology, North Carolina State University, Raleigh, North Carolina 27695-7633.

Holbrook, David J., Department of Biochemistry, University of North Carolina, Chapel Hill, North Carolina 27599.

Koper, Norbert P., National Institute of Environmental Health Sciences, Research Triangle Park, North Carolina 27709.

Kulkarni, Arun P., Environmental and Occupational Health, University of South Florida, Tampa, Florida 33612–3085.

Lawler, Cindy P., Neuroscience Center, Departments of Psychiatry and Pharmacology, University of North Carolina School of Medicine, Chapel Hill, North Carolina 27599-7250.

LeBlanc, Gerald A., Department of Environmental and Molecular Toxicology, North Carolina State University, Raleigh, North Carolina 27695-7633.

Levi, Patricia E., (Retired) Department of Environmental and Molecular Toxicology, North Carolina State University, Raleigh, North Carolina 27695–7633.

Luebke, Robert W., Immunotoxicology Branch, US Environmental Protection Agency, Research Triangle Park, North Carolina, 27711.

Mailman, Richard B., Neuroscience Center, Departments of Psychiatry and Pharmacology, University of North Carolina School of Medicine, Chapel Hill, North Carolina 27599–7250.

Meyer, Sharon A., College of Pharmacy and Health Sciences, University of Louisiana at Monnoe, Monnoe, Louisiana 71209-0400.

Monteiro-Riviere, Nancy A., Cutaneous Pharmacology and Toxicology Center, North Carolina State University, Raleigh, North Carolina 27606.

Moreland, Donald E., Department of Crop Science, North Carolina State University, Raleigh, North Carolina 27695.

Morell, Pierre, Neuroscience Center, Department of Biochemistry, University of North Carolina School of Medicine, Chapel Hill, North Carolina 27599-7250.

Nikula, Kristen K., Lovelace Respiratory Research Institute, Albuquerque, New Mexico 87185-5890.

Preston, R. Julian, Environmental Carcinogenesis Division, U.S. Environmental Protection Agency, Research Triangle Park, North Carolina 27711.

Ramos, Kenneth, Department of Physiology and Pharmacology, Texas A & M University, College Station, Texas 77843.

Reed, Donald J., Department of Biochemistry and Biophysics, Oregon State University, Corvallis, Oregon 97331.

Ronis, Martin J. J., Department of Pediatric Research, University of Arkansas Medical Sciences, Little Rock, Arkansas 72205.

Rose, Randy, L., Department of Environmental and Molecular Toxicology, North Carolina State University, Raleigh, North Carolina 27695-7633.

Sailstad, Denise M., Immunotoxicology Branch, US Environmental Protection Agency, Research Triangle Park, North Carolina 27711.

Selgrade, MaryJane K., Immunotoxicology Branch, US Environmental Protection Agency, Research Triangle Park, North Carolina 27711.

Smialowicz, Ralph J., Immunotoxicology Branch, US Environmental Protection Agency, Research Triangle Park, North Carolina 27711.

Smart, Robert C., Department of Environmental and Molecular Toxicology, North Carolina State University, Raleigh, North Carolina 27695-7633.

Stern, Mariana C., National Institute of Environmental Health Sciences, Research Triangle Park, North Carolina 27709.

Tarloff, Joan B., Philadelphia College of Pharmacy, University of the Sciences in Philadelphia, Philadelphia, Pennsylvania 19104-4495.

Taylor, Jack A., National Institute of Environmental Health Sciences, Research Triangle Park, North Carolina 27709.

Ward, Marsha D., Toxicology Curriculum, University of North Carolina, Chapel Hill, North Carolina 27599.

CHAPTER ONE

Biochemical Toxicology: Definition and Scope

ERNEST HODGSON and ROBERT C. SMART

1.1 INTRODUCTION

While the field of biochemistry has expanded to include the area of molecular biology, so has the field of biochemical toxicology expanded to include what is now referred to as molecular toxicology. Biochemical toxicology is concerned with the definition, at the molecular and cellular levels, of the cascade of events that is initiated by exposure to a toxicant and culminates in the expression of a toxic end point. Molecular techniques involving molecular cloning and the manipulation of RNA and DNA have provided a wealth of mechanistic information about the role of gene function in the interaction of xenobiotics and living organisms. The third edition of *Introduction to Biochemical Toxicology* reflects this by inclusion of molecular studies and approaches currently used in understanding the metabolic processing and mode of toxic action of xenobiotics.

Toxicology can be defined as the branch of science dealing with poisons. Having said that, attempts to define all of the various parameters lead to difficulties. The first difficulty is seen in the definition of a poison. Broadly speaking, a poison is any substance causing harmful effects in an organism to which it is administered, either deliberately or by accident. Clearly, this effect is dose related inasmuch as any substance, at a low enough dose, is without effect, while many, if not most, substances have deleterious effects at some higher dose. Much of toxicology deals with compounds exogenous to the normal metabolism of the organism, such compounds being referred to as foreign compounds or, more recently, as *xenobiotics*. However, many endogenous compounds, including metabolic intermediates such as glutamate, or hormones such as thyroxine, are toxic when administered in unnaturally high doses. Similarly, trace elements such as selenium, which are essential in the diet in low concentrations, are frequently toxic at higher levels. Such effects are properly included in toxicology, while the endogenous generation of high levels of metabolic intermediates due to disease or metabolic defect is not, although the effects on the organism may be similar. Whether the harmful effects of physical phenomena such as irradiation, sound, temperature, and humidity are included in toxicology appears to be largely dependent on the preference of the writer. It is convenient, however, to include them under the broad definition of toxicology.

The expression of toxicity, hence the assessing of toxic effects, is another parameter of considerable complexity. *Acute toxicity*, usually measured as mortality and expressed as the LD50 (the dose required to kill 50% of a population of the organism in question under specified conditions) is probably the simplest measure of toxicity. Even so, reproducibility of LD50 values is highly dependent upon many variables. These include the age, sex, and physiological condition of the animals, their diet, the environmental temperature and humidity, and the method of administering the toxicant.

Chronic toxicity may be manifested in a variety of ways—carcinomas, cataracts, peptic ulcers, and reproductive effects, to name only a few. Furthermore, compounds may have different effects at different doses. Vinyl chloride, a potent hepatotoxicant at high doses, is a carcinogen with a very long latent period at low doses. Most drugs have therapeutic effects at low doses but are toxic at higher levels. Acetylsalicylic acid (aspirin), relatively nontoxic to humans, is a useful analgesic at low doses but is, however, toxic at high doses, and may cause peptic ulcers with chronic use.

Considerable variation also exists in the toxic effects of the same compound administered to different animals, or even to the same animal by different routes. The insecticide malathion has a low toxicity to mammals, whereas it is toxic enough to insects to be a widely used commercial insecticide. The route of entry of toxicants into the animal body is frequently oral, in the food or drinking water in the case of many chronic environmental contaminants such as lead or insecticide residues, or directly as in the case of accidental or deliberate acute poisoning. Other routes for poisoning include dermal absorption and pulmonary absorption. The above routes of administration are all used experimentally, and, in addition, several types of injection are commonly used, including intravenous, intraperitoneal, intramuscular, and subcutaneous. The toxicity of many compounds varies tenfold or greater depending upon the route of administration.

1.2 TOXICOLOGY

Toxicology is clearly related to two of the applied biologies—medicine and agriculture. In the former, clinical diagnosis and treatment of poisoning as well as the management of toxic side effects are areas of significance, whereas in the latter the development of agricultural biocides such as insecticides, herbicides, nematocides, and fungicides is of great importance. The detection and management of the nontarget effects of such compounds is also an area of increasing importance that is essential to their continued used. Toxicology may also be considered an area of fundamental biology because the adaptation of organisms to toxic environments has important implications for ecology and evolution.

The tools of chemistry and biochemistry are the primary ones of toxicology, and progress in toxicology is closely related to the development of new methodology in these sciences. Those of chemistry provide analytical methods for toxic compounds, particularly for forensic toxicology and residue analysis, and those of biochemistry provide the techniques to investigate the metabolism and mode of action of toxic compounds. On the other hand, studies of the chemistry of toxic compounds have contributed to fundamental organic chemistry, and studies of the enzymes involved in detoxication and toxic action have contributed to our basic knowledge of biochemistry.

Toxicology in the most general sense may be one of the oldest practical sciences. From the earliest beginnings, humans became aware of numerous toxins such as snake venoms and those of poisonous plants, and from the earliest written records it is clear that the ancients had considerable knowledge of poisons. The Greeks made use of hemlock as a method of execution and they and, more particularly, the Romans made much use of poisons for political and other assassinations. Indeed, it was Dioscorides, a Greek at the court of Nero around 50 AD, who made the earliest known attempt to classify poisons. His *Materia Medica*, which classified poisons as plant, animal, or mineral, as toxic or therapeutic, as well as methods of treatment, remained as a major descriptive work on poisons for 15 centuries. Although poisoning has enjoyed a considerable vogue at many times and places since, the scientific study of toxicology can probably be dated from the German physician Paracelsus, who, in the sixteenth century, put forward the necessity for experimentation and included much in his range of interests that would today be classified as toxicology, especially the concept of a dose-response relationship. This statement from his *Third Defense* is indeed a cornerstone of toxicology today—"What is there that is not a poison? All things are poison and nothing without poison. Solely the dose determines that a thing is not a poison."

The modern study of toxicology is usually dated from the Spaniard, Orfila (1787–1853), who, at the University of Paris, identified toxicology as a separate science. Among his many contributions, he devised chemical methods for the detection of poisons and stressed the value of chemical analysis to provide legal evidence. He was also the author, in 1815, of the first book devoted entirely to the toxic effects of chemicals, "Traité des Poisons Tirés des Règnes Minéral, Végétal et Animal, ou, Toxicologie Général Considérée sous les Rapports de la Physiologie, de la Pathologie et de la Médecine Légale."

Toxicology can be subdivided in a variety of ways. Loomis refers to the three "basic" subdivisions as environmental, economic, and forensic. Environmental toxicology is further divided into such areas as pollution, residues, and industrial hygiene; economic toxicology is said to be devoted to the development of drugs, food additives, and pesticides; and forensic toxicology is concerned with diagnosis, therapy, and medicolegal considerations.

Environmental toxicology continues to be a rapidly growing branch of the science. Public concerns over environmental pollutants and their possible chronic effects, particularly carcinogenicity and reproductive toxicity, has given rise, in the United States, to new research and regulatory agencies and to the Toxic Substances Control Act (TSCA), the Federal Insecticide, Fungicide and Rodenticide Act (FIFRA), the Food Quality Protection Act (FQPA), etc. Similar developments are also taking place in many other countries. The range of environmental pollutants is enormous, including industrial and domestic effluents, combustion products of fossil fuels, agricultural chemicals, and many other compounds that may be found in food, air, and water. Such compounds as food additives and cosmetics are also being subjected to the same scrutiny.

Other subspecialities are frequently mentioned that do not fit readily into the above divisions. Behavioral toxicology, an area of increasing importance, is usually treated as a separate subspeciality. Analytical toxicology provides the methods of chemical analysis used in essentially every branch of the subject, while biochemical toxicology, the subject of this text, provides the fundamental basis for all branches of toxicology. Since the publication of the second edition of this text, the application of the methods of molecular and cellular biology (chapters 2–4) has brought about dramatic changes in our understanding of mechanistic toxicology, changes reflected in every chapter of the current edition.

1.3 BIOCHEMICAL TOXICOLOGY

1.3.1 General Description

Biochemical and molecular toxicology deal with events that occur at the molecular level when toxic compounds interact with living organisms. Defining these interactions is fundamental to our understanding of toxic processes, both acute and chronic, and is essential for the rational development of new therapies, for the determination of toxic hazards, and for the development of new biocides for agriculture and medicine. The poisoning process may be thought of as a series of several more or less distinct phases, and biochemical toxicology is intimately involved with all but one of these. *Exposure*, or the way in which an organism becomes part of an environment containing the toxicant, is properly the domain of epidemiology and industrial hygiene, although the development of molecular biomarkers of either uptake or effect (Chapter 17) clearly falls with the purview of biochemical toxicology. *Uptake*, however, involves the biochemistry of cell membranes in the portals of entry and the structure-function relationships of toxicant absorption. *Distribution* involves the mechanism of transport throughout the body and the distribution between tissues. *Metabolism*, which may take place at the portal of entry or in such organs as the liver, involves the study of the enzymes that detoxify and/or activate xenobiotics. Compounds with intrinsic toxicity or activated metabolites are involved in the *mode of action* phase, in which they interact with cellular components at their site of action to initiate the toxic effect. The study of mode of action is a critical area of biochemical toxicology. Although mechanisms of carcinogenesis are being intensively investigated, many of the other chronic effects of toxicants are poorly described and understood and much work is needed. A further distribution phase brings metabolic products to the organs of *excretion*, primarily the kidneys, gallbladder, and lungs, again involving processes that are proper subjects for biochemical toxicology.

This introductory chapter is a general guide to the overall process and also to the appropriate section of the book for the discussion of a particular body locus or metabolic process. The organization follows the route that a typical toxicant might take through the mammalian body (Fig. 1.1).

1.3.2 Portals of Entry

The principal portals of entry are the skin, the gastrointestinal tract, and the lungs. The toxicant must pass through a number of biological membranes before it can be

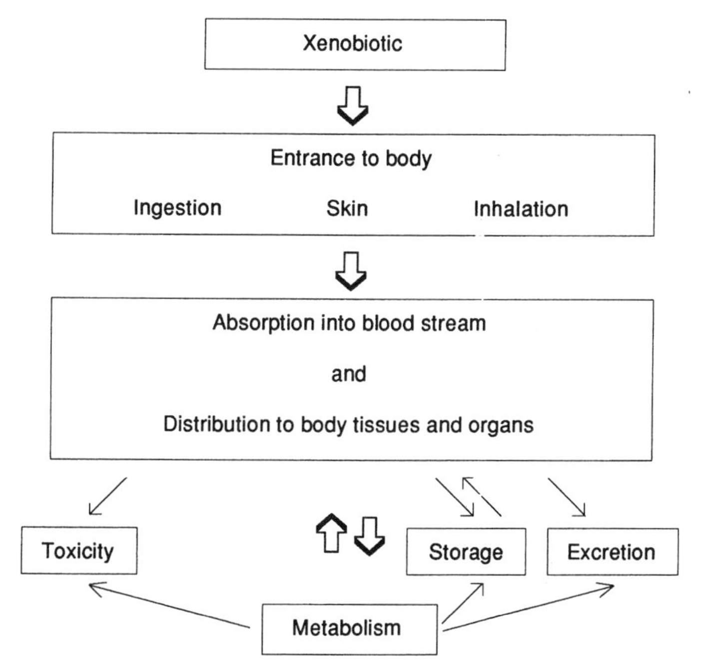

FIGURE 1.1 Entry and fate of chemicals in the body.

distributed throughout the body and that uptake depends on the nature of the membrane as well as the physical and chemical properties of the toxicant. The structure of cell membranes, basically bimolecular lipid leaflets with associated proteins, and their various modifications are important inasmuch as their nature is responsible for the fact that lipophilicity is the most important determinant of the rate of uptake of exogenous molecules. Active transport, pinocytosis, etc. are much less common than diffusion across the lipid membranes.

1.3.3 Distribution

The distribution of toxicants throughout the body is important and usually involves the binding of toxicants to blood proteins, particularly lipoproteins. Lipoproteins are an important class of protein, particularly in the vascular fluids. They vary in molecular weight from 200,000 to 10,000,000, and the lipid content varies from 4% to 95%, being composed of triglycerides, phospholipids, and free and esterified cholesterol. Although lipoproteins are classified into groups based on their flotation constants, each group is, in fact, a mixture of many similar proteins.

The nature and importance of various types of ligand-protein interactions are assessed, including covalent binding, ionic binding, hydrogen bonding, van der Waals forces, and hydrophobic interactions. Many of the same binding forces are also important in toxicant-receptor interactions, as discussed in Chapter 12.

1.3.4 Metabolism

The majority of xenobiotics that enter the body do so because they are lipophilic and can readily diffuse across body membranes. The metabolism of xenobiotics, which is carried out by a wide range of relatively nonspecific enzymes, serves to increase the water solubility of foreign chemicals and make possible their elimination from the body. This process generally consists of two phases. In phase I metabolism, considered in Chapter 5, a reactive polar group is introduced into the molecule, rendering it a suitable substrate for phase II reactions. Phase I reactions include the well-known cytochrome P450-dependent monooxygenations as well as reductions, hydrolyses, etc. Phase II reactions (Chapter 6) include all of the conjugation reactions in which a polar group on the toxicant is combined with an endogenous compound such as glucuronic acid, glutathione, sulfate, etc. to form a highly water-soluble conjugate that can be readily eliminated from the body.

It is well known that these metabolic reactions are not all detoxications because many foreign compounds are metabolized to highly reactive products that are responsible for their toxic effects (Chapter 9). Such reactions include the activation of thiophosphates to potent phosphate cholinesterase inhibitors and the activation of relatively inert chemicals to carcinogens and hepatotoxicants.

Although the liver is the most studied organ with regard to xenobiotic metabolism, several other organs are known to be active in this respect, although neither the specific activity nor the range of substrates metabolized is as large as in the liver. These organs include the lungs and the gastrointestinal tract, as one might expect of organs that are important sites for the entry of xenobiotics into the body, and to a lesser extent, the other important portal of entry, the skin. Other organs, such as the kidney, may also be important sites for xenobiotic metabolism. Activation at the site of toxic action by extrahepatic tissues, although quantitatively less than that in the liver, may be more important because of proximity of the activated metabolites to the target site. The distribution of xenobiotic-metabolizing enzymes between different organs is touched upon in several chapters, including chapters 5–7 and 18–26.

Because toxicants are both activated and inactivated metabolically, physiological factors affecting metabolic rates can have dramatic effects on the expression of toxicity. These effects, including age, sex, pregnancy, and diet, are considered in Chapter 7.

Foreign compounds can be substrates, inhibitors, or inducers of the enzymes that metabolize them and, not infrequently, serve in more than one of these roles. Since the enzymes in question are nonspecific, numerous interactions between foreign compounds are possible. These may be synergistic or antagonistic and may have a profound effect on the expression of toxicity. Depending upon the compounds and the enzymes involved in a particular interaction, the effect can be an increase or a decrease in either acute or chronic toxicity. The basic principles of such interactions are summarized in Chapter 8, while various specific effects are noted in other chapters, such as Chapter 16 on genetic toxicology, Chapter 15 on carcinogenesis, and Chapter 20 on hepatotoxicity.

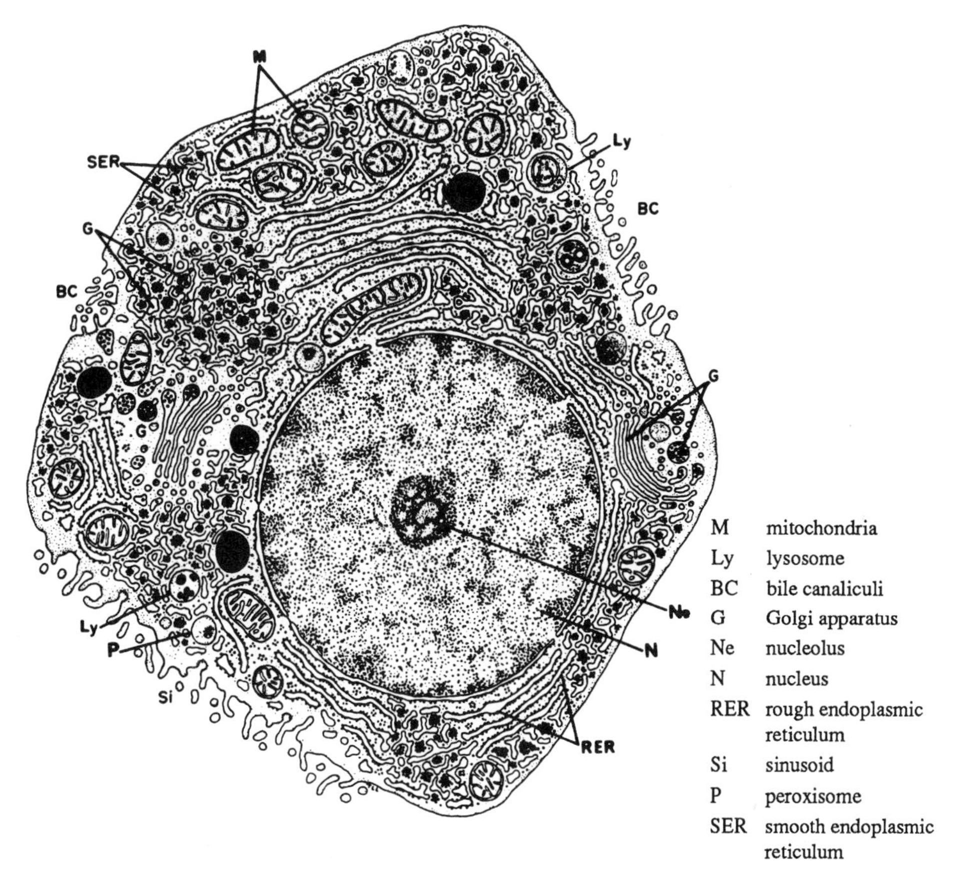

FIGURE 1.2 Diagrammatic cross section of a mammalian liver cell.
Source: Lentz, *Cell Fine Structure*, W.B. Saunders, Philadelphia, 1971.

The cell type that has been studied most intensively in biochemical toxicology is the hepatocyte, the cell that forms the bulk of the liver. A diagrammatic cross section of a liver cell is shown in Figure 1.2. These cells are highly active metabolically, both in normal intermediary metabolism and in reactions involving xenobiotics. The principal cell organelles shown in the diagram almost all play some role of importance in biochemical toxicology.

The *nucleus*, the chromosomes of which contain the DNA responsible for the code for most of the proteins synthesized in the cell (Fig. 1.3), is the site for reaction of genotoxic carcinogens with DNA, as described in Chapter 16. Depending upon the toxicant, the organ, and the cell type involved, similar reactions are involved in mutagenesis and other reproductive effects as well as in teratogenesis.

Mitochondria are the site of electron transport and oxidative phosphorylation, pathways that provide sites for the action of many acute toxicants (Chapter 13).

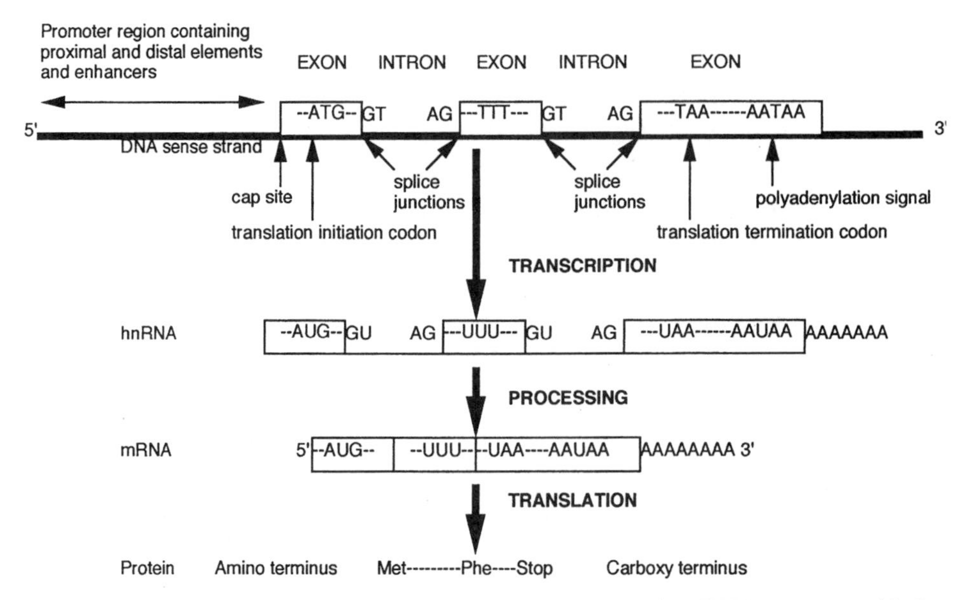

FIGURE 1.3 Transcription, m-RNA processing and translocation. DNA sense strand is designated by bold line, hnRNA and m-RNA by thinner line. Exons are shown as rectangles and introns as the intervening sequence between the exons.

Source: Adopted from The Genetic Bases of Cancer (B. Vogelstem and K.W. Kinzler. Megian Hell NY 1988.

The *endoplasmic reticulum* exists in two forms: rough endoplasmic reticulum (RER) with attached ribosomes, which has been associated with protein synthesis, and smooth endoplasmic reticulum (SER), which is devoid of ribosomes (Chapter 14). While both rough and smooth endoplasmic reticulum are active in the oxidation of xenobiotics, the latter usually has the highest specific activity. After disruption of the cells, followed by differential centrifugation, the two types of membrane are isolated as rough and smooth spherical vesicles known as microsomes.

The enzymes and metabolic pathways associated with the metabolism of toxicants are discussed in detail in chapters 5–8.

The most important advance of the last ten years in the study of xenobiotic metabolism has been the use of molecular biology techniques to characterize the enzymes involved and to study the mechanism of induction of these enzymes by exogenous compounds. These techniques are discussed in chapters 2–4 and include the use of antibodies in Western blotting to study the protein products, the use of Northern blots to characterize the mRNA involved in protein synthesis as well as the cloning and sequencing of cDNA to compare and characterize closely related isozymes. These techniques have been particularly useful in studies of the many isozymes of cytochrome P450 and their induction, permitting their functional and evolutionary relationships to be explored (Chapter 5).

Genetic susceptibility to xenobiotic insult can be the result of genetic polymorphisms in phase I and phase II enzymes and, as discussed in chapters 5 and 17, the area of molecular epidemiology has also been advanced rapidly by the use of molecular techniques in identifying these polymorphisms and determining their toxicological significance.

1.3.5 Sites of Action

Compounds of intrinsic toxicity as well as active metabolites produced in the body ultimately arrive either at a site of action or an excretory organ. Although almost any organ can show the effects of toxicity, some are more easily affected than others by particular classes of toxicants, and some have been studied in greater detail than others. In all cases, however, toxicant-receptor interactions are of importance. The fundamentals of such interactions are discussed in Chapter 12.

Acute toxicants tend to affect either oxidative metabolism, as discussed in Chapter 13, the synapses of the nervous system, or the neuromuscular junction. Toxic effects on the central nervous system are discussed in Chapter 19, and the peripheral nervous system in Chapter 18.

Chronic toxicity involving interaction with nucleic acids causing carcinogenesis (Chapter 15) or reproductive effects (Chapter 22) is a primary concern of toxicological research. Significant advances in our understanding of the mechanisms of carcinogenesis have resulted from the application of the techniques of molecular biology. The discovery of oncogenes and their activation in initiation as well as the role of second messenger systems in tumor promotion are examples (Chapter 15).

The study of organ specific toxicity has continued to advance since the publication of the second edition. Nine chapters (18–26) are now devoted to toxic events in critical organ systems.

1.3.6 Excretion

Either the unmetabolized toxicants or their metabolic products are ultimately excreted, the latter usually as conjugated products resulting from phase II reactions. The two primary routes of excretion are the urinary system and the biliary system, although minor routes of excretion such as the lungs, sweat glands, sebaceous glands, hair and nails, and sex-related routes such as milk and fetus are also known. The kidney is also a site of toxic action and renal toxicity is discussed in Chapter 25.

1.4 CONCLUSIONS

The preceding description of the nature and scope of biochemical toxicology should make it clear that the biochemistry of toxic action is a many-faceted subject, covering all aspects from the initial environmental contact with a toxicant to its ultimate excretion back into the environment. A considerable amount of material is summarized in the following chapters and, in general, the authors assume a competent knowledge of basic biochemistry on the part of the student. Much essential information is still missing—to be discovered, one might hope, by some of the students who use this book.

SUGGESTED READING

Klassen, C.D. (Ed), Casarett and Doull's *Toxicology, The Basic Science of Poisons*, 5th Ed., McGraw-Hill, New York, 1996.

Hodgson, E. and Levi, P.E. *A Textbook of Modern Toxicology*, 2nd Ed., Appleton and Lange, Stamford, Connecticut, 1997.

CHAPTER TWO

Overview of Molecular Techniques in Toxicology: Genes/Transgenes

ROBERT C. SMART

2.1 APPLICABILITY OF MOLECULAR TECHNIQUES TO TOXICOLOGY

Recombinant DNA techniques that facilitate molecular cloning, as well as techniques for the manipulation of DNA and RNA, have revolutionized the biological sciences. These techniques have allowed for the elucidation of gene function in complex biological processes such as cell growth, differentiation, development, and cancer. Many of the techniques of molecular biology can be applied to toxicology, and a subdiscipline of toxicology termed *molecular toxicology* has emerged. It is expected that the Human Genome Project will have completed the sequencing of the entire haploid human genome by 2003. It is also expected that the sequence of the mouse genome will soon be completed and plans are underway to sequence the rat genome. Genomic and bioinformatic technologies derived from these projects will have a major impact on our understanding of individual susceptibility to and mechanisms of chemical-induced toxicity. The term *toxicogenomics* has been coined to describe this new area of molecular toxicology.

Molecular techniques have wide applicability in understanding the mechanisms and the responses of a host organism to xenobiotic insult. For example, the response of a host organism to a xenobiotic often results in adaptive responses involving alterations in gene expression. Such genes can be identified using a molecular approach and their roles elucidated. DNA microarrays can be utilized to study the expression of hundreds or even thousands of genes in a single experiment. Through various molecular approaches, chemical carcinogen-induced mutations in oncogenes and tumor suppressor genes can be identified and in certain cases the mutation spectra can be used as a molecular fingerprint to aid in the identification of the responsible carcinogen. This approach is currently utilized in the emerging field of molecular epidemiology. Genetic susceptibility to xenobiotic insult can be the result of certain genetic polymorphisms. Molecular epidemiological studies can identify these genetic polymorphisms and determine their toxicological significance. Molecular approaches can be employed to understand species, age and sex differences in toxicant responses with regard to differences in gene structure, function, and/or expression.

The manipulation of cells and animals through the introduction of genes into the host DNA is a powerful approach to determine the function of a specific gene in a toxicological response. For example, this approach can be used to express xenobiotic metabolizing genes in recipient cells in culture. Such ectopically expressed genes can facilitate the determination of substrate specificity and the role of individual enzymes in the production and detoxification of toxic metabolites. Of particular importance in toxicology is the ability to make transgenic animals and gene "knockout" animals where the function of a specific gene can be studied in vivo. The creation of transgenic and nullizygous mice provides powerful and predictive tools in toxicological studies.

Molecular cloning and techniques facilitating the manipulation of DNA and RNA provide the toxicologist with opportunities to elucidate the mechanisms of toxicity. In addition, these techniques allow for the development of in vivo and in vitro transgenic models that may be predictive of human toxicity. This chapter is intended to provide a conceptual overview or primer of molecular techniques and approaches. For more comprehensive and detailed information, the reader is referred to the textbooks *Molecular Biology of the Gene* and *Genes VI*. For technical information and protocols, the reader is referred to the following laboratory manuals, *Molecular Cloning, A Laboratory Manual* and *Current Protocols in Molecular Biology*.

2.2 OVERVIEW OF THE GENETIC CODE AND FLOW OF GENETIC INFORMATION

Cells are composed primarily of proteins, nucleic acids, lipids, and carbohydrates. The genetic material of the cell, deoxyribonucleic acid (DNA), is composed of individual deoxyribonucleotide building blocks. Each deoxyribonucleotide or "nucleotide" is composed of a nitrogenous base such as cytosine, adenine, thymine, or guanine linked to a deoxyribose sugar moiety that has a phosphate group on either the 5′ or 3′ carbon of the sugar. These nucleotides are covalently linked together by phosphodiester bonds to form a linear strand that contains millions of nucleotides (Fig. 2.1A). A typical DNA molecule is composed of two antiparallel strands forming the Watson-Crick double helix with the bases of one strand forming hydrogen bonds with the bases of the other strand (Fig. 2.1B). Deoxyadenosine monophosphate always pairs, or hydrogen bonds, to thymidine monophoshate to form A-T pairs whereas deoxycytidine monophosphate always pairs with deoxyguanosine monophosphate to form G-C pairs. Within the coding sequence of a gene, the order of the nucleotides in DNA dictates the order of amino acids in protein. Three adjacent nucleotides (a codon) code for a specific amino acid. Thus, the linear sequence of DNA ultimately specifies the order of amino acids for all cellular proteins.

The flow of information from DNA to protein is transmitted through messenger ribonucleic acid (mRNA). Ribonucleic acid contains ribonucleotides composed of nitrogenous bases, cytosine, adenine, uracil, or guanine linked to a ribose sugar moiety that is phosphorylated on the 5′ or 3′ carbon of the sugar. Transcription of DNA results in a copy of the DNA sequence of a given gene in

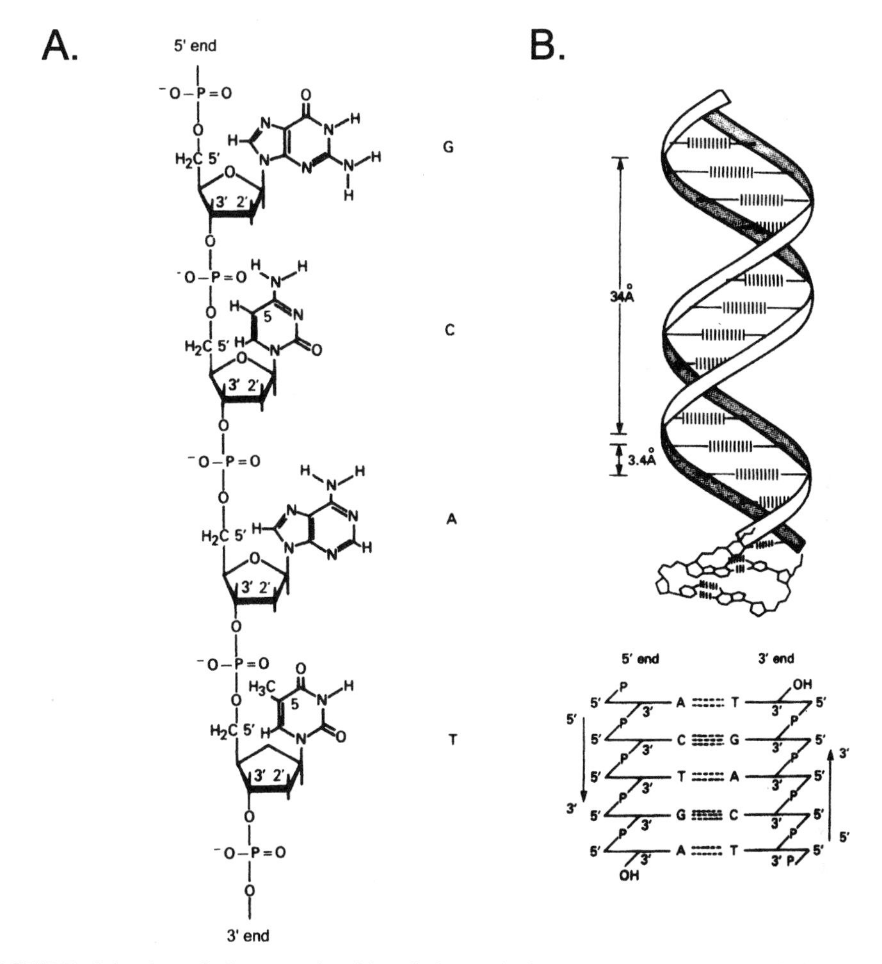

FIGURE 2.1 A polydeoxynucleotide chain and the structure of DNA. (A) A polydeoxynucleotide chain composed of four different deoxynucleotides covalently linked to one another by a phosphodiester bond (note orientation of the 5′ and 3′ end of the molecule). (B) Watson Crick double helix (top) and simple representation of the antiparallel strands (bottom).

Source: Adapted from B. Volgelstein and K.W. Kinzler (Eds), *The Genetic Basis of Human Cancer*, McGraw-Hill, New York, 1998.

the form of mRNA (Fig. 2.2). The mRNA contains the sequence of bases that is complementary to the bases of the transcribed (antisense) strand of DNA. Eukaryotic genes contain introns (interruptions) in the coding sequence (exons) of the gene. Therefore, eukaryotic DNA is first copied into a primary transcript called heterogenous nuclear RNA (hnRNA). hnRNA undergoes extensive processing in the nucleus that involves the removal of introns and the splicing together of exons, the capping of the 5′ end of mRNA with 7-methylguanosine, and polyadenenylation of the 3′ end with 100–200 adenosines. These last two modifications

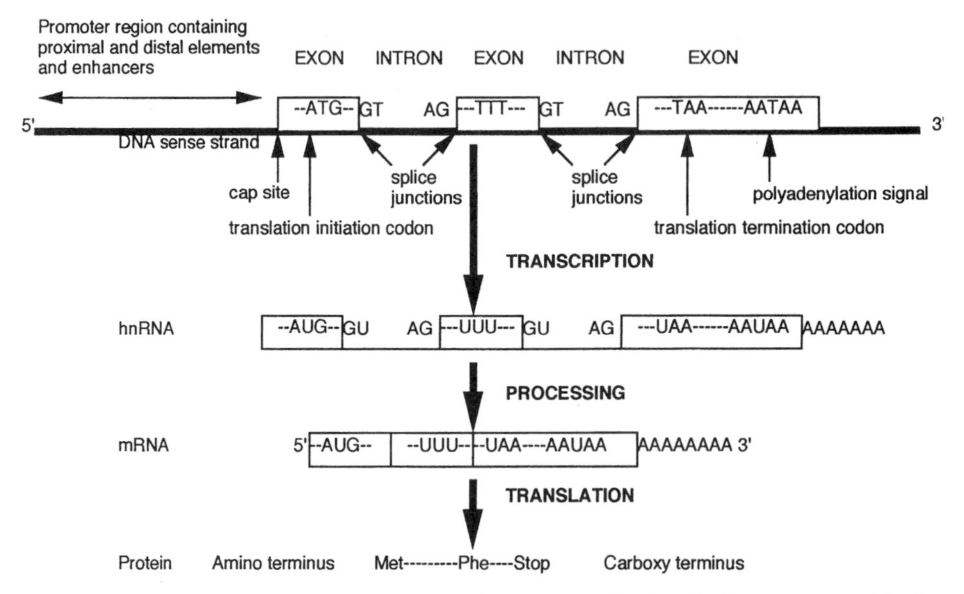

FIGURE 2.2 Transcription, mRNA processing, and translation. DNA sense strand is designated by bold line, hnRNA and mRNA by thinner lines. Exons are shown as rectangles and introns as the intervening spaces between exons.

Source: Adapted from B. Volgelstein and K.W. Kinzler (Eds), *The Genetic Basis of Human Cancer*, McGraw-Hill, New York, 1998.

tend to stabilize the mRNA and prevent its degradation. The mature mRNA is transported to the cytosol where it is translated into protein with the assistance of transfer RNA (tRNA) and ribosomal RNA (rRNA). Translation of the mRNA involves the ordered and sequential covalent incorporation of amino acids into a polypeptide chain to form a functional protein (Fig. 2.2). The order of amino acids is determined by the order of the codons present in the mRNA and the subsequent interactions with specific tRNA. Each tRNA is covalently linked to a specific amino acid and has a sequence complementary (anticodon) to the codons in the mRNA. Through the aid of complex ribosomal structures and the specific interaction of the mRNA codon and the tRNA anticodon, the correct amino acid is aligned and released from the tRNA and subsequently covalently linked to the growing polypeptide chain.

Regulation of gene expression can occur at the transcriptional and posttranscriptional levels and may involve mRNA initiation, mRNA stability, mRNA splicing, polyadenylation, transport of the mRNA from the nucleus to cytoplasm, and translation of the mRNA. The regulation of transcription is accomplished by the binding of sequence-specific DNA binding proteins, or transcription factors to specific nucleotides (cis element) within the 5′ noncoding sequence or promoter region of a gene. These transcription factors or trans-acting factors can increase or suppress gene expression depending upon the cis elements within a given promoter. Transcription factors bound to promoters interact with the RNA polymerase complex to regulate gene expression.

2.3 MOLECULAR CLONING

The human haploid genome contains approximately 3 billion base pairs encoding 50,000–100,000 individual genes. It is thought that about 1500 genes are expressed within a given cell type. In order to study the structure, function, and regulation of genes and their protein products one must have the capability to isolate the gene of interest. The general idea of molecular cloning is to insert a DNA segment (a gene or some part thereof) of interest into a vector that is an autonomously replicating DNA molecule. The vector containing the DNA is then inserted into a host, such as a bacterium or yeast. Once inside the host cell, the vector containing the insert replicates many times and produces hundreds of exact copies per cell, effectively cloning the inserted DNA.

2.3.1 Plasmid Vectors

There are many kinds of cloning vectors, including plasmids, bacteriophages, and yeast artificial chromosomes (YAC), and there are also many variations on the cloning theme. Generally plasmids, bacteriophages, cosmids, and YACs are employed to clone DNA fragments of up to 10 kb (kilobase = 1×10^3 bases), 20 kb, 40 kb, and 1 Mb (megabase = 1×10^6 bases) in length, respectively. Bacteriophages and cosmids are frequently used to make genomic libraries while YACs are frequently used in chromosomal mapping and positional cloning studies. Plasmids such as pBR322 and pUC (Fig. 2.3) are very useful for cloning smaller DNA fragments and making cDNA libraries. Bacterial plasmids are double-stranded closed circular DNA molecules that range in size from 1 kb to more than 200 kb and have an origin of DNA replication that allows vector replication independent of the bacterial chromosome. Plasmids have been engineered to contain a multiple cloning site that contains numerous unique restriction endonuclease sites. When the insert DNA and the vector have been cut with the same restriction endonuclease, both molecules have "sticky" or complementary ends that enable the molecules to anneal. The annealed molecules can then be covalently joined with DNA ligase. In this manner, the insert DNA is placed into the plasmid to form recombinant DNA (Fig. 2.4). Plasmids also contain genes encoding enzymes that give bacteria a selective advantage. For example, these genes can provide resistance to antibiotics such as ampicillin. Recombinant DNAs are introduced into the bacterium through a process referred to as transformation. This can be accomplished by creating bacteria that are "competent" to take up DNA by incubating with a mixture of divalent cations such as $CaCl_2$, which makes the cell permeable to small DNA molecules such as plasmids. The transformed bacteria are then incubated on plates containing media and the appropriate antibiotic. For example, if the recombinant plasmid contains a gene that provides resistance to ampicillin, then the media would contain ampicillin to select for bacteria that have taken up plasmid. Inside the bacterium, the vector will replicate to produce many copies and the bacterium will grow to form a colony. Colonies can subsequently be isolated and examined for the presence of the DNA insert.

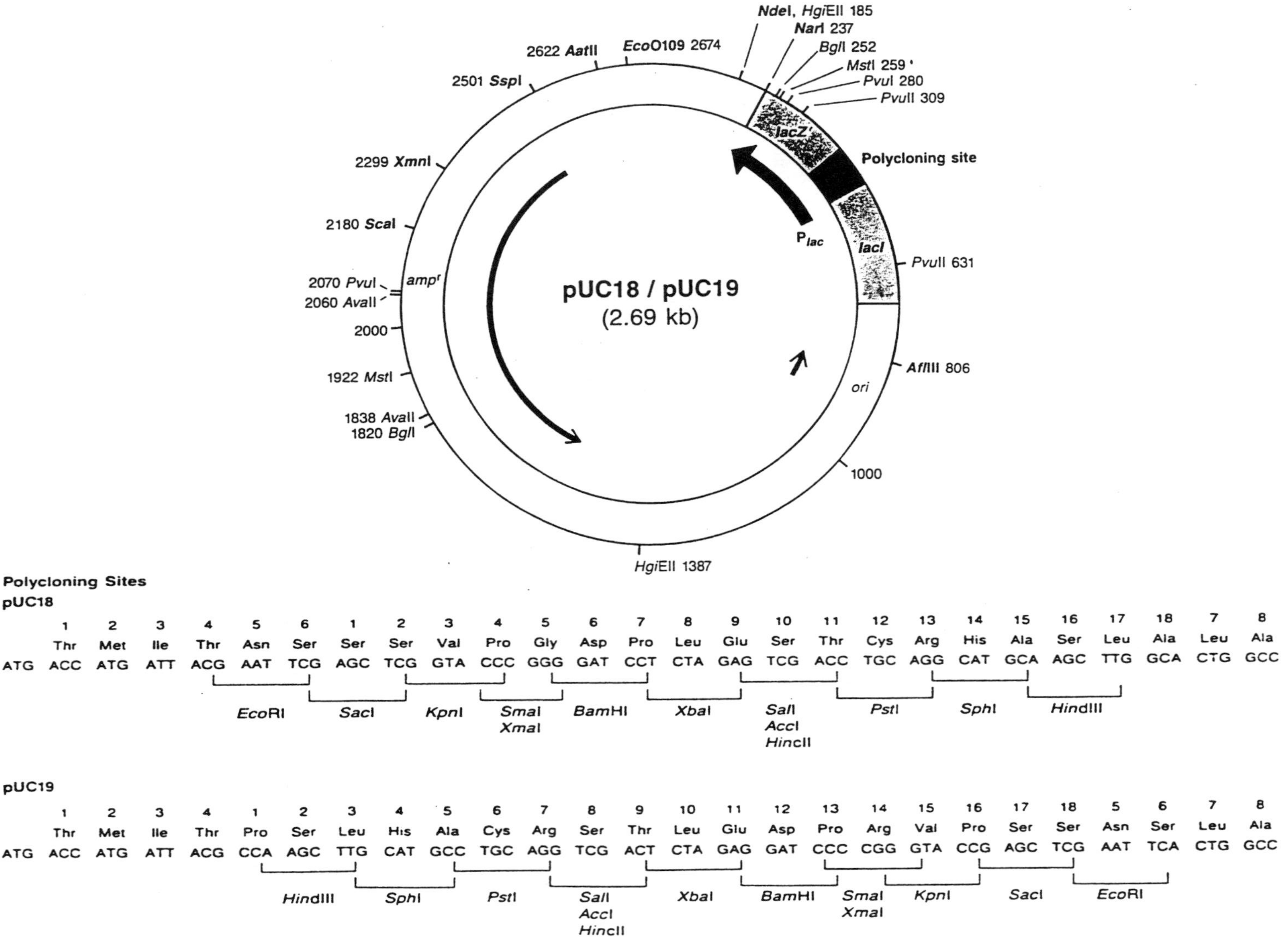

FIGURE 2.3 Map of pUC18/pUC19.

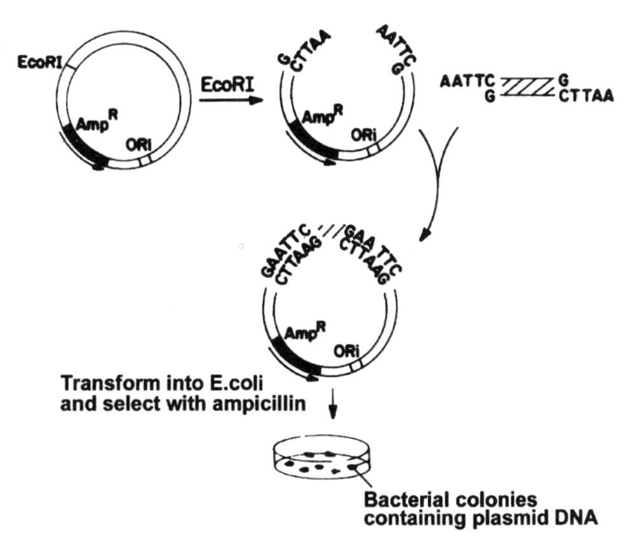

FIGURE 2.4 Molecular cloning using a plasmid vector.

2.3.2 Identification of Bacterial Colonies that Contain Recombinant DNA

There are several methods to identify bacterial colonies harboring recombinant plasmid. One of the more commonly used methods is referred to as α-complementation. The pUC series of plasmids contains the coding sequence for the first 146 amino acids of the amino terminus of the β-galactosidase gene (*lacZ*). Within this region is the polycloning site, however, the site is not large enough to disrupt the reading frame and results in the harmless incorporation of a small number of amino acids into the amino terminal fragment of β-galactosidase (Fig. 2.3). These plasmids are used in the *Escherichia coli* strain JM103, a strain that has the genetic information to produce the carboxy terminal region of β-galactosidase but not the amino terminal region. Neither the host alone or plasmid alone can form an active β-galactosidase enzyme, however, together they form an active β-galactosidase enzyme. Thus, α-complementation results in an active β-galactosidase enzyme or Lac^+ bacteria. These Lac^+ colonies are easily identified in the presence of the chromogenic substrate 5-bromo-4-chloro-3-indolyl-β-D-galactopyranoside (X-gal) because β-galactosidase metabolizes X-gal to form a blue product and Lac^+ colonies are blue. However, insertion of foreign DNA into the multiple cloning site results in the disruption of the coding sequence of β-galactosidase amino terminus and an amino terminal fragment that is not capable of α-complementation. These colonies appear white and represent the bacterial colonies that contain recombination DNA.

2.3.3 Construction of cDNA and Genomic Libraries

For the construction of a cDNA library, total RNA is isolated from the cells or tissue of interest. Total RNA contains only 1%–2% mRNA and approximately 80% rRNA. mRNA can be isolated from rRNA and tRNA using an oligo (dT)-cellulose column or coated beads. The poly (A) 3′ tail of mRNA binds to the immobilized

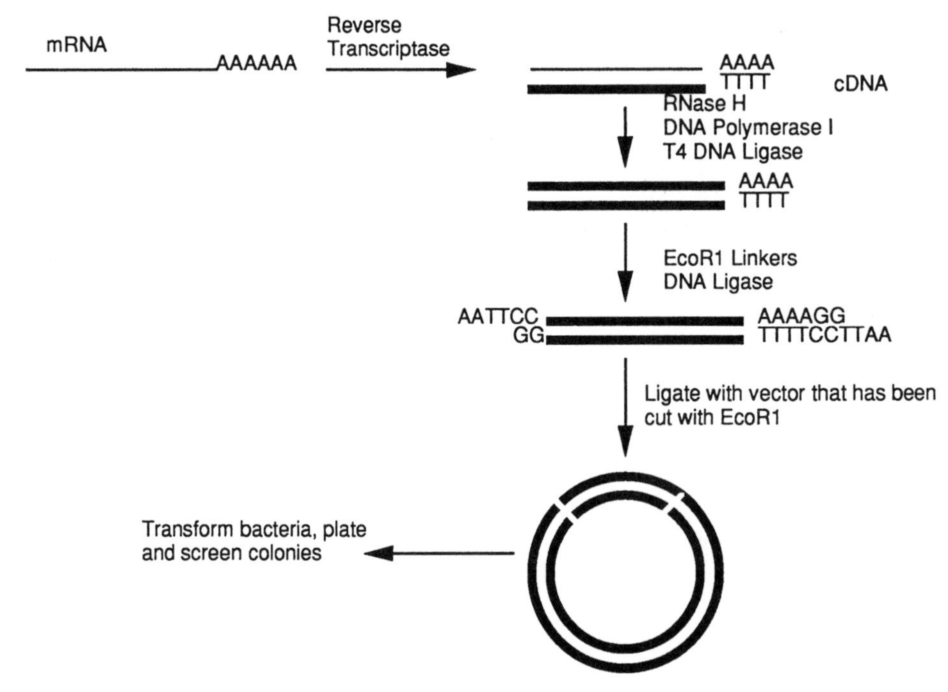

FIGURE 2.5 Generation of cDNA library using a plasmid vector. RNA is designated with a thin line and DNA with a bold line.

oligo dT, and tRNA and rRNA pass through the column or can be separated from the coated bead. The poly (A) mRNA is then eluted from the column or beads and is utilized to make complementary DNA (cDNA) with the aid of reverse transcriptase, an RNA-dependent DNA polymerase (Fig. 2.5). The cDNA is inserted into a vector and the recombinant vector is then transformed into the appropriate host and plated as described above. For the construction of a genomic library, the genomic DNA is digested with a restriction enzyme to produce an overlapping set of DNA fragments 12–20 kb in length. These fragments are cloned into a bacteriophage vector and encapsidated with viral coat protein, a process often referred to as in vitro packaging. The recombinant infectious phage are then used to infect *E. coli* as shown in Fig. 2.6.

Different methods are available to identify the recombinant clones containing the specific cDNA or genomic DNA of interest. One of the most common methods involves screening by nucleic acid hybridization. Plasmid or phage-containing bacterial colonies are replica plated from a master plate onto a nitrocellulose or nylon membrane (Fig. 2.7). The cells are lysed on the membrane and the DNA is immobilized on it. The membrane is then incubated with a radiolabeled DNA fragment or "probe" that is complementary to the desired DNA sequence. Specific annealing conditions allow for the hybridization of probe to the desired DNA sequence. The membrane is then washed and subjected to autoradiography. The probe will bind to the DNA of the colonies that contain the DNA of interest and

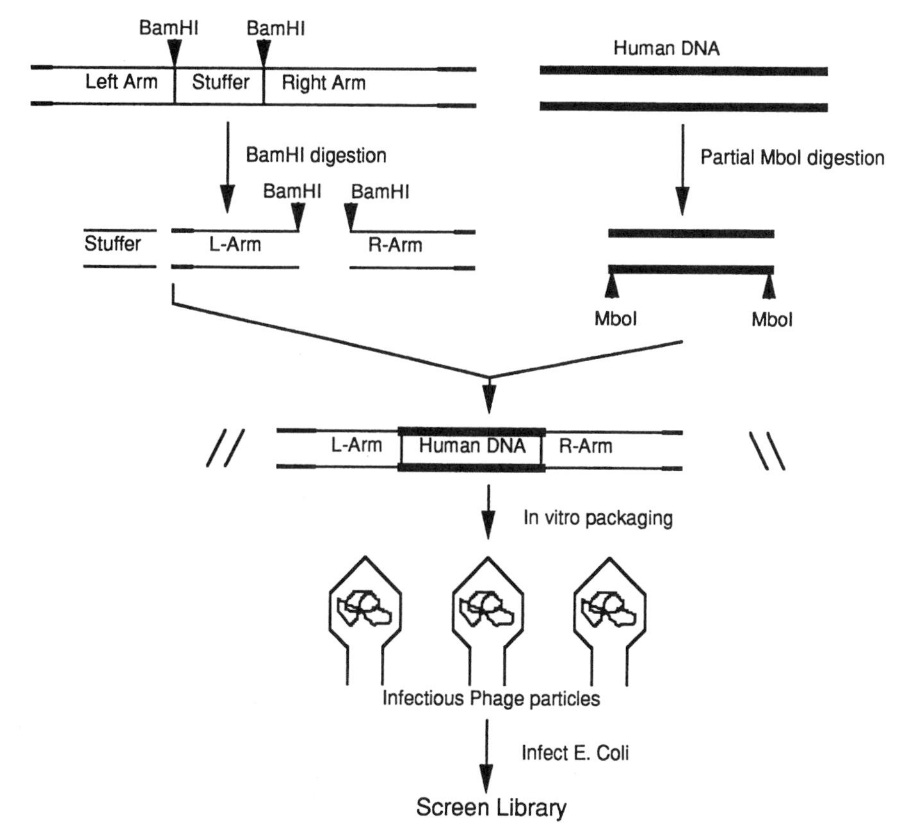

FIGURE 2.6 Generation of genomic library using a bacteriophage vector. Human DNA is digested with MboI, which can be ligated into the BamHI sites of the bacteriophage arms. The GATC overhang generated by MboI anneals to the BamHI site.

Source: Adapted from B. Volgelstein and K.W. Kinzler (Eds), *The Genetic Basis of Human Cancer*, McGraw-Hill, New York, 1998.

will expose the film only in this specific area. The corresponding clone from the master plate can then be identified, retrieved, and grown to obtain large quantities of the inserted DNA. The probe used can be the entire cDNA or a portion of the cDNA or genomic sequence if these are known. Related cDNA sequences can also be identified by reducing the stringency of hybridization of the probe so that it will bind to closely related sequences. In this way members of a gene family can be identified. If the cDNA or genomic sequences are not available but a portion of the protein sequence is known, then synthetic antisense degenerate oligonucletides encoding possible codon combinations can be synthesized and used to screen the library. If no genetic sequence information or protein structural information is known, then special methods are available to express the protein in bacterial or mammalian cells and the library can be screened with an antibody to the protein of interest.

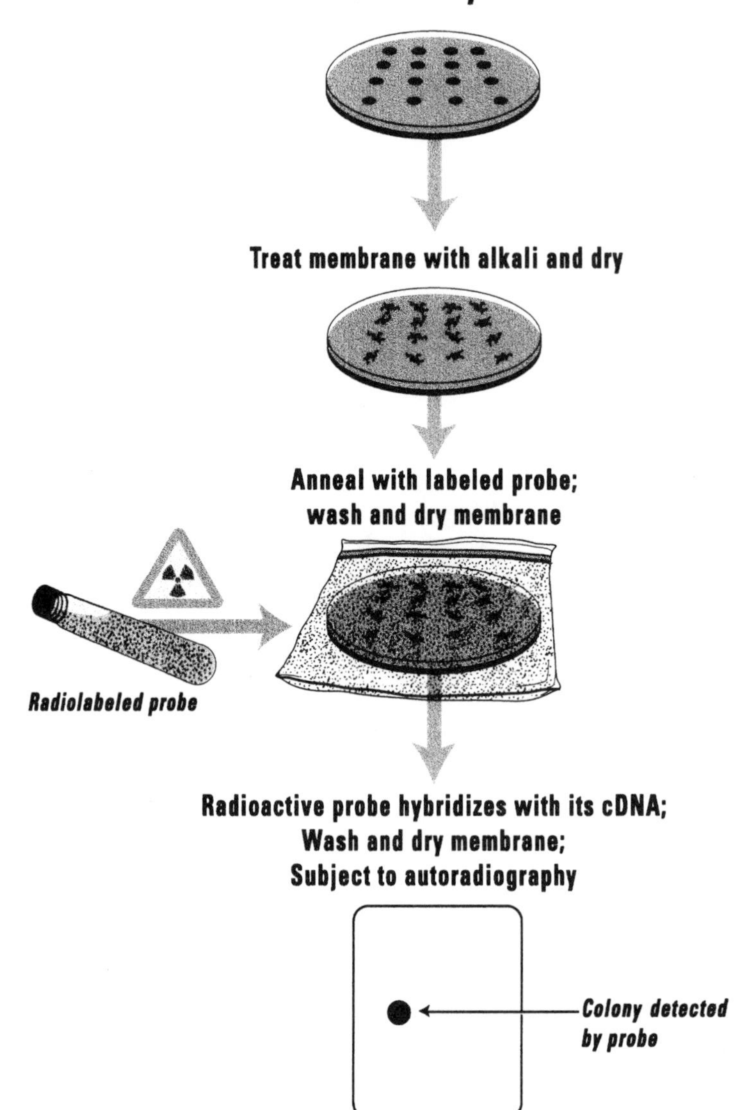

FIGURE 2.7 Replica plating and hybridization to identify colonies that contain the DNA of interest.

2.4 SOUTHERN AND NORTHERN BLOT ANALYSES

Southern analysis of DNA is frequently used to: (1) determine if a gene of interest is present, (2) determine the copy number of a particular gene (e.g., gene amplification is a common occurrence in cancer and results in additional copies of certain

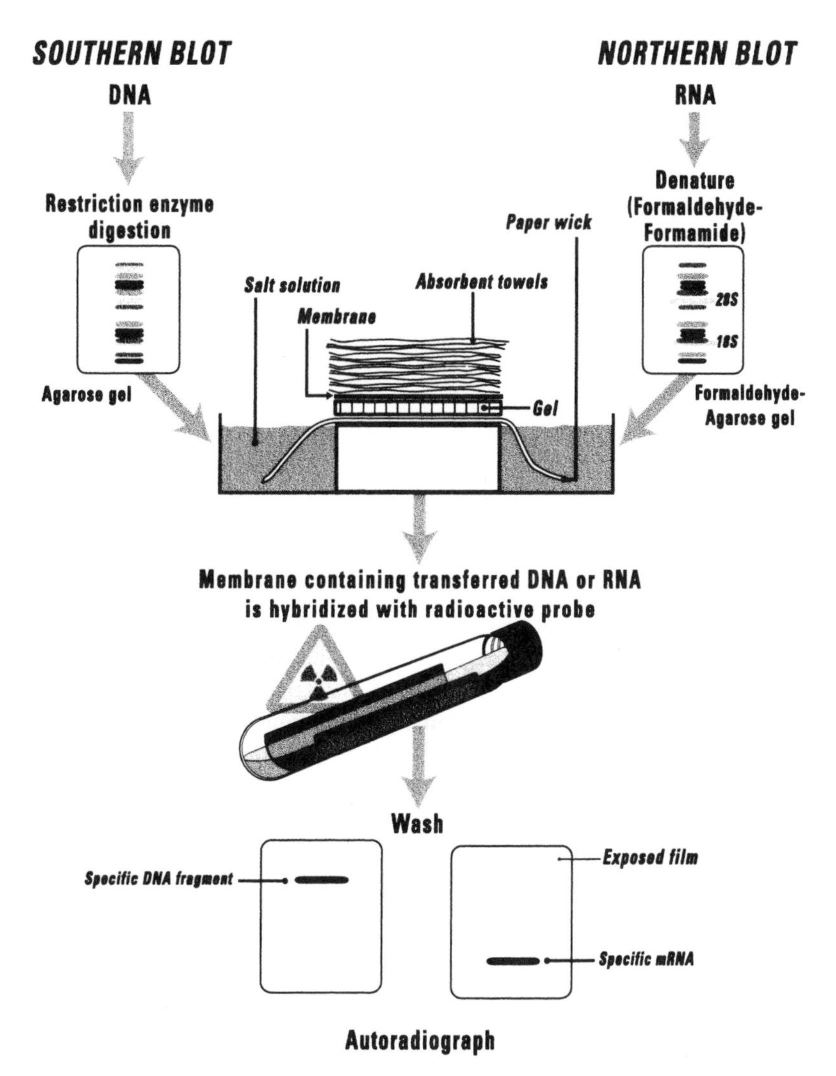

FIGURE 2.8 Generalized procedure for Southern and Northern analysis.
Source: Adapted from B. Volgelstein and K.W. Kinzler (Eds), *The Genetic Basis of Human Cancer*, McGraw-Hill, New York, 1998.

oncogenes, chromosome duplication or loss can also result in differences in gene copy number), (3) examine restriction fragment length polymorphism (RFLP) (e.g., changes in restriction sites are an indication of genetic differences between individuals or samples) and (4) determine loss of heterozygosity or reduction to homozygosity which can be important in evaluating and identifying loss of known or putative tumor suppressor genes. Northern analysis is most frequently used to examine mRNA size, splice variants, and levels.

Northern and Southern blot analyses are carried out as shown in Fig. 2.8 and as described below. Restriction digested DNA (Southern) or total RNA or poly (A) mRNA (Northern) is subjected to agarose gel electrophoresis, which separates DNA and RNA molecules based on size. The separated molecules are then transferred to a nylon or nitrocellulose membrane by capillary action transfer or electroblotting. The immobilized DNA or RNA on the membrane is incubated with a radiolabeled probe for several hours to allow the probe to hybridize to its complementary target sequence. The membrane is then gently washed to remove unbound probe and subjected to autoradiography in which an autoradiographic film is placed over the membrane. Following exposure, the film is developed and the blackened or exposed area corresponds to the complementary sequence of the probe. Many methods are currently available that do not use radioactive probes but instead use fluorescent or chemiluminescent probes.

2.5 POLYMERASE CHAIN REACTION

Polymerase chain reaction (PCR) is a powerful technique that can amplify specific DNA and RNA species. Amplification of DNA by PCR can be used for mutational analysis, RFLP analysis, allele specific hybridization, or to simply clone a gene of interest. It can be used to amplify regions of genes and these products can then be analyzed for single-strand conformational polymorphism (SSCP). SSCP analysis of PCR-amplified products is an effective screening method for detecting single base mutations. SSCP uses gel electrophoresis to separate strands of DNA on a non-denaturing polyacrylamide gel. If a mutation is present in the DNA, then the mobility of that strand of DNA will be different than the wild-type strand. The mutation within the strand can then be determined by direct DNA sequencing of the PCR product. To use PCR, the sequence of the gene of interest must be known or at the very least the sequence in the area where the primers are to bind must be known. Using the known sequence, primers can be constructed that will provide starting points for DNA synthesis on each strand. The target DNA of interest is incubated with thermostable DNA polymerase, all four dNTP and a set of primers that are complementary to the nucleotides at the end of the area of each strand to be amplified. PCR is conducted in a thermal cycler in which the temperature is raised to separate the double stranded DNA (denaturation step), the temperature is lowered to allow the primers to anneal to the complementary sequences in the target DNA (annealing step), and finally the temperature is raised to allow the polymerase to synthesize DNA (synthesis step) (Fig. 2.9). This cycle of denaturation and annealing followed by synthesis is repeated 20–40 times. Using this PCR approach the area of the gene of interest can be amplified at least 10^5 times providing large amounts of the DNA of interest. The procedure is so powerful that DNA from a single cell can be amplified, for example, cells from histological sections can be scraped from a microscope slide and DNA can be amplified for mutation analysis. Recently, the use of laser capture microdissection (LCM) has allowed for the analysis of DNA collected from just a few cells on a histological section. This is especially useful in mutational analysis of tumor cells and in studies of chemical carcinogenesis and tumor progression.

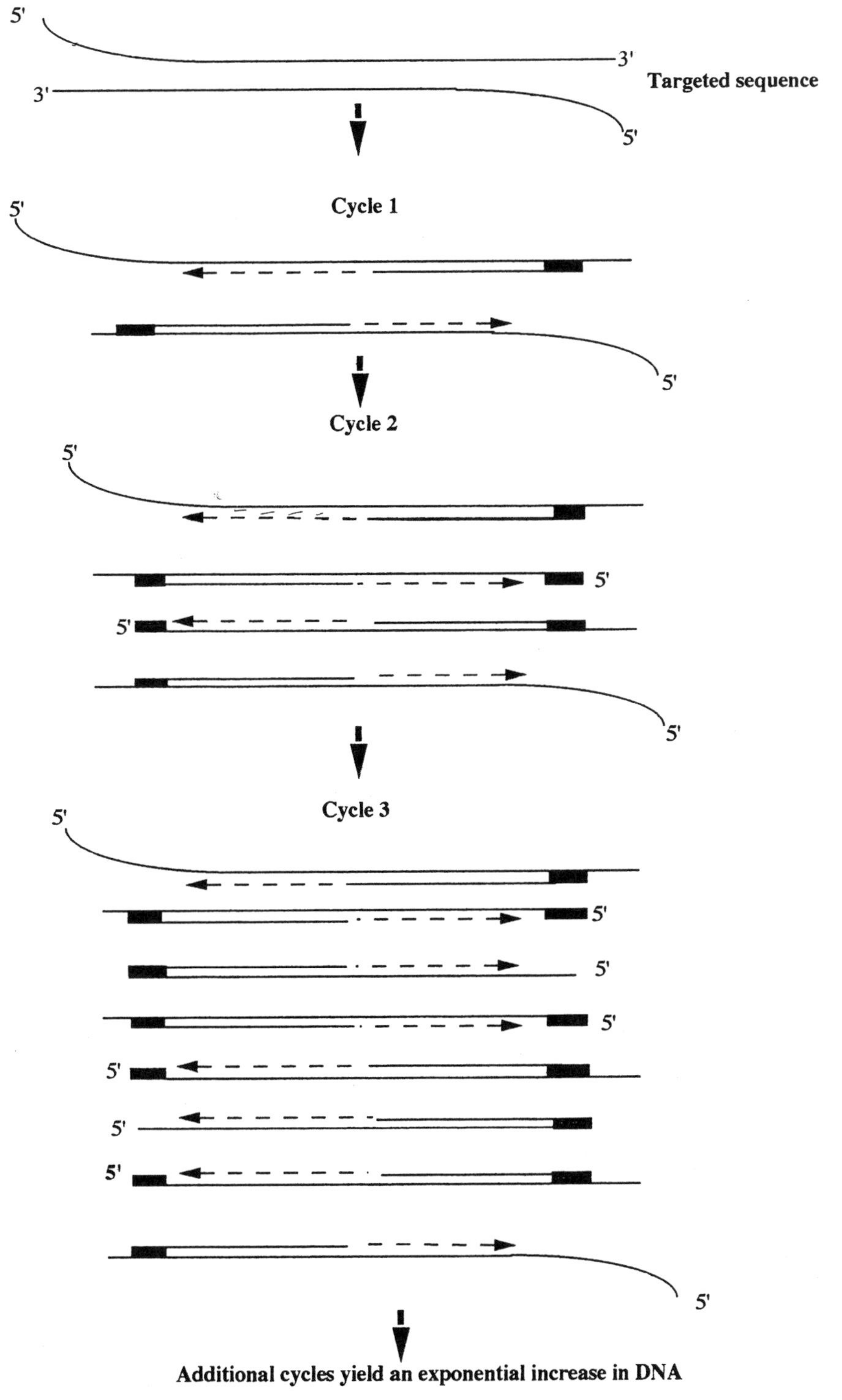

FIGURE 2.9 Polymerase chain reaction (PCR).

2.6 SOME METHODS TO EVALUATE GENE EXPRESSION AND REGULATION

2.6.1 Northern Analysis

As described above, Northern blot analysis is commonly used to determine levels of mRNA of interest. mRNA levels can be compared between experimental and control samples and should be normalized to the expression of a gene whose expression is invariant, for example, housekeeping genes such as actin or GAPDH (3-glyceraldehyde-3-phosphate dehydrogenase) are commonly used.

2.6.2 Nuclear Run On

Although Northern analysis can reveal an increase in the level of mRNA for a specific gene, this increase could be due to mRNA stabilization, an increased rate of transcription, or both. To determine if the rate of transcription is increased, a nuclear run on assay should be performed (Fig. 2.10). Basically, nuclei are isolated from the control and treated cells of interest at various times and incubated with [α-^{32}P]UTP. The ^{32}P labeled uridine will be incorporated into elongating hnRNA. At various times the hnRNA is isolated and spotted onto a nylon or nitrocellulose membrane that has the cDNA of the gene of interest immobilized to it. The newly synthesized radiolabeled hnRNA hybridizes to the immobilized cDNA, the membrane is washed to remove unincorporated [α-^{32}P]UTP and nonspecific binding of other radiolabeled hnRNA and autoradiography is conducted. The [α-^{32}P]UTP incorporation can be compared between experimental and control cells. If the experimental group demonstrates increased exposure of the film, it indicates that the rate of mRNA synthesis of the gene of interest is increased.

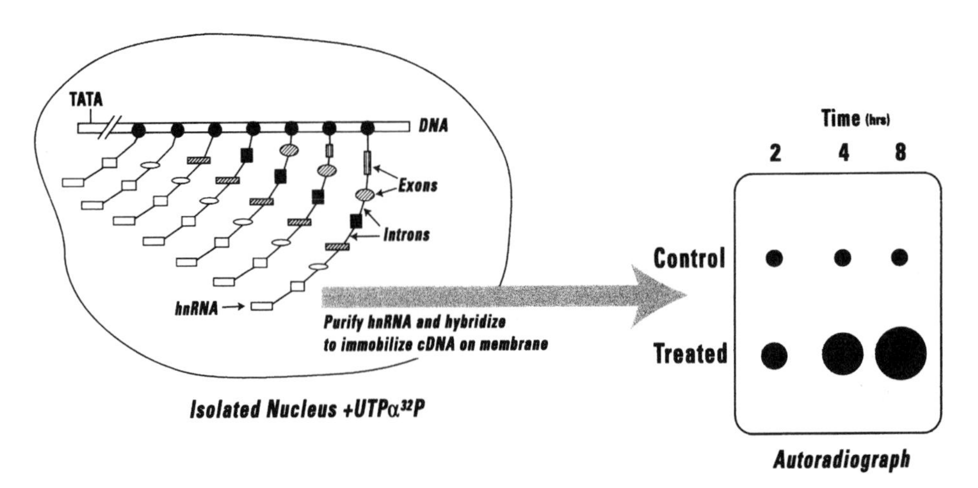

FIGURE 2.10 Nuclear run on assay.

2.6.3 Promoter Deletion Analysis

If after conducting a nuclear run on assay the rate of transcription is found to be increased, sites within a promoter that are required for transcription and corresponding transcription factors that contribute to the increased rate of transcription can be identified. In order to do this, one must have the promoter region of the gene of interest in hand. The promoter region is inserted into a plasmid such that the promoter regulates the transcription of a reporter gene such as luciferase or chloramphenicol acetyl transferase (CAT). The reporter construct is transfected into a mammalian cell line of choice and the cells are treated with an agent or conditions that increase the expression of the gene. Under these conditions, the reporter will be transcribed, mRNA translated, and the resultant luciferase activity or CAT activity can be easily quantified. Generally, cells are harvested 24–48 h after transfection, cell homogenates prepared and luciferase or CAT activity determined. Promoter deletion analysis is conducted to identify the specific cis elements within the promoter that are responsible for the increase in the expression of the reporter gene. Small areas of the promoter are deleted and these constructs are then transfected and the activity of the promoter is determined as described above. Through this type of analysis, the specific enhancer elements and transcription factors that are important in the regulation of the gene under study can be identified.

2.6.4 Microarrays and DNA Chips

Microarrays can be utilized to study the expression of hundreds or even thousands of genes in a single experiment. A microarray is a solid support to which hundreds to thousands of individual cDNA have been covalently linked. The cDNA are applied to the support in a grid pattern so that each individual cDNA has a unique position within the grid. mRNA is isolated from the experimental and control tissue or cells and cDNA is made using reverse transcriptase in the presence of fluorescent or radiolabeled nucleotides. The labeled cDNA are hybridized to the complementary DNA sequences that are immobilized on the microarray. The degree of hybridization of the cDNA is quantitiated using a specialized scanner or a phosphoimager and the expression of each gene is normalized to housekeeping genes or other genes whose expression is not altered between the experimental and control group. Through this approach one can examine the expression of thousands of genes in a single experiment. DNA chips consist of short segments of DNA (oligonucleotides of 20–30 base pairs) synthesized directly onto a solid silicon support. Like microarrays these DNA chips are used to measure changes in gene expression but they are also used to detect polymorphisms or mutations in genes.

2.6.5 Reverse Transcriptase-Polymerase Chain Reaction

PCR can also be used to amplify mRNA through a technique called reverse transcriptase-polymerase chain reaction (RT-PCR). mRNA is first converted into DNA using reverse transcriptase and then the cDNA are amplified using gene-specific primers by PCR. This technique is suitable for studies to determine if a gene is or is not expressed. To compare levels of mRNA between samples, a quantitative RT-PCR technique has been developed whereby PCR products are directly related to the amount of starting target sequence. The amount of the

PCR product is normalized to the PCR product levels of a housekeeping gene that is amplified in the same PCR reaction using its gene-specific primers. By using specific primers to multiple genes, changes in the expression of numerous genes can be determined and normalized to the expression of a housekeeping gene within a single sample.

2.6.6 RNase Protection Assay

Ribonuclease protection assay (RPA) is a sensitive method for the detection and quantitation of mRNA species as well as the detection of alternatively spliced mRNA. The assay uses high specific activity ^{32}P, ^{33}P, or ^{35}S-labeled antisense RNA probes expressed from a T7 or T3 polymerase promoter containing vector using DNA-dependent RNA polymerase. The antisense RNA probe is hybridized to the target mRNA in solution, after which the free probe and other single-stranded RNA are digested with RNases. The double-stranded RNA is protected from digestion and the RNase-protected probes are purified and resolved on a denaturing polyacrylamide gel and quantified by autoradiography or phosphorimage analysis. The quantity of the target mRNA is based on the level of intensity of the signal.

2.6.7 Electrophoretic Mobility Shift Assay

Electrophoretic mobility shift assay (EMSA) is used to determine the level of binding of a transcription factor to its specific DNA consensus sequence. It is based on the observation that the movement of a DNA molecule through a nondenaturing polyacrylamide gel is slowed when a protein is bound to it. Nuclear extracts are incubated with the radiolabeled DNA consensus sequence of the transcription factor under study and then subjected to electrophoresis on a nondenaturing polyacrylamide gel. If the nuclear extract contains the transcription factor that binds to the specific consensus sequence, then the mobility of the consensus sequence will be retarded or slowed on the gel. Autoradiography of the gel will reveal that the free DNA consensus sequence (not bound to its transcription factor) has migrated to the bottom of the gel. However, if the transcription factor specific to the DNA consensus sequence is present in the nuclear extract and it is capable of binding, then a radiolabeled protein/DNA complex with retarded mobility will appear in the top portion of the gel (a mobility shift) (Fig. 2.11). The identity of the binding proteins can be confirmed by incubating the nuclear extract, radiolabeled DNA consensus sequence with an antibody to the suspected transcription factor thought to be responsible for the binding. The samples are then subjected to the procedure described above with the end result demonstrating a "supershift" upward of the transcription factor/DNA complex due to the binding of the antibody to the transcription factor.

2.7 METHODS TO EVALUATE GENE FUNCTION

2.7.1 Eukaryotic Expression Systems

Numerous eukaryotic expression systems are used to study gene function in an intact cell and these most often use a plasmid or viral expression system. In choosing an expression system, one has to choose between (1) viral infection versus

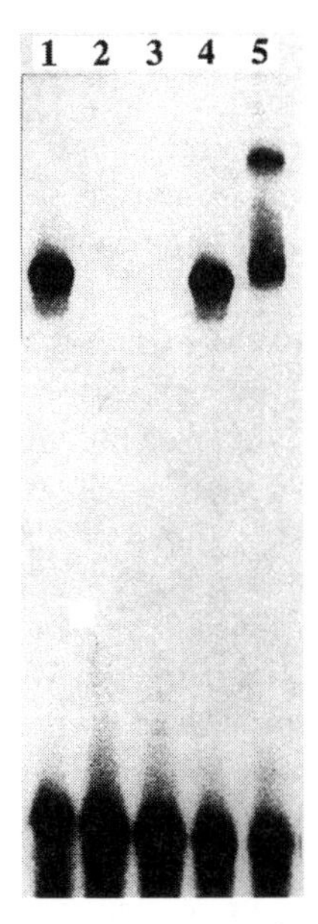

FIGURE 2.11 Electorphoretic mobility shift assay (EMSA).
Probe: an oligo carrying two binding sites for heterodimeric transcription factor E2F. Protein: Baculovirus produced E2F/DP-1 heterodimer. Lane 1—protein + radiolabeled probe; Lane 2—protein + radiolabeled probe + competition with cold (not radiolabeled) homologous probe carrying one binding site; Lane 3—protein + probe + competition with cold homologous probe carrying two binding sites; Lane 4—protein + probe + competition with cold homologous probe crrying no binding sites; Lane 5—protein + probe + anti-E2F antibody.

Source: Courtesy of J.M. Horowitz.

plasmid transfection systems, (2) a constitutively active promoter versus an inducible promoter, and (3) transient expression versus stable integration into the hosts' genome. Viral vectors are extremely useful when it is important to obtain expression in a large percentage of target cells, however, these systems require helper viruses or packaging cell lines and are more labor intensive than the plasmid based systems. Plasmid based vectors require transfection and, depending upon the cell type used, efficiencies of transfection can range from 5% to 80%. Low transfection efficiency may preclude plasmid vector use in some cells. Plasmid eukaryotic expression systems often produce very high levels of expression of the gene of interest due to their use of powerful viral promoters and their ability to autonomously replicate within the host cell.

In general, a plasmid-based expression system can be divided into two general types: constitutive expression vectors or inducible expression vectors. In both cases, the gene of interest is cloned into the expression plasmid such that it is regulated by the promoter provided by the vector. The construct is then transfected into mammalian cells using a variety of methods, including those that use cationic lipids, $CaCl_2$, or electroporation. A constitutively active promoter is fully active, driving the transcription of the gene under its regulation independent of extracellular stimuli. The cytomegalovirus (CMV) promoter is a commonly used promoter in this type of expression vector, as it is a potent promoter in many cell types. Inducible expression vectors utilize promoters that can be turned off and on. For example, the metallothionein promoter is inactive in the absence of metals such as zinc. The addition of zinc to tissue culture media triggers the promoter to become active and transcription of the inserted gene ensues. The transfection experiments described above are referred to as transient transfection or transient expression experiments in that the vector is not incorporated into the host genome and cells are generally studied within 48 hours of transfection. However, stable or genetically modified cell lines that have incorporated the expression vector into the host genome can be generated. Selection for cells that have stably incorporated expression constructs into their genome can be accomplished by the presence of a drug resistance marker gene in the vector. The neomycin resistance gene, which is expressed from a constitutive promoter, is commonly used and facilitates the selection of stable transformants in the presence of the antibiotic G418 (an aminoglycoside related to gentamicin, which is inactivated by the neomycin resistance gene product). G418 blocks translation and kills cells that do not contain the neomysin resistance gene whose product inactivates G418. Cells that have stably integrated the vector DNA into their genome survive G418 selection forming colonies of drug-resistant cells that may be isolated as clones or grown as populations. Many excellent mammalian expression systems, both constitutive and inducible, are commercially available and are constantly being refined.

2.7.2 Transgenic Mice

A transgenic organism is one that has exogenous DNA in its genome. In order to achieve inheritance of exogenous DNA or transgene, integration of the DNA must occur in the cells that can give rise to germ cells. This is accomplished by introducing the DNA into a fertilized egg cell or into embryonic stem cells. Embryonic stem cells are commonly used to make 'knock out" or nullizygous mice that are deficient in a single gene, whereas injection of the fertilized egg with DNA is most commonly used to make transgenic mice that overexpress a specific gene. To understand how transgenic and nullizygous mice are made, it is necessary to briefly review early post-fertilization and preimplantation development (Fig. 2.12).

2.7.2a Procedure for Making Transgenic Mice Using Zygote Injection

Within a few hours after mating, zygotes can be removed from a female mouse before the female pronucleus and male pronucleus fuse. The pronucleus from the male is injected (male pronucleus is larger and provides bigger target for injection) with a construct containing the gene of interest (Fig. 2.13). These constructs can be targeted to specific tissues in the adult animal by the use of specific promoters. For example, an albumin promoter will target the expression to the liver, certain keratin promoters will target the expression to the epidermis, and an actin promoter (housekeeping gene) will direct the expression of the transgene in all cells. After injection of the DNA into the pronucleus of the male, the zygotes are placed in the oviducts of a 0.5 day pseudopregnant female (a female that has mated with a sterile male and is physiologically capable of allowing for the development of the injected zygotes). About 15% of the offspring will contain the transgene and will be transgenic hemizygous animals. Transgenic mice can be identified by Southern analysis

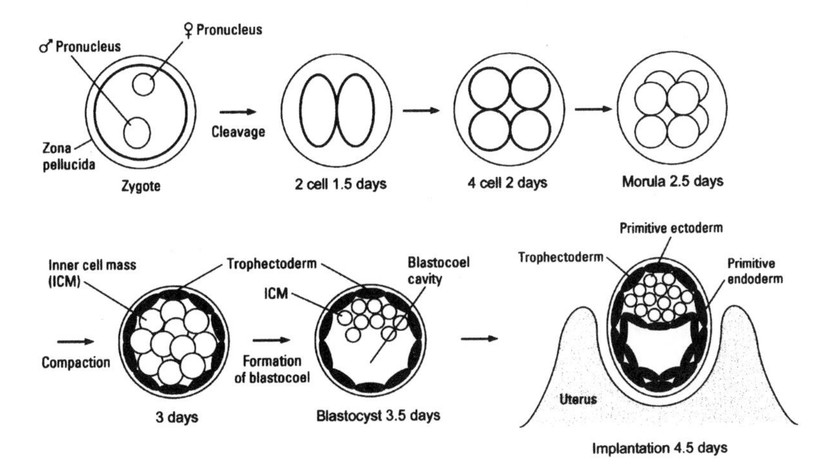

FIGURE 2.12 Fertilization and preimplantion development.

Source: Adapted from J.M. Sedivy and A.L. Joyner (Eds), *Gene Targeting*, Oxford University Press, New York, 1992.

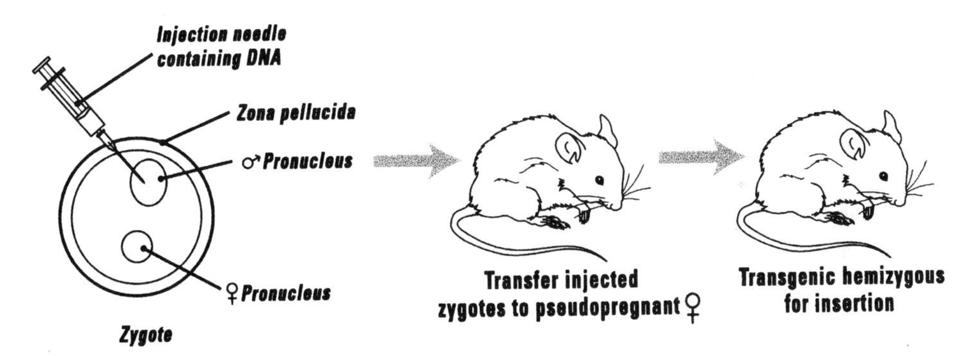

FIGURE 2.13 Procedure of making transgenic mice using zygote injection.

Source: Adapted from J.M. Sedivy and A.L. Joyner (Eds), *Gene Targeting*, Oxford University Press, New York, 1992.

of a DNA sample purified from a clipping of a small piece of their tail. Transgene positive mice are referred to as founders and can be mated with wild-type mice to produce offspring that also contain the transgene, which also confirms that the transgene is transmitted by the germ cells. Generally, the approach outlined above to make transgenic mice is used to study the overexpression of a gene (gain of function), however, antisense as well as a dominant negative approaches can be utilized and can result in a decrease in the level of the endogenous protein.

2.7.2b Procedure for Making Knockout Mice Using Embryonic Stem Cells

Embryonic stem cells (ES cells) and homologous recombination are utilized to inactivate an endogenous gene from a host's genome. ES cell lines are derived from a 3-day embryo (ICM cells) and are undifferentiated but remain totipotent. These

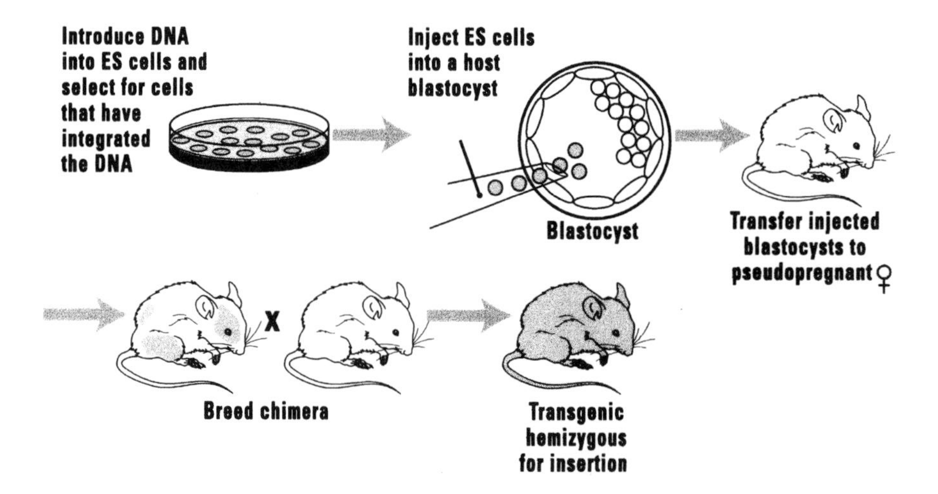

FIGURE 2.14 Procedure for making nullizygous mice using embryonic stem cells. *Source*: Adapted from J.M. Sedivy and A.L. Joyner (Eds), *Gene Targeting*, Oxford University Press, New York, 1992.

cells are transfected in cell culture with a vector that generally contains a homologous portion of the gene (not cDNA) of interest with a neo resistance gene inserted into an exon to disrupt its function. This piece of DNA can undergo homologous recombination with the ES cell gene and result in the replacement of the ES gene with the defective gene (this is an uncommon event but it does happen at low frequencies). Through positive and negative selection homologous recombinants can be identified and then characterized by PCR analysis to confirm the presence of the defective gene. These ES cells of interest can then be injected into a blastocyst and then transferred into a pseudopregnant female (Fig. 2.14). If the ES cells were derived from mice with black fur and the blastocyst is from an albino mouse, then offspring will be chimeras containing both white and black fur (cells of such chimera mice are from ES cells as well as cells derived from the blastocyst into which they were injected). These chimera can be bred with albino (recessive hair phenotype) animals and black (dominant hair phenotype) offspring will have inherited one inactive allele of the particular gene of interest. By mating heterozygous mice (one allele inactivated), mullizygous animals can be produced.

SUGGESTED READING

Davis, L.G., Dibner, M.D. and Battey, J.F. (Eds), *Basic Methods in Molecular Biology*, Elsivier Science Publishing Co., New York, 1986.

Sambrook, J.F. and Russell, D.W. *Molecular Cloning: A Laboratory Manual*, 3rd ed., Cold Spring Harbor Press, New York, 2000.

Ausubel, F. et al., *Current Protocols in Molecular Biology*, John Wiley & Sons, New York, 1988.

Joyner, A.L. (Ed), *Gene Targeting: A Practical Approach*, Oxford University Press, New York, 1993.

Sedivy, J.M. and Joyner, A.L. *Gene Targeting*, Oxford University Press, New York, 1992.

Hogan, B., Beddington, R., Costantini, I. and Lacy, E. (Eds), *Manipulating the Mouse Embryo: A Laboratory Manual*, Cold Spring Harbor Press, New York, 1994.

Lewin, B. *Genes VI*, Oxford University Press, New York, 1997.

Watson, J.D., Hopkins, N.H., Roberts, J.W., Steitz, J.A. and Weiner, A.W. (Eds), *Molecular Biology of the Gene*, 4th Ed., Benjamin Cummings Publishing, Menlo Park, CA, 1987.

CHAPTER THREE

Immunochemical Techniques in Toxicology

GERALD A. LEBLANC

3.1 INTRODUCTION

The advent of antibody technology has revolutionized every branch of biology that involves protein analyses, including toxicology. Proteins serve as both basic structural and functional components of organisms and, as such, are major target sites for the action of many toxic agents. Furthermore, proteins can modulate chemical toxicity through various processes including activation, deactivation, sequestration, and elimination. Antibody technology provides powerful tools for a variety of toxicological analyses including: (a) assessing the effects of a toxicant on intracellular protein localization, (b) precisely assessing the effects of a toxicant on specific protein levels (i.e., protein induction or suppression), (c) measuring interactions between a specific protein in a mixture of proteins and a toxicant (i.e., contribution of a specific enzyme in a complex cellular mixture to the metabolism of a toxicant), and (d) isolating individual proteins for subsequent characterization. Immunochemical techniques also have provided means for the rapid quantification of toxicants, drugs, hormones, and other small molecules in biological and environmental samples.

3.2 DEFINITIONS

It is not the intent of this chapter to provide a synopsis of the science of immunology. Rather, this chapter is intended to provide a general overview of some basic immunochemical techniques that have proven invaluable to toxicological research. The following immunological terms are integral to the discussion of immunochemical analyses and are routinely used in this chapter:

Immunogen: Any foreign molecule (usually a protein) that elicits an immune response.

Immune response: The production of antibodies by an organism in response to exposure to a foreign molecule (immunogen). The antibodies produced will recognize the foreign molecule.

Antibody: Proteins secreted by B lymphocytes that recognize and bind to a specific immunogen.

Antigen: Any foreign molecule that is recognized by an antibody. Note that a single protein is typically first an immunogen (elicits the immune response) and is then an antigen (is bound by the resulting antibodies).

Epitope: The specific site on an antigen that is recognized by an antibody.

Immunogenicity: The ability of a foreign molecule to elicit an immune response.

Antigenicity: The ability of a foreign molecule to be recognized and bound by an antibody.

Hapten: A small molecule (i.e., a hormone, drug, or toxicant) that is conjugated to a larger carrier molecule (i.e., a protein) so that an immune response is elicited and antibodies are produced that recognize epitopes on the small molecule.

Adjuvants: Nonspecific stimulators of the immune response that are administered during immunization.

Primary antibody: The antibody used in an immunoassay that recognizes and binds to the antigen.

Secondary antibody: An antibody used in an immunoassay that recognizes and binds to the primary antibody (i.e., anti-IgG). The secondary antibody is equipped with some means of detection that allows for the analyses of the immune complex.

3.3 POLYCLONAL ANTIBODIES

When a mammal is injected with a foreign molecule (immunogen), the immune response elicited includes the generation of antibodies from a variety of B lymphocytes. The antibodies produced from each lymphocyte will recognize a single epitope on the foreign molecule (antigen). However, because of their different cellular origins, the population of antibodies produced will recognize and bind to multiple epitopes on the antigen. These antibodies, collected from the serum of the immunized animal, are polyclonal antibodies (Fig. 3.1).

Advantages of polyclonal antibodies over monoclonal antibodies are: (a) they are highly reactive due to the binding of multiple epitopes on the same antigen and (b) they are easy to produce. The disadvantage of polyclonal antibodies as compared with monoclonal antibodies is that they often have a lower degree of specificity (i.e., they will bind antigens that are not of interest).

3.4 MONOCLONAL ANTIBODIES

After an immune response has been elicited in an immunized animal, the isolation and propagation of individual B lymphocytes in cell culture will provide antibodies that are all of the same clonal origin and thus recognize a single epitope on the antigen. These antibodies, collected from the cell culture medium, are monoclonal antibodies (Fig. 3.1). Two important characteristics of monoclonal antibodies are (a) their single-epitope specificity of binding and (b) their ability to be produced in

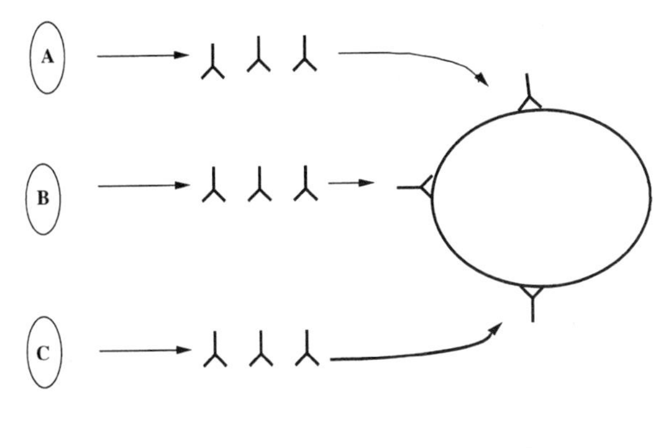

FIGURE 3.1 The generation of antibodies of several clonal origins (polyclonal antibodies) with antibodies from each clonal origin (monclonal antibodies, clones A, B, and C) recognizing a distinct epitope on the antigen.

unlimited quantities. The disadvantages of monoclonal antibodies are that (a) their single-epitope specificity often makes them unsuitable for certain immunoassays and (b) their production is time intensive and requires cell culture facilities and expertise.

3.5 POLYCLONAL ANTIBODY PRODUCTION

The generation of polyclonal antibodies involves the following steps and considerations.

3.5.1 Immunogen Preparation

The first step in generating an immune response with the intent of collecting antibodies is to prepare the immunogen to which the response will be generated. Thus, the generation of antibodies for use in protein analyses typically requires that the protein that the antibodies will recognize will be available in purified or semipurified form. The following characteristics of the immunogen must be considered to maximize success in generating useful antibodies.

3.5.1a Immunogen Purity

The antigenic specificity of the antibodies produced will only be as good as the purity of the immunogen used to immunize the animal. Partially purified immunogen may be adequate for certain immunoassays; however, fully purified immunogen will yield antibodies containing the greatest specificity for the immunogen. Partially purified immunogen can often be completely purified for immunization by electrophoretically separating the immunogen from the contaminating proteins, cutting the immunogen from the solid matrix with which it is associated (i.e., acrylamide gel, nitrocellulose membrane), and using it for immunization.

3.5.1b Immunogen Size
Immunogens typically must be greater than approximately 5000 Da to produce an immune response. If the immunogen is too small, it can be used as a hapten.

3.5.1c Immunogenicity
Some foreign compounds have high immunogenicity, while others have low immunogenicity. The following should be considered for materials having poor immunogenicity:

- Change the species or strain of animal being immunized. Immunization of a species with an immunogen derived from the same species may not elicit a sufficient immune response because the immunogen will not be recognized as foreign.
- Increase the dose of the immunogen. Within limits, the degree of immune response increases with increasing dose of immunogen. Weakly immunogenic materials may require a higher dose to stimulate a sufficient immune response.
- Alter the physical makeup of the immunogen (e.g., denature the protein, conjugate to dinitrophenol). This may reveal immunogenic sites that were not available on the native protein.

3.5.2 Immunization

Once the immunogen has been adequately prepared, it is then administered to an appropriate species with which an immune response will be generated. The following points should be considered during animal immunization.

3.5.2a Selection of Animal Species to Be Immunized
Both scientific and practical considerations contribute to the selection of the appropriate animal species to be immunized.

- *Self-recognition* Administration of an immunogen to the same species from which the immunogen was derived poses the risk that the immunogen will not be recognized as foreign and will accordingly not generate a sufficient immune response. The species selected for immunization should be different from that used to generate the immunogen.
- *Amount of antibodies required* Polyclonal antibodies are generally obtained from the immunized animal by collecting blood and preparing serum from the blood. The antibody-rich serum is used as the immunochemical reagent. The amount of immunochemical reagent derived from an immunization is therefore dependent upon the amount of blood that can be collected from the immunized animal. For example, a maximum of 2 ml of serum can generally be collected from mice, whereas up to 500 ml of serum can be derived from rabbits.
- *Amount of immunogen available* A limited amount of immunogen may dictate that a small species be used for immunization. For example, mice typically require 10 times less immunogen than that required by rabbits to elicit an equivalent immune response.

- *Animal maintenance capabilities* Larger species require more extensive animal holding facilities. While goats or horses may be used to generate large volumes of immunochemical reagent, most laboratories are not equipped with adequate facilities to accommodate such species.

3.5.2b Adjuvant Selection

An adjuvant should be selected that will adequately stimulate the immune system of the immunized animals without posing undo stress to the animals. Historically, Freund's adjuvant was most commonly used in antibody preparation. Freund's adjuvant consists of a water-oil emulsion containing heat-killed bacteria. Although very effective in stimulating a prolonged immune response, Freund's adjuvant has fallen into disfavor due to its propensity to induce persistent and aggressive lesions (granulomas) resulting in pain and distress to the immunized animals.

MPL+TDM adjuvant has proven to be a viable alternative to Freund's adjuvant. MPL+TDM adjuvant consists of a water-oil emulsion containing bacterial-derived chemicals or analogs thereof. MPL+TDM stimulates the immune system without eliciting the degree of toxicity associated with Freund's adjuvant. Aluminum salts also have been used to stimulate the immune response without severe toxic side effects to the immunized animal. Aluminum salts appear to function by trapping the immunogen and providing for a slow, prolonged release of the immunogen following administration.

3.5.2c Immunogen Dose

If the immunogen is readily available, then one strives to administer a dose that will elicit the maximum immune response (generally 0.5–1.0 mg in rabbits, 10 times lower in mice). If the immunogen is in short supply, then one strives to administer the minimum dose that will elicit a sufficient immune response (generally 10–100 μg for rabbits, 10 times lower for mice). An animal is generally injected several times with the immunogen to elicit a satisfactory immune response. Thus, total dose must be considered when determining the amount of immunogen required.

3.5.3 Serum Collection and Screening for Antibody Titer

Blood samples are generally collected and serum prepared 8–10 weeks after the first injection. Serum samples are analyzed for antibody titer. When the titer is sufficiently high, the animal is bled. Serum samples are generally screened for the presence of the antibody by enzyme-linked immunosorbent assay (ELISA). ELISA methods are discussed later in this chapter.

3.6 MONOCLONAL ANTIBODY PRODUCTION

All of the processes and considerations described for polyclonal antibody production are also relevant to monoclonal antibody production. The only variation between the two, with respect to the immunization process, is species selection. Monoclonal antibodies are most commonly produced in mice because of the commercial availability of mouse myeloma cells that are particularly suited as partner cells in hybridoma preparation (discussed below).

Monoclonal antibody production deviates from polyclonal antibody production once the immune response has been generated. During polyclonal antibody production, high titer antibodies in the serum denotes the end of the process and the antibody-rich blood is sampled from the animal. During monoclonal antibody production, a high titer of antibodies in the serum denotes the time to begin the generation of monoclonal antibodies. This encompasses the preparation of hybridomas, the production of hybridomas of monoclonal origin, and the screening of the hybridoma clones for the production of the antibodies that recognize the antigen of interest. This process is diagrammed in Figure 3.2.

The B lymphocytes of the immunized animal are the source of the antibodies that recognize the administered immunogen. Removal of the spleen and culture of individual B lymphocytes derived from this organ may seem to be a viable approach to the generation of monoclonal antibodies in culture. However, primary B lymphocytes will not survive in culture. These antibody-producing cells are therefore immortalized by fusion to myeloma cells. These cells are derived from tumors of B lymphocyte origin and will grow indefinitely in culture. B lymphocytes from the immunized mice and the myeloma cells are most commonly fused by stirring and centrifugation in polyethylene glycol.

Following fusion, hybridomas must be separated from fusion products of the same cell type (e.g., fusion of two B lymphocytes or two myeloma cells) and cells that did not undergo fusion. Unfused B lymphocytes or the fusion product of two B lymphocytes will be eliminated during continuous culture as these cells have a short life span in culture. Commercially available myeloma cells for hybridoma production have mutations in one of the enzymes of the salvage pathway of purine nucleotide biosynthesis. Hybridoma cells are cultured in medium that forces the cells to utilize the salvage pathway for nucleotide synthesis. The mutated myeloma cells or hybridization products of two myeloma cells will die in this selection medium because they are incapable of nucleotide synthesis under these propagation conditions. However, myeloma cells that have fused to the B lymphocytes derived from the spleen of the immunized animal will have an intact salvage pathway and will survive in the selection medium. Thus, only the B lymphocytes-myeloma hybridomas will survive prolonged culture in the selection medium.

Hybridomas are then cultured and the culture media assayed for the presence of antibodies to the antigen of interest. Cultures producing desirable antibodies are then aliquoted at dilutions calculated to yield 0.5 to 1.0 cell per aliquot. These aliquoted samples are then cultured. Many of these cultures will be initiated with a single cell, thus the resulting cell population will be of monoclonal origin. Media from these cultures are again assayed for the presence of the antibodies. Positive cultures are again diluted and aliquoted to ≤1 cell per aliquot. This ensures that the resulting cell populations are of monoclonal origin. Media from these cultures are again assayed for the presence of the antibody and positive clones are used in mass culture for monoclonal antibody production. The ability to produce the antibodies in culture provides for an inexhaustible supply of immunochemical reagent.

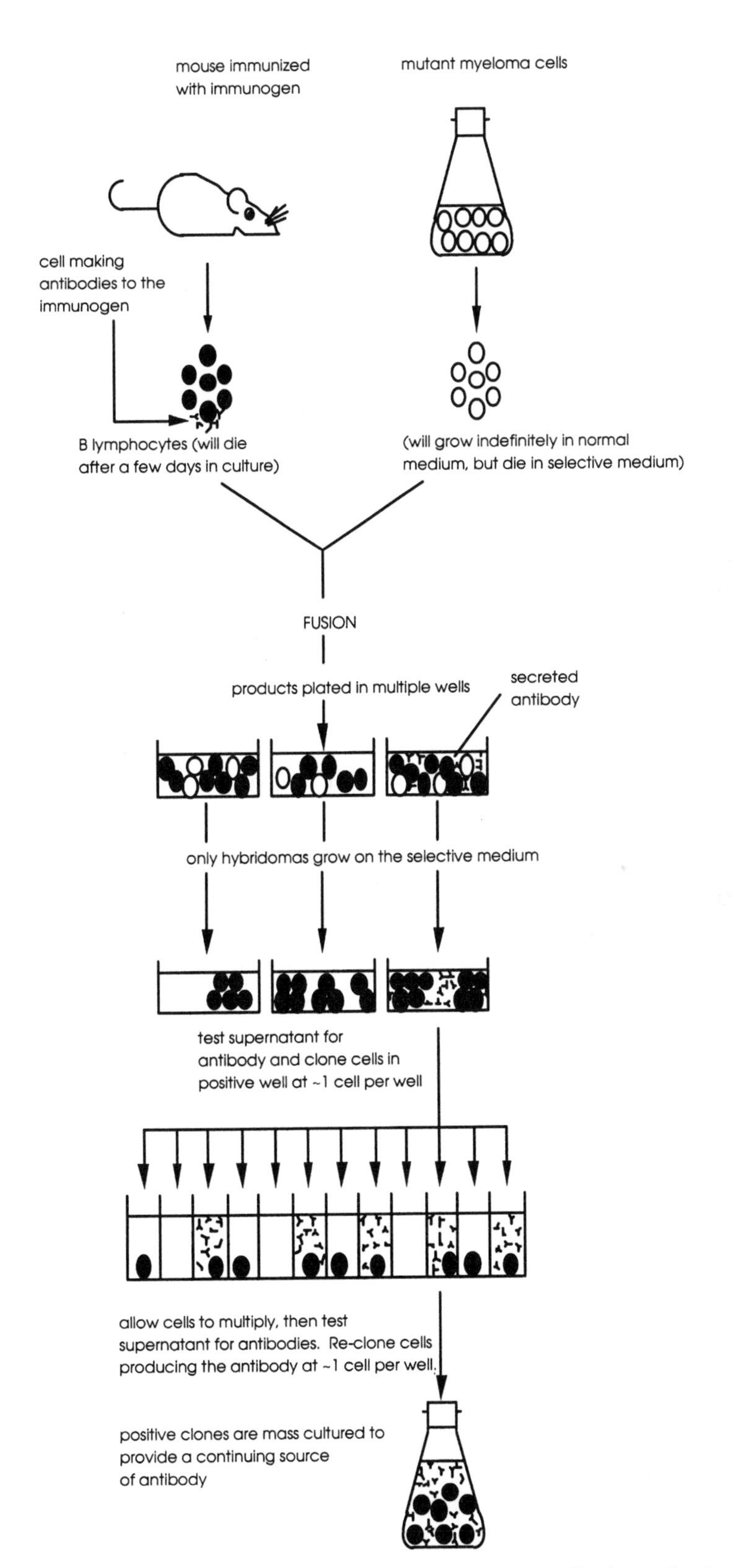

FIGURE 3.2 Diagrammatic overview of monoclonal antibody production.

3.7 IMMUNOASSAYS

The polyclonal or monoclonal antibodies produced can be used in a variety of immunoassays for the detection or characterization of an antigen. Some of the more common immunoassays used in toxicology are described below.

3.7.1 Immunolocalization

This technique is a semiquantitative means of determining the presence of a protein within a cell, its abundance, and its subcellular localization. This approach could be used in toxicology, for example, to determine if, upon treating cells with a specific toxicant, a receptor to which the toxicant binds localizes from the cell cytoplasm to the nucleus.

The general approach used to assess the intracellular localization of a protein within a cell is as follows:

- Cells under study are immobilized onto a solid support to facilitate later manipulation. Cells are generally secured to a glass slide on which they subsequently can be viewed with a microscope. This can be accomplished by growing adherent cells on a slide, drying cells onto the slide, or chemically attaching the cells to the slide.
- The cells are then fixed to preserve cellular structure and immobilize proteins in place. Fixation of the cells is often accomplished by incubating the cells in organic solvent, paraformaldehyde, or glutaraldehyde.
- Cells are incubated with the primary antibody. The primary antibody will recognize and bind to the protein of interest.
- Cells are incubated with secondary antibody. The secondary antibody will recognize and bind to the primary antibodies. This tertiary (protein-primary antibody-secondary antibody) complex can now be subjected to detection methods.
- Preparations are incubated with appropriate reagents to allow visualization based upon the detection system associated with the secondary antibody. The secondary antibody may be conjugated to a enzyme (i.e., alkaline phosphatase, horseradish peroxidase). Incubation with the appropriate substrate to the enzyme will result in the production of an insoluble, colored product that can be detected upon microscopic analyses of the cells. Secondary antibodies can also be conjugated to fluorochromes (i.e., fluorescein, rhodamine) that can be detected using a microscope equipped to detect fluorescence.

Immunolocalization has proven to be a powerful tool in biochemical toxicology allowing for in situ assessments of protein responses to toxicant exposure.

3.7.2 Immunoaffinity Purification

Immunochemical approaches allow for the purification of a protein from a complex mixture in a single chromatographic step. Because antibodies to the protein targeted for purification are required, immunoaffinity purification techniques are often used for the routine purification of a protein subsequent to its initial purification by traditional chromatographic approaches. Immunoaffinity purification of a protein involves the following general steps:

- Antibodies are bound to a solid-phase matrix. Protein A beads are commonly used. Protein A is a polypeptide located on the cell wall of some bacteria. Protein A binds the Fc (constant) region of antibodies without affecting the ability of the antibodies to bind antigen. Protein A is commercially available conjugated to sepharose, agarose, or acrylic beads. A chromatographic column is prepared with the tertiary complex (antibody-protein A-solid matrix).
- The mixture containing the protein targeted for purification is added to the column and sufficient time is allowed for the protein and antibodies to bind.
- Unbound proteins are washed from the column.
- Antibody-bound proteins are dissociated from the antibodies and eluted from the column. This is often the most difficult step because the antibodies and antigen bind with high affinity. Nondestructive approaches should be attempted to dissociate the antibodies and antigen such as by manipulating the salt concentration or pH of the elution buffer. Sometimes, more drastic approaches are required such as adding a denaturent (i.e., detergent) to the elution buffer.

3.7.3 Immunoprecipitation

Immunoprecipitation provides a means by which a protein can be selectively removed from a complex mixture. While conceptually similar to immunopurification, immunoprecipitation is typically used in the direct functional characterization of a protein, not as a means of obtaining protein. For example, a toxicologist may determine that two enzymes (designated A and B) are capable of inactivating a specific chemical carcinogen. Immunoprecipitation could then be used to determine the relative contribution of each enzyme in a tissue preparation to the metabolism of the chemical as follows:

- Incubate increasing concentrations of antibody to enzyme B with the tissue homogenate. Increasing amounts of enzyme B will be bound with increasing antibody concentration in the assay.
- Precipitate the immune complexes (antibody-antigen) by adding Protein A beads to the mixture, incubate, then centrifuge the mixture. The Protein A will bind to the Fc region of the antibodies. The beads (i.e., sepharose) associated with the Protein A will provide sufficient mass to the complex to allow its precipitation during centrifugation.
- Use the supernatant from the precipitations to measure biotransformation of the carcinogen. With increasing concentration of antibody, greater amounts of enzyme B will be removed from the supernatant. If enzyme B is responsible for the biotransformation of the carcinogen, then progressively less biotransformation will be measured with increasing concentrations of antibody added to the mixture (Fig. 3.3).

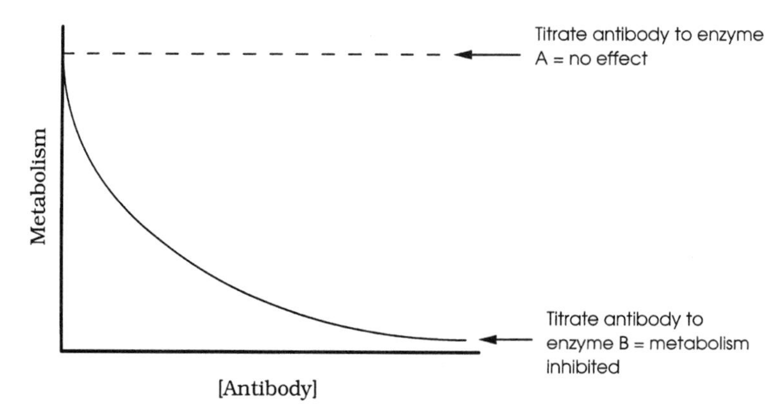

FIGURE 3.3 The use of immunoprecipitation to assess the relative contribution of enzymes A and B in a cellular preparation to the metabolism of a chemical carcinogen. Metabolism was decreased following immunoprecipitation with increasing concentration of antibodies to enzyme B. These results indicate that enzyme B and not enzyme A was responsible for the metabolism of the carcinogen.

- Repeat the above procedure but using antibodies to enzyme A instead of enzyme B. This portion of the experiment will reveal the relative contribution of enzyme A to the biotransformation of the carcinogen in the tissue preparation. Taken together, results will establish the relative contribution of each enzyme to the biotransformation of the chemical carcinogen in the tissue sample (Fig. 3.3).

3.7.4 Immunoblotting (Western Blotting)

This procedure makes use of electrophoretic and immunochemical techniques to identify a specific protein in a complex mixture. Immunoblotting can be used: (a) to determine the presence of a specific protein in a biological preparation, (b) to determine relative amounts of a specific protein in a biological preparation, (c) to estimate the molecular weight of a specific protein.

Consider the following experiment. A toxicologist wants to determine whether administration of a polychlorinated biphenyl (PCB) to rats induces a specific hepatic P450 enzyme (Fig. 3.4).

- Liver samples from both PCB treated (T) and untreated (C) rats are subjected to SDS-polyacrylamide gel electrophoresis to separate individual proteins based upon their molecular weights. While separated proteins are depicted in Figure 3.4, these proteins would not be visible without the use of some means of protein detection (i.e., protein staining).
- The separated proteins are transferred from the fragile acrylamide gel used in electrophoresis to a membrane (e.g., nitrocellulose) to which the proteins will bind.
- The remaining protein-binding sites on the membrane are blocked to prevent nonspecific binding of the antibodies to the membrane. This is generally accom-

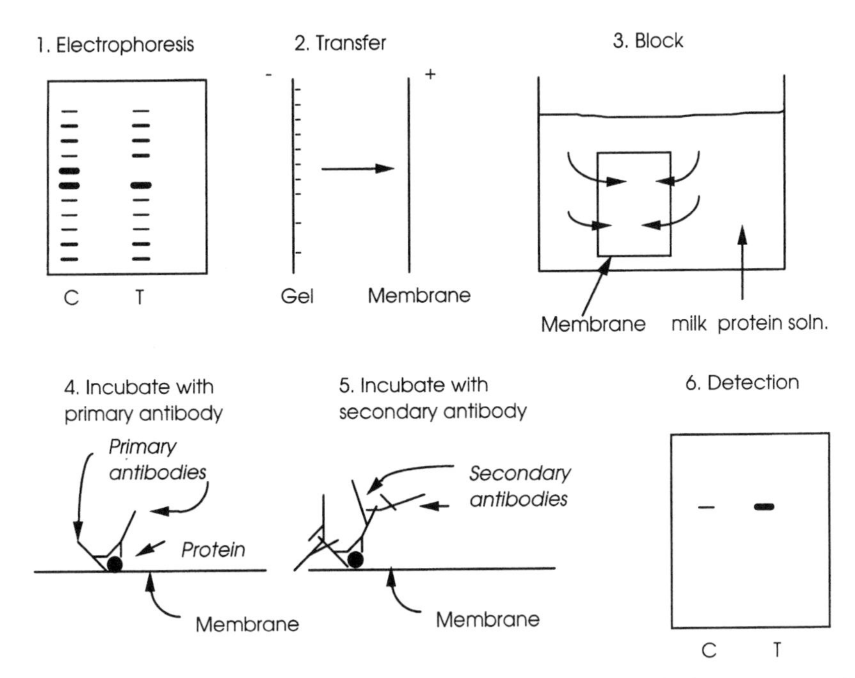

FIGURE 3.4 Diagrammatic representation of the use of immunoblotting to assess relative levels of a P450 protein following treatment of rats with a PCB. *C* = hepatic microsomal proteins from a control, untreated rat; *T* = hepatic microsomal proteins from a rat treated with PCBs.

plished by incubating the membrane in a solution containing a high concentration of proteins (solution of milk proteins, solution of serum proteins, etc.).

- Membrane is incubated with primary antibody. The binary complex consisting of the antigen on the membrane and the primary antibody forms.
- Membrane is incubated with secondary antibody. The tertiary complex consisting of the antigen-primary antibody-secondary antibody forms.
- The immune complex on the membrane is visualized using appropriate procedures (color development or chemiluminescence with alkaline phosphatase or horseradish peroxidase-conjugated secondary antibodies are most commonly used).

3.7.5 Radioimmunassay (RIA)

Radioimmunassay (RIA) is used to detect and quantify minute quantities of an antigen in biological or environmental samples. RIA is commonly used to measure nonprotein antigens such as drugs and toxicants. This is the most quantitative of the immunoassays. An example of the use of RIA in toxicology would be the analyses of dioxin levels in the blood of employees from a chlorophenol production plant.

To perform this assay, antibodies to dioxin would be required. Because the molecular weight of dioxin is less than 5000 Da, it would be necessary to use dioxin as a hapten to generate the antibodies. Various strategies can be used in RIA, although the antigen capture strategy is most commonly used (Fig. 3.5).

1. Known amount of dioxin antibodies bound to wells

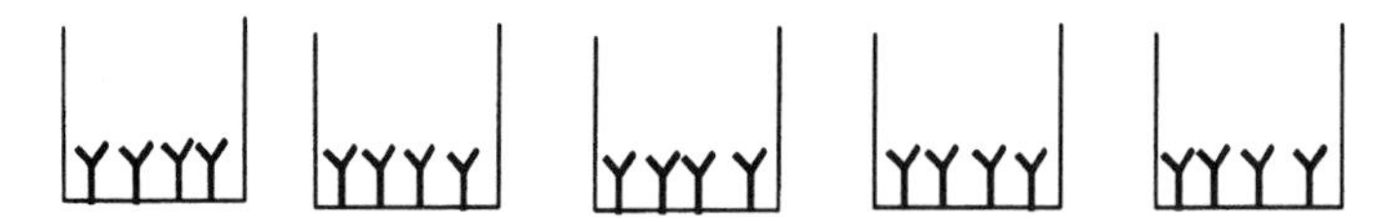

2. Fixed concentration of radiolabeled dioxin (•) and increasing concentrations of unlabeled dioxin (o) added to wells

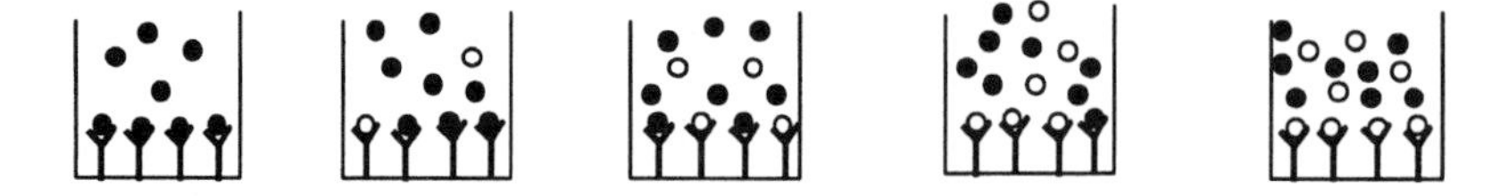

3. Wash out unbound dioxin and measure radioactivity associated with wells

4. Prepare a standard curve comparing the amount of unlabeled dioxin added to the wells and radioactivity (DPMs) measured in the wells

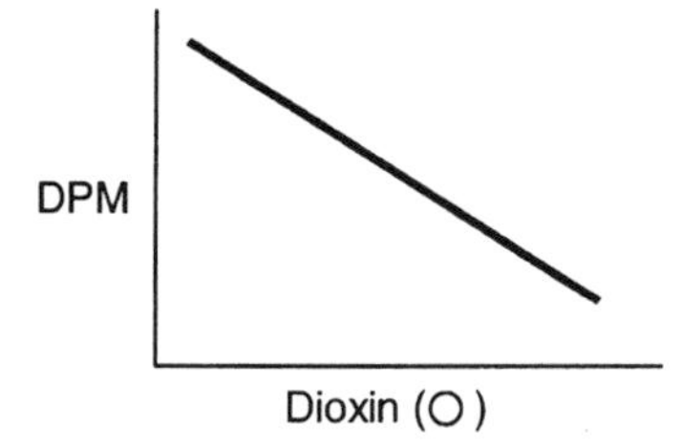

5. Incubate blood samples in wells (step 2) in place of unlabeled dioxin. Process through step 3. Determine concentration of dioxin in blood samples by using the standard curve to convert DPM to the concentration of dioxin in the sample.

FIGURE 3.5 Diagrammatic representation of the use of RIA for the quantification of dioxin levels in the blood of factory workers.

- A known amount of antibody is bound to a solid phase (i.e., PVC test tubes or wells of a microtiter plate).
- A known amount of [^{125}I]dioxin is added to a series of antibody-containing tubes. An increasing amount of unlabeled dioxin is added to each tube. The solutions are incubated during which time the iodinated dioxin and the non-iodinated dioxin will compete for binding with the immobilized antibodies.

- The solutions are removed, the tubes rinsed, and the radioactivity remaining in each tube measured. The radioactivity remaining in each tube will be equal to the amount of [^{125}I]dioxin bound by the antibodies in the tubes. The amount of bound [^{125}I]dioxin will decrease with increasing concentrations of unlabeled dioxin (due to increased competition for the antibody). The amount of bound radioactivity associated with each concentration of unlabeled dioxin can then be used to construct a standard curve.
- Incubate the blood sample in an antibody-containing tube along with the same amount of [^{125}I]dioxin used to develop the standard curve. Remove the solution and measure radioactivity associated with the tube. The level of radioactivity measured can then be applied to the standard curve to determine the amount of dioxin in the blood sample.

3.7.6 Enzyme-Linked Immunosorbant Assay (ELISA)

The ELISA encompasses a variety of assay designs for the analyses of either an antigen or antibody in a mixture using some means of immune complex immobilization and enzyme-mediated detection. The basic assay has been used as a rapid screening method for assessing antibody titer in plasma samples during antibody production, the detection of antigens in biological samples, and the quantification of chemicals (e.g., pesticides) in environmental samples. Although historically used as a qualitative measure of antigen or antibody presence, the assay has been modified to provide quantitative measures that rival RIA in some instances. Following are three examples of the use of ELISA:

Example 1: A rabbit has been injected with a protein with the intent of generating polyclonal antibodies to the immunogen. After adequate time for the generation of the immune response, a blood sample is drawn from the rabbit and serum is analyzed for the presence of the antibodies by ELISA (Fig. 3.6). The method is as follows:

- The antigen is bound to a solid phase (i.e., well of a PVC microtiter plate). This bound antigen will serve as the means of capturing the antibodies of interest from the serum. Remaining unoccupied protein binding sites are blocked with a protein-containing reagent (i.e., milk protein solution).
- Serum from the rabbit is added to the antigen-containing well and incubated. Antibody-antigen complexes will form during this time.
- Serum is removed and the well is thoroughly washed. After washing, the only serum proteins that will remain in the well are those bound to the immobilized antigen (i.e., the antibodies of interest).
- Enzyme-linked secondary antibody is added to the well. The solution is incubated for a sufficient time to allow for the tertiary complexes (antigen-primary antibody-secondary antibody) to form.
- The solution is removed, the well is washed thoroughly (to remove unbound secondary antibody), and the immune complex is assayed for using the appropriate colorimetric or luminescent substrate.
- The presence of the antibodies in the serum samples will be indicated by the generation of color or light.

1. Wells coated with antigen (●) remaining protein binding sites are blocked

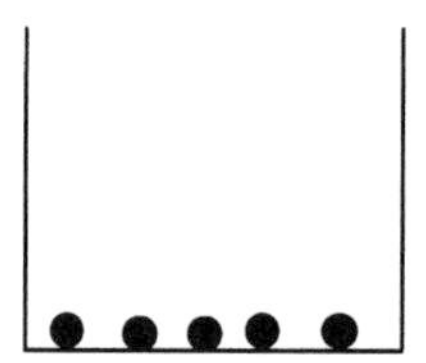

2. Serum from immunized rabbit added to well.
⅄=desired antibodies; 0 =other serum proteins

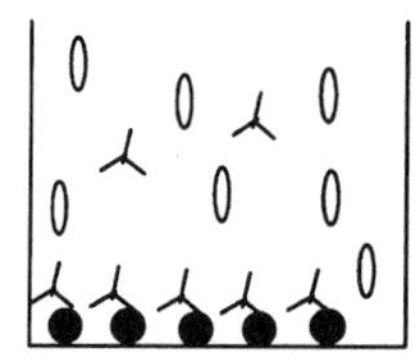

3. Wash out unbound proteins

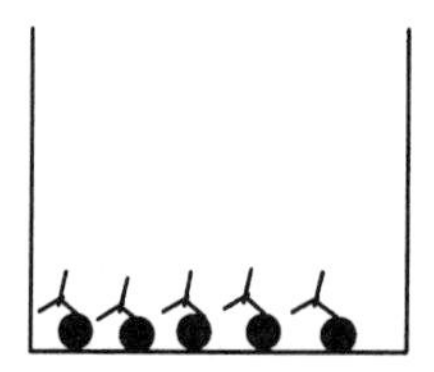

4. Add secondary antibody ( )

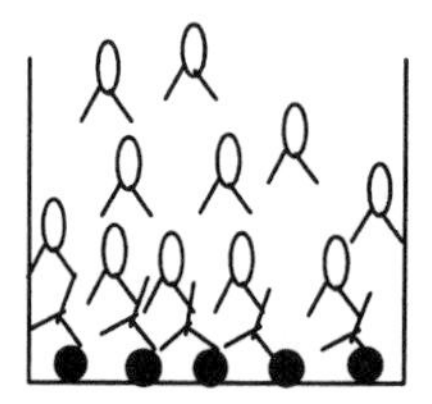

5. Wash out unbound secondary antibodies and assay for complexes

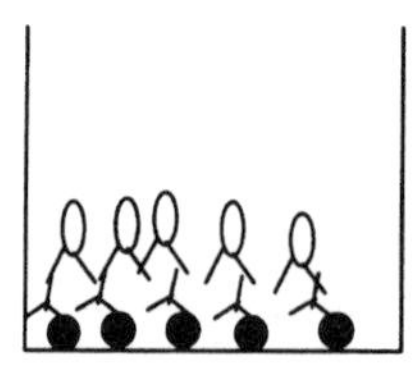

FIGURE 3.6 Diagrammatic representation of the use of ELISA to assess antibody titer in the serum of an immunized rabbit.

Example 2: Male fish have been collected from a reservoir suspected of being contaminated with an estrogenic chemical. Blood samples are drawn from the fish and are to be analyzed by ELISA for the presence of a protein that is induced by estrogenic compounds. The method is as follows:

- Sera collected from the fish is added to the wells of a PVC microtiter plate. Serum protein, including the antigen of interest, will bind to the PVC. Remaining unoccupied protein binding sites are blocked with a protein-containing reagent (i.e., milk protein solution).
- Antibodies to the antigen of interest are added to the wells and incubated. Antibody-antigen complexes will form during this time.
- Solutions are removed and the wells are washed thoroughly. After washing, the only antibodies that will remain in the wells are those bound to the immobilized antigen.
- Enzyme-linked secondary antibodies are added to the wells and incubated for a sufficient time to allow for the tertiary complexes (antigen-primary antibody-secondary antibody) to form.
- The antibody solutions are removed, the wells are washed thoroughly (to remove unbound secondary antibody), and the immune complexes are detected using the appropriate colorimetric or luminescent substrate.
- The presence of the estrogen-inducible protein in the serum samples will be indicated by the generation of color or light.

Example 3: Water samples have been collected from a farm pond that was found to contain deformed frogs. The water samples are to be analyzed by ELISA to determine the concentration of an agricultural fungicide that may be responsible for the deformities. A *competitive ELISA* in conjunction with a standard curve would be necessary due to the quantitative nature of the analyses (Fig. 3.7). The standard curve preparation is as follows:

- Increasing concentrations of the antigen (fungicide) are incubated with a fixed concentration of antibody. With increasing antigen concentration, more of the antibodies in the incubation solution will be complexed and inactivated.
- A fixed amount of antigen is bound to the wells of a microtiter plate. The remaining protein-binding sites in the wells are blocked by incubating with a protein-containing reagent (i.e., milk protein solution).
- Solutions containing the preincubated antibodies and antigen, from step 1, are transferred to the antigen-coated wells. Free (uncomplexed) antibody will bind to the antigen fixed to the well. The amount of free antibody available to bind to the fixed antigen in the well will decrease with increasing concentration of the antigen added to the preincubation solution.
- Wells are washed and secondary antibody is incubated in each well. The tertiary complex forms with the amount of complex determined by the amount of free primary antibody added to the well.
- The solutions are removed from the wells, the wells are washed thoroughly (to remove unbound secondary antibody), and the immune complexes are assayed

1. Standard curve preparation. Incubate increasing concentrations of antigen (●) with a fixed concentration of primary antibodies ( ).

2. Wells coated with antigen (●)

3. Solutions from step 1 added to wells prepared in step 2

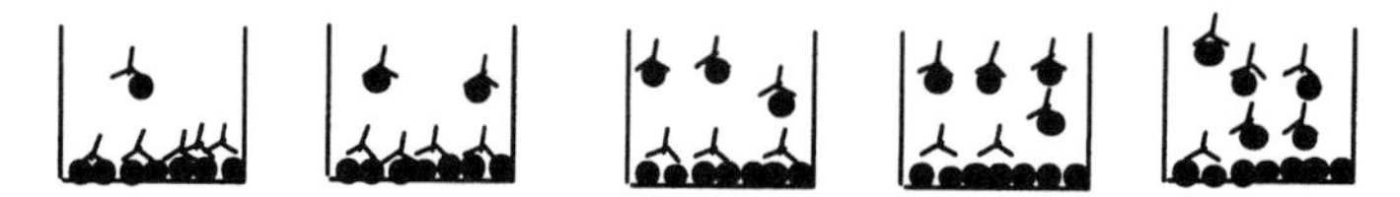

4. Wash out unbound proteins and add secondary antibodies ( ).

5. Wash out unbound secondary antibodies and assay complexes.

6. Prepare a standard curve consisting of amount of antigen added to tubes in step 1 to signal intensity in step

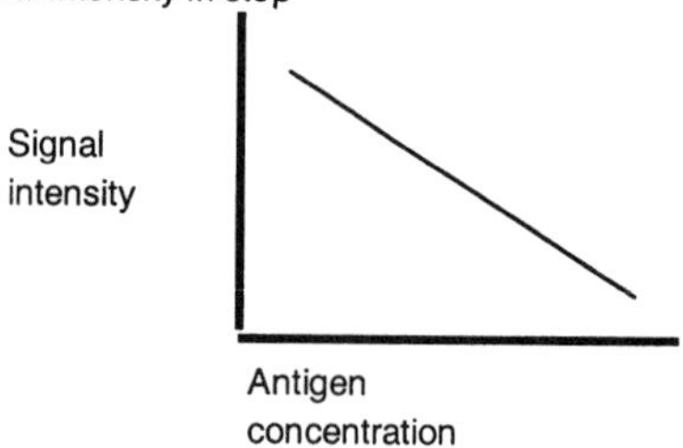

7. Incubate environmental samples in step 1 in place of antigen. Process through step 5. Determine concentration of antigen in samples by using the standard curve to convert signal intensity (step 5) to antigen concentration.

FIGURE 3.7 Diagrammatic representation of the use of a competitive ELISA for quantifying the amount of a fungicide in reservoir water samples.

using the appropriate colorimetric or luminescent substrate. The intensity of the detection signal will be inversely proportional to the original concentration of antigen used as part of the standard curve.

Following are sample analyses:

- Water samples containing an unknown amount of fungicide (antigen) are analyzed in the same manner as the standard curve with the water samples used in place of the known amount of antigen in the preincubations.
- Resulting intensity of the detection signal is compared with the standard curve and used to calculate the concentration of fungicide in the water sample.

3.8 CONCLUSIONS

A variety of immunoassays are available to toxicologists for the detection, characterization, and quantification of antigens. Antigens are most typically proteins, but through the use of haptens, antibodies can be generated to virtually any molecule. Such advances have allowed for the development of methods for the immunoquantification of toxicants, drugs, and other chemicals in biological (i.e., blood) and environmental (i.e., water) samples.

SUGGESTED READING

Harlow, E., Lane, D., *Antibodies: A Laboratory Manual*, Cold Spring Harbor Laboratory, New York, 1988.

Ferrone, S., Dierich, M.P. (Eds), *Handbook of Monoclonal Antibodies: Applications in Biology and Medicine*, Noyes Publications, Park Ridge, New Jersey, 1985.

Goers, J., *Immunochemical Techniques Laboratory Manual*, Academic Press, San Diego, California, 1993.

Price, C.P., Newman, D.J., (Eds), *Principles and Practice of Immunoassay*, Stockton Press, New York, 1997.

CHAPTER FOUR

Overview of Cellular Techniques in Toxicology

SHARON A. MEYER

4.1 INTRODUCTION

A variety of cellular techniques are useful tools for biochemical toxicology, especially in their provision of (1) context against which molecular changes associated with toxicant exposure can be related to toxic effects, (2) models for study of toxicodynamic mechanisms that are simplified, "living" systems relative to intact animals, (3) replacements for whole animal toxicity testing for some well-validated end points, and (4) human cells for comparison with target tissue effects of toxicants identified in animal testing. Techniques enabling evaluation of cellular toxic effects of molecular changes are an indispensable complement to nucleic acid-based screening techniques. Molecular changes ranging from increased expression of normal or mutated protein to ablation of protein function can be engineered into isolated cells or studied in cells isolated from engineered animals (see Chapter 2). Simplified cellular systems are amenable to relating cause and toxic effect and especially valuable for characterizing toxicant interactions within multicomponent pathways. Cellular changes that have been specifically associated with organismal toxic effects and exhibit high sensitivity may be useful surrogates allowing replacement of certain animal tests. Determination of similar toxic effects and mechanisms in human cellular systems and those of experimental animals argues relevance to human health, information important for toxicant risk assessment.

Limitations of cellular techniques are largely due to two factors (1) qualitative changes that occur in cells upon isolation from the organism and propagation and (2) difficulty in duplicating kinetic aspects of toxicant exposure as occurs in the intact animal. The most common alteration cells exhibit upon cultivation in culture is loss of differentiated function. This is a consequence of the conflicting need to enhance cell proliferation to provide adequate experimental material. Differentiated function of a particular cell type is partly determined by interaction with other cell types and noncellular tissue constituents; these are lost upon isolation and traditional cell monoculture. Also, undefined biological materials are routinely added to cell culture because they have been shown empirically to have mitogenic effects, but they also have often unappreciated effects on other cellular functions. With regard to toxicokinetics, one of the most challenging problems is relating effective toxicant concentrations in cell culture to that in vivo. Frequently, related toxic effects can be qualitatively induced in cultured cells and target cells in vivo, but toxic concentrations in culture often exceed those effective in vivo. Although sometimes interpreted as evidence for irrelevance of cell culture systems, this discrepancy can also result from the different milieus in vivo and in culture that influence the activity of toxicant at the cell. These toxicokinetic limitations can impose difficult problems for use of cell culture systems for quantitative information.

Although a major asset of cell culture is provision of context, that of the cell isolated from a multicellular organism (i.e., ex vivo) should closely mimic context within the intact organism. A modified toxic effect due to simplified, ex vivo context can be informative providing the basis for this change is understood and allows extrapolation to the in vivo response. Also, considerable progress in addressing many of the above limitations has been guided by discoveries of the fundamental properties of cell growth, differentiation, and their interaction. Many improvements have come from efforts to closely simulate the in vivo environment with the intent of maintenance of the differentiated state of cells in culture. Another approach is to propagate growth-competent cells and induce their differentiation in culture. Some tumor-derived cells and nontumorous cells transformed with inducible transforming proteins from DNA viruses, especially SV40 large T antigen, exhibit such conditional proliferation and differentiation within a specific lineage. Pluripotential embryonic stem (ES) cells offer this potential across multiple lineages. Cultured mouse ES cells have long been available and used primarily for generation of "knock-out" mice. Now that conditions for human cells have been defined, ES cells are likely to find greater application in biochemical toxicology.

The focus of this chapter is on technical aspects of isolation of cells from intact tissue, their propagation and maintenance in culture, and use in biochemical toxicology. General cellular mechanisms of toxic effect are the subject of Chapter 10. The methodology included is treated generally and is only a small collection from applied research on ex vivo cellular biology. For a more thorough description, the reader is referred to an excellent group of books by I. Fresney. Two informative web sites with practical cell culture information are maintained by the American Type Culture Collection (http://www.atcc.org) and the Industrial In Vitro Toxicology Group (http://www.invitro.org). There are also journals and serials devoted to in vitro toxicology, some of which are listed at the end of this chapter.

4.2 CELLULAR STUDIES IN INTACT TISSUE

4.2.1 Whole Animal Studies

Although the emphasis of this chapter is on isolated cells in culture, responses at the cellular level can be assessed in intact tissue after exposure of the whole animal. Observations of cellular responses are most often made with in situ detection techniques and microscopic observation, such as immunohistochemisty (Chapter 3) and nucleic acid hybridization (Chapter 2). Preparations used for these in situ techniques are generally tissue that has been fixed after toxicant treatment, then embedded and sliced thinly enough (~5 μm) to enable observation by microscopy, usually with transmission or fluorescent modes. Immunological and nucleic acid probes are available that allow observation of many proteins as modulated at transcriptional and post-transcriptional steps and of complex cellular processes such as cell proliferation and apoptosis. Probes for protein modification are available for such changes as signal transduction-induced phosphorylation, oxidant-induced reactions, and alkylation by reactive metabolites.

4.2.2 Tissue Slices

Tissue slices are an ex vivo system that offers simplicity relative to the organism, but retains some of the tissue-level complexity relative to isolated, cultured cells. Although earlier preparations suffered from extremely short viability, tissue slices have again become a popular source of cellular material since the introduction of vibrating, precision slicing instruments. Tissue slices have found greatest applicability in assessment of toxicant metabolic profiles, although loss of metabolic enzyme activity with time has limited this application to short-term studies (<24 hr). This limitation also restricts studies on toxic effects to immediate responses to acute exposures. Although the thickness of tissue slices (~200 μm) precludes observation with standard microscopic techniques at the cellular level, the newer development of confocal microscopy allows real-time observation of fluorescently tagged responses and should enable broader application of tissue slices in the study of toxic effects.

4.3 STUDIES WITH DISPERSED, ISOLATED CELLS

Suspensions of freshly isolated, dispersed cells commonly are used in short-term incubations for biochemical studies on intermediary metabolism. Similar systems are useful for biochemical toxicology, although the limited viability of these preparations restricts their application to acute exposures and measurement of relatively rapid responses. Alternately, toxicant exposure in suspension followed by monolayer culture appears to be a suitable method for study of activation-dependent toxic effects that require longer times for expression. Rodent hepatocytes are frequently used in this type of study because of the ease of isolation of large numbers of viable cells and because hepatocytes from numerous species, including humans, are readily available for comparative studies. A recent emphasis on improvement of cryopreservation techniques, driven largely by the potential for organ transplants from banked, isolated cells, will further improve availability.

4.3.1 Tissue Digestion and Cell Separation

Except for circulating blood cells and cells easily flushed by lavage, such as peritoneal and alveolar macrophages, preparation of isolated cells generally requires some means of release of individual cells from a solid tissue environment and separation from other dislodged tissue constituents. Cell-cell and cell-substratum interactions are the basic forces that maintain tissue architecture. Cell-cell interactions are several types of varying strength, but in general are mediated through protein binding, some of which is Ca^{2+}-dependent. Relatively strong intercellular connections, tight junctions and desmosomes, join cells of tight endothelia and epithelia, like duodenum. Several of the membrane proteins of these junctional complexes have been identified at the molecular level and their associations have been shown to be Ca^{2+}-dependent. In addition, adherens junctions that are dispersed along the lateral walls of tight epithelia and through the plasma membranes of nonepithelial cells mediate weaker cell-cell associations. The proteins of these structures, the cadherins and associated catenins, also require Ca^{2+} for binding.

Cell-substratum interactions occur at hemidesmosomal complexes. These structures contain the transmembrane α and β subunits of cell-type specific integrins. The extracellular ligands for the integrins are the proteins fibronectin or laminin embedded within a collagenous matrix. There are several α and β integrin, laminin, and collagen gene products and each integrin-fibronectin/laminin-collagen combination is cell-type specific. This association is Ca^{2+}-dependent and, for fibroblasts, relies on recognition by integrin of the arginine-glycine-aspartatic acid (RGD) tripeptide motif of fibronectin. Extracellular matrix also includes considerable amounts of carbohydrate polymers in glycosaminoglycan chains linked to the polypeptides of proteoglycans and as hyaluronic acid. Cellular integral membrane proteins that bind hyaluronic acid have been identified. Also, the transmembrane proteoglycan syndecans participate in cell-matrix adhesion by bridging cytoskeletal actin to various matrix components bound to their sticky glycosaminoglycans.

Because protein complex formation and Ca^{2+} are critical to cell fixation within a tissue, dissociation media usually contain some type of proteolytic enzyme and the Ca^{2+} chelator, EDTA. The proteolytic enzyme can be of general specificity, such as trypsin, or a more targeted enzyme, such as a collagenase selective for the collagen type characteristic of the tissue of interest. Hyaluronidase has been also used with matrix rich in hyaluronic acid, such as for isolation of duodenal enterocytes. In all cases, the appropriate incubation times and concentrations to achieve cell dispersal, but retain high viability, need to be determined empirically. One factor that compromises viability is the variable proteolysis of cells at different depths in the tissue, even when minced into small blocks ($\sim 1\,mm^3$). For some tissues, the inherent structure enables access of dissociation medium to deeper cells. The ease of preparation of hepatocyte suspensions in enabled by perfusion through the existing sinusoidal vasculature such that each hepatocyte is bathed on two sides with dissociation medium.

All solid tissues are composed of a variety of cell types. If the cells of interest are fibroblasts, traditional monolayer culture on plastic substratum in medium supplemented with fetal calf serum will select for this type of cell. If differential sensitivity of a tissue's cell types to any of the dispersal conditions exist, then these can be exploited to generate a suspension enriched in a given cell type. An example is the isolation of nonparenchymal cells from liver digested with pronase, which destroys parenchymal hepatocytes. However, isolation of a specific cell type more frequently requires application of separation techniques that exploit differences in physical or biological characteristics of the different cell types after dispersal. The most common method is velocity sedimentation (i.e., centrifugation at a fixed g-force and time) and separation is based upon size. Hepatocytes are approximately twice as large as nonparenchymal cells and thus can be cleanly separated by low-speed centrifugation. A better technique for cells of similar size entails centrifugation through a density gradient such that a bouyant force counters the centrifugational force. When equilibrium is achieved, cells with band at a density equivalent to the cellular density. Ficoll, Percoll, and Nicodenz are polymers that contribute density but not excessive osmotic strength, and are commonly used with cell suspensions.

Several methods can be used to achieve separation based upon cell type-specific surface markers. Surface markers can be tagged with fluorescent antibodies and cells separated with a fluorescence-activated cell sorter (FACS). Antibodies to cell surface markers bound to magnetic beads allow target cells to be pulled from suspension with a magnet. Ligands of surface receptors can be coated on tissue culture plasticware and adherent cells collected by panning. In some cases, the biology of the cell type can be used for selection. For example, hepatocytes are the only liver cells that will survive in arginine-free, ornithine-supplemented medium because arginase, a urea cycle enzyme exclusive to hepatocytes, is needed to synthesize the essential amino acid arginine.

4.3.2 Limited Maintenance in Defined Media

Once the cell type of interest is isolated, short-term culture in suspension will require choice of an appropriate incubation medium. In general, media requirements are relatively simple compared to formulations for longer-term monolayer culture. Minimally, suspended cells need salts to maintain osmolarity, a pH buffer, and an energy source. Usually NaCl, bicarbonate, and glucose serve these needs in the standard balanced salt solutions (e.g., Earle's, Hanks'). It is also necessary to provide a gas phase of oxygen and CO_2. The latter buffers the solution upon equilibration with bicarbonate anion and is the system of choice because cells independently require some bicarbonate for metabolic processes. Oxygen (95%)/5% CO_2 gas is usually provided as the head space in a gas-impermeable container, but this arrangement introduces pH instability if interim samples are to be taken from a common vessel. To compensate, media may contain other solution-based buffers such as HEPES. Also, oxygen delivery from the gaseous phase can be diffusion-limited, and thus shallow solutions and gentle mixing are required.

4.3.3 Long-Term Suspension Culture

Cell types that are normally nonadherent, such as lymphocytes, and some tumor cells are able to grow in suspension when the medium is supplemented with adequate nutrients. For small-scale cultures, cells can simply be seeded onto standard tissue culture plasticware and maintained in a shallow layer of medium in a CO_2 incubator. If necessary, cell attachment can be prevented by coating the plastic surface and reducing the medium Ca^{2+}. Suspension cultures can also be propagated in roller bottles, and larger vessels (spinner flasks) have been designed for hundreds of milliliters of stirred culture medium. Growth in suspension culture is advantageous for ease of propagation of large quantities of material. This feature is the impetus for adaptation of adherent cells to suspension culture by seeding them onto microcarriers. In this way, their growth requirement for attachment is met, yet large quantities can be efficiently cultured using suspension techniques, such as in industrial bioreactors.

4.4 MONOLAYER CELL CULTURE

Proliferation of most cells in culture requires attachment to a substratum. The most common means to achieve this is to seed cells onto the plastic base of tissue cultureware. Tissue culture plastic is usually polystyrene modified to carry a charged surface to which cells can initially attach, then elaborate an underlying matrix and spread to form a monolayer. Growth and maintenance of cultured cells for longer than a few hours also requires additional nutrients, which are added to bicarbonate- and glucose-supplemented basal salt solutions in various formulations as standard tissue culture media. Necessary nutrients are essential amino acids, vitamins, additional salts, and trace minerals. Many cells also require serum for optimal growth, although this contributes considerable variability to cell culture because serum is largely undefined and its constituents vary from sample to sample. Serum replacements have been formulated from the identified serum components known to support growth and, for those cells that will adapt to serum-free media, their use can remove some of the empiricism of cell culture. Some of the more important serum constituents are insulin, growth factors, the nutrients ethanolamine and pyruvate, selenium for glutathione peroxidase synthesis, and transferrin for provision of bioavailable iron. All standard culture media, except L15, use bicarbonate buffering; thus, a CO_2 gas phase also is needed for monolayer culture. In this application, CO_2 is mixed at ~5–10% volume with ambient air (~20% O_2) in a tissue culture incubator, a specialized cabinet that maintains humidity and temperature, as well as CO_2. Phenol red is also included in standard media to enable convenient visual assessment of pH during culture.

4.4.1 Propagation in Culture

The monolayer culture that results after seeding a cell suspension from dissociated tissue is called a primary culture. Those cells that are growth competent initially will exhibit "log-phase" growth during which they proliferate exponentially, providing that their medium is replenished every few days (Fig. 4.1). When the surface area of the vessel bottom is covered (i.e., at confluence), the closely apposed cells will then arrest their growth, a phenomenon called "contact inhibition." To resume proliferating, the cells must be replated onto a larger surface area per cell. This procedure, called subculturing, involves detachment of adherent cells from the substratum and dissociation to a single-cell suspension, then reseeding into a larger vessel or reseeding a portion of the cells in a vessel of the same size. The detached, dissociated cell suspension is generated using a limited protease digestion, usually with trypsin, and EDTA. This procedure is repeated, with each repetition referred to as a "passage" and with the ratio of the surface area per cell after subculture to that before subculture as the "split ratio."

After numerous passages, cells will no longer proliferate and cultures will begin to deteriorate. This property, "senescence," is thought to be a manifestation of a predetermined limit on the number of proliferative cycles available to a cell. During senescence of the culture, rare cells may arise that exhibit resistance, continue to proliferate, and overgrow the culture. These "immortalized" cells then become the founders of continuous cell lines. Intense research has been conducted to define the

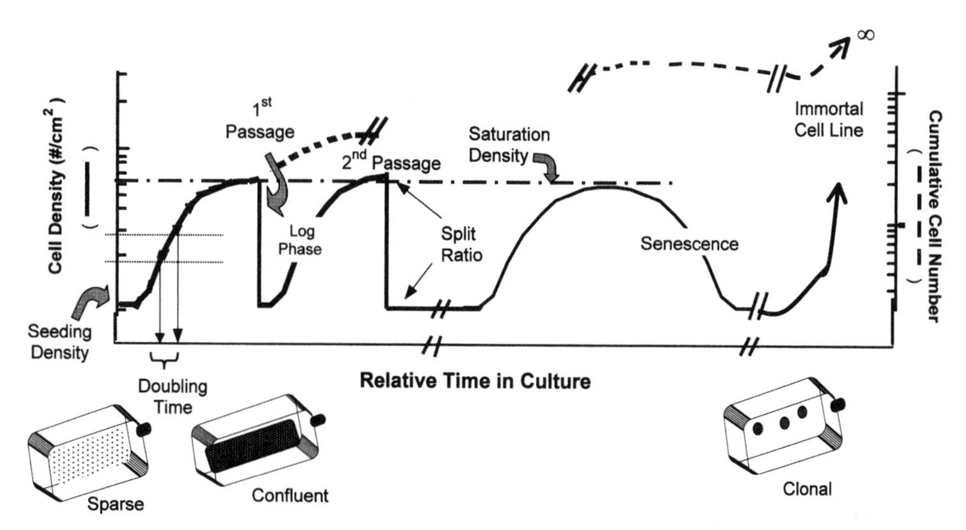

FIGURE 4.1 Propagation of cells in monolayer culture. Both cell proliferation as an increase in number of cells/cm^2 within a passage (—) and as cumulative cells from all passages (- - -) are shown. Tissue culture flasks at various stages of monolayer growth are diagrammed below the graph to indicate the corresponding stage of confluency. Note the logarithmic scaling of the ordinate.

molecular basis of senescence inasmuch as its occurrence has profound implications for aging and carcinogenesis. One of the best supported concepts is that chromosomal ends, telomeres, shorten with each replicative cycle until chromosome loss limits viability. Cells upon immortalization then derepress expression of the enzyme telomerase that restores chromosome length. Rodent cells have a low rate of spontaneous immortalization that can be substantially increased by transfection with the oncogenes of certain DNA tumor viruses. DNA tumor viral oncogenes are not sufficient to immortalize human cells, but they do condition human cell cultures to facilitate additional complementary changes in rare cells that enable them to escape senescence. The most commonly used viral oncogene for immortalization is the large T antigen of SV40.

Once an immortalized cell line has been established, it is possible to obtain derivatives or "clones" that originate from a single cell. These can be generated by plating a cell suspension diluted to theoretically yield one cell per small vessel or by sparsely plating on a larger plate and subsequently collecting isolated cell clusters with cloning cylinders. Random clones can be isolated from normal medium with the objective of generating a homogenous cell line or can be isolated from conditions that impose a selective pressure, such as resistance to a given toxicant. In the latter case, there will be genotypic changes characteristic of the clonal cell line that enabled its survival in the selective medium. The number of cells that survive in selective medium is increased by mutagenic and clastogenic events and quantitation of clonal survival is a frequently used parameter of genotoxicity (see Chapter 16).

Immortalized cells can be induced to undergo additional changes in phenotype. Transformation is a change that reflects a further loss of growth control from immortalization and is highly correlated with tumorigenicity of cells when they are transplanted into animal hosts. Transformation has been an invaluable tool for identification of carcinogenic chemicals and oncogenes and for mechanistic studies on carcinogenesis (Chapter 15). Properties of transformed cells in culture are (1) a spindle-shaped morphology, (2) reduced serum requirement, (3) loss of contact inhibition, and (4) anchorage-independent growth. Loss of contact inhibition is apparent as small (~2 mm diameter) piles of cells or foci that overlie the monolayer. Anchorage-independent growth is operationally defined as the ability to grow in soft agar, which precludes contact with a substratum. Another phenotypic change that can be induced in several immortalized cell lines is expression of differentiated properties. Examples include cell lines that express differentiated properties of neurons, hepatocytes, adipocytes, adrenocortical cells, and renal tubule epithelial cells. Differentiation can be induced by a variety of hormonal and chemical agents such as cyclic AMP, glucocorticoids, butyrate, retinoic acid, and DMSO. Selected examples of application of differentiated cell lines in biochemical toxicology are presented in Table 4.1.

4.4.2 Modifications to Monolayer Cell Culture

The tendency to lose differentiated function in culture has been one driving force for innovative improvements in traditional monolayer cell culture. Provision of substrata to mimic that in vivo has improved maintenance of differentiated function in some cases. For example, primary rodent hepatocytes retain sensitivity to phenobarbital induction of the cytochrome P4502B1 when cultured on matrix that contains mostly laminin with some fibronectin and proteoglycan. A synthetic polymer with attached repeats of the RGD integrin-binding motif has also been developed. Polarity of transporting epithelial cells, such as MDCK kidney cells, is sustained when they are cultured on permeable membranes inserted into the medium of traditional tissue culture plates. Culture in three dimensions, such as with multicellular aggregates or on alginate microcarriers, has improved differentiation of some cells. Culture in the presence of other cell types (i.e., co-culture), can also improve viability and differentiation as can culture of one cell type in medium conditioned by exposure to another cell type.

Co-culture also provides a system in which questions about cell-cell interactions can be addressed. If a toxic effect in a tissue requires cooperation between cell types, then considerable information about the mechanism of this interaction can be obtained through use of co-culture constructs. For example, an effect may be observed when different cell types are admixed, but not when cell types are physically separated and only communicate through a common medium, as can be achieved by plating one type on the vessel bottom and another on a membrane insert. This result would argue against involvement of a soluble mediator and suggest a mechanism involving direct contact of the different cell types. An elaborate three-dimensional co-culture system for human skin keratinocytes layered upon a synthetic mesh infiltrated with dermal fibroblasts, when floated to allow contact of the uppermost keratinocytes with air, exhibits stratification remarkably similar to in vivo squamous epithelia. It has been anticipated that this reconstructed epithelial model can be used as an in vitro replacement for ocular irritation testing, but thus far validation studies have yielded mixed results.

TABLE 4.1 Examples of Application of Cell Lines Retaining Differentiated Properties in the Study of Toxic Effects.

Cell Line	Source	Differentiated Cell Type	Toxicant	Measured End Point
N1E-115	Mouse neuroblastoma	Cholinergic neuron	Lead	Blockage of voltage-dependent Ca^{2+} channels
			Pyrethroid insecticide	Prolonged open time for voltage-dependent Na^{+} channels
PC12	Rat pheochromocytoma (adrenal medullary tumor)	Adrenergic neuron	Tricresyl phosphate (organophosphate)	Inhibition of neurofilament assembly and axonal growth
SK-N-SH	Human neuroblastoma	Neuron	Anesthetic N_2O	Depressed cholinergic Ca^{2+} signaling
Hepa-1	Mouse hepatoma	Hepatocyte	2,3,7,8-tetrachloro-dibenzodioxane (TCDD)	Induction of CYP 1A1 and 1B1
HII4 E	Rat hepatoma	Hepatocyte	Polychlorinated biphenyls (PCBs)	Induction of CYP 1A1
HepG2	Human hepatoblastoma	Hepatocyte	Cyclophosphamide (antineoplastic)	Cytochrome P450-dependent genotoxicity
3T3-L1	Mouse embryo fibroblasts	Adipocyte	TCDD	Inhibition of glucose transport and lipoprotein lipase
Y1	Mouse adrenocortical tumor	Adrenocortical cell	Methyl sulfone metabolites of DDT and PCBs	Inhibit corticosterone synthesis by competitive inhibition of CYP
LLC-PK1	Pig kidney	Renal tubule epithelial cell	Cadmium	Cytotoxicity, apoptosis
MDCK	Dog kidney	Renal tubule epithelial cell	Organic mercury compounds	Cytotoxicity, transepithelial leakiness

4.5 OBSERVATION OF CULTURED CELLS

Real-time observation of living cells is invaluable for monitoring the progress of cell cultures. However, living cells are translucent and thus require techniques other than standard bright-field microscopy for visualization. Phase contrast microscopy is the preferred technique for routine observation of monolayer cultures and relies upon the change in phase that occurs when white light, made coherent by a phase ring in the microscope condenser, passes through living cells due to differences in the refractive index. These phase changes can be observed as differences in brightness when the transmitted light rays constructively or destructively interfere when passed through a second phase ring in the objective lens of a phase contrast microscope. Also, because cell monolayers are viewed while attached to the bottom of a culture vessel, conventional microscopic design does not provide a large enough working distance. Thus, the inverted phase contrast microscope is used, in which the relative position of objective lenses and condenser is reversed. With this design, light enters from the top and the image is collected underneath the specimen.

The recent development of cell-permeable fluorescent tags has provided a means to observe alterations in cell function in addition to morphological changes. Inverted fluorescent microscopes are used for these studies, and various stage attachments can be added to allow control of temperature and other variables. Fluorescent tags are available for monitoring oxidant status, sulfhydryl content, intracellular Ca^{2+}, H^+, Na^+ and K^+, mitochondrial function, and membrane potential. Digital electronic imaging and computerized data analysis can enhance the sensitivity of this technique and provide information on temporal relationships between multiple responses within a single cell. A disadvantage of fluorescence imaging is poor resolution inasmuch as light emitted above and below the plane of focus is also collected in the objective lens. However, this problem can be circumvented with the use of laser-scanning confocal light microscopy, which can optically limit the image to thin slices within the depth of the monolayer. Confocal microscopy has also enabled visualization of three-dimensional structures and fluorescent imaging of systems other than cell monolayers, such as tissue slices and multicellular aggregates.

4.6 INDICATORS OF TOXICITY

4.6.1 Intact Tissue

Overt cell death in intact tissue is recognized microscopically by loss of definition of cell margins and organelles with the vacated space filled with amorphous material (coagulative necrosis) and/or lymphocyte infiltrate. Nuclear disintegration is easily recognized as karyolysis. Apoptosis is a less extensive form of cell death and can be recognized microscopically as isolated cellular residues, sometime phagocytosed within neighboring cells, or after labeling for associated internucleosomal DNA strand breaks. Biochemical parameters of cytotoxicity are used to assess viability of tissue slices. In general, leakage into the incubation medium of normally intracellular components is used. The most commonly monitored parameters are K^+, ATP, and the cytosolic enzyme lactate dehydrogenase (LDH).

4.6.2 Cell Culture

Parameters used to measure toxicity of cultured cells can be general properties related to loss of vital functions or impairment of cell type-specific functions. These indicators can be further classified as those occurring shortly after toxicant exposure and those requiring prolonged culture for expression.

Assessment of general vital functions is done within minutes to hours of toxicant exposure and is based upon monitoring performance of critical organelles. These endpoints are schematically represented in Fig. 4.2. The most definitive indicator is cell death that occurs upon rupture of the plasma membrane. However, several parameters of dysfunction can be observed prior to this terminal event. Microscopically, various membrane extrusions are observed. The formation of filopodia, pseudopodia and "blebs," that is, ballooned regions devoid of intracellular organelles, are indicative of disruption of plasma membrane interaction with the underlying cortical cytoskeleton. Damaged plasma membranes will also exhibit leakiness. This can be monitored as the egress of normally impermeant molecules, such as LDH or $^{51}Cr^{2+}$, which is preloaded in healthy cells as permeant $^{51}Cr^{3+}$ and trapped upon reduction. Alternately, influx of the exogenous markers trypan blue or propidium iodide, which concentrate in the nucleus after breaching the plasma membrane barrier, is used.

Function of mitochondria is also commonly monitored as an indicator of cellular toxicity. Mitochondrial uptake and retention of the fluorescent dye rhodamine 123 can be visualized microscopically. Biochemical measurements of mitochondrial function include the ATP:ADP ratio and dehydrogenase activity with MTT (3-(4,5-dimethylthiazol-2-yl)-2,5-diphenyltetrazolium bromide), which yields a colored formazan product upon reduction. The dye neutral red (3-amino-7-

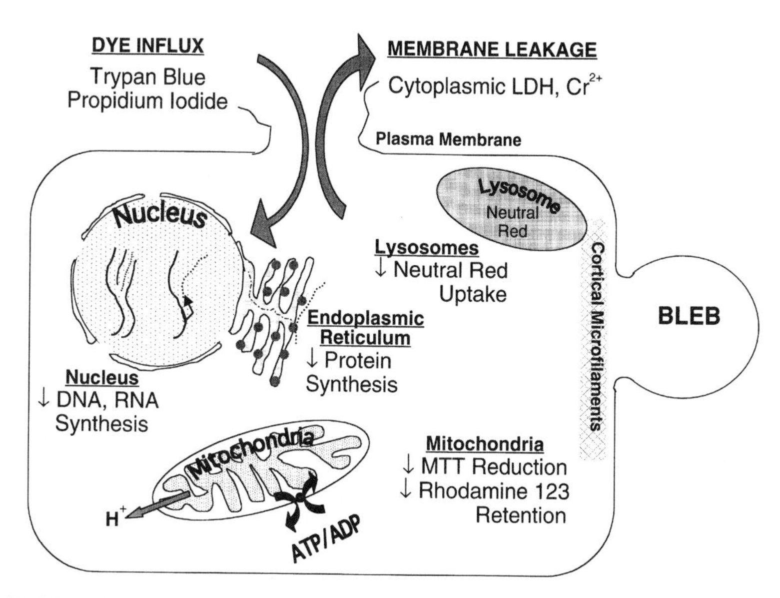

FIGURE 4.2 Idealized diagram of a cell to illustrate parameters often used to measure cytotoxicity and the corresponding affected subcellular organelle.

dimethyl-amino-2-methylphenazine hydrochloride) targets lysosomes and its retention is inversely related to cytotoxicity. Commercially available versions of the MTT and neutral red assays have been adapted to microtiter plate formats to provide highly efficient screening assays. Examples of how cell type-specific functions can be followed as indicators of cell toxicity are included in Table 4.1.

Longer-term assays of cytotoxicity include assessment of general cell function. Toxicant effects on cellular protein, RNA, and DNA synthesis can be measured as incorporation of radioactivity in macromolecules of cells prelabeled with cell permeant, radioactive precursors, e.g., [^{35}S]methionine, [^{3}H]uracil, and [^{3}H]thymidine. Cell proliferation can simply be monitored by counting cell numbers with a hemocytometer or Coulter counter. Assessment of cell proliferation data is usually determined as rate of log-phase growth, as expressed as time for cell number to double (doubling time). Number of cells per area at stationary phase (saturation density) may also be affected. These parameters are indicated in Fig. 4.1. A more stringent test of growth competency is afforded by assessment of clonal growth in which isolated colonies that survive and grow out from a known number of sparsely plated cells are scored. Apoptosis is measurable in cultured cells using DNA end-labeling or morphology. Cell culture techniques are also used for a subset of genotoxicity tests. Clonal growth in selective media is used for mutagenicity studies, and observations on cells arrested in mitosis can be used to assess larger chromosomal abnormalities. Excision repair of DNA damage is also observable in cultured cells as nonreplicative nuclear incorporation of radiolabeled deoxynucleoside precursor ("unscheduled DNA synthesis," UDS). The theoretical basis and interpretation of these types of genotoxicity assays are detailed in Chapter 16.

4.7 REPLACEMENT OF ANIMAL TESTING WITH CELL CULTURE MODELS

The above discussion has largely been focused upon techniques enabling provision of cell culture models for mechanistic study of toxic effects. The greatest priority in model development to achieve this objective is maintenance of target tissue differentiated function. However, another major thrust driving the use of cell culture models in toxicology is the promise that they hold for replacement of animals in toxicity testing. Much of the momentum for in vitro toxicology arose because of the ethical questions inherent in imposition of potential pain, distress, and lethality on nonhuman, higher species for the purpose of predicting human risks from chemical exposure. However, another equally compelling reason to incorporate cell culture models is that they can be used to address the limitations in extrapolation of toxic effects in nonhuman species to humans through the so-called "parallelogram" approach (Fig. 4.3).

Several academic, government, and industrial organizations have been formed to address the potential of "alternative" methods in toxicity testing. Alternative test methods have been defined as procedures that accomplish the "3 Rs": replacement, reduction, or refinement of use of higher animals. Several cell culture systems are being considered as alternative test methods, and comprehensive, interlaboratory studies designed to validate their results against historical animal results are in

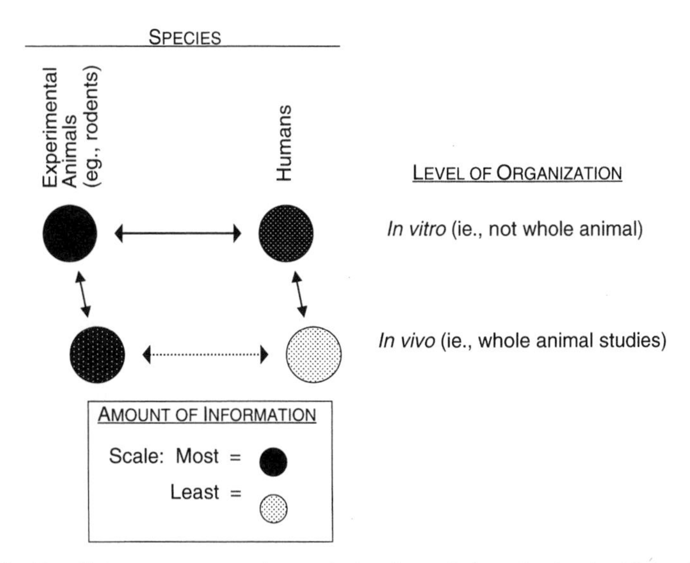

FIGURE 4.3 Parallelogram approach to relating knowledge obtained with in vitro systems from experimental animals and humans to in vivo observations with experimental animals and extrapolation to predict human in vivo responses.

Source: Modified from Sobels, *Arch. Toxicol.* 46 (1980), 21–30.

progress. In general, these efforts have revealed that qualitatively there is often good concordance between the in vitro and in vivo methodologies for chemicals at the extremes of toxicity and nontoxicity. However, ranking of chemicals with more similar toxicities is less well predicted and quantitative relationships are difficult to relate to in vivo exposures. Thus, it appears that the most beneficial application of cell culture models as an alternative will be as early screens in tiered protocols for product safety testing. These in vitro systems appear adequate to identify highly hazardous compounds for which no further animal testing may be needed to justify curtailment of new product development or withdrawal of marketed products.

Another application of cell culture expected to impact on toxicity testing is the use of engineered cell lines for hazard identification in high-throughput screening protocols. Knowledge gained from basic research identifying mechanisms and molecular mediators of certain toxic effects has been utilized in constructing reporter systems for certain classes of chemicals. A noteworthy example is the use of cell lines designed to detect estrogenic or androgenic activities of chemicals. It has recently been mandated that chemicals be tested for "endocrine disrupting activity." The magnitude of this task, which is predicted to require assessment of over 60,000 chemicals in the near future, has necessitated prioritization based upon prescreening results. For this purpose, cell lines have been engineered to contain a vector with a reporter gene whose expression is responsive to activation of a cotransfected steroid hormone receptor. Activation or inhibition of the steroid receptor activity by a test compound can then be efficiently determined by monitoring the reporter gene product. Similar engineered cell lines are available for detecting dioxin-like compounds through their interaction with the Ah receptor and others that are sensitive to heavy metals, oxidative stress, and DNA damage.

4.8 CONCLUSION

Acceptance of cell culture techniques in biochemical toxicology has been slow relative to other basic science fields. The most important limitation contributing to this hesitancy has been the loss of differentiated function in culture, primarily loss of components necessary for metabolic activation. However, these problems are resolving as discoveries about the basic biology of cell differentiation are translating into modifications to cell culture methodology that support maintenance of metabolic function. Consequently, several cell culture systems are now available for study of metabolism-dependent toxic effects. Also, the increased availability of cell culture systems from humans has substantially improved the predictability of chemical toxicity through metabolic profiling. Perhaps the last frontier will be application of cell culture systems in toxicology testing. At present, some limited applications have gained acceptance and practical and ethical issues concerning animal usage will most certainly stimulate further attempts to develop and validate in vitro toxicology models.

SUGGESTED READING

Freshney, R.I. *Culture of Animal Cells. A Manual of Basic Technique*, 3rd ed., John Wiley & Sons, New York, 1994.

Freshney, R.I. *Freshney's Culture of Animal Cells: A Multimedia Guide*, John Wiley & Sons, New York, 1999.

Freshney, R.I. *Culture of Epithelial Cells*, John Wiley & Sons, New York, 1992.

Freshney, R.I. *Culture of Hematopoietic Cells*, John Wiley & Sons, New York, 1994.

Freshney, R.I. *Culture of Immortalized Cells*, John Wiley & Sons, New York, 1996.

Jones, G.E. *Human Cell Culture Protocols*, Humana Press, Totowa, NJ, 1996.

Kedderis, G.L. Extrapolation of in vitro enzyme induction data to humans in vivo. *Chem. Biol. Interact.* 107 (1997), 109.

Lemasters, J.J., Gores, G.J., Nieminen, A.-L., Dawson, T.L., Wray, B.E. and Herman, B. Multiparameter digitized video microscopy of toxic and hypoxic injury in single cells. *Environ. Hlth. Perspec.* 84 (1990), 83.

O'Hare, S., Atterwill, C.K. *In Vitro Toxicity Testing Protocols*, Humana Press, Totowa, NJ, 1995.

Tyson, C.A., Frazier, J.M. (Eds) *In Vitro Biological Systems*, Methods in Toxicology, Vol. 1A, Academic Press, San Diego, CA, 1993.

Tyson, C.A., Frazier, J.M. (Eds) *In Vitro Toxicity Indicators*, Methods in Toxicology, Vol. 1B, San Diego, CA, 1994.

Walum, E., Stenberg, K. and Jenssen, D. *Understanding Cell Toxicology. Principles and Practice*, Ellis Horwood, New York, 1990.

Watson, R.R. *In Vitro Methods of Toxicology*, CRC Press, Boca Raton, FL, 1992.

Journals

In Vitro Cellular & Developmental Biology—Animal, journal of the Society for *In Vitro* Biology; publisher: SIVB, Largo, MD.

In Vitro Toxicology, A Journal of Molecular and Cellular Toxicology, journal of the Industrial In Vitro Toxicology Group; publisher: Mary Ann Liebert, Larchmont, NY.

Toxicology In Vitro; in association with BIBRA; publisher: Elsevier, New York.

Cell Biology and Toxicology; publisher: Princeton Scientific Publishers, Princeton, NJ.

CHAPTER FIVE

Metabolism of Toxicants: Phase I Reactions and Pharmacogenetics

ERNEST HODGSON and JOYCE A. GOLDSTEIN

5.1 INTRODUCTION

The majority of xenobiotics that enter the body tissues are lipophilic, a property that enables them to penetrate lipid membranes and to be transported by lipoproteins in body fluids. The metabolism of xenobiotics, carried out by a number of relatively nonspecific enzymes, usually consists of two phases. During phase I, a polar group is introduced into the molecule, and although this increases the molecule's water solubility, the most important effect is to render the xenobiotic a suitable substrate for phase II reactions. In phase II reactions, the altered compounds combine with an endogenous substrate to produce a water-soluble conjugation product that is readily excreted.

Although this sequence of events is generally a detoxication mechanism, in some cases the intermediates or final products are more toxic than the parent compound, and the sequence is termed an activation or intoxication mechanism. See Chapter 9 for discussion of activation and toxicity.

Although most phase I reactions are oxidations, reductions, or hydrolyses, the hydration of epoxides and dehydrohalogenations also occur. A summary of the most important reactions is shown in Table 5.1.

5.2 MICROSOMAL MONOOXYGENATIONS

Microsomal monooxygenation reactions are catalyzed by nonspecific enzymes such as the flavin-containing monooxygenases (FMOs) or the multienzyme system that has cytochrome P450s as the terminal oxidases.

5.2.1 Microsomes and Monooxygenations: General Background

Microsomes are derived from the endoplasmic reticulum of the cell as a result of homogenization of the tissue and are isolated by ultracentrifugation of the

TABLE 5.1 Phase I Xenobiotic-Metabolizing Enzymes with Examples of Substrates.

Enzymes	Examples
Cytochrome P450 (P450)	
Epoxidation and hydroxylation	Aldrin, benzo(a)pyrene, aflatoxin, bromobenzene
N-Dealkylation	Ethylmorphine, atrazine, dimethylnitrocarbamate, dimethylaniline
O-Dealkylation	*p*-Nitroanisole, chlorfenvinphos, codeine
S-Dealkylation	Methylmercaptan
S-Oxidation	Thiobenzamide, phorate, endosulfan, methiocarb, chlorpromazine
N-Oxidation	2-Acetylaminofluorene
P-Oxidation	Diethylphenylphosphine
Desulfuration	Parathion, fonofos, carbon disulfide
Dehalogenation	CCl_4, $CHCl_3$
Nitro reduction	Nitrobenzene
Azo reduction	O-Aminoazotoluene
Flavin-containing monooxygenase (FMO)	
N-Oxygenation	Nicotine, dimethylaniline, imipramine
S-Oxygenation	Thiobenzamide, phorate, thiourea
P-Oxygenation	Diethylphenylphosphine
Desulfuration	Fonofos
Prostaglandin synthetase (PGS) cooxidation	
Dehydrogenation	Acetaminophen, benzidine, DES, epinephrine
N-Demethylation	Dimethylaniline, benzphetamine, aminocarb
Hydroxylation	Benzo(a)pyrene, 2-aminofluorene, phenylbutazone
Epoxidation	7,8-dihydrobenzo(a)pyrene
Sulfoxidation	Methylphenylsulfide
Oxidations	FANFT, ANFT, bilirubin
Molybdenum hydroxylases (aldehyde oxidase, xanthine oxidase)	
Oxidations	Purines, pteridine, methrotrexate, quinolones, 6-deoxycyclovir
Reductions	Aromatic nitro compounds, azo dyes, nitrosamines, N-oxides, sulfoxides
Alcohol dehydrogenase	
Oxidations	Methanol, methanol, isopropanol, glycols, glycol ethers (2-butoxyethanol)
Reductions	Aldehydes and ketones
Aldehyde dehydrogenase	
Oxidations	Aldehydes from alcohol and glycol oxidations
Esterases and amidases	Paraoxon, dimethoate, phenyl acetate
Epoxide hydrolase	Benzo(a)pyrene epoxide, styrene oxide
DDT-dehydrochorinase	*p,p*-DDT
Glutathione reductase	disulfiram

postmitochondrial supernatant fraction. The endoplasmic reticulum is a continuous anastomosing network of lipoprotein membranes that extends from the plasma membrane to the nucleus and mitochondria. The microsomal fraction derived from the endoplasmic reticulum, however, consists of membranous vesicles contaminated with free ribosomes and fragments of mitochondria, nuclei, Golgi apparatus, etc. In the hepatocyte the surface area of the endoplasmic reticulum is 37 times that of the plasma membrane and 8.5 times that of the outer mitochondrial membrane.

The endoplasmic reticulum and, consequently, the microsomal vesicles, consist of two types, rough and smooth, the former having the outer surface studded with ribosomes, which the latter characteristically lack. Although both rough and smooth microsomes have all the components of the cytochrome P450-dependent monooxygenase system, the specific oxidative activity of the smooth fraction is usually considerably higher.

Although microsomes have been prepared by several methods, the method of choice remains homogenization followed by differential centrifugation in which a microsomal pellet is sedimented by ultracentrifugation of the post mitochondrial (10,000 g for approximately 10 minutes) supernatant at about 100,000 g for 30–120 minutes. The microsomal pellet is then resuspended in a reaction medium suitable for the measurement of oxidative reactions or for spectral examination. Rough and smooth microsomes can be separated, if necessary, by a further two-step density gradient centrifugation.

Monooxygenations are those oxidations in which one atom of molecular oxygen is reduced to water while the other is incorporated into the substrate. The term "mixed function oxidase" was formerly used for enzymes that catalyze such oxidations; "monooxygenase" is now the term of choice.

Because the electrons involved in the reduction of cytochrome P450 (P450) are derived from NADPH, the overall reaction can be written as follows (where RH is the substrate):

$$RH + O_2 \xrightarrow{NADPH + H^+ \;\rightarrow\; NAD^+} ROH + H_2O$$

or alternatively:

$$RH + O_2 \xrightarrow{\text{Reduced P450} \;\rightarrow\; \text{Oxidized P450}} ROH + H_2O$$

5.2.2 Constituent Enzymes of the Cytochrome P450-Dependent Monooxygenase System and the P450 Reaction Mechanism

Cytochrome P450, the carbon monoxide-binding pigment of microsomes, is a hemoprotein of the b cytochrome type. Unlike most cytochromes it is named, not from the absorption maximum in the visible region of the reduced cytochrome, but rather from the wavelength of the carbon monoxide derivative of the reduced cytochrome, which has an absorption maximum at 450 nm. Direct proof of the role of P450 as the terminal oxidase in monooxygenase reactions has been afforded, in recent years, by the demonstration that monooxygenase systems, reconstituted from purified or recombinant P450, NADPH-cytochrome P450 reductase, and phosphatidylcholine, can catalyze monooxygenase reactions.

As discussed later, many isozymes of P450 exist, often in the same species and organ, each of which is coded for by a different gene. Although substrate specificity, relative to either xenobiotics or endogenous substrates, may vary, the basic oxidative mechanisms, as regards oxidation states and electron transport, is the same. Individual isozymes may be highly specific, oxidizing only a single substrate, or relatively nonspecific, oxidizing a range of related compounds. The latter situation is generally the case for those isozymes oxidizing xenobiotics.

Cytochrome P450, like other hemoproteins, has a characteristic absorption in the visible region. Addition of many organic and some inorganic ligands results in a perturbation of this spectrum. Because the particulate nature of microsomes gives rise to light scattering, and the light scattering is a function of wavelength, such changes in the absolute spectrum are seen as fluctuations along a sloping baseline—an undesirable situation for quantitative measurements. In difference spectroscopy, light-scattering, nonspecific absorption and the absolute spectrum of the microsomal cytochromes are balanced out by placing microsomes in both cuvettes of a split-beam spectrophotometer; then only the difference caused by the addition of a ligand to the microsomes in the sample cuvette is seen as a difference spectrum. Although the detection and measurement of such spectra requires a high-resolution instrument, spectral studies have been of tremendous use in the characterization of P450.

The most important difference spectra of oxidized P450 are type I, with an absorption maximum at 385–390 nm and a minimum around 420 nm, and type II, with a peak at 420–435 nm and a trough at 390–410 nm.

Type I ligands are found in many different chemical classes and include many drugs, environmental compounds, insecticides, etc. They appear to be generally unsuitable on chemical grounds as ligands for the heme iron and are believed to bind at a hydrophobic site in the protein in close enough proximity to the heme iron to allow both perturbation of the absorption spectrum and interaction with the activated oxygen. Although the vast majority of type I ligands are substrates, it has not been possible to demonstrate a quantitative relationship between Ks (concentration required for half-maximal spectral development) and Km (Michaelis constant), even with purified P450 isozymes.

Type II ligands, on the other hand, interact directly with the heme iron of P450 and are associated with organic compounds having nitrogen atoms with sp^2 or sp^3 nonbonded electrons that are sterically accessible. Such ligands are frequently inhibitors of P450-dependent monooxygenase activity.

The most important difference spectra of reduced P450 are the well-known carbon monoxide spectrum, with its maximum at or about 450 nm, and the type III spectrum, with two pH-dependent peaks at approximately 430 and 455 nm. The best known and most investigated type III ligand for P450 is ethyl isocyanide. This is an unstable interaction, however, because the ligand can be readily displaced. Compounds such as the methylenedioxyphenyl synergists and SKF-525A form stable type III complexes that appear to involve covalent binding and to be related to the mechanism by which they inhibit monooxygenase reactions.

Reducing equivalents are transferred from NADPH to P450 by a flavoprotein enzyme known as NADPH-cytochrome P450 reductase (P450 reductase). The evidence that this enzyme is involved in microsomal monooxygenases was originally derived from the observations that cytochrome c, which can function as an artificial electron acceptor for the enzyme, is an inhibitor of such oxidations. Furthermore, phenobarbital, a known inducer of P450, also causes a parallel induction of the reductase. Direct evidence, including inhibition of monooxygenase activity by antibodies to NADPH-P450 reductase, has shown that P450 reductase is an essential component of monooxygenase systems reconstituted from purified or recombinant components.

The purified reductase is a flavoprotein of approximately 80,000 Da that contains 1 M each of flavin mononucleotide (FMN) and flavin adenine dinucleotide (FAD) per 1 M of enzyme. The only other component necessary for activity in reconstituted systems is a phospholipid, phosphatidylcholine. This substance is not directly involved in electron transfer but appears to be involved in the coupling of the reductase and the cytochrome and in the binding of the substrate of the cytochrome. In effect it functions as a "lipid membrane."

Although the mechanism of P450 function has not been established unequivocally, the generally recognized steps are summarized in Figure 5.1. The initial step consists of the binding of substrate to oxidized P450 followed by a one-electron reduction catalyzed by P450 reductase to form a reduced cytochrome-substrate complex. The next several steps are less well understood. They involve an initial interaction with molecular oxygen to form a ternary oxygenated complex. This ternary complex accepts a second electron, resulting in the further formation of one or more poorly understood complexes. One of these, however, is probably the equivalent of the peroxide anion derivative of the substrate-bound hemoprotein (Fig. 5.1). Under some circumstances this complex may break down to yield hydrogen peroxide and the substrate-oxidized cytochrome complex. After the transfer of one atom of oxygen to the substrate and the other to form water, dismutation reactions occur that lead to the formation of the hydroxylated product, water and the oxidized cytochrome.

The possibility that the second electron is derived from NADH through cytochrome b_5 has been the subject of discussion for some time and is still to be completely resolved. It is clear, however, that this pathway is not essential for all P450-dependent monooxygenation reactions because many occur in systems reconstituted from NADPH, O_2, phosphatidylcholine, P450 reductase, and P450. Nevertheless, evidence is available that this pathway can occur under some circumstances or with particular substrates and that it may, in fact, facilitate oxidative activity in the intact endoplasmic reticulum.

It has also been suggested that the oxygenated complex may decompose to yield the substrate-oxidized cytochrome complex and superoxide anion. If this is the case, it is of importance in such manifestations of superoxide anion toxicity as lipid peroxidation.

5.2.3 Multiplicity of P450: Purification and Reconstitution of the P450-Dependent Monooxygenase System

Even before appreciable purification of P450 had been accomplished, it was already apparent from indirect evidence that mammalian liver cells contained more than one P450 enzyme. More direct evidence was obtained when solubilized microsomal

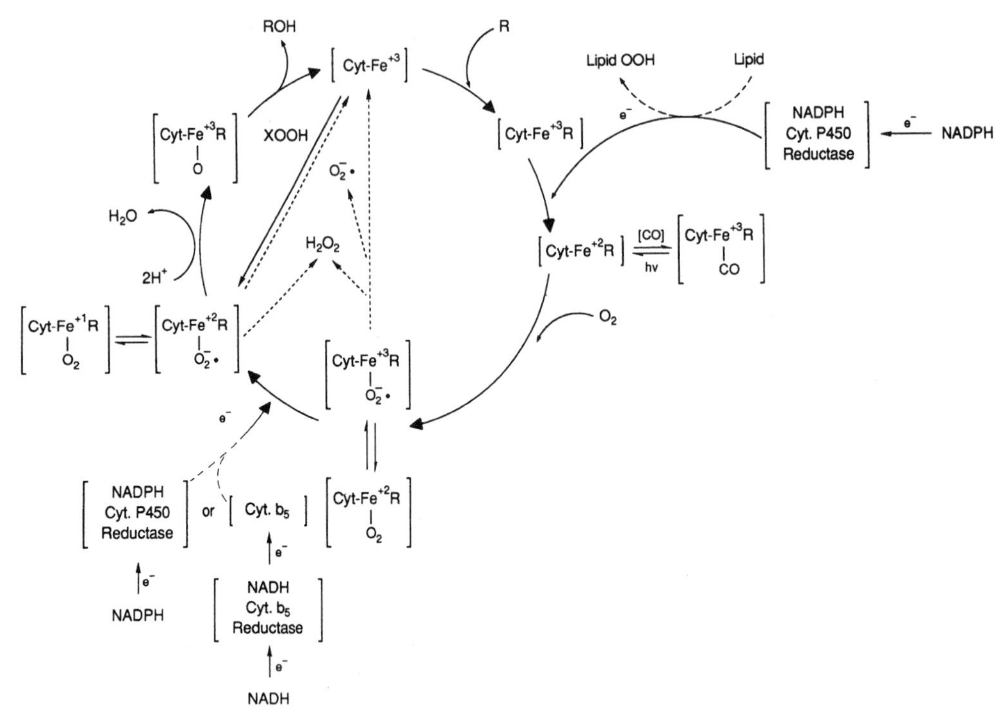

FIGURE 5.1 Generalized scheme showing the sequence of events for cytochrome P450 monooxygenations.

preparations could be separated into multiple P450s chromatographically, and could be distinguished from each other by their differing immunochemical properties, substrate specity, immunochemical characteristics, and by differences in their apparent molecular weights following sodium dodecyl sulfate polyacrylamide gel electrophoresis (SDS-PAGE).

Considerations of multiplicity may be of considerable importance in several aspects to be discussed in subsequent chapters, including developmental changes, gender differences, inhibition, and induction. Any of these effects on P450 or monooxygenase activity could involve either changes in the proportion of specific cytochromes or specificity differences between cytochromes.

Purification of P450 was, for many years, an elusive goal. The initial difficulty involved the instability of P450 on solubilization, which resulted in degradation of P450 to P420, an enzymatically inactive form. This problem was solved by the use of glycerol and dithiothreitol as protectants. As a result of the hydrophobic nature of P450, the protein has a tendency to aggregate upon solubilization from the lipid membrane. This is partially overcome by maintaining a low concentration of a suitable detergent, such as Emulgen 911, throughout the purification procedure. Multiple forms have been purified from liver and other tissues using a combination of chromatographic methods in the presence of glycerol and Emulgen 911. Detergent is then removed before catalytic activity can be assessed.

Monooxygenations of xenobiotics have been demonstrated frequently in systems reconstituted from purified P450, P450 reductase, and phosphatidylcholine.

Such systems will, in the presence of NADPH and oxygen, oxidize xenobiotics such as benzphetamine, often at rates comparable to microsomal oxidations. The fact that this minimum number of constituents is enzymatically active does not mean, however, that other microsomal constituents, such as cytochrome b_5, may not facilitate activity in vivo or that all substrates can be oxidized by this minimal system.

One important finding from such studies is that the lack of specificity of the hepatic microsomal monooxygenase system is not due to the presence of a mixture of several specific P450s because it appears that many of the cytochromes isolated are relatively nonspecific. The relative activity toward different substrates does, however, vary greatly from one isoform to another. This is illustrated in Table 5.2, a compendium of the most important human P450s and their known substrates.

As discussed in Chapter 2, the techniques of molecular biology have been applied extensively to P450. Many cDNAs have been sequenced (477 as of January 2, 2000) and the nucleotide sequences and derived amino acid sequences compared. In some cases the location of the gene on a particular chromosome has been determined and the mechanism of gene expression investigated.

A system of nomenclature based on protein sequence was proposed in 1987 and updated most recently in 1996. Under the most recent update of this system, P450 genes are designed CYP (or cyp in the case of mouse genes). This is followed by an arabic numeral designating the gene family, a letter designating the subfamily when more than one subfamily exists within the family, and finally an arabic numeral designating the individual gene. If there are no subfamilies or if there is only a single gene within a family or subfamily the letter and/or the second numeral may be omitted (e.g., CYP17).

The protein sequence of any member of a gene family is equal to or greater than 40% similar to that from any member of the same gene family. Protein sequences within subfamilies are greater than 55% similar in the case of mammalian genes and 46% in nonmammalian genes. In cases examined so far, genes from the same subfamily lie on the same chromosome within the same gene cluster and are nonsegregating, suggesting a common origin through gene duplication events. Sequences showing less than 3% divergence are arbitrarily designed allelic variants unless other evidence exists to the contrary.

The gene products, the P450 isozymes, may still be designated P450 followed by the same numbering system used for the genes. Common names may still be used (although this is not recommended) provided no use is made of hyphens, subscripts, or superscripts. An example of this classification system is shown in Table 5.3 and the rate and extent of P450 evolution is shown in Figure 5.2. The known sequences fit the scheme surprisingly well, only a few exceptions are found at the family, subfamily, or allelic variant levels, and in each case additional information is available to justify the departure from the rules set out.

In some cases a homolog of a particular P450 enzyme is found across species (CYP1A1). In other cases, the genes diverged subsequent to the divergence of the species and no exact analog is found in various species (e.g., the CYP2C subfamily). In this case, genes are numbered in the order of discovery and the gene products from a particular subfamily may even have differing substrate specificity in different species (e.g., rodent vs. human).

TABLE 5.2 Human CYP Enzymes and Their Substrates.

Subfamily	Enzyme	Substrate
CYP1A	1A1	Polycyclic aromatic hydrocarbons
	1A2	Caffeine, N-hydroxylates aromatic amines to mutagens, phenacetin
CYP2A	2A6	Coumarin, (methylnitrosamino)-1-(3-pyridyl)-1-butanone (NNK), a nitrosamine found in tobacco smoke, nicotine
CYP1B	1B1	Polycyclic aromatic hydrocarbons such as dimethyldibenzanthracene, 17 β-estradiol
CYP2B	2B6	4-hydroxylation of cyclophosphamide, *S*-mephenytoin-demethylation, certain barbiturates, 7-ethoxy-4-trifluoromethylcoumarin (7EFC)
CYP2C	2C19	*S*-Mephenytoin (anticonvulsant), omeprazole (antiulcer), proguanil (antimalarial), certain barbiturates, diazepam (valium), propranolol, (β-blocker), imipramine (antidepressant), certain herbicides
	2C9	Phenytoin (anticonvulsant), anti-inflamatory drugs, e.g., ibuprofen, warfarin (anticoagulant), tolbutamide (diabetic drug)
	2C8	Taxol (breast cancer drug), retinoic acid
CYP2D	2D6	Antihypertensive: debrisoquine β-blockers: metoprolol, propranolol, bufuralol Antidepressants: nortriptyline, desipramine, clomipramine Neuroleptics: thioridizine, perphenazine, trifluperidol, colzapine Opiates: O-demethylates codeine to morphine
CYP2E	2E1	Alcohol, carbon tetrachoride, benzene, drugs such as acetaminophen (Tylenol), activates nitrosamines to mutagens and carcinogens
CYP3	3A4	Aflatoxin B_1 aldrin, 60% of clinically used drugs including erythromycin (antibiotic), nifedipine (antihypertensive), lidocaine (anesthetic), (antihistamine terfenadine), cyclosporine (immunosuppresive drug), 17α-ethnylestradiol (replacement estrogen), Tamoxifen (used for breast cancer), lovastatin (a cholesterol-lowering drug), dapsone (leprosy), testosterone, cortisol (hormones)
	3A5	Overlapping substrate specificity with CYP3A4
	3A7	Fetal form; metabolized 3A substrates
CYP4A	4A11	Fatty acids, clofibrate (hypolipidemic drug)

5.2.4 Distribution of Cytochrome P450

In vertebrates, the liver is the richest source of P450 and is also the most active organ in the monooxygenation of xenobiotics. P450s and the other components of this monooxygenase system are found also in skin, lung, nasal mucosa, and gastroin-

TABLE 5.3 Examples of Nomenclature for Cytochrome P450 Genes and Gene Products.

Gene Symbol	Gene Product	Trivial Name	Species
CYP1A1	CYP1A1 or P4501A1	c, βNF-B	Rat
		P_1, c, form 6	Human
		form 6	Rabbit
		1A1	Trout
		Dah1	Dog
		Mkah1	Monkey
Cyp1a-1	CYP1A1 or P4501A1	P_1	Mouse
CYP1A2	CYP1A2 oe P4501A2	P-448, d, HCB	Rat
		P_3, d, form 4	Human
		LM_4	Rabbit
		MC4	Trout
		Dah2	Dog
		Mkah1	Monkey
Cyp1a-2	CYP1A2 or P4501A2	P_2, P_3	Mouse

testinal tract, presumably reflecting the evolution of defense mechanisms of portals of entry. In addition to these organs, P450s have been demonstrated in kidney, adrenal cortex and medulla, placenta, testes, ovaries, bladder, fetal and embryonic liver, nervous system, corpus luteum, aorta, and blood platelets. In humans, P450s have been demonstrated in fetal and adult liver, lung, placenta, kidney, testes, fetal and adult adrenal gland, skin, blood platelets, and lymphocytes.

Although P450s are found in many tissues and organs, their substrate specificities differ. The liver cytochromes carry out a large number of xenobiotic oxidations as well as the oxidation of some endogenous steroids and bile pigments. Figure 5.3 shows the relative distribution and importance for the metabolism of clinical drugs of the major P450 isoforms in human liver. The P450 forms in the lung appear to be concerned primarily with xenobiotic oxidation, although the range of substrates tested has been more limited than in the case of the liver. While skin and small intestine also carry out xenobiotic oxidations, very few activities have been studied to any extent. Usually, human placental microsomes display little or no ability to oxidize foreign compounds, appearing to function primarily as a steroid hormone-metabolizing system. In the placenta from women who smoke, however, CYP1A1-dependent aryl hydrocarbon hydroxylase activity is readily apparent.

The P450 of the kidney is active in the ω-oxidation of fatty acids such as lauric acid and arachidonic acid. Although the renal cytochromes bind some xenobiotics, the renal P450s are relatively inactive in the oxidation of most xenobiotics. P450s in extrahepatic organs such as the heart are involved predominantly in the metabolism of endogenous substrates such as arachidonic acid.

Distribution of P450 within the cell has been studied primarily in the mammalian liver, where it is present in greatest quantity in the smooth endoplasmic reticulum and in smaller, but still appreciable, quantities in the rough endoplasmic reticulum. Other cell membranes may also possess smaller amounts of P450. The nuclear membrane, for example, has been reported to contain P450 and to have detectable aryl hydrocarbon hydroxylase activity, an observation of great importance in studies of

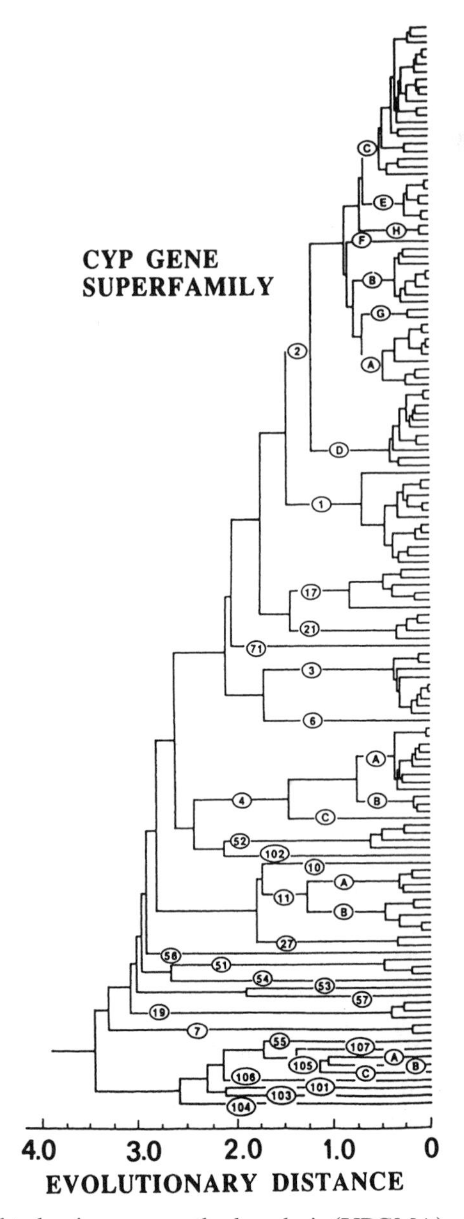

FIGURE 5.2 Unweighted-pair-group method analysis (UPGMA) of the P450 superfamily.

the activation of chemical carcinogens. Traces of P450 have been reported also in lysosomes and Golgi apparatus, while its presence is virtually undetectable in plasma membranes. However, certain steroid and vitamin hydroxylases (e.g., vitamin D) occur in hepatic mitochondria and mitochondrial P450s in the adrenal cortex and placenta have specialized functions in steroid hormone metabolism. In the brain, xenobiotic-metabolizing P450s occur in both mitochondrial and microsomal membranes.

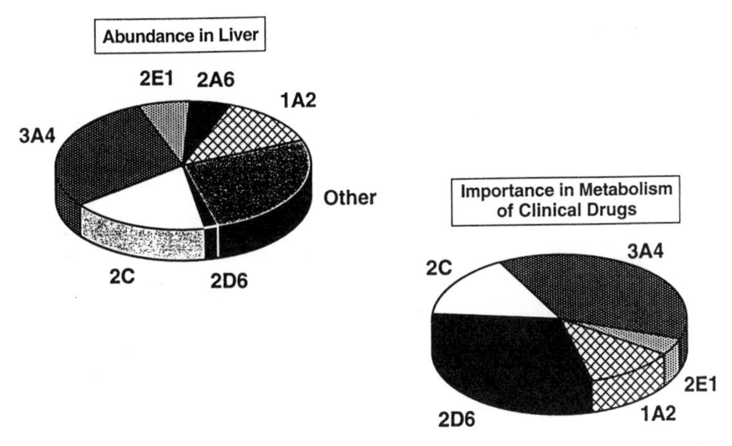

FIGURE 5.3 The relative abundance of cytochrome P450 isoforms in human liver and their importance in drug metabolism.

5.2.5 P450-Dependent Monooxygenase Reactions

Although microsomal monooxygenase reactions are basically similar with respect to the role played by molecular oxygen and in the supply of electrons, the enzymes are markedly nonspecific, both substrates and products falling into many different chemical classes. It is convenient, therefore, to classify these activities on the basis of chemical reactions, bearing in mind that, not only do the classes often overlap, but the same substrate may undergo more than one oxidative reaction.

5.2.5a Epoxidation and Aromatic Hydroxylation

Epoxidation is an extremely important microsomal reaction because not only can stable epoxides be formed but arene oxides, the epoxides of aromatic rings, are intermediates in aromatic hydroxylations. In the case of polycyclic hydrocarbons, the reactive arene oxides are known to be involved in carcinogenesis.

The epoxidation of aldrin to dieldrin is one of the best known examples of the metabolic formation of a stable epoxide (Fig. 5.4A). The oxidation of naphthalene, on the other hand, was one of the earliest understood examples of an epoxide as an intermediate in aromatic hydroxylation (Fig. 5.4B). The epoxide can rearrange nonenzymatically to yield predominantly 1-naphthol, can interact with the enzyme epoxide hydrolase to yield the dihydrodiol, or can interact with glutathione S-transferase to yield the glutathione conjugate that ultimately is metabolized to a mercapturic acid. This reaction is also of importance in the metabolism of the insecticide carbaryl, which contains the naphthalene nucleus.

The proximate carcinogens arising from the metabolic activation of benzo(a)pyrene are isomers of benzo(a)pyrene 7,8-diol-9,10-epoxide (Fig. 5.4C). These metabolites appear to arise by the prior formation of the 7,8-epoxide, which gives rise to the 7,8-dihydrodiol through the action of epoxide hydrolase. The diol is further metabolized by the microsomal monooxygenase system to the 7,8-diol-9,10-epoxides, which are both potent mutagens and unsuitable substrates for the further action of epoxide hydrolase.

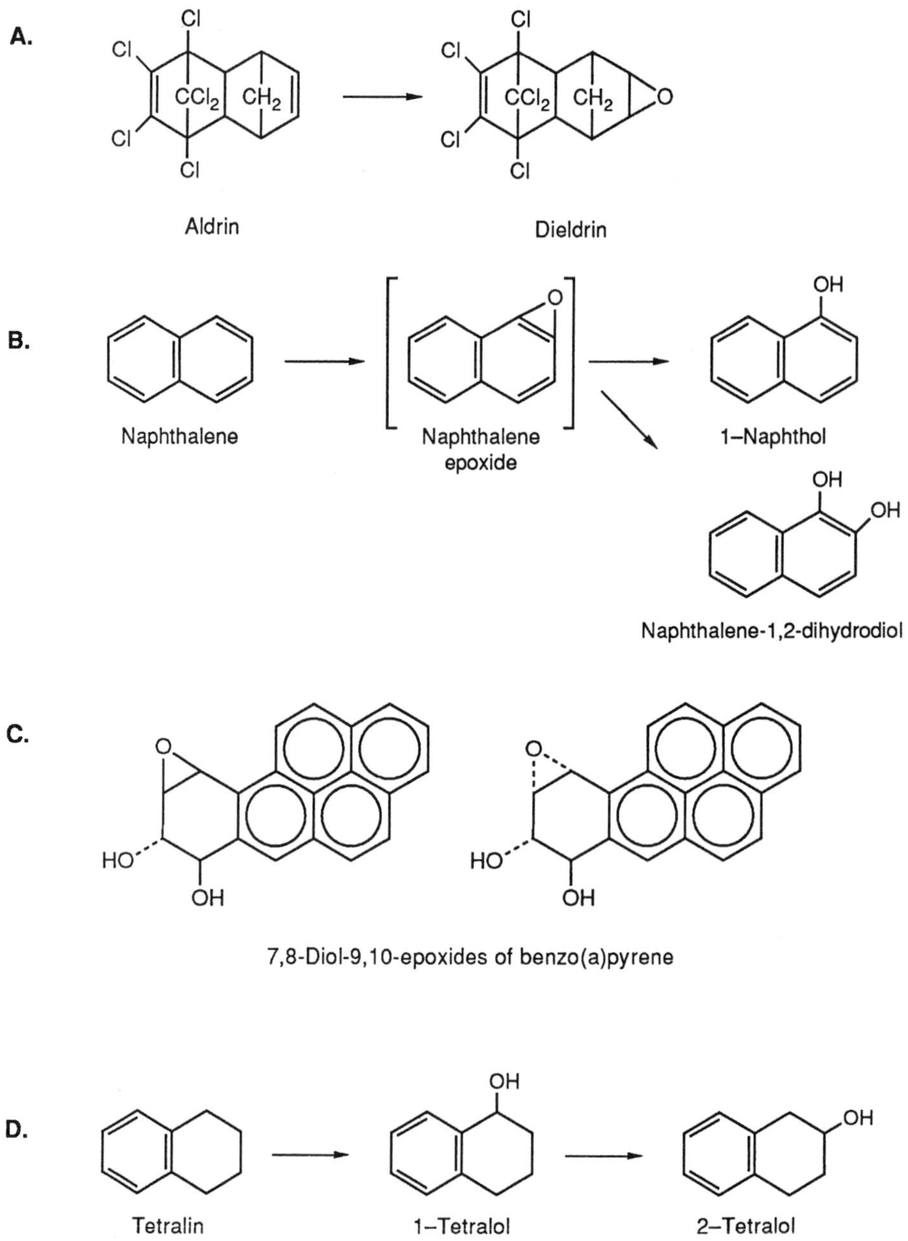

A. Aldrin → Dieldrin

B. Naphthalene → [Naphthalene epoxide] → 1–Naphthol; Naphthalene-1,2-dihydrodiol

C. 7,8-Diol-9,10-epoxides of benzo(a)pyrene

D. Tetralin → 1–Tetralol → 2–Tetralol

FIGURE 5.4 Examples of epoxidation and aromatic hydroxylation reactions.

5.2.5b Aliphatic Hydroxylations

Alkyl side chains of aromatic compounds are readily oxidized, often at more than one position, and provide good examples of this type of oxidation. In the rabbit the n-propyl side chain of n-propylbenzene can be oxidized at any of the three carbon atoms to yield 3-phenylpropan-l-ol ($C_6H_5CH_2CH_2CH_2OH$) by ω-oxidation, benzylmethylcarbinol ($C_6H_5CH_2CHOHCH_3$) by ω-1-oxidation and ethylphenylcarbinol

($C_6H_5CHOHCH_2CH_3$) by ω-2-oxidation. Further oxidation of these alcohols is also possible.

Alicyclic compounds, such as cyclohexane, are also susceptible to oxidation, in this case first to cyclohexanol and then trans-cyclohexane-1,2-diol.

In compounds with both saturated and aromatic rings, the former appears to be the most readily hydroxylated. For example, the major oxidation products of tetralin (5,6,7,8-tetrahydronaphthalene) in rabbit are 1- and 2-tetralol, whereas only a trace of the phenol, 5,6,7,8-tetrahydro-2-naphthol, is formed (Fig. 5.4D).

5.2.5c Dealkylation: O-, N-, and S-Dealkylation

Probably the best known example of O-dealkylation is the demethylation of *p*-nitroanisole. Due to the ease with which the colored product *p*-nitrophenol, can be measured, *p*-nitroanisole is frequently used for the demonstration of monooxygenase activity. The reaction is thought to proceed via an unstable methylol derivative (Fig. 5.5A). Other substrates which undergo O-dealkylation include the drugs codeine and phenacetin and the insecticide methoxychlor.

The O-dealkylation of organophosphorus triesters differs from the above reactions in that it involves the dealkylation of an ester rather than an ether. The reaction was first described for the insecticide chlorfenvinphos (Fig. 5.5B), but is now known to occur with a wide variety of vinyl, phenyl, phenylvinyl, and naphthyl phosphates and the thionophosphate triesters. At least one phosphonate, O-ethyl O-*p*-nitrophenyl phenylphosphonate (EPNO), is also metabolized by this mechanism.

N-Dealkylation is a common reaction in the metabolism of drugs, insecticides, and other N-alkyl xenobiotics. Both the N- and N, N-dialkyl carbamate insecticides are readily dealkylated and in some cases the methylol intermediates are stable enough to be isolated. N, N-dimethyl-*p*-nitrophenyl carbamate is a useful model compound for this reaction (Fig. 5.5C). Another important example is the insecticide carbaryl, which undergoes several different microsomal oxidation reactions, including an attack on the N-methyl group. In this case, the methylol compound is stable enough to be isolated or to be conjugated in vivo.

The drug aminopyrene undergoes two N-demethylations to form first monomethyl-4-aminoantipyrene and then 4-aminoantipyrene (Fig. 5.5D).

S-Dealkylation is known to occur with a number of thioethers such as methylmereaptan, 6-methylthiopurine, etc. (Fig. 5.5E).

5.2.5d N-Oxidation

N-Oxidation can occur in a number of ways, including hydroxylamine formation, oxime formation, and N-oxide formation. The latter is primarily dependent on the flavin-containing monooxygenase found also in the microsomes and will be discussed in Section 5.3.

Hydroxylamine formation occurs with a number of amines, such as aniline and many of its substituted derivatives (Fig. 5.6A). In the case of 2-acetylaminofluorene, the product is a potent carcinogen and thus the reaction, catalyzed by CYP1A2, is an activation reaction (Fig. 5.6B).

Oximes can be formed by the N-hydroxylation of imines and primary amines. Imines have, furthermore, been suggested as intermediates in the formation of oximes from primary amines (Fig. 5.6C).

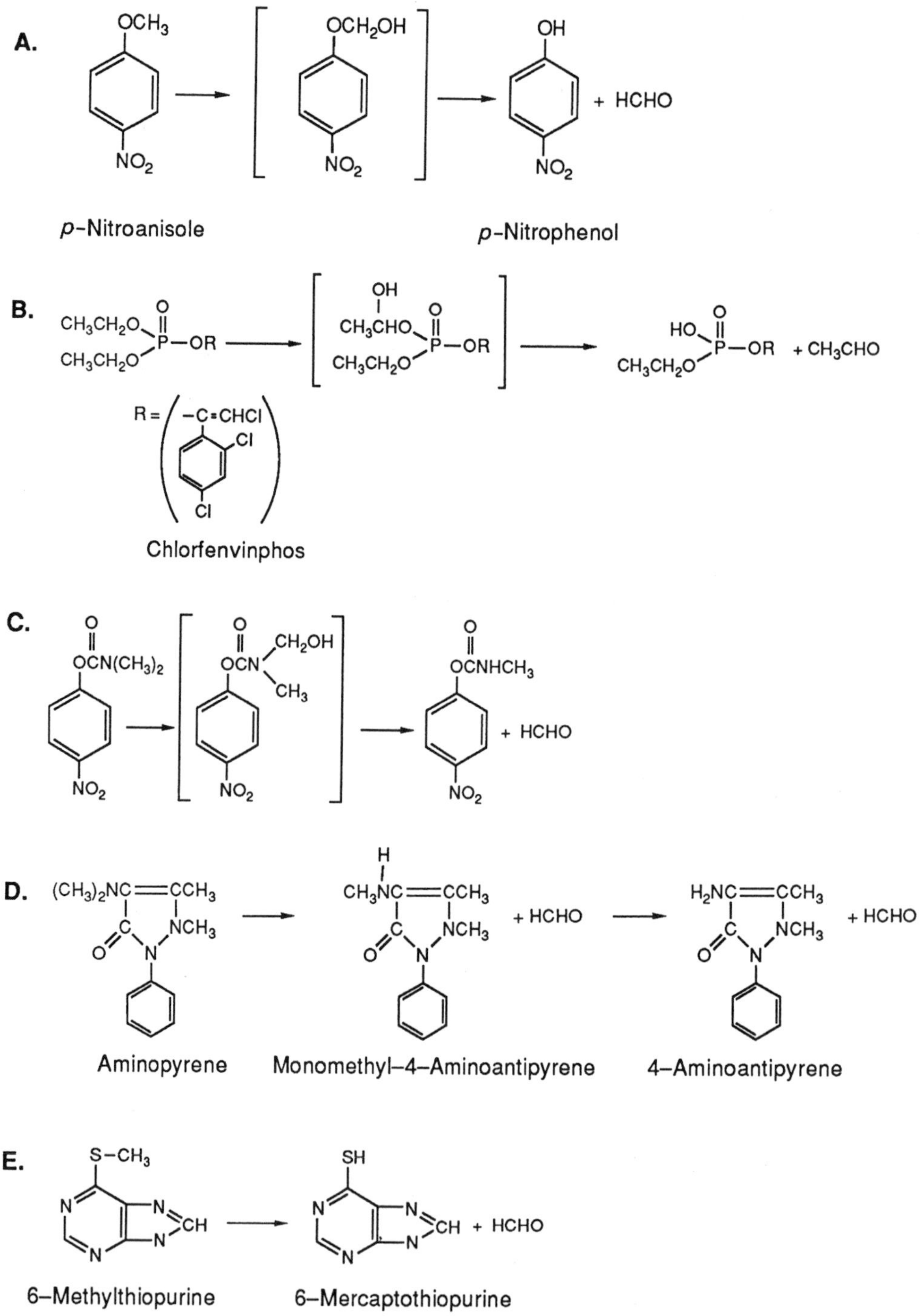

A.

OCH_3 / NO_2 → [OCH_2OH / NO_2] → OH / NO_2 + HCHO

p-Nitroanisole *p*-Nitrophenol

B.

$(CH_3CH_2O)_2P(O)$—OR → [$CH_3CH(OH)O(CH_3CH_2O)P(O)$—OR] → $HO(CH_3CH_2O)P(O)$—OR + CH_3CHO

R = (—C=CHCl, Cl, Cl)

Chlorfenvinphos

C.

$OC(O)N(CH_3)_2$ / NO_2 → [$OC(O)N(CH_2OH)CH_3$ / NO_2] → $OC(O)NHCH_3$ / NO_2 + HCHO

D.

$(CH_3)_2NC{=}CCH_3$, NCH_3 → $CH_3N(H)C{=}CCH_3$, NCH_3 + HCHO → $H_2NC{=}CCH_3$, NCH_3 + HCHO

Aminopyrene Monomethyl–4–Aminoantipyrene 4–Aminoantipyrene

E.

S–CH_3 → SH + HCHO

6–Methylthiopurine 6–Mercaptothiopurine

FIGURE 5.5 Examples of dealkylation reactions.

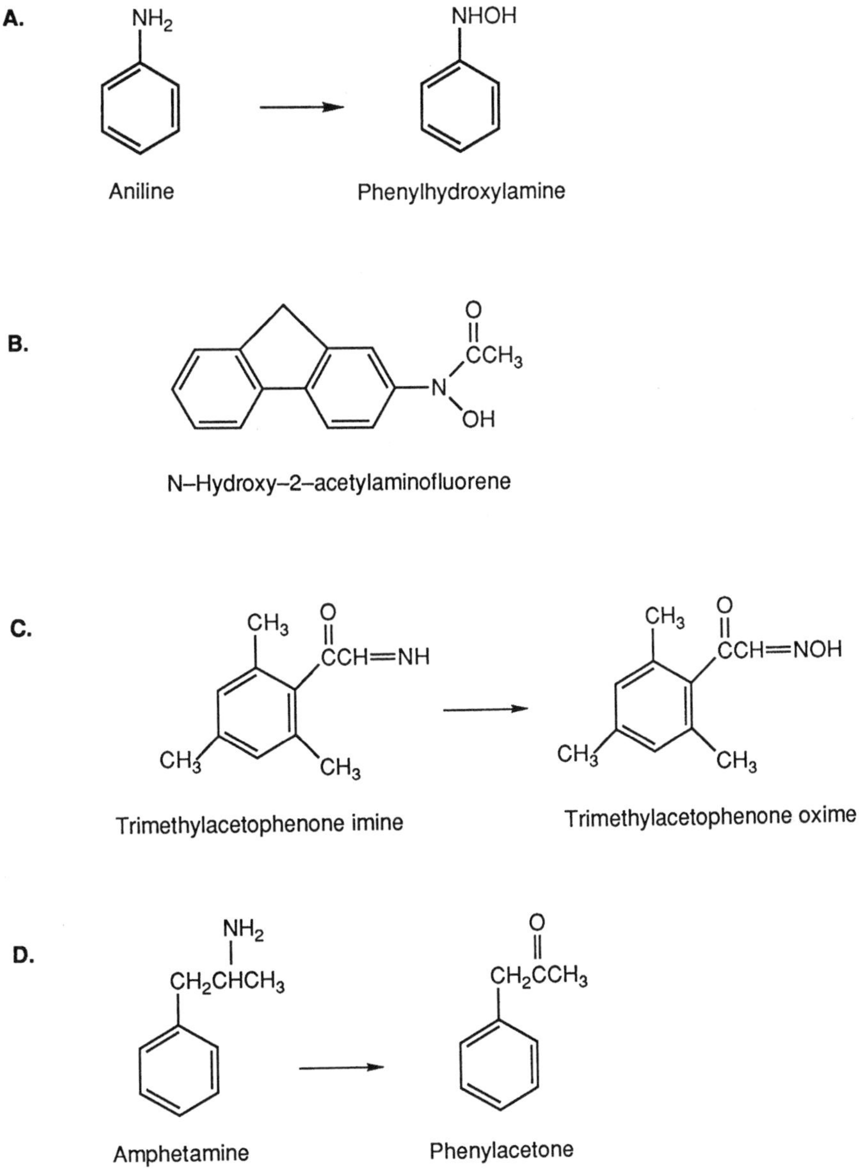

A.

NH_2 → NHOH

Aniline Phenylhydroxylamine

B.

N-Hydroxy-2-acetylaminofluorene

C.

Trimethylacetophenone imine → Trimethylacetophenone oxime

D.

Amphetamine → Phenylacetone

FIGURE 5.6 Examples of *N*-oxidations.

5.2.5e Oxidative Deamination

Oxidative deamination of amphetamine occurs in the rabbit liver but not to any extent in either the dog or rat, which tend to hydroxylate the aromatic ring (Fig. 5.6D). A close examination of this reaction indicates that it is probably not an attack on the nitrogen but rather on the adjacent carbon atom, giving rise to a carbinol amine, which produces the ketone by elimination of ammonia:

$$R_2CHNH_2 \xrightarrow{O} R_2C(OH)NH_2 \xrightarrow{-NH_3} R_2C=O$$

The carbinol, by another reaction sequence, can also give rise to an oxime:

$$R_2C(OH)NH_2 \xrightarrow{-H_2O} R_2C=NH \xrightarrow{O} R_2CNOH$$

The oxime can now be hydrolyzed to yield the ketone, which is thus formed by two different routes:

$$R_2CNOH \xrightarrow{H_2O} R_2C=O$$

5.2.5f S-Oxidation

Thioethers in general are oxidized to sulfoxides by microsomal monooxygenases, including the flavin-containing monooxygenase (Section 5.3), as well as P450. Some of the sulfoxides are further metabolized to sulfones. This reaction is very common among insecticides of several different chemical classes including carbamates, organophosphates, and chlorinated hydrocarbons (Fig. 5.7A).

Organophosphates include phorate, dimeton, and others, whereas among the chlorinated hydrocarbons, endosulfan is oxidized to endosulfan sulfate and methiochlor to a series of sulfoxides and sulfones to yield eventually the bissulfone.

S-Oxidation is also known among drugs, for example, chlorpromazine, whereas the solvent dimethylsulfoxide is further oxidized to the sulfone:

$$(CH_3)_2SO \rightarrow (CH_3)_2SO_2$$

5.2.5g P-Oxidation

This little-known reaction involves the conversion of trisubstituted phosphines to phosphine oxides. It is catalyzed not only by P450 but also by the flavin-containing monooxygenase. Known substrates are diphenylmethylphosphine and 3-dimethylaminopropyl-diphenylphosphine (Fig. 5.7B).

5.2.5h Desulfuration and Ester Cleavage

Organophosphorus insecticides containing the P = S moiety owe their insecticidal activity and their mammalian toxicity to an oxidative reaction in which the P = S group is converted to P = O, thereby converting compounds relatively inactive toward cholinesterases into potent cholinesterase inhibitors. This reaction is known for many organophosphorus compounds and has been studied most intensively in the case of parathion.

Much of the splitting of the phosphorus ester bonds in organophosphorus insecticides, which was formerly believed to be due entirely to hydrolysis, is now known also to be due to oxidative dearylation and is catalyzed by P450.

The question of whether desulfuration and dearylation occur independently of each other, possibly catalyzed by different P450s, or whether they involve common intermediates is not yet resolved with certainty. However, persuasive evidence, including investigations of purified, reconstituted systems, has been brought forward to support the hypothesis that both reactions involve a common intermediate of the "phosphooxythiiran" type (Fig. 5.7C).

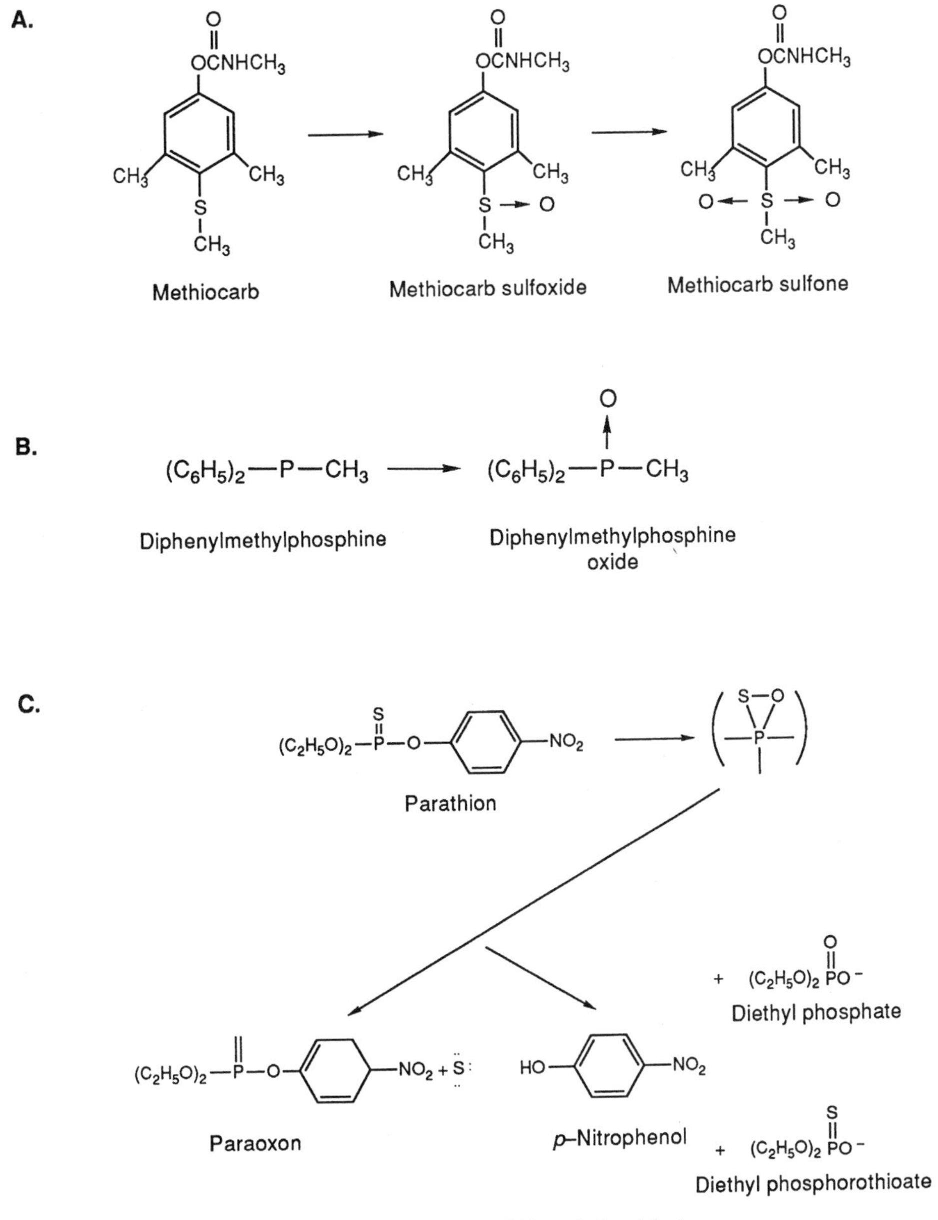

A. Methiocarb → Methiocarb sulfoxide → Methiocarb sulfone

B. $(C_6H_5)_2$—P—CH_3 → $(C_6H_5)_2$—P(→O)—CH_3

Diphenylmethylphosphine → Diphenylmethylphosphine oxide

C. Parathion → Paraoxon + S; p-Nitrophenol + Diethyl phosphate; Diethyl phosphorothioate

FIGURE 5.7 Examples of *S*- and *P*-oxidations.

5.2.5i Cyanide Release

Recently, there has been considerable evidence suggesting that the toxicity of organonitriles may occur as a result of the release of cyanide from the parent compound. Organonitriles are used in numerous manufacturing processes including production of synthetic fiber, plastic, pharmaceuticals, and dye stuffs, thus presenting the potential for extensive human exposure. Because many of these organonitriles are volatile and water-soluble, they would be absorbed to a large extent in the nasal mucosa. The nasal cavity contains a high concentration of some P450 isozymes and in fact microsomal preparations of nasal mucosa are more efficient than liver microsomes in catalyzing cyanide release from a number of organonitriles. For primary organonitriles, P450 catalyzes the oxidation of the carbon alpha to the cyano group to produce cyanohydrins that decompose to hydrogen cyanide and an aldehyde.

$$RCH_2C \equiv N \rightarrow RCHO + HCN$$

In general, substrate affinity and cyanide release increase with increasing size of the R group with benzylcyanide having a Km as low as 2.3 pm for rat nasal microsomes. Metabolism of acrylonitrile may proceed via an epoxide, formed by P450, which is then converted to the diol by epoxide hydrolase. The diol would then release cyanide by the same mechanism as other cyanohydrins.

5.3 THE FLAVIN-CONTAINING MONOOXYGENASE

Flavin-containing monooxygenases (FMOs), like P450, are located in the endoplasmic reticulum and are involved in the oxidation of numerous organic xenobiotics containing nitrogen, sulfur, or phosphorus heteroatoms as well as some inorganic ions.

The types of compounds oxidized by FMOs are shown in Table 5.4 with some examples of reactions catalyzed shown in Figure 5.8. Substrates are soft nucleophiles and compounds containing additional charged groups (anionic or cationic) are excluded as substrates. With the exception of cysteamine, there are no known endogenous substrates.

FMO was first purified to homogeneity from pig liver microsomes and subsequently from the livers of several other mammalian species. It is a highly lipophilic protein containing FAD as the only flavin, with a monomeric molecular mass of 56,000 per mole of FAD. Although it seems unlikely that there will be a large number of FMO isozymes, as there are for P450, investigations to date indicate that there are at least five forms, named FMO1 through FMO5.

Several unique features of the catalytic cycle of the FMOs are important to understanding the mechanism by which they oxidize xenobiotics. The catalytic mechanism for the FMO has been shown to involve the formation of an enzyme bound 4a-hydroperoxyl-flavin (Fig. 5.9) in an NADPH and O_2 dependent reaction. Reduction of the flavin by NADPH occurs before binding of oxygen can occur, and activation of oxygen by the enzyme occurs in the absence of substrate by oxidizing NADPH to form NADPI and peroxide. Finally, addition of the substrate to the peroxyflavin complex is the last step prior to oxygenation. This is in contrast to the P450 cycle in which the substrate binds to the oxidized enzyme which is then reduced.

The flavin hydroperoxide intermediate in the FMO enzyme forms a relatively stable, potent oxygenating species. Thus, any nucleophile that can be oxidized by an organic peroxide and can gain access to the active site is a potential substrate for the FMO. This capability accounts for the wide substrate specificity of the FMO. Although the flavin hydroperoxide of the FMO is a strong electrophile, it exhibits

TABLE 5.4 Substrates Oxidized by the Flavin-Containing Monooxygenase.

Inorganic
HS^-, S_8, I^-, IO^-, I_2, CNS^-
Organic nitrogen compounds
Sec- and tert-acyclic and cyclic amines
N-alkyl and N,N-dialkylarylamines
Hydrazines
Primary amines (FMO2)
Organic sulfur compounds
Thiols and disulfides
Cyclic and acyclic sulfides
Mercapto-purines, -pyrimidines, -imidazoles
Dithio acids and dithiocarbamides
Thiocarbamides and thioamides
Organic phosphorus compounds
Phosphines
Phosphonates
Others
Boronic acids
Selenides, selenocarbamides

a high degree of selectivity toward certain types of soft nucleophiles, primarily organic compounds with sulfur, nitrogen, and phosphorus heteroatoms. Compounds containing ionized carboxyl groups, which include most physiological sulfur compounds, are not substrates for the FMO, with the exception of cysteamine, which is an excellent substrate.

Reactivation of the enzyme is considered to be the rate-limiting step in the reaction. Because oxidation of the substrate occurs more rapidly than the regeneration of the active enzyme, the Vmax values are relatively similar for a variety of substrates even though the Km values differ.

Often the same substrate is metabolized by both P450 and FMO; this situation is especially prevalent with many N- and S-containing pesticides and drugs (e.g., phorate and nicotine). To study the relative contributions of these two enzymes with common substrates, methods have been developed to measure each separately in microsomal preparations. The most useful of these techniques is the inhibition of P450 activity by using an antibody to P450 reductase, thus permitting measurement of FMO activity alone. A second procedure is heat treatment of microsomal preparations (50°C for 1 minute) which inactivates the FMO, thus allowing determination of P450 activity, which is unchanged by heat treatment. Thermal inactivation, however, is ineffective with lung microsomes, because the lung FMO is more heat stable than the liver FMO.

Although the levels of FMO are not readily altered by classic chemical inducing and inhibiting agents, unlike the P450s, the balance of enzyme activity between P450 and FMO is easily disturbed, especially in the liver, by compounds that alter the concentration of P450 isozymes. Of special interest is a change in the balance of

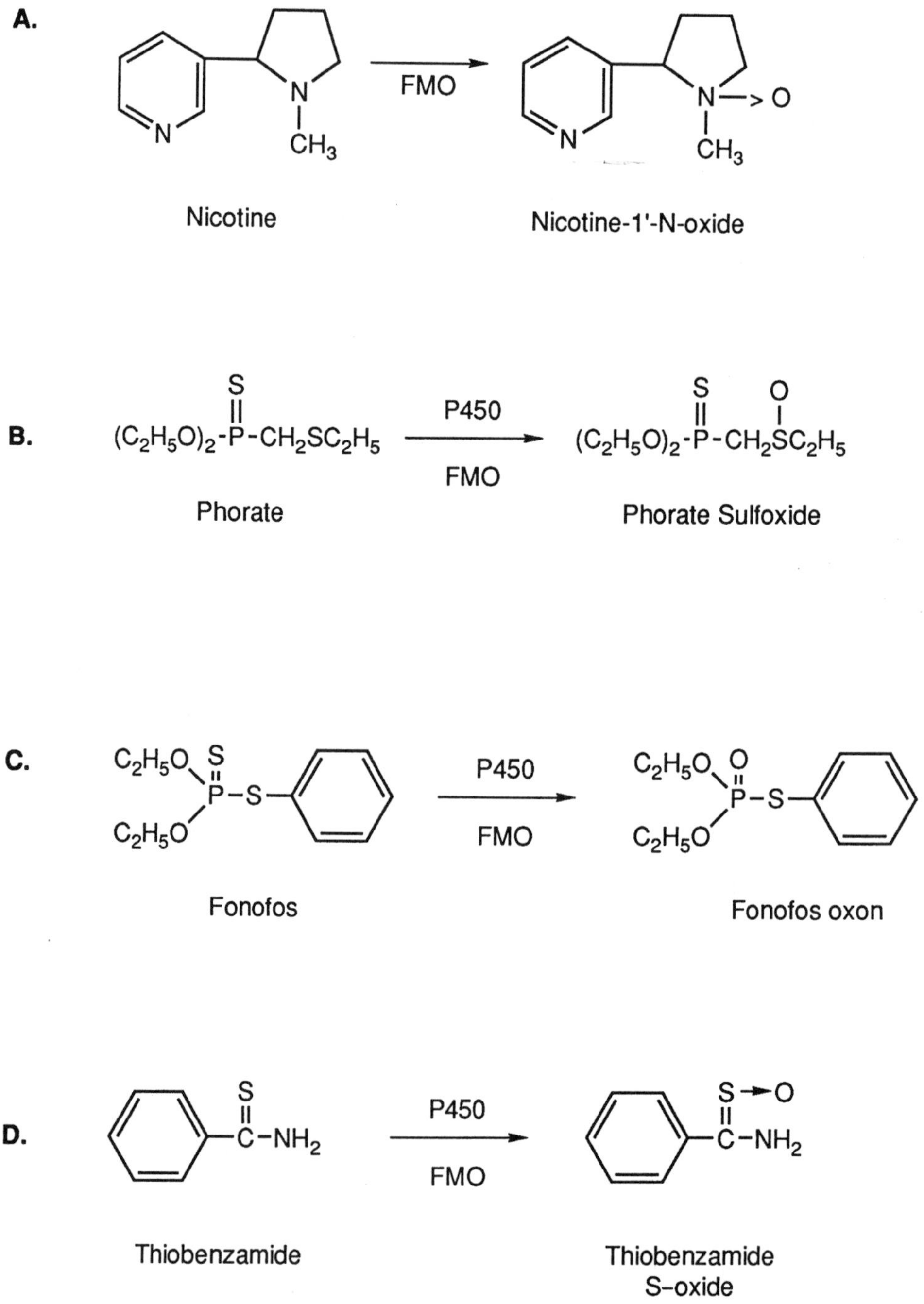

A.

Nicotine $\xrightarrow{\text{FMO}}$ Nicotine-1'-N-oxide

B.

$(C_2H_5O)_2$-P(=S)-$CH_2SC_2H_5$ $\xrightarrow[\text{FMO}]{\text{P450}}$ $(C_2H_5O)_2$-P(=S)-$CH_2S(O)C_2H_5$

Phorate → Phorate Sulfoxide

C.

$(C_2H_5O)_2$P(=S)-S-C_6H_5 $\xrightarrow[\text{FMO}]{\text{P450}}$ $(C_2H_5O)_2$P(=O)-S-C_6H_5

Fonofos → Fonofos oxon

D.

C_6H_5-C(=S)-NH_2 $\xrightarrow[\text{FMO}]{\text{P450}}$ C_6H_5-C(=S→O)-NH_2

Thiobenzamide → Thiobenzamide S-oxide

FIGURE 5.8 Examples of flavin-containing monooxygenase-catalyzed oxidations.

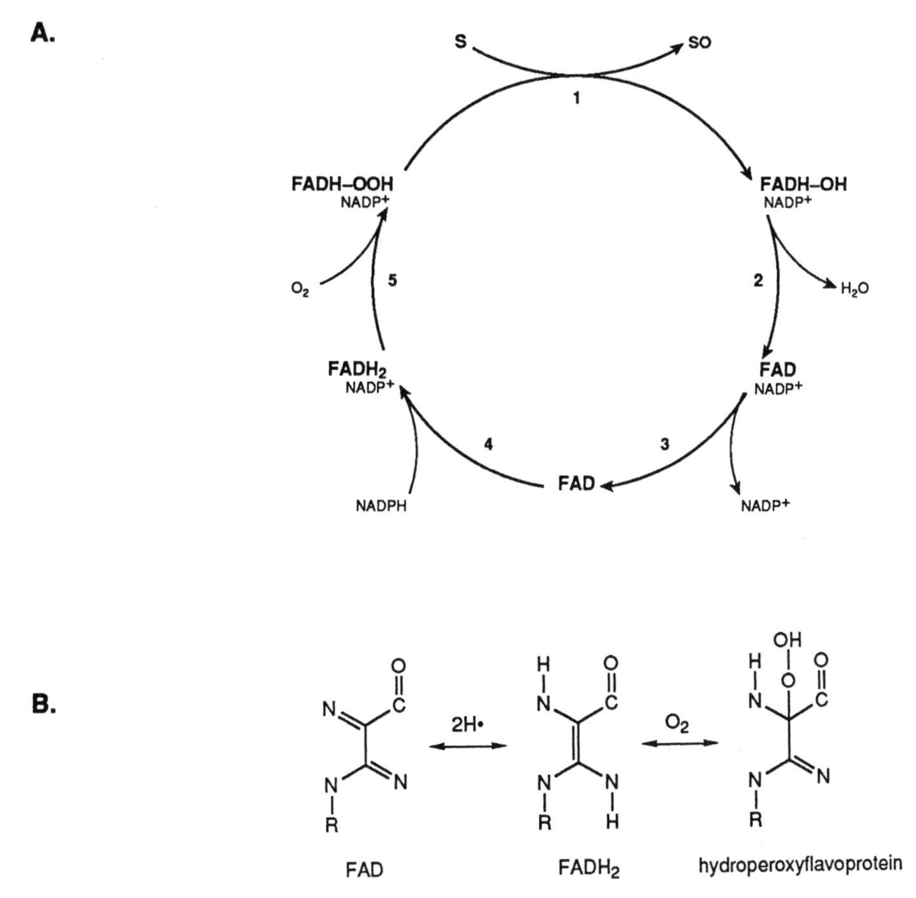

FIGURE 5.9 Catalytic cycle of the flavin-containing monooxygenase.

activity after in vivo exposure of animals to either inducers or inhibitors of P450 activity. For example, microsomes prepared from the livers of mice pretreated with phenobarbital showed not only an increase in the total rate of oxidation of the insecticide phorate, but also in the proportion metabolized by P450. As a result the percentage of products due to FMO is decreased.

The effects of xenobiotics on the relative contributions of FMO and P450 appear to be mediated primarily by P450 because the FMO does not appear to be inducible by xenobiotics. FMO levels may, however, vary with nutrition, diurnal rhythms, gender, pregnancy, and corticosteroids, although the effects appear to be both species and tissue dependent. Such alterations in the relative contributions of the two enzyme systems may assume toxicological importance when the products from the two enzymes differ, and particularly when one metabolite is more toxic than the others. Thus prior exposure of animals to environmental agents can have a significant effect on activation/detoxication pathways and the toxicity of other xenobiotics.

The five FMO isoforms are the products of distinct genes, but all fall into a single gene family. The isoforms are 50–58% identical compared with each other and are highly conserved in mammals, each isoform being over 80% identical across species. The importance of FMO polymorphisms is discussed below (Section 5.10.5). No sequences are known for FMOs from nonmammalian vertebrates or from invertebrates, thus the true extent of the gene family is not readily apparent. FMOs are expressed in several tissues and organs but have been most studied in the liver and lung and, to a lesser extent, in the kidney. Isoforms 1, 3, and 5 are generally expressed in the liver. FMO2 is expressed predominantly in the lung and the kidney. The expression of FMO2 as the sole isoform in the lung of rabbit is unusual. In other species, although FMO2 is almost always expressed in the lung, other isoforms are also present. The mRNA for FMO4, the least understood of the FMO isoforms, has been demonstrated in the rabbit nervous system, where it appears to be the only FMO message present.

Studies on the cellular localization of FMO have also been carried out. An immunohistochemical method utilizing peroxidase-labeled antibodies and diaminobenzidine revealed that in the rabbit lung the FMO is highly localized in the nonciliated bronchiolar epithelial (Clara) cells. Similar immunohistochemical studies of FMO distribution in the skin of mice and pigs revealed significant staining in epidermis, sebaceous gland cells, and hair follicles. Since the lung and skin are often a major route of entry for environmental chemicals, such as pesticides, the presence of the FMO in these tissues is of considerable interest.

FMO1, FMO2, and FMO3 are all broadly nonspecific, although substrate specificity does vary from one to another, indicating some differences in the substrate binding regions. Because FMO4 has proven difficult to express in heterologous expression sytems, it has been identified primarily from the cDNA and inferred amino acid sequences and no substrates are yet known. FMO5 has a restricted substrate specificity, oxidizing primary amines such as n-octylamine.

Although FMOs, unlike P450s, are not induced by xenobiotics, their level of expression can be dramatically affected by gender and during development. In the mouse, overall FMO activity is higher in female liver than in male. FMO1 is two to three times higher in females that males, whereas FMO3 is not expressed in the liver of male mice, being completely suppressed at puberty by testosterone. FMO5, on the other hand, is expressed to the same extent in both genders. In the rat, FMO1 activity is higher in males than females, whereas FMOs 3 and 5 show no difference between genders.

Development can also affect the level of expression of FMO isoforms. In humans, FMO1 is the predominant form in fetal liver, FMO3 not being expressed, and FMO3 is the predominant form in the adult, with FMO1 not being expressed. This situation appears to be gender-independent. In the mouse, FMO1 is expressed in the liver by gestation day 13, the gender effect becoming apparent by 4 weeks of age. Hepatic FMO3 is not expressed until about 2 weeks of age, is equivalent in males and females until puberty, and is then completely suppressed in males. Thus, the expression of FMOs in the female mouse, with FMO1 expressed in the fetal liver and FMO3 in the adult liver, resembles FMO expression in humans.

5.4 NONMICROSOMAL OXIDATIONS

In addition to the monooxygenases, there are a number of other enzymes that are involved in the oxidation of foreign compounds. The oxidoreductases are located in either the mitochondrial fraction or the 100,000g supernatant of tissue homogenates.

5.4.1 Alcohol Dehydrogenases

As a class of enzymes, the alcohol dehydrogenases catalyze the conversion of alcohols to aldehydes or ketones, for example,

$$RCH_2OH + NAD^+ \rightleftharpoons RCHO + NADH + H^+$$

This should not be confused with the P450-catalyzed monooxygenation of ethanol observed in liver microsomes. The alcohol dehydrogenase reaction is reversible and carbonyl compounds can be reduced to alcohols. Alcohol dehydrogenase is probably the most important dehydrogenase involved in the metabolism of foreign alcohols and carbonyl compounds. The enzyme is found in the soluble fraction of the liver, kidney, and lung and requires NAD or NADP as a coenzyme. The reaction proceeds at a slower rate with NADP. In the intact organism, the reaction proceeds to the right, because aldehydes are further oxidized to acids. The further oxidation of aldehydes to acids is a vital detoxication reaction because aldehydes are usually toxic and, because of their lipid solubility, are not readily excreted.

All of the human alcohol dehydrogenases are dimeric zinc metalloenzymes existing in multiple forms that may be homo- or heterodimers with subunits of aproximately Da 40,000. Five classes exist in humans and their distribution varies between organs. Details of the isoforms and their polymorphic forms are presented below (Section 5.10.3).

Alcohol dehydrogenases will metabolize primary alcohols to aldehydes, n-butanol being the substrate oxidized at the highest rate. Secondary alcohols are oxidized to ketones but at a reduced rate. Butanol-2 is oxidized at one-third the rate of n-butanol. Tertiary alcohols are not readily oxidized.

Poisoning caused by ethylene glycol (CH_2OHCH_2OH) results from activation to aldehydes, glycolate, oxalate, and lactate, resulting from an initial attack by alcohol dehydrogenase. This is similar to the activation of methanol to formaldehyde and subsequent oxidation by aldehyde dehydrogenase to formic acid.

These two enzymes also activate the solvent 2-butoxyethanol ($CH_3CH_2CH_2CH_2OCH_2CH_2OH$) to 2-butoxyacetic acid, responsible for its hematotoxicity. (See Chapter 9 for further discussion.)

A number of other dehydrogenases, described from various sources, play an important role in steroid, lipid, and carbohydrate metabolism. However, in general, their substrate specificity is narrow, and they are unlikely to be important in the metabolism of xenobiotics.

5.4.2 Aldehyde Dehydrogenase

The oxidation of aliphatic and aromatic aldehydes to their corresponding acids allows the acids to be excreted or conjugated in phase II reactions:

$$RCHO + NAD^+ \rightarrow RCOOH + NADH + H^+$$

Aldehyde dehydrogenases have been isolated from a variety of sources, with the most important enzyme being the one from the liver, which has the capability to handle a wide variety of substrates. With a series of linear aliphatic aldehydes, the rate of oxidation increases as the carbonyl carbon becomes more positive.

In addition to liver aldehyde dehydrogenase, a number of other enzymes are present in the soluble fraction of liver homogenates that will oxidize aldehydes and certain N-heterocyclic compounds. Among these are aldehyde oxidase and xanthine oxidase (see below), both flavoprotein enzymes containing molybdenum. These enzymes catalyze the oxidation of aldehydes formed by the deamination of endogenous amines by amine oxidases.

5.4.3 Amine Oxidases

The biological function of amine oxidases appears to involve the oxidation of biogenic amines formed during normal biological processes. In mammals, the monoamine oxidases are involved in the control of the serotonin: catecholamine ratios in the brain, which in turn influence sleep and EEG patterns, body temperature, and mental depression. Two groups of amine oxidases are involved in the oxidative deamination of naturally occurring amines as well as foreign compounds.

5.4.3a Monoamine Oxidases

Monoamine oxidases (MAO) are flavoprotein enzymes located in the mitochondrial fraction of liver, kidney, and brain. The enzymes have been found also in blood platelets and intestinal mucosa. The MAO exist as a large group of similar enzymes with overlapping substrate specificities and inhibition patterns. The number is difficult to assess because of the great variety and variation within each tissue as well as in different animal species.

The general reaction catalyzed is shown in Figure 5.10. MAO will deaminate primary, secondary, and tertiary aliphatic amines, the reaction rate being faster with the primary, and slower with the secondary and tertiary amines. Electron-withdrawing substituents on an aromatic ring increase the rate of deamination. Compounds that have a substituted methyl group on the α-carbon atoms are not metabolized by the MAO system (e.g., amphetamine and ephedrine).

5.4.3b Diamine Oxidases

Diamine oxidases (DAO) also oxidize diamines to the corresponding aldehydes in the presence of oxygen. The DAO are pyridoxal phosphate proteins containing copper that are found in the soluble fraction of liver, intestine, kidney, and placenta. A typical DAO reaction is the oxidation of putrescine (Fig. 5.10).

The rate of deamination is determined by the chain length, and with polymethylene diamines, NH_2-$(CH_2)_n$-NH_2, the maximum rate occurs when $n = 4$ (putrescine) and $n = 5$ (cadaverine) and decreases to zero when $n = 9$ or more. At this stage, MAO becomes active and can oxidatively deaminate the diamines. It should be noted that although MAO can deaminate both substituted and primary amines, DAO can deaminate only primary amines.

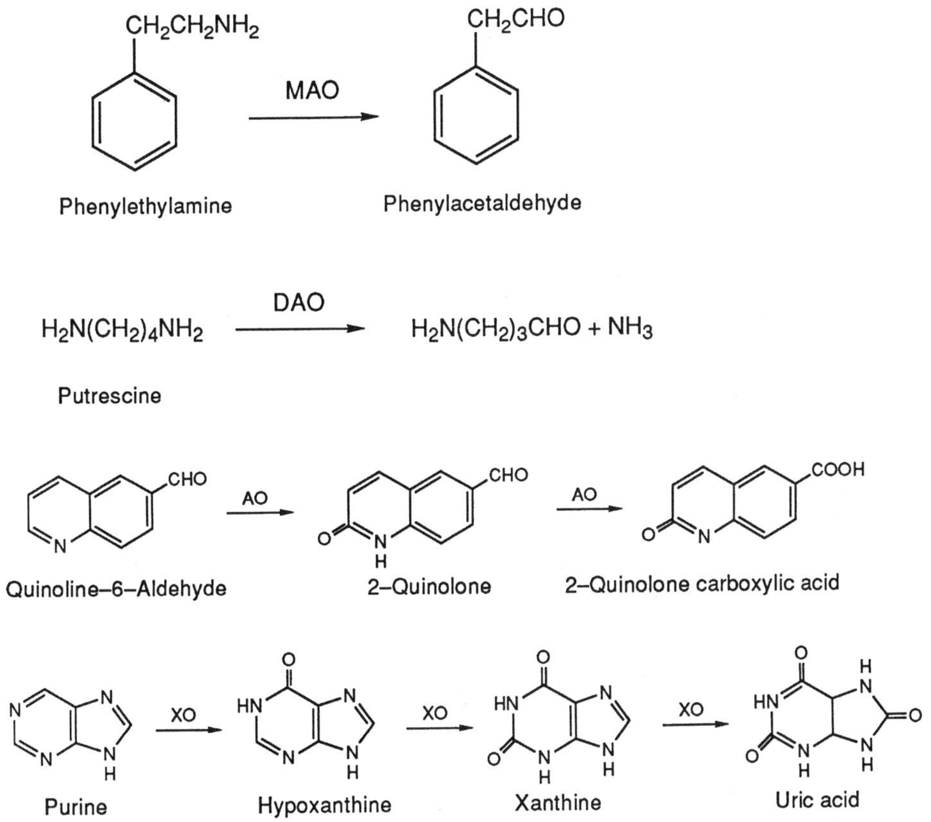

$CH_2CH_2NH_2$ —MAO→ CH_2CHO

Phenylethylamine Phenylacetaldehyde

$H_2N(CH_2)_4NH_2$ —DAO→ $H_2N(CH_2)_3CHO + NH_3$

Putrescine

Quinoline–6–Aldehyde —AO→ 2–Quinolone —AO→ 2–Quinolone carboxylic acid

Purine —XO→ Hypoxanthine —XO→ Xanthine —XO→ Uric acid

FIGURE 5.10 Examples of nonmicrosomal oxidations.

5.4.4 Molybdenum Hydroxylases

There are a number of molybdenum-containing enzymes, but those that are important in carbon oxidation of xenobiotics are aldehyde oxidase (AO) and xanthine oxidase (XO), also referred to as molybdenum hydroxylases. Both enzymes catalyze the oxidation of a wide range of aldehydes and N-heterocycles. The name aldehyde oxidase is somewhat misleading, however, because oxidation of heteroaromatics is more significant. The differences in substrate specificities between monooxygenases and molybdenum hydroxylases is partly based on chemistry since with monooxygenases, the mechanism involves an electrophilic attack on the carbon, whereas the hydroxylases catalyze nucleophilic addition at an unsaturated carbon, which can be represented as an attack by the hydroxylation.

$$RH + OH^- \rightarrow ROH + 2e^- + H^+$$

Although molecular oxygen may be involved in the overall oxidation process, the oxygen atom incorporated into the product is derived from water, whereas with monooxygenations, the oxygen comes from molecular oxygen. In N-heteroaromatic ring compounds, the most electropositive carbon is usually adjacent to a ring N atom and this is the normal position of nucleophilic molybdenum hydroxylase attack. In contrast, a P450 catalyzed attack would tend to occur distant from the N, probably in an adjacent carbocyclic ring. Uncharged carbocyclic compounds, such as benzene and naphthalene, which are P450 substrates, are not substrates for the molybdenum hydroxylases. However, as the number of N atoms increases in a ring, the affinity of the hydroxylases with the substrate increases. Thus the bases purine and pteridine are oxidized exclusively by AO and XO, although the oxidations do not necessarily occur at the same positions.

These enzymes are found predominantly in the cytosol and occur in most tissues. However, the highest levels are found in those tissues most often exposed to ingested foreign compounds; AO is most prevalent in the liver, whereas the highest levels of XO occur in the small intestine, milk, and mammary gland. Both enzymes react with endogenous substrates; XO is involved in the final stages of purine catabolism and AO catalyzes the oxidation of vitamins such as pyridoxal and N-methylnicotinamide. Aldehyde oxidase and xanthine oxidase also catalyze the oxidation of a wide range of xenobiotics including aldehydes, uncharged bases, quaternary N-heterocycles, and carbocyclics. Some examples are shown in Figure 5.10. In addition to their oxidative role, these enzymes may be involved in reductive pathways in vivo, and they are known to catalyze the in vitro reduction of aromatic nitro compounds, azo dyes, nitrosamines, N-oxides, and sulfoxides. Table 5.1 lists some of their known substrates. Because these enzymes metabolize such a wide range of xenobiotics, including many that are also substrates for the monooxygenases, the role of the molybdenum hydroxylases in detoxication cannot be discounted. In addition, they may be important also in the production as well as deactivation of toxic metabolites through reductive pathways.

5.5 COOXIDATION BY PROSTAGLANDIN SYNTHETASE

The most important peroxidase reaction involved in the metabolic oxidation of xenobiotics is probably that catalyzed by prostaglandin synthetase (PGS). PGS catalyzes the oxygenation of polyunsaturated fatty acids to hydroxy endoperoxides, with the preferential substrate in vivo being arachidonic acid (AA). PGS catalyzes two activities, the fatty acid cyclooxygenase activity that brings about the oxygenation of polyunsaturated fatty acids, such as arachidonic acid, to form a cyclic endoperoxide, prostaglandin G (PGG), which is then reduced to a hydroperoxy endoperoxide termed prostaglandin H (PGH). A number of xenobiotics may be cooxidized during this second hydroperoxidase reaction. (Fig. 5.11 and Chapter 9)

The enzyme, which is membrane bound, is present in virtually all mammalian tissues, and is especially concentrated in seminal vesicles, a rich source that has been used extensively in experimental studies. PGS is high also in platelets, lungs, skin, kidney medulla, endothelial cells, and embryonic tissue. It is a glycoprotein with a subunit molecular mass of about 70,000 Da, containing one heme per subunit. A wide variety of compounds can undergo oxidation during PGH biosynthesis; the types of cooxidations catalyzed include dehydrogenation, demethylation, epoxidation, sulfoxidation, N-oxidation, C-hydroxylation, and dioxygenation. Some of the more important compounds that can be cooxidized are listed in Table 5.1.

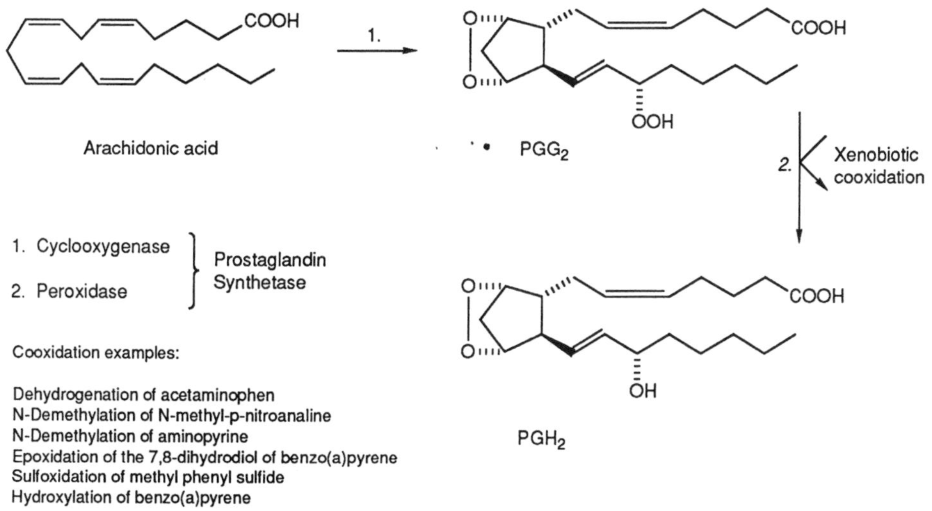

Arachidonic acid

PGG2

PGH2

1. Cyclooxygenase
2. Peroxidase
Prostaglandin Synthetase

2. Xenobiotic cooxidation

Cooxidation examples:

Dehydrogenation of acetaminophen
N-Demethylation of N-methyl-p-nitroanaline
N-Demethylation of aminopyrine
Epoxidation of the 7,8-dihydrodiol of benzo(a)pyrene
Sulfoxidation of methyl phenyl sulfide
Hydroxylation of benzo(a)pyrene

FIGURE 5.11 Cooxidation during prostaglandin biosynthesis.

Many xenobiotics can be metabolized to reactive metabolites by PGS. For a discussion of activations catalyzed by PGS, the reader is referred to Chapter 9.

It is now increasingly evident that PGS can serve as an alternate enzyme for xenobiotic metabolism, particularly in tissues with low monooxygenase activity. Consequently, the cooxidation of drugs, chemical carcinogens, and other xenobiotics by PGS is an important area of study. In addition, this enzyme system, like the monooxygenases, may be important in screening chemicals for mutagenic and teratogenic potential.

5.6 REDUCTION REACTIONS

A number of functional groups, such as nitro, diazo, carbonyls, disulfides, sulfoxides, and alkenes, are susceptible to reduction. In many cases it is difficult to determine whether these reactions proceed nonenzymatically by the action of biological reducing agents such as NADPH, NADH, and FAD, or through the mediation of functional enzyme systems. As noted above, the molybdenum hydroxylases can carry out, in vitro, a number of reduction reactions, including nitro, azo, N-oxide, and sulfoxide reduction. Although the in vivo consequences of this are not yet clear, much of the distribution of "reductases" described below may be, in whole or in part, the distribution of molybdenum hydroxylases.

5.6.1 Nitro Reduction

Aromatic amines are susceptible to reduction by both bacterial and mammalian nitroreductase systems (Fig. 5.12A).

Nitroreductase activity has been demonstrated in liver homogenates as well as in the soluble fraction, whereas other studies have reported that nitroreductase

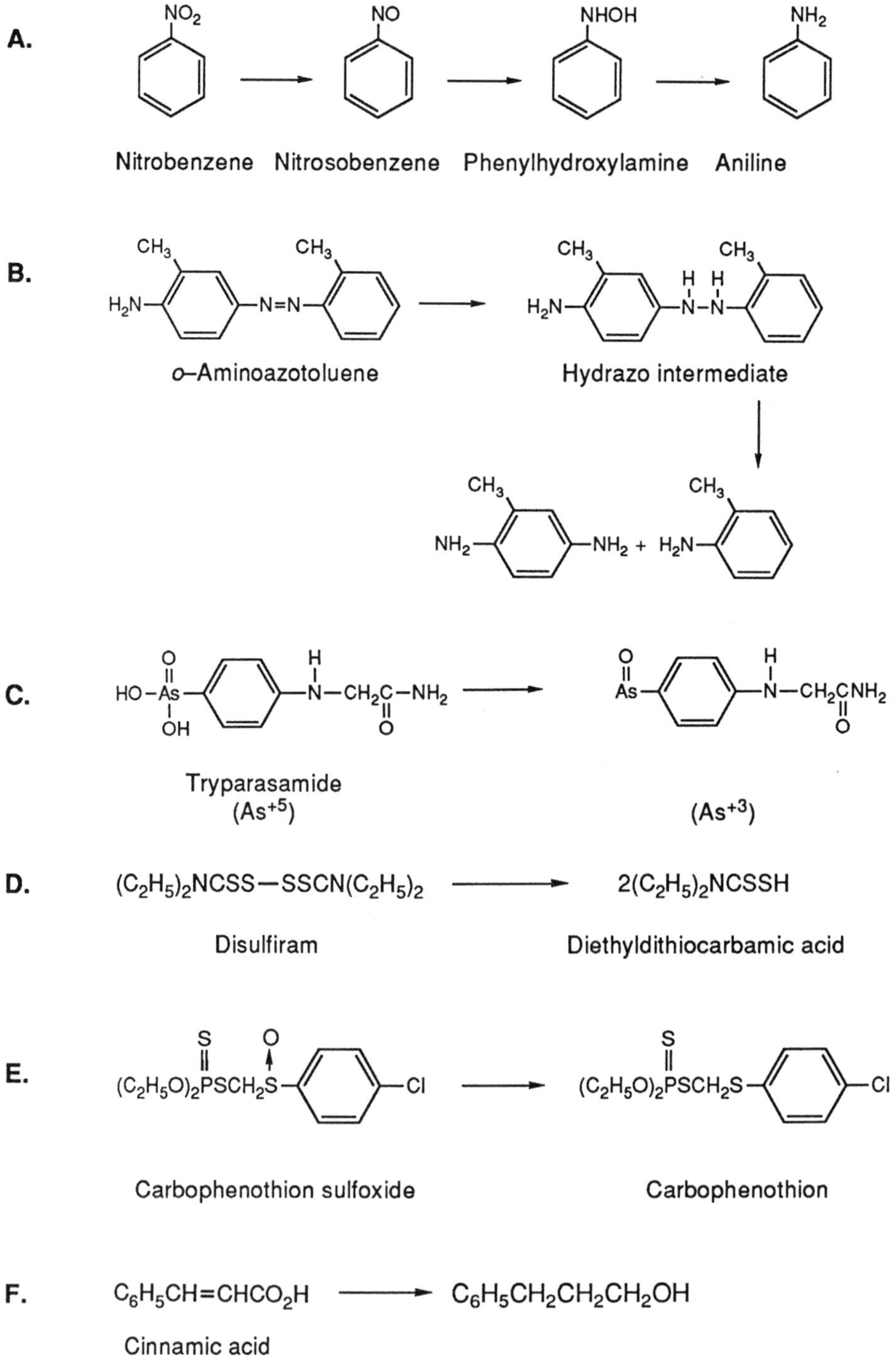

A. Nitrobenzene → Nitrosobenzene → Phenylhydroxylamine → Aniline

(NO_2, NO, NHOH, NH_2)

B. H_2N–(CH$_3$)C$_6$H$_3$–N=N–C$_6$H$_4$(CH$_3$) → H_2N–(CH$_3$)C$_6$H$_3$–NH–NH–C$_6$H$_4$(CH$_3$) → NH_2–(CH$_3$)C$_6$H$_3$–NH_2 + H_2N–C$_6$H$_4$(CH$_3$)

o–Aminoazotoluene — Hydrazo intermediate

C. HO–As(=O)(OH)–C$_6$H$_4$–NH–CH_2C(=O)–NH_2 → O=As–C$_6$H$_4$–NH–CH_2C(=O)NH_2

Tryparasamide (As^{+5}) — (As^{+3})

D. $(C_2H_5)_2NCSS{-}SSCN(C_2H_5)_2$ → $2(C_2H_5)_2NCSSH$

Disulfiram — Diethyldithiocarbamic acid

E. $(C_2H_5O)_2P(=S)SCH_2S(\rightarrow O)$–C$_6H_4$–Cl → $(C_2H_5O)_2P(=S)SCH_2S$–C$_6$H$_4$–Cl

Carbophenothion sulfoxide — Carbophenothion

F. $C_6H_5CH{=}CHCO_2H$ → $C_6H_5CH_2CH_2CH_2OH$

Cinnamic acid

FIGURE 5.12 Examples of reduction reactions.

activity has been found in all liver fractions evaluated. The reductase appears to be distributed in liver, kidney, lung, heart, and brain. The reaction utilizes both NADPH and NADH and requires anaerobic conditions. The reaction can be inhibited by the addition of oxygen. The reaction is stimulated by FMN and FAD, and at high flavin concentrations they can act simply as nonenzymatic electron donors. The reduction process probably proceeds via the nitroso and hydroxylamine intermediates as illustrated above.

5.6.2 Azo Reduction

Requirements for azoreductases are quite similar to those for nitroreductases, that is, they require anaerobic conditions and NADPH and are stimulated by reduced flavins (Fig. 5.12B).

The ability of mammalian tissue to reduce azo bonds is rather poor. With p-[2,4-(diaminophenyl)azo]benzenesulfonamide (Prontosil), in vivo reduction forms sulfanilamide. However, pretreatment with antibiotics destroys the intestinal bacteria, which results in a decrease in the formation of the amino compound. Generally, it would appear that both nitro and azo reduction as a function of specific tissues is of minor importance, and intracellular bacteria as well as intestinal bacteria are actually responsible for these reductions.

5.6.3 Reduction of Pentavalent Arsenic to Trivalent Arsenic

Pentavalent arsenic compounds such as tryparsamide appear to require reduction to the trivalent state for antiparasitic activity (Fig. 5.12C).

Various arsenicals containing pentavalent arsenic have only slight in vitro antiprotozoal activity, whereas compounds containing trivalent arsenic are highly active.

5.6.4 Reduction of Disulfides

A number of disulfides are reduced in mammals to their sulfhydryl compounds. An example is disulfiram (Antabuse), a drug used for the treatment of alcoholism (Fig. 5.12D).

5.6.5 Ketone and Aldehyde Reduction

The reduction of ketones and aldehydes occurs through the reverse reaction of alcohol dehydrogenases (Section 5.4.1):

$$R_1R_2CO \rightarrow R_1R_2CHOH$$

5.6.6 Sulfoxide and N-Oxide Reduction

The reduction of sulfoxides and N-oxides has been reported to occur in a number of mammalian system. It appears that liver enzymes reduce sulfoxide or sulfonyl compounds to thioether or sulfides under anaerobic conditions (Fig. 5.12E).

N-Oxides have also been reported to be reduced by a bacterial reductase:

$$R_3CNO \rightarrow R_3CN$$

5.6.7 Reduction of Double Bonds

Certain aromatic compounds have been reported to be reduced by intestinal flora, for example, cinnamic acid (Fig. 5.12F).

5.7 HYDROLYSIS

A large number of xenobiotics, such as esters, amides, or substituted phosphates that are composed of ester-type bonds, are susceptible to hydrolysis. Hydrolytic reactions are the only phase I reactions that do not utilize energy. Numerous hydrolases are found in blood plasma, liver, intestinal muscosa, kidney, muscle, and nervous tissue. Hydrolases are present in both soluble and microsomal fractions. The general reactions are shown in (Fig. 5.13A).

The acid and alcohol or amine formed on hydrolysis can be eliminated directly

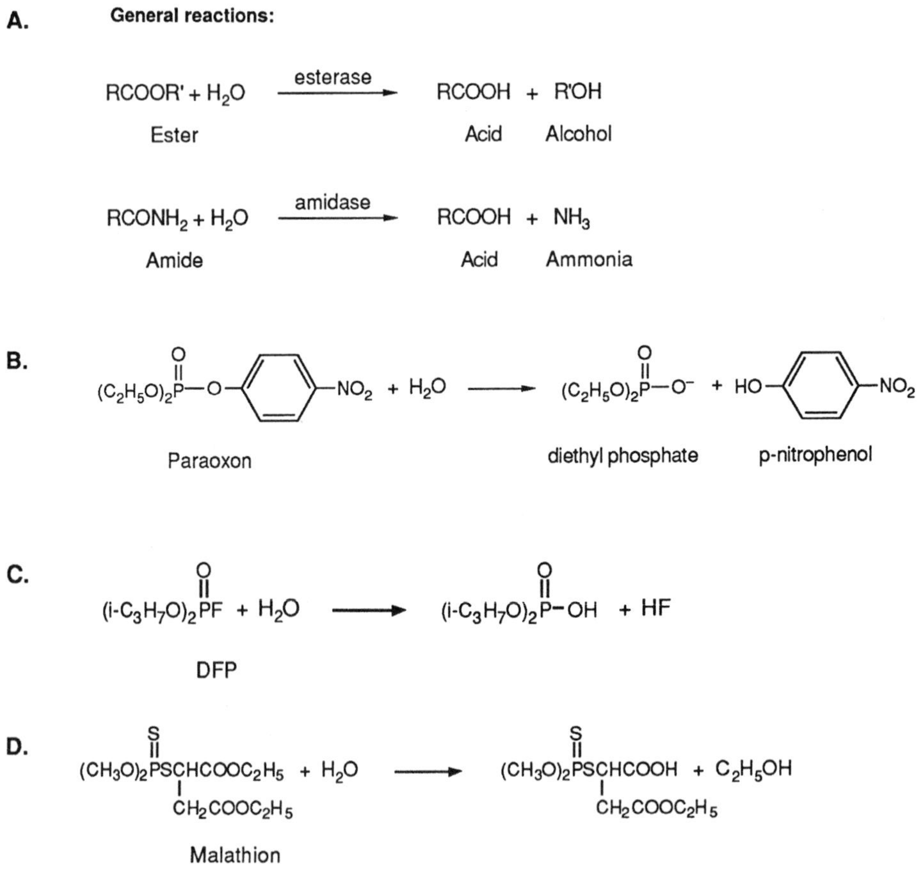

A. **General reactions:**

$$RCOOR' + H_2O \xrightarrow{\text{esterase}} RCOOH + R'OH$$

Ester → Acid, Alcohol

$$RCONH_2 + H_2O \xrightarrow{\text{amidase}} RCOOH + NH_3$$

Amide → Acid, Ammonia

B.

$$(C_2H_5O)_2P(=O){-}O{-}C_6H_4{-}NO_2 + H_2O \longrightarrow (C_2H_5O)_2P(=O){-}O^- + HO{-}C_6H_4{-}NO_2$$

Paraoxon → diethyl phosphate, p-nitrophenol

C.

$$(i\text{-}C_3H_7O)_2P(=O)F + H_2O \longrightarrow (i\text{-}C_3H_7O)_2P(=O){-}OH + HF$$

DFP

D.

$$(CH_3O)_2P(=S)SCH(CH_2COOC_2H_5)COOC_2H_5 + H_2O \longrightarrow (CH_3O)_2P(=S)SCH(CH_2COOC_2H_5)COOH + C_2H_5OH$$

Malathion

FIGURE 5.13 Examples of hydrolysis reactions.

or conjugated by phase II reactions. In general , amide analogs are hydrolyzed at slower rates than the corresponding esters.

Tissue esterases have been divided into two classes, A-type esterases, which are insensitive, and the B-type esterases, which are sensitive to inhibition by organophosphorus esters. The A esterases include the arylesterases, whereas the B esterases include cholinesterases of plasma, acetylcholinesterases of erythrocytes and nervous tissue, carboxylesterases, lipases, etc. The nonspecific arylesterases that hydrolyze short-chain aromatic esters are activated by Ca^{2+} ions and are responsible for the hydrolysis of certain organophosphate triesters such as paraoxon (Fig. 5.13B).

Certain hydrolases also are present in mammalian plasma and tissue, and these enzymes hydrolyze and subsequently detoxify chemical warfare agents such as the nerve gases tabun, sarin, and DFP (Fig. 5.13C). A variety of foreign compounds, such as phthalic acid esters (plasticizers), phenoxyacetic and picolinic acid esters (herbicides), and pyrethroids and their derivatives (insecticides), as well as a variety of ester and amide derivatives of drugs, are detoxified by hydrolases found in plants, animals, and bacteria, that is, the entire plant and animal kingdom.

In certain strains of insects that are resistant to malathion, the resistance mechanism is associated with a higher level of a carboxylesterase, which detoxifies malathion (Fig. 5.13D).

5.8 EPOXIDE HYDRATION

Epoxide rings of certain alkene and arene compounds are hydrated enzymatically by epoxide hydrolases to form the corresponding trans-dihydrodiols (Fig. 5.14). The epoxide hydrolases are a family of enzymes known to exist both in the endoplasmic reticulum and in the cytosol. In earlier studies they were named epoxide hydratase, epoxide hydrase, or epoxide hydrolase. Epoxide hydrolase, however, has been recommended by the International Union of Biochemists Nomenclature Committee and is now in general use.

5.8.1 Microsomal Epoxide Hydrolase

The microsomal epoxide hydrolase converts arene and alkene oxides to vicinal dihydrodiols by hydrolytic cleavage of the oxirane ring. This is a detoxication reaction in that it converts generally highly reactive electrophilic oxirane species to less reactive nonelectrophilic dihydrodiols. It should be borne in mind, however, that some epoxides are unreactive, both toward macromolecules and toward epoxide hydrolase (e.g., dieldrin) and that dihydrodiols may be further epoxidized to produce metabolites that are even more reactive, as is the case with benzo(a)pyrene. An example of a substrate, benzo(a)pyrene 7,8-epoxide, is shown in Figure 5.14A.

Microsomal epoxide hydrolase is widely distributed, having been described from plants, invertabrates, and vertebrates. In vertebrates it has wide organ distribution. In the case of the rat, the most studied species, the enzyme has been found in essentially every organ and tissue. Although predominantly located in the endoplasmic reticulum (microsomes), epoxide hydrolase is also found in the plasma and nuclear membranes and, to some extent, in the cytosolic fraction.

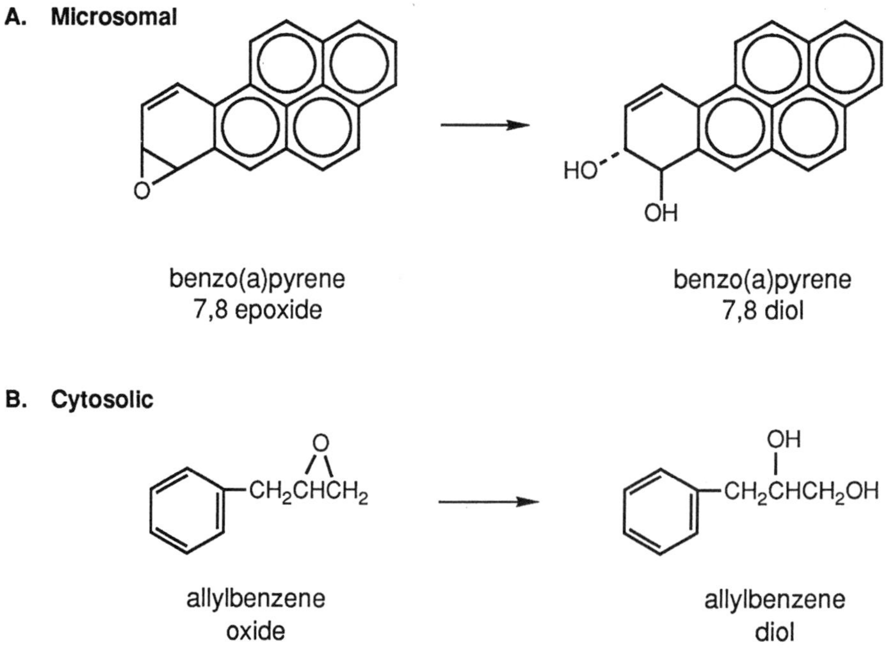

A. Microsomal

benzo(a)pyrene 7,8 epoxide → benzo(a)pyrene 7,8 diol

B. Cytosolic

allylbenzene oxide → allylbenzene diol

FIGURE 5.14 Examples of epoxide hydrolase reactions.

Microsomal epoxide hydrolase has been purified from several sources. It is a hydrophobic protein with a molecular mass of approximately 50,000 Da. The complete amino acid sequence has been determined directly and has been deduced from the cDNA sequence. It appears to be a single polypeptide of 455 amino acid residues with a high degree of homology between species. Although existence of multiple forms within a single tissue has been suggested, this may be an artifact of purification because only a single gene coding for microsomal epoxide hydrolase has been isolated from rat liver.

The reaction is highly regio- and stereospecific, always proceeding with inversion of configuration. With arene oxides, this means that only trans-dihydrodiols are formed. All of the experimental evidence points to a general base catalyzed nucleophilic (SN_2) addition of water to the oxirane ring.

A number of inhibitors of this enzyme are known. They include epoxides that are hydrolyzed by the enzymes, such as trichloropropene oxide, metal ions such as Hg^{2+}, Zn^{2+}, and Cd^{2+} and 2-bromo-4′-acetophenone, a potent inhibitor that binds to imidazole nitrogen atoms. Microsomal epoxide hydrolase can be induced by compounds such as phenobarbital, Arochlor 1254, 2(3)-t-butyl-4-hydroxyanisole (BHA) and 3, 5-di-t-butyl-hydroxytoluene (BHT). Many microsomal epoxide hydrolase inducers are inducers also of P450 and produce a general proliferation of the endoplasmic reticulum. Induction does, however, involve an increase in the mRNA specific for the hydrolase.

A second unique microsomal epoxide hydrolases has been described. This enzyme appears to have a narrow substrate specificity, being specific for choleserol 5,6-oxides and related steroid 5,6-oxides. It is not well characterized and will not be discussed further.

5.8.2 Cytosolic Epoxide Hydrolase

The existence of a cytosolic epoxide hydrolase was first indicated by its ability to hydrolyze analogs of insect juvenile hormone not readily hydrolyzed by microsomal epoxide hydrolase. Subsequent studies demonstrated a unique cytosolic enzyme catalytically and structurally distinct from the microsomal enzyme. It appears probable that the cytosolic enzyme is peroxisomal in origin. Both enzymes are broadly nonspecific and have many substrates in common. It is clear, however, that many substrates hydrolyzed well by cytosolic epoxide hydrolase are hydrolyzed poorly by microsomal epoxide hydrolase and vice versa. For example, 1-(4′-ethylphenoxy)-3,7-dimethyl-6,7-epoxy-trans-2-octene, a substituted geranyl epoxide insect juvenile hormone mimic, is hydrolyzed 10 times more rapidly by the cytosolic enzyme than by the microsomal one. In any series, such as the substituted styrene oxides, the trans configuration is hydrolyzed more rapidly by the cytosolic epoxide hydrolase than is the cis isomer. At the same time, it should remembered that in this and other series, cis as well as trans isomers are hydrolyzed, although generally at lower rates. An example of a substrate, allylbenzene oxide, is shown in Figure 5.14B.

As with substrates, the microsomal epoxide hydrolase and the cytosolic epoxide hydrolase have inhibitors in common and, again, those that are most effective toward one are less effective toward the other. Inhibitors of cytosolic epoxide hydrolase include poorly metabolized substrates, such as chalcone oxide and its substituted analogs. A group of cyclopropyl oxiranes were specifically designed as suicide inhibitors of cytosolic epoxide hydrolase, binding covalently to the active site through a transient intermediate formed during the enzymatic induced ring opening. All of the known inducers of cytosolic epoxide hydrolase are also peroxisome proliferating agents. They include the drug clofibrate and the plasticizer di-(2-ethylhexyl)phthalate and many of their analogs and related compounds.

Cytosolic epoxide hydrolase has been purified from several sources. It is a dimer consisting of two identical monomers of approximately 60,000 Da molecular mass. Its amino acid sequence is not known but antibodies to the microsomal and cytosolic enzymes do not cross-react. It occurs broadly across species; activity in the liver of those species examined is highest in the mouse, lowest in the rat, with activity in rabbit, guinea pig, and human liver being intermediate. It is also broadly distributed between organs, occurring in all major organs of the species most investigated, the mouse and rat.

Leukotriene A4 hydrolase is a unique cytosolic epoxide hydrolase, structurally dissimilar to the cytosolic enzyme described above. Its substrate specificity is narrow, being restricted to leukotriene A4, (5(S)-trans-5,6-oxido-7,9-cis-11,14-trans-eicosatetraenoic acid), and related fatty acids.

5.9 DDT-DEHYDROCHLORINASE

In the early 1950s, it was demonstrated that DDT-resistant houseflies detoxified DDT mainly to its noninsecticidal metabolite DDE. The rate of dehydrohalogenation of DDT to DDE was found to vary between various insect strains as well as between individuals. This enzyme occurs also in mammals but has been studied more intensively in insects.

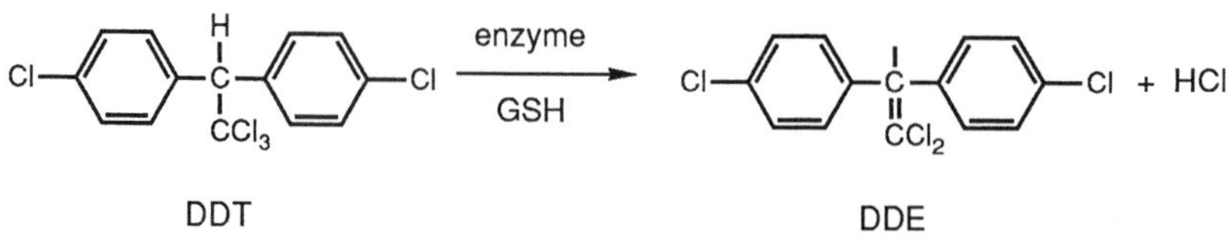

DDT DDE

FIGURE 5.15 Conversion of DDT to DDE.

DDT-dehydrochlorinase, a reduced glutathione (GSH)-dependent enzyme, has been isolated from the 100,000 g supernatant of resistant houseflies. Although the enzyme-mediated reaction requires glutathione, the glutathione levels are not altered at the end of the reaction (Fig. 5.15).

The lipoprotein enzyme has a molecular mass of 36,000 Da as a monomer and 120,000 Da as the tetramer. The Km for DDT is 5×10^{-7} M with optimum activity at pH 7.4.

The enzyme system catalyzes the degradation of *p,p*-DDT to *p,p*-DDE or the degradation of *p,p*-DDD (2,2,-bis(*p*-chlorophenyl)-1,1-dichloroethane) to the corresponding DDT ethylene TDEE (2,2-bis(*p*-chlorophenyl)-l-chloroethylene). *o,p*-DDT is not degraded by DDT-dehydrochlorinase, suggesting a *p,p*-orientation requirement for dehalogenation. In general, the DDT resistance of house fly strains is correlated with the activity of DDT-dehydrochlorinase, although other resistance mechanisms are known in certain strains.

At one time, it was believed that DDT-dehydrochlorinase, glutathione S-aryltransferase, and enzymes metabolizing the γ-isomer of hexachlorocyclohexane (γ-BHC) were one enzyme because experimental results suggested the existence of a nonspecific enzyme that catalyzed all these reactions. However, it has since been demonstrated that DDT-dehydrochlorinase is different from the glutathione S-aryltransferase that metabolizes γ-BHC based on the following: a difference in electrophoretic mobility; a difference in stability and response to inhibitors; results of genetic studies; and purification of a house fly glutathione S-aryltransferase that lacks DDT-dehydrochlorinase activity, although both are now known to be glutathione S-transferases.

5.10 PHARMACOGENETICS

5.10.1 Introduction

The way that individuals respond to drugs shows considerable interindividual variation, partially due to variations in drug metabolism. Some of this variation has been shown to be due to genetic mutations (polymorphisms) at certain gene loci. *Pharmacogenetics* is defined as the study of the hereditary basis of the differences in response to drugs (Nebert, 1997). A *polymorphism* is defined as an inherited monogenetic trait that exists in the population in at least two genotypes (two or more variant alleles) and is stably inherited. Several defective alleles may occur at the same gene locus. There are often racial differences in the incidences of polymorphisms.

Pharmacogenetics first began to be studied intensively during World War II when red cell hemolysis was seen in some soldiers after administration of the drug primaquine. The incidence of hemolysis was more frequent in African-American soldiers than Caucasian soldiers. After the war, this drug-induced hemolysis was shown to be due to a genetic deficiency of glucose-6-phosphate (G6PDH) dehydrogenase. The racial differences were explained by an association between G6PDH deficiency and the ability to survive malaria, which conveyed a biological advantage to individuals with G6PDH deficiency in malaria-infested countries (Kalow and Bertilsson, 1994). The first polymorphism in a drug metabolizing enzyme was documented in the 1950s, when it was found that there was a high incidence of peripheral neuropathy in response to the antituberculosis drug isoniazid due to slow clearance of the parent drug in some individuals. Subsequently individuals were phenotyped and found to be rapid or slow acetylators (Chapter 17). The slow metabolizer phenotype was shown to be inherited recessively as a monogenetic trait in family studies. In population studies, Caucasians could be divided into almost equal numbers of rapid and slow acetylators, while only 15% of Japanese and Chinese were slow acetylators. This polymorphism is important because it also affects the metabolism of aromatic amines to which humans have been exposed industrially, through the diet and environmentally. It therefore affects susceptibility to certain cancers. The first known polymorphisms in the phase I CYP drug metabolizing enzymes were reported in the 1970–1980s. One was a polymorphism in *CYP2D6*, which affects the metabolism of many clinically used drugs including the antihypertensive drug debrisoquine. The second was a polymorphism in the metabolism of an anticonvulsant drug mephenytoin, which was later shown to be due to polymorphisms in CYP2C19.

Polymorphisms in drug metabolizing enzymes are usually inherited as autosomal recessive traits. In many cases, polymorphisms are indicated by a bimodal distribution of metabolism in population studies (Fig. 5.16). Individuals can be divided into two populations, extensive metabolizers (EMs) and poor metabolizers (PMs) of the drug. *Phenotype* refers to the biological traits of the individual (e.g., EMs or PMs), *genotype* refers to the genetic makeup of the individual. Because of the autosomal recessive nature of most of these traits, PMs carry two mutant alleles (m_{1-n}/m_{1-n}) with $m_1 - m_n$ representing various deleterious mutants of the same enzyme, whereas EMs can be either homozygous or heterozygous wild type (wt/wt or wt/m_{1-n}).

Metabolic polymorphisms affect the rate of clearance of certain drugs. A good example can be seen in the pharmacokinetics of haloperidol in poor and extensive metabolizers of CYP2D6 (Fig. 5.17). If a drug has a narrow therapeutic index (ratio of toxic dose/therapeutic dose), the polymorphism may produce toxic levels in the PM (Fig. 5.18).

5.10.2 Polymorphisms in Cytochrome P450(CYP) Enzymes

5.10.2a CYP2D6

The first CYP enzyme shown to be polymorphic was CYP2D6. A dramatic event occurred during the course of a clinical trial of debrisoquine, which included a study of the metabolism and pharmacokinetic in volunteers. One of the investigators, Robert Smith, who used himself as a volunteer for the study, developed severe orthostatic hypotension in response to debrisoquine, with blood pressure dropping

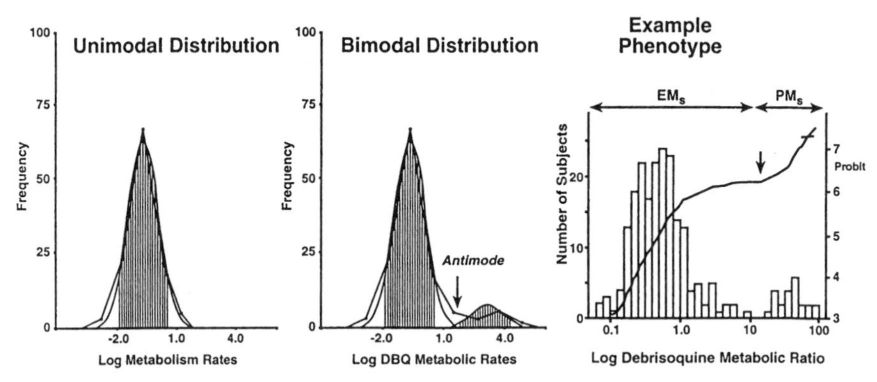

FIGURE 5.16 Demonstrates a unimodal and bimodal distribution of drug metabolism. The theoretical unimodal (single gaussian distribution) on the left is typical of a nonpolymorphic metabolism, and the bimodal distribution (mixture of two guassian distributions) on the right represents the results of phenotyping for a polymorphic substrate, debrisoquine. The number of subjects is shown on the y axis and log of the metabolic rate is shown on the x axis (debrisoquine/4-OH-debrisoquine) (middle panel). The dots connected by straight lines represent the actual frequencies whereas the smooth curves represent the guassian distributions that best describe the data. The intersection of the two bimodal distributions is the *antimode* or the estimated cutoff point between the two phenotypes used to predict the PM phenotype.

The middle panel is redrawn from Henthorn T.K., Benitez, J., Avram, M.J., Martinez, C., Llerena, A., Cobaleda, J., Krejcie, T., Gibbons, R.D. Assessment of the debrisoquin and dextromethorphan phenotyping tests by gaussian mixture distributions analysis. *Clin. Pharmacol. Ther.* 45 (1989), 328–333. The panel on the right represents an example of actual data redrawn from Küpfer, A., Presage, R. Pharmacogenetics of mephenytoin: a new drug hydrolyzation polymorphism in man. *Eur. J. Clin. Pharmacol.* 26 (1984), 753–759.

to 70/50. These symptoms persisted for two days. His other colleagues took a similar dose but showed few if any significant cardiovascular effects. Analysis of the urine showed that Smith eliminated the drug essentially unchanged, whereas other participants of the study excreted primarily the major metabolite, 4-hydroxydebrisoquine. This study prompted an immediate search for a polymorphism of debrisoquine metabolism in a study of 94 volunteers and three families of PMs. This led to the first description of a genetic polymorphism in CYP mediated drug oxidation in man in 1977. The CYP2D6 polymorphism is prevalent in Caucasians and African Americans, in which ~7% are PMs. In Asian populations, the frequency of PMs is only 1%.

CYP2D6 metabolizes a large number of clinically used drugs. These include *antihypertensive drugs* such as debrisoquine; *antiarrhythmic drugs* such as flecainide; *β-blockers*: metoprolol, propranolol, and bufuralol; *antidepressants*: nortriptyline, desipramine, and clomipramine; a number of *neuroleptics* such as halperidol and thioridazine; and *opiates* such as codeine. Many of these drugs have a narrow therapeutic index, which results in greater toxicity in PMs (Fig. 5.18). The oxytocic drug sparteine was actually withdrawn from the market because of fetal deaths in women treated with this drug. Interestingly, opiates such as codeine are actually activated by O-demethylation to morphine by CYP2D6. Therefore codeine is not effective in individuals who are PMs of CYP2D6. There are now at least 37 known alleles of

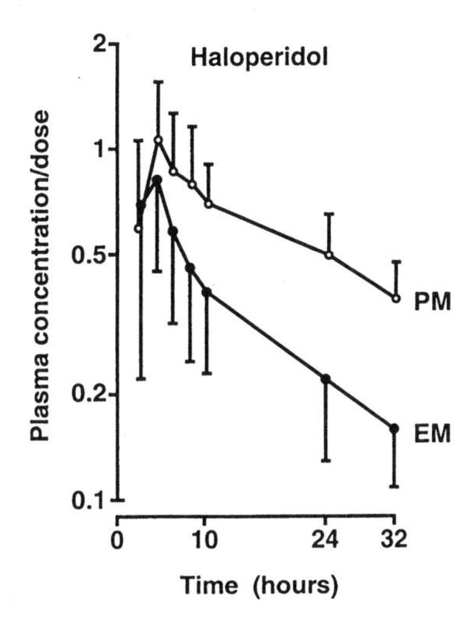

FIGURE 5.17 Pharmacokinetics of haloperidol (a CYP2D6 substrate) in six EMs and six PMS of debrisoquine. Plasma concentration of haloperidol versus time.
Redrawn from Kalow and Bertilsson, 1994 (Llerena, A., Alm, C., Dahl, M.L., Ekqvist, B. and Bertilsson, L. *Ther. Drug Monitor.* 14 (1992), 92–97).

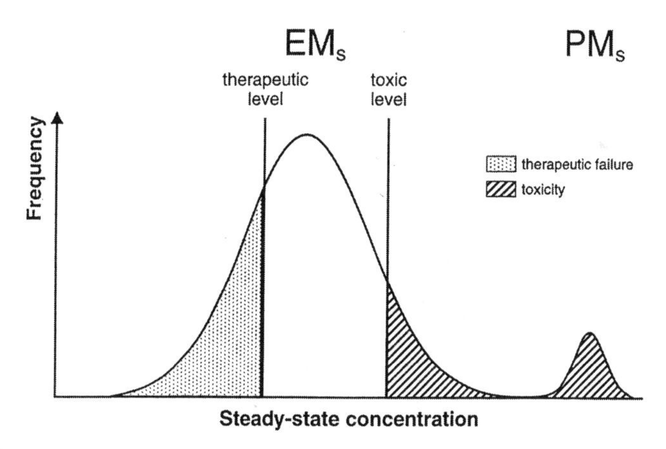

FIGURE 5.18 Steady-state plasma concentrations of a drug metabolized exclusively by CYP2D6 subsequent to a single dose of the drug. Shows the higher mode for PMS, lower mode for EMs, and the concentration at which therapeutic failure, therapeutic improvement, and toxicity occur.

(Taken with permission from Brøsen, K., Gram, L.F. Clinical significance of the sparteine/debrisoquine oxidation polymorphism. *Eur. J. Clin. Pharmacol.* 36 (1989), 537–547.)

CYP2D6. A unified nomenclature for polymorphic CYP alleles has been proposed and can be accessed from the website http://www.imm.ki.se/CYPalleles/. In the unified nomenclature, the gene and allele are separated by an asteric followed by Arabic numerals and uppercase roman letters (e.g., *CYP2D6*4A*). Although the gene is italicized, the protein is not italicized, and a period replaces the asteric (CYP2D6.4A). An allele is given a new Arabic number if it contains an inactivating mutation such one that affects splicing, transcription, deletions, or insertions that cause frame shifts, stop codons, or amino acid changes. Alleles that contain such a functional change but differ by silent base changes and nonfunctional amino acid changes have historically been differentiated by letters such as *CY2D6*4A* and **4B*. Table 5.5 summarizes a number of CYP polymorphisms, prototype drugs metabolized by these enzymes, clinical consequences, and examples of genetic defects causing these polymorphisms.

5.10.2b CYP2C19

The anticonvulsant drug mephenytoin exists as two enantiomers (*R*- and *S*-). As part of his doctoral thesis, Adrian Küpfer reported the stereoselective metabolism of *S*-mephenytoin in the dog. While later studying the metabolism of mephenytoin in man during a postdoctoral fellowship at Vanderbilt University, Küpfer found that the formation and urinary excretion of the 4′-hydroxymetabolite is rapid during the first 24 hours after administration of the drug and this metabolite is derived almost entirely from the *S*-enantiomer. Unexpectedly, one of the human subjects complained of unacceptable sedation on a dose of mephenytoin, which was without effect in several other individuals. Taking advantage of a family reunion over a Thanksgiving holiday, Küpfer examined metabolism of mephenytoin in the family of the individual who metabolized the drug poorly, and found the trait was inherited as an autosomal recessive trait. In 1984, he performed population studies and showed that ~3–5% of Caucasians were PMs. Subsequent studies showed that as many as 13–23% of Asians were PMs of mephenytoin.

It was not until the 1990s that CYP2C19 was shown to be the enzyme responsible for the 4′-hydroxylation and of *S*-mephenytoin. At least seven defective alleles of *CYP2C19* were discovered, accounting for the majority of the PM phenotype. The most frequent two polymorphisms are an aberrant splice site in exon 5 and a premature stop codon in exon 4. CYP2C19 also metabolizes the antiulcer drug omeprazole, the HIV protease inhibitor nelfinavir, certain barbiturates, a number of antidepressants, is partially responsible for metabolism of the anxiolytic drug diazepam, and activates certain antimalarial drugs, such as proguanil.

5.10.2c CYP2C9

Approximately 0.2% of Caucasians have a rare polymorphism in CYP2C9. However, despite the rarity of the genetic defects, CYP2C9 metabolizes many clinically important drugs including the anticoagulant warfarin, the diabetic agent tolbutamide, the antihypertensive losartan, the anticonvulsant phenytoin, and numerous antiinflammatory drugs such as ibuprofen. Two mutant alleles have been reported. One allele, *CYP2C9*3* contains a Ile → Leu substitution at amino acid 359. This allele has lower affinity for drugs like tolbutamide, and people who are homozygous for this allele are slow metabolizers of warfarin, tolbutamide, and phenytoin. Another allele, *CYP2C9*2* appears to have intermediate activity toward some substrates.

5.10.2d Other CYP Polymorphisms

CYP1B1 metabolizes many premutagenic polycyclic hydrocarbons. Several mutations have been found to be associated with congenital glaucoma, suggesting this enzyme has a function in maintenance of intraocular pressure. An upstream

TABLE 5.5 Examples of Polymorphisms in CYP Enzymes and Their Consequences.

ENZYME	Drug Substrates	Effect	CHANGE
CYP2D6	Debrisoquine Sparteine Codeine	Exaggerated cardiovascular effects Fetal deaths Lack of efficacy	~37 alleles Splice defects gene deletion
CYP2C19	Mephenytoin Proguanil	Exacerbated sedation Decreased effectiveness	>8 alleles, splice variants, stop codons
CYP2C9	Warfarin Tolbutamide Phenytoin	Excessive bleeding Dangerous lowering of blood sugar toxicity requires lower dose	Amino acid change alters Km
CYP1B1	Polycyclic hydrocarbons, Benzpyrene, Dibenzanthracene, 3-Methylcholanthrene	Mutations occur in primary congenital glaucoma	25 mutant alleles amino acid changes frameshifts
CYP2A6	Coumarin, Nicotine, NNK(methylnitrosamino)-1-(3-pyridyl)-1-Butanone	Reduces risk for tobacco addiction	2 inactive alleles
CYP2E1	Benzene Nitrosamines	Upstream polymorphism may affect expression, an amino acid change effects activity	Changes in upstream region; amino acid changes
CYP1A1	Polycyclic aromatic hydrocarbons	Uncertain	RFLP in 3′ end amino acid changes
CYP1A2	Caffeine, N-hydroxylates food mutagens	Uncertain	Amino acid changes, silent base changes

polymorphism in the *CYP2E1* gene has been suggested to change expression of the gene, at least in promotor constructs in vitro, and one of several amino acid changes decreases expression of the protein. A change in the downstream region of CYP1A1 and an amino acid change have been intensively studied for alterations in risk for lung cancer in epidemiology studies. However, it is questionable whether these changes affect catalytic activity. An amino acid change in *CYP1A2* has been recently identified, but its effect on catalytic activity is not yet known. Two rare inactivating alleles of *CYP2A6* have been reported. This enzyme metabolizes coumarin. It metabolizes certain procarcinogens including (methylnitrosamino)-1-(3-pyridyl)-1-butanone (NNK), a nitrosamine found in tobacco smoke, and also metabolizes nicotine to conitine. It has been suggested that individuals with defective CYP2A6 who have an impairment of nicotine metabolism decrease the risk for nicotine addition and are less likely to become tobacco-dependent smokers. Recently, polymorphisms in CYP3A have been reported. There is one SNP (single nucleotide polymorphism) in the upstream region of the gene (CYP3A4*2) but it does not appear to affect promotor activity. A second polymorphism includes a Ser → Pro amino acid change that appears to change the intrinsic clearance of CYP3A4 for nefedifine but not testerone.

5.10.3 Polymorphisms in Alcohol Dehydrogenase

Alcohol dehydrogenase exists as at least seven genetic loci. However, three genes produce the low Km class I ADHs. They produce the subunits α, β, and γ. The ADH molecule is a dimer, consisting of homodimers α/α, β/β, and γ/γ and heterodimers α/β, α/γ, and β/γ. If an individual is heterozygous for β1 and β2, for example, ADH consists of 10 dimers formed by random association with either of the B alleles substituting randomly to form α/α, β1/β1, β2/β2 and β1/β2, γ/γ, and heterodimers α/β1, α/β2 α, γ, β1/γ, β1/γ. Several of these genes, including ADH2 (determines subunit β chains) and ADH3 (determines γ chains), are polymorphic. The three β subunits are encoded by the genes ADH2*1, ADH2*2, ADH2*3, and two γ subunits are encoded by ADH3*1 and ADH3*2. ADH2*1 (β1) contains arginine, whereas the ADH2*2(β2), which contains histidine, has a 40-fold higher Vmax for ethanol than ADH2*1. The ADH2*2 allele (encoding the β2 subunit, which shows greater efficiency in metabolizing alcohol) may reduce the risk for alcoholism, whereas the ADH2*1 allele with low capacity for metabolizing alcohol is thought to be a risk factor. ADH2*3 encodes β3, which has a high Km but a high Vmax, which makes it less efficient at low ethanol concentrations but gives it greater capacity at high alcohol concentrations, which can occur during binge drinking. This allele has been suggested to be somewhat protective against alcohol related birth defects. In contrast, γ1 and γ2 encoded by the ADH3*1 and ADH3*2 alleles have similar catalytic activity toward ethanol. The ADH4 gene (encoding ADH π) has also been found to have several polymorphisms in the upstream region one of which affected the activity of promotor constructs in vitro approximately twofold and could potentially affect expression of this protein. Both the alcohol and aldehyde dehydrogenase polymorphisms are being widely studied with respect to their influence on alcoholism and alcohol related diseases.

5.10.4 Polymorphisms in Aldehyde Dehydrogenases

There are at least four forms of aldehyde dehydrogenase. ALDH2 is the major form that metabolizes acetaldehyde. A polymorphism in this enzyme, which consists of an amino acid substitution Lys487Glu results in loss of activity of the enzyme. This mutation is particularly prevalent (50%) in Asians and causes high levels of acetaldehyde after ingestion of ethanol, causing flushing and nausea in affected individuals. This phenotype generally causes an aversion to alcohol and individuals homozygous for this allele are at greatly reduced risk for alcoholism.

5.10.5 Polymorphisms in Flavin-Containing Monooxygenases

5.10.5a FMO3

Individuals with a defect in FMO3 have a condition known as fish odor syndrome, or trimethylaminurea, which is characterized by increased levels of free triethylamine. Individuals with this syndrome exhibit an objectionable body odor reminiscent of rotting fish due to increased levels of free triethylamine in sweat and urine. This syndrome can lead to social isolation, clinical depression, and even suicide. This disease is linked to deficiency in the N-oxidation of trimethyl amines obtained from foods including meat, eggs, and soybeans. A single amino acid mutation has been discovered which leads to a Pro153 $\rightarrow$ Leu substitution that abolishes the catalytic activity of the enzyme. Although FMOs oxidize a number of drugs, pesticides and other xenobiotics, the toxicological significance of this polymorphism is still not known. However, affected individuals do have a deficiency in the ability to N-oxidize nicotine.

5.10.5b FMO2

FMO2 is one of five forms identified in mammals. It is expressed primarily in the lung and can catalyze N-oxidation of certain primary alkylamines. Surprisingly, analysis of the gene for FMO2 in humans indicates that it contains a C $\rightarrow$ T nonsense mutation at codon 472, resulting in a gene that produces a peptide that is lacking 64 amino acids at its 3′-terminus compared with the FMO2 gene of closely related primates including the gorilla and chimpanzee. Heterologous expression of the recombinant protein revealed it was catalytically inactive. While this is not a polymorphism, it is an example of a gene that is inactivated by a nonsense mutation in a whole species. Therefore, data on lung metabolism by FMO2 in laboratory animals is not predictive of human pulmonary metabolism.

5.10.6 Polymorphisms in Epoxide Hydrolase

Microsomal epoxide hydrolase matabolizes xenobiotic expoxides including epoxides of promutagenic and procarcinogenic polycyclic hydrocarbons. Mutations in this enzyme could be of potential importance in carcinogenesis. A number of SNPs have been reported that cause amino acid substitutions. However, these mutations do not affect the specific activity of the enzyme or rate of synthesis of the protein. Therefore, it is not clear whether any of the mutations in microsomal epoxide hydrolase have biological importance.

5.10.7 Polymorphisms in Serum Cholinesterase

A polymorphism in serum cholinesterase is one of the oldest polymorphisms known. It leads to prolonged muscle relation or prolonged paralysis after administration of the muscle relaxant succinylcholine. Several mutants occur, the most common is a point mutation causing the substitution of glycine for aspartic acid at position 70. This variant shows defective binding of choline esters to the anionic binding site but has normal activity with neutral or positively charged esters. There also numerous other variants, many with partial or complete loss of activity.

5.10.8 Polymorphisms in Paraoxonase

Paraoxonase hydrolyzes organophosphate esters, carbamates, and aromatic carboxylic esters. There are two allozymes (the Q and R forms) containing a single amino acid difference: either a glutamine (Q allele) or an arginine (R allele) at position 191. The type Q alloenzyme has higher hydrolytic activity toward sarin, soman, and diazoxon, while the R form shows a higher turnover for some substrates such as paraoxon, and the two enzymes have similar activities for other substrates such as the pesticide metabolite chlorpyrifos-oxon. Recently, it has been suggested that Gulf War veterans with neurological symptoms were more likely to have an R allele than to be homozygous for the Q allele.

5.10.9 Mechanisms of Polymorphisms

There are many different types of defective alleles. These include (1) splicing defects or changes at intron-exon junctions that cause frame shifts, alter the coding sequence, and generally produce premature stop codons and truncated inactive proteins; (2) insertions or deletions of nucleotides that cause frame shifts; (3) gene deletions or rearrangments; (4) single nucleotide changes that can either result in amino acid changes that may affect catalytic activity, or produce premature stop codons or silent (noncoding) changes; and (5) alterations in promotor upstream regions have been reported for some genes such as *CYP3A* and *CYP2E1*.

Splicing defects occur because most eucaryotic genes contain exons that code for protein interrupted by noncoding regions or introns. The DNA is transcribed into pre-mRNA, then the introns are spliced out from the RNA to form mRNA, which is translated into the protein (Fig. 5.19). Correct splicing depends partially on conserved splice sites at intron-exon junctions. There is normally a gt at the 5′-end of the donor splice site of the intron and an ag at the 3′-end of the intron (Fig. 5.20A). Changes in splice site junctions often are responsible for incorrect splicing. For example, the most common defective *CYP2D6* allele contains a mutation in the acceptor site of a splice junction (*CYP2D6*4*) (Fig. 5.20B). There is also a null mutation in which *CYP2D6* is completely deleted. There are also ultrarapid metabolizers of CYP2D6 substrates who contain more than two copies of the *CYP2D6* gene. These individuals produce more CYP2D6 protein, which decreases the effectiveness of commonly used doses of compounds metabolized by CYP2D6 in these individuals. A single base mutation that causes a new splice site in exon 5, and a single base substitution that produces a stop codon are the most frequent defect respon-

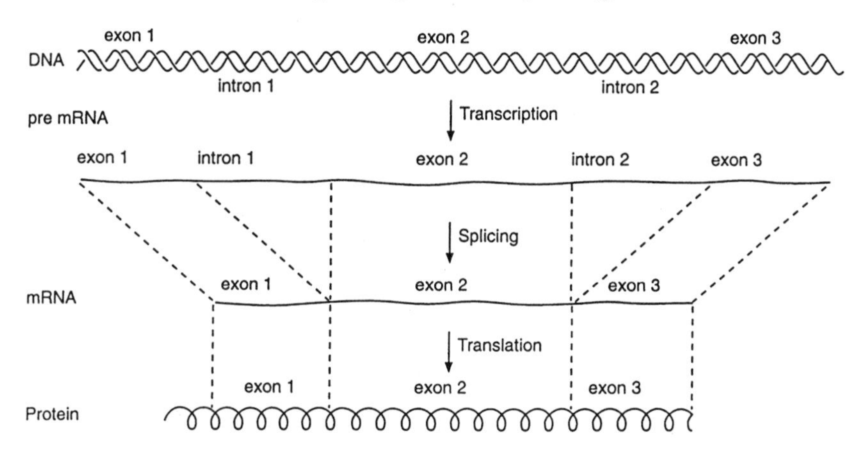

FIGURE 5.19 Normal splicing mechanism of eucaryotic genes.
(Taken from Genes, 2nd ed (B. Lewis, ed) John Wiley & Sons, NY, 1983, with permission).

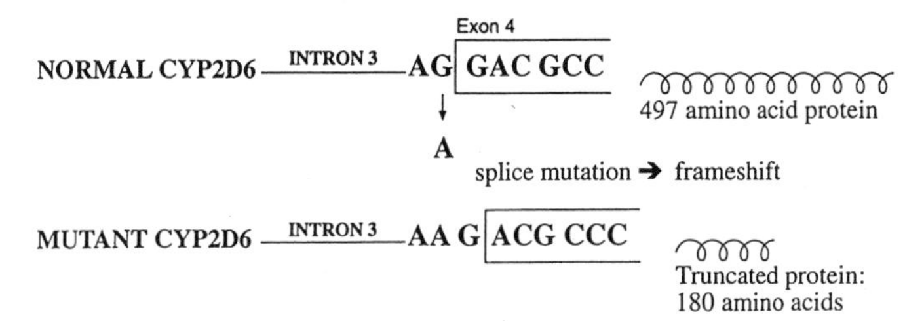

FIGURE 5.20 Mutation of intron splice sites involved in formation of a mutant allele of *CYP2D6**. (A) Normal intron consensus sequences of mammalian gene. (B) Mutation of the *CYP2D6* splice site that moves the splice site downstream to the next 3′ consensus sequence, altering the reading frame of the protein and resulting in a premature stop codon (*CYP2D6*4*) and a truncated, inactive protein.

sible for PMs of drugs metabolized by CYP2C19. Mutations that cause amino acid changes that alter activity are also responsible for additional PMs of CYP2C19 drugs. A single amino acid change in *CYP2C9* (*CYP2C9*3*) is responsible for PMs of the antidiabetic drug tolbutamide, the anticoagulant warfarin, and the anticonvulsant phenytoin. This change alters the affinity of the protein for various drugs (see Table 5.5 for examples).

5.10.10 Consequences of Polymorphisms in Drug Metabolism

The first consequence of a polymorphism in a drug metabolizing pathway is to alter the disposition and pharmacokinetics of the drug. For drugs with a narrow therapeutic index, this may increase toxicity. For example, PMs of CYP2D6 substrates exhibit exaggerated hypotension with debrisoquine; the oxytocic drug sparteine was taken off the market because of fetal deaths in PMs; perhexiline is no longer used because of increased neuropathy in PMs; and the antiarrythmic drug encainamide was removed from the market because of toxicity in PMs (see Table 5.5 for examples). The polymorphism in cholinesterase can produce apnea and prolonged paralysis after administration of the muscle relaxant succinylcholine. There is increased toxicity in CYP2C9 PMs treated with the anticonvulsant phenytoin, and much lower doses of the anticoagulant warfarin are required in CYP2C9 PMs. In contrast, some substrates are activated by drug metabolizing enzymes, and codeine is ineffective in PMs of CYP2D6.

A polymorphism may also confer an advantage to PMs. For example, omeprazole is more effective in treatment of ulcers in PMs of CYP2C19, due to greater efficacy in the eradication of Helicobacteria. Moreover, in rapid metabolizers of CYP2D6 (containing multiple copies of the gene), CYP2D6 substrates are less effective at any given dose.

Because of these differences in efficacy and toxicity, polymorphisms in drug metabolizing enzymes are of great importance to the pharmaceutical industry. The development of a new drug may be impacted negatively if its metabolism is mediated via a pathway that is known to be polymorphic, because of the chances of increased drug toxicity or decreased efficacy in a certain proportion of the population. For this reason, whether the metabolism of a drug is mediated by a known polymorphic pathway is studied with human liver preparations in vitro and recombinant enzymes during the development of a drug, and clinical trials now address the effect of any known polymorphism on pharmacokinetics of the drug.

5.10.11 Methods Used in Studying Polymorphisms

Polymorphisms are discovered by resequencing genes in known PMs or by resequencing a gene in a random set of the population. Verification of a polymorphism involves developing a genetic test for a mutation, and testing a population that has been phenotyped with a drug in vivo. If the polymorphism is inherited recessively, PMs should have two mutant (m_{1-n}) alleles, and EMs should have at least one wild-type allele. Expression of the wild-type cDNA versus the mutated cDNA in a recombinant system such as *Escherichia coli* or bacculovirus permits the investigation of the effect of that mutation on catalytic activity of the two recombinant proteins, and expression in a mammalian cell line (e.g., Cos-1 cells) is useful for determining whether the mutation affects the level of expression or stability of the protein in mammalian cells.

Phenotyping of individuals involves administering the test drug, and collecting timed urine or blood samples for analysis of parent drug and/or metabolites. This technique has the advantage of detecting unknown genetic mutations, but it is labor intensive and invasive to the patient or volunteer. This method obviously cannot be used on large populations.

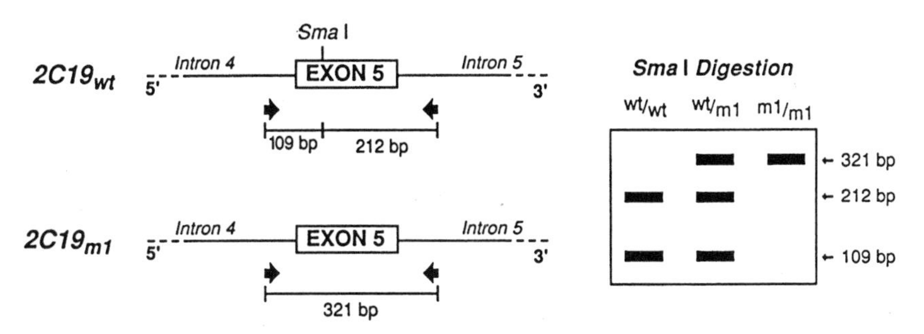

FIGURE 5.21 A typical restriction length polymorphism-polymerase chain reaction (RFLP-PCR) test for the *CYP2C19*2* mutation. Complementary primers that are specific for CYP2C19 (heavy arrows) are used to amplify a 312 bp fragment. The DNA is digested with a restriction enzyme *Sma*I and electrophoresced on an agarose gel. The wild-type DNA is cut by *Sma*I into two pieces of 109 bp and 212 bp, whereas the homozygous mutant allele is not cut. Individuals heterozygous for the two alleles show all three bands.

Genotyping involves detecting known defective alleles. This can be done from a single drop of blood, and the methodology is rapid. However, to be effective, most of the known defective alleles must be known.

A common type of genetic test for polymorphisms involves amplification of the affected region of the gene by the polymerase chain reaction (PCR). Mutations often affect the way specific restriction enzymes such as *Sma*I cut the DNA. If a mutation affects a restriction enzyme site, PCR of the region can be followed by restriction by the enzyme, the DNA run on a gel, and the cutting pattern will reveal the genotype (Fig. 5.21). The test is known as PCR-RFLP (PCR-restriction length polymorphism). Alternatively, if no restriction site is produced, a restriction site can often formed by placing a mismatch in one of the PCR primers so that only the normal or mutant DNA will be cut (mismatch PCR). *Allele specific PCR* is the use of two sets of primers, the 3′ end of one of the primers matches the wild-type DNA and the 3′ end of the other primer matches only the mutant DNA. With careful control of conditions, only the DNA will be amplified that is complementary to the correct primer. Southern blotting is still used to identify gene deletions.

Recently more *rapid throughput* procedures have been developed for studying polymorphisms. Only one such procedure is described. This procedure involves the binding of short oligonucleotides (up to 25 bp) on a solid surface such as nylon or silica. This is referred to as microarray- (or chip-) based hybridizational analysis. Probes are designed so that the questionable base is in the center position of the probe. Arrays of 400,000 nucleotides can be produced in a 20 μm region. Five million probes will currently fit on a 2 × 2 cm array. Scanning techniques such as laser confocal fluorescence scanning can be used to analyze the results. Affymetrix, Inc. (Santa Clara, CA) is one company commercializing this type of high throughput methodology (for further reading on this subject, see *Nature Genetics*, 21, No. 1, supplement, January 1999). Microarrays that analyze for the more common CYP2D6 and CYP2C19 methods have been developed for commercial use. Other types of high density methods dependent on PCR technology are being developed for genotyping.

5.10.12 Odds Ratios and Epidemiology

In epidemiology studies, the odds ratio is used to estimate relative risk. This estimates how likely you are to have a particular disease (e.g., bladder cancer or colon cancer) if you have a particular risk factor (e.g., if you are homozygous for a defective genotype vs. how likely you are to have the disease if you are not in the risk group (e.g., an EM with one wild-type copy of gene). The methods for calculating odds ratios are covered in Chapter 17 (see the example for the acetylation polymorphism). An odds ratio significantly >1.0 indicates that the PM genotype is at risk. Odds ratio of 1.0 means no effect of the PM genotype.

The effects of various CYP polymorphisms as well as other drug-metabolizing polymorphisms (e.g., the acetylation polymorphisms, GST polymorphisms, see Chapter 6) on risk to various types of cancer is an area of active study. The *CYP2D6* splice site mutation has been associated with altered risk to meningioma and astrocytoma. A polymorphism in a *Msp*1 restriction site in the 3′ region of *CYP1A1* has been widely studied with respect to different cancers and has been associated with increased lung cancer in Japanese smokers. The functional relevance of the 3′ polymorphism itself seems doubtful, but it has been associated with increased inducibility of the *CYP1A1* gene, which is one possible explanation of the association with lung cancer in Japanese studies. However, the importance of the *Msp*1 polymorphism to lung cancer in Caucasians has been debated, at least partly because the low incidence of this polymorphism in Caucasians lowers the power of statistical testing. CYP2E1 metabolizes a variety of tobacco-derived chemicals including nitrosamines. A polymorphism in a *Rsa*1 restriction site in the upstream region of *CYP2E1* has been associated with increased expression of the promotor constructs in vitro. The *Rsa*1 polymorphism and a *Dra*1 restriction site polymorphism have been widely studied with respect to different cancers. The *Dra*1 restriction site has been associated with increased risk to lung cancer in Japanese smokers. Recent preliminary studies have suggested an increased risk for lung cancer and esophageal cancer in Japanese PMs of CYP2C19.

SUGGESTED READING

Arinc, E., Schenkman, J.B. and Hodgson, E. (Eds), *Molecular and Applied Aspects of Oxidative Drug Metabolizing Enzymes*, Kluwer Academic/Plenum Publishers, New York, 1999.

Beedham, C. Molybdenum hydroxylases: Biological distribution and substrate-inhibitor specificity, *Progress in Medicinal Chemistry*, Vol. 24, G.P. Ellis and G.B. West (Eds), Elsevier Biomedical, New York, 1987, pp. 85–127.

The Chipping forecast. *Nat. Genet.*, vol. 21, no. 1. supplement (January 1999). Eichelbaum, M., Gross, A.S. The Genetic Polymorphism of Debrisoquine/Sparteine Metabolism-Clinical Aspects. Pharmacol. Ther. 46 (1990), 377–394.

Goldstein, J.A. and Blaisdell, J. Genetic Tests Which Identify the Principle Defects in CYP2C19 Responsible for the Polymorphism in Mephenytoin Metabolism. In: Johnson, E.F., Waterman, M.R. (Eds), *Methods in Enzymology*, vol. 272, Cytochrome P450, Part B, Academic Press, San Diego, CA, 1996, pp. 210–217.

Gonzalez, F.J. The molecular biology of cytochrome P450s. *Pharmacol. Rev.* 40 (1990), 243–288.

Gorrod, J.W., Oelschlager H. and Caldwell, J. (Eds), *Metabolism of Xenobiotics*, Taylor and Francis, London and Philadelphia, 1988.

Guengerich, F.P. Enzymatic oxidation of xenobiotic chemicals. *Crit. Rev. Biochem. Mol. Biol.* 25 (1990), 97–153.

Kato, R., Estabrook, R.W. and Cayen, M.N. (Eds), *Xenobiotic Metabolism and Disposition*, Taylor and Francis, London, 1989.

Kulkarni, A.P. and Hodgson, E. The metabolism of insecticides: the role of monooxygenase enzymes. *Ann. Rev. Pharmacol. Toxicol.* 24 (1984), 19–42.

Lewin, B. *Genes*, 2nd Ed., John Wiley and Sons, New York, 1985.

Nebert, D.W. Polymorphisms in drug metabolizing enzymes: What is their clinical relevance and why do they exist. *Am. J. Hum. Genet.* 60 (1997), 265–271.

Nelson, D.R., Koymans, L., Kamataki, T., Stegeman, J.J., Feyereisen, R., Waxman, D.J., Waterman, M.R., Gotoh, O., Coon, M.J., Estabrook, R.W., Gunsalus, I.C. and Nebert, D.W. The P450 superfamily: update on new sequences, gene mapping, accession numbers and nomeclature. *Pharmacogenetics* 6 (1996), 1–42.

Nelson, D.R. P450 website; http://drnelson.utmem.edu/CytochromeP450.html

Okey, A.B. Enzyme induction in the cytochrome P450 system. *Pharmacol. Ther.* 45 (1990), 242–298.

Ryan, D.E. and Levin, W. Purification and characterization of hepatic microsomal cytochrome P450. *Pharmacol. Ther.* 45 (1990), 153–239.

Ziegler, D.M. Flavin-containing monooxygenases: enzymes adapted for multisubstrate specificity. *Trends Pharmacol. Sci.* 11 (1990), 321–324.

Website: http://www.imm.ki.se/CYPalleles/ and references therein.

CHAPTER SIX

Conjugation and Elimination of Toxicants

GERALD A. LEBLANC and WALTER C. DAUTERMAN

6.1 INTRODUCTION

Lipophilic toxicants can passively diffuse across the lipid-rich surface membranes of cells. Upon entry in into the cell, the toxicant is available to interact with molecular target sites of toxicity. Phase I biotransformation processes can reduce the reactivity or toxicity of the chemical through processes such as hydroxylation (Chapter 5). Toxicants can also undergo metabolic conjugation, either directly or subsequent to phase I biotransformation. The conjugation process or phase II detoxification involves coupling of the toxicant to small, endogenous molecules that are present within the cell. Conjugation typically reduces the reactivity of the toxicant and, hence, reduces its toxicity to the cell. Conjugation also generally facilitates the elimination of toxicants from the body through aqueous routes (i.e., urine) by increasing the aqueous solubility of the molecule.

Because both phase I and II biotransformation processes can increase the polarity and aqueous solubility of the toxicant, these biotransformations can essentially trap the toxicant in the cell by compromising its ability to passively diffuse across the surface membrane of the cell. The cellular elimination of toxicants is facilitated by membrane proteins that actively transport phase I and II biotransformation products out of the cell and making them available for elimination from the body. The active cellular elimination processes are often referred to as phase III detoxification/elimination processes.

6.2 CONJUGATION REACTIONS

Conjugation reactions may be divided into two types of mechanistic reactions. The first involves the formation of a conjugate in which the xenobiotic reacts with a high-energy or reactive endogenous ligand.

Type I: xenobiotic + reactive conjugating ligand = conjugated product

TABLE 6.1 Conjugation Reactions.

Reaction	Enzyme	Functional Group
Glucuronidation	UDP-glucuronyltransferase	OH COOH NH_2 SH C—C
Glucosidation	UDP-glucosyltransferase	OH COOH SH C—C
Sulfation	Sulfotransferase	NH_2 OH
Acetylation	Acetyltransferase	NH_2 SO_2NH_2 OH
Methylation	Methyltransferase	OH NH_2 SH
Amino acid conjugation	Acyltransferase	COOH
Glutathione conjugation	Glutathione S-transferase	Epoxides Organic halides Organic nitro compounds Unsaturated compounds
Lipophilic conjugation		COOH OH

The second type of conjugation reaction involves the coupling of the endogenous conjugating ligand with a high-energy or reactive xenobiotic. The reactivity associated with the xenobiotic is sometimes the consequence of phase I biotransformation (i.e., epoxidation, Chapter 5).

Type II: reactive xenobiotic + conjugating ligand = conjugated product

Type I reactions include the formation of glycosylated, sulfated, methylated, and acetylated conjugates, whereas type II reactions include peptide and glutathione conjugation.

In order for conjugation to occur, hydroxyl, amino, carboxyl, epoxide, thiol, or halogen groups must be present in the xenobiotic molecule (Table 6.1). Conjugation reactions most commonly result in the formation of hydrophilic products. These products are generally less lipid soluble, more polar, more readily eliminated, and less toxic. However, lipophilic conjugates of xenobiotics can also be generated that result in the storage of the xenobiotic in lipid compartments of the organism and cause a delay in excretion. Although conjugation reactions have been considered generally to result in detoxification of the xenobiotics, a number of examples of conjugation products that are the ultimate toxicants have been reported (see Section 6.10).

6.3 GLYCOSIDES

Glycosylation is the most common form of conjugation found in the animal and plant kingdom. The reactive intermediates are all derived from the universal fuel, glucose, and the supply of this molecule is less likely to be depleted than that of other endogenous intermediates. Also, glycosylation is a major pathway because the endogenous conjugating molecule has the capacity to react with a wide range of molecular groupings.

In order for glycosylation to occur, a high-energy endogenous molecule must be available to donate the conjugating group to the aglycone (targeted xenobiotic). For glucosidation, the reactive intermediate is uridine diphosphate glucose (UDPG); and for glucuronidation, the reactive intermediate is uridine diphosphate glucuronic acid (UDPGA).

$$\text{UDP} + \text{D-glucose-1-Pi} \xrightarrow{\text{UDPG pyrophosphorylase}} \text{UDP-}\alpha\text{-D-glucose} + \text{PPi}$$

$$\text{UDP-}\alpha\text{-D-glucose} + 2\text{NAD} + \text{H}_2\text{O} \xrightarrow{\text{UDPG dehydrogenase}} \text{UDP-}\alpha\text{-D-glucuronic acid} + 2\text{NADH}_2$$

Glucuronidation is a major conjugative pathway in vertebrates, whereas glucosidation generally is associated with plants and invertebrates. The multistep reaction for glycoside formation involves the two reactions presented below:

1. $\text{UDPG} + \text{ROH} \xrightarrow{\text{UDP glucosyl transferase}} \text{RO-}\beta\text{-D-glucoside} + \text{UDP} + \text{H}_2\text{O}$
2. $\text{UDP} + \text{ROH} \xrightarrow{\text{UDP glucuronosyl transferase}} \text{RO-}\beta\text{-D-glucuronide} + \text{UDP} + \text{H}_2\text{O}$

The first reaction forms glucosides and the second forms glucuronides. The reactions result in the nucleophilic displacement of the UDP moiety. The acceptor group attacks the C_1 of the pyranose ring to which UDP is attached in an α-glycosidic bond and a Walden inversion occurs. The resulting glycoside has a β-configuration.

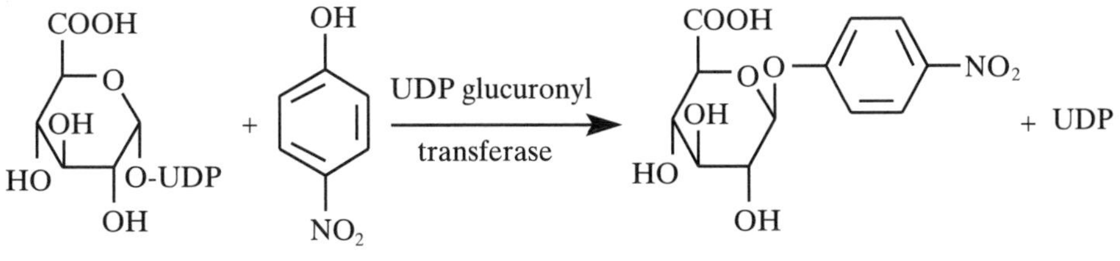

COOH, OH, HO, O-UDP, OH + OH, NO_2 —UDP glucuronyl transferase→ COOH, OH, HO, OH, O, NO_2 + UDP

For a high rate of glycosylation to occur, the reactive group of the xenobiotic must be sufficiently nucleophilic. Any molecule that is nucleophilic or capable of becoming nucleophilic is, in theory, a candidate for glycoside bond formation. The reactivity of the aglycone will depend on both steric and electronic factors.

6.3.1 Glucuronides

A wide variety of compounds can be substrates for glucuronidation. Four general categories of O-, N-, S-, and C-glycosides have been reported (Table 6.2). Both glucosides and glucuronides react to form phenolic, enolic, and ester glycosides. Similarly, thiolic and dithioic glycosides have been reported for the S-glycosides. Several types of N-glucuronides have been identified when the glucuronosyl moiety is attached to the nitrogen. They include the following functional groups: aromatic amino, sulfonamide, carbamyl, and heterocyclic nitrogen. Although C-glucosyl compounds have been known in plants for some time, only recently have C-glucuronides been demonstrated with C(4)-β glucuronides of phenylbutazone. Many factors affect glucuronidation such as age, diet, sex, species and strain differences, genetic factors, and diseases (see Chapter 7).

Xenobiotic glucuronidation occurs in the liver, the intestinal mucosa, and the kidney. Almost all other organs and tissue possess some glucuronidation activity but at very low levels. Normally, the glucuronide conjugates are excreted in the urine and bile unless they are involved in enterohepatic circulation. The conjugates may be hydrolyzed by β-glucuronidases present in the various tissues or organs or upon biliary excretion by bacterial β-glucuronidases.

Glucuronidation of many substrates is low or undetectable in fetal mammalian tissues but increases with age. The rate of development is dependent upon the species, tissue, and substrate. The inability of newborns of most mammals, except the rat, to form glucuronides is associated with deficiencies in glucuronosyl transferase activity and the cofactor, UDPGA. Slow excretion of the glucuronide conjugate may also hinder its formation. The blood serum of newborn babies may contain pregnandiol, which is an inhibitor of glucuronide formation. The inhibition as well as low levels of the conjugating system are responsible for an increase in bilirubin levels and the development of neonatal jaundice.

UDP-glucuronosyl transferases (UGT), the enzymes responsible for glucuronidation reactions, are the products of a multigene superfamily. Rat liver UGTs thus far identified are members of two gene families, UGT1 and UGT2. Each gene family consists of at least four distinct enzymes. Four enzymes constituting the UGT1 family are the products of alternate splicing of a single gene. UGT1 family members are inducible by Ah-receptor ligands (e.g., 2,3,7,8-tetrachlorodibenzo-p-dioxin, 3-methylcholanthrene) phenobarbital, clofibric acid, and pregnenolone-16α-carbonitrile (PCN). UGT1 family members are responsible for bilirubin and thyroxine conjugation as well as the conjugation of some xenobiotics (i.e., opioids). Humans afflicted with Crigler-Najjar syndrome type 1 and the Gunn strain of rat are deficient in the ability to form glucuronide conjugates of bilirubin. This deficiency is due to a mutation in a UGT1 gene and results in the accumulation of unconjugated bilirubin (unconjugated hyperbilirubinemia). Humans produce at least six distinct UGT1 gene products.

Thyroid hyperplasia, hypertrophy, or tumorigenesis has been associated with

TABLE 6.2 Types of Glucuronides Formed.

O-Glucuronides

R–(aryl)–OH + UDPGA Phenols	⟶	R–(aryl)–$O \cdot C_6H_9O_6$
RCH_2OH + UDPGA Alcohols	⟶	$RCH_2O \cdot C_6H_9O_6$
RCOOH + UDPGA Carboxylic acids	⟶	$RC(O)O \cdot C_6H_9O_6$
R—NH—OH + UDPGA Hydroxylamines	⟶	$RNHO \cdot C_6H_9O_6$
RC(O)NH—OH + UDPGA Hydroxamic acids	⟶	$RC(O)NHO \cdot C_6H_9O_6$

S-Glucuronides

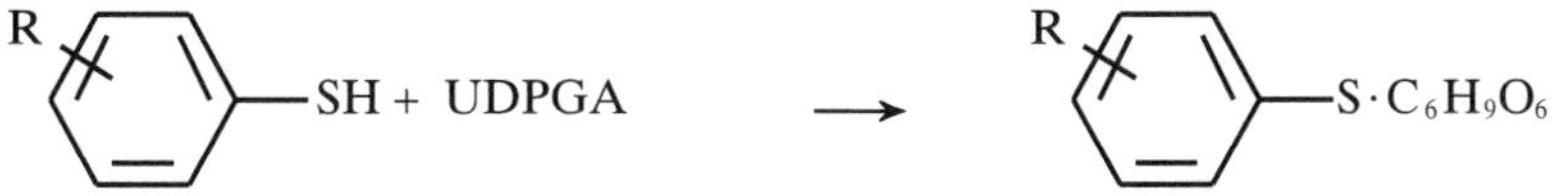

R–(aryl)–SH + UDPGA Thiophenols	⟶	R–(aryl)–$S \cdot C_6H_9O_6$
(thiazoline ring: N, S) C—SH + UDPGA Thiols	⟶	(thiazoline ring: N, S) C—S + $C_6H_9O_6$
R_2N—C(S)SH + UDPGA Thiocarbamic acids	⟶	R_2N—$C(S)S \cdot C_6H_9O_6$

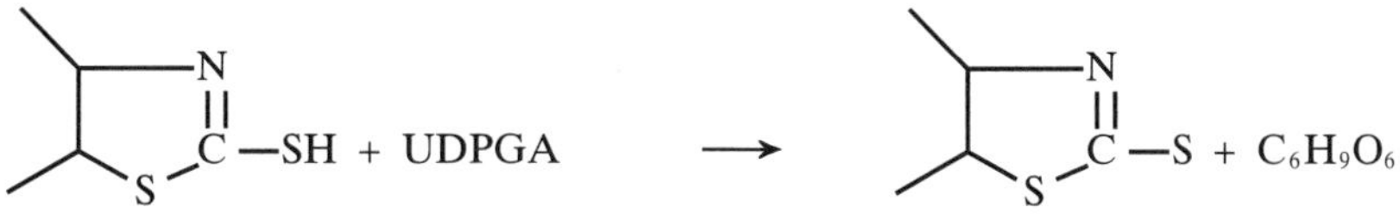

N-Glucuronides

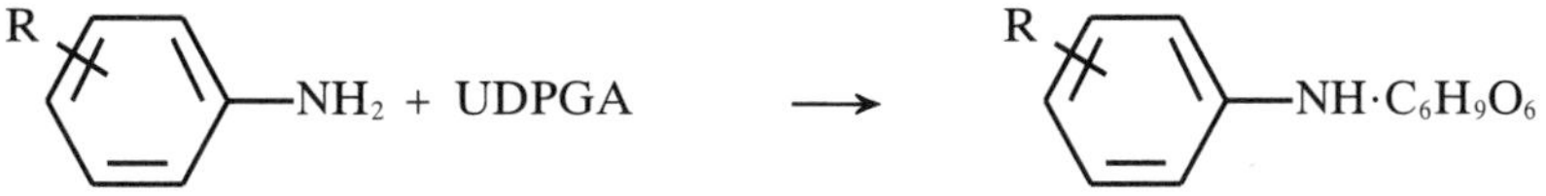 R–(aryl)–NH_2 + UDPGA Aromatic amines	⟶	R–(aryl)–$NH \cdot C_6H_9O_6$
R—NHOH + UDPGA Hydroxylamines	⟶	$RN(OH) \cdot C_6H_9O_6$

TABLE 6.2 *Continued*

$RCH_2OC(O)NH_2$ + UDPGA Carbamates	⟶	$RCH_2OC(O)NH \cdot C_6H_9O_6$
RSO_2—NH—R + UDPGA Sulfonamides	⟶	RSO_2—NR$\cdot C_6H_9O_6$
C-Glucuronides		
$R_1-\overset{\overset{O}{\|}}{C}-CH-\underset{\underset{O}{\|}}{C}-R$ + UDPGA 1,3-dicarbonyl compounds	⟶	$R_1-\overset{\overset{O}{\|}}{C}-CH(C_6H_9O_6)-\underset{\underset{O}{\|}}{C}-R$

exposure of rodent models to xenobiotics such as 2,3,7,8-tetrachlorodibenzo-p-dioxin, diniconazole, and thiazopyr. This toxicity has been attributed to the inductive effects of these compounds on hepatic UGT1. Induction of UGT1 causes increased conjugation and clearance of thyroxine and decreased circulating thyroxine levels. In response, thyroxine stimulating hormone (TSH) secretion by the pituitary increases and the thyroid gland is stimulated to produce and secrete thyroxine. Overstimulation of the thyroid gland by TSH may account for hyperplasia, hypertrophy, and ultimately tumorigenesis in this tissue. Thus, thyroid toxicity associated with these chemicals may be secondary to effects of the chemicals on conjugative processes in the liver.

Enzyme products of the UGT2 gene family share less than 50% amino acid identity with the UGT1 family members. UGT2 enzymes appear to all be products of individual genes. The UGT2 family is divided into two subfamilies, UGT2A and UGT2B, that are differentially expressed in tissues. The UGT2A subfamily members are preferentially expressed in olfactory epithelium, whereas UGT2B subfamily members are preferentially expressed in the liver. Rat and human express at least four and six UGT2B gene products, respectively. Some of the UGT2 family members are susceptible to induction by phenobarbital. These enzymes are largely responsible for glucuronide conjugation of steroid hormones and some xenobiotics (i.e., 1-naphthol, 4-nitrophenol, 4-hydroxybiphenyl).

6.4 SULFATION

Sulfate conjugation is important in the biotransformation of xenobiotics as well as many endogenous compounds. It plays a role in the biosynthesis of thyroid and steroid hormones as well as certain proteins and peptides. Sulfation has been demonstrated in vertebrates, invertebrates, plants, fungi, and bacteria. Sulfate ester

TABLE 6.3 . Examples of Sulfate Conjugation to Alcohols, Phenols, and Aromatic Amines.

Alcohol conjugation

$$C_2H_5OH + PAPS \rightleftharpoons C_2H_5OSO_3H + ADP$$

Phenol conjugation

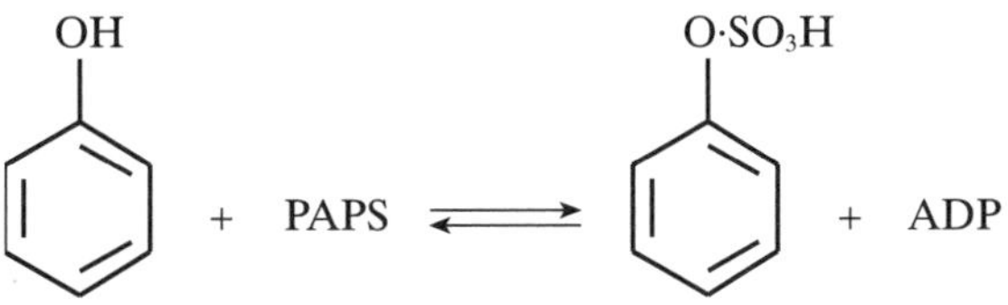

$$C_6H_5OH + PAPS \rightleftharpoons C_6H_5O\cdot SO_3H + ADP$$

Aromatic amine conjugation

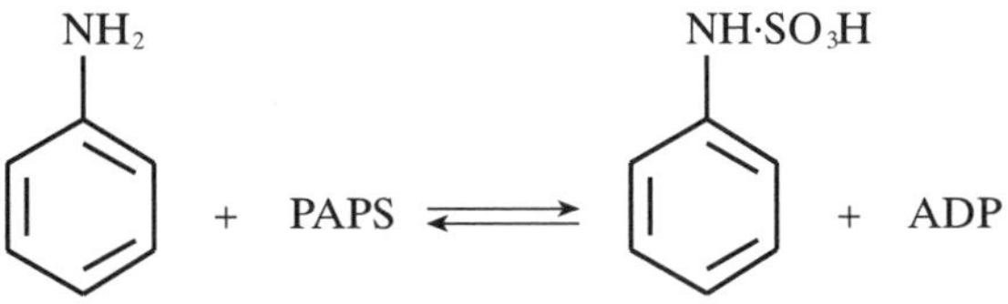

$$C_6H_5NH_2 + PAPS \rightleftharpoons C_6H_5NH\cdot SO_3H + ADP$$

formation is known to occur with primary, secondary, and tertiary alcohols, phenols, and arylamines. Sulfate esters, in terms of biological conjugations, are in reality half-esters (i.e., $ROSO_3^-$) that are completely ionized, very water soluble, and quickly eliminated from the organism. With many xenobiotics, sulfation appears to be a detoxification reaction. However, a number of compounds are converted into highly labile sulfate conjugates that form reactive intermediates that have been implicated in carcinogenesis and tissue damage (see Section 6.10).

The sulfate molecule used in conjugation reactions is derived from 3-phosphoadenosine-5′-phosphosulfate (PAPS). Efficient synthesis of PAPS requires an adequate amount of inorganic sulfate with L-cysteine, D-cysteine, or L-methionine serving as precursors for the inorganic sulfate. PAPS is likely synthesized in every vertebrate cell, with the concentration of PAPS being the highest in the liver.

The most common acceptor for sulfation is the hydroxyl group of phenols, alcohols, and N-substituted hydroxylamines (Table 6.3). In addition, sulfation of thiols and amines forming thiosulfates and sulfamates have been reported. In most species, both the sulfation and glucuronidation of the same phenolic substrates occur. Sulfation often provides for high-affinity, low-capacity catalysis and provides for efficient substrate conjugation at low substrate concentrations. Glucuronidation provides for low-affinity, high-capacity catalysis and provides for efficient substrate conjugation at high substrate concentrations. Sulfate conjugates are excreted predominately in the urine and to a lesser degree in the bile. Sulfotransferases are typically not inducible by the common inducers of the microsomal drug-metabolizing enzymes (i.e., Ah receptor ligands, phenobarbital). However, levels of some sulfotransferases can be elevated by steroid hormones or endocrine modulators.

The sulfotransferase enzymes comprise a superfamily of enzymes. These enzymes can be either membrane bound or cytosolic, though the enzymes involved in xenobiotic conjugation are primarily cytosolic. Although various nomenclatures have been used to describe the sulfotransferase, the major subfamilies of cytosolic sulfotransferases thus far identified are commonly referred to as SULT 1, SULT 2, and SULT 3.

The SULT 1 subfamily consists of the phenol sulfotransferases (formally called P-PST). Members of this subfamily are responsible for the sulfate conjugation of endogenous substrates such as thyroid hormone, catecholamines, and 17β-estradiol as well as exogenous compounds such as the drugs acetaminophen and minoxidil. Four members of this subfamily have been identified in humans (SULT 1A1, SULT 1A2, M-PST, EST). Each enzyme is the product of a distinct gene. SULT 1A1 and SULT 1A2 are involved in the conjugation of a variety of endogenous and exogenous phenolic compounds. M-PST and EST exhibit greater substrate specificity and are primarily responsible for the conjugation of monoamine neurotransmitters such as catecholamines and estrogens such as 17β-estradiol, respectively.

The SULT 2 sulfotransferase subfamily consists of the hydroxysteroid sulfotransferases. Three sulfotransferases that comprise this subfamily (SULT 2A1, SULT 2B1a, SULT 2B1b) have been identified in humans. Dehydroepiandrosterone (DHEA) is the prototypical substrate for the SULT 2 enzymes. However, other hydroxysteroids such as testosterone and its phase I hydroxylated derivatives are substrates for these enzymes. The SULT 2 sulfotransferases are also responsible for the sulfate conjugation of a variety of alcohols and xenobiotics that have undergone phase I hydroxylation. SULT 2A1 is the product of a distinct gene, whereas SULT 2B1a and SULT 2B1b are both products of the same gene. These two distinct enzymes are derived from the same gene by alternative transcription initiation and alternative splicing.

SULT 3A (also known as ST 3A or AST-RB1) is responsible for the sufation of amines such as aniline, 4-chloroaniline, and 2-naphthylamine. Thus far, a single member of the subfamily has been isolated from rabbit. Both SULT 3A and SULT 1 subfamily members are known to contribute to the sulfate conjugation of amines. Other amine sulfotransferases have been purified (i.e., HAST 1, HAST 3, HAST 4, HAST 4v, AST-RB2) that may represent additional members of these or other sulfotransferase subfamilies.

6.5 METHYLATION

Methylation is a common biochemical reaction involving the transfer of methyl groups from one of two methyl donor substrates (S-adenosylmethione or N^5-methyltetrahydrofolic acid). The transfer occurs with a wide variety of methyl-acceptor substrates (proteins, nucleic acids, phospholipids, and other molecules of diverse structures). Of the cofactors, S-adenosylmethionine (SAM) is the most important in the methylation of xenobiotics containing oxygen, nitrogen, or sulfur nucleophiles. SAM is biosynthesized from L-methionine and ATP by ATP:L-methionine S-adenosyltransferase. Methyl conjugates generally are less water soluble than the parent compound, except for the tertiary amines. Despite the decrease in water solubility, methylation generally is considered a detoxification reaction. Methylation reactions are classified according to substrates as follows.

6.5.1 N-Methylation

N-Methylation is an important reaction by which primary, secondary, and tertiary amines are substrates of methylation. Most tissues catalyze the methylation of a large variety of amines. The source of the methyl group that is transferred in each instance is SAM, and the products are secondary, tertiary or quaternary N-methylamines as well as S-adenosyl-L-homocysteine (SAH). The reaction shown below is with a primary amine as substrate and is catalyzed by an amine N-methyltransferase:

$$RCH_2NH_2 + SAM \rightarrow RCH_2NHCH_3 + SAH$$

Humans express at least three forms of N-methyltransferase that are catagorized base upon substrate specificity. Nicotinamide N-methyltransferase methylates compounds have an indole ring such as serotonin and tryptophan or a pyridine ring such as nicotinamide and nicotine. Histamine methyltransferase is a cytoplasmic enzyme characterized by its ability to methylate histamine to form 3-N-methylhistamine. The transferase is expressed in several tissues including the kidney, gastric mucosa, erythrocytes, brain, and skin. Phenylethanolamine N-methyltransferase, also known as noradrenaline methyltransferase, catalyzes the methylation of norepinephrine to epinephrine, the final biosynthetic step is formation of the hormone. The enzyme is located in the soluble fraction of the adrenal gland, retina, heart, and brain. Multiple forms of phenylethanolamine N-methyltransferase have been reported to be present in several species. At least five forms have been described in cow and rabbit, two in the dog, one in rat, and one in human. A variety of phenylethanolamines, but not phenylethylamine, serve as methyl acceptor substrates when SAM is the methyl donor. Examples of natural substrates are norepinephrine, epinephrine, octopamine, normethaneprine, metanephrine, and synephrine. In general, primary amines are better substrates than secondary amines.

6.5.2 O-Methylation

Catechol-O-methyltransferase (COMT) is widely distributed throughout the animal and plant kingdom and localized in the soluble fraction of rat liver, skin, kidney, glandular tissue, heart, and brain. However, a membrane-bound COMT also has been detected in a variety of tissues. Upon solubilization, this enzyme appears to have properties similar to the soluble COMT. Multiple forms of COMT may be expressed within a species. However in humans, COMT appears to be encoded by a single gene. COMT catalyzes the transfer of a methyl group from SAM to one of the phenolic groups of a catechol in the presence of Mg^{++}ions. With only a few exceptions, all of the methyl acceptor substrates of COMT require the catechol moiety.

HO COOH + SAM → CH_3O COOH + SAH

HO HO

3,4-Dhydroxybenzoic acid 3-Methoxy-4-hydroxybenzoic acid

Three types of compounds are known to inhibit the in vitro activity of COMT. SAH is a potent inhibitor of COMT as well as the other SAM-dependent methyltransferases. The inhibition seems to be the result of a feedback control mechanism. Certain divalent ions such as Ca^{+2} and trivalent metal ions such as the salts of lanthanides, neodymium, and europium are excellent inhibitors of COMT. A number of catechol-type substrates such as pyrogallol, flavonoids, pyrones, pyridenes, hydroxyquiolines, 3-mercaptotyramine, and tropolones act as irreversible inhibitors for COMT.

Phenol O-methyltransferase (POMT) is a membrane-bound methyltransferase that is responsible for the methylation of phenolic compounds that are more commonly recognized as substrates for glucuronic acid or sulfate conjugation (i.e., acetaminophen). Phenol O-methylation is typically regarded as a minor catalytic processes with respect to xenobiotic detoxification.

6.5.3 S-Methylation

Methyl groups from SAM are also transferred to thiol-containing drugs and xenobiotics. It appears that the only structural requirements are the presence of a free thiol and that the compound is not too hydrophilic in nature. S-Methylation has been demonstrated with mercaptoethanol, thioacetanilide, 6-mercaptopurine, as well as a variety of thiopurine and thiopyrimidine drugs. In general, aromatic thiols are the best substrates, while simple aliphatic thiols are less active.

$$\underset{\text{Mercaptoethanol}}{HSCH_2CH_2OH} + SAM \xrightarrow{\text{Thiol S-methyltransferase}} \underset{\text{S-methylthioethanol}}{CH_3SCH_2CH_2OH} + SAH$$

Thiol methyltransferase has been detected in erythrocytes, lymphocytes, lungs, and cecal and colonic mucosae. At the present time, it is not clear as to the nature and number of thiol methyltransferases. A soluble and a microsomal enzyme have been reported, with the microsomal enzyme being solubilized relatively easily. The microsomal enzyme in rat liver has been solubilized and purified to homogeneity. The enzyme is a 28,000 Da monomer with an isoelectric point of 6.2. A wide variety of xenobiotic thiols are methylated, but cysteine and glutathione are not substrates. S-Methylation is an important component in the "thiomethyl shunt." Thiomethyl conjugates are metabolized to the methylsulfoxides by oxidation (see Chapter 5) and reenter the mercapturic acid pathway as substrates for glutathione S-transferase.

6.6 ACYLATION

Foreign carboxylic acids and amines undergo biological acylation to form amide conjugates. Acylation reactions are of two types. The first involves an activated conjugating intermediate, acetyl CoA, and the xenobiotic. The reaction is referred to as acetylation.

$$\underset{\text{acetyl-CoA}}{CH_3C(O)SCoA} + \underset{\text{amine}}{RNH_2} \rightarrow \underset{\text{acetyl conjugate}}{RNHCOCH_3} + \underset{\text{CoA}}{CoASH}$$

The second type involves the activation of the xenobiotic to form an acyl CoA derivative, which then reacts with an amino acid to form an amino acid conjugate.

$$\underset{\text{acyl-CoA}}{RC(O)SCoA} + \underset{\text{glycine}}{NH_2CHCOOH} \rightarrow \underset{\text{amino acid conjugate}}{RC(O)NHCH_2COOH} + \underset{\text{CoA}}{CoASH}$$

6.6.1 Acetylation

Compounds containing amino or hydroxyl groups are quite readily acetylated in vivo. The only known example of acetylation of a sulfhydryl group occurs with the formation of acetyl-CoA.

The acetylation of amino groups is quite common. Five types of amino groups (arylamino, aliphatic amino, α-amino, hydrazino, and sulfonamido) can be biotransformed by acetylation. Besides direct acetyl transfer, inter- and intramolecular transfers of the acetyl group from the nitrogen to the oxygen of arylhydroxamic acids have been reported. Recently, direct O-acetyl transfer has been shown to form carcinogenic acetoxyarylamines. Acetylation typically results in the masking of the amine group with a nonionizable acetyl moiety. As a result, the acetylated derivatives are generally less water soluble than the parent compound.

There is a large species variation in N-acetylation. The hamster and rabbit possess relatively high N-acetylating capacities, whereas the dog has little or no acetylating capacity. Other species such as rats and mice are intermediate. N-Acetyltransferase activity has been demonstrated in liver, intestinal mucosa, colon, kidney, thymus, lung, pancreas, brain, erythrocytes and bone marrow. N-Ethylmaleimide, iodoacetate, and *p*-chloromercuribenzoate are irreversible inhibitors of the N-acetyltransferase. A variety of divalent cations such as Cu^{2+}, Zn^{2+}, Mn^{2+}, and Ni^{2+} are also inhibitory.

Two cytoplasmic N-acetyltransferases, NAT1 and NAT2, have been identified in human. NAT1 is rather ubiquitous in its expression, whereas NAT2 is associated with the liver and the gastrointestinal tract. These enzymes have different but overlapping substrate preferences. A third enzyme, NAT3, has been identified in mouse.

Genetic polymorphisms in N-acetylation have been identified in humans and other species. The human population is segregated into slow acetylators and fast acetylators based upon rates of acetylation of the drug isoniazid. The incidence of slow acetylators is high among Middle Eastern populations, intermediate in Caucasian populations, and low in Asian populations. The slow acetylator phenotype is the result of mutations in the NAT2 gene that compromises the activity of the enzyme. Slow acetylators are predisposed to toxicity of drugs that are inactivated by acetylation such as isoniazid, dapsone, sulfamethazine, procainamide, and hydralazine. This enzyme also acetylates aromatic amine dyes to which workers have been exposed industrially such as benzidine dyes, 4-aminobiphenyl, 4,4′-methylbis(2-chloroaniline)naphthylamine, and o-toluidine. Workers in the arylamine dye industry who are slow acetylators have been shown to have an increased risk for bladder cancer. The low activity of NAT2 in the liver of slow acetylators may make the aromatic amines more available for hydroxylation by CYP1A2. The resulting hydroxylamines then accumulate in the bladder where they are acetylated by NAT1.

6.6.2 Amino Acid Conjugation

The conjugation of carboxylic acid xenobiotics with amino acids occurs in both liver and kidney and is catalyzed by an enzyme system located in the mitochondria. Conjugation requires initial activation of the xenobiotic to a CoA derivative in a reaction catalyzed by acyl:CoA ligase. The acyl CoA subsequently reacts with an amino acid, giving rise to acylated amino acid conjugate and CoA.

The reactions are catalyzed by acyl-CoA:amino acid N-acyltransferase, of which two distinct N-acyltransferases exist in mammalian mitochondria. The predominant transferase conjugates medium-chained fatty acyl CoA and substituted benzoic acid derivatives with glycine and is termed an aralkyl-CoA:glycine N-acyltransferase, whereas the other enzyme conjugates arylacetic acid derivatives with glycine, glutamine, or arginine and is an arylacetyl-CoA:amino acid N-transferase.

Amino acid conjugation has been reported to occur with glycine, glutamine, arginine, and taurine in mammals and certain primates. In other organisms, different amino acid acceptors are utilized in peptide conjugation. Ornithine is utilized by reptiles and some birds, whereas ticks utilize arginine and glutamine, insects use alanine, glycine, serine, and glutamine, and fish use taurine. There is more interspecies variation with this type of conjugation than in any other phase II reaction. The specific amino acids employed in peptide conjugation within a class of animals generally depend on the bioavailability of the amino acids from endogenous and dietary sources and are those amino acids that are normally nonessential. The extent of peptide conjugation with glycine in mammals follows the order: herbivores > omnivores > carnivores.

Amino acid conjugation typically results in detoxification and elimination of xenobiotics. Amino acid conjugation is an alternative conjugation process for carboxylic acid-containing xenobiotics such as ibuprofen. Glucuronidation of carboxylic acid-containing xenobiotics can result in the generation of toxic acylglucuronides. Amino acid conjugation of these xenobiotics limits the production of these toxic metabolites. Amino acid conjugates of xenobiotics are generally cleared from the body by urinary elimination.

6.7 GLUTATHIONE S-TRANFERASES

The glutathione S-transferases (GST) are a family of isoenzymes involved in the conjugation of reduced glutathione with electrophilic compounds. GSTs mediate

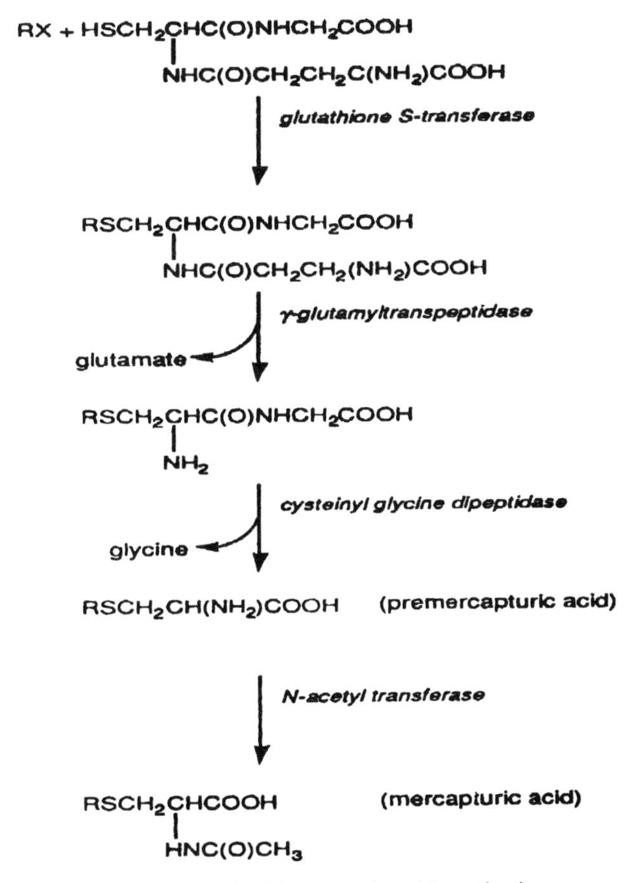

FIGURE 6.1 Mercapturic acid synthesis.

the initial reaction in the biosynthesis of mercapturic acids (Fig. 6.1). The biosynthesis of mercapturic acids occurs in several steps, initially involving the formation of a glutathione conjugate, then formation of cysteinyl glycine conjugate as the result of removal of glutamic acid by the γ-glutamyltranspeptidase. This is followed by removal of glycine by cysteinyl S-conjugate, that is, the premercapturic acid. Acetylation of the cysteine S-conjugate by N-acetyltransferase results in the formation of the N-acetyl derivative, a mercapturic acid.

The GSTs are distributed quite widely throughout the animal and plant kingdom, having been demonstrated in vertebrates, invertebrates, plants, and bacteria. GST activity is ubiquitous in mammalian tissues with high levels found in tissues associated with chemical detoxification and elimination. Soluble GSTs are localized to the cytoplasm of cells, whereas membrane-bound GSTs are associated with the endoplasmic reticulum. GSTs have also been shown to compartmentalize within interchromatinic regions of the cell's nucleus. The cytoplasmic GSTS are the most abundant forms of the enzyme and are the forms primarily associated with xenobiotic detoxification.

The glutathione S-transferases are involved in the conjugative metabolism of a wide variety of electrophilic substrates that include epoxides, haloalkanes, nitroalka-

TABLE 6.4 Glutathione S-transferase-Mediated Conjugations Involving Nucleophilic Displacement, Michael Addition, and Nucleophilic Attack on Strained Oxirange Rings.

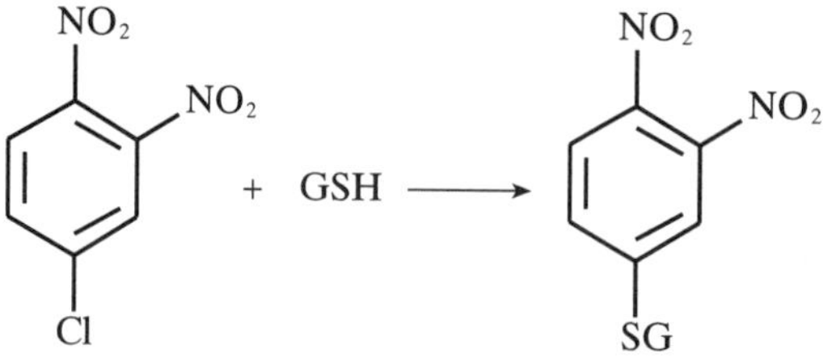

NO_2, NO_2, Cl + GSH → NO_2, NO_2, SG

CDNB
(chlorodinitrobenzene)

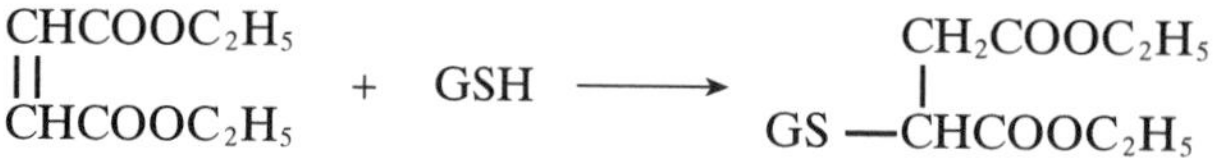

$CHCOOC_2H_5$ = $CHCOOC_2H_5$ + GSH → $CH_2COOC_2H_5$ | GS —$CHCOOC_2H_5$

Diethyl maleate

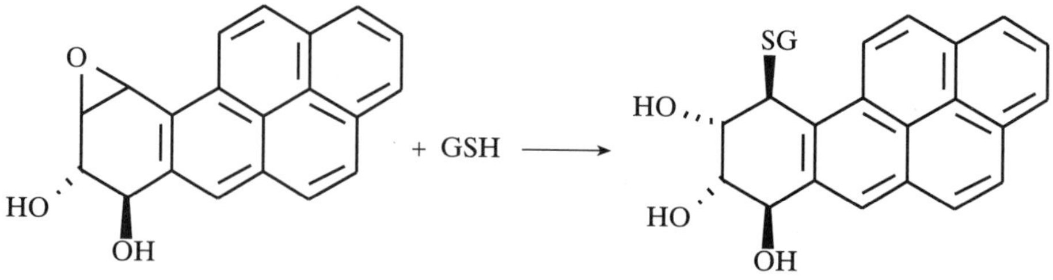

O, HO, OH + GSH → SG, HO, HO, OH

+ antibenzo(a)pyrene-7,
8-diol-9, 10-epoxide

nes, alkenes, methyl sulfoxide derivatives, organophosphates, and aromatic halo- and nitro- compounds. Among the compounds catalytically conjugated to glutathione are antibiotics, vasodilators, herbicides, insecticides, analgesics, anticancer agents, and carcinogens. The reaction involves nucleophilic displacement, Michael addition, and nucleophilic attack on strained oxirane rings. Each of these reactions is illustrated in Table 6.4.

The cytoplasmic GSTs comprise a supergene family of homo- and heterodimeric enzymes whose individual members are composed of various combinations of different monomers, that is, subunits with molecular weights ranging from ~20,000 to 25,000 Da. The monomeric proteins all appear to be products of individual genes. Mammalian cytosolic GSTs have been segregated in six classes—alpha, mu, pi, sigma, theta, and zeta—based upon amino acid sequence similarity. Subunits within a class typically share at least 70% homology, whereas subunits between classes are typically ~30% identical. Heterodimeric combinations occur only among subunits within the same class. The class alpha, mu, and pi GSTs are responsible for most of the catalytic activity associated with organs of detoxification such as the liver.

Various nomenclatures have been used to identify GSTs. Presently, individual GSTs are most commonly identified using a lowercase letter to designate species, an uppercase letter to designate class, and an Arabic number to designate the specific protein subunit within the class. Thus, rGSTM1-2 refers to the rat class mu GST consisting of a heterodimeric combination of subunits 1 and 2. The GSTs typically exhibit overlapping substrate specificities although enzymes within a class sometimes exhibit substrate preferences. For example, class alpha GSTs exhibit high organic peroxidase activity toward cumene hydroperoxide relative to members of the other GST classes. Class mu and pi GSTs exhibit higher relative glutatione conjugating activity toward 1,2-dichloro-4-nitrobenzene and ethacrynic acid, respectively.

6.7.1 Allosteric Regulation

In addition to the substrate and the glutathione-binding sites that are involved in catalysis, some class alpha GSTs contain noncatalytic ligand-binding sites. Binding of ligands to some of these sites can increase enzymatic activity. For example, interaction of class alpha GSTs with the herbicide 2,4,5-T increases catalytic activity of the GST towards 1-chloro-2,4-dinitrobenzene. Binding of the ligand to the GST appears to change the conformation of the protein in a manner that enhances its catalytic activity. Therefore, exposure to some xenobiotics may increase GST activity without affecting GST protein levels.

6.7.2 Induction by Xenobiotics

Following treatment with xenobiotics, such as some planar aromatic hydrocarbons, phenols, and quinones, transcription of some class alpha and mu GSTs is significantly elevated. Induction by planar aromatic hydrocarbons such as 2,3,7,8-tetrachlorodibenzo-p-dioxin (TCDD), polyclorinated biphenyls (PCBs), and polycyclic aromatic hydrocarbons (PAHs) is dependent upon the Ah receptor and CYP 1A1 activity and appears to be mediated by two distinct regulatory pathways. The response element known as the dioxin response element (DRE) or the xenobiotic response element (XRE) has been identified in some class alpha GSTs. These GSTs appear to be induced in direct response to the interaction of the Ah receptor and the DRE (see Chapter 8 for more detail of this induction mechanism). Planar aromatic hydrocarbons can also induce levels of CYP 1A1 in the cell via interaction with the Ah receptor. This increase in CYP 1A1 results in increased oxidative metabolism of the xenobiotic to oxidation-reduction labile metabolites, such as diphenols, aminophenols, and quinones. These reactive metabolites generate a redox signal in the cell that initiates a GST induction pathway. The antioxidant response element (ARE) or electrophile response element (EpRE) has been identified in some class alpha GSTs that may respond to such an induction pathway. According to this model, a redox signal in the cell stimulates a signal transduction pathway resulting in the generation of cfos/cjun heterodimers. These heterodimers then interact with the ARE/EpRE, which is a variant of the AP-1 binding site, and stimulate gene transcription. AP-1 sites are known to regulate a variety of genes involved in cell proliferation and other processes through interaction with fos/jun heterodimers. Xenobiotics such as some quinones and phenols may directly generate a redox signal (e.g., putting the cell under oxidative stress) causing GST gene induction via the same redox signal.

Many class alpha and mu GSTs are induced by phenobarbital and other barbiturates. Gene induction is associated with the presence of barbiturate inducible elements (BARBIE boxes) associated with the gene. BARBIE boxes have been identified in the 5′-flanking region of both mouse and rat GSTs. The glucocorticoid response element also has been identified in some class alpha GSTs. This element may mediate the induction of some GSTs by the synthetic glucocorticoid dexamethasone.

Glutathione peroxidase activity is catalyzed in the cell by both a selenium-dependent peroxidase and class alpha GSTs (selenium-independent peroxidase). Under conditions of selenium deficiency, class alpha GST protein levels are elevated in apparent compensation for the loss of selenium-dependent peroxidase activity. These observations suggest that selenium normally suppresses constitutive class alpha GST levels. In the absence of selenium, class alpha GSTs are derepressed.

GST pi expression is generally not considered to be induced by xenobiotics. However, hepatic GST pi expression is elevated 50- to 100-fold in hepatic preneoplastic tissue and in hepatocarcinomas induced by chemical carcinogenesis. Due to this high level of induction, GST pi is regarded as a marker of hepatic preneoplastic cells. Further, recent studies have demonstrated that GST pi is expressed at high levels in tumor cells from various organs that have developed resistance to the toxicity of anticancer drugs. GST pi gene expression is under the regulatory control of both enhancer and silencer elements. Deletion of the silencer region elevates gene transcription, whereas deletion of the enhancer region inhibits gene transcription. The enhancer region called GPE1 contains two response elements that, like the ARE, are capable of binding cfos/cjun heterodimers resulting in gene activation. Therefore, GPE1 could be responsible for the induction of GST pi during carcinogenesis. Some silencer regions of the GST pi gene contain the recognition sequence for the transcriptional regulator protein $NF_{\kappa}B$. As long as the silencer region of the gene is occupied by the appropriate transcriptional regulator protein, the enhancer region is nonfunctional and gene expression is suppressed.

GSTs also play an important role in resistance. They have been implicated in resistance to the effects of a variety of chemicals including antibiotics, anticancer agents, analgesics, herbicides, insecticides, and vasodilators. Preneoplastic lesions, tumors, and cultured tumor cell lines commonly overexpress GSTs of the classes alpha, mu, and pi. The overexpression of GSTs is considered partly responsible for the multidrug resistance often associated with these cells.

The expression of a class mu GST is polymorphic in the human population. The principal variant is the null variant in which the gene is not expressed. In most racial groups, the frequency of individuals that are homozygous for the null polymorphism is approximately 50%. This polymorphism has been widely studied with regard to cancer risk and is associated with increased risk for bladder cancer, colorectal cancer, head and neck cancers, various cutaneous tumors, and lung adenocarcinomas.

6.8 CYSTEINE S-CONJUGATE β-LYASE

Cysteine S-conjugate β-lyase is responsible for converting a number of cysteine S-conjugates into pyruvate, ammonia and thiols.

$$\underset{\text{premercapturic acid}}{RSCH_2CH(NH_2)COOH} + H_2O \rightarrow RSH + NH_3 + \underset{\text{pyruvic acid}}{CH_3C(O)C(O)OH}$$

This activity has been demonstrated in mammalian liver and kidney, as well as in intestinal bacteria. Glutathione S-transferases catalyze the formation of glutathione conjugates, which are processed via the mercapturic biosynthetic pathway, to acetylated cysteine S-conjugates. The unacetylated premercapturic acids are substrates for cysteine S-conjugate β-lyase, whereas, the acetylated cysteine S-conjugates, the mercapturic acids, do not function as substrates for the enzymes.

Cysteine S-conjugate β-lyase activity has been implicated in the bioactivation of certain halogenated alkenes. This bioactivation results from the generation of reactive thiols. The thiol produced by the β-lyase-dependent metabolism of S(1,2-dichlorovinyl)-L-cysteine is an unstable electrophile that binds covalently to cellular macromolecules and leads to nephrotoxicity and genotoxicity. Alternatively, certain other cysteine S-conjugates form chemically stable thiols that exhibit no toxicity. The chemically stable thiols may be excreted intact or further metabolized to S-glucuronides or methylated to yield methylthio-derivatives.

6.9 LIPOPHILIC CONJUGATION

Recent studies utilizing improved methods of isolation, purification, and characterization of metabolites have demonstrated that certain xenobiotic metabolites are not readily eliminated and that some of these conjugates have lipophilic character. While their physical properties are different from the classical hydrophilic conjugates, the mechanism of formation clearly defines them as conjugates (i.e., they are formed as the result of a union of xenobiotic metabolites with endogenous molecules). The reactions involve the coupling of xenobiotic acids and alcohols with endogenous intermediates of lipid synthesis (i.e., the acids with glycerol and cholesterol and the alcohols with the fatty acids).

With lipophilic conjugation, the xenobiotic metabolites appear to be incorporated in the lipid biosynthetic pathway similar to the normal constituents. One would expect the conjugates to have turnover times similar to those of their natural counterparts. Whether the lipophilic conjugates have any deleterious effect on the organism would depend upon the type and amount of bioactivity retained by the metabolites before and after conjugation.

6.10 PHASE II ACTIVATION

The biological activity of xenobiotics is generally decreased by conjugation. Phase II reactions normally increase the rate of elimination of most lipid-soluble xenobiotics, thereby terminating their possible effect in a biological system. However, the number of xenobiotics that are metabolized to penultimate or ultimate toxicants as the result of phase II reactions is ever-expanding. Therefore, conjugation can no longer be considered strictly a detoxification reaction.

A number of arylamines are potent bladder carcinogens such as 4-aminobiphenyl, l-naphthylamine, and benzidine. Metabolic activation of these carcinogens requires the action of UDP-glucuronosyl transferase on N-hydroxyarylamines to form N-glucuronides.

$$\text{Ar-NHO glucuronide} \xrightarrow{\text{pH} < 7} \text{Ar-NHOH} \rightleftarrows \underset{\text{arylnitrenium ion}}{\text{Ar-}\overset{+}{\text{N}}\text{-OH}}$$

The N-glucuronides of the N-hydroxyarylamines are transported to the bladder whereupon hydrolysis by β-glucuronidase and in the presence of acidic urine, N-hydroxylarylamines are formed. Spontaneously, the electrophilic arylnitrenium ion forms, which can then react with nucleophilic centers of the epithelium in the bladder to initiate tumor formation.

Metabolic activation occurs with the sulfation of 2-acetylaminofluorine (2-AAF). N-hydroxylation of the amino nitrogen by the monooxygenases followed by sulfation results in an unstable N-O-sulfate ester. The ester decomposes to an electrophilic nitrenium ion-carbonium ion, which can form covalent adducts with macromolecules. Several other xenobiotics such as mono- and dinitrotoluene, N-hydroxyphenacetin, hydroxysafrole, and other N-hydroxyarylamides are in part activated by this mechanism.

Recent studies have also implicated glutathione S-conjugate formation in the bioactivation of certain halogenated alkanes and alkenes to form potent nephrotoxicants. Glutathione conjugates are metabolized by the mercapturic acid pathway to the cysteine conjugates. Two different mechanisms have been proposed by which cysteine S-conjugates may produce nephrotoxicity. One is the spontaneous nonenzymatic formation of electrophilic episulfonium ions, which may alkylate essential renal macromolecules. The other mechanism is dependent upon β-lyase activation, forming unstable reactive thiols that react with biological macromolecules. Both mechanisms seem to be responsible for the formation of mutagenic and DNA-damaging intermediates that are responsible in the initiation of neoplastic transformations.

6.11 PHASE III ELIMINATION

Conjugation of lipophilic xenobiotics to polar cellular constituents renders the xenobiotic more water soluble. While the lipophilic xenobiotics could readily diffuse into the cells, the increase in polarity associated with conjugation greatly reduces the ability of the compound to diffuse across the lipid bilayer of the cell membrane thus trapping the compound within the cell. The polar conjugates must therefore rely upon active transport processes to facilitate efflux from the cell. Hepatocytes and other cells involved in chemical detoxification are rich with members of the ATP-binding cassette superfamily of active transport proteins. Cellular efflux of xenobiotics by these transporters is often referred to as phase III elimination because phase I or II detoxification processes often precede and are a requirement of phase III elimination. Two toxicological important ATP binding cassette transporter families are MRP/MOAT and P-glycoprotein.

6.11.1 MRP/MOAT

The multidrug resistance associated protein (MRP) and the multispecific organic anion transporter (MOAT) represent a family of active transport proteins involved in the cellular elimination of phase II conjugates from hepatocytes and other cells.

These proteins have been shown to transport various polar anionic conjugates of xenobiotics as well as some unconjugated xenobiotics. This family consists of at least seven transporters in humans that are the products of distinct genes. MRP1 was the first of the MRPs to be identified and characterized. This transporter is expressed in a variety of tissues including the lateral membrane of hepatocytes and its overexpression often has been found to be associated with multixenobiotic resistance of cells. MRP2 (also called cMOAT) is expressed on the canalicular surface of hepatocytes where it functions in elimination of transport substrates into the bile canaliculus. MRP2 is also expressed in the kidney and perhaps other organs of excretion. MRP3 (also called MOAT-D) is associated with the basolateral membrane of hepatocytes and other organs of excretion such as the colon, pancreas, and kidney. MRP4 and MRP5 (also called MOAT-B and MOAT-C, respectively) are widely distributed in extrahepatic tissues. MRP4 and MRP5 levels are particularly high in prostate and skeletal muscle, respectively. Homologs of the human MRPs 1, 2, 3, 5, and possibly 6 have been identified in mouse or rat. The MRP/MOAT proteins are capable of transporting a variety of conjugates including those of glutathione (i.e., dinitrophenyl-glutathione, bromosulfophthalein-glutathione), glucuronic acid (i.e., *p*-nitrophenyl-glucuronide, bilirubin-glucuronide), and sulfate (i.e., taurolithocholate-3-sulfate). MRP/MOAT proteins have been shown to transport a variety of conjugated anti-cancer drugs and contribute to the multidrug resistance often associated with cancer cells.

Dubin-Johnson disease in humans is caused by a defect in the MRP2 protein due to a gene mutation. Individuals afflicted with this disorder have limited capability to efflux glucuronic acid-conjugated bilirubin into the bile canaliculus. As a result, these individuals experience elevated blood levels of conjugated bilirubin (conjugated hyperbilirubinemia). Hepatic MRP3 is overexpressed in some individuals afflicted with this disorder. This suggests that conjugated bilirubin is eliminated from the hepatocyte in these individuals across the basolateral membrane into the circulation where the metabolite accumulates.

6.11.2 p-Glycoprotein

p-Glycoprotein comprises a family of ATP-binding cassette transporters that are involved in the cellular efflux of a variety of structurally-diverse compounds. The p-glycoprotein product of the *MDR1* gene is responsible for xenobiotic efflux in human. In mouse and rat, two p-glycoproteins, mdr1a and mdr1b, contribute to xenobiotic efflux. p-Glycoprotein substrates typically have one or more cyclic structures, a molecular weight of 400 or greater with moderate to low lipophilicity (log Kow <2) and high hydrogen (donor) bonding potential. Highly lipophilic compounds (log Kow >2) may bind to the transport site of the molecule but tend to be poor transport substrates. Such compounds may actually inhibit p-glycoprotein function. Moderate to low lipophilicity with high hydrogen (donor) bonding potential is characteristic of hydroxylated compounds. This suggests that phase I hydroxylation reactions may render some compounds susceptible to efflux by p-glycoprotein. This relationship between hydroxylation and active efflux has been demonstrated with steroid compounds. Progesterone, which is highly lipophilic and has low hydrogen (donor) bonding potential is not appreciably transported by p-glycoprotein, whereas its hydroxylated derivatives cortisol and aldosterone are actively transported by this protein.

p-Glycoprotein is localized to the secretory surfaces of tissues of elimination such as the brush border of the renal proximal tubules, the canalicular membrane of hepatocytes, the apical surface of mucosal cells in the intestines, and endothelial cells of the blood-brain barrier. The protective action of p-glycoprotein against chemical toxicity was demonstrated using a mouse strain in which the *madr1a* gene was disrupted and functional mdr1a p-glycoprotein was not produced. These mice were fertile and displayed a normal phenotype but were 100-fold more susceptible to the toxicity of the neurotoxicant ivermectin. Mice deficient in p-glycoprotein were also shown to accumulate significantly more ivermectin in the brain and other tissues. Administration of p-glycoprotein inhibitors to mice expressing normal levels of functional p-glycoprotein has been shown to increase the susceptibility of the mice to the neurotoxicity of ivermectin. Many xenobiotics, including pesticides, surfactants, and medicinals have been shown to inhibit p-glycoprotein function and may function to sensitize organisms to the toxicity of p-glycoprotein transport substrates.

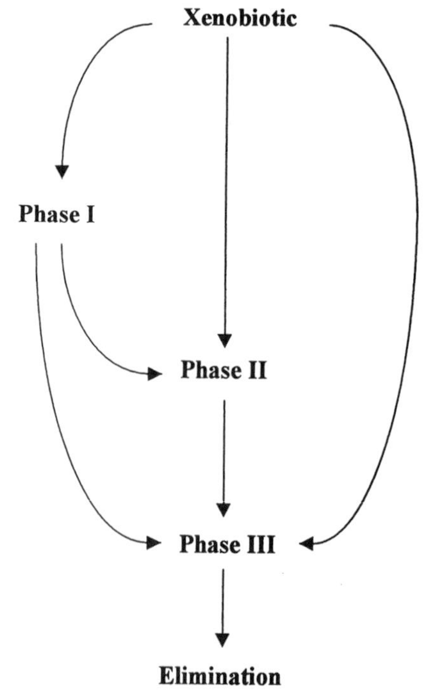

FIGURE 6.2 Interrelationships among phase I (hydroxylation), phase II (glucuronic acid, sulfate, and glutathione conjugation), and phase III (MRP/MOAT and p-glycoprotein mediated efflux) detoxication processes leading to the inactivation and elimination of xenobiotics.

6.12 CONCLUSIONS

Conjugation and active transport processes provide efficient means for both the inactivation and elimination of xenobiotics. Xenobiotics may possess the molecular properties that render them suitable for active cellular efflux with no need for biotransformation (Fig. 6.2). Alternatively, xenobiotics may first undergo phase I hydroxylation, phase II conjugation, or sequential hydroxylation and conjugation (Fig. 6.2). The resulting modifications can render the xenobiotic susceptible to active efflux from the cell, and ultimately, elimination from the body. Through these processes of inactivation and efflux, conjugation and active elimination processes confer a high degree of protection against the toxicity of chemicals. The over-expression of the proteins responsible for these processes is a common mechanism of resistance to chemical toxicity. Conversely, the inhibition of these proteins or the suppression of their expression can increase one's susceptibility to some toxicants.

SUGGESTED READING

Mulder, G.J. (Ed), *Conjugation Reactions in Drug Metabolism*, Taylor and Francis, London, 1990.

Weinshilboum, R.M., Otterness, D.M., Aksoy, I.A., Wood, T.C., Her, C. and Raftogianis, R.B. Sulfation and sulfotransferases1: sulfotransferase molecular biology: cDNAs and genes. *FASEB J.* 11 (1997), 3–14.

Hayes, J.D. and Pulford, D.J. The glutathione S-transferase supergene family: regulation of GST and the contribution of the isoenzymes to cancer chemoprotection and drug resistance. *Crit. Rev. Biochem. Mol. Biol.* 30 (1995), 445–600.

Eaton, D.L. and Bammler, T.K. Concise review of the glutathione S-transferases and their significance to toxicology. *Toxicol. Sci.* 49 (1999), 156–164.

Ambudkar, S.V., Dey, S., Hrycyna, C.A., Ramachandra, M., Pastan, I., Gottesman, M.M. Biochemical, cellular, and pharmacological aspects of the multidrug transporter. *Annu. Rev. Pharmacol. Toxicol.* 39 (1999), 361–398.

Awasthi, S. and Zimniak, P. Multiple transport proteins involved in the detoxification of endo- and xenobiotics. *Front. Biosci.* 2 (1997), 427–437.

LeBlanc, G.A. Hepatic vectorial transport of xenobiotics. *Chemico-Biol. Interact.* 90 (1994), 101–120.

CHAPTER SEVEN

Physiological Factors Affecting Xenobiotic Metabolism

MARTIN J.J. RONIS and HELEN C. CUNNY

7.1 INTRODUCTION

Xenobiotics are biotransformed by phase I enzymes and phase II conjugation reactions to form a variety metabolites that are generally more water soluble and less toxic than the parent compound. Occasionally, the enzymatic action of phase I or II systems leads to the formation of unstable intermediates or reactive metabolites that are toxic or carcinogenic; this is the subject of Chapter 9. Many factors influence the rate of xenobiotic metabolism and the relative importance of different pathways of metabolic activation or detoxication. This chapter deals with the physiological factors that affect these processes. Chemical and environmental factors are discussed in Chapter 8.

Phase I and II xenobiotic metabolizing systems exist as large gene families of enzymes, many of which have some degree of overlapping substrate specificity. The enzymes involved probably originally evolved to deal with endogenous processes such as steroid biosynthesis and inactivation, fatty acid and vitamin metabolism, and for the disposal of waste products such as bilirubin. Therefore, it is not too surprising that these enzymes are under close physiological regulation in addition to being inducible by exogenous chemical agents. A good example is the cytochrome P450 enzyme CYP2E1. This enzyme has toxicologically important roles in the metabolic activation of acetaminophen (Tylenol) and industrial solvents such as carbon tetrachloride, alcohol-induced liver damage, and activation of nitrosamine procarcinogens but is also an important enzyme in metabolism of two carbon units and gluconeogenesis. As such, in addition to being inducible by compounds such as ethanol, acetone, and the drug isoniazid, CYP2E1 is under tight physiological regulation by hormones such as glucagon, insulin, and growth hormone. It is also affected by diet and pathophysiological conditions such as diabetes, fasting, and obesity.

In vivo metabolism and clearance of xenobiotics is the sum of many different processes including (1) the action of many different phase I and II enzymes in the liver and extrahepatic tissues and (2) xenobiotic transport systems leading to excretion in urine or bile. Modern biochemical and molecular biological techniques have revealed striking differences in tissue and sex-specific expression, as well as developmental and hormonal regulation of phase I and II enzymes. In addition, xenobiotic metabolism is affected by alterations in physiology and pathology (such as found in pregnancy and infection) and undergoes circadian and seasonal cycling. In addition to changes in protein expression of phase I and II enzymes, rates of xenobiotic metabolism are also affected by the levels of essential cofactors such as NADPH, GSH, PAPS, and UDPGA, which may also be altered by physiological status. In the sections below, the effects of age, gender, and hormonal status on xenobiotic metabolism are considered. This is followed by sections on the effects of pregnancy and pathology. Finally, information is included on the molecular mechanisms underlying physiological regulation of phase I and II enzyme expression.

7.2 DEVELOPMENT AND AGING

In general, the ability to metabolize xenobiotics is low or absent in the fetus and neonate, develops rapidly after birth, is at its highest in early adulthood, and declines in old age. The development of individual phase I and II enzymes is under complex ontogenic control. Most xenobiotic-metabolizing enzymes do not appear gradually during development. Instead, clusters of different enzymes seem to develop rapidly during critical periods in an organism's life such as birth, weaning, and puberty. The timely expression of these enzymes complements the physiological needs of the developing organism in its changing environment. Before birth, metabolic capacities for handling xenobiotics are low largely because the mother's metabolism provides fetal detoxication reactions. A number of cytochrome P450 enzymes, including CYP1A1, CYP1B1, CYP2C8, CYP2D6, and CYP2E1, have been detected in human fetal tissues and some such as CYP1A1 are inducible in the feto-placental unit by environmental chemicals such as polycyclic aromatic hydrocarbons found in cigarette smoke (see Chapter 8). However, the major fetal cytochrome P450 in human liver is CYP3A7, which probably has a major physiological role in steroid inactivation within the fetus. At birth, CYP3A7 rapidly disappears and is replaced by rapid increases in the major adult CYP3A enzyme CYP3A4. Thus, relative to other species such as laboratory rodents, total levels of cytochrome P450 are high in the neonatal human liver.

In the rat, a major suite of P450 enzymes is expressed immediately after parturition. Overall, P450 concentrations increase from less than 10% to 50% of adult levels within a few days of birth. One of the P450 enzymes that is regulated in this fashion is CYP2E1, and it is one of the few for which this process has been studied at the molecular level. Increased hepatic CYP2E1 expression at birth involves gene transcription. This appears to be associated with the binding of a liver-specific transcription factor HNF1α to a response element between –100 to –120 base pairs (bp) in the promoter region of the CYP2E1 gene upstream of the transcription start site. The process that initiates this response is still unknown, but seems to involve alterations in methylation status of the CYP2E1 promoter.

In addition to P450 enzymes, many phase II enzymes are expressed soon after birth. These include the UDPGT enzymes responsible for the glucuronidation of steroids and bilirubin. This latter process is important clinically for protection against the development of neonatal jaundice.

A number of phase I and II enzymes are expressed at high levels during lactation in rodents and then decline following weaning. These include CYP2E1, CYP2B family members, and a number of steroid sulfotransferases such as hydroxysteroid sulfotransferase ST-40/41. This presumably is the result of important physiological roles during this period of high fat intake and estrogen imprinting of the brain and reproductive physiology. It is not clear if lactation-specific expression of particular phase I and II enzymes occurs in humans.

A further group of phase I and II enzymes are expressed at weaning and are involved in the detoxication of dietary xenobiotics such as flavinoids, alkaloids, and terpenes. These include CYP1A2 in humans and CYP1A1, 1A2, and CYP3A6 in the rabbit. Increased expression of CYP1 enzymes may be the result of induction by dietary factors. However, artificial alteration of the age of weaning by controlling food availability to the pups had no effect on CYP3A6 expression in the rabbit and this suggests endogenous physiological regulation of this enzyme rather than the influence of dietary factors.

Finally, major alterations in expression of hepatic phase I and II enzymes occur in rodents at puberty. This is largely the result of the appearance of sexual dimorphism in these species (see Section 7.3 below). Male-specific rat enzymes such as CYP2C11 and sulfotransferase ST1C1, which metabolizes N-hydroxy-2-aminofluorene (N-OH-2AAF), appear at puberty. So do female specific rat enzymes such as CYP2C12 and sulfotransferases ST-20/21 and ST-60. In contrast, other P450 enzymes such as CYP3A2, which is expressed at high levels prior to puberty in both sexes of rat, are selectively suppressed in female animals. Glutathione S-transferases are expressed at low levels in prepubertal rats, increase in activity, and become sexually dimorphic at puberty. In addition, UDPGTs also become sexually dimorphic at puberty. It is not clear whether puberty alters xenobiotic metabolism in humans where sexual dimorphism in hepatic metabolism is much less pronounced.

The effects of old age on xenobiotic metabolism continue to receive much less attention than earlier developmental events. In general, drug and xenobiotic metabolism rates and clearance seem to decline. In old male rat livers, male-specific P450 enzymes such as CYP2C11 and CYP3A2 decline whereas female-specific forms such as CYP2C12 begin to appear. This may reflect loss of sex steroids and sexually dimorphic patterns of growth hormone secretion (see Sections 7.3 and 7.5 below). In addition, glucuronidation activity also declines with old age whereas other enzyme systems such as monoamine oxidase have been shown to increase.

7.3 GENDER DIFFERENCES

Gender differences in hepatic biotransformation of xenobiotics have been recognized in experimental animals for many years. A classic example of this is the longer sleep time observed in female rats compared with males following administration of barbiturate anesthetics due to slower oxidative metabolism. Where metabolism is an important determinant of toxicity, many gender differences have been described in rodents. For example, phenacetin, salicylic acid, nicotine, picrotoxin, and warfarin are all more toxic to female than to male rats. In contrast, epinephrine, ergot, and monocrotaline are more toxic to the male rat. As stated in Section 7.2 above, many of the gender differences in phase I and II enzymes in rodents develop at puberty. It appears that these differences are either directly due to differences in sex steroid concentrations in male and female rodents or the indirect effect of sex steroid imprinting of sexually dimorphic patterns of growth hormone (GH) secretion by the pituitary and that GH is the endocrine mediator that maintains sexually dimorphic enzyme expression (Fig. 7.1).

In adult rats, GH is secreted in a pulsatile fashion. In males, low-frequency, high-amplitude pulses are observed with nadirs between pulses in which no GH is secreted. In contrast, in females, GH is released as high-frequency, low-amplitude pulses superimposed on a baseline of continuous GH secretion. The sexually dimorphic pattern of GH secretion by the pituitary appears at puberty and is determined by pulsatile secretion of the hypothalamic peptides GHRH and somatostatin 180 degrees out of sync with each other. The release patterns of GHRH, somatostatin, and GH are themselves imprinted and under feedback regulation by the sex steroids. Prenatal and neonatal androgen surges seem to imprint the masculine pattern of GH secretion. It is thought that aromatization of androgens to estrogens within the neonatal hypothalamus is required for the imprinting of GH because nonaromatizable androgens such as dihydrotestosterone are unable to reverse the effects of neonatal castration.

In some cases, testosterone and estrogens directly stimulate expression of xenobiotic metabolizing enzymes. An example of this is in the mouse kidney, where many cytochrome P450 enzymes including CYP2E1 are under direct androgen regulation. This accounts for the unique susceptibility of the female mouse kidney to acetaminophen nephrotoxicity compared with the male mouse kidney. In the rat liver, expression of CYP3A9 has recently been shown to be positively regulated by estrogens. In the rat kidney another male specific cytochrome P450 enzyme, CYP2C11, has been shown to be under direct androgen regulation. Castration and hypophysectomy (Hx, removal of the pituitary), which remove testosterone either directly or indirectly via loss of luteinizing hormone (LH) signaling, both significantly reduce renal CYP2C11 expression. However, whereas testosterone replacement completely restores this enzyme, GH replacement seems to have no effect. These results are in direct contrast to the expression of CYP2C11 in male rat liver. In this case, although both castration and Hx reduce hepatic CYP2C11 expression by 50% in male animals, testosterone replacement has no effect in Hx animals and intermittent administration of GH mimicking the male pattern restores levels to normal. Moreover, continuous administration of GH mimicking the female pattern completely suppresses CYP2C11 expression. In addition, passive immunization against GH in vivo also significantly suppresses hepatic CYP2C11 expression in male rats (see Fig. 1, Chen et al. in "Suggested Reading"). Regulation of the female-specific rat cytochrome P4502C12 is exactly the reciprocal of CYP2C11—the female GH pattern stimulates expression and the male pattern suppresses it. These enzymes appear at puberty in the rat and their appearance coincides with the development of the sexually dimorphic patterns of GH secretion. Neonatal castration and steroid replacement experiments have demonstrated that both GH patterns and these hepatic P450 enzymes are imprinted by neonatal androgen exposure. In conditions such as lead poisoning, where androgen surges at birth and puberty are suppressed and pubertal development is delayed, developmental expression of these enzymes is also delayed. In addition to CYP2C11 and CYP2C12, many other rat liver P450 enzymes are sexually dimorphic to some degree. In the male liver, CYP2C13,

CYP3A2, and CYP2A2 are the major forms. In female rat liver, CYP2A1 and CYP2C7 are predominant. Epoxide hydrolase, GSTs, and some UDPGT enzymes are also expressed at a higher level in male rat liver. In contrast, many flavin-containing monooxygenases (FMOs) are found at higher levels in the livers of female rodents. Testosterone has been shown to significantly suppress FMO1 and FMO3 in female mice but it is not clear whether this is a direct steroid effect or mediated via alterations in GH secretion.

Sulfotransferase enzymes are also sexually dimorphic in rat liver. ST1A1, ST1C1, and ST1E1, which metabolize phenol, N-OH-2AAF, and estrone, respectively, are male dominant whereas STs, which metabolize glucocorticoids, hydroxysteroids (STs 20/21, 40/41 and 60), and bile acids, are female dominant. As with the P450 enzymes, the primary regulator of gender-specific ST expression seems to be GH rather than the sex steroids directly.

A considerable amount of research has been conducted to examine which aspects of the sexually dimorphic GH pulse are responsible for sexually dimorphic enzyme expression. With CYP2C11 and 2C12, plasma GH concentrations and pulse amplitude seem unimportant for regulation. Administration of GH to Hx rats in different patterns and concentrations has revealed that 2C11 is only expressed if GH is completely absent for 1½ to 2 hours between pulses. In contrast, CYP2C11 is completely suppressed and CYP2C12 is expressed at normal levels in the presence of continuous GH at concentrations as low as 3% of physiological levels. Other sexually dimorphic P450 enzymes are less dependent on GH pulse pattern. Male rat-specific CYP3A2 is markedly elevated by Hx, and the enzyme is suppressed by GH in a pattern-independent fashion. Recently, the molecular events underlying GH regulation of a number of rat P450 enzymes have been examined and these findings will be discussed in Section 7.8 below.

In humans, GH secretion patterns are much less sexually dimorphic than in rodents and there are no human orthologues of the sexually dimorphic rat liver P450s CYP2C11, 2C12, 2C13, or CYP3A2. As such, genetic polymorphism and environmental factors such as diet probably play a more important role in interindividual variations in drug and xenobiotic metabolism than do gender differences. However, the major human hepatic P450 enzyme CYP3A4 does appear to be expressed overall at higher levels in women than men, suggesting that there is endogenous hormonal regulation of this enzyme. Like the human, little sexual dimorphism in xenobiotic metabolism is observed in the guinea pig, dog, pig, or rhesus monkey. These species differences in gender-specific metabolism must be taken into account when extrapolating from studies in laboratory animals to humans.

7.4 HORMONAL STATUS

Many of the effects discussed in other sections of this chapter are mediated through changes in endocrine status. Hormones display many complex interrelationships, with one hormonal cascade modulating the effects of another through synergistic interactions and multiple feedback loops. The complex interrelationships of this system are amply demonstrated in the interaction of the hypothalamic-pituitary-gonadal axis with the hypothalamic-pituitary-growth hormone axis in the regula-

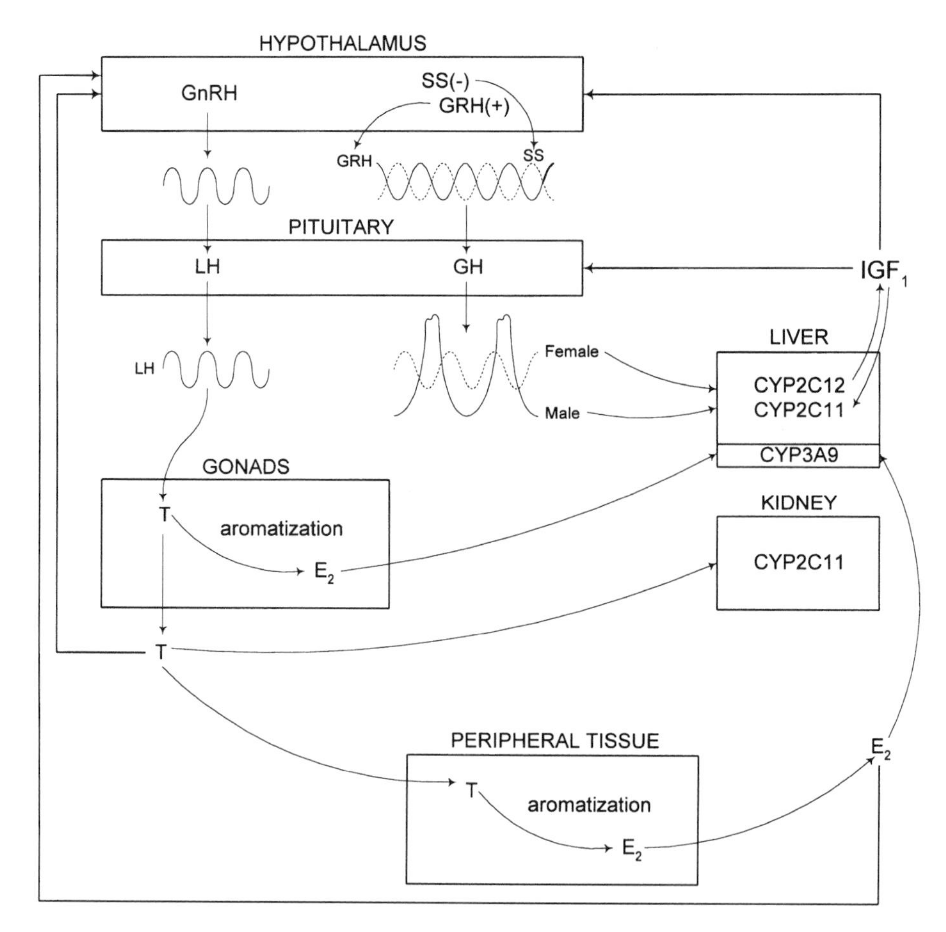

GnRH = gonadotropin releasing hormone; SS = somatostatin; GRH = growth hormone releasing hormone; GH = growth hormone; LH = lutenizing hormone; T = testosterone; E_2= 17ß-estradiol; IGF1 = insulin-like growth factor; CYP2C12 = female-specific cytochrome P450 2C12; CYP2C11 = male-specific cytochrome P450 2C11; CYP3A9 = a female predominant cytochrome P450

FIGURE 7.1 Regulation of sex-specific cytochrome P450 expression in rat liver and kidney by the hypothalamic-pituitary-gonadal axis.

tion of the ontogeny and sexual dimorphism of the rat hepatic P450 enzymes CYP2C11 and CYP2C12 discussed in the previous two sections and summarized in Figure 7.1. The current section deals with hormonal effects on xenobiotic metabolism that are not directly linked to gender differences.

7.4.1 Pituitary Hormones

The hypothalamic-pituitary axis ultimately controls many of the hormonal effects on enzymes mediating xenobiotic metabolism. Releasing and release-inhibiting factors produced by the hypothalamus control the synthesis and secretion of trophic hormones from the pituitary gland. These include the gonadotrophins, follicle-stimulating hormone (FSH), and luteinizing hormone (LH), which control gonadal biosynthesis of androgens and estrogens; prolactin, which regulates mammary gland development and milk production; adrenocorticotrophic hormone (ACTH), which regulates the synthesis and release of mineralocorticoid and glucocorticoid hormones from the adrenal cortex; and thyroid stimulating hormone (TSH), which controls the synthesis and release of triiodothyronine (T3) and thyroxine (T4) by the thyroid. In addition, the pituitary secretes GH, which has ubiquitous effects on somatic growth and development throughout the body. Many of the physiological and pathological conditions that alter xenobiotic metabolism operate through changes in the rate or pattern of secretion of these trophic hormones.

7.4.2 Growth Hormone

As discussed in Section 7.3, the sexually dimorphic pattern of GH secretion has an important role to play, together with the sex steroids, in regulating the expression of gender specific and predominant phase I and II enzymes in rodent species such as the rat, mouse, and hamster. However, GH also has effects on the expression of other phase I and II enzymes and their inducibility by environmental chemicals that are less obviously sex related. Although there are small male/female differences in expression of the rat P450 enzymes CYP2B1 (the major phenobarbital-inducible forms) and CYP2E1 (the major alcohol-inducible form), both of these enzymes are significantly elevated in rat liver following hypophysectomy (Hx); CYP2E1 expression is elevated even more in rat kidney. Replacement of GH in either pulsatile (male-like) or continuous (female-like) fashion results in suppression of these enzymes toward control levels. Thus, GH is a major negative regulator of these enzymes, in a pattern-independent fashion, in both hepatic and extrahepatic tissues. Moreover, it has been shown that this is an effect at the level of gene transcription. In addition, GH has been shown to modulate the magnitude of phenobarbital induction of CYP2B1. Therefore, it appears that there is some cross-talk between GH-activated secondary messenger pathways and those activated by phenobarbital.

GH also has effects on the expression of liver esterases. In rats, Hx increases isocarboxazid esterase activity but decreases malathion and α-naphthyl acetate esterase activities. Hx also alters the pattern of alcohol dehydrogenase (ADH) expression in rat liver. Hx decreases ADH activity and this is restored by GH replacement in a pattern-independent fashion.

Administration of many chemical compounds and some pathophysiological conditions such as diabetes can result in alterations in GH expression. For example, chronic ethanol intake has been shown to significantly reduce the pulse amplitude and demasculinize the GH pulse pattern in male rats and reduce GH secretion in humans. A secondary, indirect consequence of this endocrine effect of ethanol consumption is a significant suppression of hepatic CYP2C11 in male rats. It has also been suggested that suppression of GH secretion by ethanol may indirectly contribute to the induction of the major alcohol-inducible P450 enzyme CYP2E1.

Finally, GH may have indirect effects on xenobiotic metabolism mediated via GH-stimulated endocrine and paracrine factors such as the insulin-like growth factors IGF-I and IGF-II. IGF-I, which is produced by the liver and many other tissues in response to GH, independent of secretory pattern, has recently been shown to have effects on rat liver P450 expression in Hx animals with similar effects to GH itself on CYP2C11 and CYP2C12 and a stimulatory effect on CYP3A mediated cortisol 6β-hydroxylation.

7.4.3 Other Pituitary Hormones

Other pituitary hormones seem to play little if any role in regulation of xenobiotic metabolism. ACTH has its effects primarily on the expression of P450 enzymes connected to steroid biosynthesis in the adrenal cortex that are not really involved in xenobiotic metabolism. Similarly, FSH and LH are primarily involved in regulation of the enzymes controlling sex steroid biosynthesis in the gonads. Although prolactin and GH have similar effects on some cellular targets, it seems that prolactin has few if any effects on expression of phase I and II enzymes.

7.4.4 Thyroid Hormones

The hormones produced by the thyroid gland, triiodothyronine (T3), and thyroxine (T4) have wide-ranging and tissue-specific effects on many aspects of metabolism and organ development. They have the general effect of increasing basal metabolic rate (BMR) via effects on the sodium pump and increase overall protein synthesis. However, it has been known for many years that changes in thyroid status are associated with alterations in drug metabolism and xenobiotic toxicity. In general, hypothyroidism results in a decrease in xenobiotic metabolism and an increase in the half-life of many drugs, whereas hyperthyroidism has the opposite effect. However, there are exceptions to this—thyroid hormone treatment decreases P450 CYP2A1, epoxide hydrolase, and monoamine oxidase in rat liver.

In the rat liver at least, thyroid hormones have been implicated in the regulation of many sexually dimorphic enzymes that are also to some degree regulated by GH. For example, thyroid hormones elevate the expression of female-predominant enzymes such as the P450 CYP2C7, steroid 5α-reductase, and the hydroxysteroid sulfotransferases and- have been shown to be involved in the regulation of hydroxyarylamine sulfotranferase.

Hyperthyroidism also appears to result in induction of the ethanol-inducible P450 enzyme CYP2E1. CYP2E1 induction has been shown to be responsible for the increased hepatotoxicity seen with carbon tetrachloride and halothane in rats treated with thyroid hormones. This effect appears to be largely post-transcriptional and involves stabilization (increased half-life) of CYP2E1 mRNA.

A major target of thyroid regulation is the enzyme cytochrome P450 reductase. Hx results in a greater than 75% decrease in P450 reductase expression. Expression can be restored by T3/T4 replacement but not by other pituitary hormones. Moreover, T3 treatment appears to work at both transcriptional and post-transcriptional levels. T3 treatment of rats made hypothyroid by chronic treatment with the antithyroid drug methimazole results in a 20- to 30-fold increase in P450 reductase mRNA levels but only a 1.8-fold increase in gene transcription rates.

Transcription rates are detemined by nuclear run-on analysis. This involves isolating fresh nuclei in which transcription has been temporarily halted with inhibitors and incubating them with radiolabeled nucleotides that become incorporated, "run-on," into partially transcribed new mRNAs when transcription is reinitiated. The now labeled mRNAs are hybridized to cDNAs against the target mRNA bound to a filter and the amount of radioactivity bound is a measure of ongoing transcription.

A thyroid hormone responsive element (TRE) at –552 to –564 bp in the promoter of the P450 reductase gene has been identified. This was done using transient transfection experiments with parts of the 5′-flanking region of the P450 reductase gene linked to a reporter gene. The principle underlying these types of experiments is that plasmids of DNA are constructed containing an inserted cDNA consisting of different parts of the promoter regions of genes of interest linked to the structural gene for an easily detected product such as the enzymes catachol amino transferase (CAT), luciferase, or β-galactosidase. These plasmids are then transfected (inserted) into cell lines that respond to the regulator of interest (in this case T3/T4), and the ability of the regulator to turn on the reporter gene is measured by measuring the enzyme activity of the reporter gene product. In this way, a promoter can be chopped into smaller and smaller pieces and a minimum sequence of DNA base pairs identified that still permits a normal response to the regulator. The P450 reductase TRE sequence consists of three imperfect direct repeats of the general TRE binding motif AGGTCA and has been shown to bind heterodimers of the TRα thyroid hormone receptor and the retinoid X receptor RXRα in gel-shift experiments (see Section 7.8). These experiments involve incubating a radiolabeled oligonucleotide of the putative response element with nuclear extracts from tissues/cells treated in this case with T3/T4 or 9-cis retinoic acid that contain the transcription factors that bind to the response element. When the incubated mixtures are run out on polyacrylamide gels, unbound oligonucleotide runs with the dye front and protein-bound oligonucleotide is retained, migrating more slowly. Competition with other putative response elements and supershifting the transcription factor/oligonucleotide complex to a yet slower rate of migration by using antibodies against various transcription factors are used to characterize response element specificity and members of transcription factor complexes. The large increase in steady-state reductase mRNA levels relative to the modest increase in gene transcription suggests an additional significant stabilization of reductase mRNA by thyroid hormones in addition to transcriptional effects mediated via binding to the TRE.

Thyroid hormones have other effects on xenobiotic toxicity independent of alterations in phase I and II enzyme activity. For example, thyroidectomy protects against dioxin (TCDD) toxicity. However, the mechanism for this effect is unclear. Thyroid hormones have only minor effects on the induction profiles of liver enzymes induced by TCDD via the Ah receptor (see Chapter 8). Changes in circulating thyroid hormones may have secondary effects on xenobiotic metabolism via interactions with other hormones. For example, they have been shown to increase the expression of glucocorticoid receptors in some tissues.

7.4.5 Steroid Hormones

The importance of the sex steroids testosterone and estradiol, either directly or indirectly, in determination of developmental and sexually dimorphic expression of xenobiotic enzymes has been largely discussed above in Sections 7.2 and 7.3.

However, other steroids such as the glucocorticoids and progestins also have effects on xenobiotic metabolism. It has been known for many years that adrenalectomy results in a decrease in the rate of clearance of many drugs and that this can be prevented by treatment with glucocorticoids such as cortisone or prednisolone. Moreover, treatment of animals and humans with synthetic glucocoticoids such as dexamethasone results in significant induction of several P450 enzymes important in the metabolism of clinical medications including members of P450 families CYP3A and CYP2B. In the rat, dexamethasone treatment also increases other hepatic enzyme activities, including alcohol dehydrogenase and sulfotransferase activities, toward a number of substrates including 1-naphthol, dopamine DHEA, and N-OH-2-AAF. Other phase II enzymes are also influenced by glucocorticoids. The appearance of UDPGTs catalyzing the metabolism of phenolic substrates and bilirubin at birth has been linked to the action of corticosteroids and digitogenin monodigitoxide glucuronidation is coordinately induced with CYP3A enzymes in the livers of rats treated with dexamethasone. Finally, synthetic glucocorticoids decrease toxicity of some metals because they are also able to induce metallothionein.

The molecular mechanisms underlying the regulation of CYP3A enzymes by synthetic glucocorticoids and some xenobiotics have recently been elucidated. Interestingly, it seems that rather than also being regulated by endogenous glucocorticoids, the physiological mediators of these CYP3A responses may be metabolites of lithocholic acid. This is discussed in detail in Section 7.8. Progesterone may also be implicated in the induction of certain xenobiotic metabolizing systems such as the flavin-containing monooxygenase FMO2 during pregnancy (see Section 7.5).

7.4.6 Adrenal Catecholamines

Considerably less attention has been paid to the effects of adrenal catecholamines on xenobiotic metabolism than on glucocorticoid effects. However, epinephrine, like the pancreatic hormone glucagon (discussed in Section 7.4.7), has its effects via increases in cAMP and activation of the protein kinase A (PKA) cascade. PKA activation has been shown to significantly suppress phenobarbital induction of cytochrome P450 CYP2B1 in hepatocyte cultures, suggesting that there is cross-talk between epinephrine/glucagon-regulated signaling pathways and those involved in the phenobarbital-induction response. In addition, epinephrine also stimulates increases in the degradation rate of the cytochrome P450 CYP2E1 in rat hepatocyte cultures via PKA activation and has been shown to increase esterase turnover in the liver, a phenomenon antagonized by insulin.

7.4.7 Pancreatic Hormones

The major pancreatic hormones are glucagon and insulin. Both have been shown to have significant influences in xenobiotic metabolizing systems. Like epinephrine, glucagon stimulates the degradation of cytochrome P450 CYP2E1 in hepatocyte cultures via a PKA-dependent pathway and has been shown to significantly inhibit the induction of cytochrome P450 CYP2B1 by phenobarbital. Insulin regulates the expression of many phase I and II enzymes and many of the changes in drug metabolism and increased sensitivity of the liver to toxic injury seen in diabetics are the result of disruption of normal insulin regulation of xenobiotic metabolism.

In alloxan and streptozotocin rat models of diabetes, insufficient circulating levels of insulin result in increased sensitivity of the animals to the hepatotoxicity produced by carbon tetrachlorone, enfluorane, acetaminophen, and thiacetamide. This effect appears to be due to an increase in expression of the alcohol-inducible cytochrome P450 CYP2E1. Inhibition of CYP2E1 in vivo by the garlic-derived suicide inhibitor diallyl sulfide (DAS) results in a reversal of the increased thioacetamide hepatotoxicity produced by streptozotocin-induced diabetes. Moreover, insulin treatment reverses the increase in CYP2E1 expression. The negative regulation of CYP2E1 by insulin appears to involve decreased stability of the CYP2E1 mRNA. Nuclear run-on experiments have failed to detect increases in gene transcription even though steady-state CYP2E1 message is elevated 6- to 10-fold. Recent experiments have suggested that the effect of insulin is mediated via protein phosphorylation cascades catalyzed by serine and tyrosine kinases triggered by increases in phosphatidylinositol (PI3). The end result of insulin-dependent kinase activation appears to be binding of a protein factor to the 3′ flanking end of CYP2E1 mRNA, which may affect its stability.

In addition to increases in CYP2E1, other P450 enzymes are modified by diabetes. CYP2B1, CYP2A1, CYP2C7, CYP4A, and CYP3A enzymes are also significantly elevated and CYP2C11 is suppressed in male rats. However, the mechanisms underlying these effects remain to be elucidated.

Insulin also appears to have significant modulatory effects on cytochrome P450 induction by drugs and environmental chemicals. Insulin deprivation in hepatocyte cultures has been shown to enhance the induction of P450 CYP2B1 by phenobarbital, the induction of CYP1A1 by polycylic aromatic hydrocarbons, and CYP3A1 by dexamethasone.

Less is known about the regulation of phase II enzymes by pancreatic hormones. The bilirubin-conjugating UDPGT is known to be increased by glucagon via a PKA mediated pathway. Diabetes results in a general decrease in glucuronidation associated with a loss of UDP-glucuronic acid (UDPGA), which is reversible by insulin. In addition, significant loss of glutathione in the diabetic liver increases susceptibility to oxidative damage and impairs overall glutathione-S-transferase activity.

7.4.8 Hormonal Effects on Xenobiotic Metabolism in Nonmammalian Species

Much less is known regarding hormonal effects on xenobiotic metabolism in species other than mammals. However, some developmental and gender-specific changes in phase I and II enzymes seen in nonmammalian vertebrates and invertebrates do appear to be related to changes in hormonal status. In birds, glucocorticoids have been shown to induce UDPGTs in the chick embryo and some cytochrome P450 activities such as ethylmorphine N-demethylase are induced by dexamethasone in the adult bird. In contrast, fish appear relatively insensitive to glucocorticoid induction even though they possess CYP3A enzymes related to the forms found in mammals. However, like rodents, many fish species demonstrate sexual dimorphism in xenobiotic metabolizing enzymes that develop during sexual maturity. For example, in mature rainbow trout a significantly higher content of cytochrome P450 is observed in the livers of male fish, and the gender difference is as much as 20- to 30-fold in the male compared with the female kidney. These gender differences seem to be due to a major suppressive effect of estrogens on some male-specific fish P450 enzymes. A similar downregulation of FMO enzymes by estrogens has been reported in fish liver.

In insects, hormonal status is known to alter xenobiotic metabolism and may have important consequences for resistance and susceptibility to pesticides. In the housefly, both hormones that control moulting, ecdysone and juvenile hormone, cause increases in P450-dependent monooxygenase activity as measured by heptachlor epoxidation. In the fruit fly *Drosophila*, the CYP18 family of P450 enzymes, which structurally is most closely related to the mammalian xenobiotic-metabolizing P450s of the CYP2 family, are tightly developmentally regulated in the body wall in correlation with ecdysteroid pulses in the first, second, and third larval instars, at the time of pupation and in the pupae. It is thought that both ecdysone and juvenile hormone act via increases in gene transcription.

7.5 PREGNANCY

Pregnancy results in major physiological and biochemical changes in female animals, which often affect their ability to metabolize xenobiotics. It is well known that a large number of maternal enzymes have decreased activity during pregnancy. For example, both monoamine oxidase and catechol-O-methyltransferase activity are significantly reduced. Many of the changes in xenobiotic metabolism during this period seem to be associated with altered hormone levels such as the rise in progesterone and pregnane-diol late in gestation and the release of placental hormones such as chorionic gonadotropin into the maternal circulation.

Pregnancy has differing effects on hepatic expression of cytochrome P450 enzymes depending on the species. In the rat, there is an overall decrease in P450 content and some enzymes such as CYP2E1 have been shown to be significantly suppressed. In addition, inducibility of CYP2B1 by phenobarbital is significantly reduced. In contrast, in the pregnant rabbit liver little effect on monooxygenase activities are observed. In the rabbit lung, pregnancy selectively induces a member of P450 family 4 by more than 100-fold between days 25 and 28 of gestation. This enzyme is involved in the omega hydroxylation of prostaglandins E1, F2α, and A and it is not clear whether it plays a significant role in xenobiotic metabolism. Evidence also exists for induction of flavin-containing monooxygenase during mid- to late gestation in the rabbit. FMO2 expression increases in rabbit lung and kidney during late gestation and peaks at parturition, and FMO1 is similarly increased during mid- and late gestation in maternal rabbit liver. It has been suggested that these effects are associated with rises in plasma progesterone and glucocorticoids during this period.

Pregnanacy also results in significant changes in conjugating enzyme systems. Increasing levels of progesterone and pregnane-diol in late pregnancy may be responsible for observed decreases in glucuronide conjugation because both steroids have been shown to be inhibitors of UDPGT in vitro. The presence of pregnane-2α-20β-diol in the milk of lactating mothers has been reported to result in jaundice in some breast-feeding infants as a consequence of inhibition of neonatal bilirubin conjugation by UDPGT. Decreases in sulfotransferase activity have also been reported in pregnant rats and rabbits.

In addition to alterations in maternal metabolism during pregnancy, it must be remembered that the placenta is involved in xenobiotic metabolism in its own right. The placenta contains at least one important cytochrome P450 enzyme involved in xenobiotic metabolism and pro-carcinogen activation: CYP1A1. The enzyme has been shown to be induced in the placenta of human smokers and in placentas from individuals accidentally exposed to polychlorinated biphenyls (PCBs).

7.6 DISEASE

Disease causes many alterations in xenobiotoic metabolism. Conditions such as acute and chronic hepatitis, cirrhosis, and hepatic porphyria that directly lead to liver damage obviously result in major impairments in metabolism and clearance. However, diseases of the kidney and cardiac system also have major effects on xenobiotic clearance through changes in pharmacokinetics. Impaired kidney function results in a buildup of conjugated metabolic products that may be hydrolyzed and further metabolized prior to elimination. Circulatory disfunction results in impaired delivery of xenobiotics to the liver and other sites of metabolism.

Infection and other inflammatory stimuli have been shown to cause changes in the activity and expression of many different cytochrome P450 forms in the liver, kidney, and brain of both humans and experimental animals. This is of particular clinical importance because alterations in phase I metabolism have the potential to adversely or favorably affect the theraputic or toxic effects of drugs. In general, hepatic metabolism is compromised and the possibility of toxic drug side effects is increased. Thus, infection can itself influence the potential for toxic side effects of the drugs being used to treat it. Moreover, infections can produce toxic side effects in patients on drug therapy for chronic conditions such as epilepsy, high blood pressure, or asthma, especially in those patients on drugs with a low theraputic index.

A good clinical example is the case of theophylline. An outbreak of moderate to severe theophylline toxicity was reported in children taking the drug for asthma in Seattle in the early 1980s as the result of an influenza B outbreak. The infection decreased cytochrome P450 CYP1A2-dependent metabolism of the drug. Because this is the rate-dependent step in clearance and theophylline has a low theraputic index, dangerously high plasma levels resulted.

In the last few years, a large number of clinical and basic studies have investigated the effects of inflammation on P450 expression and there has been a tendency to lump all of these studies together. However, whereas it is clear that P450 enzymes are suppressed by various interferons (IFNs) and inflammatory cytokines such as TNFα, IL-1β, IL-6, and TGFβ, the patterns of IFNs and cytokines produced vary by the type of infection or inflammation and this in turn results in different patterns of effects on P450 expression. This probably underlies the tremendous variability of phase I responses and drug clearance studies reported clinically in diseased patients.

Animal models also differ. Administration of IFNs is probably a good model for viral infections whereas endotoxin administration better elicits the effects of bacterial infections. The most common findings in humans in vivo are suppression of CYP1A2 and sometimes CYP3A4 dependent activities. In animals, IFNα, endotoxin, IL-1, IL-6, and TNFα have most commonly been reported to suppress CYP3As, CYP1A2, and CYP2C11 and CYP2C12 expression. However, not all cytokine-mediated effects on hepatic P450 enzymes result in suppression. In human hepatocytes TNFα, IL-1, and IL-6 all suppress expression of CYP1A2, CYP2C forms, CYP2E1, and CYP3A enzymes, but IL-4 has been reported to increase CYP2E1 expression. In extrahepatic tissues, infections and interferon have been shown to suppress expression of pulmonary and renal P450s. However, endotoxin has been reported to increase CYP4A expression in the kidney and inflammation is the only thing known to induce CYP2E1 in the brain. Many signaling pathways have been implicated in the effects of inflammation, IFNs, and cytokines on cytochrome P450 expression. Induction of nitric oxide synthetase (iNOS) has been suggested to mediate the suppression observed in phenobarbital-inducible CYP2B1 expression in rats treated with endotoxin since iNOS inhibitors reverse this effect. In addition, iNOS inhibitors have been shown to reverse the suppression of CYP2C11 and CYP3A2 produced in rat hepatocyte cultures by TNFα and IL-1 treatment. Cytokine-induced breakdown of cell membrane lipids has also been suggested to play a role in supression of some rat hepatic P450 enzymes such as CYP2C11. Sphingolipid backbones such as ceramide and sphingosine have been suggested to act as secondary messengers for cytokines such as TNFα and IL-1β. It has been shown that sphingomyelinase and ceramidase are induced following treatment of hepatocytes with low concentrations of IL-1β. This results in elevations in the ceramide metabolite sphingosine, which has been shown to be a more potent inhibitor of CYP2C11 expression than ceramide in the same system. The steps leading to transcriptional downregulation of CYP2C11 by sphingosine or its metabolites are yet to be elucidated.

Carcinomas and preneoplastic lesions seem to have altered xenobiotic metabolism relative to the untransformed tissue surrounding them. In hyperplastic liver nodules induced by dietary N-OH-2-AAF, microsomal monoxygenase activities are depressed whereas epoxide hydrolase and some phase II enzymes such UDPGTs and GSTs are increased. In preneoplastic liver foci produced by treatment with N-nitrosomorpholine, a "permanent" induction of UDPGT activity of up to fivefold has been observed even 330 days after dosing with the initiator. The UDPGT enzyme affected is that involved in the conjugation of planar phenols that is inducible by polycyclic aromatic hydrocarbons. A radical alteration in the pattern of GST enzyme expression is also observed with significant expression of pi class GSTs that may be involved in resistance to chemotheraputic drugs. It is interesting to note that the phenolic UDPGT and GST pi are normally expressed in fetal tissues. GST pi expression is now being utilized as a marker in the detection of some preneoplastic lesions.

7.7 CYCLES IN XENOBIOTIC METABOLISM

In many species, xenobiotic metabolism exhibits circadian and seasonal cycles. In wild populations, these cycles may be related to many factors including day length, diet, breeding cycles, and temperature. Because most laboratory animals are kept under controlled environmental conditions, they do not display seasonal changes in metabolism. However, they do exhibit circadian rhythms in activity and endocrine function related to feeding and the light-dark cycle. In rats, several monooxygenase activities display circadian rhythms that seems to be related to changes in cytochrome P450 reductase expression. One P450 enzyme that demonstrates large circadian rhythms is CYP7A, which is involved in the 7α-hydroxylation of cholesterol and is active in bile acid synthesis. Steroid 5α-reductase also undergoes circadian variations in activity in rats. Some species of birds display diurnal variations in B-esterase activities with daily highs and nightly lows. This has been suggested to be related to feeding patterns.

Interestingly, in models where alcohol is continuously infused intragastrically into rats, blood and urine ethanol concentrations cycle from high peaks of 500–600 mg/dl to nadirs of nearly zero with a periodicity of 6–7 days. It is unclear whether this is an example of substrate-driven, time-dependent pharmakokinetics or if an underlying endocrine driven cycle in ethanol metabolism is being revealed. Cyclic changes in ethanol intake with the same periodicity have been observed in pigs fed beer *ad libitum*, and it is possible that these cycles in ethanol metabolism and consumption have a relationship to binge drinking in humans. The biochemical mechanisms underlying increased ethanol metabolism are unclear but do not involve changes in activity of the ethanol-inducible P450 CYP2E1 because P450 2E1 inhibitors do not affect the cyclic blood and urine ethanol concentrations in this model.

Seasonal variations in phase I and II enzymes are often seen in conjunction with breeding cycles. This is particularly true in amphibians, fish, and birds and probably reflect underlying endocrine changes associated with the establishment of reproductive competence. For example, in a bird, the razorbill, elevated metabolism of organochlorine insecticides such as aldrin has been reported in females collected in April and May. This correlated with increased ovarian size and may be related to increases in circulating estradiol and/or progesterone. Similarly, increases in lauric acid hydroxylase, benzo(a)pyrene hydroxylase, and ethoxycoumarin O-deethylase have been reported in spawning frog populations in April through July. In addition to reproductive state, seasonal variations in metabolism observed in marine species such as fish and mollusks may be associated with adaptations to changes in water temperature.

7.8 MOLECULAR MECHANISMS UNDERLYING REGULATION BY PHYSIOLOGICAL FACTORS

In the last few years there has been an explosion in our understanding of the molecular events that are involved in regulation of phase I and II enzyme activity. The distinction between endogenous physiological regulatory mechanisms and those regulating induction by environmental and chemical factors is becoming increasingly blurred. Many of the molecular pathways are shared by endogenous regulators and xenobiotic inducers. The most intensely studied area of cytochrome P450 research today is the search for endogenous regulators and substrates of enzymes such as CYP1 forms previously considered exclusively involved in xenobiotic metabolism.

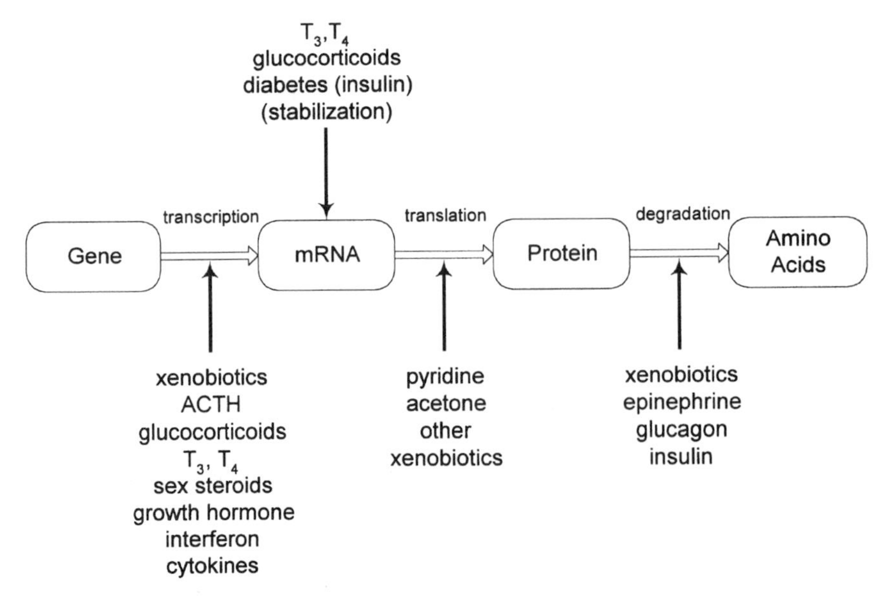

FIGURE 7.2 Factors affecting the rate of protein synthesis and degradation.

The steady-state level of any protein in the cell is determined by the balance of its synthesis and degradation. The rate of synthesis is determined by the level of gene transcription, the stability of the resulting mRNA, and the efficiency of mRNA translation. The factors controlling protein degradation rate are largely unknown. It is clear from the previous sections of this chapter that physiological regulation of xenobiotic metabolizing enzymes occurs at every level of control (Fig. 7.2). In the current section, three examples are utilized to illustrate different molecular mechanisms involved in regulation of phase I and II enzyme expression. Other pathways are probably also involved and remain to be elucidated, but we will focus first on intracellular receptor-associated mechanisms mediated via the hormone receptor superfamily such as those involved in induction of cytochrome P450 CYP3A enzyme transcription by synthetic glucocorticoids and endogenous pregnanes, second on kinase/phosphorylase cascades mediating the male GH pattern-dependent expression of rat cytochrome P450 CYP2C11, and third on the post-transcriptional mechanisms involved in the regulation of cytochrome P450 CYP2E1 by insulin and glucagon/epinephine.

7.8.1 Regulation of CYP3A via the PXR

The hormone receptor superfamily is a set of structurally related intracellular proteins that are involved in transduction of the signaling of steroid hormones and other nonsteroidal lipophilic molecules such as the thyroid hormones, retinoids, fatty acids, and ecoisinoids. The superfamily has been roughly divided into steroid- and nonsteroid-mediated pathways. Although these pathways have some features in common, they also differ from one another (Fig. 7.3). In general, steroid hormones such as the glucocorticoids cortisone and corticosterone pass through the cell

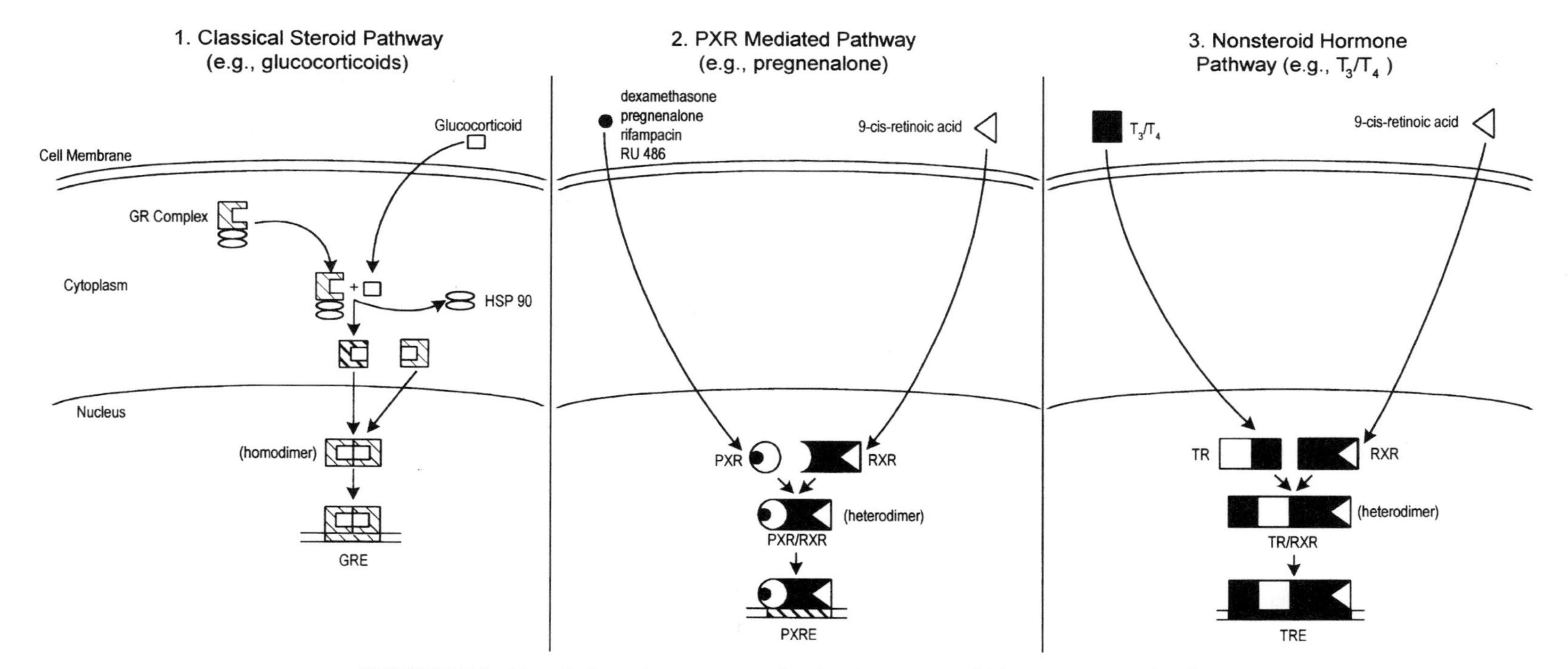

FIGURE 7.3 Regulation of gene expression by the sex steroid hormone superfamily.

membrane and bind to a cytosolic protein complex that includes the glucocorticoid receptor (GR) and the heat shock protein HSP90. When glucocorticoids bind, the conformation of the GR is altered such that HSP90 is lost and the receptor migrates into the nucleus and forms a homodimer that binds to a particular set of base pair sequences on the promoter of glucocorticoid-responsive genes known as the glucocorticoid response element (GRE). It is binding of the GR dimer complex to the GRE that increases transcription of the target gene. In general, steroid hormone response elements consist of two half sites organized as a palindrome and separated by a three nucleotide spacer. In contrast to the steroid hormone receptors, the nonsteroid hormone receptors reside in the nucleus. Hormones such as T3 bind to such a nuclear receptor (the thyroid receptor, TR), which is not associated with HSP90. Binding alters the conformation of the receptor allowing heterodimerization with the RXR receptor, so named for its high binding affinity for 9-cis retinoic acid. It is the TR/RXR heterodimer that binds a DNA response element (the TRE) and enhances gene transcription. In general, nonsteroid hormone response elements are arranged as two half sites that are a direct repeat with the number and composition of nucleotides separating the half sites determining receptor selectivity.

It was originally thought that induction of P450 enzymes in the CYP3A family in humans and rats by synthetic glucocorticoids was mediated by the glucocorticoid receptor, although not via the classical pathway inasmuch as induction required supraphysological levels of the glucocorticoids and because anti-glucocorticoids such as pregnenolone-16α-carbonitrile (PCN) and RU486 are also capable of eliciting the same response (Fig. 7.4). However, in the last year, a new intracellular receptor belonging to the hormone receptor superfamily, the PXR, has been cloned and characterized that seems to mediate the CYP3A induction response and the transduction pathway which appears to combine elements of both the steroid and nonsteroid receptor pathways (Fig. 7.3).

PXR is known as an "orphan receptor," one of about 30 or so members of the hormone receptor family that have been isolated from vertebrates and as yet lack known, defined endogenous ligands. These receptors probably modulate novel endocrine signaling pathways that will in the future impact on our understanding of normal physiology and disease. Using molecular biological techniques involving two-cell hybrid yeast systems in which the PXR was fused to the transcription factor GAL-4 and the GAL-4 DNA binding site fused to a reporter gene, both synthetic glucocorticoids such as dexamethasone and dexamethasone t-butyl acetate and anti-glucocorticoids such as PCN and RU486 have been shown to activate gene transcription via the PXR. Moreover, co-transfection of cells with the PXR and CYP3A promoters fused to reporter genes demonstrated that the presence of the PXR was necessary for induction of either human CYP3A4 or rodent CYP3A1 by these compounds. Further research has demonstrated that the PXR has a unique spectrum of ligand binding (Fig. 7.4). Although synthetic glucocorticoids and antiglucocorticoids are capable of activating the PXR, endogenous glucocorticoids such as cortisol and corticosterone are not. Moreover, a wide spectrum of xenobiotics that are known to be CYP3A inducers also bound and activated the PXR. These included macrolide antibiotics, fungicides such as clotrimazole, and organochlorine pesticides such as transnonachlor. Differences in binding affinity to the PXR seem to explain species differences in CYP3A induction. Thus, rifampicin, a human-specific 3A inducer, binds well to the human PXR and has much lower affinity for the rat PXR whereas

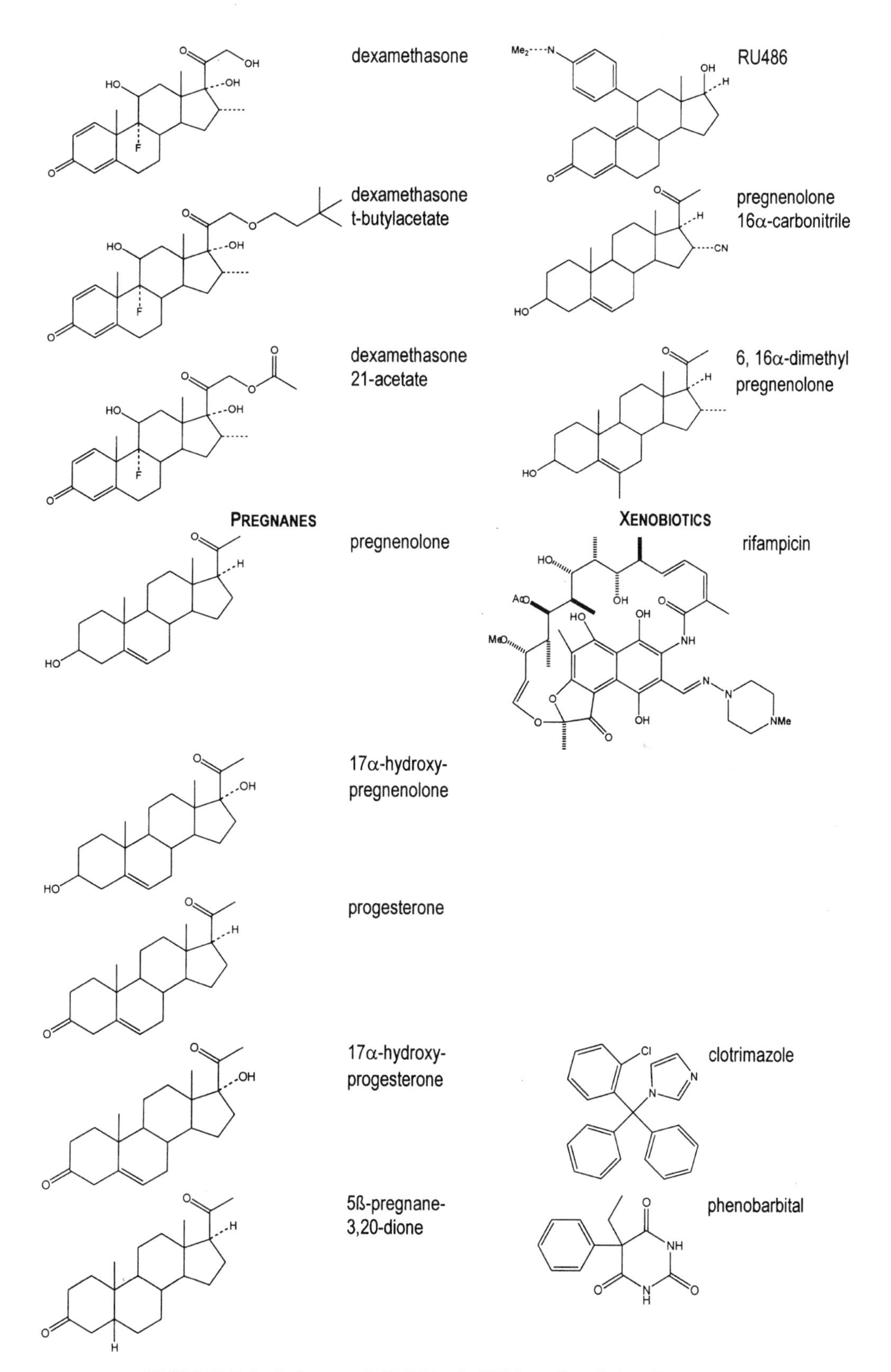

FIGURE 7.4 Inducers of CYP3A via PXR mediated signaling.

PCN, which is a better rat inducer, has higher affinity for the rat PXR. It has been shown that PXR is tissue specific, with high expression only in liver and intestine and that binding to DNA is in the form of a heterodimer with RXRα. Thus, even though the PXR binds steroids, its signal transduction path resembles that of the nonsteroid hormone receptors. The response elements for the rodent and human PXR (the PXREs) have been identified. They consist of direct half site repeats separated by three bases in the rodent CYP3A1 gene and six bases in the human CYP3A4 PXRE:

CYP3A1 PXRE aga **TGAACT** tca **TGAACT** gtc

Human CYP3A4 PXRE ata **TGAACT** caaagg **AGGTCA** gtg

Interestingly, endogenous pregnane steroids such as progesterone, pregnenolone, and their 17-OH metabolites and bile acids such as lithocholic acid have the highest known affinity for the PXR. It is likely that the true endogenous regulator is a bile acid metabolite and the pathway evolved for regulation of cholesterol metabolism and bile acid metabolism.

7.8.2 Regulation of P450 CYP2C11 by Growth Hormone via JAK/STAT Phosphorylation

In contrast to physiological regulation via intracellular hormone receptors, many endocrine and paracrine factors have membrane-bound, cell surface receptors. In these cases, signal transduction is mediated by cascades of protein kinases and phosphorylases that alter protein phosphorylation patterns and ultimately lead to increases in gene transcription. The best characterized of these pathways for a xenobiotic metabolizing enzyme is the regulation of the male-specific rat hepatic cytochrome P450 CYP2C11 by the male pattern of growth hormone secretion (Fig. 7.5). The GH receptor is present in the cell membrane and upon GH binding undergoes dimerization and a conformational change that allows the binding and activation of a member of the Janus kinase family of protein kinases known as JAK2. This kinase has been shown to autophosphorylate itself and to phosphorylate the GH receptor at specific tyrosine and serine residues. It seems that these phosphorylation events in turn allow the binding and subsequent phosphorylation of the transcription factor STAT5b. STAT5b phosphorylation releases the transcription factor from the GH receptor complex and it undergoes homodimerization with a second phosphorylated STAT5b and translocation into the nucleus where it binds its response element on the promoter of the CYP2C11 gene and initiates transcription. The reason that this signaling pathway seems to be sensitive to the pattern of GH secretion is that if GH is not removed, the GH receptor/JAK2 complex remains active and other signals are initiated that result in the permanent loss of STAT5b phosphorylation and degradation of the transcription factor. It is not clear what these other pathways are yet, but they may involve activation of other GH dependent transcription factors such as STAT 1, STAT3, the Ras/Raf kinase pathway, or hepatocyte nuclear factor 6, which in female rats exposed to a more constant pattern of GH secretion is required for expression of the female-specific P450 enzyme CYP2C12. If GH is removed from the male for a minimum of approximately 3 hours

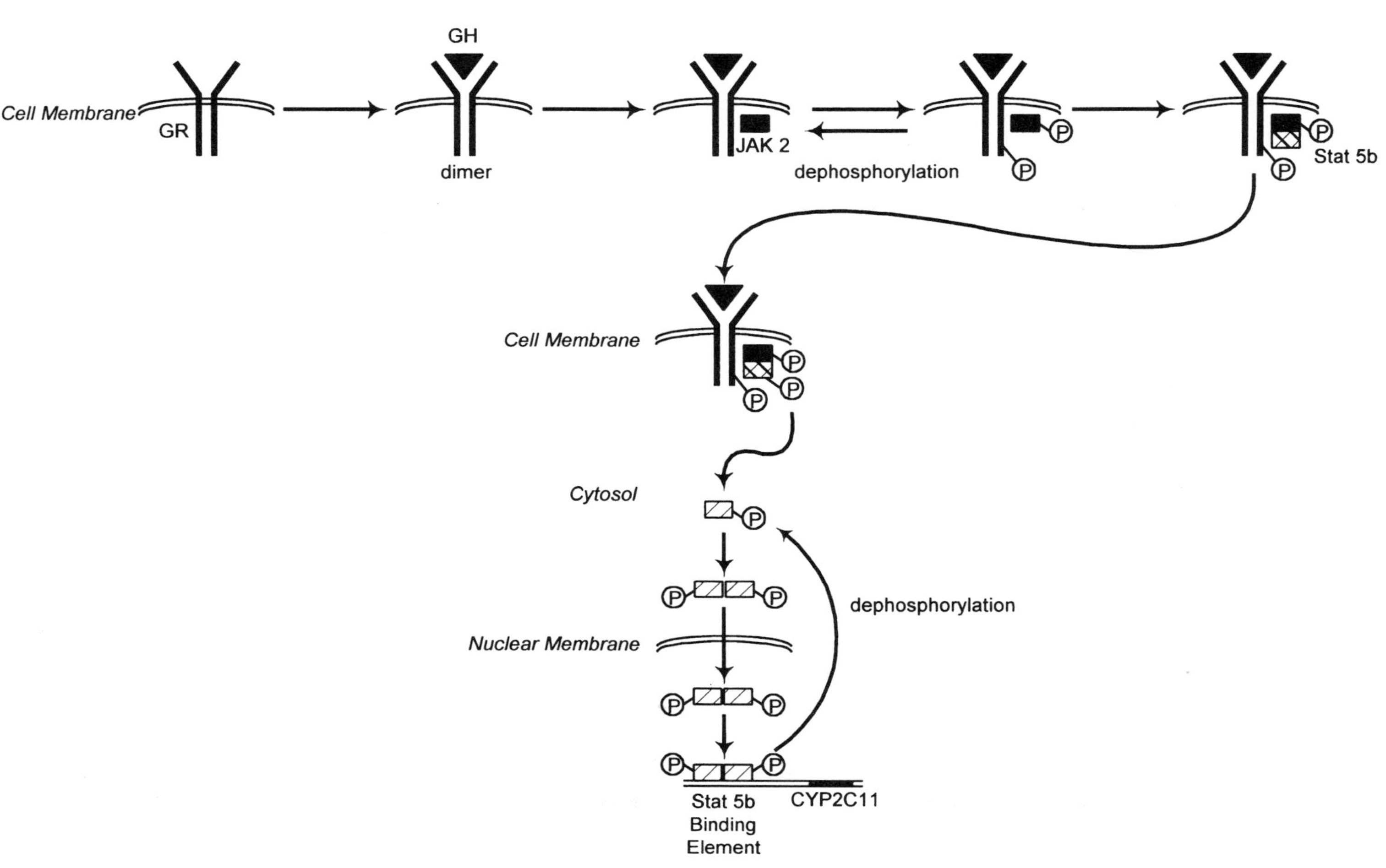

FIGURE 7.5 Growth hormone (GH) regulation of male-specific rat CYP2C11 expression.

between pulses, STAT5b in the nucleus is dephosphorylated possibly via the protein phosophatase SHP1, recycles into the cytosol, and the GH receptor/JAK2 complex also becomes dephosphorylated and inactivated. A second GH pulse initiates the signaling cycle all over again and CYP2C11 expression is maintained.

Many endocrine factors in addition to GH work through similar kinase/ phosphates pathways. For example, P450 regulation by insulin and some cytokines is probably mediated in a similar fashion.

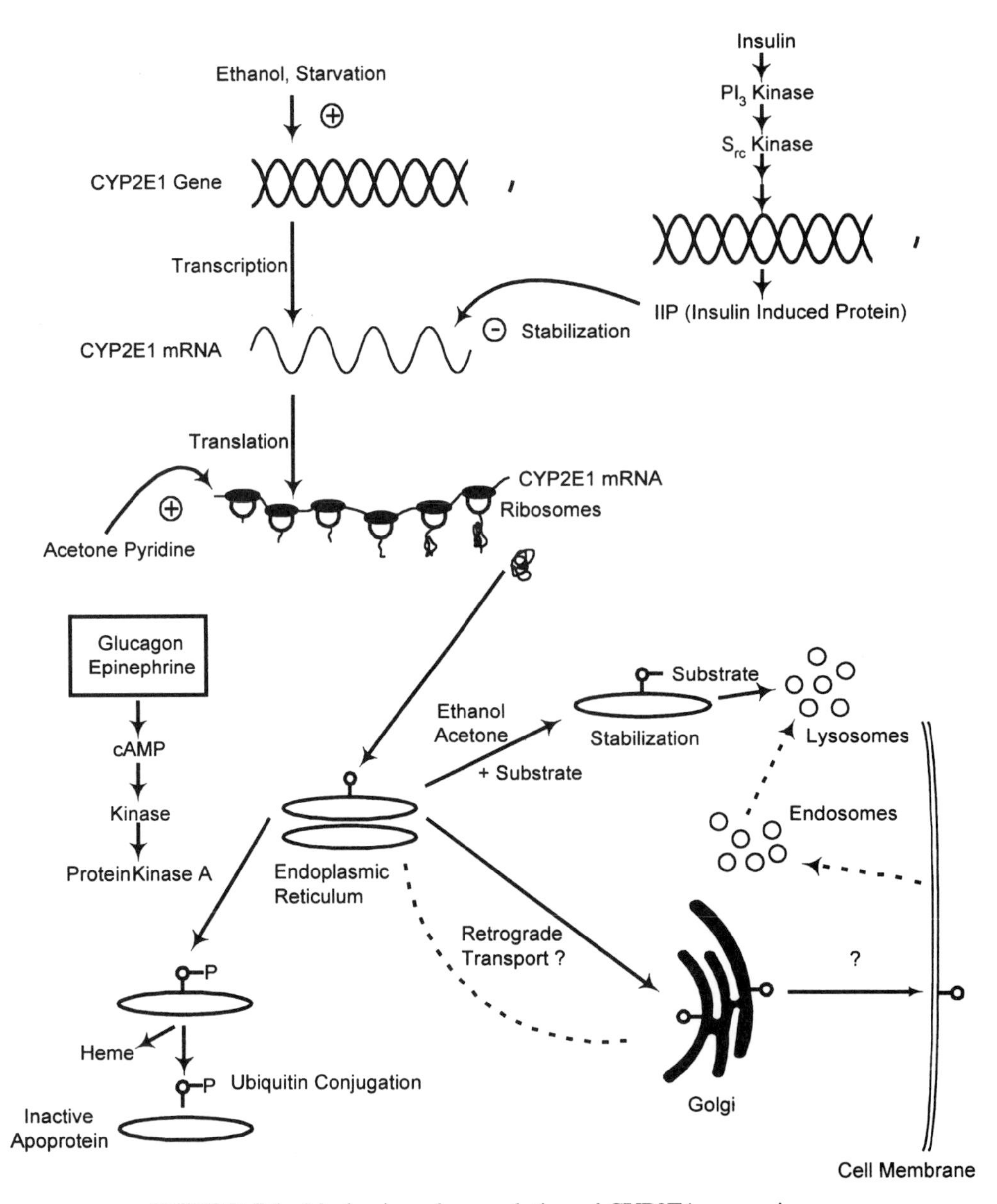

FIGURE 7.6 Mechanisms for regulation of CYP2E1 expression.

7.8.3 Post-Transcriptional Regulation of CYP2E1

Finally, as discussed above, many xenobiotic metabolizing enzymes also undergo physiological regulation by post-transcriptional mechanisms. To date, the best understood system is the regulation of the alcohol-inducible cytochrome P450 CYP2E1 by insulin, glucagon, and epinephrine (Fig. 7.6). Insulin seems to down-regulate CYP2E1 post-transcriptionaly via destabilization of the CYP2E1 mRNA. As discussed in Section 7.4.5, this seems to involve a cascade of events initiated by insulin binding to its receptor at the cell surface, which results in activation of the enzyme phosphatidylinositol 3-kinase and a buildup of inositol-3-phosphates, which in turn activate other protein kinases such as the src kinases, which ultimately produce a change in binding of proteins to the 3′-flanking region of the CYP2E1 mRNA altering its stability.

In contrast to insulin, glucagon and epinephrine seem to increase the rate of CYP2E1 degradation by a pathway involving activation of protein kinase A (PKA). Binding their receptors activates adenylate cyclase via a G protein transducer in the cell membrane resulting in increases in intracellular concentrations of cAMP. This in turn results in activation of PKA, which phosphorylates CYP2E1 at Ser 129. In vitro, this has been demonstrated to result in loss of the heme group and inactivation of the catalytically active holoenzyme. It is believed that phosphorylation results in a conformational change making the protein vulnerable to rapid degradation by either a protease within the endoplasmic reticulum or via ubiquitinization and subsequent degradation in a proteosomal complex. In vivo and in vitro CYP2E1 has a biphasic half-life with a fast phase 7-hour and slow phase 47-hour component. It is believed that PKA-mediated degradation regulates the fast phase component because CYP2E1 inducers such as ethanol and acetone block phosphorylation at Ser 129 and alter the half-life of the protein to a single, long phase 47-hour component.

SUGGESTED READING

Agrawal, A.K. and Shapiro, B.H. Gender, age and dose effects of neonatally administered aspartate on the sexually dimorphic plasma growth hormone profiles regulating expression of the rat sex-dependent hepatic CYP isoforms. *Drug Metab. Disp.* 25 (1997), 1249–1256.

Badger, T.M., Ronis, M.J.J., Ingelman-Sundberg, M. and Hakkak, R. Pulsatile blood alcohol and CYP2E1 induction during chronic alcohol infusions in rats. *Alcohol* 10 (1993), 453–457.

Chen, G.-F., Ronis, M.J.J., Thomas, P.E., Flint, D.J. and Badger, T.M. Hormonal regulation of microsomal cytochrone P450 2C11 in rat liver and kidney. *J. Pharmacol. Exp. Ther.* 283 (1997), 1486–1494.

Clarke, D.J. and Burchell, B. The uridine diphosphate glucuronosyltransferase multigene family: function and regulation. *Hanb. Exp. Pharamcol.* 112 (1994), 3–43.

Eliasson, E., Johansson, I. and Ingelman-Sundberg, M. Substrate-, hormone- and cAMP-regulated cytochrome P450 degradation. *Proc. Natl. Acad. Sci. U.S.A.* 87 (1990), 3225–3229.

Falls, J.G., Ryu, D.Y., Cao, Y., Levi, P.E. and Hodgson, E. Regulation of mouse liver flavin containing monooxygenases 1 and 3 by sex steroids. *Arch. Biochem. Biophys.* 342 (1997), 212–223.

Gebert, C.A., Park, S.-H. and Waxman, D.J. Regulation of signal transducer and activator of transcription (STST) 5b activation by the temporal pattern of growth hormone stimulation, *Mol. Endocrinol.* 11 (1997), 400–414.

Gebert, C.A., Park, S.-H. and Waxman, D.J. Termination of growth hormone pulse-induced STAT5b signaling. *Mol. Endocrinol.* 13 (1999), 38–56.

Hakkola, J., Pelkonen, O., Pasanen, M. and Raunio, H. Xenobiotic-metabolizing enzymes in the human feto-placental unit: role in intrauterine toxicity. *Crit. Rev. Toxicol.* 28 (1998), 35–72.

Hakkola, J., Tanaka, E. and Pelkonen, O. Developmental expression of cytochrome P450 enzymes in human liver. *Pharmacol. Toxicol.* 82 (1998), 209–217.

Juchau, M.R., Boutelet-Bochan, H. and Huang, Y. Cytochrome P450-dependent biotransformation of xenobiotics in human and rodent embryonic tissues. *Drug Metab. Rev.* 30 (1998), 541–568.

Klaassen, C.D., Liu, L. and Dunn, R.T. II. Regulation of sulfotransferase mRNA expression in male and female rats of various ages. *Chem-Biol. Interact.* 109 (1998), 299–313.

Kliewer, S.A., Moore, J.T., Wade, L., Staudinger, J.L., Watson, M.A., Jones, S.A., McKee, D.D., Oliver, B.B., Wilson, T.M., Zetterstrom, R.H., Perlmann, T. and Lehmann, J.M. An orphan nuclear receptor activated by pregnanes defines a novel steroid signaling pathway. *Cell* 92 (1998), 73–82.

Lacroix, D., Sonnier, M., Moncion, A., Cheron, G. and Cresteil, T. Expression of CYP3A in the human liver—evidence that the shift between CYP3A7 and CYP3A4 occurs immediately after birth. *Eur. J. Biochem.* 247 (1997), 625–634.

Lamartiniere, C.A. and Lucier, G.W. Endocrine regulation of xenobiotic conjugation enzymes. *Basic Life Sci.* 24 (1983), 295–312.

Leaman, D.W., Leung, S., Li, X. and Stark, G.R. Regulation of STAT-dependent pathways by growth factors and cytokines. *FASEB J.* 10 (1996), 1578–1588.

Lemann, J.M., McKee, D.D., Watson, M.A., Wilson, T.M., Moore, J.T. and Kliewer, S.A. The human orphan nuclear receptor PXR is activated by compounds that regulate CYP3A4 gene expression and cause drug interactions. *J. Clin. Invest.* 102 (1998), 1016–1023.

Liu, L. and Klaassen, C.D. Ontogeny and hormonal basis of male-dominant hepatic sulfotransferases. *Mol. Pharmacol.* 50 (1996), 565–572.

Liu, L. and Klaassen, C.D. Ontogeny and hormonal basis of female-dominant rat hepatic sulfotransferases. *J. Pharmacol. Exp. Ther.* 279 (1996), 386–391.

Liu, L. and Klaassen, C.D. Regulation of hepatic sulfotransferases by steroidal chemicals in rats. *Drug Metab. Disp.* 24 (1996), 854–858.

Morgan, E.T. Regulation of cytochromes P450 during inflamation and infection. *Drug Metab. Rev.* 29 (1997), 1129–1188.

Mugford, C.A. and Kedderis, G.L. Sex-dependent metabolism of xenobiotics. *Drug Metab. Rev.* 30 (1998), 441–498.

Okita, R.T. and Okita, J.R. Prostaglandin-metabolizing enzymes during pregnancy: characterization of NAD-dependent prostaglandin dehydrogenase, carbonyl reductase and P450-dependent prostaglandin omega-hydroxylase. *Crit. Rev. Biochem. Mol. Biol.* 31 (1996), 101–126.

O'Leary, K.A., Li, H.-C., Ram, P.A., McQuiddy, P., Waxman, D.J. and Kasper, C.B. Thyroid regulation of NADPH: cytochrome P450 oxidoreductase: identification of a thyroid-responsive element in the 5′-flank of the oxidoreductase gene. *Mol. Pharmacol.* 52 (1997) 46–53.

Peng, H.M. and Coon, M.J. Regulation of rabbit cytochrome P450 2E1 expression in HepG2 cells by insulin and thyroid hormone. *Mol. Pharmacol.* 54 (1998), 740–747.

Rich, K.J. and Boobis, A.R. Expression and inducibility of P450 enzymes during liver ontogeny. *Microsc. Res. Tech.* 39 (1997), 424–435.

Ronis, M.J.J. and Ingelman-Sundberg, M. The CYP2E subfamily. In: Ioannides, C. (Ed), *Cytochromes P450: Metabolic and Toxicological Aspects*, CRC Press, Boca Raton, FL, 1996, pp 211–239.

Ronis, M.J.J., Badger, T.M., Shema, S.J., Roberson, P.K., Templer, L., Ringer, D. and Thomas, P.E. Endocrine mechanisms underlying the growth effects of developmental lead exposure in the rat. *J. Toxicol. Environ. Health, Part A* 54 (1998), 101–120.

Sidhu, J.S. and Omiecinski, C.J. Insulin-mediated modulation of cytochrome P450 gene induction profiles in primary rat hepatocyte cultures. *J. Biochem. Mol. Toxicol* 13 (1999), 1–9.

Tanaka, E. In vivo age-related changes in hepatic drug-oxidizing capacity in humans. *J. Clin. Pharm. Ther.* 23 (1998), 247–255.

Viera, I., Sonnier, M. and Cresteil, T. Developmental expression of CYP2E1 in the human liver. Hypermethylation control of gene expression during the neonatal period. *Eur. J. Biochem.* 238 (1996), 476–483.

Waxman, D.J. and Chang, T.K.H. Hormonal regulation of liver cytochrome P450 enzymes. In: Oritiz de Montellano, P. (Ed), *Cytochrome P450: Structure, Mechanism and Biochemistry*, Plenum Press, New York, 1995, pp 391–417.

Waxman, D.J., Pampori, N.A., Ram, P.A., Agrawal, A.K. and Shapiro, B.H. Interpulse interval in circulating growth hormone patterns regulates sexually dimorphic expression of hepatic cytochrome P450. *Proc. Natl. Acad. Sci. U.S.A.* 88 (1991), 6868–6872.

Waxman, D.J., Ram, P.A., Park, S.-H. and Choi, H.K. Intermittent plasma growth hormone triggers tyrosine phosphorylation and nuclear translocation of a liver-expressed, STAT 5-related DNA binding protein. *J. Biol. Chem.* 270 (1995), 13262–13270.

Woodcroft, K.J. and Novak, R.F. Insulin effects on CYP2E1, 2B and 4A expression in primary cultured rat hepatocytes. *Chem-Biol. Interact.* 107 (1997), 75–91.

Yamazoe, Y., Shimada, M., Murayama, N. and Kato, R. Suppression of levels of phenobarbital-inducible rat liver cytochrome P450 by pituitary factors. *J. Biol. Chem.* 262 (1987), 7423–7428.

CHAPTER EIGHT

Adaptation to Toxicants

RANDY L. ROSE and ERNEST HODGSON

8.1 INTRODUCTION

In today's society, exposure to xenobiotics is an unavoidable aspect of life. Crops and animals are treated with pesticides to eliminate pests that compete for foodstuffs. Manufacturing processes, transportation, construction, and many of the essentials of modern living each involve the use of a multitude of chemicals. Although hazard assessments to chemicals are initially determined on an individual basis, the reality is that humans and other organisms are exposed to many xenobiotics simultaneously involving different portals of entry, modes of action, and metabolic pathways. For this reason, the effect of chemicals on the metabolism of other exogenous compounds is one of the most important areas of biochemical toxicology because it bears directly on the problem of toxicity-related interactions between different xenobiotics.

Exposure to chemicals can produce a variety of physiological adaptations. Many of these adaptations are direct responses of the biological system to the chemical and are transitory in nature. This is true for those xenobiotics that induce or inhibit enzymes involved in their own metabolism and often the metabolism of other chemicals. Some compounds that have little or no toxicity by themselves may serve to synergize or potentiate the toxicity of other compounds as a result of their ability to interact effectively with the enzymes involved in detoxication. It is important to understand these types of interactions, particularly in the pharmaceutical industry, where drug interactions have the potential to produce disastrous results.

Other adaptations include those resulting from selection pressure on populations, which causes genetic alterations that are passed on to living offspring. The term "resistance" is applied to populations where genetic changes resulting from chemical exposure have increased the number of individuals that can survive the exposure relative to the unexposed population. Because agrochemicals, pesticides, and chemotherapeutic agents are widely used, many species have acquired resistance. These include bacteria, parasites, plants, insects, rodents, normal mammalian cells, and tumor cells. Mechanisms of resistance include increased metabolism, target site insensitivity, physiological mechanisms such as decreased penetration or increased uptake and efflux, and behavioral changes. Organisms possessing a single mechanism of resistance may be cross-resistant to other chemicals as a result of the same mechanism, whereas multiple resistance denotes simultaneous resistance to several types of toxicants as a result of different mechanisms. Because of our increased reliance on chemicals in modern medicine and agriculture and their use to enhance and manipulate our environment, resistance can have important economic, social, and health-related consequences.

8.2 CHEMICAL FACTORS

From the point of view of both logistics and scientific philosophy, the study of the metabolism and toxicity of xenobiotics must be initiated by considering single compounds rather than the complex mixtures to which organisms are exposed. It is not surprising, therefore, that our knowledge of the mechanism of their chemical interactions is at a rudimentary stage. As documented below, xenobiotics, in addition to serving as substrates for a number of enzymes, may serve also as inhibitors or inducers of these or other enzymes. In fact, many examples are known of compounds that first inhibit and subsequently induce enzymes such as the microsomal monooxygenases. The situation is even further complicated by the fact that although some substances have an inherent toxicity and are detoxified in the body, others without inherent toxicity can be metabolically activated to potent toxicants. The following examples are illustrative of the situations that might occur involving only two compounds:

1. Compound A, without inherent toxicity, is metabolized to a potent toxicant. In the presence of an inhibitor of its metabolism there would be a reduction in toxic effect.
2. Compound A, in the presence of an inducer of the activating enzymes, would exhibit greater toxicity.
3. Compound B, a toxicant, is metabolically detoxified. In the presence of an inhibitor of the detoxifying enzymes there would be an increase in the toxic effect.
4. Compound B, in the presence of an inducer of the detoxifying enzymes, would exhibit less toxicity.

In addition to the above cases, the toxicity of the inhibitor or inducer as well as the time dependence of the effect must also be considered because, as previously mentioned, many xenobiotics that are initially enzyme inhibitors ultimately become inducers.

8.2.1 Inhibition

As indicated above, inhibition of xenobiotic-metabolizing enzymes can cause either an increase or a decrease in toxicity. Several well-known inhibitors of such enzymes are shown in Figure 8.1. The effects of inhibition can be demonstrated

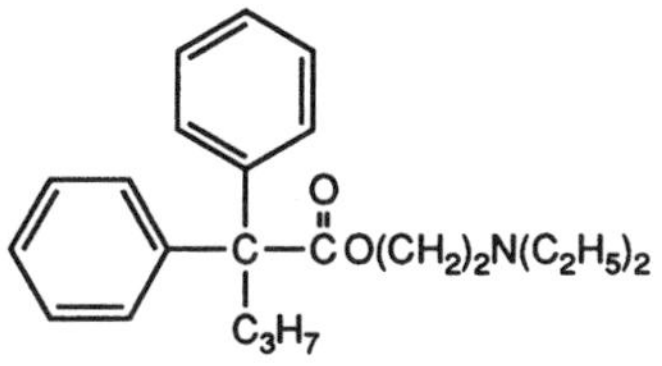

2-(Diethylamino) ethyl-
2,2-diphenylpentanoate (SKF-525A)
[P450]

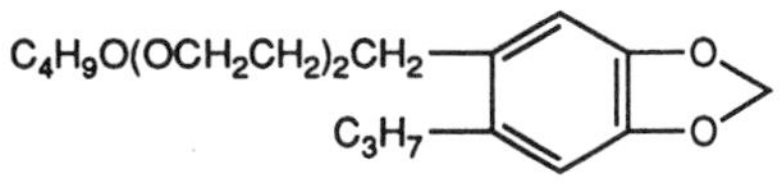

3,4-Methylenedioxy-6-propylbenzyl
n-butyl diethyleneglycol ether
(Piperonyl Butoxide) [P450]

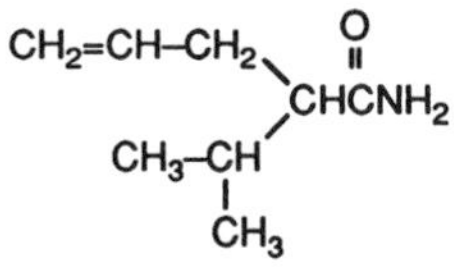

Allylisopropylacetamide
[P450]

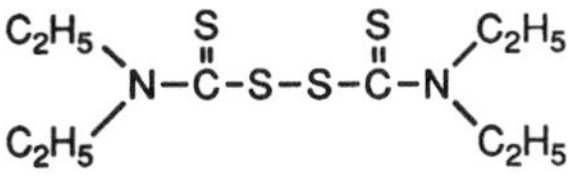

Disulfiram (Antabuse)
[aldehyde dehydrogenase]

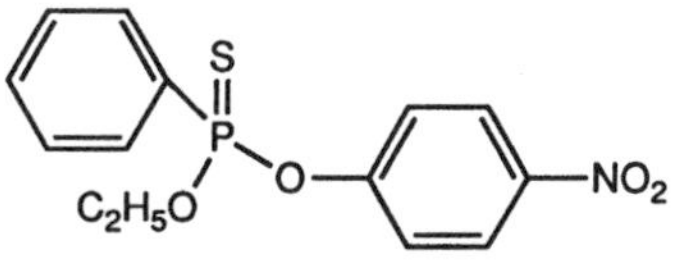

O-Ethyl-O-*p*-nitrophenyl
phenylphosphorothioate (EPN)
[esterases]

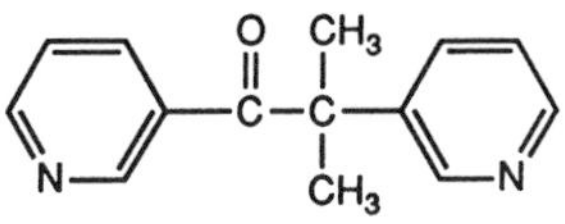

Metyrapone
[P450]

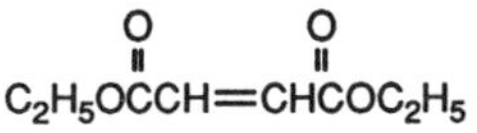

Diethyl maleate
[glutathione s-transferase]

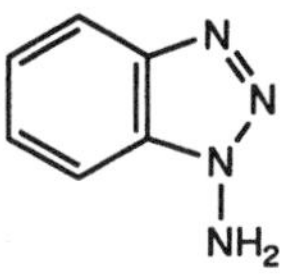

1-Aminobenzotriazole (1-ABT)
[P450]

FIGURE 8.1 Some inhibitors of various xenobiotic-metabolizing enzymes. Enzymes affected are indicated in [].

in a number of ways at different organizational levels, from the intact animal to purified enzymes.

Experimental demonstration of inhibition can be carried out by observing in vivo symptoms. The measurement of the effect of an inhibitor on the duration of drug action in vivo is the most common method of demonstrating inhibitory action. These methods are open to criticism, however, because effects on duration of action can be mediated by systems other than those involved in the metabolism of the drug. Furthermore, they cannot be used for inhibitors that have pharmacological activity similar or opposite to the compound being used. The most useful and reliable in vivo tests involve the measurement of effects on the hexobarbital sleeping time and the zoxazolamine paralysis time. Both of these drugs are fairly rapidly deactivated by the hepatic microsomal cytochrome P450 system, and thus P450 inhibitors prolong the action of these drugs.

For example, treatment of mice with chloramphenicol 0.5–1.0 hr before pentobarbital treatment prolongs the duration of the pentobarbital sleeping time in a dose-related manner; chloramphenicol is effective at low doses (<5 mg/kg) and has greater than a 10-fold effect at high doses (100–200 mg/kg). The well-known inhibitor of drug metabolism SKF-525A causes an increase in both hexobarbital sleeping time and zoxazolamine paralysis time in rats and mice, as do the insecticide synergists piperonyl butoxide and tropital, the optimum pretreatment time being about 0.5 hr before the narcotic is given. Similar results have been obtained with many other methylenedioxyphenyl compounds, including the natural product safrole and some of its derivatives. In the case of activation reactions, such as the activation of the insecticide azinphosmethyl to its potent anticholinesterase oxon derivative, a decrease in toxicity is apparent when rats are pretreated with SKF-525A.

Distribution and blood levels may also be affected by inhibition. Treatment of an animal with an inhibitor of xenobiotic metabolism may cause changes in the blood levels of an unmetabolized toxicant and/or its metabolites. This procedure may be utilized in the investigation of the inhibition of detoxication pathways; it has the advantage over in vitro methods in that it yields results of direct physiological or toxicological interest because it is carried out on the intact animal. Moreover, the time sequence of the effects can be followed in individual animals, a factor of importance when inhibition is followed by induction—a not uncommon event. For example, if animals are first treated with either SKF-525A, glutethimide, or chlorcyclizine, followed in 1 hr or less by pentobarbital, it can be shown that the serum level of pentobarbital is considerably higher in treated animals than in controls within 1 hr of its injection.

Use of specific metabolic enzyme inhibitors may often provide valuable information with respect to the metabolism of a particular drug. For example, quinidine is a potent and selective inhibitor of CYP2D6. This drug has been used in clinical studies as a pharmacological tool to mimic the lack of CYP2D6 in humans. By demonstrating that quinidine substantially slows the metabolism of trimipramine (a tricyclic antidepressant), investigators implicated CYP2D6 in its metabolism.

Effects on in vitro metabolism following in vivo treatment as a method of demonstrating inhibition is of variable utility. The preparation of enzymes from animal tissues usually involves considerable dilution with the preparative medium during homogenization, centrifugation, and resuspension. As a result, inhibitors not tightly bound to the enzyme in question are lost, either in whole or in part, during the preparative processes. Therefore, negative results can have little utility because failure to inhibit and loss of the inhibitor give identical results. Positive results, on the other hand, not only indicate that the compound administered is an inhibitor but also provide a clear indication of excellent binding to the enzyme, most probably due to the formation of an inhibitory complex. The inhibition of esterases following treatment of the animal with organophosphates such as paraoxon is a good example because the phosphorylated enzyme is stable and is still inhibited after the preparative procedures. Inhibition of the same enzymes by carbamates is greatly reduced by the same procedures, however, because the carbamylated enzyme is unstable and, in addition, residual carbamate is highly diluted.

Microsomal monooxygenase inhibitors that form stable inhibitory complexes with cytochrome P450, such as SKF-525A, piperonyl butoxide and other methylenedioxyphenyl compounds, and amphetamine and its derivatives, can be readily investigated in this way because the microsomes isolated from pretreated animals have a reduced capacity to oxidize many xenobiotics.

Another form of chemical interaction resulting from inhibition in vivo that can then be demonstrated in vitro involves those xenobiotics that function by causing destruction of the enzyme in question. Exposure of rats to vinyl chloride results in a loss of P450 and a corresponding reduction in the capacity of microsomes subsequently isolated to metabolize foreign compounds. Allyl isopropylacetamide and other allyl compounds have long been known to have a similar effect.

In vitro measurement of the effect of one xenobiotic on the metabolism of another is the most common type of investigation of interactions involving inhibition. Although it is the most useful method for the study of an inhibitory mechanism, particularly when purified enzymes are used, it is of more limited utility in assessing the toxicological implications for the intact animal. The principal reason for this is that it does not assess the effects of factors that affect absorption, distribution, and prior metabolism, all of which occur before the inhibitory event under consideration. Although the kinetics of inhibition of xenobiotic-metabolizing enzymes can be investigated in the same ways as any other enzyme mechanisms, a number of problems arise that may decrease the value of this type of investigation. They include the following:

1. Many investigations have been carried out on a particulate enzyme system, the microsomal monooxygenase system, using methods developed for single soluble enzymes. As a result, Lineweaver-Burke or other reciprocal plots are frequently curvilinear, and as a result, the same reaction may appear to have quite different characteristics from laboratory to laboratory, species to species, and organ to organ.
2. The nonspecific binding of substrate and/or inhibitor to membrane components is a further complicating factor affecting inhibition kinetics.
3. Both substrates and inhibitors are frequently lipophilic with low solubility in aqueous media.
4. Xenobiotic-metabolizing enzymes commonly exist in multiple forms (e.g., glutathione S transferases and cytochromes P450) that are all relatively nonspecific but differ from one another in relative affinities for different substrates.

The primary considerations in studies of inhibition mechanisms are reversibility and selectivity. The inhibition kinetics of reversible inhibition give considerable insight into the reaction mechanisms of enzymes, and for that reason have been well studied. Generally speaking, reversible inhibition involves no covalent binding, occurs rapidly, and can be reversed by dialysis or, more rapidly, by dilution. Reversible inhibition is usually divided into competitive inhibition, uncompetitive inhibition, and noncompetitive inhibition, although these are not rigidly separated types, and many intermediate classes have been described.

Competitive inhibition is usually due to two substrates competing for the same active site. Following classical enzyme kinetics, there should be a change in the apparent K_m but not in V_{max}. In microsomal monooxygenase reactions, type I ligands, which often appear to bind as substrates but do not bind to the heme iron, might be expected to be competitive inhibitors, and this frequently appears to be the case. Examples include the inhibition of O-demethylation of p-nitroanisole by aminopyrene, aldrin epoxidation by dihydroaldrin, and N-demethylation of aminopyrene by nicotinamide. Some of the polychlorinated biphenyls, notably dichlorobiphenyl, but also, less effectively, tetrachlorobiphenyl and hexachlorobiphenyl, have been shown to have a high affinity as type I ligands for rabbit liver cytochrome P450 and to be competitive inhibitors of the O-demethylation of p-nitroanisole. Pilocarpine, which potentiates nicotine-induced convulsions and hexobarbital hypnosis, has been shown to be a potent competitive inhibitor of the microsomal metabolism of these two compounds. Because pilocarpine and nicotine are type II ligands, whereas hexobarbital is type I, it is clear that inferences about the type of inhibition developed from a consideration of the ligand-binding spectra must be regarded as highly tentative. Competitive inhibition may also result from an inhibitor binding to a site on the enzyme other than the active site, which, nevertheless, blocks the active site to the substrate by bringing about a conformational change.

Uncompetitive inhibition has seldom been reported in studies of xenobiotic metabolism. It is seen when an inhibitor interacts with an enzyme-substrate complex but cannot interact with free enzyme. Both K_m and V_{max} change by the same ratio, giving rise to a family of parallel lines in a Lineweaver-Burke plot.

Simple noncompetitive inhibitors can bind to both the enzyme and enzyme-substrate complex to form either an enzyme-inhibitor complex or an enzyme-inhibitor-substrate complex. The net result is a decrease in V_{max} but no change in K_m. Metyrapone, a well-known inhibitor of monooxygenase reactions, can also, under some circumstances, stimulate metabolism in vitro. In either case, the effect is noncompetitive in that the K_m does not change while V_{max} does, decreasing in the case of inhibition and increasing in the case of stimulation. The characteristics of the three principle types of reversible inhibition are shown in Table 8.1.

Irreversible inhibition can arise from a variety of causes, some of which are extremely important toxicologically. In the vast majority of cases, either covalent binding or disruption of the enzyme structure is involved. In neither case can the effect be reversed in vitro by either dialysis or dilution.

Covalent binding may involve the prior formation of a metabolic intermediate that then interacts with the enzyme. An excellent example of this type of inhibition is the effect of the insecticide synergist piperonyl butoxide on hepatic microsomal monooxygenase activity. This compound can form a stable inhibitory complex that blocks CO binding to cytochrome P450 and also prevents substrate oxidation. This complex causes the appearance of a characteristic difference spectrum that has two pH-dependent peaks in the Soret region, which, apart from its stability, resembles that of ethyl isocyanide. This complex is the result of metabolite formation, which

TABLE 8.1 Comparison of Three Types of Reversible Inhibition.

Type of inhibition	Interaction	Vmax	Km	Lineweaver-Burke plot[a]
Competitive	I and E	±	+	Lines converge on y axis
Noncompetitive (simple)	I, ES, and E	–	±	Lines converge on x axis
Uncompetitive	I and ES	–	–	Parallel lines

Abbreviations: I, inhibitor; E, enzyme; ES, enzyme-substrate complex; ±, unchanged; –, decreased; +, increased.

[a] Each line referred to is a plot of l/v versus l/S in which v is the initial velocity and S the substrate concentration in the presence of a given amount of inhibitor.

By varying the amount of inhibitor in separate experiments a family of such lines is generated.

is shown by the fact that the type of inhibition changes from competitive to irreversible as metabolism, in the presence of NADPH and oxygen, proceeds (see Section 8.2.2). Piperonyl butoxide inhibits the in vitro metabolism of many substrates of the monooxygenase system, including aldrin, ethylmorphine, aniline, aminopyrene, carbaryl, biphenyl, hexobarbital, p-nitroanisole, and many others. Although most of the studies carried out on piperonyl butoxide have involved rat or mouse liver microsomes, they have also been carried out on pig, rabbit, and carp liver microsomes and in a variety of preparations from houseflies, cockroaches, and other insects.

A number of classes of monooxygenase inhibitors, in addition to methylenedioxyphenyl compounds, are now known to form "metabolic-intermediate complexes" including amphetamine and its derivatives, and SKF-525A and its derivatives.

Disulfiram (Antabuse) inhibits aldehyde dehydrogenase irreversibly, causing an increase in the level of acetaldehyde, which has been formed from ethanol by the enzyme alcohol dehydrogenase. This results in nausea, vomiting, and other symptoms in the human—hence its use as a deterrent of alcohol consumption in the treatment of alcoholism. The inhibition by disulfiram appears to be irreversible, the level returning to normal only as a result of protein synthesis.

The inhibition by other organophosphate compounds of the carboxylesterase which hydrolyses malathion is a further example of xenobiotic interaction resulting from irreversible inhibition because, in this case, the enzyme is phosphorylated by the inhibitor.

Another class of irreversible inhibitors of toxicological significance consists of those compounds that bring about the destruction of the xenobiotic-metabolizing enzyme. The drug allylisopropylacetamide, as well as a number of other allyl compounds, has long been known to cause the breakdown of cytochrome P450 and resultant release of heme. The hepatocarcinogen vinyl chloride has been shown to have a similar effect, probably also mediated through the generation of a highly reactive metabolic intermediate. A great deal of information, discussed in detail in Chapter 20, has accumulated in the past decade on the mode of action of the hepatotoxicant carbon tetrachloride, which effects a number of irreversible changes in both liver proteins and lipids, such changes being generated by active intermediates formed during its metabolism. The less specific disrupters of protein structure, such as urea, detergents, strong acids, etc., are probably of significance only in vitro experiments.

8.2.2 Activation In Vitro

There are several examples of activation of xenobiotic-metabolizing enzymes by compounds other than the substrate. This differs from induction (described in Section 8.2.5) in that it is an immediate effect on a preexisting enzyme, occurring in an enzyme preparation in vitro, that does not involve de novo protein synthesis. The occurrence and significance of such stimulation in vivo is not apparent.

Cytochrome P450-mediated microsomal oxidations are stimulated by the addition of any of a rather heterogeneous group of compounds, including ethyl isocyanide, acetone, 2,2′-bipyridyl, and, under certain conditions, the inhibitor metyrapone. The effect is not uniform, however, because acetone, for example, stimulates aniline hydroxylation but has no effect on the N-demethylation of N-methylaniline, N,N-dimethylaniline, or ethylmorphine, or the O-demethylation of p-nitroanisole. 2,2′-Bipyridine, on the other hand, stimulates both aniline hydroxylation and N-demethylation of N-methyl- and N,N-dimethylaniline but, at the same time, inhibits the N-dealkylation of ethylmorphine and aminopyrine. Presumably these differences are related to isoform specificity for the activator and/or the substrate.

The mechanism of this stimulation is unclear, and a number of suggestions have been advanced. For example, ethyl isocyanide may act via stimulation of cytochrome P450 reduction. Acetone and 2,2′-bipyridine do not appear to have a common mode of action with ethyl isocyanide and may exert their effect on the microsomal membrane, thus changing the availability of enzyme and/or substrate.

The activation of another membrane-bound enzyme of interest in biochemical toxicology, UDP glucuronyltransferase, is probably due to effects on the membrane. In this case, significant stimulation is brought about by "aging" the enzyme preparation, by sonication, and by such agents as dilute detergents and proteolytic enzymes.

8.2.3 Synergism and Potentiation

The terms "synergism" and "potentiation" have been variously used and defined but, in any case, involve a toxicity that is greater when two compounds are given simultaneously or sequentially than would be expected from a consideration of the individual toxicities of the compounds.

Generally, toxicologists use the term "synergism" for cases that fit the above definition, but only when one compound is toxic alone whereas the other has little or no intrinsic toxicity. This is the case with the toxicity of insecticides to insects and mammals and the effects on this toxicity of methylenedioxphenyl synergists such as piperonyl butoxide, sesamex, and tropital. The term "potentiation" is then reserved for those cases in which both compounds have appreciable intrinsic toxicity, such as in the case of malathion and EPN.

Thus, both synergism and potentiation involve toxicity greater than would be expected from the toxicities of the compounds administered separately, but in the case of synergism, one compound has little or no intrinsic toxicity administered

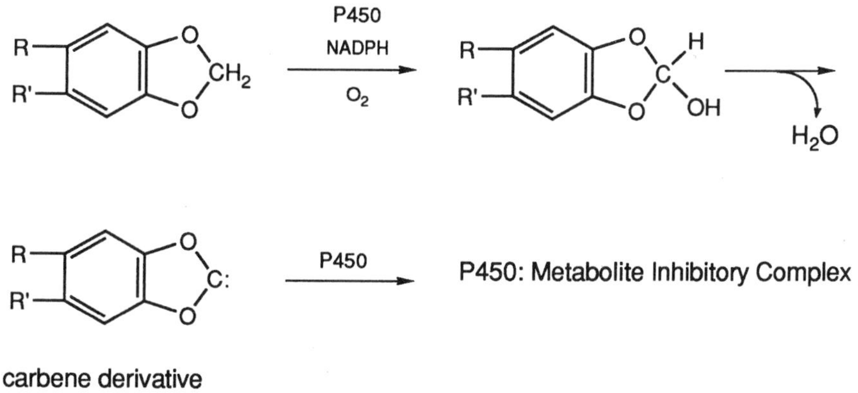

P450: Metabolite Inhibitory Complex

carbene derivative

FIGURE 8.2 Inhibition of cytochrome P450 by methylenedioxyphenyl compounds.

alone, whereas, in the case of potentiation, both compounds have appreciable toxicity when administered alone. Pharmacologists use the terms in the opposite sense, that is, synergism when both compounds are toxic and potentiation when one is nontoxic. What is even more unfortunate is the tendency to use synergism and potentiation as synonyms, leaving no distinction based on relative toxicity.

Although examples are known in which synergistic interactions take place at the receptor site, the majority of such interactions appear to involve the inhibition of xenobiotic-metabolizing enzymes. Two of the best known examples in toxicology involve the insecticide synergists, particularly the methylenedioxyphenyl synergists, and the potentiation of the insecticide malathion by a large number of other organophosphate compounds.

The first example has already been mentioned. Piperonyl butoxide, sesamex, and related compounds increase the toxicity of certain insecticides by inhibiting the insect monooxygenase system. They are of commercial importance in household aerosol formulations containing pyrethrum. This inhibition, which appears to be the same in mammals and insects, involves the formation of a metabolite-inhibitory complex with cytochrome P450. The complex probably results from the formation of a carbene (Fig. 8.2), which then reacts with the heme iron in a reaction involving n-bonding, as well as the dative σ-bond formed by the free pair of electrons, to form a complex that blocks CO (and presumably O_2) binding and inhibits the metabolism of xenobiotics.

Other insecticide synergists that interact with the monooxygenase system include aryloxyalkylamines such as SKF-525A, Lilly 18,947, and their derivatives, compounds containing acetylenic bonds such as aryl-2-propynyl ethers and oxime ethers, organothiocyanates, N-(5-pentynyl)phthalimides, phosphate esters containing propynyl functions, phosphorothionates, benzothiadiazoles, and some imidazole derivatives.

The general nature of this mode of action raises important questions concerning effects on nontarget species. The question of whether piperonyl butoxide or other methylenedioxyphenyl compounds can increase the toxicity of environmental or medicinal xenobiotics has often been asked but not yet definitely answered. It is of interest that two members of another class of insecticide synergists, 3-bromophenyl-

4(5)-imidazole and 1-naphthal-4(5)-imidazole, are potent inhibitors of the metabolism of estradiol and ethynylestradiol by rat liver microsomes.

The best known example of potentiation involving insecticides and an enzyme other than the monooxygenase system is the increase in the toxicity of malathion to mammals brought about by certain other organophosphates. Malathion has a low mammalian toxicity due primarily to its rapid hydrolysis by a carboxylesterase. EPN, a phosphonate insecticide, was shown to cause a dramatic increase in malathion toxicity to mammals at does levels that, given alone, caused essentially no inhibition of cholinesterase. In vitro studies further established the fact that the oxygen analog of EPN, as well as many other organophosphate compounds, increases the toxicity of malathion by inhibiting the carboxylesterase responsible for malathion's degradation. In a similar way, EPN and tri-o-toyl phosphate increase the toxicity of dimethoate by inhibiting the carboxylamidase necessary for its detoxication.

Synergistic action is often seen with drugs. Almost all cases of increased hexobarbital sleeping time or zoxazolamine paralysis time by other chemicals could be described as synergism or potentiation. Synergism may also result from competition for binding sites on plasma proteins or for the active secretion mechanism in the renal tubule.

8.2.4 Antagonism

In toxicology, antagonism may be defined as that situation in which the toxicity of two or more compounds administered together, or sequentially, is less than would be expected from a consideration of their toxicities when administered individually. Strictly speaking, this definition includes those cases in which the lowered toxicity results from induction of detoxifying enzymes (see Section 8.2.5). Apart from the convenience of treating such antagonist phenomena together with other aspects of induction, they are frequently considered separately because of the significant time span that must elapse between treatment with the inducer and subsequent treatment with the toxicant. The reduction of hexobarbital sleeping time and the reduction of zoxazolamine paralysis time by prior treatment with phenobarbital are obvious examples of such induction effects at the level of drug action, whereas protection from the carcinogenic action of benzo(a)pyrene, aflatoxin, and diethylnitrosamine by phenobarbital are examples at the level of chronic toxicity.

Antagonism not involving induction is a phenomenon often seen at a marginal level of detection and is consequently both difficult to explain and of less significance. In addition, several types of antagonism of importance to toxicology that do not involve xenobiotic metabolism are known but are not appropriate to discuss in this chapter. These include competition for receptor sites, such as the competition between CO and O_2 in CO poisoning, or situations in which one toxicant combines nonenzymatically with another to reduce its toxic effects, such as in the chelation of metal ions. Physiological antagonism, in which two agonists act on the same physiological system but produce opposite effects, is also of importance.

Parathion is known to inhibit monooxygenase activity, whereas DDT, aldrin, and dieldrin are all inducers. In experiments with separate and combined treatments, pretreatment of the animal with parathion inhibited O-demethylation, O-dearylation, N-demethylation, azo-reduction, and nitro reduction in the microsomes subsequently isolated, whereas the organochlorine compounds caused a stimulation of some or all of these parameters. Combined pretreatment with parathion and any one of the organochlorine compounds gave results that ranged from inhibition to induction, depending upon the particular organochlorine compound used and/or the parameter being measured. It was apparent that the results were largely antagonistic although highly unpredictable.

8.2.5 Induction

Some 40 years ago, during investigations on the N-demethylation of aminoazo dyes, it was observed that pretreatment of mammals with the substrate or, more remarkably, with other xenobiotics, caused an increase in the ability of the animal to metabolize these dyes. It was subsequently shown that this effect was the result of an increase in the microsomal enzymes involved in the metabolism of the dyes. Since that time it has become clear that this inductive phenomenon is widespread and quite nonspecific.

A landmark review by Conney (1967) gives a summary of this early work. Several hundred compounds of diverse chemical structure have been shown to induce cytochrome P450 and other enzymes. These compounds include drugs, insecticides, polycyclic hydrocarbons, and many others; the only obvious common denominator is that they are all organic and lipophilic. In addition to P450 induction, NADPH-cytochrome P450 reductase may also be increased in these situations. It is apparent that individual inducers induce one or a small number of P450 isozymes. However, because of the relatively nonspecific nature of many of these isozymes, a larger number of P450-dependent monooxygenase activities are increased.

8.2.5a Specificity of Monooxygenase Induction

Many of the inducers of monooxygenase activity fall into two principal classes, one exemplified by phenobarbital and containing many drugs, insecticides, etc., of diverse chemical classes, and the other exemplified by TCDD and polyaromatic hydrocarbons such as 3-methylcholanthrene and benzo(a)pyrene. Many inducers require either fairly high dose levels or repeated dosing to be effective, frequently greater than 10 mg/kg and some as high as 100–200 mg/kg. However, some insecticides, such as mirex, can induce at dose levels as low as 1 mg/kg, and the most potent inducer known, 2,3,7,8-tetrachlorodibenzo-p-dioxin (TCDD), is effective at 1 μg/kg in some species.

In mammalian liver, phenobarbital-type inducers cause a marked proliferation of the smooth endoplasmic reticulum concomitant with induction of P450. A wide range of oxidative is induced, including O-demethylation of p-nitroanisole, N-demethylation of benzphetamine, pentobarbital hydroxylation, aldrin epoxidation, and many others, but increases in aryl hydrocarbon hydroxylase activity are minimal.

Induction by polycyclic hydrocarbons, on the other hand, causes no increase in smooth endoplasmic reticulum and results in the appearance of P450s characterized by a shift in the λ max of the reduced cytochrome P450-CO complex to 448 nm. In contrast with phenobarbital type inducers, the range of oxidative activities induced by polycylcic hydrocarbons is primarily limited to aryl hydrocarbon hydroxylase

TABLE 8.2 Induction of Cytochrome P450 Isozymes by Xenobiotics.

Inducer Category	Examples	Isozymes Induced
Polycyclic aromatic hydrocarbon	3-Methylcholanthrene TCDD PCBs	1A1, 1A2
Phenobarbital	Phenobarbital DDT Phenytoin	2B1, 2B2
PCN/Glucocorticoid	Dexamethasone Pregnenalone-16a-carbonitrile Erythromycin	3A1
Ethanol	Ethanol Acetone Imidazole	2E1
Peroxisome proliferator	Clofibrate Tridiphane Phthalates 2,4-D and 2,4,5-T	4A1

activity. The potent inducers TCDD and 3-methylcholanthrene, mentioned above, fall into this class.

There are several other less extensively investigated classes of inducers. They include pregnenolone-16 α-carbonitrile and other steroid antagonists and synthetic and natural glucocorticoids, ethanol and other related compounds, as well as a group of compounds that also cause peroxisome proliferation. In addition, there are some chemicals that induce in more than one of the categories described above. This situations is referred to as "mixed" or "overlap" induction. Examples of xenobiotics representative of the several classes of inducer are shown in Table 8.2.

It appears reasonable that, because there are several isozymes of P450 associated with the hepatic endoplasmic reticulum, various inducers may induce one or more of them. Because each of these types has a relatively broad substrate specificity, differences in metabolic activity may be caused by variations in the extent of induction of different cytochromes.

Cytochrome P450 isozymes have been characterized in detail and immunochemical methods devised for their identification (chapters 3 and 5). The utilization of these methods has enabled relatively precise correlations to be made between the members of the various inducer classes and the isozyme(s) induced. Examples of the isozymes induced by members of the different classes are given in Table 8.2.

Although the bulk of published investigations of the induction of P450 isozymes have dealt with mammalian liver, it should be pointed out that induction has been observed in other mammalian tissues and in nonmammalian species, both vertebrate and invertebrate. Cytochrome P450 induction has been demonstrated in mammalian tissues including the lung, small intestine, adrenal gland, testis, kidney, olfactory epithelium, mammary gland, prostate, and Zymbals gland of both rats and mice. Although all of the common categories of inducers have been shown to affect one or more extrahepatic tissue, tissue-specific patterns are common. In general, however, any particular extrahepatic tissue will respond to fewer inducers than the liver and, overall, induction of CYP1A1 by TCDD-like inducers is the most common form of extrahepatic induction. Sensitivity to inducers also changes during development. For example, it has long been known that fetal rat liver is refractory to induction by phenobarbital but will respond to 3-methylcholanthrene.

8.2.5b Mechanism and Genetics of Induction in Mammals

Induction has been demonstrated following perfusion of the isolated liver with the inducer and also following treatment of isolated hepatocytes. However, it has not been duplicated in cell-free systems, presumably due to the time required for the process and the fact that several complex multienzyme processes are required simultaneously—notably the processes involved in gene expression and protein synthesis. It has been known for some time that the induction of monooxygenase activity is generally a true induction involving synthesis of new enzyme, and not the activation of enzyme already synthesized, inasmuch as it is prevented by inhibitors of protein synthesis. For example, aryl hydrocarbon hydroxylase induction is inhibited by puromycin, ethionine, and cycloheximide.

The use of suitable inhibitors of RNA and DNA metabolism has shown that inhibitors of RNA synthesis such as actinomycin D and mercapto-(pyridethyl)-benzimidazole block aryl hydrocarbon hydroxylase induction, whereas hydroxyurea, at levels that completely block the incorporation of thymidine into DNA has no effect. Thus, it appears that the inductive effect is generally at the level of transcription and that DNA synthesis is not required. This is clearly supported by the observation that increased levels of specific mRNAs can be detected soon after treatment with different inducers, including inducers of both the phenobarbital type and the 3-methylcholanthrene type. Transcription rate and increase in specific mRNAs are excellent measures of the early events in P450 induction.

As noted above, the early induction literature tended to emphasize "phenobarbital-like" and "3-methylcholanthrene-like" inducers. Not only is it now known that other classes of inducers occur, for example, ethanol, pregnenalone-16-alpha-carbonitrile and peroxisome proliferators, but in most cases the specific isozymes induced have been identified (Table 8.2). It should also be noted that not all P450 isozymes are inducible, whereas others are inducible primarily by endogenous compounds rather than xenobiotics.

Receptor-mediated induction by 3-methyleholanthrene-type inducers. Although it was clear from the outset that cells in which enzymes could be induced by xenobiotics must have a mechanism for first recognizing the inducer and then bringing about increased protein synthesis, only one receptor-based mechanism is known in detail. These mechanism studies all derive from the initial observation that 2,3,7,8-tetrachlorodibenzo-p-dioxin (TCDD) was 30,000 times more potent as an aryl hydrocarbon hydroxylase inducer than the prototypic inducer in this class, 3-methylcholanthrene. Thus, TCDD provided a ligand of high affinity for the receptor, subsequently named the AhR (for aryl hydrocarbon receptor).

The AhR is a ligand-activated transcription factor (LTF) that shares many similarities with the steroid and thyroid hormone receptor family of LTFs. The AhR has been identified in a variety of tissues from both mammalian and nonmammalian

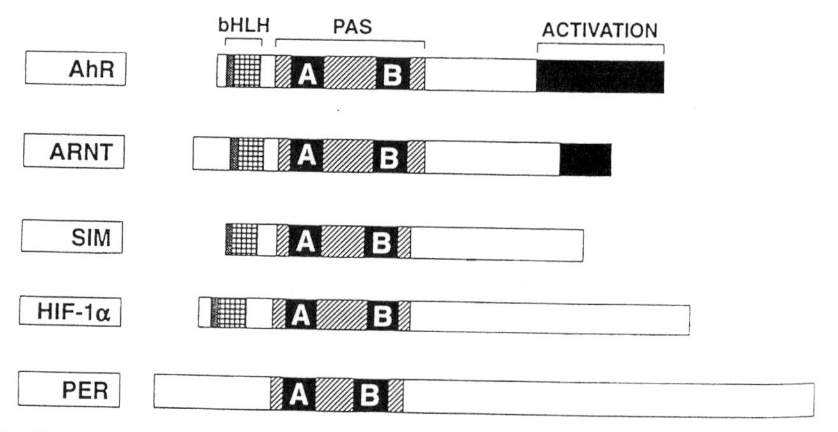

FIGURE 8.3 Structural and functional organization of the AhR. The basic helix-loop-helix (bHLH), PER, ARNT, SIM homology domain (PAS), two 50 amino acid PAS degenerate repeats (A and B), and transactivation domains (TAD) are indicated.

Source: From Rowlands, J.C. and Gustafsson, J.-A., *Crit. Rev. Toxicol.* 27 (1997), 109.

species. It has a calculated molecular weight of 90–96 kDa. Several studies have provided evidence that the AhR is active in the expression of a wide variety of genes. These include CYP isoforms 1A1 and 1A2, a glutathione S transferase of the alpha class, an NADPH-quinone-oxidoreductase, a UDP-glucuronosyltransferase, and an aldehyde dehydrogenase.

Analysis of the amino acid sequence of the AhR suggests that it belongs to a new family within the superfamily of the helix-loop-helix (HLH) proteins. Most HLH proteins are transcriptional regulators and are involved in cellular development and differentiation. Other members of this family include the AhR nuclear translocator protein (ARNT), the *Drosophila* proteins SIM and PER, and the hypoxia-inducible factor 1 alpha (HIF-1α). These proteins share several homologous regions including a basic region at the NH_2 terminal to the HLH motif (bHLH) and a region of approximately 300 amino acids termed the PAS (PER, ARNT, AhR, SIM) homology domain (Fig. 8.3), which is described in greater detail by Rowlands and Gustafsson (1997). The HLH-PAS proteins are involved in the regulation of diverse biological processes.

The unliganded AhR resides in the cytosol where it is complexed with two molecules of a 90 kDa heat shock protein (hsp-90) as well as an unidentified 43-kDa protein (p43). After ligand binding, the receptor translocates to the nucleus where it sheds hsp-90 and p43 and forms a dimer with ARNT (Fig. 8.4). In the nucleus, the transformed receptor interacts with the xenobiotic responsive element (XRE) located in the promoter of TCDD-responsive genes where interactions of the TCDD-AhR-ARNT complex with one or more XREs disrupts chromatin structure, allowing increased transcription followed by increased protein synthesis.

Several independent lines of evidence support the role of the Ah receptor in the regulation of P450 1A1 induction as well as the induction of other proteins. They include the following:

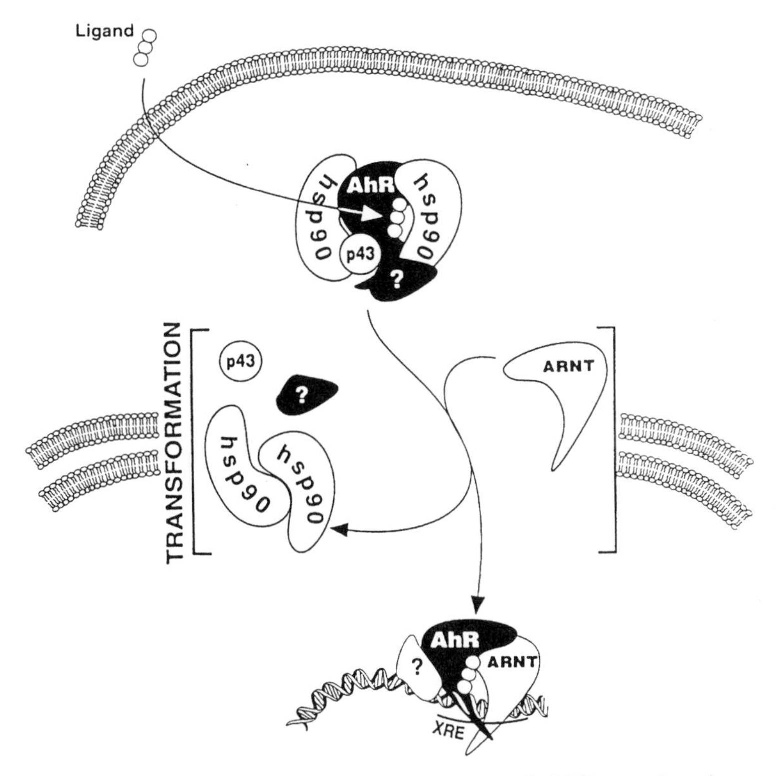

FIGURE 8.4 Proposed mechanism for ligand-activated AhR translocation and DNA binding.

Source: From Rowlands, J.C. and Gustafsson, J.-A., *Crit. Rev. Toxicol.* 27 (1997), 109.

1. *Structure-activity relationships*. The affinity with which most inducers bind to the AhR is generally well correlated with their ability to induce CYP1A1 and aryl hydrocarbon hydroxylase activity. This is particularly clear within chemically related groups of inducers, for example, dibenzo-p-dioxins or dibenzofurans or polychlorinatedbiphenyls. The most potent halogenated inducers are planar molecules with the halogens in the four lateral positions, the optimum two-dimensional size being a rectangle 3×10 A.
2. *Genetic variation*. Strains of mice have been identified that are "responsive," that is, they show a high level of induction, or "nonresponsive," showing no significant induction with polycyclic aromatic hydrocarbons. Although both are induced by TCDD, a nonresponsive strain (e.g., DBA) requires an approximately 15-fold higher dose for the same level of induction than a responsive (e.g., C57) strain. In the nonresponsive strain, the presence of a low-affinity AhR is well correlated with the reduced sensitivity of these strains to induction. The phenotypes of AhR knock-out mice further demonstrate that the AhR is necessary for TCDD induction as well as for the constitutive expression of CYP1a1, 1a2, and uridine diphosphate glucuronosyltransferase.

3. *Molecular biology*. Studies with mice treated with TCDD in vivo have shown a good correlation between the levels of Ah receptor-TCDD complex and the levels of mRNA for CYP1A1. Subsequent studies have utilized mutant mouse hepatoma cell lines that were either deficient in their ability to respond to TCDD induction of aryl hydrocarbon hydroxylase activity or were readily induced. These latter studies have revealed that several genes are required for aryl hydrocarbon hydroxylase induction. Although one of these clearly codes for the CYP1A1 protein, two or more are concerned with the AhR, coding either for structural components or for factors that modify the ligand-receptor complex into a form that can interact with specific DNA sequences. Cell lines with decreased inducibility have either decreased levels of cytosolic receptor or decreased translocation of ligand-receptor complex into the nucleus.
4. *Antagonist studies*. The development of 6-methyl-1,3,8-trichlorobenzodifuran (MCDF) as an induction antagonist has provided additional evidence for AhR involvement. MCDF is a weak CYP1A1 inducer but is a potent antagonist to induction by TCDD. This effect is paralleled by inhibition of TCDD binding to the AhR.

The AhR binds TCDD with high affinity, the apparent equilibrium dissociation constant being approximately 1 nM, and has the characteristics of a true receptor. That is, it does not serve merely as a binding protein, but rather regulates a specific biological process. The AhR-ligand complex seems to function by interaction with specific DNA sequences found 1000 bp or more upstream from the CYP1A1 transcription site. Two of these, the existence of which is supported by direct evidence, are transcriptional enhancer (induction) sites while a third, more speculative, site may function as an inhibitor or suppressor site (Fig. 8.5).

In many respects, the AhR resembles the steroid hormone receptors and is believed to belong to the steroid hormone superfamily, although no endogenous ligand has yet been identified with certainty and no known steroid competes with TCDD for binding. Recently, however, it has been shown that tryptophane photooxidation products do compete with TCDD for binding to the Ah receptor, although the functional significance of this observation is not yet clear.

Induction via the AhR can result in a variety of interactions not only with xenobiotics but also with chemicals having endogenous functions. For example, clearance rates for theophylline, a drug widely used in asthma therapy, is approximately 50% greater in smokers than in nonsmokers as a result of CYP1A2 induction. Exposure of rats to high concentrations of TCDD results in a perturbation of retinoid homeostasis, likely as a result of increased oxidation of retinoic acid by the induced CYP1A2. Increased levels of CYP1A1/1A2 in livers of TCDD-treated rats induces the formation of reactive estrogen metabolites, contributing to the genotoxic and tumorigenic effects of TCDD treatments. Similarly, serum thyroxin levels are reduced by TCDD exposure due to the induction of a hepatic UDP-glucuronosyl-transferase.

Induction mediated by the AhR pathway not only affects a variety of xenobiotic metabolizing enzymes but is also linked to the expression of a several other genes that affect basic cellular responses such as growth, differentiation, and programmed cell death. Some of the effects of TCDD observed in tissue culture

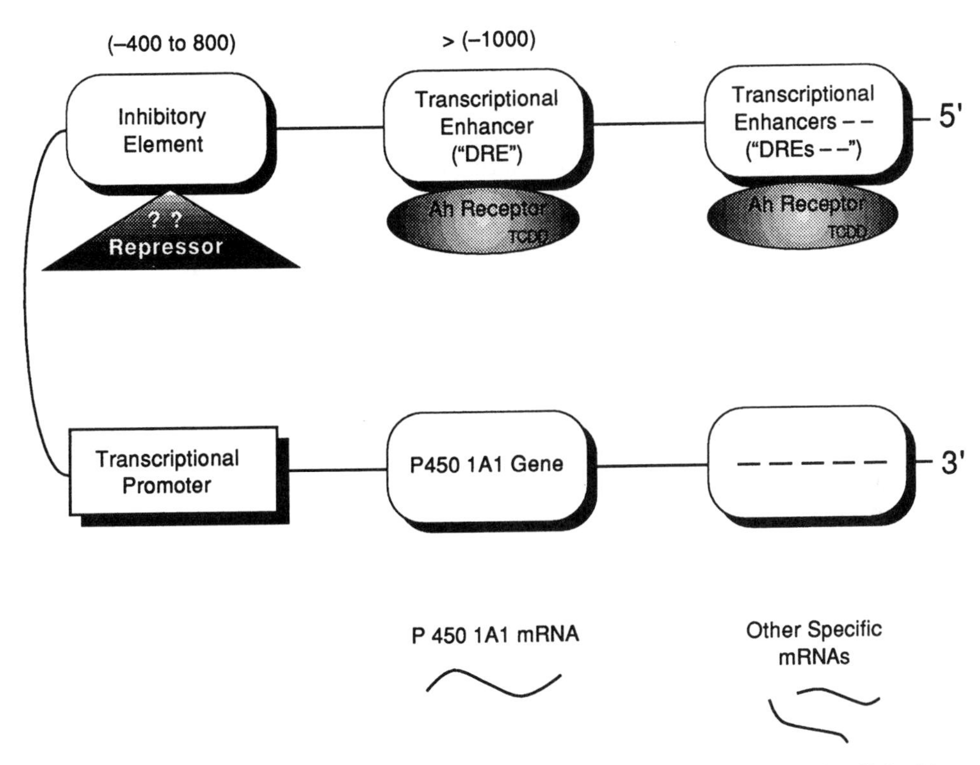

FIGURE 8.5 Interaction of the Ah receptor receptor-ligand complex with the 5′ flanking region of the P450 1A1 gene. Two dioxin responsive elements (DREs) appear to lie approximately 1000 or more base pairs upstream from the P450 1A1 transcriptional start site. These elements seem to be transcriptional enhancers, although less direct evidence indicates an inhibitory element (negative control element) between 400 and 800 base pairs upstream. The negative control element may inhibit the P450 1A1 promoter, although the conditions for this inhibition are, as yet, undefined.

Source: From Okey, A.B., *Pharmacol. Ther.* 45 (1990), 241–298.

include inhibited or accelerated differentiation of epidermal cell lines, inhibition of adipose differentiation, enhancement of growth factor-stimulated DNA synthesis, and suppression or stimulation of apoptosis in hepatocytes and thymocytes, respectively.

At least three genes are necessary for Ah receptor synthesis and function. They appear to be located on mouse chromosome 12, the genes for CYP isozymes 1A1 and 1A2 being on mouse chromosome 9.

Phenobarbital-type induction. Phenobarbital and a variety of other xenobiotics including other barbiturates, chlordane, DDT, and certain polychlorinated biphenyl compounds have been demonstrated to induce several enzyme systems including CYPs, glutathione transferases, aldehyde dehydrogenases, and UDP glucuronyl transferases. Although this induction has been known to involve gene stimulation at the transcriptional level, there has been considerable difficulty in determining the underlying molecular mechanisms, either by which the cell recognizes the inducer or how the inducer affects transcription. Phenobarbital-like inducers are of low potency, requiring concentrations several orders of magnitude higher than, for example, TCDD. In addition, no significant strain differences appear to exist in either rats or mice relative to susceptibility to induction by phenobarbital-like inducers. Induction is apparent predominantly in the liver, inducing CYP isoforms 2B1 and 2B2. As indicated earlier, this induction is concurrent with proliferation of the smooth endoplasmic reticulum and an increase in total liver weight.

Recent studies have begun to identify important structural genetic elements involved in phenobarbital induction. One such element, discovered in *Bacillus megaterium*, involves the removal of a repressor protein Bm3R1 from a 17 bp promoter regulatory sequence referred to as the "BARBIE box." Even though this mechanism for barbiturate responsiveness appears to be important in bacteria, and BARBIE box-like sequences have been characterized in several mammalian PB-inducible CYP genes, there is abundant evidence to suggest that this particular sequence is not involved in mammalian PB responsiveness. Instead, studies now suggest that a phenobarbital responsive unit (PBRE) upstream from the BARBIE box is important in modulating PB activation. In rats, this PBRE consists of a 163 bp fragment that is found –2318 to –2155 bp upstream of the transcription start site of the CYP2B2 gene. Multiple cis acting elements within this fragment cooperate to form a PBRE. These include a nuclear factor 1 site and other recognition sites for DNA binding factors present in rat liver nuclear extracts. These studies indicate that multiple regulatory proteins and their respective recognition sequences are critical for obtaining maximal PB responsiveness.

Pregnenolone 16-α-carbonitrile (PCN)-type induction. PCN induces CYP isoform 3A1. As with 3-methylcholanthrene- and phenobarbital-type induction, increased gene transcription plays an important role. However, in PCN-type induction, post-transcriptional events are also important. For example, dexamethasone appears to bring about an increase in mRNA primarily by message stabilization rather than increased transcription whereas erythromycin seems to stabilize the CYP protein. PCN-type inducers are a complex class, including synthetic and endogenous glucocorticoids, glucocorticoid and mineralocorticoid antagonists, phenobarbital-type inducers, macrolide antibiotics, and imidazoles, as shown in Table 8.2. It should be noted that, although some phenobarbital-type inducers can also induce CYP 3A1, the structure activity relationships for this process and for the induction of 2B1 and 2B2 are different, implying that different processes may be involved.

Ethanol-type induction. Although ethanol is produced by bacterial fermentation in the gut, the majority of ethanol is exogenous in origin, being readily absorbed from the gastrointestinal tract. Some ethanol can be eliminated by the kidneys or lungs, however, most is oxidized by hepatic tissues. For many years, metabolism by alcohol dehydrogenase, a cytosolic enzyme, was thought to be the primary pathway for ethanol metabolism, although catalase also is capable of oxidizing ethanol in vitro in the presence of an H_2O_2 generating system. Certain features of ethanol metabolism, such as the adaptive increases in metabolism observed after chronic alcohol consumption, could not be explained on the basis of ADH or catalase, suggesting the existence of another pathway. The observation that ethanol consumption also increased the proliferation of the smooth endoplasmic reticulum, as was commonly observed for a variety of hepatotoxicants, barbiturates, and other therapeutic agents, also suggested the possibility that CYPs (which are located within the

SER) might be involved. Subsequent studies established that ethanol is metabolized by CYP2E1 and that this metabolic system is inducible by chronic ethanol consumption. Although CYP2E1 is in the same gene family as 2B1 and 2B2, which are induced by phenobarbital-type inducers, 2E1 is not known to be induced by any of the latter. The critical role of this ethanol-induced increase in this CYP isoform and the enhanced vulnerability of the alcoholic to the hepatotoxicity of many xenobiotics has long been recognized.

Many other inducers of CYP2E1 have been discovered, some of which may be more potent than ethanol itself (Table 8.2). It is interesting that CYP2E1 is also inducible in rats by fasting, diabetes, and obesity. These physiological conditions that induce CYP2E1 are likely the result of increasing concentrations of ketones, such as acetone, which are generated by these conditions. Acetone is metabolized by CYP2E1 to acetal, which can be further metabolized to methylglyoxal, which is used in the synthesis of glucose. Indeed, both fasting and diabetic ketoacidosis result in increased utilization of acetone in the gluconeogenic pathway.

The regulation of CYP2E1 expression is complex, involving transcriptional, post-transcriptional, and post-translational events. Recent studies demonstrated that rapid increases in CYP2E1 protein levels following birth are due to stabilization of preexisting proteins by ketone bodies released at birth. Subsequent changes in protein levels and observed differences among tissues were demonstrated to be due to varying degrees of the demethylation pattern of CpG residues in the 5′ end of the gene. Others had previously demonstrated that substrates such as ethanol, imidazole, and acetone had little effects on CYP2E1 transcript content and that these substrates tend to prevent degradation of the protein. Other studies suggest that transcriptional induction resulting in increased levels of mRNA occurs at higher ethanol levels, whereas at low concentrations, post-translational mechanisms predominate.

Induction by peroxisome proliferators. Peroxisome proliferators, including clofibrate and other hypolipodemic drugs, phthalate plasticizers, and certain herbicides, bring about induction of CYP isozyme CYP4A1, a form catalyzing the ω-hydroxylation of lauric acid. It is part of a pleiotropic response in the liver, other concomitant effects including increased liver weights, peroxisome proliferation, liver hypertrophy, and hyperplasia leading to nongenotoxic hepatocarcinogenesis. Induction of the CYP4A enzymes by peroxisome proliferators is mediated by a nuclear peroxisome proliferator-activated receptor α (PPAR-α). Fatty acids, which are important substrates of the CYP4A family, were recently identified as endogenous ligands of PPAR-α, implicating involvement of this receptor in lipid homeostasis. Starvation and diabetes have also been shown to result in 3- to 17-fold increases in the mRNA levels of CYP4A isoforms that were associated with corresponding changes in protein levels and arachidonic and lauric acid ω-hydroxylase activities.

Mixed-type induction. Arochlor 1254, a complex mixture of polychlorinated biphenyls (PCBs), has been used to induce a broad array of CYP isozymes, and it might be inferred that this is possibly due to the presence, in the mixture, of individual isomers that act as 3MC-type inducers and others that act as phenobarbital-type inducers. Although this is, in part, correct, careful studies with individual, highly purified, isomers have shown that some, rather than being of one type or the other, can induce in both categories. These include 2,3,3′,4,4,′,5,5′ heptachlorobiphenyl as well as several mono-ortho and di-ortho PCB, all of which induce CYP isozymes

1A1, 1A2, 2B1, and 2B2. Similarly, with the polybrominated biphenyls several mixed-type inducers are known including 2,3′,4,4′,5,5′ hexabromobiphenyl and 2,3′,4,4′5 pentabromobiphenyl. Other possible mixed-type inducers include 2-acetylaminofluorene, hexachlorobenzene, and phenothiazines, although it should be noted that because these studies are conducted in vivo, metabolites could account for one or the other type of induction.

8.2.5c Phylogenetic Variation in Induction

Induction of CYP isozymes, although studied most intensively in mammals, is known to occur with other groups, plants and animals, vertebrates and invertebrates. Induction in birds is generally of the 3MC-, AhR-dependent type, although phenobarbital-type induction is also well documented. Despite the demonstration that a functional AhR is present in chick liver embryo cells, some differences may exist between birds and mammals inasmuch as it has been shown that aryl hydrocarbon hydroxylase activity is induced, to about the same level, by either 3MC or phenobarbital.

In fish, induction by 3MC-type inducers has been well documented and a P450 similar to 1A1 identified as the isozyme involved. Early attempts to document the presence of the AhR using traditional methods such as [3H]TCDD and velocity sedimentation on sucrose gradients were not successful, although subsequent attempts provided the first direct evidence for its existence in rainbow trout. Several AhR homologs have recently been sequenced from fish, some sharing as much as 60–80% amino acid identity with mammalian AhR. Fish appear to be unresponsive to phenobarbital-type inducers.

High levels of P450 and glutathione S-transferase are associated with resistance of insects to insecticides. These high levels, however, are usually the result of selection for genes preexisting, at low frequency, in the susceptible population that confer high, heritable, levels of constitutive P450 isozymes. In addition to this situation, however, induction by both phenobarbital and 3MC have been demonstrated. Although there have been reports of [^{3}H]TCDD binding in two insect species, the significance of such is poorly understood. For example, [^{3}H]TCDD binding in *Helicoverpa zea* was only partially displaced by unlabeled TCDD and not at all by 3MC. In *Drosphila*, three out of four strains showed specific binding to [^{3}H]TCDD in the cytosol, yet AHH activity was not inducible by TCDD nor by 2,3,7,8 tetrachlorodibenzofuran (TCDF). Further, induction of AHH activity in *Drosophila* by phenobarbital did not correspond with the presence of the TCDD binding protein in these strains. The resolution of these and other anomalies, such as the finding that in *Drosophila* the most effective inducer of aryl hydrocarbon hydroxylase activity is phenobarbital, must await sequencing of isozymes and determination of their substrate specificities.

Induction in bacteria is also well known, particularly in the case of P450cam, a highly substrate specific form induced in *Pseudomonas putida* by its substrate, camphor. CYP102, on the other hand, is induced in *Bacillus megaterium* by compounds such as phenobarbital, which are not substrates for this isozyme, a form that functions as a fatty acid monooxygenase. Induction has also been demonstrated in fungi, most notably in yeasts such as *Saccharomyces cerevisiae*, and in *Neurospora crassa*.

8.2.5d Effects of Induction

The effects of inducers are usually the opposite of those of inhibitors, and thus their effects can be demonstrated by much the same methods; that is, by their effects on pharmacological or toxicological properties in vivo, or by the effects on enzymes in vitro following prior treatment of the animal with the inducer.

In vivo effects are frequently reported, the most common being the reduction of the hexobarbital sleeping time or zoxazolamine paralysis time. These effects have been reported for numerous inducers and can be quite dramatic. For example, in the rat the paralysis time resulting from a high dose of zoxazolamine can be reduced from 1 hr to 17 min by treatment of the animal with benzo(a)pyrene 24 hr before the administration of zoxazolamine.

The induction of monooxygenase activity may also protect an animal from the effects of carcinogens by increasing the rate of detoxication. This has been demonstrated in the rat with a number of carcinogens including benzo(a)pyrene, N-2-fluorenylacetamide, and aflatoxin. Effects on carcinogenesis may be expected to be complex because some carcinogens are both activated and detoxified by monooxygenase enzymes whereas, at the same time, epoxide hydrolase, which can also be involved in both activation and detoxication, may also be induced. For example, the toxicity of the carcinogen 2-naphthylamine, the hepatotoxic alkaloid monocrotaline, and the cytotoxic cyclophosphamide are all increased by phenobarbital induction—an effect mediated as a result of the increased production of reactive intermediates.

Organochlorine insecticides are also well-known inducers. Treatment of rats with either DDT or chlordane, for example, will decrease hexobarbital sleeping time and offer protection from the toxic effect of warfarin. It may be noted that induction by DDT in rats substantially reduced the incidence of both mammary tumors and leukemia following treatment with dimethylbenzanthracene. This effect may be explained by an increase in metabolism in the liver reducing the dose reaching the mammary gland and the hematopoietic cells in the bone marrow.

Effects on xenobiotic metabolism in vivo are also widely known in humans and other animals. Cigarette smoke, as well as several of its constituent polycyclic hydrocarbons, is a potent inducer of aryl hydrocarbon hydroxylase in the rat placenta, liver, and other organs. Examination of the term placentas of smoking human mothers revealed a marked stimulation of aryl hydrocarbon hydroxylase and related activities—remarkable in an organ that, in the uninduced state, is almost inactive toward foreign chemicals. Similarly, cigarette smoking lowers the plasma levels of phenacetin by induction of the enzymes responsible for its oxidation to N-acetyl-p-aminophenol. People exposed to DDT and lindane metabolized antipyrine twice as fast as a group not exposed, whereas those exposed to DDT alone had a reduced half-life for phenylbutazone and increased excretion of 6-hydroxycortisol.

The effects of inducers on the metabolic activity of hepatic microsomes subsequently isolated from treated animals have been reported numerous times. Whereas the polycyclic hydrocarbons primarily induce aryl hydrocarbon hydroxylase activity and a few related activities that are probably all catalyzed by CYP1A1 and CYP1A2, inducers such as phenobarbital, DDT, etc. have been shown to induce many oxidative reactions, including benzphetamine N-demethylation, p-nitroanisole

O-demethylation, ethylmorphine N-demethylation, aldrin expoxidation, and many others, a reflection of the broad substrate specificity of CYP2B1, the main isozyme induced by phenobarbital.

There is a continuing debate on whether induction of one or more CYP isozymes is likely to increase or decrease the toxicity of a chemical subsequently administered. The ultimate toxicity of any specific xenobiotic is related to relative increases in activation and detoxication reactions and to the relative induction of phase II as well as phase I enzymes. Because the induction of phase II enzymes is characteristically lower than that of CYPs, the rate of removal of active metabolites by conjugation may be reduced relative to the rate of their formation by oxidative reactions.

In terms of human health, the inducibility of CYP1A1 in humans relative to cancer risk and the predictability of individual risk has been of interest. Several workers, using an assay in which human lympocytes were exposed to 3MC-type inducers, concluded that the population distribution of inducibility of CYP1A1 was trimodal and, furthermore, that the distribution was consistent with a genetic model of two alleles at a single locus. They estimated that an individual in the "intermediate-inducibility" class was 16 times more likely to develop a smoking-related bronchogenic carcinoma than a member of the "low inducibility" class. Similarly, a member of the "high inducibility" class was 36 times more likely. As observed by Okey (see "Suggested Reading"), this may be the case only when the carcinogen is applied directly to the cancer site, as in smoking, but less likely when the tumor site is remote from the site of application.

Okey generalized that when high cytochrome P450 levels are tightly coupled to high levels of phase II enzymes and the precarcinogen is given orally or intraperitoneally, then "first pass effects" in the liver are likely to produce beneficial results. This would protect peripheral tissues such as mammary gland and bone marrow. Similar reasoning would predict that high cytochrome P450 in the placenta may protect the fetus. On the other hand, if high cytochrome P450 is not coupled to high phase II activity and the precarcinogen is applied directly to the site where it is activated, the effect is likely to be deleterious, as in smoking-related lung tumors and skin tumors.

8.2.5e Induction of Xenobiotic-Metabolizing Enzymes Other than Cytochrome P450 Isozymes

Xenobiotic-metabolizing enzymes other than cytochrome P450 may also be induced, including both phase I and phase II enzymes. The epoxide hydrolase of rat liver is induced by phenobarbital, TCDD-type inducers, pregnenolone-16 carbonitrile, and polychlorinated biphenyls. Transferases such as UDP-glucuronosyl transferase and glutathione S-transferase are also induced, the former by both phenobarbital and 3MC and the latter by phenobarbital, stilbene oxide, and 2-acetylaminofluorene. Induction of P450 isozymes, however, is generally more pronounced than that of phase II enzymes.

8.2.6 Biphasic Effects: Inhibition and Induction

Many inhibitors of mammalian monooxygenase activity can act also as inducers. Inhibition of microsomal monooxygenase activity is fairly rapid and involves a

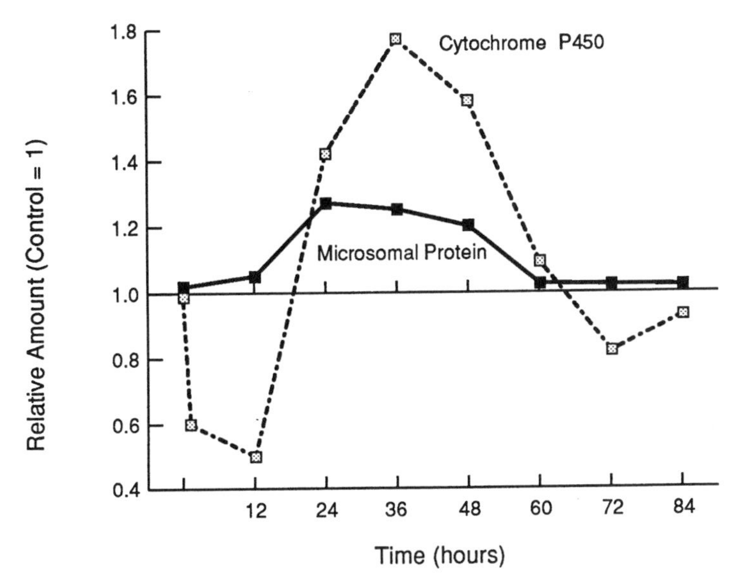

FIGURE 8.6 Effect of a single injection of piperonyl butoxide on cytochrome P450 and microsomal protein in mouse liver.

Source: From Philpot, R.M. and Hodgson, E. *Chem. Biol. Interact.* 4 (1971), 185–196.

direct interaction with the cytochrome, whereas induction is a slower process. Therefore, following a single injection of a suitable compound, an initial decrease due to inhibition is followed by an inductive phase. As the compound and its metabolites are eliminated, the levels would be expected to return to control values. Some of the best examples of such compounds are the methylenedioxyphenyl synergists including piperonyl butoxide and propyl isome (Fig. 8.6). Because P450 is combined with the methylenedioxyphenyl compound in a type III metabolite-inhibitory complex and cannot interact with CO, the CYP titer, as determined by the method of Omura and Sato, appears to follow the same curve. The biphasic effect on the magnitude of the CO spectrum was confirmed in the case of both synergists, the initial reduction of the spectrum reaching a minimum after 2–12 hr and the subsequent induction reaching a maximum after 36–48 hr. It is of interest that the two methylenedioxyphenyl compounds that have essentially identical interactions with both oxidized and reduced microsomes both in vivo and in vitro have different inductive effects, apparently inducing different isoforms. The difference also extends to more general effects: piperonyl butoxide causes a significant increase in liver weight and microsomal protein, whereas propyl isome has no such effect. In both cases, however, the biphasic effect is apparent.

At least three isozymes are known to be induced by piperonyl butoxide. CYPs 2B1 and 1A2 can be induced at moderate doses, whereas 1A1 is induced only at high dose levels. Studies conducted in mice suggested that piperonyl butoxide was marginally effective as an Ah-dependent inducer. Direct evidence that 1A2 was induced by an Ah-independent mechanism was not obtained until it was demonstrated in AhR-knockout mice that there was no induction of 1A1 whereas both 1A2 and 2B10 were both induced in treated animals.

It is apparent from extensive reviews of the induction of microsomal monooxygenase activity by xenobiotics that many compounds other than methylenedioxyphenyl compounds have the same effect. It might be expected that any synergist that functions by inhibiting microsomal cytochrome P450 activity could also induce this activity on longer exposure. This would result in a biphasic curve as described above for methylenedioxyphenyl compounds. This curve has been demonstrated for NIA 16824 (2-methylpropy1-2-propynyl phenylphosphonate) and WL 19255 (5,6-dichloro-1,2,3-benzothiadiazole), although the results were less marked with R05-8019 [2-(2,4,5-trichloro-phenyl)-propynyl ether] and MGK 264] [N-(2-ethylhexyl)-5-norbornene-2,3-dicarboximide].

8.3 RESISTANCE VERSUS TOLERANCE

The terms "resistance" and "tolerance" are closely related and are often used to mean different things (Table 8.3). Resistance denotes the situation in which the genetic constitution of the population is changed in response to the stressor chemical so that a greater number of individuals are able to survive the chemical than were able to survive it in the unexposed population. Thus, resistance involves selection of a population by a chemical followed by inheritance by the generations that follow. In higher organisms it usually involves selection for genes already present in the population at low frequency; in microorganisms this frequently involves mutations and induction of enzymes by the toxicant. Tolerance (sometimes referred to

TABLE 8.3 Differences between Resistance and Tolerance.

RESISTANCE
- related to the genetic constitution of a population
- involves selection of genes
- level of adaptation usually high (100- to 1000-fold)
- multigeneration
- usually high concentration of chemical needed
- must kill susceptible individuals to select
- months to years (several generations) to reverse
- common in insects, microbes and plants
- relatively rare in vertebrates

TOLERANCE
- related to the genetic constitution of an individual
- does not involve selection of genes
- level of adaptation usually low (2- to 10-fold)
- single generation
- usually low concentration and does not kill
- hours to days to reverse
- common in mammals, insects, and plants

as "intrinsic resistance") is reserved for those situations in which individual organisms, during their lifetime, acquire the ability to resist the effect of a toxicant, usually, but not always, as a result of prior exposure. Tolerance may also be used for populations that have the genes for resistance at a high frequency before exposure. However, this is also known as natural resistance, in contrast to acquired resistance, which is derived by selection as described above.

"Cross-resistance" is a term used to describe resistance to chemicals other than the selecting agent. For example, resistance to organophosphates may be conferred either by increased metabolism or by altered acetylcholinesterase activity. Because carbamates also inhibit acetylcholinesterase, cross-resistance to carbamates may occur by either mechanism. However, if cross-resistance was present to pyrethroids, one might deduce that metabolism was the mechanism of resistance, inasmuch as pyrethroids do not interact with acetylcholinesterase. Multiresistance is the case where a resistant population may be resistant to multiple chemicals with differing modes of action, likely as the result of several different resistance mechanisms.

With microorganisms, resistance is frequently initiated by an alteration in a gene caused by the stressor chemical In addition, selection of naturally occurring favorable genetic mutations may take place. In higher organisms resistance occurs by selection of favorable genes already present in the population at low frequencies. If the gene for resistance to the xenobiotic was present in the original population at low levels and the xenobiotic was required for selection of the adaptive manifestations, resistance is then acquired.

Changes at the molecular level that may be responsible for resistance include gene amplification, gene transfer, gene deletion, point mutation, and hypomethylation. Loss of cis-acting or trans-acting elements may also result in resistance.

Resistance is evident only in generations that follow selection pressure and not throughout the life of single individuals. Thus, it is necessary to the development of resistance that a substantial part of the population (i.e., the susceptible individuals) be destroyed for selection to occur. This is not the case with tolerance. Resistance is a multigeneration phenomenon, whereas tolerance is considered a single-generation phenomenon. Induction of enzymes is a common mechanism for tolerance.

For a population of resistant organisms to revert to a susceptible population requires the absence of the stressor chemical for several to many generations, this process may take several weeks for rapidly reproducing organisms to years with other organisms. The process depends on the fitness of the resistant genotype compared with the susceptible genotype and involves the return of the gene pool to a more biologically fit, but susceptible condition.

8.3.1 Mechanisms Responsible for Adaptation

There are many reports of resistance or tolerance where the mechanism(s) is unknown and has not been explored. The number of examples of resistance in vertebrates is remarkably low. Populations of mosquito fish and other fish species have developed resistance to chlorinated hydrocarbon pesticides. Also, populations of two frog species may have developed resistance to DDT. Resistance to coumarin anticoagulants in rodents is the most widespread pesticide resistance in vertebrates. Tolerance in humans to coumarin anticoagulant drugs has also been reported.

Rodents may develop true tolerance (as distinguished from bait avoidance) to a number of rodenticides such as arsenic oxide, zinc phosphide, and sodium fluoroacetate. Pine voles have developed 12-fold resistance to endrin. However, when one thinks of resistance, one immediately thinks of insects and a variety of other pests. In 1984, populations of 447 species of insects, 100 species of plant pathogens, 55 species of plants, 2 species of nematodes, and 5 species of rodents were reported to be resistant to insecticides, fungicides, herbicides, nematocides, and rodenticides, respectively.

Specific mechanisms that appear to provide an adequate scientific explanation for resistance or tolerance can be grouped into biochemical, physiological, and behavioral categories. Examples of resistance discussed in this chapter will be taken primarily from microorganisms and insects in which selection by drugs and pesticides coupled with rapid development, many life cycles, and large populations has been more evident than in other organisms.

8.3.2 Biochemical Mechanisms

Changes in the normal biochemistry of an organism with regard to increased titers of enzymes, altered enzymes, or binding proteins may affect the metabolic rate of a xenobiotic and thus alter the response to the chemical stress.

8.3.2a Metabolism

Detoxication. An organism or cell that acquires the ability to detoxify a xenobiotic more rapidly is able to tolerate amounts of the toxicant that would otherwise be toxic. Resistant organisms have utilized a variety of enzymatic detoxication reactions to overcome the effect of xenobiotics. Some examples of well-documented reactions to overcome the effect of xenobiotics are presented in Table 8.4. They represent both phase I and phase II reactions occurring in a variety of organisms (chapters 5 and 6).

Increased oxidative metabolism by the family of enzymes known as monooxygenases, or cytochrome P450s (CYPs), is a major mechanism of insecticide resistance for most insecticide classes. During the past decade, a number of studies have identified several specific CYP isoforms involved in insecticide resistance. These include CYP6A1 and 6D1 from houseflies, 6A2 from DDT-resistant *Drosophila melanogaster*, 6B7 and 4G8 from *Helicoverpa armigera*, and 9A1 from *Heliothis virescens*. In all cases examined thus far, increased gene transcription appears to be the predominant resistance mechanism, often involving loci on different chromosomes than those the P450s themselves.

Many cases of insecticide resistance are associated with hydrolases that cleave carboxyester and phosphorotriester bonds. Carboxyester bonds are found in some organophosphates, such as malathion, and in most pyrethroids. Phosphorotriester bonds are invariably found in organophosphates. In most cases described, increases in carboxyesterase expression have been the result of gene amplification. The resulting overproduced esterases may act primarily as high-affinity "sponges" that preferentially bind to the insecticide, not allowing it to interact at the active site. In some cases, esterases produced by gene amplification may account for as much as 6% of the total body protein, resulting in decreased fitness costs to the organism in the absence of selection. There are also examples of single base substitutions in esterase

TABLE 8.4 Examples of Metabolic Reactions Responsible for Resistance.

Substrate	Organism / conditions / enzyme	Product
OXIDATION		
$(C_2H_5O)_2P(S)O$–C_6H_4–NO_2 ethyl parathion	housefly O_2,NADPH ⟶ P450	$(C_2H_5O)_2P(S)H$ + HO–C_6H_4–NO_2 increased formation
$(C_2H_5O)_2P(S)H$ + HO–C_6H_4–NO_2 methyl parathion	tobacco budworm O_2, NADPH ⟶ P450	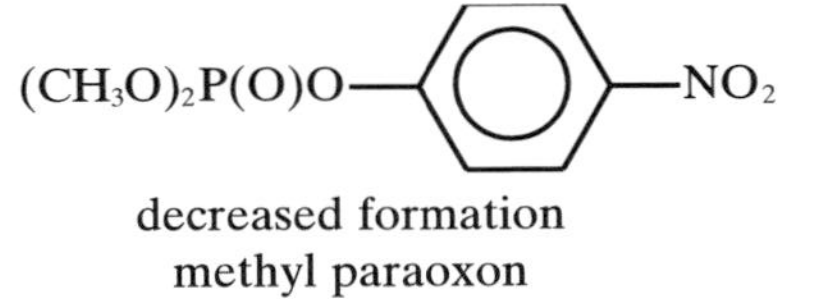 $(CH_3O)_2P(O)O$–C_6H_4–NO_2 decreased formation methyl paraoxon
HYDROLYSIS		
$(CH_3O)_2P(O)O$–C_6H_4–NO_2 + H_2O methyl paraoxon	tobacco budworm ⟶ phosphotriester hydrolase	$(CH_3O)_2P(O)OH$ + HO–C_6H_4–NO_2
R–CH–CH–S–C–$(CH_3)_2$; O=C–N–CHCOOH penicillins	*Staphlococci* ⟶ penicillinase	R–CH–CH–S–C–$(CH_3)_2$; HOOC–N–CHCOOH penicilloic acid
$(CH_3O)_2P(S)S\ CHCOOC_2H_5$ | $CH_2COOC_2H_5$ malathion	housefly mosquito ⟶ carboxylesterase	$(CH_3O)_2P(S)SCHOOH$ | $CH_2COOC_2H_5$ malathion monoacid

TABLE 8.4 ***Continued***

Substrate	Enzyme / organism	Product
REDUCTION		
$NO_2-C_6H_4-CH(OH)-R$ chloramphenicol	*E. coil* ⟶ reductase	$H_2N-C_6H_4-CH(OH)-R$ amino derivative
CONJUGATION		
$Cl-C_6H_4-C(H)(CCl_3)-C_6H_4-Cl$ DDT	DDT dehydrochlorinase ⟶ i.e., insect glutathione S-transferase	$Cl-C_6H_4-C(=CCl_2)-C_6H_4-Cl$ DDE
Triazine ring with Cl, C_2H_5HN and $NHC_3H_7{-}i$ substituents $+$ GSH atrazine	weeds plants ⟶ glutathione S-transferase	Triazine ring with SG, C_2H_5HN and $NHC_3H_7{-}i$ substituents
$(CH_3O)_2P(S)O-C_6H_4-NO_2 + GSH$ methyl parathion	insects ⟶ glutathione S-transferase	$CH_3O(OH)P(S)O-C_6H_4-NO_2 + CH_3SG$

genes that have been linked with increased carboxyesterase metabolism of insecticides in resistant insect populations. In some cases, low carboxyesterase activity has been associated with increases in organophosphate hydrolysis resulting in resistance. In one of these examples, a single amino acid substitution in the resistant population eliminated activity toward esterase substrates while conferring increased insecticide hydrolysis. The same amino acid substitution in an orthologous esterase from a different species also resulted in organophosphate resistance.

Glutathione S-transferases are a group of enzymes possessing a broad substrate specificity which have also been associated with insecticide and herbicide resistance. In insects, glutathione S-transferases appear to be a major mechanism of DDT and organophosphate resistance. In the few cases examined at the molecular level, insecticide resistance can occur as a result of increased transcription and gene amplification, as well as sequence divergence.

Besides enzymatic reactions, nonenzymatic reactions may also account for the detoxication of certain xenobiotics and thus mechanistically be involved in resistance. A strain of granary weevil is reported to be resistant to the fumigant methyl bromide. The amount of reduced glutathione in the resistant strain is twice as high as that in the susceptible strain. Because no difference in glutathione S-transferase activity was found between the two strains, it is possible that the reduced glutathione is serving as a binding sink, reacting nonenzymatically with methyl bromide and protecting the strain from the toxic effects of the fumigant.

Activation. Many insecticides, particularly the organophosphates, require metabolic activation for toxicity. For example, the substitution of the P = S group with P = O on parathion, resulting in the formation of the much more toxic paraoxon, results in greater than a 10,000-fold increase in the inhibition of acetylcholinesterase. Indeed, most organophosphates rely upon this activation reaction in vivo for their toxicity to insects.

In some cases, insects have developed the ability to actually decrease the rate of metabolic activation, resulting in the development of resistance. In one example, a 50-fold resistance level to methyl parathion was associated with a fourfold decrease in the ability to form methyl paraoxon. Because in this case the amount of cytochrome P450 remained the same in both populations, differences in the regulation of a specific parathion metabolizing isoform must be responsible for the differing levels of paraoxon product.

Another example of reduced activation as a mechanism of resistance has been reported for 5-fluorouracil. This compound is an anticancer agent that mimics thymine and prevents the replication of DNA. The 5-fluorouracil must first be activated to fluorodeoxyuridylic acid in order to be incorporated into nucleic acids. In resistant mouse tumor cells, the kinase activity responsible for the phosphorylation of fluorouridine to fluorouridylic acid is markedly decreased in comparison with nonresistant cells. Thus, the formation of the activated product is reduced and only a limited amount is incorporated into DNA.

8.3.2b Change in Target Site

An organism or cell may resist the action of a toxicant because the target site has been altered. The target site may be an enzyme whose inhibition is subsequently responsible for the toxic action or it may be a binding site or receptor protein to which the toxicant becomes attached. Changes in amino acid sequence, alteration of distances between critical amino acid residues, and altered enzyme topography may be responsible for resistance.

Resistance to organophosphate and carbamate insecticides as a result of an altered target enzyme, acetylcholinesterase, has been reported in numerous strains of mites, ticks, and insects. Acetylcholinesterase is an enzyme located on the postsynaptic membrane that normally rapidly hydrolyzes acetylcholine following its release from the presynaptic vesicle. The inhibition of the hydrolysis by organophosphates and carbamates causes accumulation of acetylcholine at the postsynaptic membrane, resulting in continuous excitation of the membrane. Single point mutations in the structure of the acetylcholinesterase gene have been shown to significantly change the affinity of the enzyme to organophosphate and other ligands. In some cases, high levels of resistance due to this mechanism may be the result of the combination of several mutations in the same gene. In at least one insect, the lesser grain borer, *Rhyzopertha dominica*, altered acetylcholinesterase resulting in resistance to parathion and malathion produced negative cross-insensitivity to carbaryl, chlorpyrifos-methyl oxon, and carbofuran.

For many years, reduced neuronal sensitivity has been proposed as a mechanism for cyclodiene resistance in several insect species. Several studies during the late 1980s provided evidence that cyclodienes exert their effects by blocking γ-aminobutyric acid (GABA)-dependent chloride flux at the $GABA_A$ receptor-chloride ionophore complex. Subsequent studies in *Drosophila melanogaster* demonstrated that resistance is always associated with replacements of the same amino acid (alanine 302) in the GABA-gated chloride channel gene Rdl (Resistance to dieldrin). Interestingly, in at least four other insect species possessing cyclodiene resistance, the same amino acid substitution in the orthologous gene has been shown to be responsible for resistance.

Pyrethroid insecticides disrupt normal sodium channel kinetics, leading to repetitive neural discharge, convulsive activity, and eventually paralysis and death. Insects with pyrethroid resistance have the ability to resist the rapid knockdown potential of this class of pesticides and thus are called knockdown resistant (kdr). Experiments to characterize resistance mechanisms related to the sodium channel suggest that there are no differences in the number of sodium channels between susceptible and resistant populations, but rather that resistance is due to reduced binding affinity of the pyrethroids to the channel. Several DNA sequences of sodium channel genes have been sequenced from several insects including para and DSC1 from *Drosophila melanogaster*, Vssc1 from the housefly *Musca domestica*, and the para-type sodium channel genes from the German cockroach *Blattella germanica*, and the tobacco budworm *Heliothis virescens*. In the majority of these orthologous genes, the presence of a single leucine to phenylalanine mutation (L1014F) has been responsible for kdr resistance, although in the tobacco budworm the mutation was from leucine to histidine (L1029H). An additional methionine-to-threonine (M918T) mutation associated with super-kdr resistance has also been characterized in some species. Although other mutations have been observed, their function in resistance has yet to be completely characterized.

The site of toxic action of the sulfonylurea herbicides is inhibition of the enzyme acetolactate synthase, the initial enzymatic step in the synthesis of branched-chain amino acids in plants and bacteria. A mutation in the acetolactate synthase gene in higher plants resistant to sulfonylureas has been identified, and it seems that the altered enzyme in resistant plants is less sensitive to inhibition by the sulfonylurea herbicides than the enzyme in the susceptible plants.

4-Hydroxycoumarin anticoagulants and vitamin K are antagonistic in their effects on the synthesis of blood-clotting factors. Studies on vitamin K metabolism in warfarin-resistant wild Welsh rats indicate an altered microsomal vitamin K epoxide reductase with reduced sensitivity to warfarin. It appears that this mechanism is associated only with rodents whose resistance is linked with an increased susceptibility to vitamin K deficiency.

8.3.2c Alteration of Receptor Proteins

Only a limited number of alterations of protein receptors (target sites) are known. Resistance of certain weeds to the herbicide atrazine involves two known mechanisms. The first is an increase in detoxication of atrazine by glutathione S-transferase in resistant plants (Table 8.4). The second involves interference with the electron carrier of the photosynthetic electron transport chain in the photosystem II complex. A 32 kDa protein has been implicated as being responsible for resistance to triazine herbicides. The protein is highly conserved from species to species and has some interesting properties. It lacks lysine, turns over rapidly in the presence of light (half-time of 6–9 hr), and is synthesized as a larger protein (33.5 kDa), which is degraded by removal of the carboxy terminal portion to form the 32 kDa size. Nucleotide sequences showed a difference in resistant and susceptible DNAs with the codon that is serine at amino acid 264 in the susceptible DNAs being altered into glycine in the resistant higher plants. Whether changes in the 32 kDa protein are responsible for resistance to other PS II-inhibiting herbicides is not clear.

8.3.2d Increased Amount of Target Molecule

A toxicological lesion may be overcome by an increase in the amount of the target molecule, thus compensating for the portion of the target that is occupied by the toxicant. In this manner the organism can maintain physiological functionality.

Studies on occupationally exposed groups in relatively polluted traffic situations have shown that an increased amount of hemoglobin provides a partial explanation for tolerance to the adverse effects of CO binding to hemoglobin and blocking of O_2 transport. When dogs were exposed to 0.08–0.1% CO for 6–8 hr daily for 36 weeks and compared with controls, a 67% increase in blood hemoglobin was observed. This created a greater reserve of the hemoglobin for O_2 transport at a given concentration of hemoglobin-CO. In addition, the number of erythrocytes and the hematocrit also increased more than 50%. This mechanism is probably an important factor in tolerance to CO by humans, but other mechanisms may also be involved.

The target of the anticancer drug methotrexate is the enzyme dihydrofolate reductase, which catalyzes the reduction of dihydrofolate to the active tefthydrofolate. In a 67-fold methotrexate-resistant cell line, the enzyme activity increased 65-fold. The degree of resistance to methotrexate was accounted for by the comparable increase in the amount of the target enzyme per cell.

8.3.2e Bypass of Target

Cases of resistance have been noted in which organisms have evolved alternate mechanisms to bypass the affected site of action. The first reported case occurred with citrus scale insects that acquired resistance to hydrogen cyanide. Utilizing an HCN-insensitive flavoprotein as the electron carrier, the insect avoids using the terminal cytochrome oxidase, which is the site of cyanide inhibition.

Electron transport in some plants and certain helminths can bypass both cytochrome c and cytochrome oxidase and electron flow proceeds directly from cytochrome b to molecular oxygen.

8.3.2f Repair of Damaged Site of Action

Although repair of genetic material is well known, an appreciable increase in such repair activities has resulted in partial resistance to genetic attack in some cases. The repair of damaged DNA has been established as a mechanism of resistance to bifunctional alkylating agents by *E. coli.* Agents such as mustard gas cross-link at N-7 of guanine to form a diguanine joined by the alkylating moiety. Resistant strains are capable of excising some of the diguanyl residues and restoring the cross-linked DNA to functional DNA. Mutant strains possess greater repair activity than the sensitive strains.

8.3.3 Physiological Mechanisms

Physiological mechanisms are for the most part associated with biochemical changes that affect the normal function of a cell or organism. Therefore, any alteration in function, that is, decreased uptake, increased efflux, increased sequestration of the xenobiotic, or increased elimination of the chemical, will permit the cell or organism to tolerate quantities of the toxicant far in excess of the lethal dose.

8.3.3a Uptake and Efflux

A cell or organism may survive the lethal effect of a toxicant as a result of reduced penetration, that is, uptake of the toxicant. A decrease in the amount of toxicant that enters the cell or organism reduces the concentration of the toxicant and also provides a greater opportunity for detoxication of the toxicant over a given time period. Similarly, the efflux of the toxicant would have the same effect—lowering the concentration of the toxicant and again providing an opportunity for detoxication.

Decreased penetration or permeability has been postulated to explain antibiotic resistance differences between Gram positive and Gram negative bacteria as well as for the resistance to certain anticancer drugs. In some cases, differences in passive diffusion processes account for differences in drug susceptibility. For example, many hydrophobic antibiotics are more effective against Gram positive bacteria than against Gram negative bacteria due to structural differences in the lipid bilayers of the bacterial cell walls. Differences in penetration can also result from mutations in active transport processes. Many mutations affecting active transport processes are implicated in the resistance to anticancer drugs including nitrogen mustard (choline transport), melphalan (L-amino acid carrier), 5-fluorouracil (purine and pyrimidine uptake system), and methotrexate (folate transport carrier).

A decrease in the penetration of arsenite in certain strains of *Pseudomonas* has been reported to be responsible for resistance. Arsenite inhibits essential sulfhydryl groups in enzymes such as dehydrogenases. *Pseudomonas* cells were grown either in the presence or the absence of 10^{-2} m arsenite and dehydrogenase activities determined. A large difference in dehydrogenase activities was found in the intact cells, whereas no difference in activity was found with the cell-free extracts. Resistant and sensitive cells were grown in the presence of different concentrations of radioactive arsenite. At a given arsenite concentration, the uptake of radioactivity by the sensitive cells was far greater than in the resistant cells, thus implicating penetration as a resistance mechanism.

Decreased penetration has been shown to be a mechanism partially responsible for insecticide resistance in a number of insect populations. The advantage provided a population with this mechanism is that the insect has more time for detoxication and the induction of enzymes involved in the detoxication of the pesticide. The relative importance of this mechanism of resistance has been difficult to document because it is seldom noted in populations without other mechanisms of resistance.

Although decreased penetration or uptake can be responsible for resistance, so can mechanisms that utilize a "pump" to extrude the chemical from the cell and thus prevent the toxicant from reaching and maintaining a lethal concentration. The observation that resistance to anticancer drugs such as vinblastine often resulted in resistance to a broad range of structurally diverse drugs led to many studies of a phenomenon now known as multidrug resistance (MDR). The resistance mechanism is based on the efflux of the drug from the cell. The drugs most associated with MDR are alkaloids or antibiotics of plant and fungal origin such as the vinca alkaloids, anthracyclines, dactinomycin, and podophyllotoxins. Cross-resistance also occurs with alkylating agents such as nitrogen mustard, melphalan, and mitomycin C. A number of compounds reverse resistance and they include calcium channel blockers, calmodulin inhibitors, and lysosomotropic agents.

Increased expression of a 170kDa P-glycoprotein (MDR-associated glycoprotein) in the plasma membrane is the most consistent change found in MDR cells. p-Glycoprotein expression correlates with both a decreased intracellular drug-concentration and the observed level of drug resistance in many MDR cell lines. A range of molecular sizes (130–200kDa) based on mobility in gel electrophoresis has been reported for p-glycoprotein. Full-length p-glycoprotein cDNA sequences from mouse and human cells predict a molecular size of approximately 140kDa with post-translational modification to form the mature glycoprotein. The modifications include N-glycosylation as well as phosphorylation and may account for 30–40kDa.

The proposed mechanism for MDR is that the drug binds directly to the p-glycoprotein and then is actively effluxed from the cell through an opening (pore or channel) formed by multiple transmembrane domains utilizing energy derived from p-glycoprotein-mediated hydrolysis of ATP. Studies with p-glycoprotein-specific monoclonal antibodies demonstrate that the protein is highly conserved in size and cross-reacts immunologically with MDR cells across species. This membrane transport protein appears to be conserved by evolution from bacteria to humans. In fact, the p-glycoprotein in higher eukaryotes is a family of highly homologous proteins.

Two examples of p-glycoprotein-mediated resistance that have been intensively studied in invertebrates involve the malarial parasites *Plasmodium falciparum* and *Leishmania donovani*. In these organisms, MDR-like genes Pfmdr1 and Ldmdr1 have been shown to cause resistance to antimalarial drugs chloroquinine and mefloquine among others. Variations in both gene copy number as well as expression have been implicated as a basis for drug resistance in these organisms.

In insects, recent evidence of the existence of p-glycoproteins have raised the question of their potential involvement in resistance mechanisms. Three p-glycoprotein homologs have been isolated and characterized in *Drosophila*, possessing 50% similarity to mammalian p-glycoprotein genes. Antibodies to p-glycoprotein have detected homologs in the blood brain barrier and malpighian tubules of *Manduca sexta*, in anal papillae of *Chironomus riparius* larvae, and in tissues of *Heliothis virescens*. In *C. riparius* larvae, ATPase activity of homogenate fractions increased in the presence of p-glycoprotein substrates and was inhibited by known p-glycoprotein inhibitors. Sublethal concentrations of p-glycoprotein inhibitors were also demonstrated to synergize pesticide treatments in *C. riparius* and *H. virescens* larvae. Insecticide-resistant *H. virescens* larvae were also shown by Western blotting to have increased p-glycoprotein levels relative to susceptible populations.

8.3.3b Storage

When a toxicant is sequestered in a tissue (fat, bone, etc.), organ (liver, kidney, etc.), or organelle (plant vacuole), the amount of reactive toxicant available to act at the site of action is reduced. No clear evidence has been presented, however, to demonstrate that increased storage of a toxicant is a mechanism that is responsible for resistance. However, one would expect certain individuals in a population to have a greater capacity to sequester certain chemicals than other individuals in the same population and thus an increased tolerance to these chemicals.

Storage of metals in liver and kidney has been well documented, with metallothionein playing an important role in this process following de novo synthesis of the protein in the presence of the appropriate metal (cadmium, mercury, zinc, copper, and possibly platinum). Bone is another site of storage. In humans, approximately 90% of the total body burden of lead is found in bone. When stored in bone, lead is not readily available, and thus has no adverse effect on soft tissues. Besides lead, strontium, fluorine, and, to a lesser extent, barium are also stored in bone. Chlorinated hydrocarbons (DDT, DDE, PCBS, dieldrin hexachlorophene) tend to be stored in fatty tissues for long periods of time. In lactating organisms, substantial quantities of these lipid-soluble compounds, if stored in the fat, can be mobilized and excreted in milk.

8.3.3c Elimination

Rapid elimination (i.e., excretion) of a chemical may allow an animal to tolerate quantities of a toxicant far in excess of the lethal dose. There are no reports of excretion of a toxicant being directly responsible for resistance, however, increased elimination is one mechanism associated with adaptation of the tobacco hornworm to nicotine. The leaves of tobacco contain 2–5% nicotine, which is nontoxic to the tobacco hornworm but extremely toxic to susceptible insects. The hornworm survives ingestion of tobacco by rapid excretion of nicotine in the feces and decreased absorption of the toxicant.

8.3.3d Behavioral Mechanisms

Acquired resistance as the result of a behavioral mechanism was first reported for a population of mosquitoes that was previously suppressed by the insecticides used in the malaria control program of the World Health Organization. In this vector control program insecticides were applied to internal surfaces of houses. Normal populations of mosquitoes resting on wall surfaces following blood meals died from pesticide exposure. The resistant strain was selected because some individuals showed a modified behavior pattern, leaving the treated houses following blood meals and thus avoiding the treated surfaces. In time, they became the dominant population even though they had not developed a biochemical or physiological mechanism of resistance to the insecticide.

The horn fly (*Haematobia irritans*) and the tobacco budworm (*H. virescens*) provide two more recently described examples of behavioral resisance. The horn fly acquired its name as a result of its propensity to gather around the horns of cattle. Since the 1980s, pyrethroid impregnated cattle ear tags have been used to control this pest. Studies of pyrethroid resistance in this pest indicated that the irritancy/repellency of pyrethroids was greater in resistant than in susceptible flies. In addition, there was a dose-dependent shift in the distribution of the resistant horn flies to the belly of the animals which was independent from dosing. An opposite effect was found in the tobacco budworm *H. virescens*. Rather than displaying increased irritancy, the resistant *H. virescens* larva move about less in the presence of the pyrethroid, leading to less pesticide exposure.

SUGGESTED READING

Conney, A.H. Pharmacological implications of microsomal enzyme induction. *Pharmacol. Rev.* 19 (1967), 317.

Conney, A.H., Pantuck, E.J., Hsiao, K.-C., Kuntzman, R., Alvares, A.P. and Kappas, A. Regulation of drug metabolism in man by environmental chemicals and diet. *Fed. Proc.* 36 (1977), 1647.

Engel, P.C., *Enzyme Kinetics*, Wiley, New York, 1977.

Evered, D. and Collins, G.M., *Origins and Development of Adaptation*, Ciba Foundation Symposium 102, Pitman, London, 1984.

Ffrench-Constant, R.H., Pittendrigh, B., Vaughan, A. and Anthony, N. Why are there are so few resistance-associated mutations in insecticide target genes? *Phil. Trans. R. Soc. Lond. B* 353 (1998), 1685.

Georghiou, G.P. and Saito, T. (Eds), *Pest Resistance to Pesticides*, Plenum Press, New York, 1983.

Guengerich, F.P. Molecular advances for the cytochrome P-450 superfamily. *TIPS* 12 (1991) 281.

Guengerich, F.P. Oxidation of toxic and carcinogenic chemicals by human cytochrome P450 enzymes. *Chem. Res. Toxicol.* 4 (1991), 391–407.

Hayes, J.D., Pickett, C.B. and Mantle, T.J. (Eds), *Glutathione S-transferase and Drug Resistance*, Taylor and Francis, London, 1990.

Hodgson, E. and Philpot, R.M. Interaction of methylenedioxyphenyt (1,3-benzodioxole) compounds with enzymes and their effects on mammals. *Drug Metab. Rev.* 3 (1974), 231.

Hodgson, E. Induction and inhibition of pesticide-metabolizing enzymes: roles in synergism of pesticides and pesticide action. *Toxicol. Ind. Health* 15 (1999), 6.

Honkakoski, P. and Negishi, M. Regulatory DNA elements of phenobarbital-responsive cytochrome P450 CYP2B genes. *J. Biochem. Molec. Toxicol.* 12 (1998), 3.

Mannering, G.J. Inhibition of drug metabolism. In: Brodie, B.B., Gillette, J.R. and Ackerman, H.S. (Eds), *Handbook of Experimental Pharmacology*, Vol. 28, Part 2, Concepts in Biochemical Pharmacology, Springer, Berlin, 1971, Chapter 49.

Murray, M. and Reidy, G.F. Selectivity in the inhibition of mammalian cytochromes P-450 by chemical agents. *Pharmacol. Rev.* 42 (1990), 85.

Okey, A.B. Enzyme induction in the cytochrome P-450 system. *Pharmacol Ther.* 45 (1990), 241.

Paine, A.J. Current status review: the cytochrome P450 gene superfamily. *Int. J. Exp. Pathol.* 72 (1991), 349.

Pesticide Resistance, Strategies and Tactics for Management, National Academy Press, Washington, D.C., July 1986.

Rowlands, J.C. and Gustafsson, J.-A. Aryl hydrocarbon receptor-mediated signal transduction. *Crit. Rev. Toxicol.* 27 (1997), 109.

Sanvordeker, D.R. and Lambert, H.J. Environmental modification of mammalian drug metabolism and biological response. *Drug Metab. Rev.* 3 (1974), 201.

Wilkinson, C.F. Insecticide interactions. In: Wilkinson, C.F. (Ed), *Insecticide Biochemistry and Physiology*, Plenum Press, New York, 1976, Chapter 15.

CHAPTER NINE

Reactive Metabolites and Toxicity

PATRICIA E. LEVI and ERNEST HODGSON

9.1 INTRODUCTION

The toxic effects of exogenous compounds (xenobiotics) often result not from the parent compound per se but from reactive metabolites formed inside the cell. This biotransformation of relatively inert chemicals into highly reactive metabolites is commonly referred to as *metabolic activation*, and is a well-recognized, essential event in numerous chemically induced toxicities. Since the publication of the previous edition of this work, the role of metabolic activation in the cascade of events leading to several toxic end points, including various organ-specific toxic end points, has been widely investigated and is summarized at appropriate points in several chapters. This chapter is intended to serve as a framework for discussion of metabolic activation with cross references to other chapters for further discussion and further examples.

In the 1940s and 1950s, the pioneering studies of James and Elizabeth Miller using the aminoazo dye N,N-dimethyl-4-aminoazobenzene (DAB), a hepatocarcinogen in rats, provided early evidence for conversion, in vivo, of chemical carcinogens to protein and nucleic acid bound derivatives. The later discoveries by Boyland that 8-naphthylamine had first to be metabolized to cause bladder cancer and the Millers that 2-acetylaminofluorene (2-AAF) was carcinogenic only after metabolism to N-hydroxy-AAF firmly established the importance of metabolic activation in chemically induced carcinogenesis.

These discoveries laid the groundwork for newer approaches to studying the toxicity and carcinogenicity of chemicals which sought to elucidate underlying mechanisms rather than relying on empirical studies in experimental animals. If reactive metabolites are responsible for the toxic effects of the parent compound, then toxicity is a consequence, not only of the chemical nature of the parent compound, but of the enzymes that metabolize the parent compound in the exposed animal. This enzyme composition in turn is determined by the genetic makeup of the animal as well as environmental and physiological factors that modulate enzyme activity, such as nutrition, disease, hormones, gender, and exposure to other chemicals.

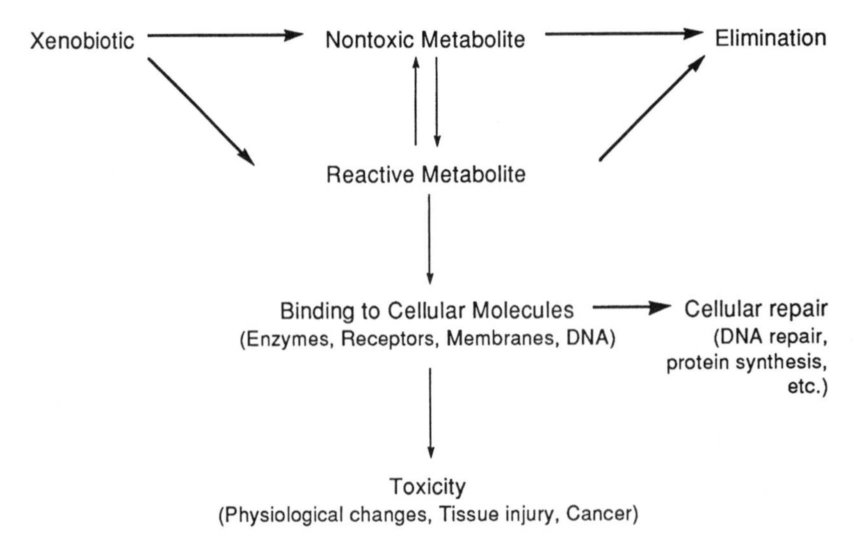

FIGURE 9.1 The relationship between metabolism, activation, detoxication, and toxicity of a chemical. The balance between activation and detoxication determines the toxicity of the chemical.

The overall metabolic scheme for the production of potentially toxic xenobiotics is outlined in Figure 9.1. Phase I metabolism of a chemical may produce inactive metabolites that are more polar and are readily conjugated and excreted (detoxication), but may also produce highly reactive metabolites that can interact with vital intracellular macromolecules resulting in toxicity. In addition, the active metabolites can also be detoxified, for example, by reaction with glutathione. In fact, glutathione, which is present rather abundantly in most cells (approximately 5 mM), readily absorbs reactive metabolites by binding either nonenzymatically, in the case of very reactive intermediates, or enzymatically by the action of the glutathione S-transferases. Generally, reactive metabolites are *electrophiles* (molecules containing positive centers), which can bind covalently to tissue *nucleophiles* (molecules containing negative centers), such as glutathione, proteins, and nucleic acids. Alternatively, reactive intermediates may be *free radicals* or act as radical generators and through interaction with oxygen produce *reactive oxygen* species, which are also capable of causing damage to membranes, DNA, and other macromolecules.

Although a chemical can be metabolized by several routes, the activation pathway is often a minor route with the remainder of the pathways resulting in detoxication. Activation, however, may become a more dominant pathway in certain situations (see Section 9.4). Thus, the amount of reactive metabolites and the degree of toxicity depends on the balance between activation and detoxication reactions.

Several important terms that are often used when discussing bioactivations include: *parent compound*, sometimes referred to as *procarcinogen* (in the case of a carcinogen); *proximate toxic metabolite* and/or *proximate carcinogen* for one or more of the intermediates; and *ultimate toxic metabolite* and/or *ultimate carcinogen* for the reactive species that binds to macromolecules and DNA.

TABLE 9.1 Enzymes Important in Catalyzing Metabolic Activations.

Type of Reaction	Enzyme
Oxidation	Cytochrome P450 Prostaglandin synthetase (PGS) Flavin-containing monooxygenase Alcohol and aldehyde dehydrogenases
Conjugation	Glutathione transferase Sulfotransferase Glucuronidases
Deconjugation	Cysteine S-conjugate β-lyase
Gut microflora	Hydrolases Reductases

9.2 ENZYMES INVOLVED IN BIOACTIVATION

9.2.1 Phase I Oxidations

9.2.1a Cytochrome P450

Although most, if not all, of the enzymes involved in xenobiotic metabolism can form reactive metabolites (Table 9.1), the enzyme systems most frequently involved in the activation of xenobiotics catalyze oxidation reactions. (See chapters 5 through 8 for discussion of xenobiotic metabolizing enzymes). The cytochrome P450 monooxygenases (P450) are by far the most important enzymes involved in oxidation of xenobiotics. This is a result of the abundance of P450, especially in the liver, the numerous xenobiotic-metabolizing isoforms of P450, each with a broad substrate specificity, and the ability of certain P450 isoforms to be induced by xenobiotics, either substrates or nonsubstrates of the isoform induced. In addition to metabolism by the liver, P450 isoforms are known to be involved in xenobiotic metabolism and activation reactions in extrahepatic organs such as the lung, kidney, and skin, sometimes resulting in toxicity. Several examples of extrahepatic activation will be discussed later in the chapter.

This enzyme system consists of a number of families, each with one or more distinct proteins that have different as well as overlapping substrate specificities. Sometimes a chemical may be preferentially metabolized by only one P450 isoform; frequently, however, a chemical serves as substrate for more than one P450, but with oxidation by different P450s occurring at different rates, at different sites on the chemical (*regioselectivity*), or with different steric configuration (*stereoselectivity*) of the products. These differences in metabolism may have profound effects on the toxicity of the chemical. One of the best characterized examples is the activation of the carcinogen benzo(a)pyrene to the ultimate carcinogen (+)-benzo(a)pyrene 7,8 diol-9,10-epoxide which binds to DNA. (This pathway is shown in Chapter 15, Figure 15.7). Formation of the benzo(a)pyrene reactive metabolites is characteristic of oxidation by P450 I Al, whereas other P450s hydroxylate the molecule in positions resulting in nontoxic metabolites.

Another class of carcinogenic compounds to which humans may be exposed from industrial, environmental, and dietary sources are the primary aromatic amines. Numerous studies have demonstrated that N-hydroxylation of these compounds is a necessary step in their activation to electrophiles that covalently bind to DNA, induce mutations, and initiate carcinogenesis. Ring hydroxylations of these compounds, on the other hand, are usually detoxication pathways. Several forms of P450, as well as the flavin-containing monooxygenase (discussed in the next section), have been shown to catalyze the N-hydroxylation of 2-aminofluorene (2-AF) to N-hydroxy-2-aminofluorene, the proximate carcinogen (Fig. 9.2). Subsequent conjugation reactions of the N-hydroxy intermediate to sulfates or glucuronides yield the

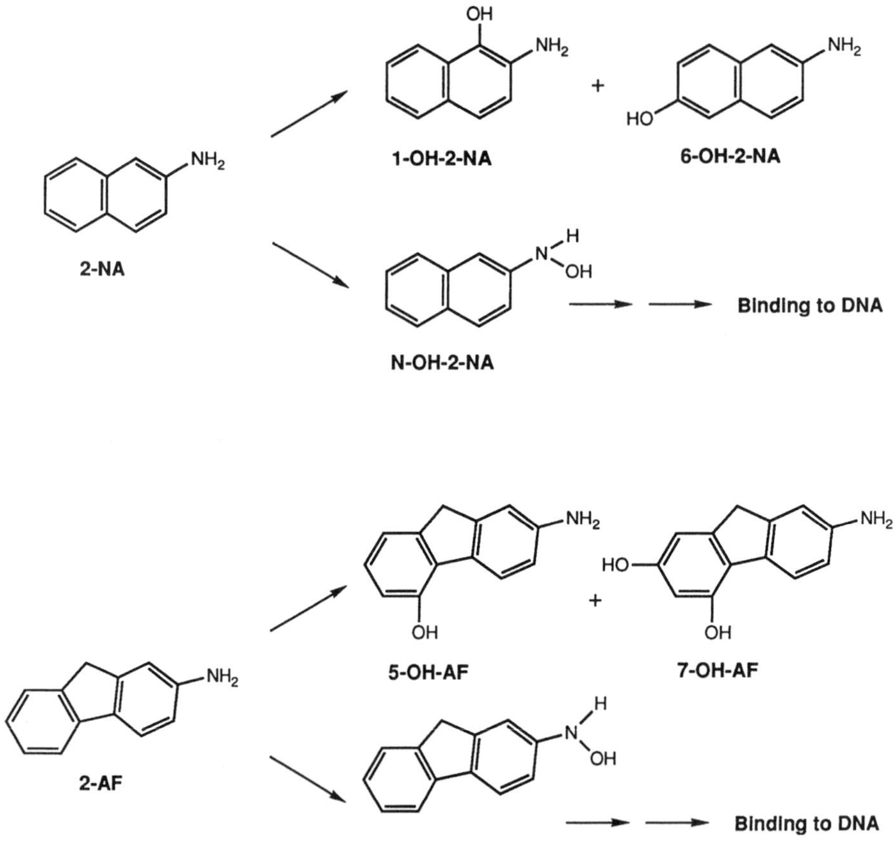

2-NA

1-OH-2-NA

6-OH-2-NA

N-OH-2-NA

Binding to DNA

2-AF

5-OH-AF

7-OH-AF

N-OH-AF

Binding to DNA

FIGURE 9.2 Structures of the primary oxidative metabolites of the carcinogenic aromatic amines 2-aminofluorene (2-AF) and 2-naphthylamine (2-NA). At least three pathways have been shown from in vitro studies to catalyze the oxidation of arylamines: specific inducible P450s, microsomal flavin-containing monooxygenase, and peroxidative enzymes such as prostaglandin synthetase.

TABLE 9.2 Examples of Chemicals Activated by P450 Enzymes.

Chemical	P450 (CYP)
Benzo(a)pyrene	1A1
3,4,3′,4′-Tetrachlorobiphenyl	1A1
Aflatoxim B_1	2C11, 1A1, 1A2
α-Naphthoflavone	1A1
Acetaminophen	2C11, 1A1, 2E1
2-Naphthylamine	1A2
2-Aminofluorene	1A2
2-Acetylaminofluorene	1A1, 1A2, 2B1
Cyclophosphamide	2B1
Fluoroxene	2B1
Bromobenzene	2B1
Benzene	2E1
Dimethylnitrosamine	2E1

ultimate carcinogen. In addition to N-hydroxylation and ring hydroxylations, both AF and N-hydroxy-AF can by acetylated in vivo in certain species, and these N-acetylated derivatives can undergo further metabolic conversions that ultimately lead to DNA binding.

Similarly, the N-hydroxylation of 2-naphthylamine (activation) is primarily catalyzed by P450 IA2, whereas ring hydroxylations (detoxication) are catalyzed by other P450 proteins (Fig. 9.2).

The isozymes in the P450 I family (IAI and IA2) are generally considered to be among the most important P450s in the activation of important groups of chemical carcinogens such as polycyclic aromatic hydrocarbons, aromatic and heterocyclic amines, azobenzene derivatives, planar polyhalogenated biphenyls, as well as several drugs (Table 9.2). Substrates for this family tend to be large, planar molecules; 1A1 shows a preference for arene oxidation and 1A2 readily catalyzes N-hydroxylation of planar aromatic and heterocyclic arnines. Although 1A2 is present constitutively in significant amounts in the livers of many laboratory animals, 1A1 is present only at extremely low levels. Both enzymes, however, can be readily induced by most of their substrates, making the metabolism (and often activation) by these P450s even more significant. All P450s, however, have the potential to be involved in activation reactions.

9.2.1b Flavin-Containing Monooxygenase

Like P450, the flavin-containing monooxygenase (FMO) catalyzes the oxygenation of N, S, and P atoms in a wide variety of xenobiotics (Chapter 5). Unlike P450, however, the FMO does not catalyze carbon oxidations and hydroxylations. This enzyme is also located in the endoplasmic reticulum, and a number of chemical compounds are common substrates for both P450 and FMO, although the oxidation products or stereochemistry of the products may be different. The FMO has been demonstrated in most tissues and is known to consist of several enzymatic forms with similar, overlapping substrate specificities.

Although the total amount of FMO is highest in the liver, the specific activity of the enzyme in some extrahepatic tissues (e.g., kidney and lung) is as high as that in the liver. In tissues where P450 levels are low or where P450s are primarily involved in metabolism of endogenous substrates, the contribution of FMO relative to P450 may be greater for the oxidation of certain xenobiotics. Because many xenobiotic substrates for the FMO are also substrates for one or more of the P450 isozymes, it is often desirable to delineate the activity of FMO and P450 in various tissues and conditions. This requires measurements in the presence and absence of selective inhibitors for the two enzyme systems. The techniques used include use of an inhibitory antibody to P450 reductase eliminating P450 activity, heat inactivation of the FMO leaving P450 activity intact, use of a P450 inhibitor such as aminobenzotriazole (ABT), or a competitive substrate for the FMO, such as methimazole, which essentially eliminates FMO activity.

Although most FMO oxidations are detoxication reactions, the enzyme is known to participate in the activation of a number of xenobiotics including several pesticides and carcinogens. As mentioned in the previous section, FMO from the liver has been shown to catalyze the N-hydroxylation of 2-aminofluorene, an activation reaction. Some sulfoxidations, such as oxidation of the hepatotoxicants thiobenzamide and thioacetamide to the corresponding sulfoxide are also activation reactions. FMO also catalyzes the P = S to P = O activation of several organophosphorus insecticides, although the reaction is restricted to compounds containing a C–P bond, such as fonofos. P450, however, has a less restrictive substrate specificity and will form oxons with a variety of organophosphorus compounds.

9.2.1c Prostaglandin Synthetase

The biosynthesis of prostaglandins occurs in two steps, and is catalyzed by the enzyme system prostaglandin (PGH) synthetase. First, arachidonic acid is oxygenated by a cyclooxygenase to form a hydroperoxide, prostaglandin G (PGG_2), which is then reduced to the alcohol prostaglandin H (PGH_2) by a hydroperoxidase, releasing oxygen. A number of compounds structurally unrelated to polyunsaturated fatty acids can undergo oxidation during PGH biosynthesis. This process, known as cooxidation, occurs when PGG_2 is reduced to PGH_2 and the exogenous compound, acting as a cofactor, is oxidized (Fig. 9.3). The substrate specificity of this reaction is extremely broad, and a number of the xenobiotics that are cooxidized in this reaction are converted to reactive electrophilic metabolites. The bladder carcinogens benzidine, ANFT (2-amino-4-[5-nitro-2-furyll-thiazole), and FANFT (N-[4-[5-nitro-2-furyll-2-thiazoyll-formamide) are converted by PGH synthetase in the presence of arachidonic acid to metabolites that covalently bind to protein and DNA. The widely used analgesic acetaminophen can also be converted in the presence of arachidonic acid and PGH synthetase to a reactive metabolite that binds covalently to protein. Other examples are shown in Table 9.3.

PGH synthetase, localized in the endoplasmic reticulum and nuclear membrane, is present in most mammalian cells, with the levels being higher in some tissues than others. Seminal vesicles are extremely rich sources, and extracts, especially from rams, are often used as a source for PGH synthetase. Embryonic tissue is especially high in this enzyme, as are platelets, endothelial cells, lungs, kidneys (with the medulla being the highest), and urinary bladder epithelium. As would be

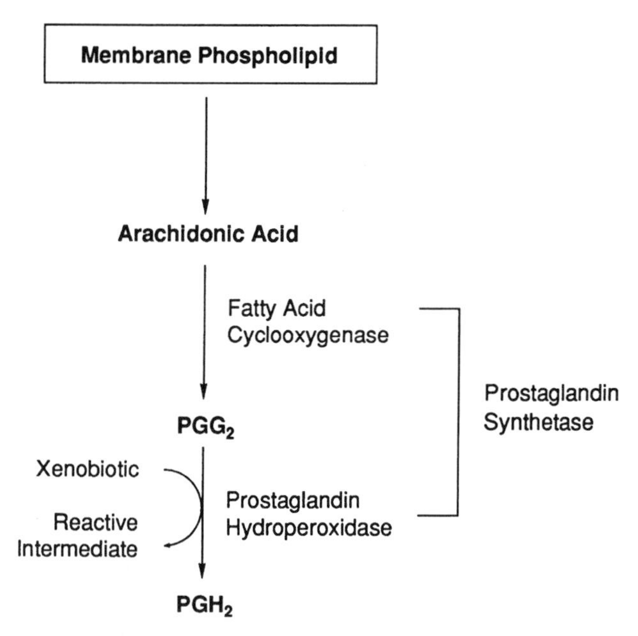

FIGURE 9.3 Mechanism of activation of xenobiotics involving cooxidation by prostaglandin synthetase in the presence of arachidonic acid.

TABLE 9.3 Examples of Chemicals Activated by the Prostaglandin Synthetase Pathway.

Chemical Group	Example
Polycyclic aromatic hydrocarbons	Benzo(a)pyrene-7,8-diol
Aromatic amines	Benzidine 2-Aminofluorene 2-Naphthylamine 4-Aminobiphenyl
Aromatic amides	2-Acetylaminofluorene FANFT
Mycotoxins	Aflatoxin B_1
Drugs	Acetaminophen Diethylstilbestrol

expected, factors that modulate the levels of arachidonic acid or PGH synthetase will influence cooxidation and activation of xenobiotics. A number of factors increase the amount of membrane bound arachidonic acid released, including cell cycle events, hormones, peptides, phospholipase, and chemicals (e.g., TPA). Inhibitors of PGH synthetase include acetylsalicyclic acid, indomethacin, and certain steroids.

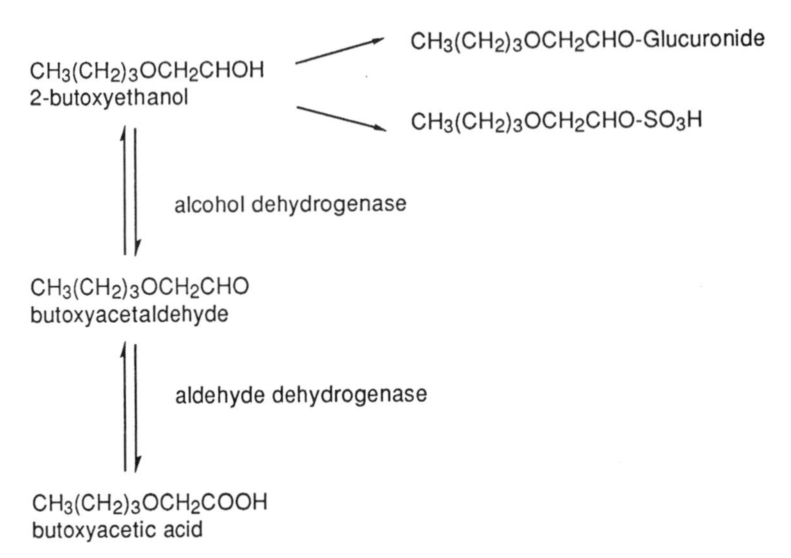

FIGURE 9.4 Metabolism of 2-butoxyethanol. Conjugation reactions with glucuronic acid and sulfate are detoxication reactions whereas alcohol and aldehyde dehydrogenases catalyze activation reactions to the ultimate toxic metabolite, butoxyacetic acid.

9.2.1d Alcohol and Aldehyde Dehydrogenase

These cytosolic enzymes can participate in the metabolism of alcohols and glycols to more toxic aldehydes and acids. A well-known example is the sequential oxidation of methanol to formaldehyde and formaldehyde to formic acid. The solvent 2-butoxyethanol (BE) is a major industrial chemical extensively used in aerosols and cleaning agents, many of which are intended for household use, thus creating a high potential for human exposure and environmental contamination. The hematopoietic system in rats is particularly sensitive to acute exposure of BE, resulting in dose-dependent decreases in circulating red blood cells and hemoglobin concentration. In addition to the hemolytic effects of BE, there is an increase in the concentration of free hemoglobin in the plasma and urine (hemoglobinuria) as well as liver and kidney damage. Metabolism of this compound is shown in Figure 9.4. Conjugation with glucuronide and sulfate are detoxication pathways whereas oxidation to butoxyacetaldehyde and butoxyacetic acid are activation reactions. Recent studies have demonstrated that the hematotoxicity of BE can be attributed to the metabolite butoxyacetic acid, and it has been suggested that therapy for glycol ether poisoning include use of alcohol dehydrogenase inhibitors and/or administration of ethanol as a competitive substrate for the enzyme.

9.2.1e Monoamine Oxidase

Other oxidative enzymes like monoamine oxidases have been found to be involved in the metabolism of xenobiotics and in the formation of toxic metabolites like those derived from the oxidation of MPTP (1-methyl-4-phenyl-1,2,3,6-tetrahydropyridine), which results in Parkinson-like brain toxicity.

9.2.2 Phase II Conjugations

Conjugation by the phase II enzymes, such as glutathione S-transferases (GST), glucuronyl transferases, and sulfotransferases is generally considered to be a detoxication process. It is now apparent, however, that there are exceptions to this and some conjugates are capable also of forming reactive intermediates by deconjugation. (See Chapter 6 for detailed discussion of phase II enzymes and additional examples of activation reactions.)

9.2.2a Sulfate Conjugation

Sulfate particularly, and the acetyl group to a lesser extent, are good leaving groups, and the resulting electrophiles are potent mutagens and carcinogens. Several sulfate conjugates are implicated as proximate or ultimate carcinogens. The first sulfate conjugate recognized as being chemically reactive and possibly involved in tumor formation was the O-sulfonate of N-hydroxy-2-acetylaminofluorene (N-hydroxy-AAF; see Fig. 15.4). When it was found from in vitro experiments that N-hydroxy-AAF itself did not react with DNA, it became apparent that additional activation must occur in vivo. Subsequently, the sulfate conjugate (see Fig. 15.4), which is extremely labile at pH 7.4, was implicated. Sulfate conjugation is now known to generate reactive intermediates from a number of compounds. Carcinogens such as safrole, 7,12-dimethylbenz(a)anthracene, and 2-aminofluorene are activated by sulfate conjugation of their hydroxylated metabolites.

9.2.2b Glucuronide Conjugation

Glucuronide conjugates, in general, are more chemically stable than are the sulfate conjugates. The glucuronide conjugate of N-hydroxy-AAF, unlike the sulfate ester, is reactive only at alkaline pH, and in vitro binding to DNA at pH 7.4 is extremely low. Such conjugates, however, might leave the cell of origin and be broken down elsewhere, resulting in toxicity. Other glucuronides, such as the conjugates of N-hydroxy-2-aminofluorene, are more reactive. The N-glucuronides of the N-hydroxy derivatives of aromatic amines are thought to be proximate carcinogens in human urinary bladder carcinogenesis. The N-hydroxy metabolites are formed in the liver by P450 oxidation followed by N-glucuronidation. The N-glucuronides are then transported to the urinary bladder by the circulatory system where, under acidic conditions, they are hydrolyzed to N-hydroxyarylamines, which nonenzymatically form highly reactive nitrenium ions.

9.2.2c Glutathione Conjugation

Glutathione (GSH) can act as an alternative (competing) nucleophile for nucleophilic portions of proteins and DNA, and thus afford protection against toxic electrophiles within the cell. As has been emphasized, the great majority of the reactions of electrophiles with GSH are detoxication processes. Several important examples are known, however, where conjugation with GSH is an activation reaction.

Ethylene dibromide reacts with GSH to form the half sulfur mustard 1-bromo-2-S-glutathionyl ethane (Fig. 9.5). This intermediate rearranges nonenzymatically to form an episulfonium (thiiranium) ion, which reacts with nucleophiles such as water,

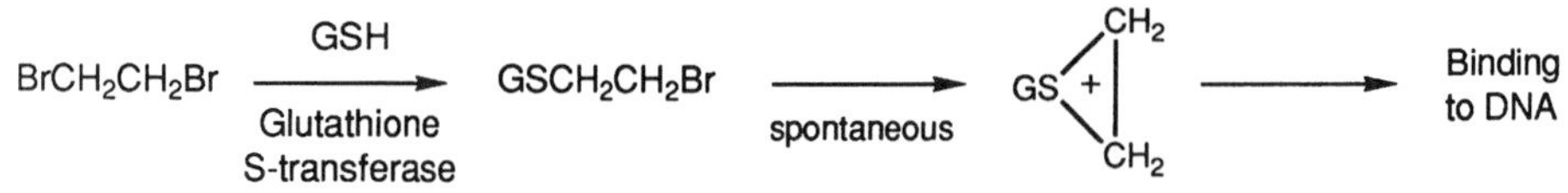

$$BrCH_2CH_2Br \xrightarrow[\text{Glutathione S-transferase}]{GSH} GSCH_2CH_2Br \xrightarrow[\text{spontaneous}]{} GS^{+}\langle CH_2, CH_2 \rangle \longrightarrow \text{Binding to DNA}$$

FIGURE 9.5 Activation of ethylene dibromide by conjugation with glutathione.

GSH, and the N-7 position of guanine in DNA. This reaction with DNA is thought to be responsible for the genotoxicity of ethylene dibromide. Although P450 metabolizes ethylene dibromide to bromoacetaldehyde, which apparently is responsible for the protein binding of ethylene dibromide, the GSH-dependent metabolism is responsible for binding to DNA.

Several haloalkenes are known to be selectively nephrotoxic and nephrocarcinogenic as a result of activation by the multistep, GSH-dependent, cysteine conjugate β-lyase pathway. For example, hexachlorobutadiene forms the GSH conjugate, S-(1,2,3,4,4-pentachlorobutadienyl)-glutathione, which is further metabolized by the mercapturic acid pathway to a cysteine conjugate that is then cleaved to a reactive intermediate by the action of the kidney β-lyase. Although glutathione S-conjugates may be predominantly biosynthesized in the liver, selective toxicity to the kidney is easily explained as a result of concentration of the S-conjugates by the kidney and/or selective biosynthesis of toxic S-conjugates at the site of toxicity, the proximal tubular cells (see also Section 9.5.3).

9.2.2d Cysteine S-Conjugate β-Lyase

This enzymes converts some cysteine S-conjugates (premercapturic acids) into thiols, ammonia, and pyruvate and has been implicated in the activation of certain halogenated alkenes (see section 6.8), leading to nephrotoxicity.

9.2.3 Intestinal Microflora

A number of metabolic transformations can be catalyzed by intestinal microflora. These include reductive reactions, hydrolytic reactions, and other reactions such as N-acetylation, aromatization, O- and N-dealkylation, decarboxylation, and nitrosamine formation. The plant glucoside cycasin, found in cycad nuts, is cleaved by the action of β-glucosidases present in intestinal microflora to the carcinogen methylazoxymethanol. This compound is carcinogenic only if administered orally because β-glucosidases are not present in mammalian tissues. Similarly, amygdalin, a cyanide-containing glucoside found in certain seed such as apricots, is cleaved by β-glucosidases in gut microflora to an intermediate that can release toxic cyanide. Of concern in recent years are nitrates that can be reduced by nitrate reductases to nitrites.

Some of the potential consequences of metabolism of xenobiotics by intestinal microflora are summarized in Table 9.4. A number of factors may affect the microfloral composition and thus the metabolism of the xenobiotic. These include species, strain, and individual variation, age, GI disease, pH effects, drugs, antibiotics, diet (especially fiber), and enzyme induction. Because glucuronides are eliminated mainly through biliary excretion, further metabolism by the gut flora, especially

TABLE 9.4 Some Consequences of Microfloral Metabolism of Xenobiotics.

1. Enterohepatic circulation resulting in delayed excretion
2. Production of toxic metabolites
3. Formation of active carcinogens
4. Detoxication
5. Pharmacologically active metabolites
6. Species variation in metabolism and toxicity
7. Individual variation in metabolism and toxicity
8. Metabolites not formed in tissues
9. Differences in toxicity resulting from route of administration

hydrolysis by glucuronidases, may yield ultimate carcinogens with the potential to initiate colon cancer.

9.3 STABILITY OF REACTIVE METABOLITES

As a result of their high reactivity, reactive metabolites are often considered to be short-lived. This is not always true, however, because reactive intermediates can be transported from one tissue to another where they may exert their deleterious effects. Thus, reactive intermediates can be divided into several categories depending on how far they are transported from the site of activation.

9.3.1 Ultrashort-Lived Metabolites

9.3.1a Reactive Oxygen Species

Reactive oxygen species (ROS) are continously produced during normal metabolic processes, and are detoxified by a variety of protective mechanisms. It is only when the balance between oxygen activation and the detoxication of ROS is shifted in the direction of activation that these reactive oxygen species inflict damage on cellular macromolecules and/or bring about lipid peroxidation, a condition refered to as oxidative stress. (Oxidative stress, its effects, and cellular defense mechanisms are discussed in detail in Chapter 10). Typically, during the one-electron reduction of oxygen, superoxide anion radicals ($O_2^{\bar{\bullet}}$) are produced. These superoxide anions radicals undergo dismutation, either spontaneously or via the enzyme superoxide dismutase to form hydrogen peroxide. Hydrogen peroxide can undergo a variety of transformations including the formation of water and oxygen catalyzed by catalase, the formation of oxidized glutathione and water in the presence of reduced glutathione and glutathione peroxidase, or reduction in the presence of additional superoxide anion radicals or transition metals such as iron or copper to yield the highly reactive hydroxylradical ($OH^{\bullet}$).

ROS are generated in a number of metabolic reactions. Cytochrome P450 (see section 5.2.2), particularly if substrate oxidation is not tightly coupled to electron transfer during the catalytic cycle, may produce superoxide anion radicals after the introduction of the first electron or hydrogen peroxide after the introduction of the second electron. Similarly, oxygen radicals are produced during the reduction of quinones to hydroquinones catalyzed by DT-diaphorase (NAD(P)H-quinone oxidoreductase).

9.3.1b Other Ultrashort-Lived Metabolites

Ultrashort-lived metabolites are typically the products of xenobiotic metabolism that bind primarily to the enzyme responsible for their production. This category includes substrates that form enzyme-bound intermediates that react only with the active site of the enzyme ("suicide substrates"). A number of chemicals are known to react in this manner with P450. Olefins and acetylenes become irreversibly bound to a pyrrolic nitrogen of P450 heme from which iron is subsequently released, destroying the heme moiety of P450. 1-Aminobenzotriazole (I-ABT), often used as a P450 inhibitor experimentally, is metabolized to an intermediate that complexes with the heme of P450. Many methylenedioxyphenyl compounds, including the carcinogens safrole and isosafrole and the insecticide synergist piperonyl butoxide, are metabolized to an intermediate, presumably a carbene, which forms a stable complex with the heme iron resulting in inhibition. This inhibition of P450, in fact, is responsible for the synergistic action of piperonyl butoxide with insecticides that are detoxified as a result of oxidation by P450. Other substrates, although not true "suicide substrates," produce reactive products that are released from the active site and bind primarily to the activating enzyme. During the metabolism of the insecticide parathion to paraoxon, highly reactive sulfur is released that binds irreversibly to the heme iron of P450, resulting in inhibition of the enzyme as well as binding to certain amino acid residues in P450.

9.3.2 Short-Lived Metabolites

These metabolites remain in the cell or travel only to adjacent cells. In this case, covalent binding is restricted to the cell of origin and nearby cells. Many metabolites fall into this group and give rise to localized tissue damage. In the liver, P450 concentration is unevenly distributed across the liver lobule, with the highest concentration being in the centrolobular region, then the midzonal, and the lowest concentration in the periportal region. Thus, it is not surprising that xenobiotics that produce highly reactive metabolites as a result of metabolism by P450 in the liver often cause centrolobular necrosis. For example, carbon tetrachloride, a toxicant that is activated to a $\bullet CCl_3$ radical by P450, causes centrolobular necrosis because this region high in P450 but has a lower oxygen tension than the other zones. The centrolobular necrosis caused by high doses of acetaminophen is explained in part by the imbalence between high P450 activation and lower glutathione-dependent detoxication.

9.3.3 Longer-Lived Metabolites

These metabolites may be transported to other cells and tissues. The primary site of benzo(a)pyrene tumorigenicity is the lung, although the liver is the most active site for formation of the ultimate carcinogen. This suggess, that the reactive intermediate may be transported from the liver by the bloodstream to the lung, and in fact i.p. administration of the reactive 7,8-diol-9, 10-epoxide results in lung tumors, confirming the stability of this metabolite. In addition, however, activation of benzo(a)pyrene by the lung may be important, especially if the route of exposure is inhalation. Another example of reactive metabolite transport occurs with benzene, a bone marrow toxicant activated by the liver.

Reactive intermediates may also be transported to other tissues, not in their original form but as conjugates, which then release the reactive intermediate under the conditions prevailing in the target tissue. For example, carcinogenic aromatic amines are metabolized in the liver to the N-hydroxylated derivatives, which, following glucuronide conjugation, are transported to the bladder where the N-hydroxy derivative is released under the prevailing acidic conditions. Moreover, additional reactive metabolites can be generated in situ from the parent compound by P450 and prostaglandin synthetase activity.

9.4 FACTORS AFFECTING ACTIVATION AND TOXICITY

In animals, detoxication of exogenous chemicals is far more prevalent than activation pathways. Under certain conditions, however, such as exposure to high concentrations of environmental chemicals, starvation, disease states, and genetic situations, activation pathways may be enhanced and become the dominant routes of metabolism.

9.4.1 Saturation of Deactivation Pathways

At high dose levels of the xenobiotic, detoxication pathways may be overwhelmed resulting in a higher concentration of the reactive metabolite being formed. Saturation of detoxication pathways may arise from saturation of the enzyme or unavailability of an essential cofactor used in the conjugation reaction. Acetaminophen is a drug that has been studied extensively and provides an excellent example of such a situation. At therapeutic doses acetaminophen is readily detoxified by glucuronide and sulfate conjugation with only a small amount being metabolized by P450 to the active quinoneimine, which is effectively conjugated with glutathione and excreted (Fig. 9.6). At very high dose levels, however, the sulfate and glucuronide cofactors (PAPS and UDPGA) are depleted, and more acetaminophen is metabolized to a reactive intermediate. As long as glutathione (GSH) is available, most of the reactive intermediate is detoxified. As GSH is depleted, however, covalent binding to -SH groups of cellular proteins will increase resulting in hepatic necrosis. If damage to the liver is extensive, death may occur.

A similar situation can occur when two drugs that use the same detoxication pathways are given simultaneously. For example, salicylamide, like acetaminophen, is conjugated with glucuronide and sulfate. Thus, when both compounds are given together, they compete for enzymes and cofactors, resulting in activation and toxicity of acetaminophen at a lower dose level than if acetaminophen alone were administered.

Most toxic chemicals follow a sequence of events comparable to those with acetaminophen in that some critical pathway must become overwhelmed by continued production of reactive intermediates in order for toxicity to occur.

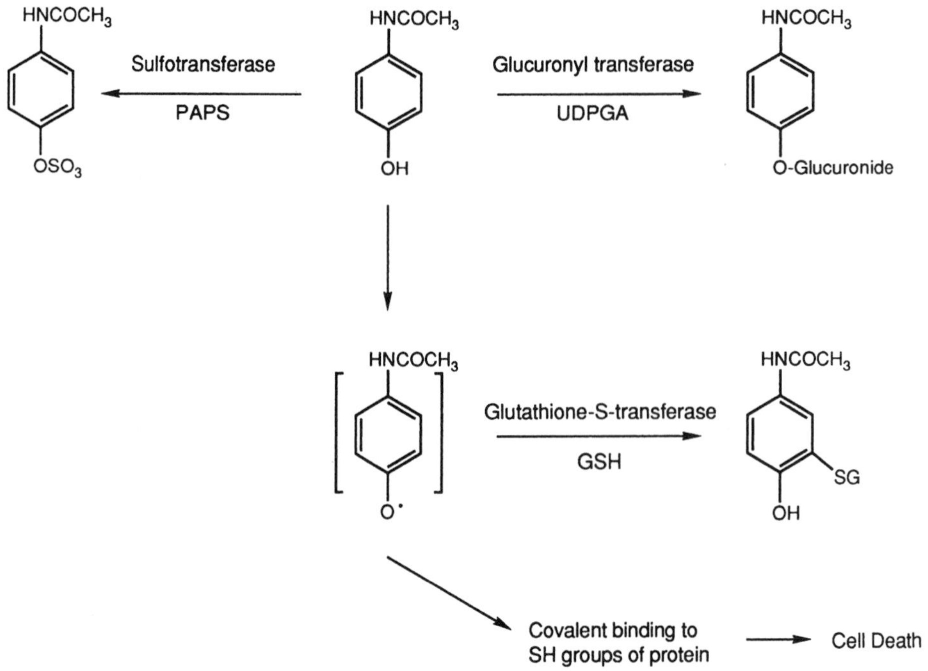

FIGURE 9.6 Metabolism of acetaminophen showing activation and detoxication pathways.

9.4.2 Enzyme Induction

In the case of the activation of acetaminophen discussed above, this compound is an excellent substrate for P450 2E1, an isozyme that can be induced by ethanol, acetone, and certain drugs such as isoniazid. Thus, animals pretreated with compounds that induce P450 2E1 are more susceptible to acetaminophen hepatotoxicity than untreated animals because more of the parent compound is metabolized by the P450 activation pathway. Interestingly enough, if ethanol and acetaminophen are given simultaneously, the toxicity of acetaminophen is diminished. In this situation induction has not yet occurred and, because both ethanol and acetaminophen are substrates for P450, the amount of acetaminophen metabolized to the reactive intermediate is reduced because ethanol acts as a competitive inhibitor.

A characteristic of many xenobiotics is that they can act as inducers of the P450 enzymes, often selectively inducing the forms involved in the metabolism of the particular xenobiotic. Undoubtedly this is an evolutionary adaptation to allow the metabolic systems to be activated as required in response to various types of chemicals and be turned off, thus saving energy, when not needed. Frequently, however, the increased oxidations are activation reactions rather than detoxication reactions. This phenomenon is especially prevalent with the polycyclic aromatic hydrocarbons and aromatic amines that induce P4501A forms that are involved in pathways activating these chemicals to mutagens (e.g., benzo(a)pyrene). In fact, S9 fractions from livers of untreated animals often fail to activate known mutagens using in vitro tests, and for this reason most tests utilize S9 preparations from animals pretreated with various inducing agents in order to maximize the potential for activation.

Induction, however, is not always associated with increased toxicity, as detoxication pathways may also be induced or be selectively induced. Often, epoxide hydrolase and one or more GSH- or UDPGA-transferases are induced so that increased activation is accompanied by increased detoxication. In addition, certain inducing agents may induce forms of P450 that do not catalyze activations of other toxicants or carcinogens. For example, activation of 2-acetylaminofluorene (2-AAF) is decreased in liver fractions from animals pretreated with phenobarbital. This agent selectively induces P450 2B1, which catalyzes the formation of the inactive ring hydroxylated products. On the other hand, pretreatment of animals with 3-methylcholanthrene induces P450 1A2, which catalyzes the N-hydroxylation of 2-AAF, the initial step in its activation to the ultimate carcinogen (Fig. 9.2). Therefore, the P450 composition of a particular animal can determine metabolism and affect the balance between activation and detoxication.

9.4.3 Genetic and Physiological Factors

Most species differences in chemical toxicity and carcinogenicity can be attributed to differences in metabolism, both qualitative and quantitative, often reflecting species differences in P450 composition. In fact, the P450 complement of any species is unique to that species and represents an evolutionary adaptation to that organism's environment. This adaptive mechanism often manifests itself in variability in metabolic pathways to xenobiotic chemicals, including pesticide selectivity and resistance as well as susceptibility to chemical carcinogenesis.

For example, 2-AAF is a carcinogen in most animal species, being activated by N-hydroxylation as discussed earlier, but is not carcinogenic to the guinea pig, an animal species that does not catalyze this hydroxylation. Insects are highly susceptible to the neurotoxic effects of the organophosphorus pesticide malathion, although this pesticide has especially low mammalian toxicity. Both insect and mammalian species catalyze the P450-mediated oxidation of malathion to malaoxon, the reactive metabolite responsible for inhibition of acetylcholinesterase. In mammals, however, the main route of malathion metabolism is hydrolysis by a carboxylesterase to the nontoxic malathion monoacid. This enzyme, which is absent or present only in low levels in most insect species, provides the basis for malathion's selective action. The effects of various physiological factors on metabolism of xenobiotics are discussed in Chapter 7.

9.5 TARGET ORGAN TOXICITY—SPECIFIC EXAMPLES

Many organ toxicities involve activation of toxicants, either in the target organ or in the liver followed by transport to the target site. Some specific examples are described below, and detailed descriptions of target organ toxicity are to be found in Chapter 20, Hepatotoxicity; Chapter 21, Dermatotoxicity; Chapter 22, Reproductive and Developmental Toxicity; Chapter 23, Immunotoxicity; Chapter 24, Respiratory Tract Toxicity; Chapter 25, Biochemical Mechanisms of Renal Toxicity; and Chapter 26, Cardiovascular Toxicity. Although metabolite activation of neurotoxicants is known to occur in the nervous system, this aspect has not been extensively investigated.

9.5.1 Liver

As mentioned previously in this chapter, the liver is often the main target for chemically induced toxicities, and several factors contribute to its being particularly susceptible. First, it is the organ with the highest complement of P450, in terms of quantity as well as numbers of isozymes and inducibility. Second, the liver is the first site of metabolism (known as "first-pass" metabolism) for xenobioties absorbed from the gastrointestinal tract, the major route of absorption for most xenobiotics. Thus, it is not surprising that the liver is the target organ for many toxicities that result from metabolic activation. Moreover, the liver may activate chemicals that can then be transported to distant tissues to affect toxicity in that organ. Details of metabolic activation in the liver are provided in Chapter 20. Although the liver has been recognized as the major site of drug or xenobiotic metabolism, recently much attention has also been paid to the metabolism of xenobiotics by extrahepatic tissues, in particular at portals of entry such as the lung and skin.

9.5.2 Lung

The lung is a very complex organ composed of at least 40 different cell types and has the potential to metabolize many foreign chemicals. Most of the xenobiotic metabolizing enzyme activities present in the liver are also present in the lung, although, in most cases, activities of the lung are significantly lower than those of the liver. Moreover, the individual enzymes are not necessarily identical in the two tissues, and some chemicals may be specifically metabolized in the lung. The cellular distribution of the metabolizing enzymes in this heterogeneous tissue as well as the balance of activation and detoxication in any particular cell are key factors in determining cellular specificity for many pulmonary toxicants. Recently, immunohistochemical methods have shown that much of the total lung content of P450 is localized in the nonciliated bronchiolar epithelial (Clara) cells, thus explaining the metabolic basis for the susceptibility of this cell type to many pulmonary toxicants such as 4-ipomeanol, 3-methylfuran, and naphthalene. Other cell types, including the type II cells and the ciliated bronchiolar epithelial cell, also contain many of the xenobiotic metabolizing enzymes. Table 9.5 lists some xenobiotics that are known to be converted to toxic metabolites by lung tissue. Three chemicals are discussed in the following section, two of which can be activated in the lung (benzo(a)pyrene and 4-ipomeanol) and one of which is activated in the liver and transported to the lung where it exerts its toxicity (monocrotaline).

9.5.2a Benzo(a)pyrene

Polycyclic aromatic hydrocarbons such as benzo(a)pyrene arise from incomplete combustion of plant and animal materials and are ubiquitous environmental contaminants. Although the lung may be systemically exposed as a result of the ingestion of smoked foodstuffs, the most important source of lung exposure is through inhalation of fuel-combustion products or cigarette smoke. The aromatic hydrocarbons such as benzo(a)pyrene have little inherent biological activity, their toxicity resulting from the formation of reactive metabolites (see Fig. 15.7). In addition to activation by the liver and possible transport to the lung (see Section 10.3.3), all of the major metabolites of benzo(a)pyrene can be formed by respiratory tract

TABLE 9.5 Examples of Xenobiotics Converted to Toxic Metabolites in the Lung.

Xenobiotic	Activating Enzymes	Type of Toxicity
4-Ipomeanol	P450	Clara cell necrosis
3-Methylfuran	P450	Clara cell necrosis
Benzo(a)pyrene	P450, epoxide hydrolase	Carcinoma
Thiourea	P450, FMO	Type I cell damage
Naphthalene	P450	Clara cell necrosis
Nitrofurantoin	P450 reductase, xanthine oxidase	Fibrosis
Paraquat	P450 reductase	Alveolar necrosis, edema
Parathion	P450	Edema
Carbon tetrachloride	P450	Clara cell and type II damage
Hydrazines	Monoamine oxidase	Carcinoma

tissues. Furthermore, the area of the respiratory tract most susceptible to developing lung cancer, the bronchial epithelium, is capable of metabolically activating benzo(a)pyrene to a metabolite that covalently binds to DNA. Thus, the lung, as well as other tissues, may produce the benzo(a)pyrene metabolites that ultimately become covalently bound to lung tissue, the relative contributions of the individual tissues depending on factors such as routes of exposure, genetic variability, and chemical and physiological modulators.

9.5.2b 4-Ipomeanol

This naturally occurring furan is produced by the mold *Fusarium solani*, which infects sweet potatoes. Lung edema in cattle appears to be associated with ingestion of mold-damaged sweet potatoes. A similar pulmonary lesion can be produced in a number of species regardless of the route of administration—oral, intraperitoneal, or intravenous. The toxicity is characterized by necrosis of the nonciliated bronchiolar epithelial (Clara) cells. Evidence points to the crucial role of a reactive metabolite formed in the Clara cells. For example, following in vivo administration of 4-ipomeanol, significantly more covalently bound radioactivity is seen in lungs than any other tissue, and this radioactivity is predominantly associated with the Clara cells. Although both liver and lung tissue are capable of activating 4-ipomeanol to a covalently binding species, when activity is expressed on the basis of covalent binding per molecule of P450, the lung is approximately eightfold more active than the liver. Moreover, two P450 isozymes purified from rabbit lung have been shown to be very active in transforming 4-ipomeanol to metabolites that bind covalently to protein. Because these two isozymes, which are major lung forms, constitute only a very small percentage of the total P450 in rabbit liver, this isozymic distribution provides a possible explanation for the target organ selectivity of 4-ipomeanol.

The reactive metabolite formed from 4-ipomeanol has not been unambiguously identified. Although much of the early data suggested that the reactive metabolite was the furan epoxide, studies with 2- and 3-methyl furan have identified the unsaturated aldehydes, acetylacrolein and methylbutenedial, respectively, as the principal reactive metabolites in the toxicity of these two compounds.

9.5.2c Pyrrolizidine Alkaloids

The pyrrolizidine alkaloids (PA), found in the genus *Senecio* and a number of other plant genera, are plant toxicants of environmental interest that have been implicated in a number of livestock and human poisonings. Grazing animals may be poisoned by feeding on PA-containing pastures, and human exposure may occur through consumption of herbal teas and contaminated grains and milk.

Monocrotaline (MCT), found in the leaves and seeds of the plant *Crotalaria spectabilis*, has been the most extensively studied of the pyrrolizidine alkaloids. When MCT is given to rats and other animals at high doses, pronounced liver injury occurs, and animals usually die early, presumably of liver failure. Lower doses, however, that are only mildly hepatotoxic result in lung injury that is associated with pulmonary hypertension and usually death in several weeks. Thus, the primary target organ in MCT toxicity depends on the dose. MCT pneumotoxicity is of particular interest because it has been used as an animal model for certain human chronic pulmonary diseases. MCT requires metabolic activation in order to produce either hepatotoxicity or pneumotoxicity. Pyrrolizidine alkaloids are metabolized in the liver by three major routes. One of these leads to an N-oxide derivative, which is relatively nontoxic. A second involves hydrolysis of the ester linkages and is also a detoxication pathway. The third pathway involves oxidation to pyrrolic metabolites that are toxic and capable of binding to tissue macromolecules. Activation of MCT to pyrrolic metabolites occurs in the liver by the action of P450 enzymes. Even though monocrotaline acts as a pneumotoxicant, several lines of evidence indicate that the lung itself is incapable of metabolizing MCT to pyrrolic derivatives or can do so only at very low levels. Although these metabolites are fairly reactive, they are apparently stable enough that a fraction is capable of leaving the hepatocytes, entering the venous circulation, and surviving passage to the lung, which comprises the next downstream capillary bed in the circulatory system. Indeed, a main site of pulmonary injury occurs in the endothelial cells, a target site consistent with a reactive metabolite being absorbed from the circulatory system.

Chemically synthesized monocrotaline pyrrole (MCTP, dehydromonocrotaline) is more toxic than MCT itself, both to liver and to lung. Although MCTP has never been isolated from animal tissues pretreated with MCT, it is thought to be the proximate toxic metabolite. When MCTP is injected intravenously into rats, a lung injury is produced that is comparable to that produced by moderate doses of MCT. Furthermore, MCT perfused directly through isolated lung tissue does not significantly damage lung tissue whereas either MCTP or the perfusate from liver treated with MCT does result in damage.

9.5.3 Teratogenesis

9.5.3a Phenytoin

The widely used anticonvulsant drug phenytoin (diphenylhydantoin, Dilantin) is known to be teratogenic in laboratory animals including mice, rats, and rabbits. Various fetal malformations are produced in these species, with cleft palate, micromelia, renal defects, and hydrocephalus being the most frequent anomalies. Phenytoin is also implicated as a human teratogen resulting in craniofacial malformations, appendicular defects, and other abnormalities such as cleft palate and cardiac defects. Collectively, the pattern of defects is termed the fetal hydantoin syndrome (FHS).

Several mechanisms have been postulated to explain how phenytoin exerts its teratogenic effect. Early studies suggested that the fetal anomalies may result from the hepatic P450-catalyzed activation of phenytoin to the reactive arene oxide intermediate, which, if not detoxified, could bind covalently to essential fetal cellular macromolecules, thereby interfering with fetal development. Several observations are difficult to explain, however, based only on the hepatic formation of arene oxides. For example, induction of P450-catalyzed arene oxide formation decreases phenytoin teratogenicity in pregnant mice, and in vivo use of P450 inhibitors increases the teratogenic effect. Additionally, the teratogenic anticonvulsants trimethadione and dimethadione, which, although they are structurally similar to phenytoin, lack the phenyt substituents and therefore do not form the reactive arene oxide intermediate.

An alternative bioactivating pathway is cooxidation by prostaglandin synthetase. This enzyme system is particularly high in embryonic tissue, whereas P450 activity, on the other hand, is extremely low. Therefore, P450-generated reactive metabolites would most likely be produced in the liver and transported to the embryo whereas PGS-generated metabolites could be formed in situ.

A number of studies have suggested an activating mechanism involving cooxidation by prostaglandin synthetase of phenytoin and similar compounds to a reactive free radical species. In vitro studies using purified prostaglandin synthetase demonstrated arachidonate-dependent covalent binding of phenytoin that was significantly reduced by coincubation with the free radical spin trapping agent PBN (a-phenyl-N-t-butylnitrone), glutathione, the antioxidants caffeic acid and butylated hydroxytoluene (BHT), or the prostaglandin synthetase inhibitor indomethacin. Coadministration of these agents in vivo reduced the fetal covalent binding of phenytoin and the incidence of fetal cleft palates. In other studies when pregnant mice were pretreated with acetylsalic acid (ASA), a potent and irreversible inhibitor of the cyclooxygenase component of prostaglandin synthetase, the incidence of fetal cleft palates induced by phenytoin, trimethadione, and dimethadione as well as the covalent binding to embryonic protein was reduced. In mice treated with both phenytoin and the tumor promoter TPA (12-O-tetradecanoylphorbol-13-acetate), which activates the release of arachidonic acid, the embryopathic effects of phenytoin were increased in a TPA dose-related manner, further substantiating the involvement of prostaglandin synthetase in activation.

9.5.4 Kidney

The kidney is a frequent target of toxic chemicals. This toxicity may be manifested as immediate organ failure or as a more general alteration of function that becomes apparent after many years. Because normal kidney function is necessary to maintain life, the consequences of such chemical insults may be fatal. There are many reasons why this organ is so susceptible to chemically induced damage. First, the kidneys receive approximately 25% of the cardiac output. This high blood flow is in part responsible for the large amount of xenobiotics in the systemic circulation delivered to the kidneys—particularly the renal cortex, which receives over 90% of the renal blood flow. Secondly, as a consequence of its normal function of filtration, reabsorption, and secretion, the kidney may also concentrate xenobiotics to toxic levels. Thus, because of its structure and physiology, the cells of the kidney are often exposed to more chemicals and at higher concentrations than are the cells of most other organs, resulting in toxic effects. In addition, although many nephrotoxicants are not toxic when reaching the kidney, they may be activated by the renal P450 system, by prostaglandin synthetase, or by reactions associated with glutathione conjugations. Several examples of activations have been discussed in previous sections of this chapter: acetaminophen activation by the prostaglandin synthetase pathway in kidney cortex and glutathione (GSH) conjugation associated toxicities.

9.5.4a Glutathione Conjugates

Although the glutathione S-transferases and other enzymes involved in formation of mercapturic acids are primarily associated with the detoxication of xenobiotics, recent evidence indicates a role for these enzymes in bioactivation reactions. Several haloalkanes and alkenes are known nephrotoxicants (e.g., hexachloro-1,3-butadiene and chlorotriflouoroethylene). Hepatic glutathione S-transferases catalyze the biosynthesis of glutathione S-conjugates, which are then metabolized by renal or intestinal γ-glutamyltransferases and dipeptidases to the corresponding cysteine S-conjugates. Depending on the chemical, the cysteine S-conjugates may be direct-acting nephrotoxicants (e.g., dibromoethane) or may undergo bioactivation catalyzed by cysteine conjugate β-lyase (see Chapter 6). Although cysteine conjugate β-lyase is found in both cytosolic and mitochondrial fractions of rat kidney cortex, the mitochondrial β-lyase is more active in activation, and in fact cytotoxicity studies show that renal mitochondria are the primary targets for nephrotoxic cysteine S-conjugates.

9.5.4b Chloroform

There is considerable evidence that the nephrotoxicity of chloroform is due to the production of a reactive metabolite, phosgene ($COCl_2$), by cytochrome P450. Although both liver and kidney microsomes can produce phosgene, it seems that chloroform nephrotoxicity is caused by phosgene generated in the kidney. (See Section 25.5.5 for further details of chloroform activation and nephrotoxicity.)

9.6 FUTURE DEVELOPMENTS

The current procedures for assessing safety and carcinogenic potential of chemicals using whole animal studies are expensive and time consuming as well as becoming less socially acceptable. More importantly, their scientific validity is also being questioned. Currently, a battery of short-term mutagenicity tests are used extensively as early predictors of mutagenicity and possible carcinogenicity. Most of these test systems use target organisms that lack a suitable enzyme system to bioactivate chemicals and therefore an exogenous activating system is used. Usually the post-mitochondrial fraction (S9) from rats pretreated with Aroclor 1254, a mixture of polychlorinated biphenyls, which selectively induces P4501A and 2B isozymes is used. The critical question is to what extent does this system represent the true in vivo situation, especially in humans. If not this system, then what is a better alternative? Many chemical carcinogens selectively induce the P450 enzymes involved in their activation, and repeated administration of the chemical will result in its metabolism by the complement of P450s which is itself induced. Thus, the ability of a chemical to induce its own pathways of activation may be as critical a factor in determining carcinogenic potential as the ability to form genotoxic metabolites. In any case, both of these factors are important points to consider in evaluating carcinogenic potential of a chemical.

In order for a chemical to manifest toxicity, the chemical must either possess intrinsic properties that enable it to interact directly with macromolecules to mediate toxicity or be transformed to a reactive intermediate that reacts with tissue molecules. In addition, the biological organism in question must have enzyme systems that are able to metabolize the chemical to a reactive intermediate and at the same time be unable to effectively detoxify the chemical and/or reactive metabolites. Because toxicity is a consequence of the nature of the chemical in question as well the enzyme complement of the exposed organism, it is crucial to evaluate toxicity in relationship to the exposed organism. As we have seen in numerous examples, a chemical that is toxic or carcinogenic to one species or gender may be inactive in another, and factors such as nutrition, disease, and exposure to environmental chemicals may affect enzyme composition and thus toxicity. An understanding of the specific enzymes, especially P450s, with which a particular chemical will interact will more accurately facilitate prediction of the metabolic pathways of the chemical and the generation of reactive intermediates that are likely to produce toxicity.

In studies of chemical toxicity, the pathways of metabolism, the rates of metabolism, as well as effects resulting from toxicokinetic factors and receptor affinities, are critical in the choice of animal species and experimental design. It is thus crucial that the animal species chosen as a model for humans in safety evaluation studies should metabolize the chemical by the same routes and, furthermore, that quantitative differences be considered in interpretation of animal toxicity data. Current methodologies of risk assessment by extrapolation of the toxicity or carcinogenic potential of a chemical from one species to another without consideration of metabolic and toxicokinetic characteristics, in addition to having little scientific validity, often lead to inaccurate conclusions. This in turn undermines confidence in the scientific process. For these reasons continued research to extend our understanding of the biochemical mechanisms underlying toxicity is of paramount importance.

SUGGESTED READING

Cohen, G.M. Pulmonary metabolism of foreign compounds: its role in metabolic activation. *Environ. Health Perspect.* 85 (1990), 31–41.

Coles, B. and Ketterer, B. The role of glutathione and glutathione transferases in chemical carcinogenesis. *Crit. Rev. Biochem. Mol. Biol.* 25:1 (1990), 47–70.

Guengerich, F.P. Enzymatic oxidation of xenobiotic chemicals. *Crit. Rev. Biochem. Mol. Biol.* 25:2 (1990), 97–153.

Koob, M. and Dekant, W. Bioactivation of xenobiotisc by formation of toxic glutathione conjugates. *Chem. Biol. Interact.* 77 (1991), 107–136.

Marnett, L.J. and Eling, T.E. Cooxidation during prostaglandin biosynthesis: a pathway for metabolic activation of xenobiotics. In: Hodgson, E., Bend, J.R. and Philpot, R.M. (Eds), *Reviews in Biochemical Toxicology*, Vol. 5, Elsevier Biomedical, New York, 1983, pp 135–172.

Miller, E.C. and Miller, J.A. Some historical perspectives on the metabolism of xenobiotic chemicals to reactive electrophiles. In: Anders, M.W. (Ed), *Bioactivation of Foreign Compounds*, Academic Press, New York, 1985, pp 3–28.

Monks, T.J. and Lau, S.S. Reactive intermediates and their toxicological significance. *Toxicol.* 52 (1988), 1–53.

Monks, T.J., Anders, M.W., Dekant, W., Stevens, J.L., Lau, S.S. and Van Bladeren, P.J. Glutathione conjugate mediated toxicities. *Toxicol. Appl. Pharmacol.* 106 (1990), 1–19.

Nelson, D.S. and Harvison, P.J. Roles of cytochromes P450 in chemically induced cytotoxicity. In: Guengerich, F.P. (Ed), *Mammalian Cytochromes P-450*, Vol. 11, CRC Press, Baca Raton, FL, 1987, pp 19–79.

Wells, P.G., Nagai, M.K. and Greco, G.S. Inhibition of trimethadione and dimethadione teratogenicity by the cyclooxygenase inhibitor acetylsalicyclic acid: a unifying hypothesis for the teratogenic effects of hydantoin anticonvulsants and structurally related compounds. *Toxicol. Appl. Pharmacol.* 97 (1989), 406–414.

Wells, P.G., Zubovits, J.T., Wong, S.T., Molinari, L.M. and Ali, W. Modulation of phenytoin teratogenicity and embryonic covalent binding by acetylsalicylic acid, caffeic acid, and a-phenyl-N-t-butylnitrone: implications for bioactivation by prostaglandin synthetase. *Toxicol. Appl. Pharmacol.* 97 (1989), 192–202.

Yost, G.S., Buckpitt, A.R., Roth, R.A. and McLemore, T.L. Mechanisms of lung injury by systemically administered chemicals. *Toxicol. Appl. Pharmacol.* 101 (1989), 179–195.

CHAPTER TEN

Mechanisms of Chemically Induced Cell Injury and Cellular Protection Mechanisms

DONALD J. REED

10.1 INTRODUCTION

Mechanisms of chemically induced cell death and the systems that protect against cell injury and death are important components of biochemical toxicology. Previously, necrotic death was essentially the only type of cell death considered when toxic levels of one or more chemicals were thought to be the cause of death. Recently, research findings indicate that necrosis and apoptosis, as presently defined, may represent the extremes of a wide spectrum of mechanisms that can contribute to chemically induced cell death. In this chapter, the mechanisms of cell injury and death are discussed along with the cellular protective systems that depend on glutathione. These protective systems are thought to be highly important in the detoxication of products from the bioactivation of chemicals and oxygen metabolism that produces a constant supply of reactive oxygen species. Without protection, chemically induced injury and death of cells and tissues with concomitant loss of organ functions would be greater as a result of chemical exposure.

10.2 CELL INJURY EVENTS

Although several hypotheses exist as the basis for mechanisms of chemical effects on cells, two will be described in some detail in this chapter: receptor-mediated and chemical-mediated effects. One hypothesis suggests that chemically induced cell injury is the result of covalent binding of biological reactive intermediates to cellular molecules, including macromolecules. Although this is a predominant mechanism for loss of cellular functions and expression of cell injury, oxidative stress, as a by-product of the metabolism of molecular oxygen, can be coupled to the effects of covalent binding and, in some instances, have a major role in the total injury and death. Oxygen is metabolized to oxygen-containing reactive intermediates, many of which can undergo a cycling process between oxidation and reduction. This process is known as redox cycling and can cause oxidative stress in cells. Such oxidative stress can result in oxidation of cellular constituents such as low molecular weight thiol-containing compounds including glutathione, protein thiols as well as other functional groups on macromolecules, and peroxidation of lipids and other susceptible cellular constituents. It now known that many chemicals can undergo both bioactivation to form biological reactive intermediates that bind to macromolecules and also enhance the formation of oxygen radicals with a concomitant oxidative stress.

10.2.1 Cell Blebbing and Morphological Changes During Chemically-Induced Injury

Cells can actively extrude and retract membrane blebbs as a normal physiological function as well as the result of chemically induced injury. Studies of experimentally induced membrane wounds have provided evidence that the opening and closing of ruptures or wounds in the cell plasma membrane offers a route for the release of molecules thought to be important in the maintenance and repair of cells and tissues. Cells are very adept at surviving experimentally induced breaks in their membranes that are so large that large molecules are introduced into or released from the cell's cytoplasm. Survival requires rapid resealing of plasma membrane wounds, some of which are so large as to perform microinjection. Because the breaks are rapidly resealed, the brief opening of the plasma membrane is thought to produce important changes to cytoplasmic constituents. For this reason, leakage of lactate dehydrogenase is often used as a measure of cellular integrity and has even been equated with cell death. Because the lipid molecules are held together by non-covalent interactions with one another and by their insolubility in the surrounding water, the bilayer structure is thermodynamically the most favorable structure. Thus, any loss of membrane integrity results in a rapid and spontaneous reversion to a bilayer. In addition, active cellular processes may also be involved in the formation of a dense network of filaments in the cytoplasm directly beneath the rupture in the membrane. Dysfunction in the formation of these protein filaments, which are composed of actin, a normal constituent in the cell's internal skeletal structure, has been linked to chemically induced injury. The network of newly formed actin filaments is very dynamic and closure of a wound leads to reversal of the actin network to pre-wound levels of actin in the vicinity of the wound.

All models of cell injury include possible roles for calcium ions beyond calcium serving as a second messenger. In general, free calcium levels are considered to be a major factor in both injury and death mechanisms as cellular integrity and various cellular components are being altered. Whether calcium ions are important initiators of cell injury is uncertain and the models differ considerably on this point. Assuming that chemically induced changes in calcium sequestration may be an early event, then the potential for alteration of calcium ion homeostasis to contribute to cell injury is very great. Almost every aspect of cell function is associated in some manner with the status and functions of calcium ions. Again, what is much less certain is the role of the various major compartments for calcium ion sequestration in cell injury. For example, with normal physiological conditions, liver mitochondria do not appear to have a major role in the control of total intracellular calcium ion concentrations. However, under pathological conditions, the loss of calcium storage ability by other sites in the cell will cause the mitochondria to become highly involved in the uptake of calcium ions and may even initiate calcium recycling. Whether calcium ion recycling occurs or not, the ability of mitochondria to take up calcium ions is enormous and has the potential of major alteration in not only inner membrane potential but also the availability of a protonmotive force for ATP synthesis, ion homeostasis, and other functions. Of course, loss of cellular integrity has the potential of allowing even greater amounts of calcium to be available from exogenous sources and can eliminate the ability of the cell to cope with the total amount of calcium ion that must be regulated by sequestration. Consideration of any mechanism for cell injury or death due to exposure to chemicals should include questions about the possible role(s) of calcium.

Calcium ion is a membrane component that is critical to the closure of ruptures in plasma membranes. It is hypothesized that, because calcium regulates the ability of cytoskeletal proteins to contract, the presence of calcium in the extracellular environment is essential for closure of membrane wounds. In the absence of extracellular calcium, frogs' eggs fail to repair wounds and the cell loses a large amount of cytoplasmic contents prior to death. Thus, calcium ions may assist the process of contraction of the cytoskeletal proteins involved for the purpose.

Blebbing of the cell surface has been described as an early consequence of hypoxic and toxic injury to cells. Because a rise in cytosolic free calcium has been suggested as the stimulus for bleb formation and the final common pathway to irreversible cell injury, individual hepatocytes in culture have been monitored for "chemical hypoxia" induced by cyanide and iodoacetate. During chemical hypoxia, free calcium in the cytosol compartment does not change with bleb formation or before loss of cell viability. Necrotic cell death appears to be precipitated by a sudden breakdown of the permeability of the plasma membrane barrier as would occur by rupture of a surface bleb. Accompanying bleb formation is a fall of the mitochondrial membrane potential. In contrast to chemical hypoxia, there is a persistence of the mitochondrial membrane potential during anoxia in suspensions of isolated hepatocytes, which may be essential for the maintenance of calcium homeostasis. An important question remaining is whether there is a "trigger" in calcium homeostasis that determines when the mitochondria will begin to become actively involved in regulation of intracellular calcium. Elucidation of the pathological conditions that can initiate similar responses may be central to resolving the nature of the proximate mechanism of cell injury and death.

Changes in cell morphology and in the cytoskeleton occur during cell injury when that injury is related to oxidative stress. An example of such injury is that generated by stimulated polymorphonuclear leukocytes in areas of inflammation associated with tissue injury. These cells produce superoxide anion radicals and hydrogen peroxide as major oxidants and cause cell lysis within a few hours after exposure. Oxidant production causes (1) oxidation of glutathione (GSH); (2) loss of NAD concomitant with activation of poly-ADP ribose polymerase, and single-stranded breaks in DNA; (3) loss of cellular ATP; and (4) elevation of intracellular free calcium. Cells that are injured and are dying display morphological changes that include swelling of the volume of cytoplasm and blebbing of the plasma membrane (Table 10.1). Membrane blebbing is associated with alteration in the redox state of GSH and intracellular homeostasis of calcium, which perturbs normal cytoskeletal organization. Because such alterations can be caused in hepatocytes as well as in other cell types by two well-known cytoskeletal toxins, cytochalasin B and phalloidin, considerable effort is being focused on determination of the mechanisms of such cell injury. A variety of chemicals, including alkylating agents, induce the formation of blebbing in different cell types. One mechanism implicates redistribution of cellular filamentous (F) actin in cells with blebs and changes in the G-actin/F-actin ratio. These effects include considerable swelling of mitochondria and subcellular organelles within 2 hours of oxidant injury. Side-to-side aggregation of F-actin bundles (microfilaments) develop during this time. The injury also produces a marked increase in F-actin, associated rearrangement of the microfilaments, and simultaneous changes in the plasma membrane prior to cell death. Thus, cytoskeletal changes during oxidant injury may have considerable importance for both the organization of subcellular organelles as well as the plasma membrane.

10.2.2 Mitochondrial Permeability Transition—A Membrane Event

Inner membrane of isolated mitochondria from a variety of tissues, which is normally impermeable to solutes, can become permeable. Although the mechanism(s) how this occurs remains controversial, this process is frequently referred to as the mitochondrial membrane permeability transition (MPT). MPT is characterized as a nonspecific and as a specific Ca^{2+}-dependent inner membrane pore that presents itself in mitochondria following treatment with Ca^{2+} and a second agent, termed an inducing agent. Many inducing agents have been identified and vary greatly in both structure and function. It is thought that these agents, in the presence of Ca^{2+}, act through a common mechanism during MPT. Examples of inducing agents include inorganic phosphate, cytosolic factor, fatty acids, heavy metals, organic sulfhydryl reagents, and oxidants such as t-butyl hydroperoxide.

Several identifiable events of MPT that have been characterized include inner membrane permeability to small ions and solutes with molecular weights <1500 Da, large amplitude swelling, loss of coupled functions, and sensitivity to cyclosporin A (CsA). In addition, loss of matrix proteins via the inner membrane pore has also been reported. During MPT, Ca^{2+} as well as other ions, are rapidly released from the mitochondrial matrix, presumably via the inner membrane pore. Following inner membrane permeability and the release of matrix solutes, a colloidal osmotic pressure arises that is slow to equilibrate in the mitochondrial matrix due to the high concentration of proteins. In order to correct the osmotic imbalance, the entrance of H_2O results in massive swelling of the mitochondria.

Mitochondrial swelling under these conditions is termed large amplitude swelling. Although secondary to inner membrane permeability and solute release, large amplitude swelling occurs within a short time (3–10 min) and is easily detected by monitoring light scattering by mitochondrial suspensions at 540 nm. Monitoring of mitochondrial swelling is a simple assay and is often utilized as an indicator of MPT. Different experimental conditions result, however, in different patterns and times of swelling, some with shorter or longer lag periods. Treatment of isolated mitochondria with CsA promotes retention of accumulated Ca^2. CsA is a potent inhibitor of the MPT when the inducing agents inorganic phosphate or t-butyl hydroperoxide were added in the presence of Ca^{2+}. When CsA was tested as an inhibitor of the inner membrane pore in the presence of several different inducing agents it was found that very low concentrations of CsA protected against MPT. These studies demonstrated the potency of CsA in preventing the MPT, as inhibition of inner membrane permeability is observed with concentrations as low as 100 pmol CsA/mg mitochondrial protein.

It has been demonstrated that the mechanism of pore formation involves the adenine nucleotide translocase as the pore structure. The "m" conformation of this inner membrane protein is modified to the "c" conformation by the binding of negative effectors such as Ca^{2+}, cyclophilin, and inducing agents resulting in a protein conformation that can no longer function as an adenine nucleotide translocase but as a nonspecific pore. Atractyloside and bongkrekic acid are known modifiers of the translocase and have been helpful in proposing this mechanism. This study also demonstrates that both ATP and ADP are positive effectors of the putative MPT pore.

The present understanding of MPT is that the pore structure is an allosteric inner membrane protein, probably the adenine nucleotide translocase, with several different regulatory sites affected by Ca^{2+}, Pi, oxidants, sulfhydryl reagents, heavy metals, cyclophilin, ADP, and ATP. Depending upon which sites are modified may determine whether MPT occurs. It is hypothesized that inhibition of MPT with CsA results from the binding of cyclophilin, a matrix protein cis-trans isomerase thought to be involved with protein folding. Whether the process of MPT is physiological remains an intriguing question but evidence is growing that it is involved in apoptotic cell death.

The oxidation-reduction state of vicinal thiols has a role in the opening of the MPT pore. Oxidants including menadione, diamide, arsenite, and tert-butyl hydroperoxide can cause cell death by one or more mechanisms that are responsive to the effects of CsA to prevent such cell death presumably via mitochondrial MPT. These workers have concluded that oxidants cause vicinal thiols to form a protein disulfide to enhance pore opening, which can be reversed by thiols such as DTT. The recruitment of mitochondrial cyclophilin to the mitochondrial inner membrane under conditions of oxidative stress that enhance the opening of a Ca^{2+}-sensitive nonspecific channel provides strong evidence that the adenine nucleotide translocase is a component of the permeability pore. Evidence for the participation of MPT in loss of cell viability during exposure to toxic agents is numerous.

Several methods are available to detect/measure mitochondrial MPT in isolated mitochondria. One method that exploits two aspects of the MPT: (1) the entry of radiolabled sucrose into the mitochondrial matrix, a normally impermeable solute to the mitochondrial inner membrane and (2) the reversible nature of the pore. By adding a chelator of Ca^{2+}, such as EDTA, or CsA, a potent pore inhibitor, the inner membrane nonspecific pore can be closed. This allows for the entrapment of sucrose into the matrix which can then be quantified. Release of added Ca^{2+} or other ions from mitochondria and its inhibition by CsA are also indicative of MPT. As previously discussed, loss of absorbance at 540 nm of mitochondrial suspensions is frequently used as an indicator of MPT in isolated mitochondria. However, because large amplitude swelling is a secondary event that does not always accompany MPT, monitoring of solute movement may be a more sensitive parameter of permeability changes. It is important to point out that recent studies strongly indicate that the MPT participates in the mechanism of toxicity of many chemicals as the collape of the inner mitochondrial membrane potential causes the loss of ATP generation and calcium homeostasis.

10.3 MECHANISMS OF CELL DEATH

These pathways are being elucidated at a rapid pace and have been a major component in development of new chemotherapeutic drugs for cancer treatment. In addition, we now know that each cell has the so-called "death" genes that may participate in chemically induced cell death. Necrotic death has been described as a toxic process in which the cell is not a targeted cell but is a passive victim that does not sustain the energy level for an orderly process of cell destruction and assimilation. In contrast, apoptosis or cell suicide is a regulated stepwise process during which the targeted cell takes an active role in its own orderly shut-down, packaging and dismantling (Fig. 10.1).

A major contribution to this increase in complexity is that receptor-mediated effects of chemicals are an important feature of apoptotic cell death and have led to a new area of scientific inquiry, namely, signal transduction pathways. Characteristic features of apoptosis include cell shrinking, surface blebbing, and chromatin condensation (Table 10.1).

Although many aspects are understood, new information concerning both the stepwise nature and regulation of cell death is being reported at a very high rate and probably will continue to do so for some time, for example, the continuous generation of reactive oxygen species within the mitochondria, the recent finding of direct involvement of mitochondria in one of the "stages" of apoptotic cell death, the release of cytochrome c, a mitochondrial electron transport protein. Subsequent binding and activation by cytochrome c of one of more than a dozen members of the caspase family of proteolytic enzymes capable of translocation and disrupting the structure and function of nuclear DNA are now thought to be important aspects of one type of apoptotic cell death and possibly having some role in necrotic cell death. Another type of apoptosis appears to occur independent of mitochondrial involvement.

The following scheme for receptor and chemically induced apoptosis involves the triggering of death receptors in distinct caspase cascades. It has been proposed that caspases form an integral part of the cell death-inducing mechanism in receptor-mediated apoptosis, whereas in chemically induced apoptosis they act solely as executioners of apoptosis (Fig. 10.2). Apoptotic cell death has been described as occurring in three phases: first a receptor-mediated commitment to cell death, which is termed an induction phase, followed by an effector phase characterized by distinct morphological changes in cell structure with modification of mitochondrial functions including a change in the status of ATP and other intracellular constituents, and finally a degradation phase with extensive break up of the targeted cell (Fig. 10.2).

These changes include condensation and fragmentation of nuclear chromatin, shrinking cytoplasmic organelles that remain intact, swelling of endoplasmic reticulum accompanied by a decrease in cell volume, and plasma membrane alterations that permit recognition and phagocytosis of apoptotic cells without an inflammatory response. The absence of inflammation is thought to be due to integrity of the subcellular vesicles that undergo phagocytosis without release of chemotactic factors. The hallmark of apoptotic cell death is the nuclear alterations are often characterized by internucleosomal cleavage of DNA that are recognized as a "DNA ladder" on gel electrophoresis.

SEQUENCE OF ULTRASTRUCTURAL CHANGES IN APOPTOSIS (RIGHT) AND NECROSIS (LEFT)

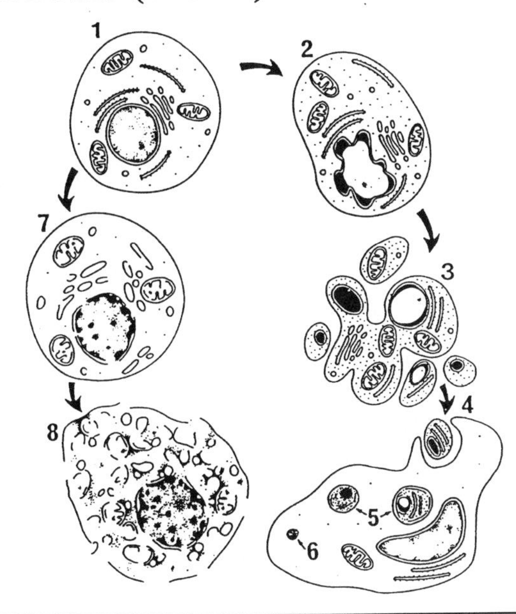

(1) normal cell

APOPTOSIS

(2) earliest recognizable stage of apoptosis

(3) compaction and margination of nuclear chromatin, condensation of cytoplasm and variable convolution of nuclear and cell outlines

(4) nuclear fragmentation and budding of the cell as a whole to produce membrane-bounded apoptotic bodies

(5) phagocytosis by nearby cells and degradation within lysosomes

(6) lysosomal residual bodies

NECROSIS

(7) irregular clumping of chromatin, swelling of all cytoplasmic compartments, the appearance of densities in the matrix of mitochondria and focal disruption of membranes

(8) cell degeneration to form a mass of debris

KERR (1993)

FIGURE 10.1 Scheme illustrating phases of morphological changes in apoptosis and necrosis.

Source: Adapted from Kerr (1993).

TABLE 10.1 Morphological Characteristics of Cell Death by Apoptosis or Necrosis.

Apoptosis	Necrosis
Death of individual cells	Contiguous cells are involved
Cytoplasm and cell volume decrease	Cytoplasm and cell volume increase
Chromatin compacted and segregate	Chromatin marginates as small aggregates
DNA fragments and nucleosome ladders form	Organelles swell (mitochondria, etc.)
Cell vesicles form with membrane integrity	Cell plasma membrane blebbs and ruptures
Cell vesicles are phagocytized	Cell contents lost and triggers chemotatics

Chemically induced necrotic cell death is characterized by cell swelling, loss of plasma membrane integrity and subsequent loss of cellular integrity resulting in recruitment of cells that cause inflammation and phagocytes cells and cellular fragments. Nuclear DNA changes in necrotic cell death occur but in general do not have the same ultrastructural characteristics as apoptotic cell death.

Although apoptosis and necrosis are considered distinct forms of cell death, it has been suggested that classical apoptosis and necrosis represent only the extreme of a wide range of possible cell deaths. Many of the same biochemical intermediates, most notably by the levels of cellular ATP, Ca^{2+}, reactive oxygen species and thiol antioxidants, regulate these processes. A key observation is that intracellular energy levels and mitochondrial functions are rapidly compromised in necrosis but not in apoptosis. Even though mitochondria are essential in one type of apoptosis, intracellular ATP has been proposed as a switch in the decision between apoptosis and necrosis. Another key finding is that oxidant-induced caspase inactivation can lead to necrosis. Thus, the oxidant, menadione, is thought to inhibit caspase activity by the generation of hydrogen peroxide through redox cycling, preventing cell death by apoptosis resulting in necrosis. A dual role for reactive oxygen species in apoptosis has been proposed; induction and inhibition of caspases depending on the extent of oxidative stress with induction caused by mild oxidative stress and prolonged or excessive oxidative stress leading to inhibition of apoptosis and induction of necrosis. A main question that remains unanswered is whether the central executioners of apoptosis are caspases or mitochondria. Now we know that this may depend on the type of apoptosis that is initiated and involves the cell's mitochondria.

The concept of dual roles for cytochrome c in oxidative phosphorylation and activation of caspases has been described as exaptation. Exaptation in evolutionary terms is the utilization of a molecule, which has a defined purpose, in a second role that has an entirely different purpose. This is in contrast to adaptation, which is thought of as a seemingly natural extension of preexisting functions. Now, in addition to cytochrome c, other molecules have been found to lead a double life associated with apoptosis. For example, human tyrosyl-transfer RNA synthethase, the enzyme that catalyzes covalent attachment of the amino acid tyrosine to the corresponding tRNA molecule prior to protein synthesis, upon proteolytic cleavage is converted into not one but two cytokines. Recruitment of tyrosyl-tRNA synthetase after secretion makes it an extracellular death messenger in contrast to cytochrome c, which is an intracellular death messenger. Normally residing in the cytoplasm,

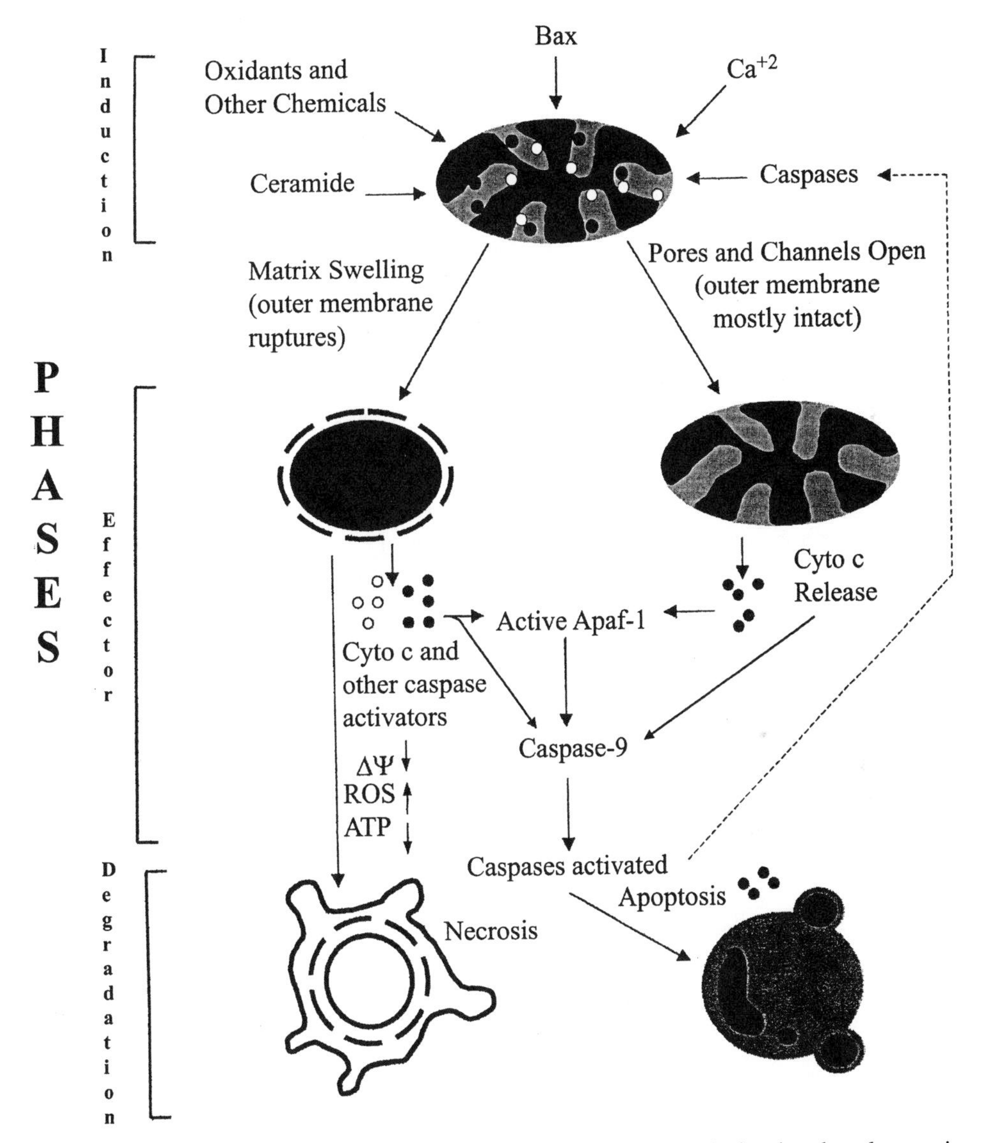

FIGURE 10.2 Signal transduction pathways leading to cell death that involve changes in mitochondrial structure and function.

tyrosyl-tRNA synthetase is secreted intact in a selective manner and then undergoes cleavage by extracellular proteolytic activity giving rise to cytokines that participate in apoptosis with IL-8 and EMAPII-like characteristics. The cytokines provide a link between protein synthesis and signal transduction. Acceleration of apoptosis may arise from the secretion of an essential component of the translational apparatus as an early event. This finding leads to the suggestion that there are more molecules with exaptation features that participate in apoptosis.

There is evidence that many cancers arise as a result of failure of abnormal cells to undergo apoptosis and that many chemical agents used for cancer therapy have been shown to kill cells by inducing apoptosis. Earlier work had focused on chemotherapeutic agents inducing necrotic cell death but today much more effort is being directed toward elucidation of both necrotic and apoptotic features of drug-induced cell death.

10.3.1 Mechanism of Chemically Induced Apoptotic Cell Death

Apoptosis, or programmed cell death, is a mode of cell death with a defined set of characteristics that distinguish apoptotic cells from cells dying by other means: shrinkage, membrane blebbing, chromatin condensation, internucleosomal DNA cleavage, cytoplasmic compaction with preservation of mitochondrial structure, and cell fragmentation. Protease activation during apoptosis causes events that are better described as a collapse of the cell. As an analogy, loss of intracellular structure has been compared to the manner in which a large tent collapses if lines are cut and the tent remains erect until a sudden crashing down of the structure. One of the main questions is whether this sudden collapse is closely associated with the loss of mitochondria structure and function—loss of energy production, protein synthesis, and the status of cytoplasmic and/or mitochondrial glutathione. Therefore, the sequence of mitochondrial events associated with death of hepatocytes that have different concentrations of glutathione due to exposure to chemicals that undergo bioactivation could be the key to chemically induced cell death.

Because the thiol status of cells is closely associated with bioactivation of chemicals, Ca^{2+} homeostasis, and reactive oxygen species (ROS) generation, it is important to understand better the role of chemical exposure to the possible effects of glutathione modulation on MPT and the sequential reduction of membrane potential ($\Delta\psi m$) and generation of reaction oxygen species in cell death. A sequential disregulation of mitochondrial function precedes cell shrinkage and nuclear fragmentation. This disregulation consists of an initial CsA-inhibitable step of ongoing apoptosis that is characterized as a reduction of $\Delta\psi m$ followed by generation of ROS upon a short-term in vitro culture. ROS generation is inhibitable by rotenone and ruthenium red. Inhibitors of MPT have been shown to interfere with the disruption of the $\Delta\psi m$ during apoptosis. As evidence, CsA, N-methyl-Val-4-CsA, and bongkrekic acid have been shown to inhibit both the disruption of MPT-mediated-$\Delta\psi m$ and prevent apoptotic chromatinolysis. The self-amplificatory properties of MPT are expected to be enhanced by a decrease in glutathione levels. Experimental evidence exists to support the suggestion that a sequential mitochondrial dysfunction expressed as loss of $\Delta\psi m$ followed by enhanced superoxide generation is implicated in apoptosis (Fig. 10.2). It is known that MPT has self-amplificatory properties including loss of matrix Ca^{2+}, glutathione, depolarization of the inner membrane and increased oxidation of thiols that are contributors to a point of no return in the apoptosis process.

Alterations of mitochondrial functions could be reflected in changes in the mitochondrial glutathione status, especially during the ROS generation phase. It is important to point out that measurement of total glutathione does not reflect change in glutathione status in the mitochondria because mitochondria are expected to contain only about 10% of the total intracellular glutathione. Because of the importance of the status of mitochondrial glutathione in hepatocytes, the depletion of hepatocyte GSH with DEM is of interest. It was important to know the amount of GSH that could be depleted prior to cell injury. Diethyl maleate under the conditions used is known not to deplete hepatocyte mitochondrial GSH. Rat hepatocytes were depleted of 50% of their total GSH by 0.4 mM diethyl maleate in 135 min without observable injury. Additional GSH depletion initiated cell injury that increased with increasing depletion of GSH. These findings suggest that less than 50% of the total GSH was required to prevent cell injury during in vitro incubation.

Mitochondrial Ca^{2+} transport may have evolved to protect the cytosol against damage from high Ca^{2+} concentration. Questions about the dynamics of thiol homeostasis, especially with respect to glutathione, the rate of energy consumption for cellular processes related to such homeostasis during pathological conditions, and the fate of glutathione during MPT and apoptosis are major areas of interest as they relate very closely to many human disease.

Unfortunately, many difficulties remain concerning distinguishing the differences between apoptotic and necrotic cell death. The actual time of death of a cell is difficult to assess, but, in general terms, it can include death being at the time that loss of cellular integrity occurred to such a degree that free exchange between the intracellular constituents and the surrounding milieu prevents cell survival or by dismemberment of the cell into intact vesicules that undergo phagocytosis. Thus, it is appropriate to continue to seek to understand better those events that are characteristic of either necrotic or apoptotic or both types of cell death.

10.3.2 Bioactivation and Covalent Binding of Biological Reactive Intermediates

Bioactivation, the consequence of metabolism of a chemical to a biological reactive intermediate, may occur in a metabolic pathway that ranges from a minor metabolic pathway to a major one when compared to the overall metabolism of a specific chemical. Biological reactive intermediates are electrophiles (electron-deficient molecules), also called alkylating agents, that react with cellular nucleophiles (electron-rich molecules), which generally results in a detoxifying event. However, because toxic injury also occurs via formation of biological reactive intermediates, a major focus of this chapter is to examine these processes in some detail as to when they lead to toxic injury rather than detoxication. The bioactivation and metabolism of many classes of xenobiotics have been well documented and reviewed. A few chemicals within these classes, which have been identified as carcinogens or cancer-causing chemicals, are known to interact with cellular constituents, particularly DNA, leading to cell transformation. The cytochrome P450 enzymes, of which more than 700 have been characterized, play pivotal roles in xenobiotic bioactivation and metabolism.

Cellular injury processes have reversible and irreversible features that are a consequence of the interplay of bioactivation and oxidative stress with the cellular protective systems that prevent or limit cellular injury and possibly repair. Reversible injury occurs even with chemicals known as "safe" chemicals. Dose is the determining factor of whether any chemical causes irreversible injury. Such irreversible injury may cause cell death regardless of what antidotal or preventive measures are

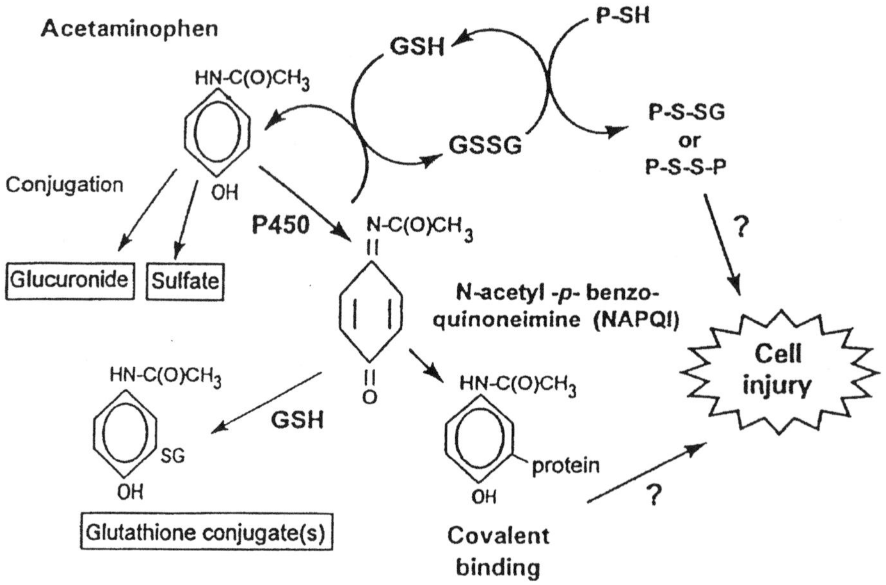

FIGURE 10.3 Structures of acetaminophen and *N*-acetyl-p-benzoquinone imine (NAPQI), metabolites and events leading to cell injury.
Source: Adapted from Cohen et al., 1998.

taken after exposure of cell or tissues to a sufficient dose of the toxic chemical. We review here the pathways by which chemicals are metabolized to proximate toxins, the mechanisms by which biological reactive intermediates are toxic to and damage cells, and mechanisms by which cells are protected by preventing or limiting cellular damage. The dynamics of in vivo chemical or xenobiotic metabolism, for example, competition between detoxication pathways, pool sizes, and rates of conjugate formation, are not well understood for most xenobiotics and drugs (Fig. 10.3). As might be expected, the viability and survival of cells that are exposed to toxic xenobiotics depend on the status and maintenance of such detoxication pathways that generally are "built-in" protective systems. Chemically induced cell injury can occur because vital constituents of systems for protection are depleted. Even molecular oxygen, during the course of normal physiologic functions, is transformed to some extent into chemically reactive intermediates including superoxide radical, hydroxyl radical, and hydrogen peroxide, which contribute to the depletion of components of cellular protective systems. GSH, a unique tripeptide and major protective cellular constituent, has a central role to provide antioxidant reducing equivalents and also serves as a substrate for formation of GSH conjugates that are excreted in the urine as mercapturic acids after further metabolism of many xenobiotics.

It should be noted that certain chemicals, including some cancer chemotherapeutic agents such as nitrogen mustards, are so chemically reactive that they are direct-acting electrophiles and therefore do not require bioactivation for either pharmacologic action or toxicity. Of course, because these agents are so very reactive chemically, they have the ability to challenge and defeat the cellular protective systems, especially the GSH-dependent systems. Depletion of protective constituents of the cell can result in extensive damage to both normal and cancer cells. For example, often a cancer patient undergoing chemotherapy with a direct-acting alkylating agent as a drug loses most if not all of his or her hair due to cellular damage of the hair follicles, which generally undergo repair and replacement subsequent to such cancer treatments.

Both the extent and the molecular targets of covalent binding by biological reactive intermediates vary with the properties of the intermediates, especially their chemical reactivity and, to some extent, sites of formation. The liver, because it is a primary site of xenobiotic metabolism, is also a major source of biological reactive intermediates derived from xenobiotics. The half-life of a biological reactive intermediate determines, in part, the site of the molecular targets with which it can react. Some biological reactive intermediates possess sufficient stability to be transported throughout the body. Also, there are threshold doses for toxicity of certain chemicals that relate to the rate of formation of biological reactive intermediates. Above such a threshold level, covalent binding of a biological reactive intermediate generally increases greatly for relatively small increases in dose of the chemical. The concept of a threshold dose, which is very important for safe use of drugs, has been considered in detail for many drugs and now is being considered for limitation of exposures to many xenobiotics such as dioxin, a potent inducer of drug-metabolizing enzymes and a carcinogen.

A noncarcinogenic chemical, bromobenzene, displays a dose response for toxicity that illustrates the threshold for an observable toxic effect. The rate of bromobenzene metabolism is dependent on the level of specific cytochrome P450s. The concentration of GSH in liver markedly decreases after the administration of toxic doses of bromobenzene, especially after induction of cytochrome P450s by prior administration of phenobarbital. The basis for this depletion is the conjugation of a biological reactive intermediate, an epoxide of bromobenzene, with GSH. Eventually, the rate of formation of the conjugate of the reactive epoxide becomes limited by the availability of GSH, which in turn is limited by its rate of synthesis. Therefore, the rate of reaction between biological reactive intermediates of bromobenzene and macromolecules in the liver increases markedly to a level that causes extensive GSH depletion. Even the rate of biosynthesis of GSH is not adequate for conjugation and cellular protection. Obviously, many factors can contribute to the rate of biosynthesis of GSH, including fasting or starvation, and can make a substantial difference in the tolerance to certain chemicals. Thus, both the rate of replenishment of GSH, as well as the rate of depletion, are critical factors in acute hepatotoxicity by bromobenzene. The enhancement of hepatotoxicity by depletion of GSH with diethyl maleate has been noted during the metabolism of bromobenzene. When hepatocytes were pretreated with diethyl maleate, which reduced intracellular levels of GSH by about 70%, added bromobenzene caused levels of GSH to fall to 5% of initial levels, and cell death (75% by 5 hr) was noted. Addition of cysteine, methionine, or N-acetyl-cysteine prevented bromobenzene-induced toxicity, but did not prevent depletion of GSH. Bromobenzene, in the presence of a cysteine source, reduced initial levels of GSH to about 40% of control. The presence of metyrapone, an inhibitor of cytochrome P450-dependent monooxygenase reactions, eliminated bromobenzene-induced toxicity. These data are consistent with a requirement for bromobenzene activation prior to GSH conjugation.

Acetaminophen (paracetamol, 4-hydroxyacetanilide), a commonly used analgesic drug, causes centrilobular hepatic necrosis upon overdosage. The drug mainly undergoes sulfation and glucuronidation in most species at normal dose levels (Fig. 10.3). At dose levels that begin to saturate the pathways of sulfation and glucuronidation, however, bioactivation occurs to an increasing extent to form the reactive intermediate *N*-acetyl-p-benzoquinone imine (NAPQI) (Fig. 10.3). As the normal conjugation pathways of sulfation and glucuronidation become saturated due to the overdose, the hepatic cytochrome P450 systems catalyze the bioactivation of acetaminophen to an increasing extent. NAPQI is a highly reactive compound. It reacts with cell components if not detoxified by conjugation with glutathione, which depletes glutathione and diminishes the efficiency of cytoprotective mechanisms. Thus, overdosage with acetaminophen causes depletion of cellular glutathione followed by oxidation and/or alkylation (covalent binding) of cysteinyl residues in cellular proteins. NAPQI undergoes a nonenzymatic two-electron reduction in the presence of glutathione to yield stoichiometric amounts of acetaminophen and GSSG. BCNU (1,3-bis-(2-chloroethyl)-1-nitrosourea)-induced inactivation of glutathione reductase prevents the reduction of NAPQI-generated GSSG and increases cytotoxicity but has no effect on the covalent binding of NAPQI to cellular proteins. A competing reaction at physiological pH is the formation of an acetaminophen-GSH conjugate. Incubation of NAPQI with hepatocytes yields the same reaction products in control and in BCNU-treated hepatocytes. The thiol, dithiothreitol, protects against cytotoxicity even after covalent binding has occurred. It has been speculated that the toxicity of NAPQI for isolated hepatocytes may result primarily from its oxidative effects on cellular proteins rather than the formation of acetaminophen-GSH conjugate.

Another possible mechanism of acetaminophen-induced cell injury is oxidative stress. NAPQI-induced cytotoxicity is accompanied by oxidation of thiol groups in proteins, and cytotoxicity of NAPQI can be prevented by a reducing agent for disulfides, for example, dithiothreitol (DTT). There is uncertainty, however, about the relative contributions of oxidative stress and injury mediated by covalent binding of NAPQI during overdosage of acetaminophen (Fig. 10.4). Recent studies have implicated the influence of acetaminophen on cell replacement and organ repair by the modulation of signal transduction and inhibition of entrance of cells into the cell cycle. Acetaminophen inhibition of passage of cells through G1 and S phases has been proposed to interfere with organ regeneration and exacerbate acute liver injury damage caused by acetaminophen bioactivation products including NAPQI.

Large doses of morphine in rats also deplete hepatic GSH and are associated with threefold elevation in activities of glutamic oxaloacetic transaminase (SGOT) and glutamic pyruvic transaminase (SGPT) in serum. Prior treatment of the rats with phenobarbital, which increases production of certain cytochrome P450 enzymes, enhances the morphine-elicited rise in SGOT and SGPT. The bioactivation of morphine involves a cytochrome P450-dependent oxidation at the benzylic C-10 position to form an electrophilic species that can react with nucleophilic thiols such as glutathione. Thus, metabolism of morphine can deplete thiols and cause liver damage. Overdosage or use of morphine in combination with other drugs that

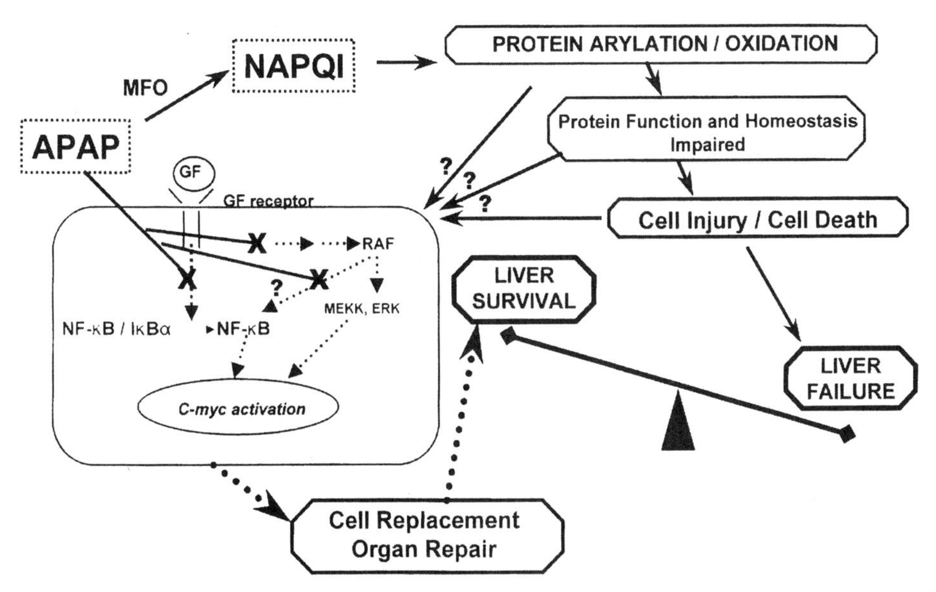

FIGURE 10.4 Proposed mechanisms for the hepatotoxicity induced by acetaminophen and its reactive metabolite, NAPQI via protein arylation and receptor modulation.

Source: Adapted from Boulares et al., 1999.

require glutathione for detoxification will therefore increase the risk of toxicity due to either agent. Alkylation of critical cellular targets by the metabolic intermediate seems to be important in the hepatotoxicity of morphine.

Furosemide, a diuretic, shows negligible covalent binding and little nephro- or hepatotoxicity below a threshold dose of about 150 mg/kg. Metabolism of furosemide dose not deplete the liver of GSH nor apparently of any other protective nucleophiles. These results suggest that the relative reactivity of nucleophiles in the cell, including GSH and thiol groups of proteins, is important in the unusually selective covalent binding of furosemide to proteins.

Because of their chemical reactivity, biological reactive intermediates can bind covalently to cellular proteins including those present in the cytoplasm, endoplasmic reticulum, nucleus, as well as nucleic acids. Such covalent binding to DNA is used as a quantitative indicator of genotoxicity of many xenobiotics. Because there is only a general understanding of how adduct formation alters the function of specific proteins, the relationship between covalent binding to proteins and the cytotoxic properties of xenobiotics remains uncertain, Also, little attention has been given to the significance of damage to either proteins or DNA as related to acute cellular toxicity.

10.4 OXIDATIVE STRESS

A major concept developed during the past 10 years is that a mechanism for chemical-induced cell death could be based upon oxidative stress. This advance in our knowledge is in part due to the many new findings that have been reported

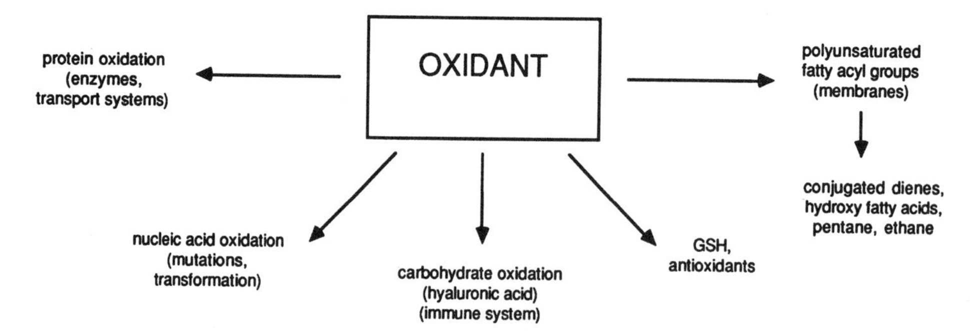

FIGURE 10.5 Intracellular glutathione homeostasis.

about reactive oxygen intermediates (ROIs) and cell defenses. Targets of oxidants have been identified and intracellular events described in detail. Inflammation that is based on cytokines and stress responses has added additional dimensions to various types of liver injury, nonhepatic diseases, immune function, and HIV. A major discovery has been the oxidative events associated with ischemia and reperfusion. Liver diseases including alcoholic liver disease, metal storage diseases, and chloestatic liver disease have been described in terms of an oxidative stress component. The occurrence of reactive oxygen species in normal metabolism of mammalian organs was first demonstrated in the liver. Rates of hydrogen peroxide production of 50–100 nmol/min/g liver under normal conditions could be enhanced up to sevenfold in the presence of substrates for peroxisomal oxidases. Subsequently, it was shown that oxidative stress-induced hepatocyte injury may accompany metabolism of chemicals to biological reactive intermediates. From these studies came the concept that a misbalance in the prooxidant/antioxidant steady state, potentially leading to damage, is termed "oxidative stress" (Fig. 10.5).

The loss of control of endogenous oxidative events related to the use of molecular oxygen by the cell is the major factor in oxidative stress injury. Such a process, which is known as chemical-induced oxidative stress, may occur to an extent that ranges from a minor to a major contribution to overall toxicity. For example, chemicals that are known to undergo redox cycling cause exogenous oxidative stress to such a degree that they play a major role in chemically induced cell injury. In addition, some of these chemicals are known to form adducts with cellular constituents, particularly proteins that are necessary for protection against oxygen toxicity. However, as indicated above, all tissues and cells contain systems for detoxification of biological reactive intermediates and antioxidants and antioxidant enzyme systems to prevent or limit cellular damage due to oxidative stress events.

Normal fluxes of proxidants serve useful functions at the cellular and whole organism levels and therefore need to be balanced with sufficient antioxidant defense to maintain a biological steady state. For example, the continual formation of reactive oxygen species is a physiological necessity and an unavoidable consequence of oxygen metabolism. When reactive oxygen species are generated in excess, they can be toxic, particularly in the presence of transition metal ions such as iron or copper. However, defense systems are present and functioning under normal conditions, Therefore, endogenous free radicals do not necessarily place biological tissues and cells at risk. Chemical exposures can cause these defense systems to be overwhelmed during various pathological conditions including, anoxia, radiation, and loss of control of both intracellular and extracellular constituents such as calcium.

It is estimated that nearly 90% of the total O_2 consumed by mammalian species is delivered to mitochondria where a four electron reduction to H_2O by the respiratory chain is coupled to ATP synthesis. Nearly 4% of mitochondrial O_2 is incompletely reduced by leakage of electrons along the respiratory chain, especially at ubiquinone, forming ROS such as superoxide (O_2), hydrogen peroxide (H_2O_2), singlet oxygen, and hydroxyl radical ($HO^\bullet$). It has been calculated that during normal metabolism, one rat liver mitochondrion produces 3×10^7 superoxide radicals per day. It is estimated that superoxide and hydrogen peroxide steady-state concentrations are in the picomolar and nanomolar range, respectively. However, an estimated hepatocyte steady state H_2O_2 concentration may be up to 25 μM. These events are thought to contribute over 85% of the free radical production in mammalian species.

10.4.1 Aerobic Metabolism: Endogenous Oxidative Stress

If endogenously formed H_2O_2 is not detoxified to H_2O, formation of the $HO^\bullet$ radical by a metal (iron) catalyzed Haber Weiss or Fenton reaction can occur (Eq. 1A,B). The $HO^\bullet$ species is one of the most reactive and short-lived biological radicals and has the potential to initiate lipid peroxidation of biological membranes although not as effectively as other radicals including the $ROO^\bullet$ radical. Unless termination reactions occur, the process of lipid peroxidation will propagate, resulting in potentially high levels of oxidative stress. Therefore, detoxification of endogenously produced H_2O_2 is critical for redox maintenance of mitochondrial as well as cellular homeostasis.

$$\text{Haber Weiss Reaction} \quad O_2^{\bullet -} + H_2O_2 + Fe^{2+} \rightarrow Fe^{3+} + O_2 + OH^- + HO^\bullet \quad \text{Eq. 1A}$$

$$\text{Fenton Reaction} \quad H_2O_2 + Fe^{2+} \rightarrow Fe^{+3} + HO^\bullet + OH^- \quad \text{Eq. 1B}$$

Other reactive oxygen species of biological relevance include singlet oxygen ($^1O_2$), hypochlorous acid (HOCl), and nitric oxide (NO). Thus, several reactive oxygen species can be formed by one electron step reductions of molecular oxygen and spontaneous reactions with nitric oxide including the formation of the oxidant peroxynitrite (Fig. 10.6).

10.5 THE ORGANIZATION OF ANTIOXIDANT DEFENSE AGAINST OXIDATIVE STRESS

The production of oxygen free radicals was not considered to be of biological consequence until 1969, when McCord and Fridovich discovered that superoxide dismutase exists in essentially every mammalian cell, suggesting ubiquitous superoxide formation in vivo (Fig. 10.6). Later, oxygen free radicals were linked to inflammatory disease states and the imbalance between proxidants and antioxidants was

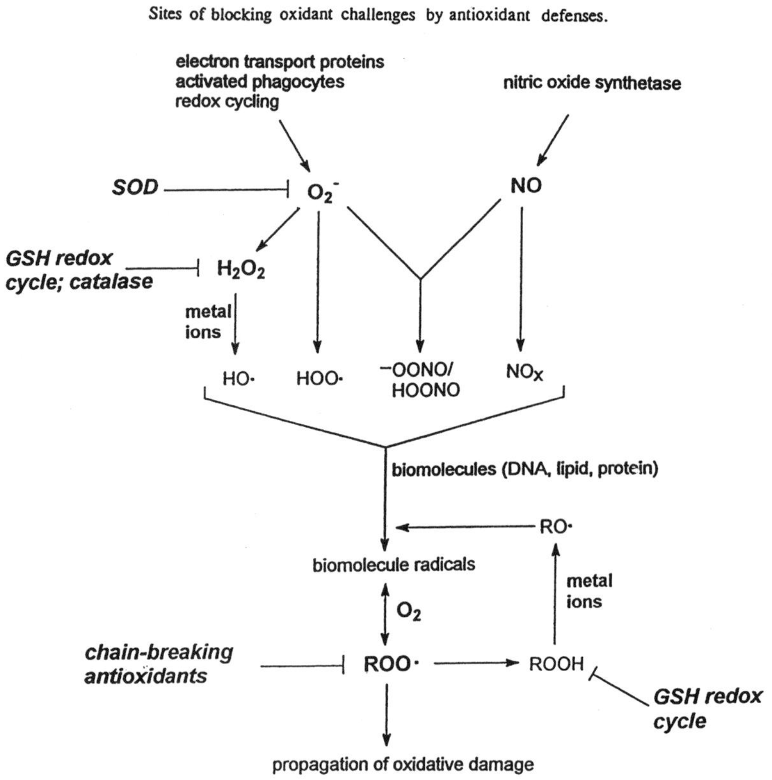

Sites of blocking oxidant challenges by antioxidant defenses.

electron transport proteins
activated phagocytes
redox cycling
nitric oxide synthetase
SOD
O_2^-
NO
GSH redox cycle; catalase
H_2O_2
metal ions
HO·
HOO·
$^-$OONO/ HOONO
NO_x
biomolecules (DNA, lipid, protein)
RO·
biomolecule radicals
metal ions
O_2
chain-breaking antioxidants
ROO·
ROOH
GSH redox cycle
propagation of oxidative damage

FIGURE 10.6 Molecular targets of oxidative injury. GSH, reduced glutathione.

defined as oxidative stress. With the more recent discovery of the second messenger functions of nitric oxide and the possible role of peroxynitrite in both toxicity and signaling, the physiological and pathological roles of nitric oxide and superoxide are being investigated vigorously as well as the defense systems that must balance between protection and providing a suitable environment for signal transduction and other biological functions based on these molecules. The evolution of bioactivation processes that form biological reactive intermediates is thought to have necessitated the concomitant evolution of cellular protection systems for cell survival in an oxygen-containing atmosphere. All tissues and cells contain systems for detoxification of biological reactive intermediates and to prevent or limit cellular damage due to oxidative stress events.

10.5.1 SOD Defense

The importance of the defense enzymes Se-glutathione peroxidase, catalase, and Cu/Zn-SOD for cell survival against oxidative stress is well established, with each enzyme having a specific as well as an irreplaceable function (Fig. 10.7). In part, the function is associated with these enzymes based on studies that have utilized specific chemicals being used as enzyme inhibitors including aminotriazole for catalase,

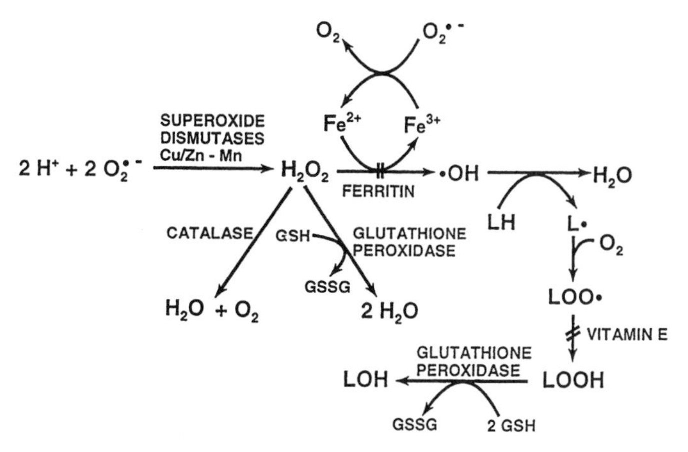

FIGURE 10.7 Origin of reactive oxygen and nitrogen species and sites of blocking their oxidant challenges by antioxidant defenses.

diethyldithiocarbamate for Cu/Zn-SOD, mercaptosuccinate for SeGpx, BCNU for glutathione reductase, and BSO for GSH synthesis. Further, such studies have as a part of the basal defense that is provided by other antioxidant enzymes including GSHP-1 and Mn-SOD. In transfection studies in which overexpression of a specific enzyme is studied, the physiological responses indicate that the metabolism of reactive oxygen species may have a critical balance. For example, Mn-SOD transfected mouse cells overexpressing SOD activity are more resistant to hyperoxia and paraquat as the levels of intracellular ROS and calcium are reduced, the mitochondrial membrane potential maintained, and apoptotic cell death prevented. Yet, SOD-enriched bacteria display increased sensitivity to hyperoxia and paraquat. SOD has an important role in defense against degenerative disease, but much remains to be understood concerning effects of manipulation of SOD expression as an intervention. An explanation for hydrogen peroxide toxicity and adaptive defense by exposure to low levels of hydrogen peroxide is that the toxicity is maximized by optimizing the concentrations of superoxide and hydrogen peroxide for additional free radical formation. Transfection experiments could therefore cause increased or decreased toxicity depending on concentrations of hydrogen peroxide and superoxide. In addition, iron (or copper) has the potential for a major influence by being optimized in its redox ratio (Fe^{2+}/Fe^{3+}) for increasing the reactivity of hydrogen peroxide. For a review see "Suggested Reading" at the end of this chapter.

10.6 ROLE OF GLUTATHIONE

All living organisms have evolved protective systems to minimize injurious events that result from bioactivation of chemicals including xenobiotics and oxidative products of cellular metabolism of molecular oxygen (Figs 10.6 and 10.7). The major protective system is dependent upon the unique tripeptide, GSH. GSH acts as both a nucleophilic "scavenger" of numerous compounds and their metabolites, via enzymatic and chemical mechanisms, converting electrophilic centers to thioether bonds, and as a cofactor in the GSH peroxidase-mediated destruction of hydroperoxides. GSH depletion to about 15–20% of total glutathione levels can impair the cell's defense against the toxic actions of such compounds and may lead to cell injury and death. Oxidative stress as a mechanism for chemical-induced hepatocyte death is based on understanding that the consumption of oxygen even during normal environmental conditions requires a considerable amount of cellular resources on a constant basis to detoxify toxic oxygen metabolites. If not reduced, these metabolites can lead to the formation of very reactive radicals including the hydroxyl radical and cause the formation of lipid hydroperoxides that can damage membranes, nucleic acids, proteins, and their functions. Failure to provide or maintain the cellular protective systems is now known to cause serious human diseases that can be greatly exacerbated by exposures to toxic chemicals.

Depending on the cell type, the intracellular concentration of glutathione is maintained in the range of 0.5–10 mM. Concentrations in the liver are 4–8 mM. Nearly all the glutathione is present as reduced GSH. Less than 5% of the total is present as glutathione disulfide (GSSG). This is so because of the redox status of GSH that is maintained by intracellular GSH reductase and NADPH (Fig. 10.7). Continual endogenous production of reduced oxygen species, including hydrogen peroxide and lipid hydroperoxides, causes constant production of some GSSG, however. The GSH content of various organs and tissues represents at least 90% of the total non-protein, low molecular weight thiols. The liver content of GSH is nearly twice that found in kidney and testes and over threefold greater than in the lung. The importance of hepatic GSH for protection against reactive intermediates has been reviewed extensively.

The unidirectional process of trans-sulfuration in which methionine sulfur and serine carbon are utilized in cysteine biosynthesis via the cystathionine pathway is essential for the maintenance of GSH stores in the body. Thus, the cystathionine pathway is of major importance to pathways of drug metabolism that involve GSH or cysteine, or both. Depletion of GSH by rapid conjugation can increase synthesis of GSH to rates as high as 2–3 μmol/hr/g wet liver tissue. The cysteine pool in the liver, which is about 0.2 μmol/g, has an estimated half-life of 2–3 min at such high rates of synthesis of GSH. Although the cystathionine pathway appears to be highly responsive to the need for cysteine biosynthesis in the liver, the organ distribution of the pathway is limited. Evidence indicates that in mammals, such as rats, the liver is the main site of cysteine biosynthesis, which occurs via the cystathionine pathway. Maintenance of high concentrations of GSH in the liver, in association with high rates of secretion into plasma and extensive extracellular degradation of GSH and GSSG, supports the concept that liver GSH is a physiological reservoir of cysteine.

The concentration of GSH in the liver is altered in rats by diurnal or circadian variations, and starvation. The diurnal variation in hepatic GSH results in the highest levels of GSH at night and early morning and the lowest levels in the late afternoon. The maximum variation is as much as 25–30%. Starvation limits the availability of methionine for synthesis of GSH in the liver, and decreases the concentration of GSH by about 50% of the level in fed animals. Assuming GSH is a physiological reservoir for plasma cysteine, efflux of GSH from liver will continue during starvation, and the released cysteine will help maintain levels of GSH in other organs including the kidney.

In vivo treatment of rats with an inhibitor of γ-glutamyl transpeptidase AT-125 (L-αS,5S)-α-amino-3-chloro-4,5-dihydro-5-isoxazoleacetic acid) prevents degradation of GSH in plasma leading to massive urinary excretion of GSH. This treatment also lowers the hepatic content of GSH because it inhibits recycling of cysteine to the liver. A physiologic decrease in interorgan recycling of cysteine to the liver for synthesis of GSH also may account in part for the decrease of hepatic GSH during starvation and for the marked diurnal variation in concentration of GSH in liver. As mentioned, the nadir occurs in the late afternoon, whereas the early morning peak occurs shortly after the animals are fed.

The efflux of liver GSH and metabolism of the resulting plasma GSH and GSSG appears to help insure a continuous supply of plasma cysteine. This cysteine pool should in turn minimize the degree of fluctuation of GSH concentrations within the various body organs and cell types that require cysteine or cystine, or both, rather than methionine for synthesis of GSH.

10.6.1 Organelle Glutathione

Compartmentation of glutathione has been demonstrated in that separate pools of glutathione exist in the cytoplasm from that in the mitochondria (Fig. 10.8). The cytosolic pool of glutathione has been characterized in terms of cellular protection (Table 10.2). A separate pool has been proposed for the nucleus, however, that finding is in dispute. Mitochondrial glutathione is a separate physiological pool of glutathione in agreement with the observation that the liver has two pools of GSH. One has a fast ((2 hr) and the other a slow (30 hr) turnover. In freshly isolated rat hepatocytes the mitochondrial pool of GSH (about 10% of the total cellular pool) had a half-life of 30 hr whereas the half-life of the cytoplasmic pool was 2 hr and concluded that the mitochondrial pool might represent the stable pool of GSH observed in whole animals.

Mitochondrial GSH functions as a discrete pool separate from cytosolic GSH. Mitochondrial GSH is mostly retained and therefore largely impermeable to the inner membrane following isolation of mitochondria, where the concentration of mitochondrial GSH (10 mM) is higher than cytosolic GSH (7 mM). The mitochondrial pool of glutathione has features that are very important for the protection of mitochondrial functions (Table 10.3).

Different rates of GSH turnover in the cytosol and mitochondria exist confirming the existence of separate intracellular GSH pools. The ratio of GSH:GSSG in mitochondria is approximately 10:1 under normal (untreated) conditions. Unlike cytosolic GSSG, mitochondrial GSSG is not effluxed from the soluble compartment. During oxidative stress induced with t-butyl hydroperoxide, GSSG is accumulated in the mitochondrial matrix and eventually reduced back to GSH. However, as the redox state of the mitochondria increased, an increase in protein mixed disulfides is also observed. Thus, mitochondria are more sensitive to redox changes in GSH: GSSG than the cytosol and therefore mitochondria may be more susceptible to the damaging effects of oxidative stress. These findings suggest that under certain experimental conditions, irreversible cell injury due to oxidative challenge may result from irreversible changes in mitochondrial function.

CELLULAR PROTECTIVE SYSTEMS

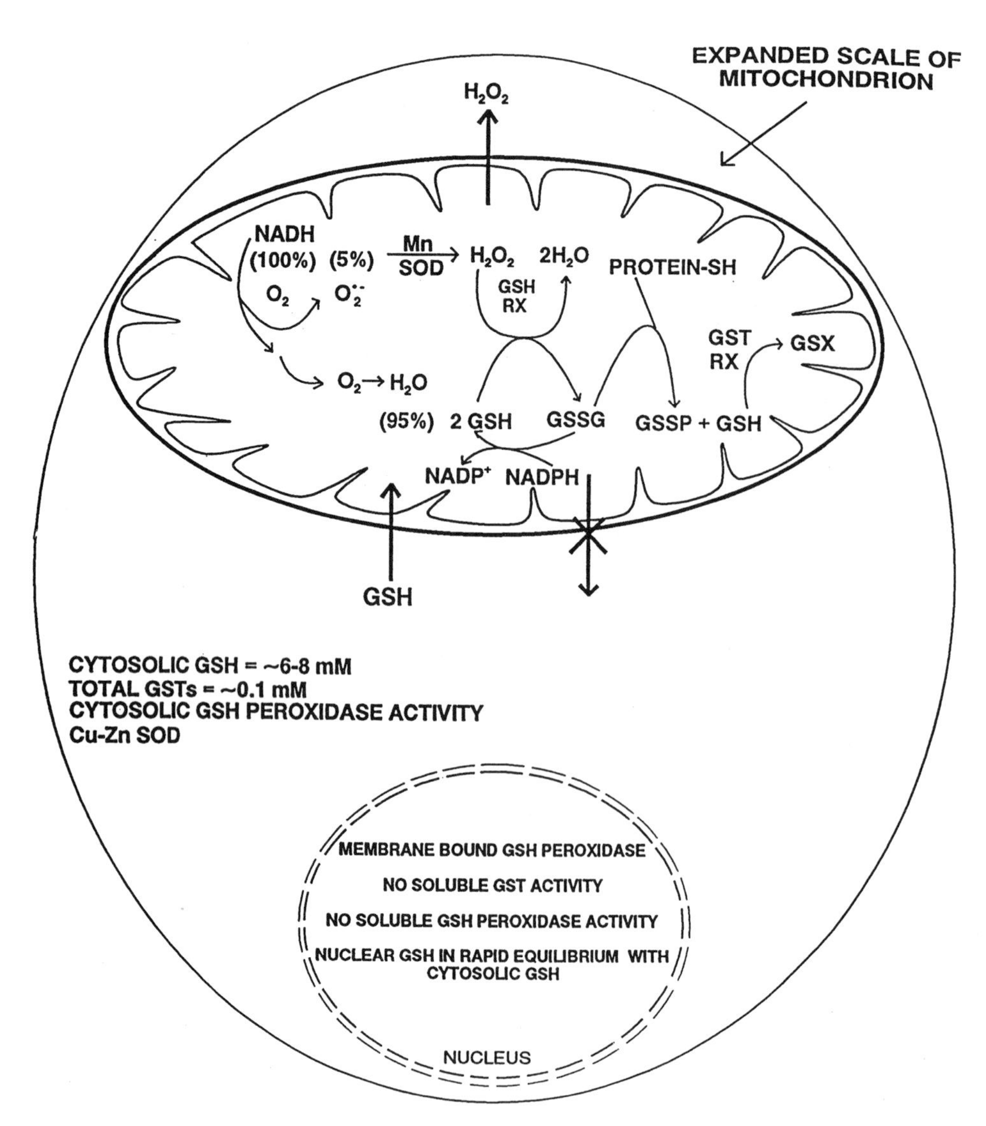

FIGURE 10.8 Cellular defense systems for the metabolism and inactivation of reactive oxygen species.

Because rat liver mitochondria contain the enzymes and cofactors necessary for the GSH/GSSG redox cycle (Figs 10.8 and 10.9) but do not contain catalase, we may assume that a primary function of mitochondrial GSH is for the detoxification of endogenously produced H_2O_2. This redox cycle requires the enzymes GSH peroxidase (selenium containing) and GSSG reductase along with the cofactors GSH and NADPH. In addition to detoxifying H_2O_2, GSH also protects protein sulfhydryls from oxidation.

TABLE 10.2 Features of Cytosolic Glutathione That Are Important in Protection Against Chemical Toxicity.

- High intracellular concentration (1–8 mM)
- Rapid synthesis by cytosolic enzymes in response to chemical depletion
- GSH, glutathione conjugate, and GSSG transport out of most cell types via plasma membrane transporters
- Very dynamic glutathione redox cycle activity to maintain GSH/GSSG ratio of about 50–100
- Protein thiol/GSH ratio is about 2–4 : 1
- High concentration of glutathione transferases (about 0.1 mM)

TABLE 10.3 Features of Mitochondrial Glutathione That Are Important in Protection Against Chemical Toxicity.

- Separate physiological pool of glutathione from cytosolic glutathione
- High concentration in the matrix (estimated at 10–12 mM)
- Only extramitochondrial synthesis of glutathione
- GSH transport; lack of GSSG transport in and out mitochondrial inner membrane
- Active glutathione redox cycle that maintains a GSH/GSSG ratio of about 25–50
- Protein thiol/GSH ratio is about 10
- Glutathione transferase activity

10.6.2 Mitochondria as Targets of Oxidants

Studies addressing mitochondria as target organelles of certain types of irreversible cell injury, as related to chemically induced effects, oxidative stress, and disrupted Ca^{2+} homeostasis, represent an area of intense investigation. GSH-dependent protection against lipid peroxidation has been demonstrated in mitochondria, nuclei, microsomes, and cytosol of rat liver. Lipid peroxidation induced in mitochondria is inhibited by respiratory substrates such as succinate, which leads indirectly to reduction of ubiquinone to ubiquinol. The latter is a potent antioxidant. The essential factor in preventing accumulation of lipid hydroperoxides and lysis of membranes in mitochondria, however, is glutathione peroxidase. Although prevention of free radical attach on membrane lipids may occur by an electron shuttle that utilizes vitamin E and GSH in microsomes, similar activity may be limited in mitochondria. Instead, mitochondrial GSH transferase(s) may prevent lipid peroxidation in mitochondria by a nonselenium GSH-dependent peroxidase activity. Three GSH transferases have been isolated from the mitochondrial matrix, and nearly 5% of the mitochondrial outer membrane protein consists of microsomal gulathione transferase. GSH transferase in the outer mitochondrial membrane could provide the GSH-dependent protection of mitochondria by scavenging lipid radicals by a mechanism that requires vitamin E and is abolished by bromosulfophthalein. Recently, mitochondrial phospholipid hydroperoxide glutathione peroxidase (PHGPx) has been suggested as having a major role in preventing oxidative injury to cells. PHGPx is synthesized in the cytoplasm as a long form (23 kDa) and a short

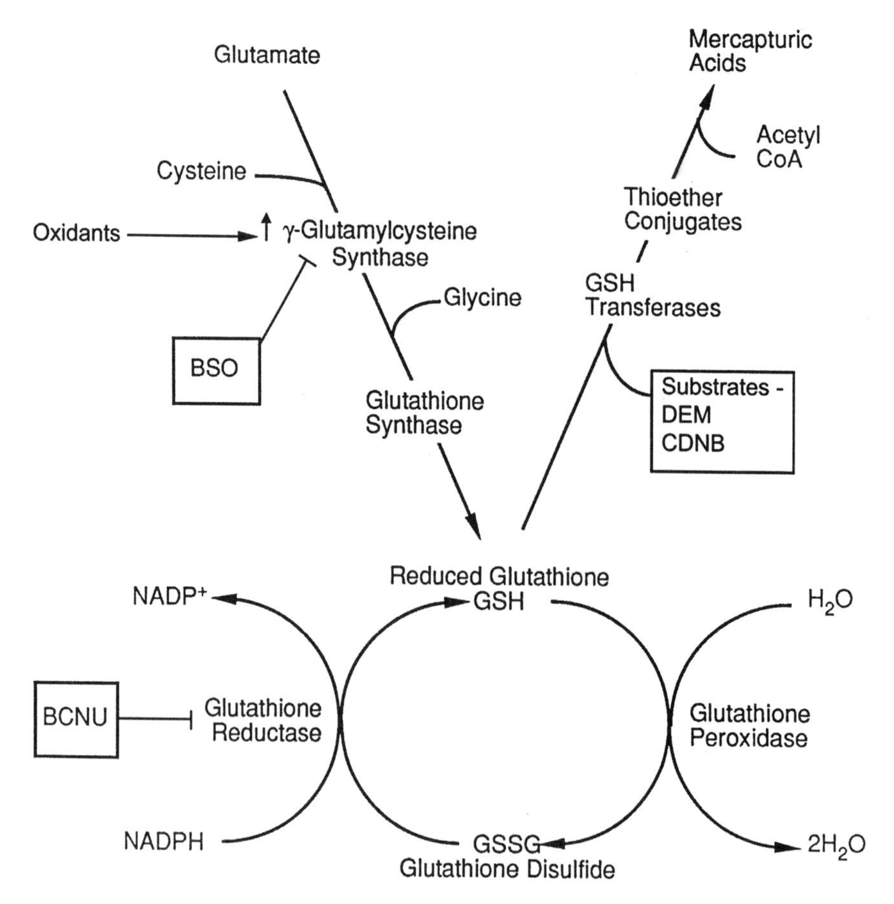

FIGURE 10.9 Cellular defense systems associated with the mitochondria.
Source: Adapted from Harlan et al., 1984.

form (20kDa). The long form contains a leader sequence that is utilized for transport to mitochondria. PHGPx in the mitochondria protects against the lipid hydroperoxides generated in the mitochondria and is thought to participate in the regulation of signal transduction pathways that are triggered by reactive oxygen species in mitochondria. Phospholipid monohydroperoxide has been shown to cause dissipation of mitochondrial membrane potential. Protection against that effect was achieved by transfection of cells with PHGPx gene. Overexpression of PHGPx has been shown to suppress cell death due to oxidative damage.

An important discovery is a previously unrecognized effect of mitochondrial oxidative stress and mitochondrial GSH defense on transcription factor activation. Not only can oxidant stress in mitochondria promote the loss of mitochondrial GSH and mitochondrial functions, it can promote extramitochondrial activation of NF-κB and therefore may affect nuclear gene expression. Mitochondria are targets of cytokines leading to the overproduction of reactive oxygen species induced by ceramide, a lipid intermediate of cytokine action and closely associated with apoptosis. Chronic ethanol intake depletes liver mitochondrial glutathione due to an ethanol induced defect in the transport of GSH from cytosol into the mitochondrial matrix. This sensitizes liver cells to the prooxidatant effects of cytokines and prooxidants generated by the oxidative metabolism of ethanol.

Because there is growing experimental evidence that oxidative stress is a factor in the neuropathology of several adult neurodegenerative disorders, many studies are being directed toward understanding of the mitochondrial oxidative stress associated with these diseases. For example, brain oxidative stress induced by t-butylhydroperoxide in 2- and 8-month-old mice suggests that aging makes the brain more susceptible to oxidative damage. It is becoming more apparent that the antioxidant status of mitochondria has a critical role in the fate of cells under oxidative stress.

10.6.3 Depletion of Glutathione by Chemicals and Fasting

Depletion of cellular glutathione has been widely studied with hundreds of chemicals including acetaminophen and bromobenzene. These studies demonstrated very clearly that bioactivation followed by glutathione adduct formation causes depletion of cytosolic glutathione and oxidative stress as indicated by indicators including enhanced levels of GSSG, lipid peroxidation, and loss of membrane integrity. The mechanism of acetaminophen-induced hepatotoxicity involves glutathione depletion followed by covalent binding and oxidative stress as indicated by the protective role of glutathione. The bioactivation of acetaminophen in vitro and in vivo is associated with GSH depletion and formation of GSSG due to the formation of the reactive metabolite N-acetyl-p-benzoquinone imine that is both an electrophile and an oxidizing agent. Acetaminophen treatment of mice increased the mitochondrial GSSG content more than 10-fold with 2% of the total GSH in controls to greater than 20% in treated animals. Loss of GSH and protein thiol homeostasis and in mitochondria could contribute to mitochondrial dysfunction and increase the potential for cell death via a mechanism involving oxidative stress.

Fasting enhances the toxicity of many chemicals. One of the earliest studies of this phenomenon compared the effects of fasting and various diets on chloroform-induced hepatotoxicity. Increased hepatotoxicity in association with fasting occurs with chemicals that are capable of depleting GSH, including carbon tetrachloride, 1,1-dichloroethylene, acetaminophen, bromobenzene, and many others. Because fasting decreases the hepatic concentration of GSH in mice and rats, such a decrease could account for the enhanced toxicity of many of these chemicals in fasted animals. In several instances, as a result of a depletion of GSH in the liver after pretreatment with diethyl maleate, acetaminophen, bromobenzene, carbon tetrachloride, and anthracycline, showed increased hepatotoxicity. Interestingly, the hepatotoxicity of thioacetamide, a substrate for the flavin-dependent monooxygenase present in microsomes, is enhanced after fasting but not after pretreatment with diethyl maleate.

The enhancement of hepatotoxicity by depletion of GSH has been noted during the metabolism of bromobenzene. When hepatocytes were pretreated with diethyl maleate, which reduced intracellular levels of GSH by about 70%, added bromobenzene caused levels of GSH to fall to 5% of initial levels, and cell death (75%

by 5 hr) was noted. Addition of cysteine, methionine, or N-acetyl-cysteine to the incubation medium prevented bromobenzene-induced toxicity, but did not prevent depletion of GSH. Bromobenzene, in the presence of a cysteine source, reduced initial levels of GSH to about 40% of control. The presence of metyrapone, an inhibitor of cytochrome P450-dependent monooxygenase reactions, eliminated bromobenzene-induced toxicity. These data are consistent with a requirement for bromobenzene activation prior to GSH conjugation. Redox cycling of a metabolite of bromobenzene derived from a GSH conjugation has been shown to cause kidney toxicity.

10.6.4 Glutathione Compartmentation and Chemical-Induced Injury

Since the depletion of hepatocellular GSH by diethyl maleate (DEM) does not exceed 90–95% of total cell GSH except at very high doses of DEM, a pool of GSH appears unavailable to conjugate with DEM. Hepatocytes, thus depleted, can remain viable for several hours, eventually resythesizing the original complement of GSH. Without additional stress, such as acetaminophen or ADP·Fe^{3+}, lipid peroxidation and cell death are not observed. Studies have shown that no formation of malondialdehyde above control levels was noted during 5 hr of incubation in hepatocytes partially depleted of GSH by BCNU. If adriamycin were included to enhance oxidative stress by redox cycling of adriamycin, however, GSH levels were decreased to less than 10% by 3 hr, coincident with a dramatic increase in production of malondialdehyde and leakage of lactate dehydrogenase from the cells. Additional studies have shown that there is a pool of GSH that is unaffected by treatment with diethyl maleate or BCNU or both, and that this pool is sequestered within the mitochondria. However, accessibility of BCNU to the mitochondrial matrix was shown by inhibition of the mitochondrial glutathione reductase. In contrast to the effects of BCNU on mitochondrial GSH depletion of the mitochondrial pool of GSH, a GSH S-transferase-dependent reaction with ethacrynic acid occurred in association with a rapid leakage of lactate dehydrogenase. Thus, alterations of chemical hepatotoxicity (or the lack thereof) by either fasting or pretreatment with diethyl maleate and related changes in the amounts of GSH should be considered from the standpoint of compartmentalization of intracellular GSH. Chemical-induced lipid peroxidation, in some instances, correlates better with depletion of GSH in the mitochondrial compartment that with cytosolic GSH. Time course studies of oxidative stress and tissue damage to zonal liver during ischemia-reperfusion in rat liver in vivo provides additional evidence that mitochondria are the major source of hydrogen peroxide in control and in reperfused liver. The uncoupler carbonylcyanide p-(trifluoromethoxy) phenylhydrazone provided almost complete inhibition of hydrogen peroxide production whereas antimycin in liver slices increased the production. These workers concluded that increased rates of oxyradical production by inhibited mitochondria appears as the initial cause of oxidative stress and liver damage during early reperfusion in rat liver. These findings support the conclusion that alterations in mitochondrial functions greatly enhances cell injury and death by a mechanism involving MPT.

The mitochondrial pool of GSH, as stated already, has a half-life of about 30 h. It is expected, therefore, that fasting will not deplete this pool of GSH. Fasting, in fact, does not increase the spontaneous rates or carbon tetrachloride-induced rates of lipid peroxidation. Hence, it may be that lipid peroxidation events are related to the size of the mitochondrial pool of GSH in liver. Moreover, perhaps certain "antioxidant" proteins in the mitochondria participate with GSH in preventing lipid peroxidation.

A controversial approach to assessing the potential for chemicals to cause lipid peroxidation in vivo is to treat an intact animal with a chemical or chemical combination and subsequently to measure products of lipid peroxidation in microsomes. In this manner, the depletion of glutathione in vivo with agents that form GSH conjugates enhances subsequent lipid peroxidation in vitro. Results from such experiments show consistently that an in vivo threshold of 1 μmol GSH/g liver is associated with spontaneous lipid peroxidation in microsomes. This critical value of GSH is about 20% of the initial concentration of GSH. Addition of exogenous GSH inhibited the lipid peroxidation in vitro in a concentration-dependent manner; 1 μM GSH yielded 50% inhibition. There also is observed a strong enhancement of spontaneous lipid peroxidation in phenobarbital-induced rats.

10.6.5 Glutathione Reductase and Glutathione Peroxidase

Glutathione reductase which is important in the regulation of the bioreductive activation of chemicals by GSH, is itself regulated by the redox status of the cell. Being similar to other reductases such as nitrate, nitrite, and $NADP^+$ reductase, GSH reductase is inactivated upon reduction by its own electron donor, NADPH. It has been proposed that this autoinactivation of glutathione reductase by NADPH and the protection as well as reactivation by GSSG regulate the enzyme in vivo. The activity of glutathione reductase may reflect the physiological needs of the cell especially during oxidative stress. For example, 40–50 μM intracellular NADPH inactivates glutathione reductase in the absence of GSSG and decreases glucose metabolism via the hexose monohposphate pathway. The physiological ratio of GSSG : GSH should provide sufficient GSSG at this level of NADPH to permit retention of significant glutathione reductase activity by preventing inactivation.

10.6.6 Glutathione Redox Cycle

It is apparent that a major protective role against the reactive chemical intermediates, which are generated by bioreduction and cause oxidative stress by redox cycling, is provided by the ubiquitous GSH redox cycle. This cycle utilizes NADPH and, indirectly, NADH-reducing equivalents in the mitochondrial matrix as well as the cytoplasm to provide GSH by the glutathione reductase-catalyzed reduction of GSSG (Fig. 10.10).

The rates of NADPH consumption in liver by the various NADPH-dependent enzymes indicate that glutatione reducatase has by far the highest rate to detoxify the reactive oxygen species generated by the various sources.

With a rat liver perfusion system, extra production of 70 nmol/min/g liver of GSSG can occur before GSSG is excreted into the bile as a result of t-butylhydroperoxide in the perfusate. These findings suggest that liver GSH could be oxidized to GSSG and reduced back to GSH over 10 times per min prior to exceeding the reducing capacity of the liver. The liver possesses high resistance to intracellular reactive oxygen formation that is dependent upon the glutathione

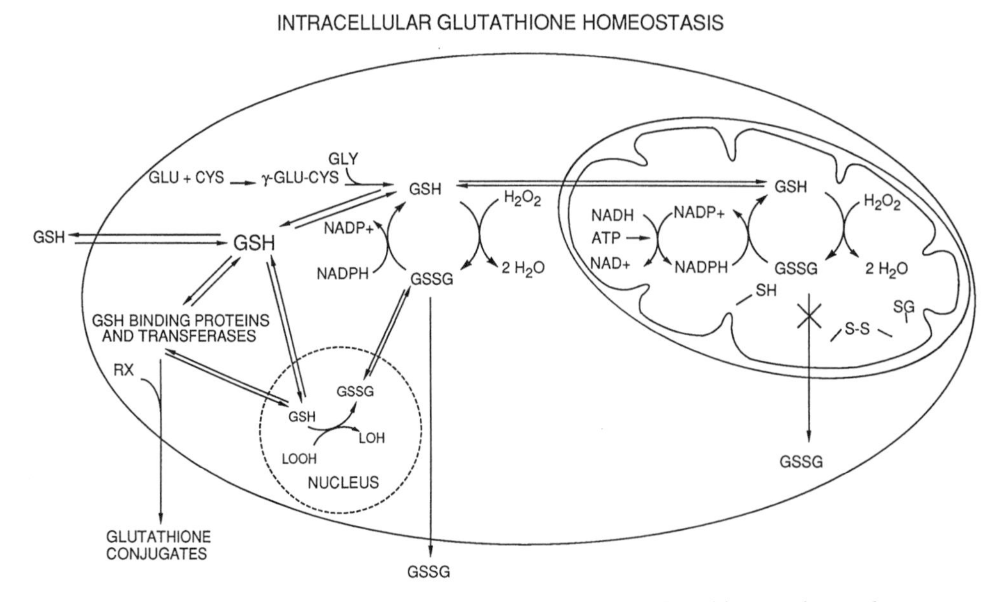

FIGURE 10.10 Modulation of glutathione and the glutathione redox cycle.

redox cycle system. Therefore, when the glutatione redox cycle is functioning at maximum capacity to eliminate hydrogen peroxide, a major regulatory effect is imposed on other NADPH-dependent pathways. The ability of the glutathione redox cycle to consume major quantities of reducing equivalents (NADPH) is further evidence that oxidative stress is a mechanism for injury and cell death since loss of this redox cycle potentiates hepatocyte death.

The mitochondrial GSH redox cycle has a role in regulating mitochondrial oxidations in liver. Various oxidants decrease O_2 uptake by isolated mitochondria and cause a complete turnover of GSH via glutathione peroxidase every 10 min. It appears that a continuous flow of reducing equivalents through the GSH redox cycle is balanced by a continuous formation of mitochondrial NADPH, which is needed for glutathione reductase activity. In addition, metabolism of hydrogen peroxide in mitochondria poses a regulatory function in regard to the oxidation of substrates by lipoamide-dependent ketoacid oxidases, which generate NADP-reducing equivalents. The entire NADPH:$NADP^+$ pool may turn over at least once every minute during a maximum oxidant challenge.

10.7 REDOX CYCLING OF CHEMICALS AND CELLULAR INJURY

Examples of chemicals that are associated with injury involving redox cycling include bromobenzene, carbon tetrachloride, ethanol, paraquat, diquat, menadione, and iron loading. Drugs that can undergo redox cycling include anthracyclines like adriamycin, mitomycin C, bleomycin, β-lapachone, alloxan acetaminophen, and the radiosensitizers (e.g., misonidazole and metronidazole). It is now known that activation of oxygen by reduction can play an important role in the toxicity of these drugs and chemicals. In general, they are enzymatically reduced or oxidized by a one-electron transfer. After a one-electron redox reaction, the intermediate formed then transfers the extra one electron to molecular oxygen to yield the superoxide anion radical and the parent drug or chemical, which can undergo repeated one electron reduction and/or oxidation to then provide one electron to molecular oxygen. This process of generation of reactive oxygen species, which is termed redox cycling, is involved in the toxicity of many hydroquinones, quinones, metal chelates, nitro compounds, amines, and azo compounds (Fig. 10.11).

In general, chemical toxicity compromises the ability of cellular processes to detoxify endogenous oxidative events related to oxygen metabolism. Further, the inherent toxicity of many chemicals includes the ability to produce additional oxidative stress, which can then become a dominant component of the overall toxicity. As mentioned above, all tissues and cells contain defense systems for detoxification of biological reactive intermediates of oxygen to prevent or limit cellular damage. Formation of hydrogen peroxide leads to the consumption of GSH as glutathione peroxidase activity transforms the hydrogen peroxide to water and GSH to GSSG, which is then reduced by an NADPH-dependent enzyme, glutathione reductase. Whether certain chemicals may cause both reactive intermediate effects such as covalent binding and redox cycling stress is being examined in some detail with the classes of compounds mentioned above.

Diquat (DQ), a bipyridyl herbicide is a model compound for redox cycling (Fig. 10.12). It generates large amounts of superoxide anion radical and hydrogen peroxide within cells. Isolated perfused rat liver when treated with diquat (200 μM) in the perfusate can generate superoxide at the rate of 1 μmol/min/g of liver. Only after depletion of GSH (10% of control) with phorone pretreatment (200 mg/kg body weight) and either ferrous sulfate (100 mg/kg) or BCNU (40 mg/kg) is it possible to significantly increase the diquat-induced liver injury. These findings support other evidence that the protective role of the GSH redox cycle in the toxicity of diquat in vivo and in vitro. Because diquat toxicity is mediated by redox cycling, it is greatly enhanced by prior treatment with BCNU; an inhibitor of glutathione reductase. Ebselen, a synthetic compound possessing glutathione peroxidase-like activity protects against diquat cytotoxicity when extracellular glutathione is present in the medium. Superoxide can reduce ferric iron to ferrous iron which is produced during diquat toxicity. Desferrioxamine, which chelates intracellular iron in the ferric state with an affinity constant of 10^{31}, provides considerable protection against diquat-induced toxicity. Therefore, hydrogen peroxide and transition metals have been suggested as major contributors to diquat toxicity. Even though the hydroxyl radical or a related species seems the most likely ultimate toxic product of the hydrogen peroxide/ferrous iron interaction, scavengers of hydroxyl radical afford only minimal protection. The high degree of reactivity of hydroxyl radicals assures, however, that the site of interaction with cellular components is within close proximity (a few angstroms) of the site of generation of the radical. Much remains to be understood about the mechanism of cytotoxicity by the various redox cycling agents including the quinones, menadione, and adriamycin, which are discussed in more detail below. Menadione has been studied as a redox cycling agent for the generation of oxidative stress in vivo and in vitro. Isolated rat hepatocytes undergo profound glutathione depletion with 25–400 μM menadione but the oxidative stress induced DNA fragmentation as demonstrated by a ladder of DNA fragments on agarose gel

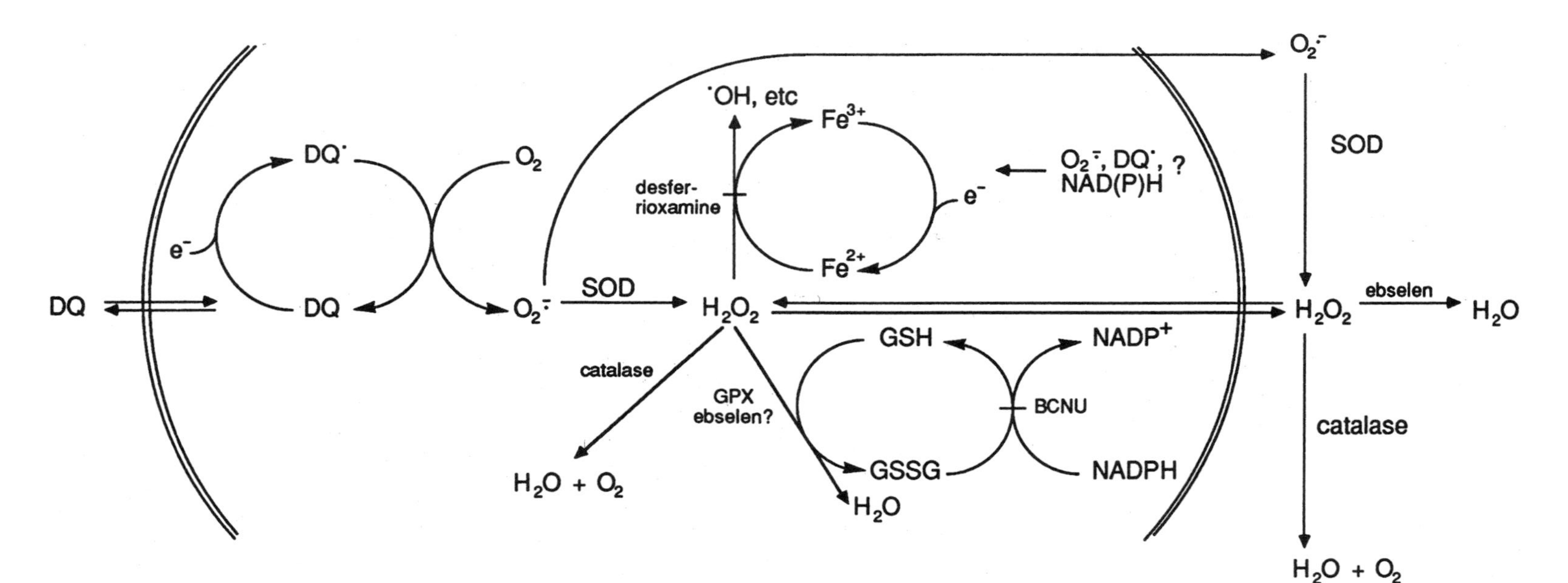

FIGURE 10.11 Scheme of postulated interactions between diquat, diquat-derived active oxygen species, internal and external enzymes, and transition metals.
DQ, diquat; DQ˙, diquat radical; GPX, glutathione peroxidase.

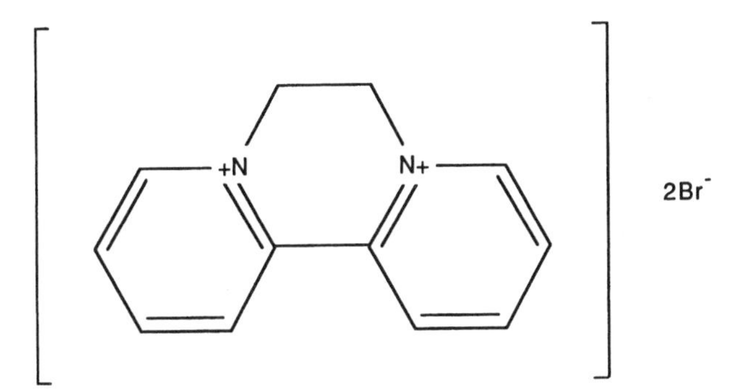

FIGURE 10.12 Structure of diquat.

electrophoresis is not accompanied by increased 8-oxodG formation. These results support apoptotic cell injury and death with the depletion of glutathione as an early event.

During their metabolism, many drugs and other chemicals involve the cellular production of radical oxygen species that participate in lipid peroxidation. For example, nonsteroidal anti-inflammatory drugs including naproxen and acetaminophen have been shown to undergo metabolism that causes a decrease in GSH, enhanced lipid peroxidation, and alteration of cellular proteins. A major question is whether the events of cell injury and death include apoptotic processes that limit the necrotic features of cell death.

t-Butylhydroperoxide (TBH) is widely utilized to investigate the mechanisms of cell injury initiated by oxidative stress. Although the mechanism remains unknown, lipid peroxidation and altered thiol status are implicated as components. Evidence has been presented that supports the hypothesis that the mitochondrial MPT can occur under conditions of oxidative stress and contribute to the mechanism of cell death. Importantly, CsA, a high-activity inhibitor of the transition, prolongs survival of hepatocytes treated with TBH under conditions where extensive lipid peroxidation is prevented. During oxidative stress induced by TBH treatment, cultured hepatocytes display the characteristics of MPT which quickly lead to mitochondrial depolarization and cell death.

The mechanism of acetaminophen-induced hepatotoxicity involves contributions by covalent binding and oxidative stress as indicated by the protective role of glutathione (Fig. 10.3). The toxicity of acetaminophen in vitro and in vivo is associated with GSH depletion and formation of GSSG due to the bioactivation of acetaminophen to the reactive metabolite N-acetyl-p-benzoquinone imine that is both an electrophile and an oxidizing agent. Although recent evidence indicates acetaminophen expressing a receptor-based mechanism along with covalent binding as primary mechanisms of toxicity, some evidence supports an oxidative stress mechanism that is associated with hepatic injury as mentioned, acetaminophen (500 mg/kg i.p.) increased the mitochondrial GSSG content from 2% in controls to greater than 20% in treated animals. It is known that GSSG does not efflux from mitochondria and must be reduced back to GSH by mitochondrial-derived reducing equivalents. Therefore, if levels of GSSG approach 20% in treated animals, then the ability to maintain mitochondrial integrity is in question as loss of calcium regulation potentiates the mitochondria for undergoing MPT. BCNU increases the susceptibility of hepatocytes to oxidative stress and increased both the formation of GSSG and cell killing produced by acetaminophen. Interestingly, acetaminophen administered by repeat exposures of incremental doses provides protection against acetaminophen-induced lethality in mice. Such autoprotection is thought to be dependent upon the ability to sustain repair processes associated with cell proliferation.

10.8 CONCLUSIONS

Much remains to be elucidated about the mechanisms of chemical-induced cell injury and death. However, there is an increasing amount of evidence that receptor-based cell injury and repair processes are critical components to the outcome of either chronic or acute exposures to toxic chemicals. Cell death mechanisms, particularly apoptotic cell death events, are being understood to a much greater extent and should provide a much better understanding of the processes that lead to both reversible and irreversible injury prior to survival or cell death.

ACKNOWLEDGEMENT

Support for this chapter was from NIH grants ES-00210, ES-00040 and ES-019001.

SUGGESTED READING

Babson, J.R. and Reed, D.J. Inactivation of glutathione reductase by 2-chloroethyl nitrosourea-derived isocyanates. *Biochem. Biophys. Acta.* 83 (1978), 754–762.

Babson, J.R., Abell, N.S. and Reed, D.J. Protective role of the glutathione redox cycle against adriamycin-mediated toxicity in isolated hepatocytes. *Biochem. Pharmacol.* 30 (1981), 2299–2304.

Beckman, J.S. and Crow, J.P. Pathological implications of nitric oxide, superoxide, and peroxynitrite formation. *Biochem. Soc. Transact.* 21 (1993), 330–334.

Boulares, H.A., Giardina, C., Navarro, C.L., Khairalla, E.A. and Cohen, S.D. Modulation of serum growth factor signal transduction in Hepa 1–6 cells by acetaminophen: An inhibition of c-myc expression, NF-kappa B activation and Raf-1 kinase activity. *Toxicol. Sciences* 48 (1999), 264–274.

Fridovich, I. Superoxide dismutases. *Adv. Enzymol. Relat. Areas Mol. Biol.* 41 (1974), 35–97.

Green, D.R. and Reed, J.C. Mitochondria and apoptosis. *Science* 281 (1998), 1309–1312.

Kerr, J.F.R. Definition of apoptosis and overview of its incidence. In: Lavin, M. and Watters, D. (Eds), *Programmed Cell Death: The Cellular and Molecular Biology of Apoptosis.* Harwood Academic Publishers, Longghome, PA, United States of America 1993, pp 1–15.

Kosower, N.S., Kosower, E.M. Glutathione status of cells. *Int. Rev. Cytol.* 54 (1978), 109–160.

Meredith, M.J. and Reed, D.J. Status of the mitochondrial pool of glutathione in the isolated hepatocyte. *J. Biol. Chem.* 257 (1982), 3747–3753.

Miller, L.J. and Marx, J. Apoptosis. *Science* 281 (1998), 1301.

Reed, D.J. Cellular defense mechanisms against reactive metabolites. In: Anders, M.W. (Ed), *Bioactivation of Foreign Compounds*. Academic Press, Orlando, FL, 1985, pp 71–108.

Reed, D.J. Glutathione: toxicological implications. *Annu. Rev. Pharmacol. Toxicol.* 30 (1990), 603–631.

Reed, D.J. Toxicity of oxygen. In: De Matteis, F. and Smith, L.L. (Eds), *Molecular and Cellular Mechanisms of Toxicity*. CRC Press Inc., Boca Raton, FL, 1995, pp 35–68.

Reed, D.J. Regulation of reductive processes by glutathione. *Biochem. Pharmacol.* 35 (1986), 7–13.

Reed, D.J. and Beatty, P.W. Biosynthesis and regulation of glutathione. Toxicological implications. In: Hodgson, E, Bend, J.R. and Philpot, R.M. (Eds), *Reviews in Biochemical Toxicology*. Elsevier Press, New York, 1980, pp 213–241.

Reed, D.J. and Ellis, W.W. Influence of γ-glutamyl transpeptidase inactivation on the status of extracellular glutathione and glutathione conjugates. In: Snyder, R., Parke, C.V., Kocsis, J.J., et al. (Eds), *Biological Reactive Intermediates, IIA*, Plenum Press, New York, 1982, pp 75–86.

Shayiq, R.M., Roberts, D.W., Rothstein, K., Snawder, J.E., Benson, W., Xiang, M.A. and Black, M. Repeat exposure to incremental doses of acetaminophen provides protection against acetaminophen-induced lethality in mice: an explanation for high acetaminophen dosage in humans without hepatic injury. *Hepatology* 29 (1999), 451–463.

St. Clair, D.K., Oberley, T.D., Ho, Y.-S. and Wheeler, K.T. Overproduction of human MnSOD modulates paraquat-mediated toxicity in mammalian cells. *FEBS Lett* 293 (1991), 199–203.

CHAPTER ELEVEN

Nutritional Factors

WILLIAM E. DONALDSON

11.1 INTRODUCTION AND SCOPE

All animals must eat to live. Humans choose their diet from among the available foods, and when they choose wisely, they are well nourished. If they eat too little or too much of a balanced diet, or if they eat an unbalanced diet, they are malnourished. Toxicologists must not only see to their own nutrition, but they have the additional burden of choosing the proper diet for their experimental animals. The science of nutrition can provide the knowledge necessary to make such choices. Hopefully the following discussion will stimulate students of toxicology to increase their awareness of nutrition and of how nutrition and toxic substances interact.

This chapter will describe how the various nutrients function in normal physiological processes in animals. Also, there will be a brief discussion of nutrient requirements of humans and laboratory animals in terms of normal function. Nutrient requirements in toxicological research will be discussed, and two theses will be presented: first, that the level of nutrients supplied to an organism can affect toxicant action, metabolism, and disposition, and second, that toxicants can affect apparent nutrient requirements in both upward and downward directions. Next, the feeding of animals in toxicological experiments will be examined with respect to types and sources of diets. The importance of the need to define, as completely as possible, the experimental diets used by toxicologists will be stressed. Finally, some of the recent literature on nutrient interactions with cancer will be discussed.

11.2 THE NUTRIENTS AND THEIR FUNCTION

Nutrients are those chemical substances that are required for the proper functioning of the body and that must be consumed by an organism on a periodic schedule. Nutrition is the science that deals with the provision, utilization, and metabolism of the nutrients. This section of the chapter provides a brief description of the nutrients and their function.

11.2.1 Water

Although one does not usually think of water as a nutrient, it is clearly a substance that fits the above definition. Among the important functions of water in the body are: (1) transport of other nutrients and wastes, (2) maintenance of pH and osmotic relationships and therefore cell turgidity, (3) regulation of body temperature, (4) cushioning and lubrication of joints, and (5) serving as a solvent (and sometimes as a reactant) for chemical reactions.

Animals can derive approximately 12g of water for each 100kcal of energy metabolized. Additionally, the water bound in consumed food can supply a portion of an animal's water requirement. By far the most important source of water for all but a few desert animals is drinking water. In most toxicological experiments, fresh, clean tap water is adequate for the needs of the animals. However, it should be borne in mind that municipal water supplies can contain significant quantities of dissolved minerals, hydrocarbons, and water treatment chemicals containing chlorine and fluorine. Given the vulnerability of groundwater supplies to environmental contamination, the use of well water in sensitive toxicological research should be considered a last resort. In any event, it is incumbent upon toxicologists to determine whether or not animal drinking water from any source is affecting experimental results and, if so, to treat the water so as to remove any compromising substances.

11.2.2 Carbohydrates

Carbohydrates serve numerous functions in the body, but the bulk of the carbohydrate ingested is used to derive energy for metabolism. The principal carbohydrate used by animals is glucose. The level of glucose in blood of normal animals, and consequently the amounts available to body organs, is closely regulated through the action of several hormones. When carbohydrate intake is high, excess glucose is converted to glycogen, a glucose polymer, for storage, but the amount of glycogen storage is limited. Any further glucose excess is converted to fat, the storage of which is essentially unlimited. In times of food or energy deprivation, glycogen stores are mobilized and when nearly exhausted, a process known as gluconeogenesis is initiated. During gluconeogenesis, the oxidation of glucose by most organs is curtailed and amino acids from body proteins are mobilized to form glucose. As a result, certain organs such as the brain, for which glucose oxidation is obligatory, can be maintained during extended periods of fasting or starvation. Concomitant with gluconeogenesis, fat stores are mobilized to provide energy for those tissues and organs, such as skeletal muscle, which no longer utilize glucose.

Other important functions of carbohydrates are as part of the structure of the genetic material, DNA (deoxyribose) and RNA (ribose). Additionally, carbohydrates make up a portion of the structure of certain biological membranes and can be part of the structure of some receptor sites located in these membranes. Of particular interest to toxicologists is that glucose is the precursor of glucuronic acid, a compound important in some phase II detoxification reactions. Finally, in those species for which vitamin C (ascorbic acid) is not a dietary essential, glucose is the precursor.

Because the amount of carbohydrate found in foods of animal origin is small, most animals obtain the bulk of their dietary carbohydrate from plant sources. These carbohydrates include fructose or fruit sugar, sucrose or table sugar (derived from sugar cane and sugar beets), or starch, a polymer of glucose, which is the primary storage form of glucose in plants. One other source of carbohydrate is lactose or milk sugar. Plants also contain a variety of substances known as fiber. Many of these fibrous substances are polymers of glucose, but unlike starch and glycogen, in which the glucose is readily available, the glucose in plant fiber is available only after extensive bacterial degradation of the fiber. Thus, while the glucose in fiber contributes to the carbohydrate nutrition of ruminants, and to a lesser degree animals such as rodents which possess a large cecum, for most monogastric animals, plant fiber is non-nutritive.

11.2.3 Lipids

Lipids are hydrophobic compounds consisting either of sterols, glycerides, or modified glycerides called phospholipids. The principal sterol found in animals is cholesterol, which functions as a major membrane component as well as a precursor of several important compounds in the body such as the sex hormones, the adrenal steroid hormones, and vitamin D. Animals can synthesize cholesterol, and, to some extent, this biosynthesis is reduced by dietary cholesterol intake. Foods of plant origin do not contain cholesterol but rather other sterols which are less readily absorbed than cholesterol and which cannot be converted to cholesterol.

Glycerides are esters of the three carbon alcohol, glycerol, with one, two, or three fatty acids forming mono-, di-, or triglycerides, respectively. Fatty acids usually found in glycerides are long-chain (14–22 carbons) and are either saturated or contain one or more double bonds (unsaturated). The predominant glyceride class is the triglycerides, which serve primarily for energy storage. In animals, most of the triglycerides are located in adipocytes, which are aggregated in several fat depots in the body. In plants, triglycerides are found mainly in the seeds. The fatty acids in glycerides can be mobilized and oxidized to yield acetyl-coenzyme A and energy for metabolic processes.

The principal phospholipids are phosphatidylcholine, phosphatidylinositol, and phosphatidylserine. These compounds contain fatty acids in the one and two position of glycerol whereas phosphoric acid and either choline, inositol, or serine are esterified in the three position. In cells, the phospholipids are oriented in a bilayer with the hydrophilic polar groups facing the exterior and the hydrophobic fatty acids facing the interior of the bilayer. Along with imbedded protein molecules, the bilayers constitute biological membranes. The fluidity, and therefore the function, of the membranes is determined to a large extent by the fatty acid composition of the phospholipids. Membrane fluidity increases with an increasing degree of unsaturation of the component fatty acids and can be altered by feeding diets containing differing levels of unsaturated fat.

Animals can synthesize a wide array of saturated and unsaturated fatty acids, however, they are unable to synthesize fatty acids containing double bonds in either the omega-3 (n-3) or the omega-6 (n-6) position (where n = the carbon number of the fatty acid chain starting from the methyl group end). Two such fatty acids are required by animals (but produced only by plants) and thus are dietary essentials. These acids are linoleic acid (containing 18 carbons and 2 double bonds— or 18:2, n-6) and linolenic acid (18:3, n-3). Animals can convert linoleic acid to arachidonic acid (20:4, n-6) and linolenic acid to eicosapentaenoic acid (EPA or 20:5, n-3).

$H_3C(CH_2)_4CH=CHCH_2CH=CH(CH_2)_7C00H$	Linoleic acid
$H_3CCH_2(CH=CHCH_2)_2CH=CH(CH_2)_7C00H$	Linolenic acid
$H_3C(CH_2)_4(CH=CHCH_2)_3CH=CH(CH_2)_3C00H$	Arachidonic acid
$H_3CCH_2(CH=CHCH_2)_4CH=CH(CH_2)_3C00H$	Eicosapentaenoic acid

The 20:4 and 20:5 acids can then be converted by tissue lipoxygenases and cyclooxygenases to a series of compounds known as eicosanoids. The eicosanoids, which include prostaglandins, prostacyclins, thromboxanes, and leukotrienes, possess potent pharmacological properties. There is rapidly growing evidence that both the levels and relative proportions of eicosanoids produced by animals can be influenced by the level and type of dietary fat.

11.2.4 Proteins

Proteins consist of amino acids, 20 of which are found in nature, linked together by peptide bonds formed between the amino and carboxyl groups. Certain of the amino acids, called essential or indispensable, must be supplied in the diet because animals either cannot synthesize them at all or cannot synthesize them in sufficient quantity to meet their needs. When protein is ingested, the component amino acids are then resynthesized into a complement of proteins that is unique for that species. Amino acids that are not incorporated into proteins and peptides by the animal are not stored. The carbon skeletons of the excess amino acids are converted to intermediates of glucose metabolism or fat metabolism while the amino moiety is eliminated in the form of urea (mammals), uric acid (birds and reptiles), or ammonia (simple aquatic animals and some fish). Animals fed insufficient quantities of the energy-bearing nutrients, carbohydrates and lipids, are forced to use a portion of their dietary protein intake to derive energy.

The proteins and peptides synthesized by animals have several diverse functions. There is the structural function of proteins found in membranes and muscle fibers. Metabolism is dependent upon metabolic and digestive enzymes all of which are proteins. Several hormones, for example, pituitary growth hormone and insulin, are proteins whereas other hormones such as enkephalins, renin, and enterokinin are peptides. Thyroxine and epinephrine are derivatives of the amino acid tyrosine. Protein-containing leukocytes, macrophages, T cells, and B cells constitute cellular immune function whereas humoral immunity is provided by proteins called antibodies. One needs only to read a list of the enzymes involved with phase I and II reactions to realize the importance of proteins to the detoxication process. Glutathione, a tripeptide composed of glutamic acid, cysteine, and glycine, is important in conjugation reactions.

11.2.5 Vitamins

Vitamins are a chemically diverse group of compounds that serve as metabolic cofactors. No energy is derived from their metabolism, and they serve no structural function in the body.

11.2.5a Fat-Soluble Vitamins

The fat-soluble vitamins, A, D, E, and K, are isoprenoid compounds that can be stored in appreciable quantities in the body. Excessive intakes, especially of vitamins A and D, can cause toxicity due to accumulation in the tissues.

Vitamin A, or retinol, is the form of the vitamin found in animals. Plants contain pigmented compounds (principally carotenes) called provitamin A because they can be converted to retinol by animal tissues. Retinol is involved in membrane transport of calcium as well as serving as the prosthetic group for the visual pigment protein, rhodopsin. A slight deficiency of retinol results in night blindness whereas a severe deficiency can result in irreversible eye damage, xeropthalmia. A deficiency leads also to increased susceptibility to infections because of damage to epithelial tissue.

Vitamin D, or cholecalciferol, can be ingested in food or formed in animals by ultraviolet irradiation (sunlight) of 7-dehydrocholesterol found in skin. Vitamin D is hydroxylated in the 25-position by liver and in the 1-position by kidney to form 1,25-dihydroxycholecalciferol. The latter compound induces a carrier protein in the gut which facilitates the absorption of calcium. A similar form of vitamin D, ergocalciferol, is found in plants. A deficiency of vitamin D results in bone deformation (rickets) in the young and bone weakness (osteoporosis) in adults.

Vitamin E, or tocopherol, is a potent antioxidant. The vitamin has no clearly established biological function except for its role in detoxifying peroxides and free radicals produced from unsaturated fatty acids in membranes. The vitamin is ubiquitous in foods, and consequently, spontaneous deficiency is not observed in humans or animals. Experimental deficiency can be produced by using fats from which vitamin E has been removed or by prolonged storage of feed containing highly unsaturated fats such as fish oil which results in destruction of the vitamin.

Vitamin K exists in several forms. Menadione is the biologically active form of the vitamin and is required in the process that converts prothrombin to thrombin. A deficiency leads to delayed blood clotting. Prolonged administration of certain drugs, such as sulfas, can provoke a deficiency by suppressing bacterial synthesis.

11.2.5b Water-Soluble Vitamins

The water-soluble vitamins are vitamin C and the B-complex vitamins. The quantities of water-soluble vitamins found in tissues are low compared to the fat soluble vitamins. Excessive intakes of water soluble vitamins are quickly excreted, therefore, toxicity is rare.

Vitamin C, or ascorbic acid, can be synthesized from glucose by most species. Among the exceptions are humans, monkeys, guinea pigs, and certain fishes and birds. These species require a dietary source. The vitamin is required for red blood cell formation, wound healing (collagen formation), and capillary integrity. No specific biochemical function is known. Physiological effects of this vitamin have been reported from ingestion by species that possess synthetic capacity. There is considerable controversy over the efficacy of megadoses (up to several hundred-fold the requirement) of ascorbic acid for the prevention and treatment of the common cold and cancer in humans. Definitive evidence to support such claims is not apparent.

Thiamin, or vitamin B_1, is part of the structure of the cofactor thiamin pyrophosphate, required of the decarboxylation of pyruvic acid. Among the symptoms of a deficiency is a nervous disorder characterized by accumulation of pyruvate and lactate in brain and decreased glucose utilization.

Riboflavin, or vitamin B_2, is a component of two coenzymes, FMN and FAD, which function as the prosthetic group of flavin-linked dehydrogenases. Deficiency symptoms include dermatitis, cataracts, and myelin degeneration.

Niacin, or nicotinamide, is a component of two coenzymes, NAD and NADP, which function as the prosthetic group of pyridine-linked dehydrogenases. NAD and NADP are associated with pathways of aerobic respiration and biosynthesis, respectively. Niacin deficiency results in dermatitis, diarrhea, and nervous disorders. Niacin can be formed in most species (except cats) from the amino acid tryptophan.

Pyridoxine, or vitamin B_6, is converted to pyridoxal phosphate and pyridoxamine phosphate. These coenzymes are important in amino acid metabolism because they catalyze transamination, decarboxylation, and racemization reactions. Deficiency symptoms include dermatitis, anemia, and nerve disorder. Human intakes >500 mg/day have been reported to cause sensory nerve damage.

Pantothenic acid is part of the structure of coenzyme A, and as such, serves as a carrier and donor of acyl groups. The vitamin also is part of the prosthetic group of some enzymes, for example, fatty acid synthase. Deficiency symptoms include dermatitis, diarrhea, and myelin degeneration with attendant nerve disorders.

Biotin serves as the prosthetic group for enzymes that catalyze carboxylation reactions. The biotin prosthetic group serves as a carrier of CO_2 in these reactions. Biotin deficiency symptoms include dermatitis and nervous disorders. Bacterial synthesis in the gut usually provides enough biotin to meet an animal's requirement. A protein, avidin, found in raw egg white binds biotin and makes it unavailable. A biotin deficiency can result from consumption of large quantities of uncooked eggs or egg white.

Folic acid is converted to the coenzyme, tetrahydrofolate, which serves as a carrier of one-carbon units (formyl or methyl) in several enzymatic reactions. Deficiency symptoms include anemia, sore mouth, gastrointestinal inflammation, and diarrhea. Some drugs (e.g., methotrexate) used in chemotherapy for cancer, are folic acid antagonists. Typical side effects for such drugs are consistent with signs of folic acid deficiency.

Cyanocobalamin, or vitamin B_{12}, is converted to a coenzyme involved in methyl transfer reactions and in reactions in which a hydrogen is transferred to an adjacent carbon in exchange for some other group. A deficiency causes severe anemia and muscle weakness, which can lead to paralysis.

Choline is considered to be a vitamin by some, but not all, nutritionists. Choline is not a dietary requirement for all species (e.g., the rat). In species that require choline, for example, the chick, the amount required is large in comparison with the requirements for other vitamins. Choline synthesis requires an adequate supply of the amino acid, methionine. The best known functions for choline are as part of the structures of acetylcholine, a neurotransmitter, and phosphatidyl choline, a phospholipid. A deficiency in growing chickens results in fatty livers and a leg-crippling condition called perosis.

11.2.6 Minerals

Several mineral elements are required in the diets of animals, and these elements serve a wide range of physiological and biochemical functions. There are two categories of elements based on the size of the dietary requirement: those required in large amounts are called major or macro elements, whereas those required in small amounts are called minor or micro or trace elements. Excessive intake of any of these elements will cause toxicity, and conversely, inadequate intake will result in specific deficiency signs.

11.2.6a Macro Elements

Calcium and phosphorus exhibit the highest requirements because of their importance in bone and tooth structure. Excess or deficiency will result in bone abnormalities such as rickets and osteoporosis (see vitamin D above). Additional functions include storage of chemical energy in cells as ATP (phosphorus) and as mediators of intracellular signals (calcium).

Magnesium is found in significant amounts in bone and acts as an activator for several enzymes. Excess intake can provoke signs of calcium deficiency. The principal sign of magnesium deficiency is muscle spasms (tetany).

Sodium, potassium, and chlorine are the major electrolytes of blood and of extracellular (sodium and chlorine) and intracellular (potassium and chlorine) fluids. The major functions of these elements are maintenance of pH and osmotic balance. Both deficiencies and excesses result in abnormal water distribution between cellular and extracellular spaces.

The requirement for organic sulfur, as part of the structure of the amino acids methionine and cysteine, is relatively large in comparison to the requirement for inorganic sulfur. Protein synthesis in animals is dependent upon adequate intake of sulfur-containing amino acids. Sulfur is also part of the structure of the vitamins thiamin and biotin. Conjugation with inorganic sulfur is an important detoxication mechanism as is organic sulfur conjugation via the cysteine moiety of glutathione. Sulfur deficiency results in poor growth, due to insufficient protein synthesis, and in lower glutathione content of several tissues.

11.2.6b Micro Elements

Iron is part of the structure of hemoglobin and myoglobin, and as such, has a major function in oxygen transport. Iron is also part of the structure of the cytochromes including cytochrome P450. The principal deficiency sign is anemia. Iron overload from ingestion is rare, except in young organisms, because the body possesses an elaborate mechanism to exclude iron from intestinal absorption. Excessive transfusions of whole blood can cause abnormal accumulation of iron in tissues (hemochromatosis), which in turn can result in enlarged liver, diabetes, and cardiac failure.

The major need for copper is as part of the structure of the enzymes in the ferroxidase system which is required for iron utilization. Consequently, anemia is the most common sign of copper deficiency. Copper also serves as part of the structure of several enzymes including tyrosinase, cytochrome c oxidase, ascorbic acid oxidase, and amine oxidases. Toxicity results from accumulation of copper in the liver which impairs hepatic function. A genetic disorder, Wilson's disease, if not recognized and treated promptly, can cause death from excessive copper deposition in liver.

Manganese is a component of one family of superoxide dismutases that is important in the metabolism of superoxide radicals. Manganese deficiency results in reproductive disorders and impaired bone formation. Excess dietary manganese interferes with iron absorption.

Iodine is important as a component of thyroxine, a hormone that is involved in regulation of metabolic rate. Iodine deficiency leads to enlarged thyroid gland (goiter). Prolonged administration of high doses can also produce goiter by reducing iodine uptake by the thyroid.

Zinc is a structural component of follicle stimulating hormone (FSH) and leutinizing hormone (LH) and of a number of enzymes including alcohol dehydrogenase, carbonic anhydrase, lactic dehydrogenase, ribonuclease, DNA polymerase, and a family of superoxide dismutases. Deficiency leads to dermatitis and impaired reproduction. Toxicity symptoms include anemia and reduced bone mineralization.

In higher organisms, selenium is as a component of glutathione peroxidase, an enzyme involved with the destruction of organic peroxides. Selenium is also a component of heme oxidase. A deficiency results in membrane damage, the manifestation of which is species dependent and can range from cellular necrosis to increased capillary permeability. Lambs and calves suffer from a muscle disease called stiff-lamb disease when raised in ranges of selenium-deficient plants. Embryo mortality in ewes from selenium-deficient areas can be reversed with selenium supplement. In humans, an endemic cardiomyopathy in China (Keshan's disease) can be alleviated by oral treatment with sodium selenite. Acute selenium poisoning produces central nervous system effects, which include nervousness, drowsiness, and sometimes convulsions. Symptoms of chronic exposure may include pallor, coated tongue, "garlic" breath, liver and spleen damage, and gastrointestinal disorders.

Fluorine is a constituent of bones and teeth. It imparts resistance in animals to formation of dental caries and may retard bone demineralization in older organisms. A deficiency is difficult to produce. Toxicity results in deranged bone and tooth formation.

The only known function of cobalt is as a component of vitamin B_{12}. Relatively high levels of cobalt have been used therapeutically in treatment of certain anemias without toxic effect. However, lower levels, used as antifoaming agents in beer, have been implicated as causative factors in cardiac failure.

Molybdenum is a constituent of the enzyme xanthine oxidase. A dietary requirement for this element has not been demonstrated in mammals. The element appears to be essential in birds, which excrete nitrogen as uric acid, because of the role of xanthine oxidase in purine metabolism. The principal symptoms of toxicity are severe diarrhea and reduced growth.

Chromium is considered a dietary essential in mammals. Deficient diets can result in hyperglycemia and glucosuria. Toxicity results in dermatitis and ulceration of the respiratory and digestive tracts.

Other trace elements may be essential, but strong evidence for their necessity is lacking. Improved growth has been reported in rats for dietary additions of lead, nickel, silicon, tin, and vanadium.

11.3 NUTRIENT REQUIREMENTS

11.3.1 Sources of Nutrient Requirement Data

The arbiter of nutritional requirements in the United States is the National Research Council of the National Academy of Sciences. The Academy's recommendations on dietary requirements and allowances are found in a series of publications available from the National Academy Press (2101 Constitution Avenue, NW, Washington, DC 20418, or http://www.nap.edu/).

In the case of humans, nutritional requirements are expressed as Recommended Dietary Allowances (RDAs), which are defined in the 10th edition (1989) of the publication titled, *Recommended Dietary Allowances*, as "the levels of intake of essential nutrients considered, in the judgement of the Food and Nutrition Board on the basis of available scientific knowledge, to be adequate to meet the known nutritional needs of practically all healthy persons."

The Academy also publishes a series on Nutrient Requirements of Domestic Animals. Separate publications are available for beef cattle, cats, coldwater fish, dairy cattle, dogs, goats, horses, laboratory animals (rats, mice, gerbils, guinea pigs, hamsters, voles, and fishes), mink and foxes, nonhuman primates, poultry, rabbits, sheep, swine, and warmwater fish and shellfish.

The nutritional recommendations of the Academy are under constant review in light of new scientific evidence. Therefore, revisions of each work are published periodically.

11.3.2 Nutritional Requirements in Toxicological Research

11.3.2a Paired Feeding

A common result of exposure of an organism to a toxicant is a reduction or cessation of food intake. Under such conditions, it is imperative to separate the effect of the reduced food intake from other effects of the toxicant. Consider the following hypothetical example. You are studying the chronic effects of a toxicant that reduces food intake and appears to alter membrane function. Your initial results suggest that the toxicant interfered with the synthesis of fatty acid precursors of membrane formation. Your next step may be to determine the effect of the toxicant on the hepatic activity of several lipogenic enzymes. The activities of enzymes such as acetyl-CoA carboxylase, fatty acid synthetase, microsomal desaturases, and microsomal systems of fatty acid elongation are depressed by reduction of food intake. To gain maximum information from the experiment, your protocol must allow you to quantitate not only the total effect of the toxicant on enzyme activities but the effect related to food intake as well. One approach is to use a technique known as pair feeding.

The control animals have food available ad libitum as do the animals treated with toxicant. Obviously, the treated animals will consume less feed. Now consider maintaining a third group of animals which, while not administered the toxicant, are allowed to eat only the amount of feed consumed by the toxicant-treated animals (pair-fed). If the depression of enzyme activities in the treated group is the same as in the pair-fed group, you conclude that the effect of the toxin on lipogenesis is mediated through reduced food intake. If the depression is greater in the treated group than in the pair-fed group, you conclude that reduced food intake is not solely responsible for the effect. Such knowledge is crucial to the design of subsequent experiments.

Pair feeding is also useful in studying the interactions of toxicants with vitamin or other nutrient deficiencies. Vitamin deficiencies usually result in reduced food intake and slower growth. Pair feeding the control diet to food intake levels of the vitamin-deficient diet with or without toxicant administration or to food intake levels of the control diet with toxicants administration can yield meaningful insights.

11.3.1b Nutritional State

The term *nutritional state* implies an all or none phenomenon, but, in reality, nutritional state can be a continuum between full satiation and prolonged starvation. In most toxicological research, nutritional state means either that feed is available ad libitum or that the animal has been fasted for some period, usually 24 or 48 hours. Nutritional state is particularly important when dealing with drugs or toxicants that are administered orally. The hydrophobicity of some compounds (e.g., griseofluvin) is so great that absorption is nil unless the compound is administered with food. Conversely, gastric absorption of aspirin and other acidic drugs is markedly reduced when consumed with food. There are numerous reports in the literature of differences in oral toxicity of chemicals between fed and fasted animals. In some cases, the differences are great enough to change the classification (slight, moderate, severe) of the toxicity.

Alteration of nutritional state can occur also with ad libitum consumption of an imbalanced feed. With inadequate intake of protein or of specific amino acids, the ability of animals to metabolize certain chemicals is reduced. The probable cause for the reduction is decreased availability of amino acids for synthesis of enzymes or peptides that are either directly or indirectly involved in metabolism of the chemicals. Deficient intake of the energy-bearing nutrients, carbohydrates and fats, can have the same effect despite adequate protein intake. This is because the energy deficiency forces the animal to utilize amino acids for energy, thereby reducing the availability of amino acids for protein and peptide synthesis.

11.3.2c Nutrient Requirements for Phase I Reactions

Phase I reactions involve either oxidation, reduction, or hydrolysis. Because the reactions are enzymatic, amino acids are required for enzyme synthesis, and metals serve as structural components or activators for some of these enzymes. For those reaction schemes that are membrane-bound, the lipid and protein of the membranes could be considered part of the nutrient requirement for detoxication.

A simplified reaction scheme and the nutrient components for phase I oxidation

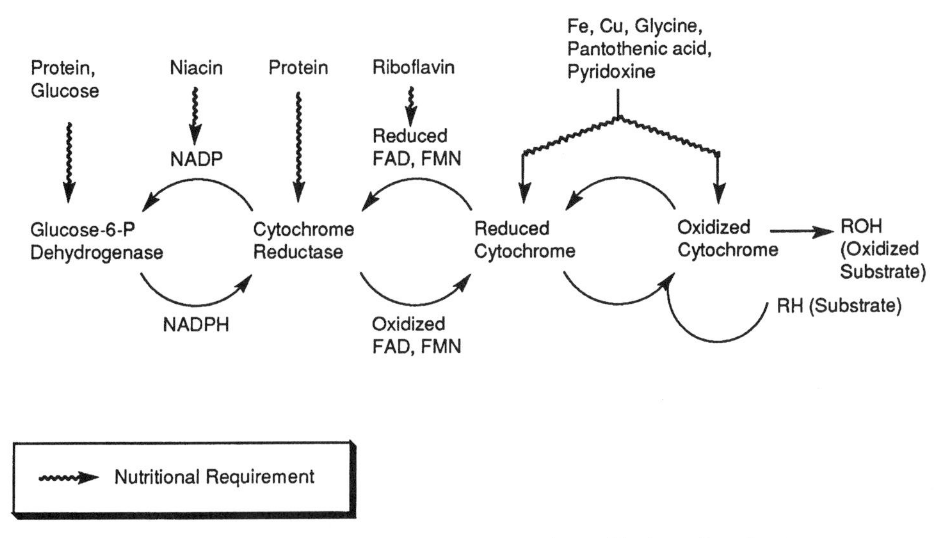

FIGURE 11.1 Monooxygenase system for oxidation of toxicants.

are shown in Figure 11.1. Phase I reduction of azo-, nitro-, or keto-compounds can proceed by reversal of the oxidation sequence. The B-complex vitamin niacin is part of NADPH which can be generated from oxidation of glucose. FMN and FAD contain another B-complex vitamin, riboflavin. Cytochrome b_5, c, or P450 contain heme moities that are synthesized from Fe, glycine, and acetyl-coenzyme A. The latter compound contains 4-phosphopantothenine, which is formed from the B-complex vitamin pantothenic acid. Pyridoxine (vitamin B_6) is also a cofactor in heme synthesis. Copper is required in the ferroxidase system responsible for conversion of ferric iron to ferrous iron prior to incorporation into heme. From the above it is obvious that deficient dietary intakes of carbohydrates, protein, minerals, vitamins or essential fatty acids could compromise the ability of the monooxygenase system to metabolize a toxicant.

Another redox reaction scheme, for dealing with peroxides, is shown in Figure 11.2. In this scheme, reducing equivalents are supplied, via reduced glutathione (GSH), for reduction of lipid hydroperoxides by glutathione peroxidase. If allowed to accumulate in tissues, the peroxides will form free radicals which can induce a cascade effect on production of additional peroxides from fatty acids in membranes. By virtue of its antioxidant properties, vitamin E can block the cascade effect of the free radicals. Several nutrients besides vitamin E are required. Niacin is part of the structure of NADPH which supplies reducing equivalents for glutathione reductase. As noted above, NADPH is formed as a result of glucose metabolism. Glutathione reductase requires riboflavin (FAD) as a prosthetic group. Methionine can be converted to cysteine, which, along with glycine and glutamic acid, is a component of glutathione. Glutathione peroxidase requires selenium as part of its structure. Finally, polyunsaturated fatty acids are the substrates for production of lipid hydroperoxides and free radicals. Thus, as in the previous example, all of the nutrient classes are involved in the reaction scheme depicted in Figure 11.2.

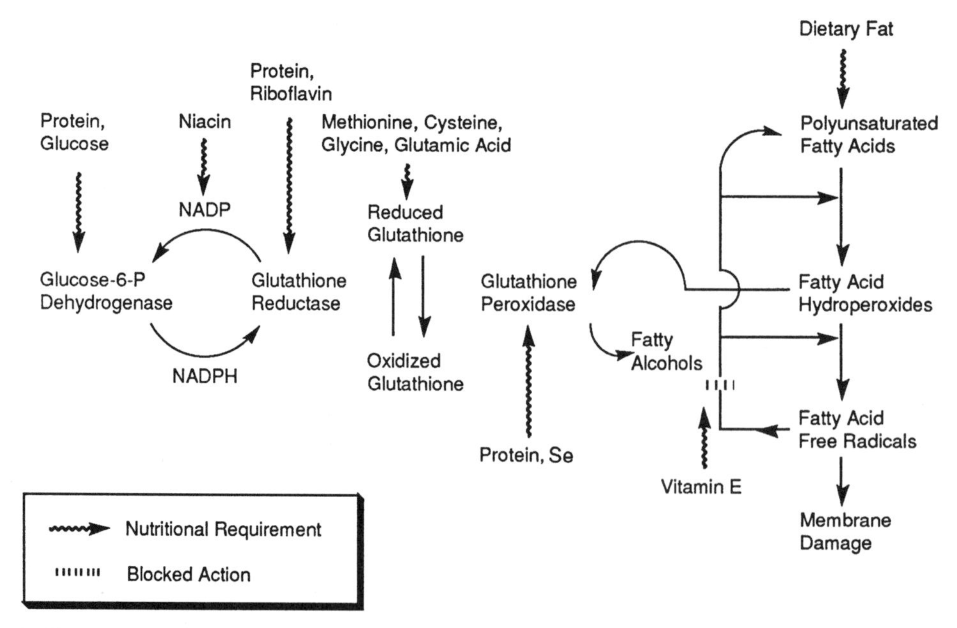

FIGURE 11.2 Glutathione redox reactions for reduction of peroxides and involvement of vitamin E in blocking the action of peroxides.

11.3.2d Nutrient Requirements for Phase II Reaction

Phase II reactions involve coupling (conjugation) of foreign compounds, foreign compound metabolites or naturally occurring compounds with endogenous compounds (conjugating agents) formed from nutrients. The major phase II reactions are described in Chapter 5, therefore only the nutrient requirements for these reactions will be listed. As with phase I reactions, the enzymatic nature of phase II reactions implies the nutritional requirement for amino acids in enzyme synthesis.

Glucuronic acid conjugation is preceded by a series of reactions in which glucose is converted to uridine diphosphate glucuronic acid (UDPGA). An intermediate step, conversion of UDP-glucose to UDPGA, requires NAD as a hydrogen acceptor. The UDPGA is available for conjugation of a variety of compounds containing oxygen, nitrogen, or sulfur groups. The nutrients required are glucose and niacin (for NAD).

Sulfate conjugation results in sulfate ester formation and proceeds only after activation of the sulfate ion with ATP. Magnesium ion is a cofactor in the conversion of sulfate to adenosine 3′-phosphate-5′-phosphosulfate (PAPS), the active conjugating agent. The PAPS is enzymatically conjugated with phenolic OH groups. Minor conjugations occur also with aliphatic OH and aromatic NH_2 groups. Nutrient requirements include magnesium and sulfur. The sulfur must be in the form of sulfate and can be obtained both by sulfate ingestion and by conversion to sulfate of the sulfur in the amino acids cysteine and methionine. The ATP requirement for PAPS formation implies a dietary requirement for phosphorus.

Methyl group conjugation occurs primarily via transfer of the methyl group from methionine to nitrogen, oxygen, or sulfur groups of foreign compounds. Methionine must be activated by ATP to form S-adenosylmethionine prior to methyl group transfer. Magnesium is required for activity of certain of the methyltransferase enzymes. Additionally, vitamin B_{12} can act as a carrier in the transfer of methyl groups from N5-methyltetrahydrofolate. Both choline and methionine can serve as sources of methyl groups for the latter reactions. It appears that 5-adenoxylmethionine is the major source of methyl groups for detoxication whereas the vitamin B_{12}-folic acid pathway plays a minor role. Nutrient requirements for methylation consist of methionine, choline, folic acid, vitamin B_{12}, magnesium, and phosphorus (for ATP).

Acylation reactions involve the conjugation either of acetic acid with a foreign amine (acetylation) or of an amino acid with a foreign carboxylic acid (amino acid conjugation). Acetylation requires the activation of acetate with coenzyme A (CoA). The acetyl group then is transferred from acetyl-CoA to the amine by acetyltransferase enzymes. Amino acids used in conjugation differ with species and include glycine, glutamine, ornithine, arginine, alanine, serine, and taurine. Amino acid conjugation requires the activation of the foreign carboxylic acid with CoA prior to conjugation. The CoA activations require ATP. Nutrient requirements for acylation are the amino acids listed above, pantothenic acid (as part of CoA), phosphorus (for ATP), and acetate, which is derived from the catabolism of glucose, fatty acids, or amino acids.

Glutathione conjugation is a primary detoxication mechanism for a variety of electrophilic compounds. Glutathione S-transferases conjugate foreign compounds to the sulfhydryl group of the cysteine moiety of glutathione as a first step in the process. Successive steps involve removal of glutamic acid and then the glycine residues from the conjugate. Finally, the amino group of the cysteine portion of the conjugate is acetylated to form a mercapturic acid derivative of the foreign compound. Both GSH synthesis and the acetylation step require ATP. Nutrients required in this system are the component amino acids of GSH, acetate and pantothenic acid for acetyl-CoA formation (acetylation step), and phosphorus for ATP formation.

11.4 NUTRIENT-TOXICANT INTERACTIONS

There is voluminous scientific literature dealing with the interactions of nutrition with toxicology. Adding to the problem of classifying and assimilating this literature is the fact that the reports appear not only in the core areas of nutrition and toxicology/pharmacology, but in the literatures of biochemistry, physiology, cell biology, immunology, clinical medicine, genetics, and general biology as well. Obviously, sheer volume prevents cataloging the available literature in a single chapter. Rather, some examples will be presented of nutrient-toxicant interactions that support the theses, first, that nutrient supply can alter toxicity, and second, that exposure to toxicants can alter nutritional requirements. From these examples, the reader should understand that nutrient-toxicant interactions cannot be predicted without knowledge of uptake, distribution, metabolism, disposition, and mode of action of a toxicant or without knowledge of nutrient requirements and nutritional state of the test organism.

11.4.1 Nutrient Effects on Toxicants

Nutrient effects on absorption of chemicals were mentioned in Section 11.3.1b. Suffice it to say here that, depending upon the properties of a xenobiotic, the presence or absence of nutrients in the gastrointestinal tract can either increase or decrease absorption. Once absorbed, the transport and distribution of a toxicant can be markedly affected by nutrients. Blood proteins, such as albumin, are able to bind exogenous lipophilic substances and selectively release them at specific tissue sites. Free fatty acids also bind to albumin, and if present in high concentration, can alter the binding and distribution of toxicants. Additionally, low dietary protein intake can reduce the plasma concentration of albumin with a corresponding reduction in binding capacity for xenobiotics.

The metabolism of xenobiotics can give rise to many complex interactions with nutrients. A dietary deficiency of several nutrients (e.g., riboflavin and protein) will elicit a general decrease in monooxygenase activities. If barbiturates are administered to deficient animals, sleeping time is prolonged because of slower conversion of the barbiturates to less active products and slower clearance of these products and the parent compound from the blood. In contrast, because the metabolism of heptachlor results in formation of products that are more toxic than the parent compound, riboflavin or protein deficiency decreases the toxicity of heptachlor. The effects of riboflavin and protein deficiencies on monooxygenase activity are considered to result from decreased cytochrome reductase activity (riboflavin) and decrease in general enzyme activity (protein).

Exactly the opposite effects on barbiturate and heptachlor toxicities are observed with a dietary deficiency of thiamin because this deficiency increases monooxygenase activity. The mechanism by which thiamin deficiency stimulates monooxygenase activity is unclear. The effects of a deficiency are to alter glucose metabolism by reducing glycolysis and increasing pentose shunt activity. The effect of the alteration would be a net increase of NADPH production, which would provide additional reducing equivalents for cytochrome reductase.

From the foregoing discussion, it is clear that four different results are possible depending upon which nutrient is deficient and upon metabolic characteristics of the toxicant to which an organism is exposed. The four possibilities are shown in Table 11.1. The situation can be further complicated by secondary factors. For example, absorbed boric acid complexes with riboflavin and enhances excretion of the vitamin. Thus, an organism fed a riboflavin-adequate diet, when exposed to borates, can become riboflavin deficient. Consider a hypothetical case of two experiments in which heptachlor toxicity is being examined using a diet with adequate, but not excess, riboflavin. In the first experiment, a dose-response curve for heptachlor toxicity is established. During the second experiment, a roach infestation of the animal quarters necessitates treating the premises with boric acid, which is inadvertently sprinkled on the feed. Now the animals are riboflavin deficient and the toxicity of the heptachlor is decreased. If the researchers are unaware of the boric acid variable, what do they conclude about the toxicity of heptachlor?

Conjugation reactions also can be affected by nutrient intake in an apparently paradoxical manner. Protein deficiency increases glucuronyltransferase activity, and therefore tends to promote metabolism and excretion and lessen toxicity of com-

TABLE 11.1 Differing Effects of Nutrient Status and Monooxygenase Activity on Toxicity of Parent Compound vs. Metabolites.

Nutrient Status	Monooxygenase Activity	Greater Toxicity With
Deficit	Decreased	Parent compound
Excess	Increased	Metabolites
Deficit	Increased	Metabolites
Excess	Decreased	Parent compound

pounds such as aniline, phenol, sulfathiazole, and thiophenol. Conversely, protein deficiency reduces glutathione conjugation and mercapturic acid formation presumably because it limits the availability of amino acids for glutathione synthesis. Methionine is of particular interest in this regard because it is an essential amino acid as well as being the precursor of cysteine. Both protein and methionine deficiency reduce excretion and increase toxicity of compounds such as dichloronitrobenzene and diethyl maleate. Lead (Pb) also forms a conjugate with glutathione, and methionine deficient diets exacerbate Pb toxicity. Surprisingly, a deficiency of lysine, another essential amino acid, protects against Pb toxicity. The lysine effect probably reflects a decreased utilization of amino acids for protein synthesis with a concomitant greater availability of methionine for glutathione synthesis.

Perhaps the most complex toxicant-nutrient interactions are in the field of chemical carcinogenesis. The series of steps by which a potentially carcinogenic compound (procarcinogen) can lead to cancer are discussed in detail in chapters 10 and 17. First, the procarcinogen is converted by single- or multi-step reactions to the ultimate carcinogen. Next, the ultimate carcinogen can form an adduct with DNA which, in turn, can initiate tumor formation. When the tumor increases in volume and its cells metastasize, cancer results. At each of these steps there are one or more protective mechanisms, all of which can be influenced by nutrition. Conversion of procarcinogens to carcinogens is accomplished primarily by monooxygenase enzymes, but in some case (e.g., N-2-fluorenylacetamide) transferase enzymes are involved. The monooxygenase and transferase enzymes also can result in detoxification of carcinogens. Nutritional alterations of these enzymes (described in sections 11.3.2c and 11.3.2d) could have a significant influence on the total dose of carcinogen available for binding with DNA. Nutritional factors, such as dietary protein and lipid, could affect membrane structure and function and thereby alter the ability of the carcinogen to gain proximity to the DNA. Genome repair is enzymatic and thus may be controlled to some extent by dietary protein. Finally, nutrition profoundly affects the immune system and could alter the immunological response to tumors and metastatic cells.

There is good experimental evidence that diet can affect the induction of certain types of cancers in laboratory animals. Elevated caloric intake and high levels of dietary fat have been shown to increase the incidence of breast and colon cancer in rats. Compared with low-fat diets, high-fat diets containing corn oil, safflower oil, and animal fat promote breast and colon cancer in animals whereas diets high in coconut oil, fish oil, and olive oil do not. Diets high in certain types of non-nutritive fiber reduce the incidence of breast and colon cancer in animals. There is convincing epidemiological evidence that the incidence of breast and colon cancer in humans is correlated with caloric intake, fat level, and fiber intake.

11.4.2 Toxicant Effects on Nutrients

When a toxicant and a nutrient interact, the usual expectation is that by supplying a higher level of the nutrient in the diet, toxic effects are lessened. Hence, the toxicant would appear to produce a higher nutritional requirement. However, as pointed out in the previous section, lower dietary levels of certain nutrients can afford protection from the effects of some toxicants. In some cases, a single toxicant can produce both higher and lower apparent requirements for different nutrients. Pb is an example of the latter phenomenon.

The toxic response to dietary Pb in animals is manifested by impaired growth, lower delta-aminolevulinic acid dehydratase activity, anemia, hemolysis, alterations of biogenic amine levels in brain, reduced food intake, tremors, apparent learning deficits, and increased mortality. Pb is a cumulative poison, and therefore the effects are both dose- and time-dependent. In avian species, one of the most reproducible effects of Pb toxicity is a marked increase of liver, plasma, and erythrocyte membrane levels of arachidonic acid (20:4). Consequently, total fatty acid unsaturation of tissues is higher and presumably, lipid membranes are more fluid and more susceptible to peroxidation. There is a concomitant, marked increase of tissue GSH that provides protection against peroxidative tissue injury. Additionally, GSH can form a conjugate with Pb, and thus can enhance Pb excretion. Supplementation of a methionine-deficient diet to a normally adequate methionine level affords protection against the growth-inhibitory effects of Pb. The methionine effect appears to be related to conversion of methionine to cysteine with subsequent incorporation into GSH. If the Pb dose is great enough, dietary methionine levels beyond the nutritionally adequate level are needed for protection. The latter results suggest an increased methionine requirement.

It is conceivable that the methionine effect could result from methionine serving as a methyl group donor. In experiments to test that hypothesis utilizing another methyl donor, choline, it was found that addition of choline to a choline-deficient, methionine-adequate diet fed to Pb-exposed animals stimulated growth. The same choline addition to a diet deficient in both choline and methionine depressed growth of Pb-exposed animals. Choline additions to diets of animals not exposed to Pb were growth stimulatory regardless of the dietary methionine level employed. The results of the experiments led to two conclusions: first, that the protective effect of methionine does not reside in its ability to donate methyl groups, and second, that the choline requirement for growth of animals fed a low-methionine diet appears to decrease with exposure to Pb. Methionine conversion to cysteine is known to be depressed by choline. Thus, it may be that with a methionine-deficient diet, addition of choline during Pb exposure further reduces cysteine supply for GSH synthesis beyond the reduction already provoked by methionine deficiency. Whatever the mechanism, it is clear the Pb increases the methionine and decreases the choline requirements. Another example, in birds, of both reduction and increase of nutritional requirements by a toxicant involves aflatoxin. Although a potent carcinogen for many species, aflatoxin is not carcinogenic in chickens. Aflatoxin produces a toxicosis in chickens that is characterized primarily by marked accumulation of liver fat with associated decreased hepatic function. The deleterious effects of aflatoxin are ameliorated by feeding a thiamin-deficient diet. The effect of the deficiency may be mediated through an increase of monooxygenase activity leading to increased metabolism and excretion of aflatoxin. Excess dietary protein or lipid has a protective effect in aflatoxicosis. Thus, while aflatoxin decreases the thiamin requirement, the requirements for protein and lipid are increased.

There are numerous additional circumstances in which xenobiotics have been shown to increase nutrient requirements. Some representative examples for the various nutrient classes follow.

11.4.2a Water

There are several types of effects of xenobiotics that can increase water requirements. Any compound with a diuretic action would increase water losses via the kidney. The fever resulting from exposure to pyrogenic compounds can increase water losses through perspiration. Any compound that induces vomiting or diarrhea increases water requirement as will any compound that promotes water retention by tissues.

11.4.2b Carbohydrates

Increases of carbohydrate requirements by toxicants can occur principally in three ways. First, the xenobiotic can interfere with absorption of carbohydrate from the intestines. This interference can be overcome by increasing carbohydrate intake. Efficacy of increased carbohydrate intake in overcoming malabsorption is suggestive of increased requirement. Second, there is an inferred increase of carbohydrate requirement when there is exposure to any chemical that is excreted after glucuronide conjugation. Third, non-nutritive carbohydrates, which are components of the fiber fraction of diets, can bind xenobiotics so that they pass through the gastrointestinal tract without producing toxicity. The fact that increased dietary fiber levels confer increased protection is an argument for the view that toxicants can increase the fiber requirement.

11.4.2c Lipids

The protective effect of increased dietary fat on aflatoxicosis in chickens was mentioned above. One effect of aflatoxin in chickens is a reduction of lipid transport. Because approximately 90% of the total fatty acid synthesis occurs in the liver of chickens, impaired transport appears to be responsible for the lipid accumulation in liver. Because one effect of dietary fat is to depress hepatic fatty acid synthesis, less fat may accumulate and consequently less liver damage would occur.

There is controversy and contradictory evidence on the effects of dietary fat on detoxication mechanisms in general. Definitive statements on effects of toxicants on dietary lipid requirements will require more research.

11.4.2d Proteins

As with dietary fats, the effects of dietary protein levels on xenobiotic metabolism can vary. In general, as the protein level in the diet is increased from deficiency through excess, activities of phase I enzymes and glutathione S-transferases are increased while activities of glucuronyltransferases and sulfotransferases are decreased. Depending upon the route of metabolism and the nature of the toxicities of the parent compound and its metabolites, increasing dietary protein can either increase or decrease the toxicity of a xenobiotic. Again, the argument can be made that in those cases where toxicity is reduced by higher protein intake, the toxicant exposure results in a higher protein requirement.

11.4.2e Minerals

Phytate, which is a natural constituent of whole grains such as wheat, becomes a xenobiotic when consumed by animals. Animals consuming diets high in whole grains can experience a zinc deficiency as a result of binding of zinc to phytate which is not absorbed. The deficiency can be counteracted by additional dietary zinc.

Iron-deficiency anemia can be provoked either by inadequate iron intake or by significant iron losses resulting from hemorrhages. Because aspirin can produce occult gastrointestinal bleeding, iron requirements are increased in susceptible individuals by prolonged aspirin administration. Diuretics can enhance excretion and thereby increase requirements for several minerals. Furosemide increases the excretion of calcium whereas chlorothiazide increases excretion of magnesium, calcium, and potassium. Thiazide treatment for hypertension can result in potassium depletion to the extent that potassium supplements or a high potassium diet are required. In contrast, spironolactone spares potassium even when used in conjunction with thiazides. Consequently, spironolactone diuretics can be fatal if the diet is supplemented with potassium.

11.4.2f Vitamins

Methotrexate is a folic acid antagonist that has been used in the treatment of several types of malignancies and psoriasis. Side effects of the drug produce the typical symptomology of folic acid deficiency, which includes soreness of the mouth, diarrhea, abdominal pain, hair loss, and delayed wound healing. Methotrexate binds to dihydrofolate reductase. The binding, and therefore the side effects, can be reduced by additional dietary folic acid, however, such therapy would defeat the purpose of the drug treatment.

Pyridoxine (B_6) requirements are markedly influenced by several xenobiotics. Some examples are: (1) L-dopa, which increases urinary excretion of B_6, (2) cycloserine, which inhibits conversion of B_6 vitamins to the active coenzyme, pyridoxal phosphate, and (3) penicillamine, which competes with pyridoxal phosphate for binding sites on B_6-requiring apoenzymes. All of these xenobiotic effects can be reversed by additional dietary B_6.

Vitamin K is available to animals either as vitamin K_1 in food or as vitamin K_2 from synthesis by intestinal bacteria. As with B_6, xenobiotics can affect vitamin K utilization at several steps. Examples are: (1) broad spectrum antibiotics that can inhibit intestinal synthesis of K_2, (2) cholestyramine, which sequesters bile salts and thereby interferes with vitamin K absorption, and (3) coumarin anticoagulants (antimetabolites of vitamin K) which inhibit vitamin K-dependent prothrombin synthesis. The end result of the actions of these xenobiotics is decreased production of blood clotting factors and proneness to hemorrhages. Their actions can be ameliorated by increased dietary vitamin K.

11.5 FEEDING OF ANIMALS USED IN TOXICOLOGICAL RESEARCH

No competent toxicologist would attempt to measure the toxicity of a substance without some knowledge of its composition; nor would a competent nutritionist attempt to measure the nutritional value of a diet without similar knowledge about the diet. Given the nature of the interactions between toxicology and nutrition, it seems imperative that toxicological research should require the benefit of both types of knowledge. Yet, a majority of studies in toxicology are performed with little or no knowledge of diet composition. The remainder of this chapter will deal with why this is so and with alternatives.

11.5.1 Closed and Open Formulas

Two types of animal diets are available commercially, the closed formula and the open formula. With the closed formulas, the purchaser knows what ingredients are included in the diet and has a manufacturer's guarantee that the diet has been formulated to meet all known nutrient requirements of the animals for which it is intended. The purchaser receives no information as to the relative proportions of the ingredients of the diet. In fact, these proportions can vary from lot to lot because the diets are formulated not only to meet nutrient requirements, but to do so as cheaply as possible. This practice is called least cost formulation and is almost universal in the feed industry. Such diets have been shown to vary in nutrient content from deficient to greatly in excess of requirements in different lots from the same manufacturer. Additionally, one can never be certain that the ingredients are not contaminated with compounds that could affect experimental results (e.g., aflatoxins, nitrates, herbicides, or pesticides).

With open formulas, the purchaser knows the level of ingredient inclusion in the diet. Open formulas can be of three types. The first is a cereal-based diet in which the majority of the diet consists of cereal grains and grain by-products. The variations in nutrient content encountered with open formulas are not as great as those encountered with closed formulas because the level of inclusion of each ingredient is the same from lot to lot. There is, however, some variability because of differences in ingredient quality among lots. The same dangers of possible contamination of ingredients exist as with ingredients used in closed formulas.

The second type of open formula is the purified diet. This type is formulated with refined carbohydrates, fats, and proteins and contains crystalline vitamins and pure (though not necessarily reagent grade) minerals. The chief advantages of this type of diet over a cereal-based diet are uniformity of nutrient content of ingredients and lower probability of contaminated ingredients. However, contamination problems are not completely eliminated, and the researcher should exert due caution.

The third type of open formula is the chemically defined diet. The principal difference between this type and the purified diet is that crystalline amino acids replace the protein source and the minerals are reagent grade. The expense of such diets limits their usefulness in toxicological research.

One final admonition about diets is appropriate. There are papers in the toxicology literature in which the effect of a vitamin deficiency on the action of a toxicant are reported. In one, and there may be others like it, the authors used a closed formula diet as the control and an open formula diet to produce the vitamin deficiency. Almost certainly the authors observed more than just the effect of the vitamin on toxicant action, but neither they nor anyone else will ever know exactly what the "more" was.

11.5.2 Prepared Versus Mix-it-Yourself Diets

There are advantages and disadvantages to both buying prepared diets and to buying ingredients and preparing the diets in the laboratory. The principal advantages of premixed diets are that extensive inventories of ingredients are not needed; there is no investment in diet preparation equipment; and there are considerable savings in labor. The principal disadvantages are that particular diets may not be available commercially; diet modifications may not be feasible; and quality control of single ingredients is not possible. Obviously, the advantages and disadvantages of one system become the disadvantages and advantages, respectively, of the other.

11.5.3 What Diet Do You Choose?

Ideally, all toxicologists working with a particular species would use the same diet, however, it would be impossible to get all nutritionists to agree on the composition of a standard diet for toxicological research with that species. The American Institute of Nutrition (now renamed the American Society for Nutritional Sciences) has proposed diets for toxicological research with rodents. These diets may be found in the *Journal of Nutrition*, 123 (1993), 1939–1951.

There is no set of rules for choosing a diet. The choice depends largely upon the type of research to be done and the intended application of the research results obtained. To illustrate, consider one nutrient, protein, and one species, the rat. Most commercially available laboratory diets for rats contain 22–24% protein. However, the National Research Council lists the protein requirement of a weanling rat as 12% with an ideal protein source and 13.6% with casein as the protein source. In view of the earlier discussion of the effect of dietary protein level on toxicological responses, the selection of a dietary protein level is crucial to the assessment of those responses. At this point, the intended use of the data becomes important to the choice of protein level. Extrapolation of data from rats to humans requires knowledge of the protein status of the human population. In some developing countries, protein deficit is common and results in a syndrome in children known as kwashiorkor. To extrapolate to such a population from data obtained using rats fed 22% protein could be misleading. Conversely in developed countries, much of the population consumes an excess of protein, and thus extrapolation from data obtained with a protein level above 13% for rats can be justified. But even here, not all persons in a developed country consume excessive or even adequate amounts of protein. Protein intake is a continuum in a population as are intakes for all other nutrients. Toxicologists must be aware of such considerations when choosing a diet, and the choice must be appropriate to the question being asked by the researcher.

A summary of the advice on diet choice will close this section. If at all possible, toxicologists should feed an open formula diet to their animals. The diet should be composed of purified ingredients to minimize nutrient variability and possible contamination. Nutrient levels of the diet should be selected based upon the aims of the research.

11.6 NUTRIENT-CANCER INTERACTIONS

Cancerous cells require a variety of nutrients for their metabolism and reproduction, just as do normal cells. Thus, it is no suprise that many cancer researchers have chosen to work in the general area of cancer-nutrient interactions. A majority of this work has centered around prevention, but there have also been studies involving putative effects of nutrients in cancer treatment. Although much is known about diet and cancer prevention, some long-held theories have recently come into question.

Epidemiological studies and animal experiments have suggested that diets high in various types of dietary fiber are beneficial in the prevention of colorectal cancers. The theory is that dietary fiber promotes regularity of bowel function and binds carcinogens to reduce their toxicity or enhance their excretion. Thus, the exposure of cells in the lower digestive tract to carcinogens of dietary or gut microflora origin would be lessened. However, a recent, widely quoted, epidemiological study suggests that inclusion of additional fiber in an otherwise normal diet is of no value in the prevention of rectal cancer.

Even more perplexing is the fact that several human intervention trials, in which the efficacy of beta-carotene was being investigated, had to be terminated because of the association of beta-carotene with *increased* incidence of lung cancer in smokers. Beta-carotene is a provitamin A that possesses strong antioxidant properties and is often added to over-the-counter vitamin preparations. This practice arose because of convincing evidence that high intakes of fruits and vegetables, which elevate blood levels of beta-carotene, are predictive of lower incidence of lung cancer in smokers. It is thought that nutrients with antioxidant properties prevent the induction of cancer by protecting the genetic material from free radicals and other oxidants. That pharmacological beta-carotene increases lung cancer whereas the dietary form decreases it may be related to the fact that the beta-carotene in food is consumed with other protective nutrients. Or, there may be factors other than beta-carotene present in fruits and vegetables that are protective.

In contrast to beta-carotene, lycopene, the red pigment in tomatoes, has been shown to protect against prostate cancer regardless of whether it is administered alone or via tomatoes. Such inconsistencies suggest that although much is known concerning cancer-diet interactions, much remains to be discovered.

Another insight into the complexity of using diet to prevent cancer can be gained by reading the short, incisive comment on brussels sprouts [Paolini, M. Brussels sprouts: an exceptionally rich source of ambiguity for anticancer strategies. *Toxicol. Appl. Pharmacol.* **152** (1998), 293–294] published recently. Members of the cabbage (Brassica) family, including brussels sprouts, contain several glucosinolates that can stimulate phase II enzymes while depressing phase I enzymes. These enzymatic changes tend to either prevent or slow the progress of some cancers. Paradoxically, these same compounds can have the opposite effect. Paolini suggests that if there is a beneficial ratio of phase I/phase II effects, genetic selection of a vegetable for a higher content of a specific compound could be dangerous.

One can also find in the literature studies that show that restricted food intake in animals is anticarcinogenic. Exactly where the optimum line is in humans between "all you can eat" and anorexia remains in question. Given the uncertainties concerning the effects of diet on cancer and other chronic diseases, probably the best nutritional advice is: eat a balanced diet (a variety of foods) and eat it in moderation.

SUGGESTED READING

American Society for Nutritional Sciences, Symposium: Animal Diets for Nutritional and Toxicological Research. *J. Nutr.* 127 (1997), 824S–856S.

Conner, M.W. and Newberne, P.M. Drug-nutrient interactions and their implications for safety evaluations. *Fund. Appl. Toxicol.* 4 (1984), 5341–5356.

Finley, J.W. and Schwass, D.E. (Eds) *Xenobiotic Metabolism: Nutritional Effects*. American Chemical Society, Washington, DC, 1985.

Hart, R.W., Neumann, D. and Robertson, R. (Eds) *Dietary Restriction: Implications for the Design and Interpretation of Toxicity and Carcinogenicity Studies*. ILSI Press, Washington, DC, 1995.

Hathcock, J.N. (Ed) *Nutritional Toxicology*. Academic Press, New York, 1982.

Heber, D., Blackburn, G.L. and Go, V.L.W. (Eds) *Nutritional Oncology*. Academic Press, New York, 1998.

ILSI Europe Concise Monographs, *Dietary Fibre: Nutritional Function in Health and Disease*. ILSI Press, Washington, DC, 1994.

Kotsonis, F.K., Mackey, M.A. and Hjelle, J. (Eds) *Nutritional Toxicology*. Taylor & Francis, Philadelphia, PA, 1994.

National Academy of Sciences. *Carcinogens and Anticarcinogens in the Human Diet: A Comparison of Naturally Occurring and Synthetic Substances*. National Academy Press, Washington, DC, 1996.

Omaye, S.T., Krinsky, N.I., Kagan, V.E., Mayne, S.T., Liebler, D.D. and Bidlack, W.R. Beta-carotene, friend or foe? *Fund. Appl. Toxicol.* 40 (1997), 163–174.

Reddy, B.S. Dietary fat and cancer: specific action or caloric effect. *J. Nutr.* 116 (1986), 1132–1135.

Roe, D.A. and Campbell, T.C. *Drugs and Nutrients*. Marcel Dekker, Inc., New York, 1984.

CHAPTER TWELVE

Toxicant-Receptor Interactions: Fundamental Principles

RICHARD B. MAILMAN and CINDY P. LAWLER

12.1 DEFINITION OF A "RECEPTOR"

Although the concept of a receptor as a selective locus of action was pioneered more than a century ago, receptor theory remains a critical concept in toxicology and in many other branches of biology. In one sense, receptors may be considered as macromolecules that bind small molecules (commonly termed ligands) with high affinity, and thereby initiate a characteristic biochemical effect. In some fields of biology, the term *receptor* has specific constraints. For example, in a cell biological context, the term refers to macromolecules (intracellular or cell surface) that recognize endogenous ligands; these ligands may be small (e.g., neurotransmitters, hormones, autacoids) or large (e.g., proteins involved in protein sorting or in intracellular scaffolding). In toxicology and pharmacology, the term *receptor* is often used to refer to the high-affinity binding site that initiates the functional change(s) induced by a xenobiotic (e.g., toxicity). It is important to note that the interaction of two macromolecules can be studied in much the same way as a ligand and a single macromolecule, and such macromolecular interactions have great importance to biological systems in general, and toxicology in particular. The use of fundamental structural precepts (both steric and electrostatic) to understand the "fit" between receptor and ligand has become more feasible during the past few years based on advances in both molecular biology, and in computational chemistry and molecular modeling.

12.1.1 A Brief History of the Concept of Receptors

The concept of receptors is credited to the independent work of Paul Ehrlich (1845–1915) and J.N. Langley (1852–1926). With Ehrlich, the concept appeared to originate from his immunochemical studies on antibody-antigen interactions. Based on the high degree of specificity of antibodies for antigens, Ehrlich postulated the existence of stereospecific, complementary sites on the two molecules. Similar reasoning led Emil Fisher to formulate the idea of a lock-and-key fit between enzymes and their substrates. In later studies with arsenicals to be used against syphilis and trypanosomes, Ehrlich observed that slight modification in chemical structure could dramatically affect the potency of a compound. This high degree of stereospecificity suggested to Ehrlich that his receptor theory was also relevant to drugs, and also led to the concept of structure-activity relationships. Although this idea seems straightforward today, it was decades before physical evidence for this idea could be gleaned.

The physiological importance of receptors was shown by Langley. Studying the denervated frog neuromuscular junction, Langley observed that muscle contraction could be elicited when nicotine was applied to the denervated muscle, and that this effect could be blocked by curare. He postulated the existence of "the receptive substance" that could bind both nicotine and curare; this was the first formulation of what is now called the neurotransmitter receptor. Langley also speculated that the formation of complexes between drug and receptive substance could be described lawfully through consideration of the relative concentration of a drug and its affinity for the receptive substance. This latter speculation predated the formal mathematical description of drug-receptor interactions according to mass action principles.

12.1.2 Toxicants and Drugs as Ligands for Receptors

The rationale for applying principles derived from the study of drug-receptor interactions to questions concerning possible toxicant-receptor interactions becomes clear when one considers that a drug is defined broadly as any chemical agent that affects living organisms; thus, a poison (toxicant) can be considered as a drug that has been administered at a dose level that results in detrimental effects, even death. Indeed, most clinically useful drugs have beneficial effects at one dose, yet become toxic (or even lethal) at higher doses. Clinical pharmacologists and toxicologists have quantified these observations in the term therapeutic index, that can be defined:

$$\text{Therapeutic Index} = \frac{\text{TD50}}{\text{ED50}}$$

The TD50 is the dose of a drug that would be toxic (or lethal) to one-half of a given population; the ED50 (ED for effective dose) is the dose that produces the beneficial effect in one-half of a comparable population. The therapeutic index is large for drugs that cause toxicity only at doses much higher than needed for therapy, and small for drugs that have a narrow margin of safety.

Although drugs are of considerable interest to toxicologists, other chemical agents are of at least equal importance. These include agricultural and industrial chemicals (e.g., herbicides, insecticides, chemical wastes, industrial solvents, etc.), as well as environmental contaminants. Whatever the source of a xenobiotic—drug or pollutant—its interaction with receptors can be described by the same general principles.

12.1.3 Is There a Normal Function for Toxicant Receptors?

As noted, a toxicant or drug receptor may be defined loosely as "a macromolecule with which a drug (or toxicant) interacts with high affinity to produce its characteristic effect." Inherent in this is the fact that binding of the toxicant (or drug) to the receptor causes a specific and predictable biological response. The obvious question, then, is whether the receptor of interest is one to which endogenous ligands bind, or rather, an opportunistic site for which the toxicant has affinity. Because this distinction is of general importance, several examples may be useful.

The pharmacological and toxicological properties of opium alkaloids (e.g., morphine and codeine) were known for centuries, and the framework provided by Langley, Ehrlich, Dale, and others led to the hypothesis that a specific receptor for these drugs existed. By the early 1970s, a receptor (actually, several related receptors) was identified that binds these opium alkaloids. The question then was whether there was an endogenous opium-like material in the nervous system that normally bound to these sites. It is now known that there are families of nerve cells that release peptides having morphine-like actions (see Chapter 19); these opioid peptides are the endogenous ligands for the morphine receptor. Another similar example relates to the alkaloids derived from the "deadly nightshade" (*Belladonna* sp.). It is now known that atropine and related compounds bind to a subtype of receptor for the endogenous neurotransmitter acetylcholine. These are only two early and well-studied examples.

In other cases, however, endogenous ligands have not been found (and may not exist) for some toxicant receptors. The binding of certain classes of toxicants to the $GABA_A$ receptor provides one excellent example (see legend for Fig. 12.1). This receptor is one of the two major subclasses of recognition sites for the inhibitory neurotransmitter GABA (gamma-aminobutyric acid). It is known that the receptor is a heteropolymer that functions as an ion channel for the chloride anion. The recognition sites for GABA on the $GABA_A$ complex also bind certain toxicants (like the mushroom alkaloid muscimol)). On the other hand, there are other distinct sites on the $GABA_A$ complex that can bind small molecules, and mediate changes in receptor function. For example, picrotoxin, barbiturates, and drugs of the benzodiazepine class bind to some of these other sites. With no known endogenous ligands for these sites, their interaction with toxicants may be adventitious.

There are many other examples (e.g., pyrethroid insecticides; tetrodotoxin; dioxin) of toxicologically important xenobiotics for which an endogenous ligand is unknown, and for which the normal physiological role of the "receptor" is unclear. In such cases, scientists often name the receptor after the toxicant of interest (e.g., the "morphine receptor" or "dioxin receptor"). Sometimes these unknown xenobiotic binding sites later are found to have roles as important physiological receptors (e.g., the "morphine receptor"). In other cases, however, the receptor (while a high-affinity binding site for the toxicant) may not subserve normal physiological processes. Exploring and resolving such issues is often at the heart of modern toxicological studies.

12.1.4 Functional Biochemistry of Toxicant Receptors

From the viewpoint of biochemistry and neuroscience, "receptor" refers to a distinct class of molecules that have unique and characteristic functions and modes of action that involve endogenous ligands. Enzymes are, in one sense, receptors with respect to their substrates (and inhibitors, too). As will be demonstrated later in this

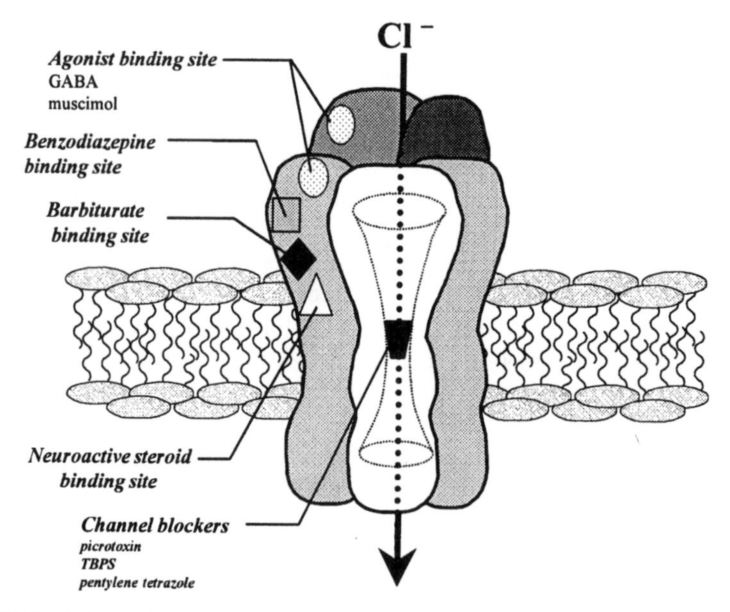

FIGURE 12.1 Schematic representation of $GABA_A$ receptor.
This polymeric receptor regulates the influx of chloride, thus providing an inhibitory neuronal influence. It consists of four or five subunits of three major types (α, β, and γ), each of which has several subforms. There are differences in the subunit composition of the receptor depending on where it is expressed. There are many sites at which xenobiotics may bind, thereby affecting the function of this receptor complex. The neurotransmitter GABA (the endogenous ligand), as well as the mushroom alkaloid muscimol, bind to the site shown as the agonist binding site in the figure. Commonly used benzodiazepine drugs, such as diazepam (Valium®), bind to a distinct site called the benzodiazepine site. A third site is where channel blockers (e.g., picrotoxinin, the active constituent of picrotoxin; the convulsant pentylenetetrazole; the pesticide derivative TBPS; the drug of abuse PCP) directly affect the chloride channel. At least two other important sites exist, including the barbiturate site (selective for drugs of the barbiturate class) and the neuroactive steroid binding site.

chapter, the interaction of receptors with their ligands is similar conceptually to the interaction of enzymes and substrates or inhibitors. In practice, however, the interaction of ligands with receptors often differs from enzyme-substrate interactions in two qualitative, but important, ways. First, the affinity of ligands for receptors is usually two or more orders of magnitude higher than the affinity of enzymes for their substrates; second, enzymes act by biochemically altering substrate molecules, whereas receptors do not function typically in this manner.

Toxicant receptors, per se, may serve a physiological role as receptors for a diverse group of endogenous ligands such as neurotransmitters, hormones, and autacoids. A common characteristic of these receptors is that they transduce chemical information of one type (the concentration of the ligand) into a variety of other secondary chemical events (including synthesis of second messengers, changes in ion flux or transport, and the like). For example, when the receptors are present on the extracellular surface of neurons, the ultimate effect of ligand binding is a change in the electrical excitability of the cell. In addition to postsynaptic receptors, cell surface receptors on neurons may be present on either presynaptic membranes or cell bodies. In these cases, they are termed autoreceptors and play a self-regulatory role (see Chapter 19 for specific examples).

Some ligands do not interact principally with cell surface receptors, but diffuse into cells and bind to intracellular receptors in the cytoplasm. For example, ligand binding to cytoplasmic steroid receptors initiates a process that is not well understood, but that involves the movement of steroid-bound receptor into the cell nucleus, where the receptor molecule interacts with genomic material, resulting in alterations in gene expression and protein synthesis.

As will become clear, this chapter is focused on toxicants for which the receptor is a high-affinity recognition site of the type discussed in the previous paragraphs. It should be noted explicitly that other toxicants have "receptors," but fall into more complex situations not appropriate for this chapter. For example, some toxicants inhibit enzymes, or are themselves enzymes. Such interesting compounds include the organophosphate and carbamate insecticides (acetylcholine esterase inhibitors) and diphtheria toxin (an enzyme).

12.1.5 Types of Interactions Between Toxicants and Receptors

Toxicant molecules that gain access to receptor sites in various tissues may then interact with those receptors in several ways. First, the interaction may mimic the endogenous ligand and cause agonist-like actions. Second, the molecule may bind to the receptor without causing resultant activation, thereby blocking access of an endogenous ligand (antagonist actions). Third, the toxicant may produce allosteric effects on receptors; for example, it has been shown that some toxicants, rather than binding to the same site as endogenous ligands, bind instead to an adjacent part of the macromolecule (see Section 12.2.2 and Fig. 12.1 for examples). This interaction causes allosteric changes that affect the function of the complex, and sometimes the binding of the neurotransmitter itself. Finally, some macromolecules may have no endogenous ligands, yet bind specific toxicants that cause physiological changes via this interaction.

The four mechanisms listed above involve direct interaction of a toxicant with a receptor; in such cases, the toxicant-receptor interaction is likely to be involved in the mechanism of action. In many cases, toxicants may affect receptor function indirectly. For example, in the nervous system, decreases in synaptic transmission (by receptor blockade or damage to a neuron) may lead to increases in the number of receptors on the target neuron (so called up-regulation). This is often felt to be one of the compensatory mechanisms by which the nervous system responds to such perturbation. Conversely, increases in synaptic transmission (e.g., by long-term receptor activation) may lead to compensatory decreases (down-regulation) in receptor number. The techniques by which such mechanisms are studied are described later in this chapter. It should be noted that the availability of molecular probes now permits evaluation not only of the characteristic of the binding sites, but also of the expression of the mRNA for the receptor(s) under study.

12.1.6 Goals and Definitions

The object of this chapter is to provide the foundation necessary for understanding the interactions of small molecules (ligands, be they toxicants or drugs) with receptors, and also the fundamentals of the techniques commonly employed for such analysis. It should be noted that the same theory is also useful for understanding the interaction of two large molecules, and the application of similar theory to such problems has increased as the importance of macromolecular interactions has been realized in recent years.

12.1.6a Definitions

There are several terms in common use that should be defined:

- *Affinity*: this is related to the "tenacity" by which a drug binds to its receptor, and reflects the difference between the rates of association and dissociation of the ligand. Thus, a ligand that binds tightly to a receptor has a small equilibrium dissociation constant, or a large equilibrium association constant. A toxicant with high affinity for a receptor will bind to that receptor even at low concentrations, whereas binding of a low-affinity toxicant will be evident only at higher concentrations.
- *Potency*: this term refers to the ability of a toxicant to cause a measured biological change (any one of interest). If the change occurs at low concentrations or doses, the compound is said to have high potency. If high concentrations or doses are required, the compound is said to have low potency.
- *Intrinsic activity* (*efficacy*): the maximal response caused by a compound (i.e., toxicant or drug) in any given test preparation. Intrinsic activity (efficacy) is always defined relative to a standard compound such as the endogenous ligand for a receptor or the prototypical toxicant for a specific system. For drugs, a *full agonist* causes a maximal effect equal to that of the endogenous ligand or reference compound; a *partial agonist* causes less than a maximal response. An *antagonist* binds to the same site but cause no functional response.

12.2 RECEPTOR SUPERFAMILIES

Receptors can be divided into several distinct classes based on the effector mechanisms evoked by ligand binding (Fig. 12.2). Three major classes of receptors that have endogenous ligands and also bind toxicants include the G protein-coupled receptors, the ionotropic receptors (also known as ligand-gated ion channels), and intracellular steroid receptors. In addition, a superfamily of enzyme-linked receptors binds a variety of polypeptide hormones and growth factors, and exert pleiotropic effects on cell physiology. Although these receptors are not known to bind xenobiotic toxicants, their central role in controlling cell proliferation and cell death makes it likely that they participate ultimately in the perturbations and adaptations induced by exposure to a variety of toxicants. Finally, voltage-gated ion channels are macromolecules that span the cell membrane and provide a pathway for particular ions. Although no endogenous ligands have been identified, a variety of xenobiotic toxicants act on these sites. Recent progress in the molecular cloning of receptors has demonstrated considerable amino acid sequence homology among receptors that share a common effector mode; such observations provide an additional rational basis for classification, but more importantly, they enable a search

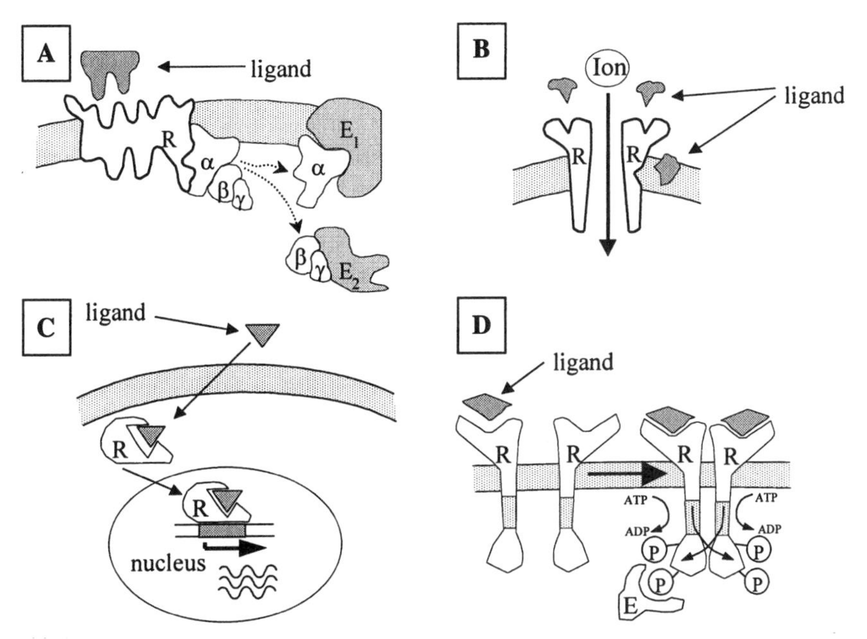

FIGURE 12.2 Schematic representation of some receptor superfamilies.
This figure represents the common mechanisms of the four major receptor families discussed in Section 12.2. (A) shows the G protein-coupled receptor. (B) represents the ionotropic receptor (ligand-gated ion channel) superfamily, on which multiple binding sites are known to exist. (C) represents the steroid receptor superfamily. (D) depicts the enzyme-linked receptor superfamily.

for characteristics (i.e., structural motifs) that must ultimately confer specificity of receptor-effector coupling.

12.2.1 G Protein-Coupled Receptors (GPCRs)

After binding of ligands, many neurotransmitter receptors (e.g., the dopamine, adrenergic, $GABA_B$, metabotropic glutamate, and muscarinic cholinergic receptors) use large guanine nucleotide binding proteins (G proteins) as their transduction systems. Thus, binding of the ligand to the receptor ultimately changes the rate of synthesis or degradation of various cytoplasmic effectors that are modulated by these G proteins (Fig. 12.2A). The G protein coupled receptors contain a single polypeptide chain with seven transmembrane spanning regions. There is an extracellular/intramembrane domain for ligand recognition, and cytoplasmic domains involved in coupling to G proteins. Interaction of the ligand with the receptor initiates a cascade of secondary events that are initiated by G protein activation. This, in turn results in the activation or inhibition of specific enzymes (e.g., adenylate cyclase or phospholipase C), with subsequent changes in the turnover of intracellular "second messengers" such as cAMP, diacylglycerol, phosphoinositols, the direct regulation of ion channels, etc. Together, these biochemical events determine the ultimate cellular consequences of ligand binding.

The best characterized model for receptor-G protein interaction is derived from studies of receptors linked to the stimulation of adenylate cyclase (e.g., β_2-adrenergic receptors). In the absence of ligand, the receptor is presumed to exist in a high-affinity state stabilized by the heterotrimer G protein that has GDP tightly bound to the α-subunit of the G protein. Agonist binding to the receptor induces a conformational change in the complex of receptor and G protein, promoting the exchange of GDP for GTP. This results in the subsequent dissociation of the G protein subunits into α and $\beta\gamma$ subunits. Both the α and $\beta\gamma$ subunits may elicit secondary events such as the activation or inhibition of the catalytic unit of adenylate cyclase.

Adenylate cyclase is considered as a second messenger that catalyzes the formation of cAMP (cyclic adenosine monophosphate) from ATP; this results in alterations in intracellular cAMP levels that change the activity of certain enzymes—enzymes that ultimately mediate many of the changes caused by the neurotransmitter. For example, there are protein kinases in the brain whose activity is dependent upon these cyclic nucleotides; the presence or absence of cAMP alters the rate at which these kinases phosphorylate other proteins (using ATP as substrate). The phosphorylated products of these protein kinases are enzymes whose activity to effect certain reactions is thereby altered. One example of a reaction that is altered is the transport of cations (e.g., Na^+, K^+) by the enzyme adenosine triphosphatase (ATPase).

These processes are all possible loci for biochemical attack by toxicants. For instance, compounds like ephedrine or mescaline are alkaloids that act as ligands at GPCRs. Many drugs have been made to act at these sites. Another possible site of attack is at the G protein level. Diphtheria and cholera toxin act on guanine nucleotide binding proteins linked to the stimulation (G_s) or inhibition (G_i) of adenylate cyclase. Both toxins act by enzymatically transferring an ADP-ribosyl group to a G protein subunit. In the case of G_s, the result is an enduring activation of adenylate cyclase, while in the case of G_i, ribosylation disrupts hormonal inhibition of adenylate cyclase.

12.2.2 Ionotropic Receptors (Ligand-Gated Ion Channels)

A number of receptors are transmembrane proteins whose structure incorporates an ion channel through the cell membrane (Fig. 12.2B). Ligand binding presumably induces conformational changes in the receptor protein that open the channel and allow ions of particular size and charge to pass through, thus altering the ionic concentrations (Na^+, K^+, Cl^-, and/or Ca^{2+}) across the membrane. In neuronal cells, the resultant shifts in membrane potential are either depolarizing or hyperpolarizing, making the cell more or less excitable, respectively. Two prototypical examples of ionotropic receptors are the nicotinic acetylcholine receptor and the $GABA_A$ receptor (Fig. 12.1 for the latter). In both cases, molecular cloning indicates that these receptors are formed from multiple subunits that are transcribed separately and then assembled after translation. For example, the $GABA_A$ receptor, one of the two major subclasses of receptor for the neurotransmitter GABA (γ-aminobutyric acid), is known to be a polymer with four or five subunits that together form a chloride channel. Besides the loci to which GABA binds, there are several other binding sites that have been characterized and are of interest to toxicologists (Fig. 12.1).

Several receptors that recognize the excitatory amino acid glutamate (e.g., the NMDA, kainate, and quisqualate classes) represent another group of ligand-gated ion channels. The NMDA receptor has been the focus of intense interest in recent years. This receptor appears to be activated primarily by membrane depolarizations that occur during periods of heightened neuronal activity. The activated NMDA receptor channel allows both Na^+ and Ca^{2+} to flow into the cell. The detrimental effects of massive Ca^{2+} influx have been implicated in the selective cytotoxicity that occurs in brain regions rich in NMDA-receptor (e.g., hippocampus) during episodes of disinhibited neuronal activity, such as during seizures. Some years ago a major poisoning episode in Canada involving domoic acid, a marine toxin that has sometimes contaminated commercial seafood. Domoic acid works as an ionotropic glutamate receptor ligand, apparently causing toxicity by permitting calcium influx through some members of these receptors.

12.2.3 Intracellular Steroid Receptors

In contrast to the preceding three categories of cell-surface receptors, binding sites for diffusible steroids can be found intracellularly, although the exact anatomical locus (cytoplasm vs. nucleus) remains controversial (Fig. 12.2C). Cloning of several of these intracellular receptors has verified the presence of distinct steroid binding and DNA binding subunits. Endogenous steroid ligands include the estrogens, androgens, progestins, glucocorticoids, mineralocorticoids, and vitamin D, as well as some active metabolites of these compounds. The proximal effect of ligand occupation of steroid receptors is the activation (or disinhibition) of a DNA binding domain, enabling the interaction of this domain with regulatory DNA sequences (promoter regions) that influence the rate of transcription of particular genes. Compared with the previously described receptor-mediated events, the changes in gene expression that are mediated by steroid binding are both slower and more enduring. One fascinating line of recent research has examined the role of glucocorticoids in promoting cellular damage during episodes of heightened neural excitability (e.g., during ischemia).

12.2.4 Enzyme-Linked Receptors

A large and heterogeneous superfamily of cell surface receptors is defined by the presence of a single transmembrane segment and a direct or indirect association with enzyme activity (Fig. 12.2D). Ligands for these receptors include cytokines, interferons, and many growth factors. Some members of this receptor superfamily possess intrinsic catalytic activity within their cytoplasmic domain (e.g., tyrosine kinase, serine-threonine kinase, guanylyl cyclase, and phosphotyrosine phosphatase). In other cases, receptors are devoid of intrinsic enzyme activity but associate directly with cytoplasmic enzymes. The intracellular "signal" generated by ligand binding of enzyme-linked receptors is conveyed by phosphorylation/dephosphorylation of various intracellular targets, or by generation of cGMP.

One important subclass of enzyme-linked receptors whose signaling pathway has been delineated recently is receptor tyrosine kinases (RTK). Two of the prototypical RTKs bind insulin and epidermal growth factor. RTK receptor structure is defined by three elements, an extracellular domain with a ligand binding site, a single hydrophobic alpha helix that spans the membrane, and an intracellular domain with intrinsic protein tyrosine kinase activity. In most cases, RTK receptors exist as monomers in the ligand-unbound state. The binding of soluble or membrane-bound peptide or protein ligands promotes the formation of dimers. This ligand-triggered dimerization elicits autophosphorylation of specific intracellular tyrosine residues. These phosphotyrosines serve as binding sites for distinct cytosolic proteins that contain a polypeptide domain called the Src homology 2 (SH2) domain. Slight differences in the amino acid sequence among SH2-containing proteins confer receptor specificity by enabling recognition of distinct amino acid sequences surrounding a phosphotyrosine residue in the receptor. SH2-containing proteins include a number of cytosolic enzymes and specific adapter proteins that link the RTK receptor to other signaling molecules. Ras proteins have been shown to be involved in the signal transduction pathway elicited by ligand binding to a number of RTKs, whose ultimate effects include regulation of cell growth, differentiation, and metabolism. The demonstration of RTK and Ras involvement in human cancers has sparked great interest in further elucidation of this important signaling cascade.

12.2.5 Additional Cell Surface Targets for Toxicants

Electrical potentials in excitable cells arise from the differential localization of a number of ions (e.g., Na^+, K^+, Ca^{2+}, and Cl^-) across the semipermeable plasma membrane. Selective changes in the permeability of the membrane to certain ions produce alterations in the membrane potential that ultimately are responsible for the propagation of electrical signals between cells. Voltage-gated ion channels are macromolecules that span the cell membrane and provide a pathway for particular ions (similar Fig. 12.2B). The mechanism for voltage dependency of these channels currently is unclear, although it probably involves a conformational change induced by altered interactions among charged amino acid residues of the channel protein. The best-characterized voltage-gated conductance channel is the Na^+ channel. It is a heavily glycosylated protein formed of a membrane spanning α-subunit and one or two polypeptide β-units that are situated on the extracellular membrane face. There are a number of binding sites present on voltage-gated channels that alter their function. Although no endogenous ligands for these sites have been identified, a variety of xenobiotic toxicants act on these sites. The puffer fish toxin tetrodotoxin and the paralytic shellfish toxin saxitoxin bind to the same site near the extracellular opening of the Na^+ channel, thereby blocking Na^+ transport through the channel. Several lipid-soluble anesthetics (e.g., lidocaine) appear to bind to a hydrophobic site on the protein. Other toxicants act to increase spontaneous channel opening (e.g., aconitine) or channel closing (e.g., scorpion toxins).

12.2.6 The Importance of Understanding Toxicant-Receptor Interactions

The approaches outlined in this chapter are important for several distinct reasons. First, they can be applied to answer mechanistic questions about the proximal actions of a toxicant. The methods and strategies of such approaches, and appropriate examples, will be provided later in this chapter. A second distinct rationale also is of importance. Toxic insult (from nervous system damage or chronic drug administration) may cause compensatory changes that include alterations in receptor characteristics. Such changes may be essential to both understanding the toxicant-induced physiological changes, and possibly, to the chain of events initiated by the toxicant. (see Chapter 19, Central Nervous System Toxicology, for examples). Finally, finding specific receptors for toxicants often provides important information that can be useful in understanding not only the toxicant, but also cellular function. Consider that finding a receptor for morphine certainly provided information about its mode of action, but more importantly, it opened a new horizon in understanding aspects of nervous system function. Similarly, the TCDD receptor, although of unknown function at the time this chapter was written, is likely to be a part of an important cell regulatory system. It is for these reasons that toxicant-receptor interactions are of great importance.

It will become apparent in the material that follows that much of the present discussion bears a great similarity to elementary enzyme kinetics. In fact, one can often understand a great deal about relevant toxicological phenomena by applying simple mass action considerations. The emphasis in this chapter shall be on the recognition of a high affinity toxicant with a macromolecule, under the assumption that one is dealing with a simple reversible phenomenon. In fact, this is often not the case. With enzyme substrates, the reaction is often not just bimolecular, and a variety of mechanisms can be invoked to yield product. Similarly, in many cases where the reaction is between a neurotransmitter receptor and a ligand, various phenomena such as internalization or trapping can complicate this picture.

12.3 THE STUDY OF RECEPTOR-TOXICANT INTERACTIONS

12.3.1 Development of Radioligand Binding Assays

As noted above, many neurotoxicants bind specifically to sites on neurons or glia, and consequently, receptor binding assays have become a useful method to study toxicants that have high-affinity interactions with a binding site. Prior to the 1970s, the study of receptors was limited largely to indirect inferences from measurement of physiological responses suspected of being mediated by the receptor of interest. The availability of radiolabeled drugs with high affinity for specific receptors enabled the development of binding assays to assess receptor-ligand interactions in isolated tissue preparations. In the two decades since these methods were introduced, in vitro binding assays have become routine. Advances in molecular biology have paved the way for new generations of techniques toward this same end, yet the radioreceptor methods described below are still used widely, and remain a powerful tool when one wishes to elucidate the mechanism of action of small molecules.

The essential ingredient in this technique is the availability of radiolabeled ligand—this radioligand is a molecule in which one or more radioactive atoms has been incorporated into the structure without significantly changing its receptor recognition characteristics. Often, scientists can choose among several radiolabeled agents specific for a particular receptor type. The choice of the type of radioactive isotope is governed by the half-life of the isotope, the amount of receptor in a tissue preparation, and the affinity of the receptor for the radioligand. In practice, specific activities of at least 10 Ci/mmol are required, meaning that ^{3}H, ^{125}I, ^{35}S, and ^{32}P can be used, whereas ^{14}C cannot. As an example, a tritiated ligand has a much lower molar specific activities (ca. 30 Ci/mmol) compared with a similar ligand with a single ^{125}I atom (ca. 2200 Ci/mmol); the former thus affords lower sensitivity. On the other hand, the replacement of hydrogen with tritium is unlikely to alter receptor recognition characteristics, whereas addition of a bulky iodine atom may have profound effects. Nonetheless, whereas ligands using ^{3}H are used commonly, ^{125}I, ^{35}S, and ^{32}P containing molecules also are sometimes employed.

Radioligand binding assays are performed typically in solutions of extensively washed membranes prepared from tissue homogenates or cultured cells. Assays of receptor binding to slide-mounted thin tissue sections (quantitative receptor autoradiography) also have become popular, and are useful particularly for the study of the anatomical localization of binding sites. In both cases, tissue is incubated with radioligand for a specific duration, followed by measurement of the amount of ligand-receptor complex. Accurate assessment of the amount of radioactivity bound to receptors requires a method to separate bound ligand from unbound and a nonbiased estimate of the amount of ligand bound specifically to the receptor of interest.

There are three criteria that are important to determine if a suspected toxicant binding site is biologically meaningful, and also whether it may be a receptor with a known physiological function. (These issues are discussed in greater detail later in the chapter). There should be a finite number of receptors in a given amount of tissue; thus, the binding should be saturable. As increasing amounts of the radioligand are added to the tissue, the amount of specific binding should plateau [see Fig. 12.6 (top) and discussion in Section 12.3.2C]. If the binding site is a known or suspected physiological receptor, other drugs or toxicants that bind to the same receptor should compete with the radioligand for receptor occupancy. The selectivity and potency of this binding should parallel the way they enhance or block the physiological effects of the toxicant believed to be mediated through the receptor. This is usually ascertained by use of competition assays (described in detail in Section 12.3.4). Finally, the kinetics of binding often will be consistent with the time course of the biological effect elicited by the toxicant.

12.3.2 Equilibrium Determination of Affinity (Equilibrium Constants)

12.3.2a Law of Mass Action and Fractional Occupancy

Although toxicant-receptor interactions may involve very complex models, they often follow an elementary mass-action model that results in algebraic expressions similar to those seen in Michaelis-Menten enzyme kinetics. Essentially, this model (Fig. 12.3) is the reversible interaction of a toxicant with a binding site (i.e., "receptor"). Although the interaction may generate physiological responses, for receptor analyses it can be modeled without making some of the assumptions needed in enzyme kinetics (e.g., initial velocity conditions). Thus, we can illustrate this model as follows.

Thus, for a toxicant (ligand) interacting with a single (or homogeneous population) of receptor(s), mass action principles require:

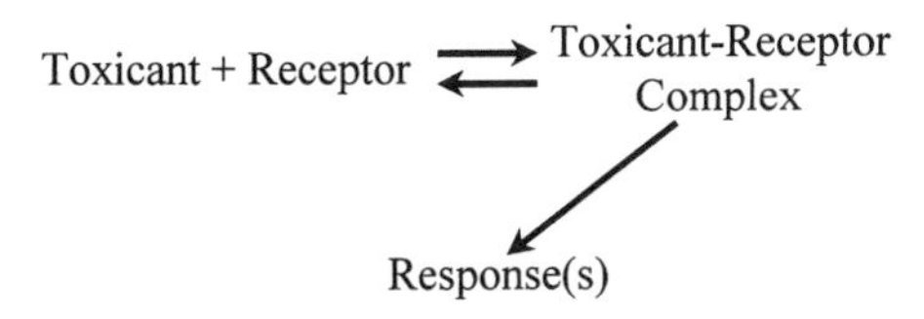

FIGURE 12.3 Simplest model of how toxicant-receptor interactions can lead to functional consequences.

$$\text{Ligand} + \text{Receptor} \underset{k_{off}}{\overset{k_{on}}{\rightleftarrows}} \text{Ligand-Receptor} \qquad \text{(Eq. 12.1)}$$

For any given system, there is a tendency for ligand and receptor to remain associated (the association rate constant k_{on}), and for the ligand-receptor complex to dissociate (the dissociation rate constant k_{off}). Equilibrium is reached when the rate of formation of new ligand-receptor complexes equals the rate at which existing the ligand-receptor complexes dissociate. Thus, equilibrium is reached when the rate of complex formation equals the rate of complex dissociation, as shown algebraically in Eq. 12.2.

$$[\text{Ligand}] \cdot [\text{Receptor}] \cdot k_{on} = [\text{Ligand} \cdot \text{Receptor}] \cdot k_{off} \qquad \text{(Eq. 12.2)}$$

From basic principles of mass action, we know the relationship between these rate constants and the association (K_A) and dissociation (K_D) equilibrium constants:

$$\frac{1}{K_A} = \frac{k_{off}}{k_{on}} = K_D \qquad \text{(Eq. 12.3)}$$

Thus, we can rearrange Eq. 12.2 to define the equilibrium dissociation constant K_D:

$$\frac{[\text{Ligand}] \cdot [\text{Receptor}]}{[\text{Ligand} \cdot \text{Receptor}]} = \frac{k_{off}}{k_{on}} = K_D \qquad \text{(Eq. 12.4)}$$

From this, it is clear that the K_D is the concentration of ligand that, at equilibrium, will cause binding to half the receptors. The K_D is the equilibrium dissociation constant, whereas the k_{off} is the dissociation *rate* constant. A summary of these definitions (and their chemical units) is shown in Table 12.1.

A term that is sometimes useful for biological scientists is *fractional occupancy*. Based on the law of mass action, this term describes (at equilibrium) receptor occupancy as a function of ligand concentration. Specifically:

$$\text{Fractional occupancy} = \frac{[\text{Ligand} \cdot \text{Receptor}]}{[\text{Total Receptor}]} \qquad \text{(Eq. 12.5)}$$

Because [Total Receptor] = [Receptor] + [Ligand · Receptor], then

TABLE 12.1 Chemical Constants and Their Units.

Variable	Name	Units
k_{on}	Association rate constant or "on" rate constant	$M^{-1}\,min^{-1}$
k_{off}	Dissociation rate constant or "off" rate constant	min^{-1}
K_D	Equilibrium dissociation constant	M

$$\text{Fractional occupancy} = \frac{[\text{Ligand}\cdot\text{Receptor}]}{[\text{Receptor}] + [\text{Ligand}\cdot\text{Receptor}]} \qquad \text{(Eq. 12.5A)}$$

From the equation for K_D derived above, it is seen that:

$$[\text{Receptor}] = \frac{K_D \cdot [\text{Ligand}\cdot\text{Receptor}]}{[\text{Ligand}]} \qquad \text{(Eq. 12.6)}$$

One can substitute this value for [Receptor] in both the numerator and the denominator of the equality for fractional occupancy and by simplifying obtain the following. (Try to derive this yourself on a piece of paper.)

$$\text{Fractional occupancy} = \frac{[\text{Ligand}]}{[\text{Ligand}] + K_D} \qquad \text{(Eq. 12.7)}$$

Eq. 12.7 assumes the system is at equilibrium. To make sense of it, think about a few different values for [ligand]. When [ligand] = 0, the fractional occupancy equals zero. When [ligand] is very, very high (i.e., many times the K_D), the fractional occupancy approaches 100%. When [ligand] = K_D, fractional occupancy is 50% (just as the Michaelis-Menten constant K_m describes the concentration of enzyme substrate that gives half-maximal velocity).

Fractional Occupancy is an important concept for several reasons. For example, if one knows the K_D of a ligand for a target site, one can estimate the occupancy, and hence the effects, that might be caused by a given concentration of the toxicant. Similarly, in experimental toxicology, it permits one to choose concentrations of a test compound that are likely to cause the desired effects without being unnecessarily high and causing undesired secondary effects. Finally, this equation makes clear why semilog dose response curves are so useful in toxicology. Table 12.2 illustrates the Fractional Occupancy (FO) values calculated for a hypothetical toxicant with a K_D = 1 nM (an arbitrary value chosen as an easy example). (Be sure that you can calculate the FO values as shown in this table.)

One can then plot these data as shown in Figure 12.4.

12.3.2b The Theoretical Basis for Characterizing Receptors Using Saturation Radioligand Assays

Radioreceptor assays were first developed in the early 1970s. They were based on two simple, but very elegant concepts:

TABLE 12.2 Values for Fractional Occupancy (FO) Versus Concentration or Log Concentration for the Bimolecular Model.

[ligand] (nM)	log[ligand]	FO
0.01	−2	0.009901
0.03	−1.52288	0.029126
0.1	−1	0.090909
0.3	−0.52288	0.230769
1	0	0.5
3	0.477121	0.75
10	1	0.909091
30	1.477121	0.967742
100	2	0.990099

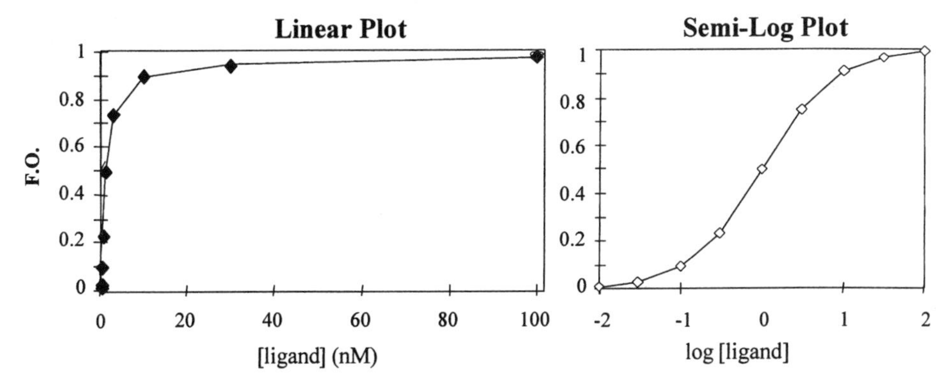

FIGURE 12.4 Plot of fractional occupancy (FO) versus the [ligand] and the log[ligand]. This provides the theoretical basis for the utility of semilog plots; a sigmoidal curve results that allows maximal visual extrapolation of information from the most biologically meaningful part of the dose response curve.

1. If a ligand had high affinity for a macromolecular target (as had been shown by classical pharmacological studies over many decades), it should be thermodynamically possible to measure the binding of the ligand to the receptor without the need to perform equilibrium dialysis (the only method then used) as long as one could separate the ligand-receptor complex from the free ligand.
2. By labeling ligands with appropriate radioactive atoms, one could detect the ligand-receptor sensitively and rapidly. (This was the key point, since chemical methods were neither sufficiently sensitive, nor inexpensive, for this use.)

If we begin with the simple law of mass action shown in Eq. 12.1:

$$\text{Ligand} + \text{Receptor} \underset{k_{off}}{\overset{k_{on}}{\rightleftarrows}} \text{Ligand-Receptor} \qquad \text{(v.i., Eq. 12.1)}$$

We can develop the following equation:

$$\frac{[\text{Ligand}]\cdot[\text{Receptor}]}{[\text{Ligand}\cdot\text{Receptor}]} = \frac{k_{off}}{k_{on}} = K_D \qquad \text{(v.i., Eq. 12.4)}$$

It is useful to replace some of these terms with equivalent ones that are both common laboratory jargon in the field and reflect the measurements that are actually made in characterizing receptors. These names are as follows:

The [Ligand-Receptor] complex is called "Bound" or "B."

The unbound ligand is called "Free" or "F."

As noted, both B and F are measurable experimentally. We can then substitute in the above equation (originally Eq. 12.4) with these terms and obtain the following:

$$\frac{F\cdot[\text{Receptor}]}{B} = K_D \qquad \text{(Eq. 12.8)}$$

Because F and B are independent variables, and we wish to solve for K_D, it is necessary to be able to quantify the unbound receptor. Yet this is almost always technically impossible or impractical. On the other hand we know the following:

$$[\text{Total Receptor}] = [\text{Receptor}] + [\text{Ligand} - \text{Receptor}] \qquad \text{(Eq. 12.9)}$$

The [Total Receptor] present is commonly termed the "$\mathbf{B_{max}}$" (i.e., the maximal number of binding sites). Thus,

$$B_{max} = [\text{Receptor}] + B \qquad \text{(Eq. 12.10)}$$

Equation 12.10 can be rearranged:

$$[\text{Receptor}] = B_{max} - B \qquad \text{(Eq. 12.11)}$$

If we now take Eq. 12.11 and substitute the right hand side of Eq. 12.11 for the [Receptor] in Eq. 12.8 as follows:

$$\frac{F\cdot[\text{Receptor}]}{B} = K_D \qquad \text{(v.i., Eq. 12.8)}$$

$$\frac{F\cdot[B_{max} - B]}{B} = K_D \qquad \text{(Eq. 12.12)}$$

We can simplify this by multiplying both sides by B and expanding the left-hand parenthetical expression to yield:

$$F\cdot B_{max} - F\cdot B = K_D * B \qquad \text{(Eq. 12.13)}$$

With simple rearrangement, we get:

$$K_D \cdot B + F \cdot B = F \cdot B_{max}$$

This can be factored to:

$$B(K_D + F) = F \cdot B_{max}$$

resulting in the following equation:

$$B = \frac{F \cdot B_{max}}{K_D + F} \qquad \text{(Eq. 12.14)}$$

This equation should look familiar, as it is functionally identical to the Michaelis-Menten equation of enzyme kinetics. This equation also should make clear the experimental design to be used in determining K_D and B_{max} using saturation isotherms. We have as the independent variable [F] and as dependent variable B. A successful experiment should permit the estimation of the two biologically meaningful constants: K_D and B_{max}.

The K_D and B_{max} are of interest to toxicologists and other biological scientists because of the information they convey. The B_{max} defines the number of available binding sites. This can provide clues about whether or not a particular tissue may bind the toxicant, or whether a toxicant alters the available number of binding sites (e.g., through killing cells or causing compensatory responses). The K_D defines the affinity of the toxicant for the binding site. This provides clues about the likelihood of the toxicant working through the mechanism being tested, and also whether intoxication alters the binding site through some compensatory or toxic response.

12.3.2c Experimental Forms of Data from Saturation Radioligand Assays

As we did with fractional occupancy, we can predict the data that would result from a toxicant-receptor interaction that was modeled by Eq. 12.14. Let's see what theoretical data would look like if we assume (for the sake of easy calculations) that the B_{max} = 100 and the K_D = 1 nM. (This is a similar exercise to what we did with fractional occupancy). Picking arbitrary [F] concentrations, Table 12.3 shows the derived values (try this yourself using Eq. 12.14).

TABLE 12.3 Derived Values for Theoretical Saturation Plot.

F (nM)	log[F]	B
0.01	−2	0.99
0.03	−1.52	2.91
0.1	−1	9.09
0.3	−0.52	23.08
1	0	50.00
3	0.48	75.00
10	1	90.91
30	1.48	96.77
100	2	99.01

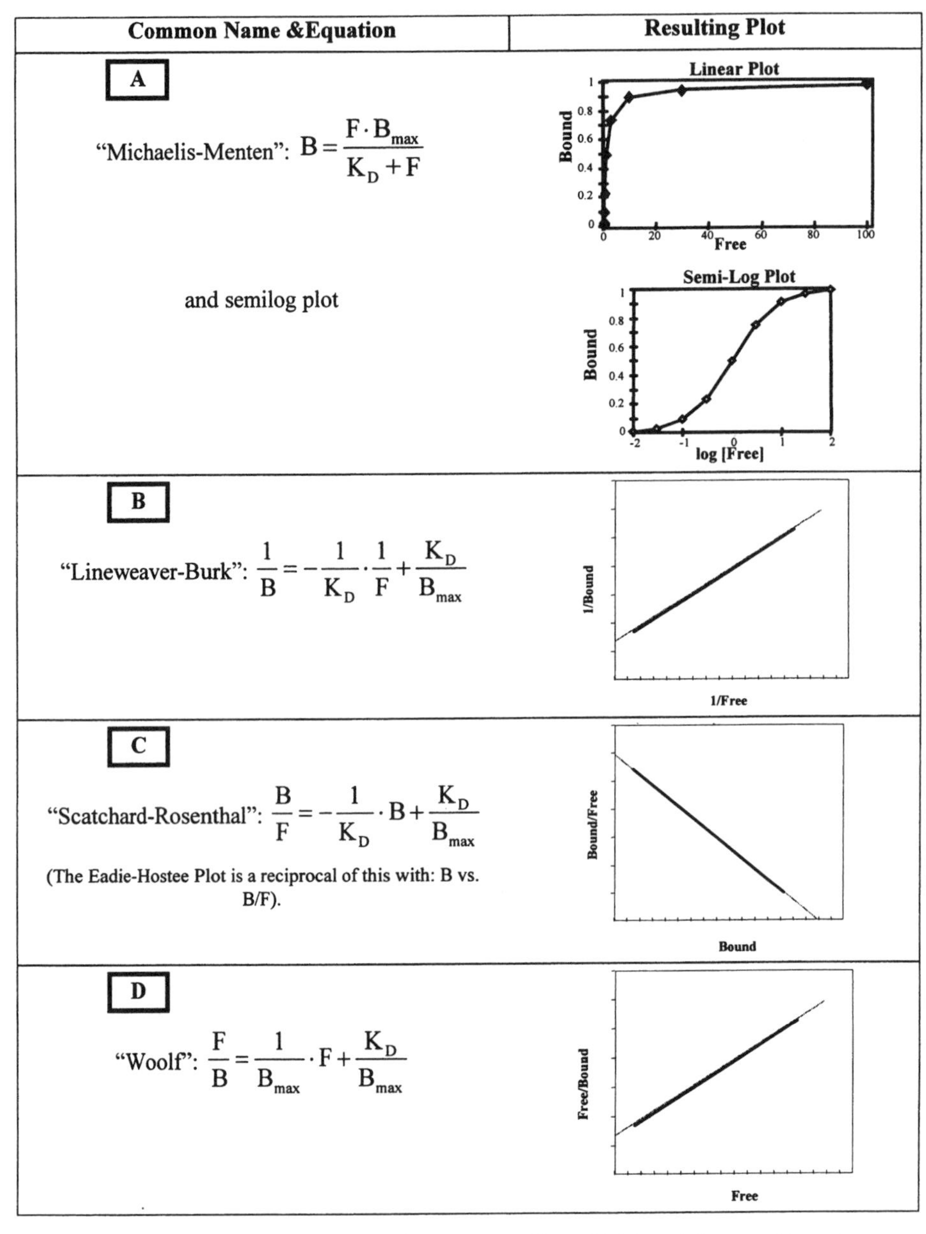

FIGURE 12.5 Some linear transformations of Eq. 12.14 and the resulting plots.

If we plot these data in either a linear or semilogarithmic format, a nonlinear plot results as shown in Figure 12.5 (panel A). In the era when computers were not readily available, it was recognized that one could linearize Eq. 12.14 by several simple algebraic rearrangements. Some of these are shown in Figure 12.5 with both their common name, and how a typical plot of the linearized data would appear. If data are near-ideal (e.g., the calculated data in Table 12.3), then all of these plots will yield *identical* results for K_D and B_{max} when their slope and intercept(s) are calculated. In practice, they each are affected differently by sources of experimental variance. In any event, direct calculations should now be done using computerized nonlinear regression, e.g., based on Eq. 12.14. This can be done with specialized programs such as LIGAND or PRISM (GraphPad, Inc.), or even with general purpose plotting or statistical software (e.g., SigmaPlot or Systat).

One of the reasons, however, that Scatchard plots still are used is that they permit quick visual estimates of the B_{max} (i.e., the B_{max} is the x-intercept of the extrapolated line), and of relative toxicant affinities (reflected as the negative reciprocal of the slope). The steeper the slope, the smaller the K_D, and the greater the affinity of the toxicant for the receptor.

Figure 12.6 (bottom) shows the Scatchard transformation of the saturation data shown in Figure 12.6 (top). In this case, a linear regression analysis adequately fits these data. As noted above, the slope of the regressed line is the negative reciprocal of the K_D, and the x-intercept is the B_{max}. Curvilinear Scatchard plots can indicate the presence of multiple populations of receptors with differing affinities for the toxicant. Alternatively, they may indicate positive or negative cooperativity (producing upwardly or downwardly 'concave" plots), where the binding of one toxicant molecule influences the binding of subsequent molecules.

A potential problem associated with Scatchard analysis occurs when the range of radiolabeled toxicant concentrations used is too narrow, such that the highest concentration does not equal or exceed the K_D (that concentration of toxicant which results in the occupation of one half of the total binding sites). Although not apparent from a Scatchard plot of such data, estimates of B_{max} under such conditions are prone to significant error.

Two other possible transformations of saturation binding data mentioned earlier are the Hofstee (B vs. B/F) and Woolf Plots (F/B vs. F). As with the Scatchard equation, these equations can be derived from algebraic manipulations of the equations listed above.

12.3.3 Kinetic Determinations of Equilibrium Constants

Although equilibrium methods are used most commonly to estimate parameters of receptor binding, separate kinetic analyses should be employed during the initial characterization of toxicant binding, for example, when a novel toxicant radioligand becomes available or an established radioligand is planned for use in a new tissue type or species. Several of the general references cited at the end of this chapter provide an excellent detailed discussion of kinetic binding methods and the derivation of the associated equations. Briefly, kinetic experiments to determine association and dissociation rate constants are conducted separately. Association rate is estimated by measuring the amount of specific toxicant bound as a function of time; assuming that the amount of free radioligand does not change appreciably over time, then a pseudo-first order rate equation is used to estimate k_1. The rate of dissociation of the toxicant is determined by incubating tissue with toxicant, allowing the reaction to proceed to equilibrium, and then stopping it either by infinite dilution or by the addition of an excess of nonradiolabeled ligand. This stops the forward association reaction, so that measurement of the amount that remains bound as a function of time can be attributed solely to the dissociation between receptor and

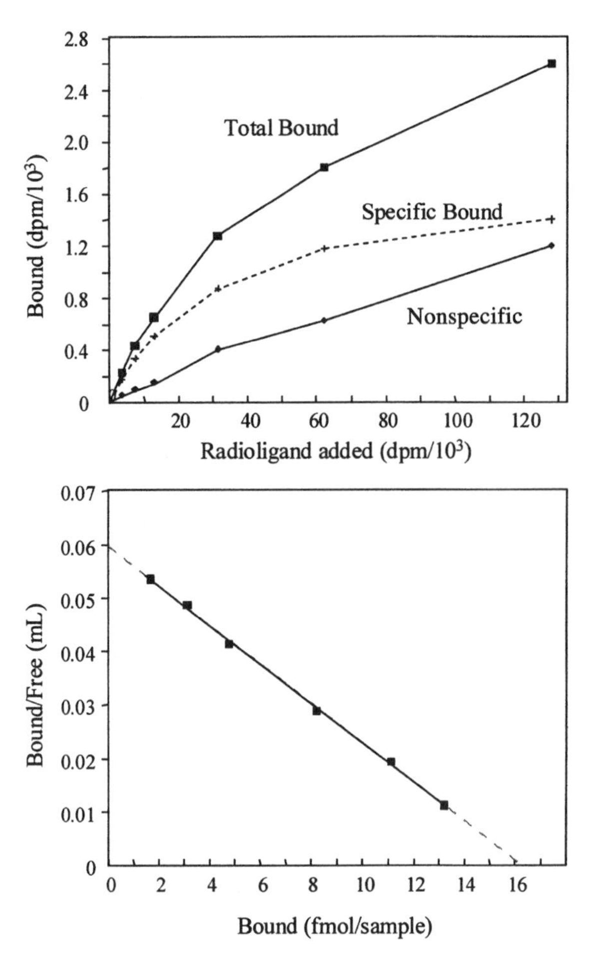

FIGURE 12.6 Characterization of a receptor using a saturation isotherm with a radioligand.
Top: In this experiment, increasing amounts of radioligand are added to tubes containing a constant amount of tissue. A duplicate series of tubes also includes an excess (>100 fold) of a known competitor for the receptor being studied. The solid lines indicate the total amount of radioligand bound to the receptor (Total Bound) and the nonspecific binding (Nonspecific). The nonspecific binding is due to physico-chemical interactions of the radiologiand with proteins or lipid (e.g., dissolving into membrane lipids). The specific binding (the term of interest) is obtained by subtracting the Nonspecific from the Total Bound. As predicted by Eq. 12.14, a plot of specific binding vs. ligand added yields a rectangular hyperbola. *Bottom*: A Scatchard plot of these data (based on equation shown in Figure 12.5, panel B). With these data, the extrapolated density of receptors is approximately 16 fmol. The KD could be calculated based on −1/slope.

toxicant. The rate of change of the receptor ligand complex over time is used to estimate k_{-1}, the dissociation rate constant.

In theory, k_D (i.e., k_{-1}/k_1) should be the same whether determined by kinetic or equilibrium approaches. In practice, however, moderate differences arise that are often attributed to technical problems associated with separating bound from free rapidly without losing a significant proportion of receptor-toxicant complex. This problem is troublesome particularly when estimating the amount bound at early time points in association or dissociation experiments, when the amount of bound ligand is changing rapidly. Large differences between the K_D as determined in saturation and kinetic experiments, however, may indicate that the reaction is more complex than a simple bimolecular reversible reaction.

12.3.4 Competition Binding Assays

Competition binding assays, also referred to as indirect binding studies, involve measuring the competition between a radioligand and an unlabeled drug for a specific receptor site. The concentration of the radioligand is held constant, while the concentration of the unlabeled drug is varied over a wide range of concentrations. If both drugs compete for the same receptor site, then the amount of radioligand bound will decrease systematically as the concentration of the unlabeled drug increases (i.e., as more of the unlabeled drug binds). Competition studies are particularly attractive because they allow the study of many drugs that are not suitable for radiolabeling and direct binding measurements due to their low affinity. Applications of this technique include determining the binding specificity of a radiolabeled toxicant, comparing the potencies of a series of unlabeled toxicants for the radiolabeled receptor site, and providing a reliable estimate of nonspecific binding.

Data from competition binding experiments are plotted routinely as the percent inhibition of specific radioligand binding (total – nonspecific) as a function of $\log_{10}$ of the concentration of competitor. The 100% binding value is defined as the amount of specific radioligand binding in the absence of competitor. When comparing the potency of several compounds, it is useful to plot several competition ("displacement") curves together because the relative position of the competition curves for various ligands will be related to their potency in displacing the radioligand, with the leftmost curves representing the most potent competitors.

Figure 12.7 (top) illustrates competition binding data of two compounds that can compete for radioligands that label the receptor for the neurotransmitter dopamine (specifically, the D_1 receptor subclass). It can be seen that Compound A can eliminate specific binding at concentrations about 100-fold lower than required for Compound B. In toxicological (and pharmacological) terms, one would say that Compound A is more potent than Compound B in this assay (or more rigorously, A has higher affinity). Specifically, higher concentrations of Compound B are required to cause 50% inhibition of binding than are required of Compound B. One other point should be obvious in these data in Figure 12.7. The slopes in the top curves are clearly not parallel for Compound A and Compound B, with the latter being more shallow. The shape of the competition curves also provides important information concerning the nature of the interaction between receptor and ligand, as will be described in sections 12.3.4b–12.3.4d.

The relevant kinetic model for competition experiments with a radiolabeled drug [D] and an unlabeled competitor [I] is shown in the two equations in Eq. 12.15. When both sets of reactions have proceeded to equilibrium, the net rate of formation of both (DR) and (IR) are zero, and the following Eq. 12.16 can be derived from mass action principles.

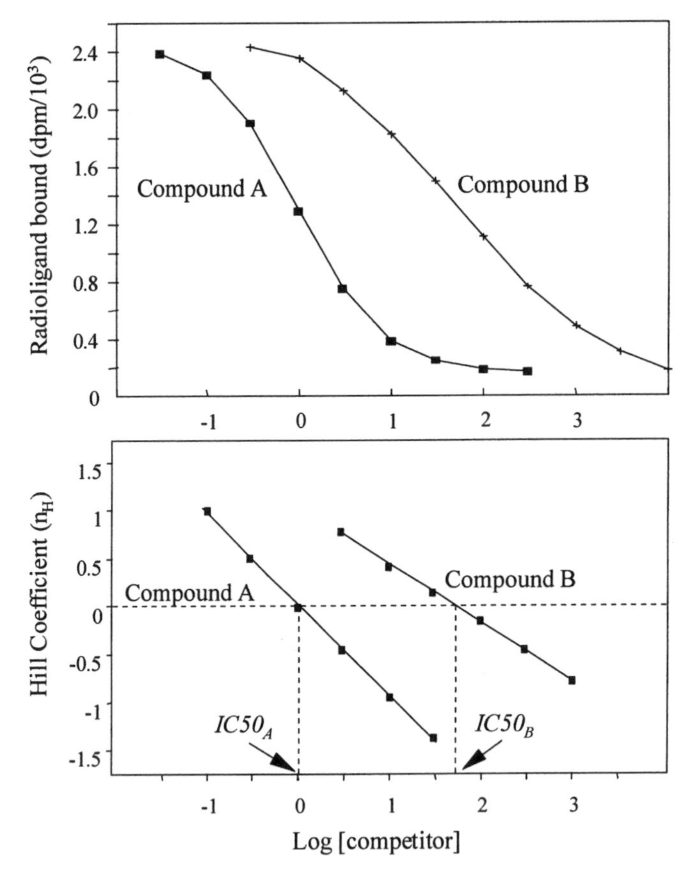

FIGURE 12.7 Competition curves for two compounds versus a known radioligand. *Top*: These data represent the competition of two compounds with a known radioligand (in this case a radioligand that labels the dopamine D_1 receptor, a member of the G protein coupled superfamily). It is important to note not only the left-right difference between Compound **A** and Compound **B**, but also the difference in the shape of their competition curves. *Bottom*: A Hill plot (based on Eq. 12.20) of the competition curves shown in the top figure provides two pieces of data. First, the slopes of the lines are different (Compound **A** = ca. 1.0; Compound **B** = ca. 0.6), which has important mechanistic meaning that is discussed in the text Section 12.3.3e. Hill plots also allow more precise estimation of IC50s. By definition, at 50% inhibition, the Hill coefficient = 0. As shown, one can estimate the IC50 for each compound from this plot.

$$D + R \underset{k_{off}}{\overset{k_{on}}{\rightleftarrows}} D \bullet R \qquad \text{(Eq. 12.15)}$$

$$I + R \underset{k_{off}}{\overset{k_{on}}{\rightleftarrows}} I \bullet R$$

$$I \cdot R = \frac{B_{max} * I}{I + K_1 * \left(1 + \frac{D}{K_D}\right)} \qquad \text{(Eq. 12.16)}$$

In Eq. 12.16, K_I is the equilibrium dissociation constant of the unlabeled competitor, while K_D is the dissociation constant of the radioligand. K_I values of different toxicants can be compared to determine the rank order of potency of toxicants in competing for the radiolabeled receptor site, where a low K_I indicates a high affinity. K_I values are routinely determined from experimentally derived IC50 values (the concentration of inhibitor that produces a 50% reduction in total specific binding of the radiolabeled drug). Although IC50 values may be estimated (with considerable error) from visual inspection of plots such as in Figure 12.7, more rigorous techniques are applied typically, including iterative nonlinear regression analysis and indirect Hill plots (see Section 12.3.4b).

12.3.4a Cheng-Prusoff Equation

For purely competitive interactions (that fit the model shown in Fig. 12.1), the relation between the IC50 and the K_I is described by the Cheng-Prusoff equation:

$$K_I = \frac{IC50}{1 + \frac{[L]}{K_D}} \qquad \text{(Eq. 12.17)}$$

where L is the concentration of the radioligand having affinity K_D. If the concentration of the radioligand is equal to its K_D, then the denominator becomes 2 and the K_I of the competitor is equal to one-half of the IC50. In contrast, if the concentration of radioligand is extremely low, then the denominator becomes unity, and the $K_I \Rightarrow$ IC50.

It is important to note that the application of the Cheng-Prusoff equation requires several assumptions: (1) both radioactive and competitor ligands are interacting reversibly with a single population of sites (bimolecular reactions); (2) the reaction is at steady state; (3) nonspecific binding is accurately estimated; and (4) the concentration of receptor is much less than the K_D for the drug or the competitor. Cheng and Prusoff (1973) also derived equations for situations when the competition is not competitive in nature.

12.3.4b The Hill Plot

This Hill plot was used frequently to determine whether obtained binding data deviate significantly from what would be expected from a simple reversible bimolecular reaction. The Hill transformation was developed originally to describe the cooperativity exhibited when O_2 bound to hemoglobin. Hill plots can be direct or indirect, depending on whether the data are obtained from saturation or competition experiments, respectively. The Hill equation can be derived from elementary enzyme kinetic principles:

The more generalized from of the Michaelis-Menten Equation is shown in Eq. 12.18:

$$B = \frac{B_{max} \cdot [D]^n}{K_D + [D]^n} \qquad \text{(Eq. 12.18)}$$

where *n* reflects whether or not the binding sites behave as a single homogeneous population rather than multiple populations or interactive sites.

If we do a series of algebraic transformations, we can arrive at the commonly used Hill equation:

$$\log\left[\frac{B}{B_{max} - B}\right] = n \cdot \log[D] - \log K_D \qquad \text{(Eq. 12.19)}$$

As can be seen by inspection, the Hill Equation (12.19) can be plotted as a straight line in the form y = m·x + b where y is the "pseudo-logits" term on the left side of Eq. 12.19, and $x = \log[D]$. In such a plot, the slope is equal to n (often called n_H; the Hill number) and the y-intercept is equal to $-\log K_D$. In the case of saturation binding data, $K_{D'}$ is the apparent equilibrium dissociation constant whose value reflects both the intrinsic K_D and the factors that determine how the binding of one ligand molecule affects the binding of subsequent molecules. If such sequential binding effects are not present, the $n_H = 1$, and $K_{D'}$ will equal the K_D. Alternatively, when n_H is significantly different from unity, then $K_{D'}$ is not equivalent to K_D. An equation of similar form (but opposite slope) can be derived for competition experiments (see below).

12.3.4c Visual Inspection of Binding Data

There are several important inferences that scientists make from examination of Hill plots. If the interaction of a ligand and its receptor follows the simple bimolecular reaction shown in Eq. 12.1, the value of the Hill coefficient n (or more commonly often n_H) will equal 1. (Please note, however, that it is usually impossible experimentally to distinguish between a single population of receptors vs. several receptors having nearly similar or identical K_Ds for the ligand.) A Hill coefficient significantly less than 1 suggests a more complex situation. These possibilities include interaction of the ligand with multiple populations of binding sites (e.g., different receptors), a single class of receptor that exists in different affinity states, or even negative cooperativity among the same receptors. Alternatively, a Hill coefficient significantly greater than 1 might be obtained when the binding sites exhibit positive cooperativity.

In toxicology, an even more common use of Hill plots is for competition assays. In contrast to the direct Hill plot (from Eq. 12.19), competition assays use indirect Hill plots. There have essentially the same form and interpretation, but have negative rather than positive slope, Again, an indirect Hill coefficient (n_H) of –1 is consistent with (although it does not prove) the notion that both competitor and radioligand are interacting with one population of receptors via a simple reversible bimolecular reaction. Similarly, an n_H significantly less in magnitude than –1 (e.g., –0.6) is consistent with multiple receptors or negatively cooperative receptors.

Figure 12.7 (bottom) shows the Hill plots from two of the competition curves illustrated in Figure 12.7 (top). In one case, $n_H = -1$, whereas in the second case it is significantly lower (ca. –0.6). The most parsimonious explanation for these data is that the binding of Compound A is occurring to a homogeneous population of binding sites, whereas that of Compound B is to multiple or negatively cooperative binding sites.

Practicing scientists are able to make reliable estimations by visual inspecting competition curves and looking at the steepness of those curves. Such a visual inspection is based on the following line of reasoning between the apparent steepness of a competition curve and the derived Hill coefficient. First, consider that the "indirect" Hill equation can take the following form:

$$\underbrace{\log\left[\frac{B}{100-B}\right]}_{y\ value} = \underbrace{n}_{slope} \cdot \underbrace{\log[D]}_{x\ value} - \underbrace{n \cdot \log IC50}_{intercept} \qquad \text{(Eq. 12.20)}$$

where 100 represents the total binding (i.e., 100%) of the radioligand in the absence of competitor and B represents the percent of total binding found at each concentration of competitor [D]. For our purposes, we shall ignore the intercept term in this equation.

From this equation, we can construct an example that is extremely useful in visually examining receptor data, indeed any data fitting the simple mass action model of Eq. 12.1. This means that a toxicant under study is competing with the radioligand for one, and only one, population of sites. By definition, the Hill slope (often called n_H) must equal 1. It turns out that by memorizing the numbers "9" and "91," one can do a very useful preliminary analysis of data fit to such a mass action model. The reason for this is as follows:

At 91% of total binding, the value for the left hand $\left(\text{i.e., } \log\left[\frac{B}{100-B}\right]\right)$ part of equation 12.20 can be calculated as follows:

$$\log\left(\frac{91}{100-91}\right) = \log\left(\frac{91}{9}\right) = \log(10.111) = 1.005$$

One can perform the same calculations for 9% of total binding:

$$\log\left(\frac{9}{100-9}\right) = \log\left(\frac{9}{91}\right) = \log(0.989) = -1.005$$

This means that the Δ (change) in the left side of the equation when going from 91% to 9% of total binding will be 1.005 – (–1.005) = 2.01 Hill units. Remember, as shown in Eq. 12.20, that the Hill number (n_H) is a slope function, in which the denominator is the log of the concentration range. If $n_H = 1$ (i.e., a system that meets the constraints of the model of Eq. 12.1), then the log of the concentration range to cause this change from 91% to 9% total binding must also have a value of 2.01. Of course, the antilog of 2.01 is ca. 100. Thus, if a curve has "normal" steepness (i.e., it fits Eq. 12.1), then a change from 91% to 9% of binding must occur with a change in concentration of 100-fold. This means that one can visually inspect dose-response data to estimate whether the 91%–9% range spans a two log order (i.e., 100-fold) concentration range. This is a very useful way to estimate of whether data is consistent with the bimolecular model.

If the change from 9% to 91% requires more than a 100-fold concentration change, than the curve is "shallow" and cannot be explained by the model in Eq. 12.1. If the change from 9% to 91% requires less than a 100-fold concentration change, than the curve is "steep" and also does not meet the model of Eq. 12.1. Of these two situations, the former is more common. One common example of shallow competition curves involves the binding of ligands to G protein coupled receptors (GPCRs; see Section 12.2.1). In such cases, the shallow steepness of the curve often is attributed to the existence of a single receptor population that exists in two interconvertible affinity states. According to this model, GPCRs exist in a complex with G proteins, and have GDP bound to the α-subunit of the G protein. These receptors are in a high-affinity state for agonists. Some of the receptors are not associated with G proteins, and exist in a low-affinity state for agonists. Thus, normally an agonist competition curve looks shallow because it detects two populations of binding sites having very different affinity.

With these GPCRs, the addition of GTP (or related guanine nucleotide analogs) to an assay system can convert those receptors in a high-affinity to a low-affinity state by displacing the GDP. Thus, one goes from having two sites (one with high- and one with low-affinity for agonists) to one site (all low affinity). Experimentally, the competition curve changes from a shallow curve (due to two different populations of binding sites) to a curve of normal steepness (representing only the low-affinity site). Moreover, the curve shifts rightward. Thus, one can, even by visual inspection, test hypotheses about the underlying model. Since dose-response relationships are at the heart of toxicology, it is important to be able to make such estimations facilely.

Finally, two practical points relative to Hill plots also should be made. As can be seen in Eq. 12.20, at the IC50 of a competitor, the left hand term will have a value equal to 0. Hill plots are used frequently to calculate the IC50 graphically; this value can then be substituted into the Cheng-Prusoff equation to determine a K_I value (see Section 12.3.3d). Experimentally, those data representing less than 5%, or more than 95%, radioligand occupancy are of little use since at these extremes, systematic or random sources of experimental error (e.g., counting error; pipeting error) make use of these points problematic.

12.3.4d Complex Binding Phenomena

In many cases, the assumption that a toxicant is interacting with a single population of binding sites is not tenable, as indicated by curvilinear Scatchard plots or Hill coefficients different from unity. A lengthy discussion of the alternative mathematical models used to describe possible receptor-toxicant interactions in such cases is beyond the scope of this chapter. Briefly, such analyses rely on powerful statistical computer programs (i.e., Prism or LIGAND) that can determine whether binding data can be modeled with greater precision by assuming the existence of two (or more) receptor sites, rather than one. If so, separate parameter estimates can be obtained for the two sites. Although two-site analysis is used frequently, the ability to resolve more than two receptor sites currently is limited. In all cases, it is important to emphasize that such analyses should be used to generate testable hypotheses about the multiple sites, rather than as proof for their existence. Finally, it should be recognized that a number of technical problems (failure to reach equilibrium at low drug concentrations) could produce data consistent with multiple sites; thus, a careful consideration of potential artifacts is wise prior to entertaining the idea of multiple receptor populations.

12.3.5 How to Conduct a Radioreceptor Assay

12.3.5a Methods to Separate Bound from Free Radioligand

Three methods have been used traditionally to separate bound from free ligand in radioreceptor assays. Dialysis, originally widely used to study enzyme-substrate interactions, has several technical problems that prevent its wider use. These include degradation or sticking of receptor or ligand, the cumbersome nature of assays when large numbers of samples are needed, and the long time to obtain equilibrium. Centrifugation to separate ligand bound to tissue receptors from free ligand in solution is limited because the ligand dissociates from the receptor during pelleting and a large amount of drug receptor complex is therefore lost. There is an inverse relationship between the affinity of the ligand for the receptor (i.e., $1/K_D$) and the time allowable for centrifugation. One can calculate this time based on the K_D of the radioligand and the derived rate constants; the relationship will be logarithmic. For example, assuming that separation must be complete in $0.15\,t_{1/2}$ to avoid losing more than 10% of ligand-receptor complex, then centrifugation must be accomplished in 10 sec for a ligand with a K_D of 10 nM, and 1000 sec for a ligand with a K_D of 0.1 nM.

The most widely used method to separate bound from free radioactivity is vacuum filtration. The suspension of tissue homogenate is diluted, and rapidly aspirated through filters that retain the tissue membranes but not the solvent. With a sufficiently high ligand affinity for the receptor, several washes of the filter can be used to achieve low levels of nonspecific binding. The use of ice-cold buffer for these washes decreases the dissociation rate of the receptor-ligand complex, permitting more extensive washing with lower nonspecific binding. The ability to prepare large numbers of samples simultaneously through the use of commercially available filtration systems is a major advantage of this method. An excellent summary of the factors that should be considered in choosing a separation method is provided by Limbrid (1996).

12.3.5b Methods to Estimate Nonspecifically Bound Radioligand

It is important to recognize that the total amount of radioligand bound to tissue represents that bound not only to high affinity receptors, but also ligand bound "nonspecifically" to other tissue components as well as to assay materials (filter paper, glass or plastic tubes, etc.) In contrast to receptor-bound ligand, nonspecifically bound ligand is assumed to be nonsaturable; nonspecific binding continues to increase as radioligand concentration increases. Most nonspecific binding often is believed to be due to hydrophobic interactions between the radioligand and lipophilic tissue constituents, such as membrane phospholipids. One may conceptualize this process as a sort of "dissolving" of the radioligand into biological membranes. Thus, while this process will be slow ("low affinity"), in an aqueous environment it will both be essentially irreversible and of high capacity relative to the limited number of specific receptors.

A method used commonly to estimate the nonspecific binding component involves measuring the amount of radioligand bound in the presence of a large excess of unlabeled ligand ($>100 * K_D$). This large excess should compete for receptor sites preventing almost all binding of radiolabeled ligand. Conversely, the model discussed in the previous paragraph would predict that radioligand bound to nonspecific components should be unaffected; a nearly infinite number of sites for nonspecific binding exist relative to specific receptor(s). Thus, the amount of radioactivity remaining under these conditions is taken as a measure of nonspecific binding. It is calculated as follows:

$$\text{specific binding} = \text{total binding} - \text{nonspecific binding}$$

where total binding is the amount of radioligand bound in the absence of a competitor, and nonspecific binding is that bound in the presence of an excess of competitor. The theory that is applied to the analyses of such data was discussed earlier.

12.3.6 The Relation Between Receptor Occupancy and Biological Response

The above discussion has focused on methods for describing the proximal events in receptor-mediated responses to toxicants, namely the binding of a toxicant ligand to a biochemically defined receptor. A critical assumption is that such an approach will elucidate the mechanisms by which receptor-toxicant interactions produce their ultimate toxic effects. The validity of this assumption rests on demonstrating that the binding of a ligand is related directly to some biological response. For instance, a straightforward prediction might be that the binding affinity of a toxicant for a receptor should correlate with its propensity to either stimulate or block a functional response. More often than not, however, this relation is complex rather than simple.

Stephenson is credited with introducing the concept of a complex relation between receptor occupancy and tissue response. His formulations were designed to explain prior anomalous data that suggested that drug effects were often not linearly related to receptor occupancy, as shown by the frequent finding that the maximal drug effect achieved was not linearly related to the concentration of drug applied. Stephenson introduced three postulates to explain such aberrant data. First, he stated that some agonists could produce a maximal response by occupying a small fraction of a receptor population. Second, he stated formally that the tissue response is not related linearly to receptor occupation. Finally, he introduced the concept of differing efficacies of agonists, such that an agonist with high efficacy may elicit a maximal effect when only a small percentage of receptors are occupied, while an agonist with low efficacy may require occupation of a substantially greater percentage of receptors. A partial agonist is a special case of a low-efficacy agonist. That is, the efficacy of a partial agonist is so low that it fails to elicit a maximal response even at saturating concentrations, when all receptors presumably are occupied. The potential antagonistic effects of partial agonists present at high concentrations can now be appreciated; partial agonists can compete effectively for receptor binding sites with more efficacious ligands, although partial agonists have only minimal biological effects. Figure 12.8 provides an example of the functional response produced by a partial agonist and a full agonist. It should be noted that the compounds in Figure 12.8 both can completely compete for the receptor believed to mediate this functional response. Several distinct mechanisms may explain partial functional efficacy, and it is not always clear which are operative. It

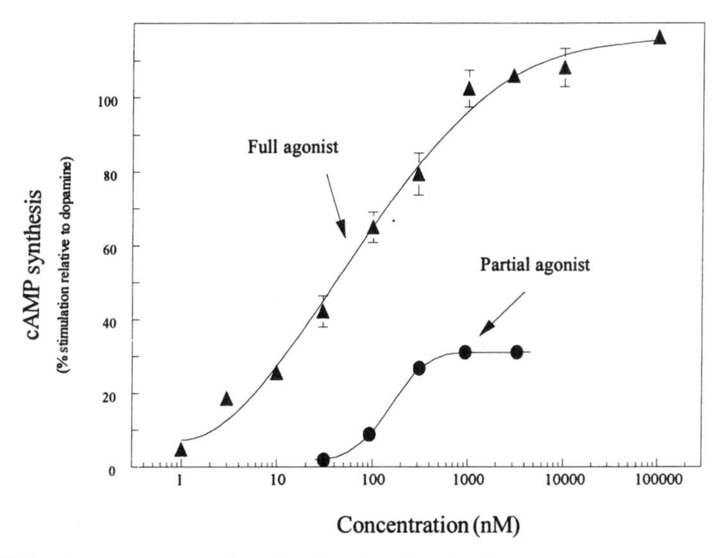

FIGURE 12.8 Dose response plot showing functional differences between partial and full agonist.
These data illustrate that two compounds of similar potency may differ in functional efficacy. In this case, the full agonist cause the same degree of stimulation as does the endogenous ligand. Conversely, the partial agonist, although of similar potency, is only partially efficacious.

Source: These data are modified from actual published studies from the authors' laboratory (Brewster et al., J. Med. Chem. 1990).

is important to remember that functional efficacy and affinity for the receptor (potency) need not correlate.

The phenomenon of partial agonism is but one example of the complexities encountered when relating binding data to biological response. Another noteworthy example is provided by the finding that many receptor populations appear to be larger than would be required to produce a maximal response. This phenomenon is referred to as 'spare receptors" or, alternately, as 'receptor reserve." The strongest empirical data supporting this notion comes from studies using irreversible receptor inactivation which demonstrate that the maximum tissue response is not depressed until a large percentage of the receptor binding sites are eliminated,

Finally, it should be noted that a comparison of receptor binding data with functional response is inherently difficult. In the case of binding assays prepared from tissue homogenates, receptors are isolated from components that are involved in the ultimate cellular response; thus, the assays are conducted in artificially pure preparations. In contrast, functional assays typically require greater preservation of cellular integrity. This also means that receptor regulatory processes are more likely to be evident during functional assays. For instance, a common finding is that the response of an agonist decreases as a function of prior exposure to the same or a similar agonist; this phenomenon is referred to a homologous or heterologous desensitization, respectively. In some model systems (e.g., β-adrenergic receptor), one mechanism for desensitization involves internalization and sequestration of receptor molecules. Additionally, an intracellular enzyme (β-adrenergic receptor kinase; BARK), has been found to be activated by agonist stimulation; this kinase phosphorylates particular amoni acid residues on the receptor which are involved in some aspects of the desensitized responses. Clearly, such dynamic receptor regulation is a critical aspect of receptor-toxicant interactions, yet such processes are not likely to be captured in receptor binding assays using extensively washed membranes.

This preceding discussion was not intended to convince one that the study of receptor binding is of limited utility, but rather that it should be accompanied by an examination of functional end points, with a realization of the potential challenges (and rewards) involved in relating the two.

12.3.7 Toxicant-Receptor Principles and Risk Assessment: The Dioxin Example

Dioxin is a potent carcinogen that has been the subject of intense political and scientific debate for the past two decades, particularly with respect to establishing safe exposure limits. Dioxin is known to bind reversibly to an intracellular protein receptor [the aromatic hydrocarbon (Ah) receptor]; the activated receptor-dioxin complex associates with a translocator protein, then enters the cell nucleus and interacts with DNA to alter gene expression. Most available evidence suggests that dioxin toxicity is receptor-mediated; thus, it has been possible recently to outline a rational alternative to the established linear extrapolation risk model. Briefly, the proposed model states that dioxin effects can be assumed to increase in a manner predicted by mass-action principles governing bimolecular reversible reactions, such as those described earlier in this chapter. Specifically, dioxin effects should increase more slowly than predicted by linear extrapolation at low levels of exposure (where a negligible percentage of receptors are presumed occupied). Conversely, as the concentration of dioxin increases further, a critical point should be reached where significant receptor occupation begins to occur. Beyond this point, receptor occupation can be expected to increase sharply, and correspondingly, so should the effects of dioxin. In contrast to the current linear extrapolation risk model (where there is no safe exposure level), the alternative model suggests that low levels of dioxin are not toxic. Although the question of potential toxic effects of low doses of dioxin may remain debatable, the application of receptor-ligand kinetic principles to models of toxicant risk assessment undoubtedly represents a significant advance.

12.4 CONCLUSIONS

An essential aspect of the mechanism of action of many (but not all) toxicants is an interaction with specific binding sites. In such cases, the approaches outlined in this chapter may provide powerful tools for understanding the biological actions of toxicants. As was illustrated in this chapter, the use of classic radioreceptor (binding) methods based on simple mass-action principles permits quantitative comparisons of toxicant potency and specificity. Moreover, an understanding of the different effector mechanisms evoked by ligand binding can provide important clues concerning toxic sequelae. For instance, the finding that a toxicant binds to dopamine receptors suggests that it may perturb specific functions linked to G proteins (e.g., adenylate cyclase activity).

At present, receptor methods can help elucidate the locus of toxicant action (even to the molecular level), and the relationship of such a mechanism to cellular physiology. Two new types of tools will make such methods even more powerful in the future. Molecular biological techniques have provided less ambiguous ways to study the interaction of toxicants with specific binding sites, and the effect of toxicants on such sites. Increasingly sophisticated computers and computer-modeling programs also will have an increasing impact. Such approaches (presently used primarily for study of chemical mechanisms or drug design) can be used to predict the stereochemical requirements for recognition of a toxicant to a specific, known site.

SUGGESTED READING

Cheng, Y.-C. and Prusoff, W.H. Relationship between the inhibition constant (Ki) and the concentration of inhibitor which causes 50 percent inhibition (I50) of an enzymatic reaction, *Biochem. Pharmacol.* 22 (1973), 3099–3108.

Kenakin, T. *Pharmacologic Analysis of Drug Receptor Interaction*, (2nd Ed.) Raven Press, New York, 1993.

Limbrid, L.E. *Cell Surface receptors: A Short Course on Theory and Methods*, 2nd Ed. City: Kluwer Academic Publishers, 1996.

Yamamura H.I., et al. *Methods in Neurotransmitter Receptor Analysis*. Raven Press, New York, 1990.

CHAPTER THIRTEEN

Effects of Toxicants on Electron Transport and Oxidative Phosphorylation

DONALD E. MORELAND

13.1 INTRODUCTION

Mitochondria are able to convert energy released by electron transport and to store it as bond energy (ATP). The energy so stored is utilized by plant and animal cells to drive or power all of the mechanical, transport, and biosynthetic work done by the cell. Hence, ATP is a pivotal metabolic compound and interference with its production or utilization constitutes one mechanism through which xenobiotics can express acute or chronic toxicity.

Mitochondrial oxidative phosphorylation is the primary process in which aerobic cells produce ATP by esterification of adenosine diphosphate (ADP) with inorganic phosphate (Pi). Mitochondria are found in all aerobic eukaryotic cells. The general structure and function of plant, mammalian, and insect mitochondria appear to be alike. The number of mitochondria per cell range from 20 up to several thousand, depending on the organism and cell type concerned. The more metabolically active cells possess the greater number. The number also varies with the type of tissue or organ and the maturation state of the tissue. For example, liver cells average 1000–2500 mitochondria per cell, paramecia about 1000, and Euglena 15–20. Mitochondria are typically about 3 μm long and 1 μm in diameter and are spear shaped, but size may vary and transient changes in shape are exhibited.

A diagrammatic representation of the membrane systems and spaces of mitochondria is shown in Figure 13.1A. The two bilayer membrane systems are the outer membrane and the inner membrane. The inner membrane is made up of two sectors—the inner boundary membrane, which parallels the outer membrane, and the cristal membrane, which consists of invaginations of the inner membrane. The space between the two membrane systems is called the intracristal space; the space in the interior of the mitochondrion is called the matrix space. The membranes are 5–7 nm thick. The outer membrane is freely permeable to both charged and uncharged molecules with molecular weights up to about 5060. Consequently, the contents of the intracristal space are in equilibrium with the cytosol with respect to low molecular weight substances. However, the inner mitochondrial membrane has limited permeability properties. It is freely permeable to H_2O, CO_2, O_2, NH_3, and acetate, but is selectively permeable to protons and most other small ions. Metabolites and nonpermeable ions are moved across the membrane under the strict control of special carriers or translocating components, that is, porters.

There is considerable variation in the number, size, and shape of cristae among mitochondria of different cell types. A large number of cristae per mitochondrion is associated with the capacity for high metabolic activity. When mitochondria are negatively stained and viewed under transmission electron microscopy, dense granules are seen to line the inner surface of the cristal or peripheral inner membrane (Fig. 13.1B). The structures appear to be attached to the membrane by means of a short neck, and protrude into the matrix. The granules and their stalks are called inner membrane subunits, and they constitute a portion of the tripartite unit. This tripartite structure consists of three bonded sectors: basepiece, stalk, and headpiece (Fig. 13.1C). The stalk and headpiece can be detached without compromising the integrity of the membrane.

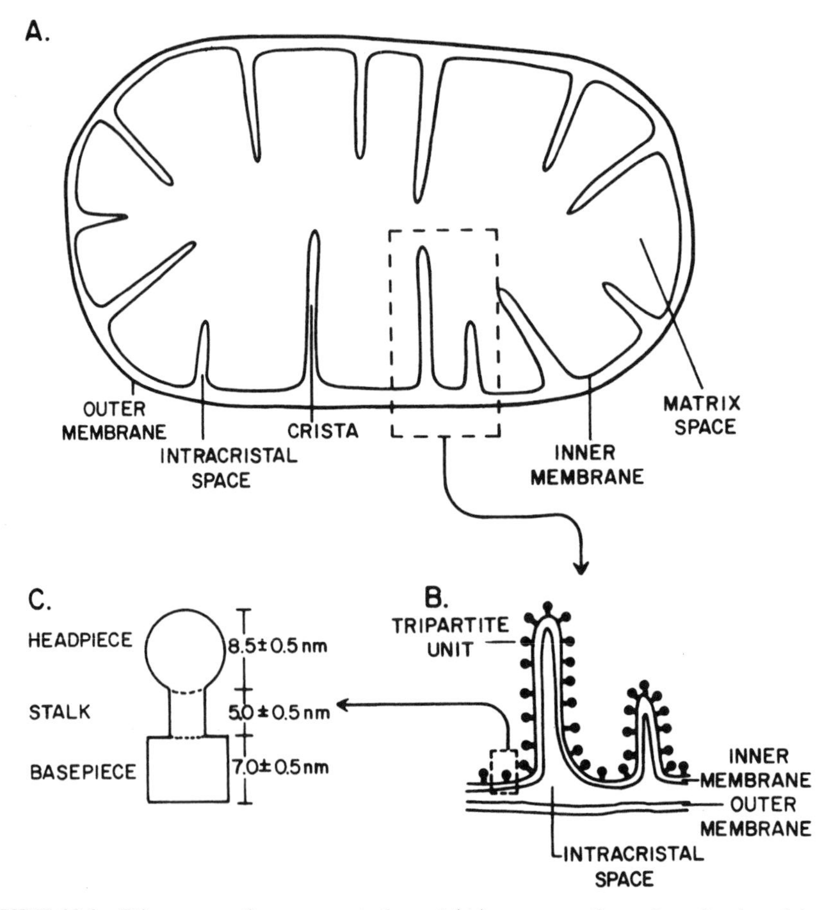

FIGURE 13.1 Diagrammatic representation of (A), cross section of a mitochondrion, (B), detailed view of cristael morphology and (C), tripartite repeating units.

Phospholipids account for approximately 40% of the weight of the inner membrane and the remainder is protein. In bovine heart mitochondria, about 40–45% of the phospholipid is phosphatidylcholine and 30% is phosphatidylethanolamine. The phospholipid is distributed asymmetrically in the lipid bilayer, with phosphatidylethanolamine predominating on the matrix (M) side and phosphatidylcholine on the cytoplasmic (C) side.

A few enzymes are associated with the outer membrane of mammalian mitochondria (monoamine oxidase, phospholipase A, nucleoside diphosphokinase) and the intracristal space (adenylate kinase, Cu^{2+}-Zn^{2+} superoxide dismutase, nucleoside diphosphokinase). The inner membrane contains the enzymes and cofactors that participate in electron transport and energy generation. Located in the matrix are the tricarboxylic acid (Krebs) cycle enzymes, NAD-linked dehydrogenases, and enzymes associated with the elongation of fatty acids, the urea cycle, and the β-oxidation of fatty acids. The matrix also contains circular DNA and ribosomes. Most of the membrane polypeptides are encoded by nuclear DNA and are synthesized in the cytoplasm. A small number of polypeptides (about 13) are encoded by mitochondrial DNA.

13.2 ELECTRON TRANSPORT AND PHOSPHORYLATION

A generalized scheme that shows components of the electron transport and coupled phosphorylation pathways as they are considered to occur in the inner membrane of mitochondria is presented in Figure 13.2. The scheme shows four complexes (customarily designated by Roman numerals). Each complex contains several of the components of the electron transport chain. Each complex also is the smallest unit in which a sector of the electron transfer chain can be isolated without a loss of native characteristics, such as the ability to react with natural electron acceptors, to show susceptibility to specific inhibitors, and to maintain the appropriate redox values of the component proteins. Components of the complexes undergo oxidation and reduction, and participate in the transfer of electrons down the chain.

The transfer of electrons from substrate to O_2 through the cytochrome pathway is coupled to an electrogenic translocation of protons from the matrix across the inner membrane to the C-side. Proton translocation occurs at the level of complexes I, III, and IV. Because the membrane is impermeable to the passive diffusion of protons, an electrochemical gradient results. The gradient has two components, a chemical component (ΔpH) and a membrane potential (ΔΨ). The M-side is negatively, and the C-side is positively, charged. The two parameters are additive and contribute to a proton motive force (pmf) differential across the membrane: $\Delta\mu H^+/F = \Delta p = \Delta\Psi - 2.3RT/F(\Delta pH)$. At 25°C, 2.3 RT/F is approximately 60 mV. Literature estimates of Δp and ΔΨ for rat liver mitochondria average 220 and 180 mV, respectively, under state 4 conditions, as measured by ion distribution and altered spectral properties of optical probes. Corresponding values obtained with plant mitochondria are somewhat lower (160–180 mV for Δp and 120–140 mV for ΔΨ). For both animal and plant mitochondria, ΔpH comprises about 20% of Δp. Values vary with the techniques used for the measurements. When mitochondria undergo transition from state 4 to state 3 respiration, ΔΨ decreases by 10–15%.

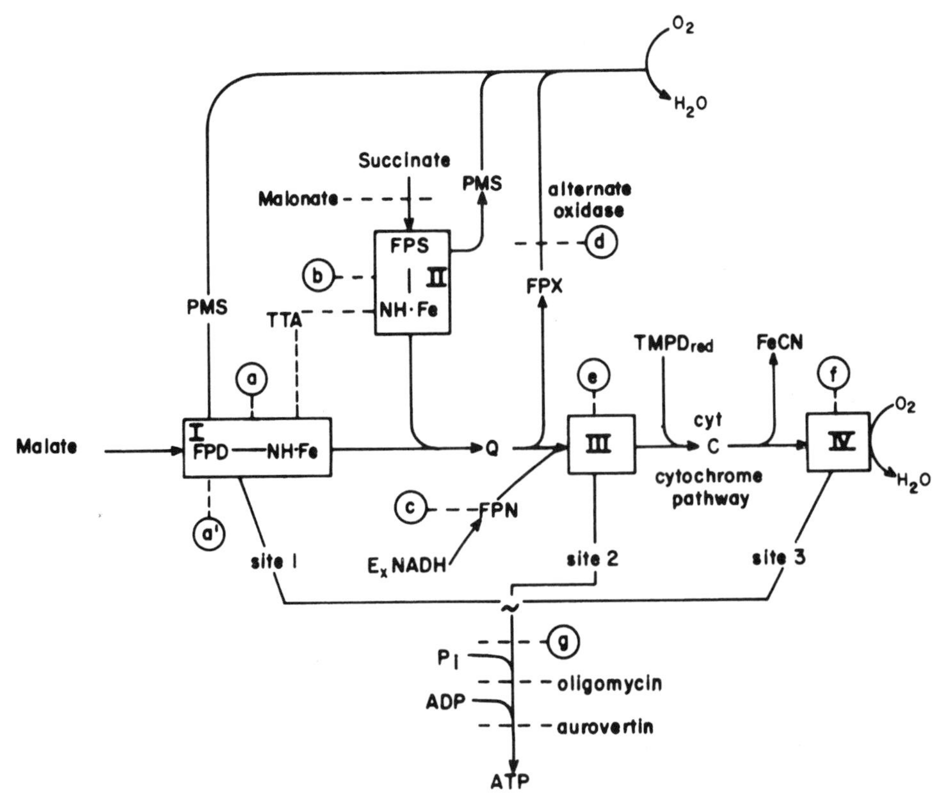

FIGURE 13.2 Schematic representation of electron transport and phosphorylation in mitochondria showing postulated sites of action of inhibitors.

Complex I (NADP-CoQ reductase) is the most complex and least understood of the four units. It is thought to consist of 26 polypeptides, has a monomeric molecular weight of 700,000, traverses the membrane, and exists as a dimer. The complex mediates the oxidation of intramitochondrial NADH generated by the Krebs cycle and transfers electrons from NADH to coenzyme Q (ubiquinone). FMN is the prosthetic moiety that is noncovalently bound to one of the polypeptides that also binds two of the nonheme iron-sulfur (Fe-S) centers. The complex contains additional Fe-S centers with differing redox potentials. FMN is located extrinsically to the membrane bilayer and is exposed to the matrix.

Succinate-CoQ reductase (complex II) mediates the transfer of electrons from succinate generated by the Krebs cycle to oxidized CoQ. Complex II exists as a monomer with a molecular weight of 200,000. The prosthetic groups consist of FAD, which is covalently linked through a histidine residue to the dehydrogenase, and three Fe-S centers. FAD and one or more of the Fe-S centers are exposed to the matrix.

Complex III (reduced CoQ-cytochrome c reductase) mediates the transfer of electrons from reduced CoQ to cytochrome c. It consists of 10 polypeptides, has a monomeric molecular weight of 300,000, and exists as a dimer. The prosthetic groups include an Fe-S center, two b-cytochromes (b-556, b-560), and cytochrome c (c-552). Cytochrome c and the Fe-S center are exposed to the C-side.

Cytochrome c oxidase (complex IV) has a monomeric molecular weight of 160,000, exists as a dimer, contains seven to eight subunits in addition to four or five other polypeptides, and mediates the transfer of electrons from reduced cytochrome c to molecular O_2. Two copper ligands and cytochromes a and a_3 function as the prosthetic groups. O_2 is considered to bind to the iron and copper centers associated with cytochrome a_3, which are exposed to the C-side of the membrane. CoQ and cytochrome c are considered to be freely mobile membrane components and to mediate transfer of electrons between the complexes.

Proton extrusion is associated with complexes I, III, and IV. Hence, each provides a site for energy conservation. The number of ATP molecules synthesized during the oxidation of substrates via the respiratory chain through molecular O_2 is defined as the P/O or ADP/O ratio. The theoretical ratio of the oxidation of NAD-linked substrates such as malate and glutamate is 3 and for succinate is 2.

Plant mitochondria in contrast to mammalian mitochondria have a somewhat more complex respiratory chain, more matrix dehydrogenases, and a more complex anion carrier system. In contrast to animal mitochondria, intact plant mitochondria readily oxidize malate without the addition of pyruvate. In addition to the oxidation of endogenous NADH through the rotenone-sensitive, complex I route with three energy transducing sites, plant mitochondria readily oxidize exogenously supplied NADH through a dehydrogenase located externally on the C-side of the inner membrane (Fig. 13.2). Complex I is bypassed and electrons are passed directly to complex III. Only two sites of energy transduction are involved. In animal mitochondria, the inner membrane is impermeable to NAD and NADH, and there is no externally located dehydrogenase. Consequently, mammalian mitochondria do not oxidize exogenous NADH unless the permeability barrier of the membrane has been altered.

Plant mitochondria isolated from many sources also possess residual respiration in the presence of inhibitors of complexes III and IV. This cyanide- and antimycin-A insensitive (alternate) pathway branches from the conventional electron transport system that begins with CoQ and terminates with an alternate oxidase (Fig. 13.2). It is inhibited by cyclic hydroxamates such as SHAM (salicylhydroxamic acid). The alternate oxidase is absent in mammalian mitochondria, but is found in millipedes, trypanosomes, and some species of bacteria and fungi, in addition to higher plants. The pathway bypasses two energy conservation sites on the cytochrome chain and is not considered to be involved in energy conservation. Some investigators have ascribed a thermogenic role to the alternate pathway. This nonphosphorylating pathway would also provide for the continued functioning of the Krebs cycle and the provision of carbon skeletons when the need for ATP is diminished.

13.3 ATP GENERATION

Mitochondrial electron transport generates a proton electrochemical gradient and these proton currents couple electron transport to ATP synthesis. According to the chemiosmotic hypothesis of Mitchell, ATP synthesis is driven by the delocalized, bulk phase, transmembrane electrochemical gradient (Δp) produced by the redox proton pumps of the electron transport system. Some investigators suggest that Δp may not be a kinetically competent intermediate between electron transport and ATP synthesis. However, whereas the causal relationship between Δp and ATP synthesis is somewhat controversial, a correlation between the two does exist.

Proton movement back across the inner membrane from the C- to the M-side together with the generation of ATP from ADP and Pi is catalyzed by the coupling ATPase (ATP synthase) or F_0-F_1 coupling factor that is sometimes referred to as complex V. The complex exists as a monomer, has a molecular weight of 500,000, and consists of 18 polypeptides. The complex is divided into two fractions, F_0 and F_1. F_0 consists of 13 subunits, with at least some of the components represented by possibly two copies. F_0 stretches across the membrane and forms a hydrophobic proton channel.

F_1 consists of five subunits with more than a single copy of the subunits being present. It catalyzes the reversible synthesis and hydrolysis of ATP. In electron micrographs, F_1 appears to be spherical is shape and extends into the matrix space (Fig. 13.1B). If separated from F_0, F_1 retains the ability to hydrolyze ATP, but loses the ability to synthesize ATP. ATP is synthesized within the matrix, but is primarily used externally. Transport of ATP out of, and ADP into, the matrix involves an adenylate porter.

13.4 ISOLATION AND CHARACTERIZATION OF MITOCHONDRIA

Well-coupled mitochondria have been isolated and characterized from several mammalian tissues. Rat livers and beef hearts have been favored sources. The isolation procedure involves gentle homogenization of the tissue with a loose fitting pestle in an isotonic medium that is buffered around pH 7.2 followed by differential centrifugation. As opposed to mammalian tissue, plant tissues have tough cell walls that have to be broken to release the cytoplasmic contents. Rough and/or prolonged homogenization and isolation procedures damage the mitochondria. In addition, plant cells have vacuoles that contain acidic constituents. Hence, when plant tissue is homogenized, vacuolar contents are released and the pH of the homogenization medium is lowered. Consequently, the extracting medium must be buffered to maintain a pH from 7.1 to 7.4. Successful isolation of intact, coupled mitochondria has been accomplished from a variety of plant tissues. Further purification can be accomplished by gradient centrifugation in sucrose or Percoll. The latter method has been utilized to isolate foliar mitochondria that are essentially free of chloroplast fragments.

O_2 utilization by isolated mitochondria is routinely measured with a Clark-type (oxygen) electrode. The Clark is a Ag^+/Pt^- electrode with a saturated KCl bridge. A potential difference, usually of 800 mV, is imposed across the junction. Dissolved oxygen reacts with the Pt^- and a flow of current results that is proportional to the concentration of oxygen. Changes in potential are monitored with a strip-chart recorder.

Measurements of mitochondrial activity are usually made in a small-volume glass reaction vessel, in which the electrode is inserted, that is accurately thermostated and magnetically stirred. Various substrates, ADP, inhibitors, and other reagents can be added to the mitochondria as desired. With this arrangement, five respiratory states have been identified. Of the five states, only states 3 and 4 are of immediate interest in studies concerned with the action of inhibitors. In state 3, all required components are present in excess and the respiratory chain itself is the rate-limiting factor. State 4 is the condition in which ADP availability has limited the respiratory rate. In state 1, both ADP and substrate are lacking. State 2 is the condition in which only substrate is rate-limiting, whereas in state 5, oxygen is lacking.

A polarographic trace of oxygen utilization obtained with isolated mitochondria as measured with a Clark-type electrode is shown in Figure 13.3, Trace A, for the oxidation of succinate. The trace reflects the stimulation of oxygen uptake (state 3 respiration) obtained by the addition of a small amount of ADP, followed by a decrease in respiration upon exhaustion of the added ADP (state 4 respiration). This sequence can be repeated by further addition of small amounts of ADP until anaerobiosis occurs (state 5). The cycling between states 3 and 4 demonstrates the presence of respiratory control that is required for the evaluation of the action of inhibitors on the energy transfer sequence. The control manifested by ADP will only be evident when mitochondria are tightly coupled.

The quality and intactness of isolated mitochondria can be evaluated by calculation of respiratory control (RC) and ADP/O ratios. The RC value is obtained by dividing the fast state 3 rate by the slow state 4 rate. It provides an indication of the relative efficiency of energy transduction of a mitochondrial preparation. RC values obtained with mammalian mitochondria, as reported in the literature, typically exceed 4, but vary somewhat with the substrate and the tissue being oxidized. The ADP/O ratio is defined as the number of ATP molecules synthesized during the oxidation of a substrate via the respiratory chain by molecular

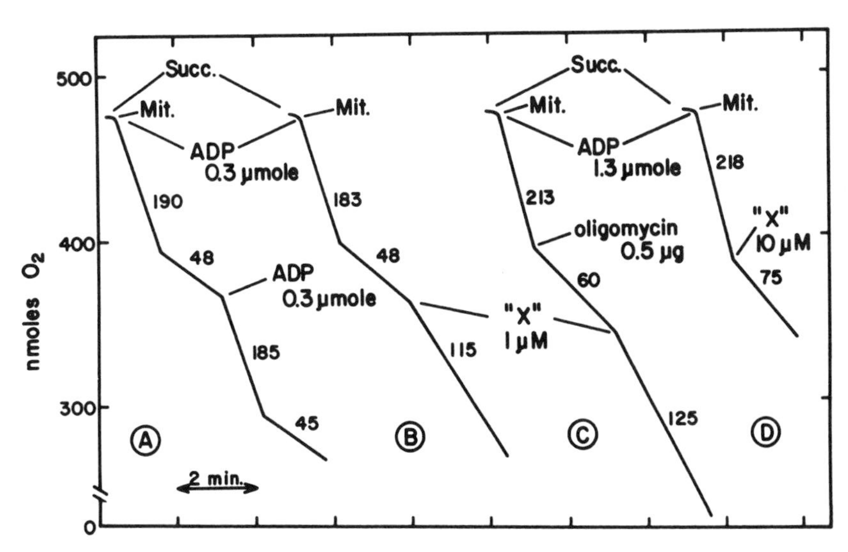

FIGURE 13.3 Representative polarographic traces depicting oxygen utilization obtained with plant mitochondria for succinate oxidation.

oxygen. The ADP/O ratio can be calculated directly from polarographic traces because the concentration of the added ADP is known and the moles of atomic oxygen consumed can be measured. The ADP/O ratio for the oxidation of NAD-linked substrates such as glutamate and malate is 3, whereas the ratio for the oxidation of succinate is 2. These values are theoretical and are seldom achieved.

In addition to studying whole-chain electron transport, it is possible to monitor partial reactions with the use of appropriate electron mediators (Fig. 13.2). Phenazine methosulfate (PMS) can be used to mediate electron flow from the oxidation of malate and succinate through complexes I and II, respectively. Ferricyanide can accept electrons in the region of cytochrome c, and reduced N,N,N′,N′-tetramethyl-p-phenylenediamine (TMPD) can be oxidized through cytochrome c and complex IV.

13.5 CLASSIFICATION OF INHIBITORS

13.5.1 General Considerations

Consideration will be extended primarily to compounds that interfere with the electron transport and phosphorylation pathways. Inhibitors of transport processes associated with the inner membrane and the many metabolic activities associated with the matrix enzymes will not be considered.

Inhibitors have played, and will continue to play, important roles in the formulation of postulates that explain oxidative phosphorylation, and in the elucidation of components of the electron transport and ATP generation pathways. Classically, chemicals that interfere with the mitochondrial system have been classified as (a) electron-transport inhibitors, (b) uncoupling agents, (c) energy transfer inhibitors, and (d) inhibitory uncouplers. Classes (a), (b), and (c) are identified and defined in the mammalian mitochondrial literature. Class (d), inhibitory uncouplers, has been added and is used here to identify the multiple types of responses that are observed with some of the nonclassical inhibitors of oxidative phosphorylation.

The effects of xenobiotics on mitochondrial reactions are usually compared with those manifested by classical agents. Unfortunately, many times the investigators fail to adequately characterize the mitochondria being used, that is, RC values and ADP/O ratios are not given. In studies with many insect preparations, homogenates of certain segments or, in the case of mites, the entire organism are necessarily used. Some investigators report results obtained for partial reactions only. Other investigators use submitochondrial particles (electron-transporting vesicles prepared from the inner membrane). During the preparation of the particles, their inner membranes are turned inside out, that is, the M-side is exposed to the toxicant as opposed to the C-side, as is the situation with intact mitochondria. It is not always possible for different investigators to reproduce responses reported by other researchers even with isolated mitochondria. Some of the reasons for discrepancies may be related to: source (species, tissue, and age of animal); composition of the reaction mixture; substrate oxidized; solvent in which water-insoluble test compounds are dissolved; final concentration of the solvent in the reaction mixture; and the extent to which the mitochondria are and remain coupled. Regardless of the source, animal or plant, isolated mitochondria generally respond similarly to a given inhibitor.

13.5.2 Electron-Transport Inhibitors

Experimentally, electron-transport inhibitors are characterized by their ability to interrupt electron flow by combining with one of the proteinaceous electron carriers of the complex and to prevent the formation of a redox couple. This action is measured most readily as inhibition of state 3 respiration as shown in Figure 13.3, Trace D, for reference inhibitor "X." When electron flow is interrupted, the coupled phosphorylating reaction also is inhibited. Amytal, rotenone, piericidin A, and steroids such as progesterone, interact with complex I (Fig. 13.2, site a), and prevent the oxidation of malate and other NAD-linked substrates. Most appear to bind to the same Fe-S protein. TTA (thenoyltrifluoroacetone) inhibits by binding to one of the nonheme iron-sulfur centers of complex 1. Malonate competitively interferes with the oxidation of succinate by succinic dehydrogenase, and TTA interacts with one of the nonheme iron sulfur centers of complex II and blocks the transfer of reducing equivalents from FAD (Fig. 13.2, site b). The antibiotic antimycin A and its synthetic analogs and homologs and HOQNO (2-n-heptyl-4-hydroxyquinoline-N-oxide) affect complex III (Fig. 13.2, site e). The functional site of inhibition is postulated to lie between cytochromes b and c. Cyanide, azide, and CO are presumed to bind to the copper ligands of complex IV (Fig. 13.2, site f). Cyclic hydroxamates, such as SHAM (salicylhydroxamic acid), inhibit the cyanide insensitive pathway (Fig. 13.2, site d). Inhibitors that act on the complexes cause $\Delta\Psi$ to decrease.

Inhibition of state 3 respiration by compound X at a concentration of 10 μM is shown in Figure 13.3, trace D. Percent inhibition can be calculated from polarographic traces by comparing the inhibited state 3 rate with the no-inhibitor control rate. I_{50} values can be obtained by titration of state 3 respiration with increasing concentrations of an inhibitor. Graphs in which percent inhibition is plotted versus the logarithm of the molar concentration are usually linear except at the extreme concentrations.

Not many pesticides or pharmaceuticals, other than those mentioned previously are pure electron-transport inhibitors. However, the cis-crotonanilide fungicides such as carboxin, oxycarboxin, and fenfuran inhibit succinate oxidation possibly by interacting with one of the iron-sulfur centers of complex II, much like TTA. The fungicide fenaminosulf inhibits NADH oxidation by interacting with both complex I and the externally located dehydrogenase in plant mitochondria. The site of action has been postulated to reside prior to the site affected by rotenone. The DDT and cyclodiene insecticides inhibit state 3 respiration and do not stimulate state 4 respiration. However, the action of these insecticides is complex and will be considered in a later section.

A number of herbicides (chlorproham, propanil, dinoseb, and ioxynil) inhibit state 3 respiration, but they also uncouple state 4 respiration, usually at slightly lower molar concentrations. Hence, these herbicides are not considered to be pure electron-transport inhibitors, but instead are classified as inhibitory uncouplers. Some of the dinitroaniline and diphenylether herbicides, however, do act as multisite electron-transport inhibitors of mitochondrial respiration without having a marked uncoupling effect.

13.5.3 Uncouplers

At appropriate concentrations, classical uncouplers such as carbonyl cyanide p-trifluoromethoxyphenythydrazone (FCCP) prevent the phosphorylation of ADP without interfering with electron transport (Fig. 13.2, site g). In general, any compound that promotes the dissipation of the proton motive force generated by electron transport, other than for the production of ATP, may be regarded as an uncoupler. Uncouplers collapse $\Delta\Psi$ to the Donnan potential of about 60 mV. When added in vitro to tightly coupled mitochondria, uncouplers stimulate oxygen uptake by eliminating the regulatory influence exerted by Pi and ADP. Stimulation of state 4 respiration by compound X at a concentration of 1 μM is evident in Figure 13.3, Trace B. As shown, the control exerted by ADP is lost. Electron transport and oxygen utilization continue at accelerated rates, but no ATP is formed. Percent stimulation can be calculated by comparing the stimulation induced by a test compound to the stimulation obtained with ADP, that is, 100% stimulation equals that obtained with ADP. The titration of state 4 respiration with increasing concentrations of an uncoupler, when plotted as percent stimulation versus the logarithm of the molar concentration, gives a characteristic parabolic or bell-shaped curve.

Stimulation of state 4 respiration in media deficient in ADP or Pi is one of several criteria that uncouplers should meet. Other criteria include (a) induction of ATPase activity, with the induction process being sensitive to oligomycin; (b) circumvention of oligomycin-imposed inhibition of state 3 respiration (as shown in Fig. 13.3, Trace C) and (c) inhibition of the various exchange reactions catalyzed by mitochondria in the absence of substrates (Pi-ATP, ADP-ATP, and O_2 of water with Pi and ATP).

Several hypotheses have been proposed to explain uncoupling of oxidative phosphorylation. These include the classical uncoupler-induced hydrolysis of a high-energy intermediate; conformational changes and altered functions of coupling proteins upon interaction with uncouplers; energy linked transport of the uncoupler anion into and passive diffusion of the protonated species out of the mitochondria; acid- or base-catalyzed hydrolysis by uncouplers of an intermediate reaction of oxidative phosphorylation in a lipophilic membrane region; increase of membrane conductance; collapse of the transmembrane proton gradient with uncoupters acting as protonophores and ionophores in transporting cations across the mitochondrial membrane; and variations of the above. Covalent labeling of specific mitochondrial proteins by reactive radio- and photo-affinity labeled uncouplers (2-azido-4-nitrophenol and 2-nitro-4-azidocarbonyl cyanide phenylhydrazone) have identified specific uncoupler-binding proteins associated with the energy conservation pathway.

Pentachlorophenol (PCP) and 2,4-dinitrophenol (DNP) are classical examples of uncouplers that have been used as pesticides for many years. Most halogenated and nitrophenols possess uncoupling activity. Their activities vary, depending upon the substituents present on the benzene ring, in that some are insecticides, ovicides, fungicides, or herbicides. Substituted 2-trifluoromethylbenzimidazoles are also uncouplers with a wide range of biological activity. Some are insecticides, acaricides, molluscicides, or herbicides. Others possess antiviral and antibacterial activities. Additional compounds identified as being uncouplers of oxidative phosphorylation include dicoumarol (an anticoagulant), salicylanilides, atebrin (the antimalarial drug), and various anesthetic gases such as halothane.

When intact mitochondria are probed with classical uncouplers, $\Delta\Psi$ is partially collapsed. The partial collapse correlates with inhibition of ATP generation and occurs at concentrations that do not suppress O_2 utilization. However, general anesthetics, free fatty acids, detergents, and some alcohols uncouple phosphorylation without significantly affecting either $\Delta\Psi$ or O_2 uptake.

Chlordecone and its derivatives stimulate state 4 respiration, but also strongly inhibit state 3 respiration. A number of herbicides (dinitrophenols, benzimidazoles, benzonitriles, thiadiazoles, acylanilides, bromofenoxim, and perfluidone) act similarly, that is, stimulate state 4 respiration at a low concentration and inhibit state 3 respiration at slightly higher concentrations. The above compounds have been classified as inhibitory uncouplers and their action will be discussed separately.

Many pesticides are detoxified microsomally by hydroxylation of their phenyl rings. Whereas the unaltered parent molecule may not be an uncoupler, the hydroxylated degradation forms are potential uncouplers if they are not rapidly complexed as glucosides or glycosides in plants or as glucuronides in mammals.

13.5.4 Energy Transfer Inhibitors

Compounds in this group, with the antibiotic oligomycin being the prototype, inhibit phosphorlyating electron transport (state 3 respiration) when the energy conserving apparatus of mitochondria is intact (Fig. 13.3, Trace C, upper part). The inhibition is circumvented by uncouplers (Trace C, lower part). Energy transfer inhibitors interact with either F_0 and F_1 of the energy coupling factor complex (F_0-F_1) and, hence, block the phosphorylation sequence that leads to the production of ATP (Fig. 13.2). They cause state 3 $\Delta\Psi$ to increase to the state 4 value. Oligomycin, N′N′-dicyclohexylcarbodiimide (DCCD), and chlorotrialkyltins inhibit the activity of F_0 by blocking the proton channel. Phlorizine, quercetin, Nbf-Cl (4-chloro-7-nitrobenzofurazan), and the aurovertin antibiotics inhibit ATPase activity by binding to one of the subunits of F_1. Activity of the adenylate porter is inhibited by atractylosides.

Cyhexatin (a chlorotriaryltin) also acts as an F inhibitor. With intact rat liver mitochondria, like oligomycin, the compound inhibits state 3 respiration and the coupled phosphorylation, and the inhibition is circumvented by uncoupters. In studies with housefly and spider mite homogenates, where intact mitochondria cannot be isolated, interference with the F_0-F_1 complex is measured as inhibition of the membrane-bound oligomycin-sensitive Mg^{2+}-ATPase. Cyhexatin is an extremely potent inhibitor. I_{50} values of 0.6 nM for partially purified preparations have been reported.

13.5.5 Inhibitory Uncouplers

The action of chemicals other than the reference standards that interfere with oxidative phosphorylation is frequently complex with a diverse spectrum of responses being manifested. Such chemicals have been classified as inhibitory uncouplers. At low molar concentrations, the chemicals satisfy most, if not all, of the criteria established for uncouplers, but at higher concentrations, they act like electron transport inhibitors.

Herbicides classified as inhibitory uncouplers, in so far as they have been tested, stimulate state 4 respiration. All produce the characteristic bell-shaped curve when percent stimulation is plotted as a function of the logarithm of the concentration of the herbicide. Within the stimulatory range, inhibition of state 3 respiration becomes evident. Hence, the higher concentrations of inhibitory uncouplers both stimulate state 4 and inhibit state 3 respiration. As explained earlier, inhibition of state 3 respiration is generally considered to reflect interference with electron transport, whereas the state 4 response is considered to measure an effect imposed on a component of the ATP-generating sequence. The occurrence of "overlapping" effects, that is, inhibition of state 3 and stimulation of state 4 respiration at the same molar concentration, can be expected to reduce the actual or apparent uncoupling capabilities of a compound because the capacity for utilizing oxygen has been reduced by a competing reaction. Additionally, the compounds induce oligomycin-sensitive ATPase activity and circumvent oligomycin-imposed inhibition of state 3 respiration associated with complexes I and II.

In addition to meeting all of the requirements for uncouplers, the herbicides classified as inhibitory uncouplers affect several sites on the electron transport pathway. That inhibition of state 3 respiration associated with the oxidation of malate, succinate, and NADH (Fig. 13.2, site c) does not occur at a single site common to the three substrates is suggested by the widely differing I_{50} values that are obtained. The herbicides also inhibit malate-PMS oxidoreductase (Fig. 13.2, site a). Malate-PMS oxidoreductase is not inhibited by rotenone, hence, action prior to the rotenone-sensitive site in complex I is suggested. Additional evidence for interference at or near complex II (Fig. 13.2, site b) is provided by inhibition of succinate-PMS oxidoreductase. Succinate-PMS oxidoreductase is not inhibited by TTA, SHAM, diphenylethers, or the dinitroanilines. Inhibitory uncouplers also inhibit malate and succinate-mediated cyanide-resistant respiration (Fig. 13.2, site d). No evidence has been obtained to date for interference at or near complex III and IV by any of the herbicides.

The herbicides can be separated into two groups according to the dose-response relationships exhibited with respect to $\Delta\Psi$ and oxygen utilization. The first group, designated as dinoseb-types (dinitrophenols, benzimidazoles, benzonitriles, thiadiazoles, and bromofenoxim), uncouple phosphorylation and collapse $\Delta\Psi$ to the Donnan level before oxygen utilization is inhibited. These compounds possess dissociable protons and are postulated to act as protonophores, much like FCCP. With the second group, termed dicryl-types (acylanilides, dinitroanilines, diphenylethers, bis-carbamates, and perfluidone), collapse of $\Delta\Psi$ parallels uncoupling of phosphorylation and inhibition of oxygen utilization. However, phosphorylation is inhibited to a greater extent than is respiration. The dicryl-type herbicides are not classical-type protonophores. Some of their action can be attributed to interference with the redox pumps. The complete collapse of $\Delta\Psi$ to the Donnan potential is associated with alterations and perturbations induced in the membranes by classical uncouplers and by both types of herbicides. The perturbations are postulated to increases the permeability of the membranes to protons and other cations, and induce unfavorable conformational changes that impede interactions between redox enzymes. Conceivably, the combined responses collapse $\Delta\Psi$ and inhibit electron transport.

DDT and cyclodiene insecticides also induce a complicated set of alterations to mitochondrial activity including interaction with the F_0-F_1 coupling factor. The diversity of the alterations preclude the strict classification of their actions. With submitochondrial particles obtained from various mammalian tissues and homogenates of insect tissue, DDT is a strong inhibitor (1 to 10 μM) of oligomycin-sensitive Mg^{2+}-ATPase. When tested with lipid-free, purified coupling factors, the water-soluble F_1 Mg^{2+}-ATPase preparations were insensitive to DDT. Inhibition occurred only when the oligomycin-sensitivity conferring protein and phospholipids from the F_0 fraction were added back to the purified F_1, preparations. In these studies, DDT acts like an F_1 inhibitor. The site of inhibitory action of DDT may be quite different from that of oligomycin and DCCD in that a specific binding protein may not be involved. In studies with intact mammalian mitochondria, the energy transfer action of the DDT and cyclodiene types of insecticides is masked. Variable results have been reported by different investigators. At low concentrations, investigators unanimously report inhibition of NADH-linked and succinate-dependent state 3 respiration. There are suggestions that DDT interrupts electron transport by acting between cytochrome b of complex III and cytochrome c. Some, but not all, investigators have observed uncoupling action measured as stimulation of state 4 respiration. Other observations reported, at overlapping or slightly higher concentrations, include inhibition of uncoupler-stimulated respiration, induction of mitochondrial swelling with KCl and/or sucrose as osmotica, inhibition of Ca^{2+} uptake driven by ATP hydrolysis, inhibition of valinomycin-mediated K^+ uptake, and inhibition of cytochrome oxidase. Some of the observations indicate that the DDTs alter the permeability properties and integrity of the inner membrane. Electrophysiologists consider DDT and related compounds to be neurotoxicants. Other investigators, however, postulate that a reduction in ATP availability, because of interference with the generation of mitochondrially produced ATP, limits the availability of energy required for operation of neural ion pumps.

In studies with intact rat liver mitochondria, chlordecone also manifests a number of alterations. Swelling is induced by concentrations as low as 2 μM. The permeability properties of the inner membrane are altered as evidenced by inhibition of valinomycin-induced swelling, induction of passive swelling, oxidation of exogenous NADH, and induction of lysis, which leads to the leakage of matrix enzymes. Associated with the alteration in membrane properties, are stimulation of state 4 and inhibition of state 3 respiration for the oxidation of both succinate and NADH-linked substrates.

With intact mammalian mitochondria, cyhexatin at low concentrations, acted as a classical energy transfer inhibitor. However, at slightly higher concentrations (>1.0 μM) cyhexatin altered the permeability properties of the inner membrane as evidenced by the induction of swelling, leakage of matrix enzymes, collapse of $\Delta\Psi$, and inhibition of uncoupled electron transport.

13.6 ATP AVAILABILITY AND METABOLISM

ATP has a ubiquitous and dominant role in cellular metabolism. ATP provides the energy, directly or indirectly, to drive most biosynthetic reactions. The several nucleotide and deoxynucleotide triphosphates (CTP, GTP, UTP, and dATP, dCTP, dGTP, and dTTP) are formed by transfer of the terminal phosphate group of ATP to the 5′-diphosphates of the respective nucleotides. Through these transfers, catalyzed by specific diphosphokinases, ATP energy is channeled into the major biosynthetic pathways involved in the synthesis of polysaccharides, porphyrins, cellulose, proteins, lipids, and the nucleic acids. None of the triphosphates can substitute for ATP in processes energized by ATP.

Membrane functions including active transport and the effects induced by membrane-bound hormones are energy dependent. The structural organization, contraction, and orientation of chromosomes and microtubules of the spindle apparatus during mitosis depend on ATP energy. The intracellular concentrations and stoichiometric relations of ATP, ADP, and AMP modulate cellular metabolism.

13.7 PERTURBATIONS OF ENERGY TRANSDUCING MEMBRANES

The multisite inhibitory action on mitochondrial reactions by xenobiotics is considerably more complex than that of the site-specific classical inhibitors. The xenobiotics are lipophilic and partition into the nonpolar (lipid) regions of all cellular membranes. Consequently, they can be expected to perturb lipid-lipid, lipid-protein, and protein-protein interactions that are required for membrane functions such as electron transport, ATP formation, and active transport. Interference with oxidative phosphorylation may involve interactions with both the lipoidal and proteinaceous components of the membranes. Partitioning also can be expected to produce alterations to the fluidity and permeability properties of the mitochondrial membrane that result in perturbational and conformational shifts to the constituent redox components. These alterations, then, could be responsible for the multiple effects imposed on energy transducing membranes.

When interfering xenobiotics partition into membranes, the state and/or interactions of the constituent proteins and lipids, and the distribution of ions are affected. These effects depend qualitatively on the type of membrane encountered. Conceivably, ion fluxes and metabolite distributions between the cytosol and mitochondria, may also be modified by changes in the proton gradient and membrane potential together with inhibition of ATP production.

Whereas, membrane perturbations can be induced very rapidly, techniques needed to measure directly these rapid alterations in membrane fluidity and permeability are not available. If a xenobiotic perturbs a particular cellular membrane, it is likely that all membranes could be affected. The consequences of perturbations can be measured readily in those membranes that involve electron transport and phosphorylation. Perturbations and subsequent alterations in the plasmalemma, nuclear membranes, and endoplasmic reticulum may be just as dramatic as in mitochondria, but the changes may be more subtle.

The production of alterations in the fluidity and permeability properties of the mitochondrial inner membrane by certain insecticides (DDT-types, cyclodienes, cyhexatin, chlordane) and some herbicides is evidenced by the induction of swelling of rat liver and plant mitochondria. Inhibitory pesticides have also been shown to inhibit the rate of valinomycin-induced swelling in the same organelles. Valinomycin is a mobile carrier of potassium, and a restriction in its movement across the membrane suggests that the fluidity of the membrane has been decreased.

The well-documented effects of the DDT, cyclodiene, and synthetic pyrethroid insecticides with excitable membranes and the effects induced on mitochondrial electron transport and phosphorylation (a nonexcitable organelle) can be explained by interference with general membrane activities. However, the existence of specific recognition membrane structures must be considered. If, however, the recognition features are absent, multiple nonspecific inhibitory effects can be expected. Inhibition of mitochondrial responses may serve as an example of such nonspecific interactions and provides a basis for the chronic toxicity of some xenobiotics.

If the experimental and therapeutic responses induced by xenobiotics are considered, it is evident that many compounds induce a large number of metabolic effects. The diverse responses would be difficult to explain if there were an equally large number of individual and specific effects on various enzymes. Many xenobiotics require the presence of membranes to exert their inhibitory activities.

Many biologically active chemicals, such as anesthetics and drugs, induce expansion when they partition into membranes. The expansion is postulated to produce multiple effects on membrane-related processes. According to this postulate, anesthetics and other neuroactive agents adsorb to hydrophobic regions (lipid and protein) of the excitable membrane, expand the hydrophobic regions of proteins, and subsequently block the ionic conductance channels that underlie nerve cell action potentials. Most of the pesticides that interfere with mitochondrial activity induce expansion of the inner membrane. Some of the inhibitory responses may be a consequence of membrane expansion.

Conceivably, for many xenobiotics, chronic toxicity could be expressed by a combination of alterations to the properties of various cellular membranes and the limitations imposed on the availability of ATP. Changes induced to the proton gradient and membrane potential of energy transducing membranes can alter ion fluxes and metabolite distribution between organelles and the cytosol. A large number of alterations can be expected to result which would have a drastic effect on the metabolism of an organism.

SUGGESTED READING

Douce, R. *Mitochondria in Higher Plants*, Academic Press, Orlando, FL, 1985.

Erecinska, M. and Wilson, D.F. Inhibitors of mitochondrial functions. In: *International Encyclopedia of Pharmacology and Therapeutics*, Section 107, Pergamon Press, Oxford, 1981.

Green, D.E. and Baum, H. *Energy and the Mitochondrion*, Academic Press, New York, 1970.

Lehninger, A.L. *The Mitochondrion*, Benjamin, New York, 1964.

Lehninger, A.L. *Bioenergetics*, Benjamin, Menlo Park, NJ, 1971.

Levings, C.S. III and Vasil, I.K. *The Molecular Biology of Plant Mitochondria*. Kluwer Academic Publishers, Dordrecht, 1995.

Papa, S., Guerrier, F. and Tager, J.M. *Frontiers of Cellular Bioenergetics*. Kluwer Academic/Plenum Publishers, New York, 1999.

Scheffler, I.E. *Mitochondria*, Wiley-Liss, New York, 1999.

Slater, E.C. Uncouplers and inhibitors of oxidative phosphorylation. In: Hochester, R.M. and Quaste, J.J. (Eds), *Metabolic Inhibitors*, Vol. 2. Academic Press, New York, 1963, pp 503–516.

Slater, E.C. Application of inhibitors and uncouplers for a study of oxidative phosphorylation. *Meth. Enzymol.* 10 (1970), 48.

Tzagoloff, A. *Mitochondria*. Plenum Press, New York, 1982.

CHAPTER FOURTEEN

Effects of Toxicants on Nucleic Acid and Protein Metabolism

DAVID J. HOLBROOK

14.1 INTRODUCTION

Toxicants may exert primary effects (e.g., inhibition of a polymerase) or secondary effects (inhibition of oxidative phosphorylation or other energy production) on various aspects of nucleic acid or protein metabolism. It is of interest to identify the direct and indirect effects of toxicants and the specific biochemical processes involved. The topics to be presented include some of the basic principles of the metabolism of nucleic acids and proteins, the methodology that may be used to examine the effects of toxicants, and examples of the application of this information to specific toxicants.

14.2 DIFFERENCES IN NUCLEIC ACID AND PROTEIN METABOLISM IN MULTIPLE CELL TYPES

The cell population in any mammalian tissue is heterogeneous, and nucleic acid and protein metabolism of a tissue represents the differing contributions from each of the multiple cell types in that tissue. The various cell types differ in characteristics of macromolecular synthesis such as the activities of DNA and RNA polymerases, the susceptibility to certain toxicants (which in part may be due to the ability of the cell to convert a nonreactive toxicant to a reactive metabolite), or the concentration of protective systems within the cell (e.g., cellular glutathione concentration). The liver includes parenchymal cells (or hepatocytes), bile duct cells, Kupffer cells, and cells lining the blood vessels. The parenchymal and nonparenchymal nuclei differ in their properties related to nucleic acid synthesis, such as incorporation of radioactive precursors into nucleic acids, in normal and in toxicant-treated animals. Furthermore, the parenchymal tetraploid and nonparenchymal diploid nuclei isolated from adult rodent liver differ in the activities of the DNA polymerases and RNA polymerases retained in the isolated nuclei.

14.3 PRECURSOR INCORPORATION

The most commonly used technique to study the effect of a toxicant on nucleic acid and protein synthesis is to determine the effect of the toxicant on the relative rates of the incorporation of a radioactive precursor into the macromolecule. The radioactive precursors most often selected include: thymidine incorporation into DNA; orotic acid, uridine, or orthophosphate incorporation into RNA; and an amino acid incorporation into protein. The convenience of the technique for measurement of the incorporation of the radioactive precursor into cold-acid-precipitable macromolecules has led to its extensive use. The DNA, RNA, and protein (including the newly synthesized macromolecule containing the incorporated radioactive precursor) are precipitated by cold acidic solutions (e.g., 0.5–1 M perchloric acid or 5–10% trichloroacetic acid), which permits separation from the radioactive precursors (nucleotides and amino acids) that are soluble in the acid solution.

In spite of the extensive use of the technique, the results are only indicative of an effect of the toxicant on the synthesis of the macromolecule. In addition to the direct effect of a toxicant on one of the steps or components involved in macromolecular synthesis, alterations in precursor incorporation into a macromolecule may be secondary to some indirect effect such as (a) alteration in the movement of the radioactive precursor across the cell membrane or from the site of injection to the tissue under study; (b) alteration in energy metabolism (oxidative phosphorylation or mitochondrial structure) such that all energy-requiring processes (including DNA, RNA, and protein synthesis) are inhibited; or (c) a change in the specific radioactivity (disintegrations per minute per micromole) of the radioactive precursor by dilution due to different pool sizes of the precursor (e.g., nucleoside triphosphate) or to the release of nonradioactive precursor from neighboring cells undergoing necrosis. A more definitive study relates the incorporation into the macromolecule (e.g., DNA) to the specific radioactivity of the precursor (e.g., deoxyribonucleoside triphosphate).

The precursor incorporation into total RNA has been generally replaced by studies of precursor incorporation into specific mRNAs. Likewise amino acid incorporation into total cellular protein has been commonly replaced by amino acid incorporation into specific proteins.

14.4 DNA METABOLISM

14.4.1 Cell Cycle and Nuclear DNA Replication

The effect of a toxicant on DNA metabolism (or replication) is markedly dependent on the stage of the cell cycle of the exposed cell. The cell cycle is divisible into at least four distinct stages in mammalian cells. In moderately rapidly dividing mammalian cells, the entire cell cycle (i.e., from one mitosis to the next mitosis) requires approximately 18–24 hours. The G1 (gap 1 or ground 1) stage requires approximately 8–10 hours and the activities during this phase include the synthesis of enzymes which will be required for the DNA replication stage, such as synthesis of enzymes involved in nucleotide synthesis and of DNA polymerases which are required for DNA replication. The next stage, approximately 8–12 hours, is the DNA replication stage, or S phase, during which all of the DNA replicative synthesis occurs. The S phase is followed by the G2 (gap 2) phase, approximately 6–8 hours, when proteins involved in mitosis are synthesized. The next stage to complete the cell cycle is mitosis, lasting approximately one hour. Mammalian cells that are not actively undergoing mitosis are referred to as being in a G-zero or an extended G1 phase.

Although DNA replicative synthesis is limited to the S phase, DNA repair (or repair synthesis) occurs throughout the cell cycle.

14.4.2 DNA Polymerases: Multiple Forms in Mammalian Cells

There are five classes of mammalian DNA polymerases, and the enzymes differ in their structure, functions, and sensitivities to certain types of inhibitors (Table 14.1). The mammalian enzymes are named with Greek letters in the order in which the enzymes were discovered.

14.4.3 Functions of the Mammalian DNA Polymerases

DNA Replicative Synthesis. The synthesis of DNA is semidiscontinuous. One strand of the newly synthesized DNA is synthesized as one continuous, long strand, that is, the leading strand. The second strand, the lagging strand, is synthesized as multiple discontinuous strands (Okazaki fragments) that must be ligated to produce the second newly synthesized strand. The mammalian DNA polymerases delta and (probably) epsilon are responsible for the nucleotide polymerization on the leading strand of the DNA after synthesis of the primer by polymerase alpha-primase. Part of the evidence for the role of polymerase delta is the requirement for PCNA (proliferating cell nuclear antigen) which is required by this polymerase. Both of these polymerases have an associated 3′ to 5′ exonuclease (as part of the polymerase molecule) that permits proofreading of the product DNA and increases the accuracy of the polymerization. The lagging strand of DNA is synthesized as units of approximately 1000 nucleotides, and most of the polymerization is catalyzed by DNA polymerase alpha. The latter enzyme also has a primase activity that has the ability to synthesize the short RNA-like primer for each of the Okazaki fragments. After removal of the RNA primer and completion of the deoxyribonucleotide fragments by polymerase delta, the fragments are then ligated by DNA ligase to produce the high-molecular-weight, complete DNA strand.

DNA Repair. During the repair of DNA by nucleotide excision repair, fragments of 24–29 nucleotides are removed by the appropriate nucleases. DNA polymerases delta and epsilon catalyze the polymerization of nucleotides to replace the damaged region of DNA. The nucleotide polymerization during nucleotide excision repair requires PCNA and provides evidence that the polymerization is catalyzed by DNA polymerase delta and (possibly) epsilon. In contrast, during base excision repair, in instances when the damaged region consists of only one to four nucleotides, the nucleotide polymerization is catalyzed primarily by DNA polymerase beta. This repair is the only primary function of this polymerase.

The attribution of a function to a specific DNA polymerase does not exclude a limited participation by other polymerases in that function. For example, although DNA polymerase beta is generally considered the primary polymerase involved in

TABLE 14.1 Properties of Mammalian DNA Polymerases.

Cellular Location	Alpha Nucleus	Beta Nucleus	Delta Nucleus	Epsilon Nucleus	Gamma Mitochondria
Function in replication	Nuclear DNA replication (lagging strand) (a)	None	Nuclear DNA replication (leading strand and lagging strand completion)	Nuclear DNA replication (leading strand)	Mitochondrial DNA replication
Function in repair		Nuclear DNA repair (base excision repair)	Nuclear DNA repair (nucleotide excision repair and mismatch repair)	Nuclear DNA repair (nucleotide excision repair and mismatch repair)	
Nuclease: $3' \rightarrow 5'$	None (b)	None	$3' \rightarrow 5'$ Exonuclease (proof-reading function)	$3' \rightarrow 5'$ Exonuclease (proof-reading function)	$3' \rightarrow 5'$ Exonuclease (proof-reading function)
Requirement for PCNA (proliferating cell nuclear antigen)	No	No	Yes	No	No
Inhibition by -SH reactive compounds (e.g., N-ethylmaleimide)	Yes	No	Yes	Yes	Yes
Inhibition by aphidicolin (d)	Yes	No	Yes	Yes	No
Inhibition by dideoxyribo-nucleoside 5′-triphosphate	No	Yes	No	No	

(a) DNA ploymerase alpha has a low processivity (number of nucleotides polymerized per attachment to the growing DNA chain) and an associated primase that are consistent with its role in the synthesis of the discontinuous strand (lagging strand) during the DNA replication.

(b) Mammalian DNA ploymerase alpha does not have a 3′ to 5′ exonuclease activity. Therefore, polymerase alpha is not capable of proof-reading the nascent DNA chain and has a lower accuracy than the polymerases that have an associated 3′ to 5′ exonuclease.

DNA base excision repair, polymerases delta and epsilon also may be involved when the repair patch is up to 10 nucleotides in length.

14.4.4 Relative Cellular Content of the DNA Polymerases

DNA polymerase beta is present as a relatively large fraction of the DNA polymerases in nondividing or slowly dividing cells. Such cells include the liver of adult rats or mice and cultured cells that are quiescent and have grown to confluence. If the same cells are stimulated to divide, the relative content of the other DNA polymerases involved in nuclear DNA replication markedly increase. In the liver, stimulation of cell division may be accomplished by partial surgical removal of the liver (partial hepatectomy) which is followed in approximately 24–26 hours by a peak in replicative DNA synthesis. Cultured cells can be stimulated to cell division and logarithmic growth by dilution of confluent cells and enrichment of the culture medium. An increase in cell division may also occur following exposure of certain cells to toxicants that result in the death of some but not all cells. For example, treatment of rodents with carbon tetrachloride kills some hepatic cells. The surviving cells initiate rapid DNA synthesis (with a peak at approximately 36–40 hours after treatment with carbon tetrachloride) and mitotic activity (with the first peak at 50–60 hours after treatment) as the surviving cells restore the original mass of liver tissue.

14.4.5 Inhibitors of DNA Replication and DNA Polymerases

The molecular structures of the polymerases obviously affect the differential susceptibility of the DNA polymerases to inhibition by certain toxicants. Various synthetic chemicals and natural products are sulfhydryl-reactive compounds. DNA polymerases alpha, delta, and epsilon contain one or more critical sulfhydryl groups, thus these enzymes are very sensitive to sulfhydryl-reactive inhibitors that react with or modify protein sulfhydryl groups. For example, N-ethylmaleimide, iodoacetamide, para-chloromercuribenzoate that form covalent adducts with protein sulfhydryl groups, are potent inhibitors of DNA polymerases alpha, delta, and epsilon. In contrast, DNA polymerase beta does not contain critical sulfhydryl groups, and polymerase beta is relatively resistant to inhibition by N-ethylmaleimide and other sulfhydryl-reactive agents.

The antibiotic aphidicolin inhibits DNA polymerases alpha, delta, and epilson but does not affect DNA polymerases beta or gamma. This property has been an important experimental tool in the establishment of the roles of the various polymerases in replication and repair.

14.5 DNA REPAIR AND MAINTENANCE OF GENOMIC STABILITY

The DNA is subject to constant damage on the basis of its chemical instability and exposures to carcinogens and mutagens. Consequently, there must be both very active and very accurate repair systems to maintain the normal genomic stability of the cell and to prevent mutations leading either to abnormal cell growth (cancer) or cell death.

14.6 BASE EXCISION REPAIR

Overall Process. Base excision repair is the removal (if necessary) of a damaged DNA base and repair of a missing base at an abasic site. The process is initiated by the absence or removal of either a normal base or an abnormal base (i.e., a purine or a pyrimidine, not a nucleotide). The process is also characterized as a short-patch repair with the damage usually being repaired by the removal of the damaged base and the replacement of one to four nucleotides in the process.

Production of Abasic Sites. The initiation of base excision repair is characterized by the loss of a base to produce an abasic site (also called an AP, for apurinic site or apyrimidinic, site that lacks a purine or pyrimidine, respectively).

The loss of a base from the DNA may involve three mechanisms:

1. The beta-glycosyl bond between a purine and the deoxyribose is inherently unstable at physiological conditions and appreciably less stable than the beta-glycosyl bond between a pyrimidine base and deoxyribose. The result is that normal, undamaged purines are spontaneous lost from the DNA structure to produce an abasic site, or more specifically, an apurinic site. It has been estimated that the DNA (as a deoxyribonucleoprotein) of a mammalian diploid cell loses one base per minute per cell or in another estimate of 2000–8000 bases per 24 hours. Such a rate of loss emphasizes the rate of base excision repair that is necessary to maintain the normal DNA structure.
2. The beta-glycosyl bond between the deoxyribose and a methylated or ethylated purine base is even less stable than the bond to a normal purine base. Thus, DNA purines that have been exposed to methylating or ethylating carcinogens or mutagens are even more susceptible to depurination than normal purine bases. Methylating and ethylating carcinogens include direct acting carcinogens such as N-methyl-nitrosourea and the metabolites of metabolically activated carcinogens such as N,N-dimethylnitrosamine and N,N-diethylnitrosamine.
3. Glycosylases are an active component in the repair of abnormal bases. These enzymes recognize an abnormal base and hydrolyze the beta-glycosyl bond between the abnormal base and its attached deoxyribose, and the free base is released as a product of the reaction. The glycosylases are generally highly specific for the abnormal base substrate.

Glycosylases. Glycosylases in mammalian cells include:

1. *Uracil-DNA Glycosylase.* Uracil can be produced in DNA either by spontaneous deamination of cytosine or by exposure of the DNA to nitrous acid (i.e., nitrite ion in the presence of the hydrogen ion even in a mildly acidic medium). The glycosylase removes uracil from the DNA and produces an abasic (specifically an apyrimidinic) site.
2. *Thymine (Uracil) Mismatch DNA Glycosylase.* The deamination of 5-methylcytosine creates a thymine-guanine mismatch that is recognized by the glycosylase, and the thymine is removed. In humans, this glycosylase also removes uracil from a uracil-guanine mismatch (produced by deamination of cytosine) in double stranded DNA.

3. *N-Methylpurine DNA Glycosylase*. This enzyme has also been referred to as 3-methyladenine DNA glycosylase or hypoxanthine DNA glycosylase, and it is also active on both 3-methyl and 7-methylpurines. N^3-methyladenine is a stable and toxic metabolite produced by various methylating carcinogens. Hypoxanthine is produced by spontaneous deamination of adenine or by exposure of adenine to nitrous acid. As the names imply, the glycosylase has a broad range of specificity and catalyzes the removal of a number of abnormal purines from DNA. Substrates for the enzyme include DNA containing 3-methyladenine, 7-methyladenine, O^4-methylthymine, various ethylated bases, and the four cyclic ethenoadenines and ethenoguanines. Several of the etheno substrates occur both naturally in DNA and all four are produced by metabolites of vinyl chloride and vinyl fluoride. 3-Methyladenine is the most rapidly repaired base in DNA; the relatively large amount of 3-methyladenine caused by exposure to methylating carcinogens and the high activity of this glycosylase account for the rapid removal of 3-methyladenine from damaged DNA.
4. *Thymine Glycol DNA Glycosylase*. This glycosylase acts on oxidatively damaged pyrimidines. Substrates, and the naming of the glycosylase, include pyrimidines with saturated 5,6 double bonds (thymine glycol and cytosine hydrate) and other ring-opened and degraded pyrimidines.
5. *8-Oxoguanine Glycosylase (8-Hydroxyguanine DNA Glycosylase)*. 8-Oxoguanine (specifically 7,8-dihydro-8-oxyguanine) is a common product produced by oxidative damage (hydroxyl free radical) or by ionizing radiation. The quantity of 8-oxyguanine in cellular DNA is commonly used as a measure of oxidative stress to the cells. The enzyme also removes formamidinopyrimidine (formed from guanine by ionizing radiation) or by ring scission of N^7-methylguanine. N^7 methylguanine is the most common product produced in DNA exposed to methylating carcinogens.

Abasic Endonucleases and Phosphodiesterase. Abasic sites are cytotoxic, block DNA replication and RNA transcription, and are mutagenic. The phosphodiester bonds to the 3′- and 5′-side of the abasic site are broken by nucleases (or lyases) and the sugar-phosphate residue is removed.

Polymerase. DNA polymerase beta is at least predominantly, and perhaps exclusively, the polymerase activity that replaces the gap that contains the damaged nucleotide (and several adjacent nucleotides). The region of the DNA replaced during base excision repair is usually 1 to 4 nucleotides in length, and thus constitutes a "short-patch" repair. A second type of repair may remove up to 10 nucleotides during base excision repair.

Ligase. Of the four DNA ligases identified in eucaryotic cells, DNA ligase I, and perhaps DNA ligase III, participate in DNA repair.

14.7 NUCLEOTIDE EXCISION REPAIR

Nucleotide excision repair is usually initiated by the production of large bulky adducts to the DNA, and is the process for the repair of adducts produced by the carcinogenic metabolites of N-acetylaminofluorene, aflatoxin and benzo[a]pyrene, ultraviolet radiation products such as thymine dimers and 6-4 photoproducts, and the anticancer therapeutic agent cisplatin. In addition, nucleotide excision repair may also repair small adducts that are predominantly repaired by other mechanisms, for example O^6-methylguanine and O^6-methyl transferase, and small adducts normally repaired by base excision repair. "There is no known covalent base modification that is not a substrate for the nucleotide excision repair system" (Sancar, 1995).

Nuclease Activity. The excision nuclease (also called excinuclease) involves as many as 16 polypeptide subunits and makes two incisions in the damaged DNA to remove the damaged site. Seven of the polypeptides that function in the excision nuclease are those that reflect the seven complementation groups of the human disease xeroderma pigmentosum. The disease is characterized by various defects in DNA repair and leads to an increased propensity for skin cancer.

In humans, the fragment of DNA containing the damaged site that is removed by the excision nuclease contains 24–29 nucleotides. One aggregate of polypeptides makes one incision at approximately the fifth phosphodiester bond on the 3′ region of the damaged site, and a different aggregate of polypeptides makes the second incision at approximately the 24th phosphodiester bond on the 5′ region of the damaged site, to produce a 29-nucleotide fragment, or 29-mer.

Polymerase. The nucleotide fragment of approximately 29 nucleotides is replaced by DNA polymerase activity, and the newly synthesized fragment of DNA exactly matches the length of the fragment removed by the excision nucleases. The polymerase activities involved in nucleotide excision repair require PCNA (proliferating cell nuclear antigen), a protein that is required only by polymerase delta. The requirement for PCNA and experiments on antibody inhibition studies indicate that polymerase delta carries out the nucleotide polymerization.

Ligase. The formation of the last phosphodiester bond is catalyzed by one of the four mammalian DNA ligases, probably DNA ligase I and perhaps DNA ligase III.

14.8 MISMATCH REPAIR

Noncomplementary base pairs in double-stranded DNA can be produced by replication errors by DNA polymerase, during recombination, or by base deaminations. Several mismatch repair systems exist in mammalian cells. A short patch repair results in the replacement of up to 10 nucleotides, and the short patch repairs are initiated by glycosylases that recognize the mismatch bases. Alternatively, there exists a long patch mismatch repair that is directed at the repair of replicative errors in the newly synthesized, nascent DNA strand.

The DNA polymerases alpha, delta, and epsilon may participate in the replicative step to replace the nucleotides removed during the repair.

14.9 REPAIR OF O^6-GUANINE AND O^4-THYMINE ALKYLATIONS AND PHOSPHOTRIESTERS

Carcinogens such as N-methyl-N′-nitro-N-nitrosoguanidine (MNNG), N-methyl-N-nitrosourea and methylmethane sulfonate, and the ethyl derivatives of the latter two react with DNA to produce N-alkylated (usually the predominant products) and O-alkylated purines and pyrimidines. Two of the more common O-alkylated products, O^6-alkylated guanine and O^4-methylthymine are important because they are promutagenic lesions and cause mispairing during DNA replication.

14.9.1 O^6 Methylguanine Methyltransferase

The repair of O^6-methylguanine and O^4-methylthymine, promutagenic sites, is catalyzed by O^6-methylguanine methyltransferase. The methyl group from the methylated base in DNA is transferred to a cysteine-sulfhydryl group of the enzyme. Transfer of the methyl group from the methylated guanine to the single receptor sulfhydryl group on the protein inactivates the enzyme. The limited quantity of methyltransferase in mammalian cells and the inactivation of the enzyme by the transfer of a single methyl group explain the saturation in the repair of O^6-methylguanine in damaged DNA, because new protein synthesis is necessary to replace the inactivated methyltransferase.

The transferase protein is referred to as an enzyme although it can catalyze the transfer of only one methyl group. Others emphasize that the protein acts only stoichiometrically (i.e., transfers only one methyl group) rather than catalytically (i.e., necessary for enzymatic activity, by definition), and they have preferred to refer to the protein as a substrate rather than an enzyme.

14.10 RNA METABOLISM

14.10.1 Types of RNA

Mammalian ribosomal RNA, messenger RNA, and transfer RNA are all produced as primary transcripts that then undergo extensive processing in the nucleus and transport to the cytoplasm. Ribosomal RNA is synthesized as a higher molecular weight polymer (45S precursor rRNA) and undergoes processing and methylation to yield the cytoplasmic 18S and 28S ribosomal RNA molecules. Messenger RNA undergoes a number of modifications from the primary transcript during maturation to the functional mRNA. Transfer RNA also undergoes a processing that includes the introduction of trace amounts of unusual nucleosides in the functional tRNA.

Although a toxicant may inhibit the synthesis of all the RNA species, the inhibition of mRNA synthesis exhibits the earliest manifestations of toxicity. The short half-life of most mRNAs and its role in regulating the synthesis of various proteins (including enzyme induction) results in acute changes in the levels of enzymes and other proteins which have short half-lives (or fast turnover) whenever there is an interruption in mRNA synthesis.

14.10.2 Methodology to Study RNA Metabolism

There are a number of techniques used to measure the synthesis of mRNAs for specific proteins after exposure to toxicants: (1) Northern slot blots, the most widely used technique, that permit the measurement of the mRNA for a specific protein;

(2) quantitative reverse transcriptase-polymerase chain reaction (QRT-PCR); (3) in situ hybridization which can localize the specific mRNA within a specific cell type and avoids dilution of mRNAs in the multiple cell types in a tissue; and (4) others such as the newer microarray-based analysis. (5) The technique to measure *total* messenger RNA synthesis is isolation of total cellular or polysomal RNA and purification of the poly A-containing mRNA by chromatography on oligo(dT)-cellulose. (For more details on these techniques, see Chapter 2.) The major disadvantage is that under some circumstances an indefinite portion of the total mRNA may not contain the 3′-poly A fragment on messenger RNA and that a portion of the mRNA may be discarded by the preparative method, although others claim that all the mRNA for a specific protein contains the poly A fragment and can be isolated.

14.10.3 RNA Polymerases: Multiple Forms in Mammalian Cells

Three classes of RNA polymerase occur in mammalian cells, unlike the single form of the RNA polymerase that occurs in bacteria like *Escherchia coli*. Each of the RNA polymerases has distinct functions in relation to the product RNA that is produced. Mammalian and other eukaryotic RNA polymerases are very complex with usually 9–11 polypeptide subunits. Some of the properties of RNA polymerases from other eucaryotic species (e.g., yeast) are very different from the properties of mammalian RNA polymerases.

The properties of the mammalian RNA polymerases are presented in the Table 14.2. The class I RNA polymerases are present in the nucleolus and synthesize the 45S precursor rRNA (pre-rRNA) that undergoes maturation to yield eventually the two large (18S and 28S) fragments of polysomal rRNA. The class II RNA polymerases synthesize the heterogeneous (in size) nuclear RNA (hnRNA). A small amount (less than 10%) of the hnRNA eventually forms mRNA, and the remainder of the hnRNA is degraded. The class III RNA polymerase synthesize two distinct products, the 4.5S precursor tRNA (pre-tRNA), that is processed to yield tRNA, and the 5S rRNA, the small rRNA molecule.

TABLE 14.2 Properties of Mammalian RNA Polymerases.

Class	I	II	III
Nuclear site	Nucleolus	Nucleoplasm	Nucleoplasm
Product	Initial product is 45S pre-rRNA that yields 18S and 28S cytoplasmic rRNA	Initial product is hnRNA, and less than 10% yields cytoplasmic mRNA	Initial product is 4.5S pre-tRNA that yields tRNA. A second product is 5S rRNA
Inhibition by alpha-amanitin	Resistant (>400 micrograms/ml)	Most sensitive (50% inhibition at 0.03–0.1 micrograms/ml)	Moderately sensitive (50% inhibition at 10–25 micrograms/ml)

Abbreviations: pre-rRNA, precursor ribosomal RNA; rRNA, ribosomal RNA; hnRNA, heterogeneous (in size) nuclear RNA; mRNA, messenger RNA; pre-tRNA, precursor transfer RNA; tRNA, transfer RNA

Alpha-amanitin is the toxin from the poisonous Amanita mushroom.

In addition to the three classes of RNA polymerases, each class also occurs in two functional states: (a) In the chromatin-bound state the enzyme is engaged in transcription (initiation, chain elongation, or termination) at the time the nuclei are isolated. In this state the enzyme is relatively tightly bound to the chromatin. (b) In the free state the polymerase is easily extractable from isolated nuclei with solutions of low ionic strength and is not engaged in transcription.

14.10.4 Effects of Toxicants and Other Substances on RNA Metabolism

Actinomycin D is a specific inhibitor of RNA synthesis. The compound binds to the deoxyguanosine moieties of DNA and blocks the movement of any of the three RNA polymerases along the template DNA. Because actinomycin D binds to the template DNA and not to a protein subunit of the polymerase molecule, actinomycin D can inhibit the synthesis of all three RNA classes. The specificity of actinomycin D as an inhibitor of RNA synthesis has resulted in its extensive use in biochemical studies to demonstrate the need for RNA synthesis (usually presumably mRNA synthesis) for further biochemical changes such as the induced synthesis of an enzyme (e.g., heme oxygenase) or other protein (e.g., metallothionein). It is observed in mammalian cells, however, that the synthesis of ribosomal RNA is inhibited preferentially at the lowest dosages of actinomycin D, and the synthesis of mRNA is affected at somewhat higher dosages of actinomycin D.

The various mammalian RNA polymerases exhibit marked differences in sensitivities to inhibition by alpha-amanitin (Table 14.2), one of the toxins in the poisonous mushroom *Amanita phalloids*. The class II polymerases are most susceptible to inhibition, and 50% inhibition of the mammalian class II polymerases occurs at concentrations of alpha-amanitin of 0.03–0.1 μg/ml and nearly complete inhibition occurs at approximately 1 μg/ml. The class III mammalian polymerases are moderately sensitive to alpha-amanitin, and they exhibit 50% inhibition at concentrations of 10–25 μg/ml and nearly complete inhibition at approximately 150–200 μg/ml. The class I polymerases are very resistant to inhibition by alpha-amanitin, and negligible inhibition occurs at concentrations of alpha-amanitin of less than 250–400 μg/ml.

The different sensitivities to alpha-amanitin can be used experimentally to assay the activities of the three classes of polymerases in a crude mixture, for example, an extract of nuclei without the separation or purification of the individual polymerases. A typical assay of the bound polymerases measures incorporation of radioactive GTP into perchloric acid- or trichloroacetic acid-precipitable material. In the absence of alpha-amanitin the sum of the activities of all three RNA polymerases is measured. In the presence of 1 μg/ml of alpha-amanitin that causes complete inhibition of the class II RNA polymerases, the sum of class I and III is measured. In the presence of 150 μg/ml of alpha-amanitin, that inhibits completely both the class II and III polymerases, only the activity of class I polymerase is measured.

A large number of toxicants inhibit the synthesis of mRNA, and in many cases this inhibition is secondary to disruption of energy production (e.g., inhibition of oxidative phosphorylation or alterations in mitochondrial structure).

The induced synthesis of metallothionein and various cytochrome P450s are preceded by the induced synthesis of the appropriate mRNA. There are also examples of the induced synthesis of rRNA. Thioacetamide administration causes the enlargement of the nucleolus and enhances the synthesis of rRNA. When the activities of the various polymerases were assayed in hepatic nuclei after the administration of thioacetamide to rats, the chromatin-bound polymerase I was increased in activity to 300% of control (i.e., increased by 200%) whereas other polymerase activities were increased only moderately. Thus, the increased synthesis of rRNA in thioacetamide-treated animals was attributable predominantly to the increase in the activity of bound polymerase I, the enzyme that is known to be responsible for the synthesis of 45S precursor ribosomal RNA.

14.11 PROTEIN METABOLISM

14.11.1 Inhibitors of Protein Synthesis and Destruction of Polysomes

A toxicant-induced decrease in the incorporation of radioactive amino acids into protein may be secondary to effects of the toxicant on energy production or on the synthesis of mRNA.

Several specific inhibitors of eucaryotic protein synthesis exist. The most noted inhibitor is cycloheximide that blocks the peptidyl transferase on the large eucaryotic ribosomal subunit. This block prevents the translocation of peptidyl-tRNA and inhibits protein synthesis. The specific nature of the cycloheximide action has resulted in extensive use of this chemical in studies to demonstrate the need for protein synthesis during enzyme induction studies. In various experimental systems, cycloheximide inhibits the induction of cytochromes P450, metallothionein, and heme oxygenase. The conclusion in each case is that new protein synthesis is required to bring about the increased activity (or content) of the appropriate protein.

Puromycin is also a specific inhibitor of protein synthesis and acts as an aminoacyl-tRNA analog and peptidyl acceptor. The latter causes the premature chain termination of the protein and the release of nascent or growing polypeptide chains. Puromycin is rarely used in studies of enzyme induction because of its other toxic effects.

Diphtheria toxin and ricin are naturally occurring toxins that are inhibitors of eucaryotic protein synthesis. The diphtheria toxin catalyzes the covalent attachment of an adenosine diphosphate-ribosyl moiety, derived from NAD+, to eEF-2, the eucaryotic elongatiion factor 2 that normally catalyzes translocation. The ADP-ribosylation of a single side chain of eEF-2 blocks its ability to catalyze translocation of the growing polypeptide chain, and the altered eEF-2 is unable to function in protein synthesis. Ricin is an extremely toxic protein produced by the plant *Ricinus communis* (castor bean). In the cytosol, one peptide from the ricin protein acts as an N-glycosylase that removes one adenine from the 28S ribosomal RNA but leaves the rRNA phosphodiester chain intact. Removal of the adenine inactivates the 60S ribosomal subunit, and it can no longer function in protein synthesis. In one estimate, a single molecule of the enzymatic polypeptide of ricin in the cytosol can inactivate 50,000 ribosomes.

Various inhibitors (including carbon tetrachloride and toluene) of protein synthesis bring about the disaggregation of polysomes into monoribosomes and, in some cases, into ribosomal subunits. The increased distribution of ribosomes as monoribosomes or ribosomal subunits, that are not active in polypeptide synthesis, is accompanied by the corresponding decrease in polysomes that are associated with mRNA and participate in polypeptide synthesis. The distribution of ribosomes into the various sized aggregates can be measured by the centrifugation of the postmitochondrial supernatant from a tissue homogenate on sucrose gradient solutions.

14.11.2 Enzyme Induction and Induced Synthesis of Proteins by Toxicants

The term *enzyme induction* refers to the specific synthesis of new enzymatic protein. An increase in enzyme activity (or the content of a nonenzymatic protein) following administration of a toxicant does not always imply an enzyme induction or the synthesis of new protein. As increase in enzyme activity (or the increase in a nonenzymatic protein) may arise by three mechanisms: (1) the synthesis of new protein, "true" enzyme induction; that may result from the synthesis of new mRNA or the stabilization of preexisting mRNA; (2) the conversion of an inactive protein to an active enzyme by either (a) removal of a shielding group, for example, conversion of pepsinogen to pepsin; or (b) complex formation with a cofactor, for example, the possible association of heme with apocytochrome P450 to yield the active holocytochrome P450; or (3) the decreased degradation of the enzyme or changes in activity with enzyme phosphorylation. In the latter case, the activity of an enzyme is a summation of the turnover process, the rate of the new synthesis of the enzyme and the rate of degradation of that protein.

The most extensive studies have examined the induction of the various cytochrome P450 isozymes by toxicants and other active agents. The induced synthesis of the cytochromes P450 is discussed elsewhere in this text (Chapter 8).

14.11.3 Induced Synthesis of Metallothionein by Metallic Cations and Other Agents

Metallothionein is a low-molecular-weight (molecular weight of 6000 to 7000), metal-binding protein. Metallothionein consists of a multigene family containing four classes (MT-I through-IV) in both mouse and human, although there are multiple genes that occur in one of the human classes. The ability of metallothionein to bind metallic cations is due to its high content of cysteine (approximately 33 mole per cent, 20 of approximately 61 amino acids). The cysteine sulfhydryl groups are capable of binding Zn^{2+}, Cd^{2+}, Hg^{2+}, methyl-Hg^{+}, and a number of other heavy metallic cations that bind to sulfhydryl groups. Each molecule of metallothionein is capable of binding up to seven Cd^{2+} or Zn^{2+} ions.

Metallothionein has several probable functions. (a) The function in normal cells is probably the formation of a complex with Zn^{2+} and as an intracellular storage form of zinc, and perhaps other ions such as Cu^{2+}. (b) Metallothionein also appears to act as a free-radical scavenger. (c) A clearly established role of metallothionein is its protective function against exposure to acute toxic doses of Cd^{2+} and other heavy metallic cations in rodents or in cultured mammalian cells.

The induced synthesis of metallothionein (or, more specifically, treatments that are known to induce metallothionein synthesis) protects against various measures of toxicity, especially hepatic toxicity: (a) lethality in rodents of acute toxic doses of Cd^{2+}, (b) hepatotoxicity of Cd^{2+} as measured by the release of hepatic enzymes into the serum, and (c) the Cd^{2+}-induced destruction of cytochrome P450. In representative experiments (of (a), above), the prior treatment (48 hours) with doses of Cd^{2+} (1 or 3 mg Cd^{2+}/kg) in mice induced the synthesis of metallothionein and markedly increased the LD50 dose (i.e., decreased the toxicity) of Cd^{2+} from 4.5 mg Cd^{2+} in the nonpretreated mice to 8.2 mg Cd/kg in the Cd^{2+} pretreated mice. Likewise, several pretreatments with Zn^{2+} increased the LD50 from 5.7 mg/kg to 10.3 in Zn^{2+} pretreated mice. In the second representative experiment ((b), above), injection with Cd^{2+} (2 mg/kg) 24 hours prior to toxic, challenge doses of 3, 4, or 5 mg Cd^{2+}/kg protected rats against hepatic toxicity as measured by the release of sorbitol dehydrogenase into the serum (an established marker of hepatic toxicity), and a similar protection was noted when measured by the release of aspartate aminotransferase. In a representative experiment of (c), above, the prior treatment of rats with lower doses of Cd^{2+} prevent the destruction of hepatic microsomal cytochrome P450 by a subsequent, higher dose of Cd^{2+} (Table 14.3).

In rodents or cultured mammalian cells, the induced metallothionein synthesis also protects against the toxicity of Hg^{2+}, carbon tetrachloride, the anticancer agent cisplatin, lipopolysaccharide, and oxidative stress inducers such as tert-butylhydroperoxide, hydrogen peroxide, paraquat, and adriamycin.

Metallothionein normally exists at low levels in the cytosol (and complexed with metal in lysosomes) in hepatic and other cells. The pattern of results obtained in studies with inhibitors of RNA and protein syntheses demonstrate the initial induced synthesis of the metallothionein mRNA and the subsequent induced synthesis of metallothionein. In a representative study, the synthesis of Cd-thionein (metallothionein bound to Cd^{2+}) was measured by the incorporation of radioactive cysteine and radioactive Cd^{2+} into a soluble low molecular weight hepatic protein isolated by gel chromatography. The injection into rats of a trace amount of radioactive Cd^{2+} (at a molar level that does not induce Cd-thionein synthesis) resulted at 42 hours in negligible cysteine incorporation into the Cd-thionein (implying no significant new synthesis of metallothionein) and a low level of labeling of Cd-thionein with radioactive Cd^{2+} (perhaps by displacement of endogenous Zn^{2+} from the protein). In contrast, if a mixture of radioactive Cd^{2+} and nonradioactive Cd^{2+} (at a molar dosage sufficient to induce the synthesis of metallothionein) was injected, there was incorporation of cysteine into the chromatographyically separated Cd-thionein and a 16,000-fold increase in the total Cd^{2+} present in the fraction. The administration of cycloheximide (an inhibitor of protein synthesis) or actinomycin D (an inhibitor of RNA synthesis) prior to the Cd-thionein-inducing dose of Cd^{2+} markedly decreased cysteine incorporation and appreciably decreased Cd^{2+} uptake into Cd-thionein. In contrast, when actinomycin D was administered 3 hr after the Cd-thionein-inducing dose of Cd^{2+} (presumably after an interval that permitted the synthesis of the appropriate mRNA during the induction), the incorporation of cysteine into Cd-thionein was not appreciably altered. Thus, the results imply that Cd^{2+} induces the formation of the mRNA of metallothionein and the subsequent use of the mRNA in the synthesis of the protein.

TABLE 14.3 Induced Synthesis of Metallothionein and Protection Against Cadmium Acute Hepatic Toxiciy.

Group	Day 1 Treatment (mg/kg)	Day 4 Treatment (mg/kg)	Day 7 Hepatic Microsomal Hexobarbital Metabolism	Interpretation
1	Sodium acetate 1.23	Sodium acetate 1.23	1.00	Control
2	Sodium acetate 1.23	Cadmium acetate 0.84	0.27	On day 4, no MT available to bind the toxic dose of Cd^{2+}, and the Cd^{2+} caused destruction of cytochrome P450 as measured by hexobarbital metabolism in vitro
3	Cadmium acetate 0.42	Sodium acetate 1.23	0.87	Although the day 1 Cd^{2+} caused destruction of cytochrome P450, resynthesis of cytochrome P450 occurred by day 7
4	Cadmium acetate 0.42	Cadmium acetate 0.84	0.93	Between day 1 and day 4, there was accumulation of MT (or its mRNA) in sufficient amounts to bind the toxic day 4 dose of Cd^{2+} and prevent destruction of cytochrome P450
5	Cadmium acetate 0.21	Cadmium acetate 0.84	0.96	Treatment on day 1 with one-fourth of the toxic dose of 0.84 was sufficient to induce MT (or its mRNA) to bind the day 4 toxic dose of Cd^{2+} and prevent cytochrome P450 destruction

On days 1 or 4, rats were treated with either cadmium acetate or sodium acetate (control). On day 7, hepatic microsomes were isolated, and hexobarbital metabolism (micromole hexobarbital/mg of protein/hr) was measured in vitro.

Source: Adapted from Roberts, S. A., Miya, T. S., and Schell R. C. *Res. Commun. Chem. Pathol. Pharmacol.* 14 (1976), 197–200.

The response of cells to the need for metallothionein upon exposure to toxic metallic cations (and other inducers) is moderately rapid (Fig. 14.1). Treatment of rats or mice with Cd^{2+} results in a 3- to 5-fold increase in the concentration of mRNA for metallothionein (specifically metallothionein-1, MT-1) by 3 hours and a peak of 12-fold in mice and a 7-fold increase in rats at 6 hours after treatment with Cd^{2+}. The concentration of MT-1 mRNA had decreased to approximately 2-fold above control by 12 hours in mice and by 18 hours in rats. Lipopolysaccharide, which also induces the synthesis of metallothionein, also caused a peak concentration at 6 hours after treatment with a similar-fold increase in the MT-1 mRNA to that induced by Cd^{2+}.

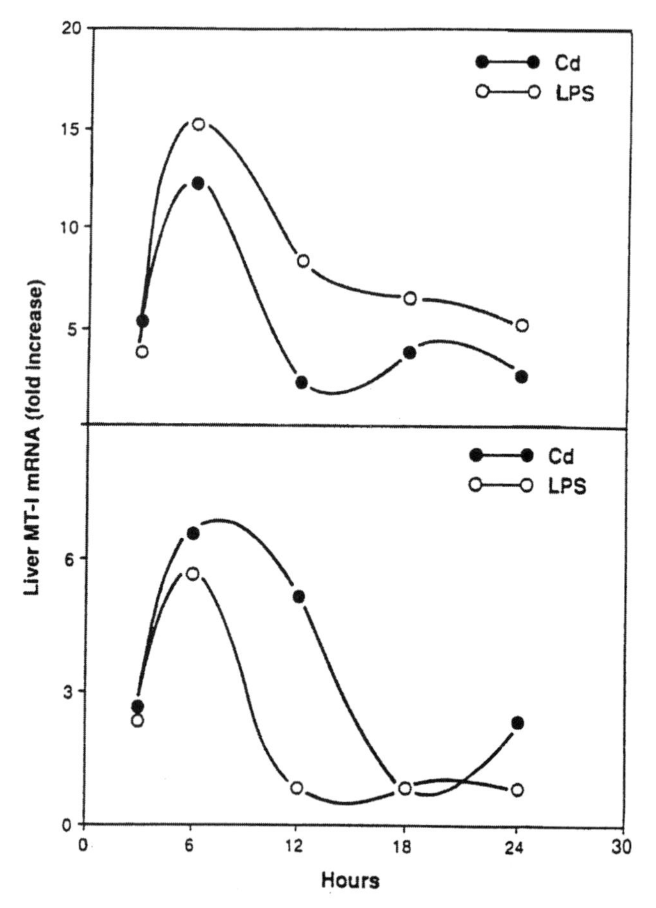

FIGURE 14.1 Time-course of metallothionein-I mRNA induction of livers of mice (*top*) and rats (*bottom*) by slot-blot analysis. At the 6-hour time point, Cd^{2+} and LPS (lipopolysaccharide) increased MT-I mRNA about 12- and 15-fold respectively over control mice. Cd^{2+} and LPS increased MT-I mRNA about seven- and six-fold, respectively over control rats. Choudhuri et al., *Toxicol. App. Pharmacol.* 119 (1993), 1–10.

14.11.4 Collagen Synthesis as a Measure of Pulmonary Toxicity

There are at least 19 proteins that are classified as collagens. Collagen I and III are the predominant collagens in lung and account for about 95% of the collagen in the lung. Collagens are synthesized by fibroblasts, by stellate cells (also called Ito cells) in the liver, and by many other cells types in cultured cells. Collagens are synthesized as larger precursor molecules known as procollagens. During the synthesis, there is an intracellular, post-translational hydroxylation of proline and lysine residues, and the former yield 4-hydroxyproline. The amino acid hydroxyproline is absent in most other tissue proteins. Collagen is secreted eventually into the extracellular space where it is polymerized. Deposition and accumulation of excessive collagen and fibrosis impairs the normal function of tissues such as lung, liver and kidney.

The release of cytokines has an important role in the induced synthesis of collagen. In the liver, transforming growth factor-beta (TGF-beta) from hepatocytes stimulates stellate cells and, in the kidney, TGF-beta in renal epithelia stimulates interstitial mesenchymal cells to produce collagen.

One parameter of exposure to various pulmonary toxicants is fibrosis characterized by the increased deposition of collagen in the alveolar interstitium, the alveolar ducts, and the respiratory bronchioles of the lung. A variety of agents, including ozone, the antibiotic/anticancer agent bleomycin, and the herbicide paraquat, are known to induce the synthesis and accumulation of collagen in the lung. Pulmonary toxicity can be evaluated either by measurements of collagen deposition (or accumulation) or by estimation of the synthesis of new collagen. The accumulation can be evaluated either by the use of stains that are moderately specific for collagen or biochemically by the measurement of hydroxyproline in the lung. The basis for the use of hydroxyproline as a measure of lung collagen is that collagen contains more than 90% of the total lung hydroxyproline. The total protein in the lung is hydrolyzed in a strong acid solution (6N HCl), and the released hydroxyproline is measured. A representative study is shown in Figure 14.2. The intratracheal

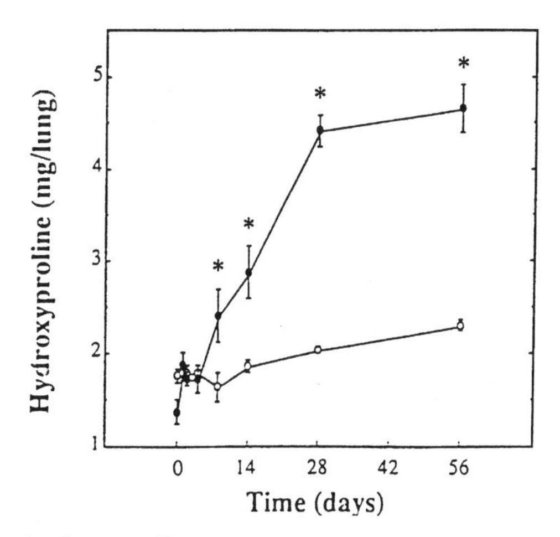

FIGURE 14.2 Lung hydroxyproline content following a single intratracheal instillation of 1.3 mg indium chloride/kg (*closed circles*) or saline (*open circles*). *Significantly greater ($P < 0.05$) than the saline treated group at this time point. Blazka et al., *Fund. Appl. Toxicol.* 22 (1994), 231–239.

instillation of indium trichloride caused an increase in collagen in rat lung as early as 8 days after a single treatment, and the lung content of collagen (measured as total hydroxyproline) was more than two times the control level at 28 days after the single treatment.

The relative rate of collagen synthesis can be measured by the incorporation of radioactive proline into lung protein-hydroxyproline. For example, after a lung section is incubated with radioactive proline, the protein is precipitated and hydrolyzed to its amino acids, the amino acids separated by chromatography, and the radioactivity in the hydroxyproline is measured. The incorporation of radioactive proline into collagen-hydroxyproline has been used to demonstrate increased collagen synthesis in the lungs of rats treated with paraquat, a pronounced pulmonary toxicant.

SUGGESTED READING

DNA Repair

Sancar, A. and Thompson, C. Multiple topics on DNA repair as listed. In: *The Encyclopedia of Molecular Biology*. New York: John Wiley and Sons, 1999.

Sancar, A. DNA repair in humans. *Annu. Rev. Genet.* 29 (1995), 69–105.

Sancar, A. Excision repair in mammalian cells. *J. Biol. Chem.* 270 (1995), 15915–15918.

Friedberg, E.C., Walker, G.C. and Siede, W. *DNA Repair and Mutagenesis*. Washington, DC: ASM Press, 1995.

Nickoloff, J.A. and Hoekstra, M.F. (Eds) *DNA Damage and Repair*, Vol. 2, *DNA Repair in Higher Eukaryotes*. Totowa, NJ: Humana Press, 1995.

Metallothionein

Klaassen, C.D., Liu, J. and Choudhuri, S. Metallothionein: an intracellular protein to protect against cadmium toxicity. *Annu. Rev. Pharmacol. Toxicol.* 39 (1999), 267–294.

Choudhuri, S., McKim, J.M., Jr. and Klaassen, C.D. Differential expression of the metallothionein gene in liver and brain of mice and rats. *Toxicol. Appl. Pharmacol.* 119 (1993), 1–10.

Garrett, S.H., Somji, S., Todd, J.H., Sens, M.A. and Sens, D.A. Differential expression of human metallothionein isoform I mRNA in human proximal tubule cells exposed to metals. *Environ. Health Perspect.* 106 (1998), 825–831.

Collagen

Last, J.A. and Reiser, K.M. Collagen biosynthesis. *Environ. Health Persp.* 55 (1984), 169–177.

CHAPTER FIFTEEN

Carcinogenesis

ROBERT C. SMART and JACQUELINE K. AKUNDA

15.1 INTRODUCTION AND HISTORICAL PERSPECTIVE

Two of the earliest observations that exposure of humans to certain chemicals or substances is related to an increased incidence of cancer were made independently by two English physicians, John Hill in 1571 and Sir Percival Pott in 1576. Hill observed an increased incidence of nasal cancer among snuff users and Pott observed that chimney sweeps had an increased incidence of scrotal cancer. Pott attributed this to topical exposure to soot and coal tar. It was not until a century and a half later in 1915 when two Japanese scientists, Yamagiwa and Itchikawa, substantiated Pott's observation by demonstrating that multiple topical applications of coal tar to rabbit skin produced skin carcinomas. This experiment is important for two major reasons: (1) it was the first demonstration that a chemical or substance could produce cancer in animals; and (2) it confirmed Pott's initial observation and established a relationship between human epidemiology studies and animal carcinogenicity. Because of these important findings, Yamagiwa and Itchikawa are considered the fathers of experimental chemical carcinogenesis. In the 1930s, Kennaway and co-workers isolated a single active carcinogenic chemical from coal tar and identified it as benzo[a]pyrene, a polycyclic aromatic hydrocarbon that results from the incomplete combustion of organic molecules. Benzo[a]pyrene has also been identified as one of the carcinogens in cigarette smoke.

The concept that cancer involves an alteration in the genetic material of the somatic cell (somatic mutation theory) was first introduced by Theodor Boveri in 1914. Boveri suggested that cancer was related to chromosome abnormalities. In support of this theory, Furth and Kahn in 1934 isolated single cell clones from a tumor and found that injection of these cells into a healthy host could reproduce the original disease, demonstrating that cancer involves a stable cellular alteration. In terms of chemical carcinogenesis, James and Elizabeth Miller in the 1950s observed that a wide variety of structurally diverse chemicals could produce cancer in animals. They suggested that each of these diverse chemicals require metabolic activation to electrophilic reactive intermediates that covalently bind to nucleophilic centers on proteins, RNA or DNA. It is now appreciated that specific genes within the DNA represent the critical sites for chemical carcinogen-induced alterations. The Millers termed this the electrophilic theory of chemical carcinogenesis. Many carcinogens, including benzo[a]pyrene have been demonstrated to be metabolized by cytochromes P-450 to produce reactive electrophilic intermediates capable of covalently binding to DNA and producing mutations in prokaryotic and eukaryotic cells.

Specific genes, found in normal cells, termed proto-oncogenes, have been identified as target sites for chemical carcinogens. Proto-oncogenes are involved in the positive regulation of cell growth and differentiation and certain chemical carcinogens can produce mutations in these proto-oncogenes, thereby activating them to dominant transforming oncogenes. For example, members of the *ras* proto-oncogene family are frequently mutated in carcinogen-induced rodent tumors as well as in human tumors. The function of the protein products encoded by proto-oncogenes and oncogenes and the interactions of chemical carcinogens with proto-oncogenes are active areas of current investigation. Another family of genes, known as tumor suppressor genes can be mutationally inactivated during carcinogenesis. Tumor suppressor genes and the proteins they encode often function as negative regulators of cell growth. The *p53* tumor suppressor gene is a target gene for and can be inactivated by numerous carcinogens, including a carcinogenic metabolite of benzo[a]pyrene.

Although the mechanism of action of many carcinogens is consistent with their mutagenicity, electrophilicity, and the somatic mutation theory, other carcinogens do not have these characteristics. It is generally accepted that many of these agents, such as tumor promoters, function through an epigenetic mechanism, one in which the primary sequence of DNA is not altered, but rather the expression or repression of certain genes or cellular functions such as cell proliferation, differentiation, or apoptosis may be altered. Experiments conducted by Berenblum in 1941 suggested that carcinogenesis is a multistage process and these experiments described a two stage mechanism in the development of a tumor and are now referred to as the initiation-promotion model. This model has been expanded to numerous organs and experimental systems; epidemiological evidence also suggest that human carcinogenesis can involve a similar process.

It is now appreciated that cancer does not arise from a single mutation but from the accumulation of sequential mutations in a single cell and that the genetic background of an individual can play an important role in terms of cancer susceptibility. In addition, certain genetic alterations may lead to genetic instability and result in the accumulation of mutations, this condition is often referred to as a mutator phenotype. Lifestyle factors such as diet, smoking, sexual and cultural behavior, and occupation all influence the carcinogenic process in humans and some of these factors are thought to function at the promotion level. The interaction between environmental factors and an individual genetic background with regard to an individuals genetic susceptibility to cancer is an important emerging area in toxicology and epidemiology.

This abbreviated description of some important events in carcinogenesis is intended to provide a historical perspective and insight into the evolution of our current thinking about the subject. It is not intended to suggest that these are the only important events in the history of chemical carcinogenesis.

15.2 HUMAN CANCER

An important aspect of toxicology is the identification of potential human carcinogens and the protection of human health. To begin to appreciate the complexity of this subject it is important to have some understanding of human cancer and its etiologies. Major insights into the etiologies of cancer can be attained through epidemiological studies that relate the role of hereditary, environmental, and cultural influences on cancer incidence.

Cancer is not a single disease but a large group of diseases, all of which can be characterized by the uncontrolled growth of an abnormal cell to produce a population of cells that have acquired the ability to multiply and invade surrounding and distant tissues. It is this invasive characteristic that imparts its lethality on the host. Epidemiology studies have revealed that the incidence of cancer increases expotentially with age. From these studies it has been suggested that three to seven mutations or "hits" are necessary for cancer development. Most cancers are monoclonal (derived from a single cell) in origin and do not arise from a single mutation but from the accumulation of sequential mutations in a single cell. Each additional mutation provides a further growth advantage for the developing tumor. These mutations can be the result of imperfect DNA replication, oxidative DNA damage, or DNA damage caused by environmental carcinogens. There is a selection for cells with mutations in oncogenes and tumor suppressor genes as these genes regulate cell proliferation, differentiation, and apoptosis and when mutated provide a growth advantage to the cell. It often requires decades for such mutations to accumulate and for mutated cells to produce a clinically detectable tumor. Molecular analyses of human tumors have documented the accumulation of mutations in critical genes in the evolution of cancer. These studies have confirmed on a molecular basis the findings of the epidemiology studies which, as indicated above, have long suggested that three to seven mutations are necessary for cancer development.

Cancer is a type of a neoplasm or tumor. Although technically a tumor is defined as only a tissue swelling, it is now used as a synonym for a neoplasm. A neoplasm or tumor is an abnormal mass of tissue, the growth of which exceeds and is uncoordinated with the normal tissue, and persists after cessation of the stimuli which evoked it. There are two basic types of neoplasms, termed benign and malignant. The general characteristics of these tumors are defined in Table 15.1. Cancer is the common name for a malignant neoplasm. Neoplasms are composed of two

TABLE 15.1 Some General Characteristics of Malignant and Benign Neoplasms.

Benign	Malignant
Generally slow growing	May be slow to rapid growing
Few mitotic figures	Numerous mitotic figures
Well differentiated and architecture resembles that of parent tissue	Some lack differentiation; disorganized; loss of parent tissue architecture
Sharply demarcated mass that does not invade surrounding tissue	Locally invasive, infiltrating into surrounding normal tissue
No metastases	Metastases

components, the parenchyma, which contains proliferating neoplastic cells, and the stroma, which provides support and is composed of connective tissue and blood vessels. In terms of cancer nomenclature, most adult cancers are carcinomas that are derived from epithelial cells. Sarcomas are derived from mesenchymal tissues, whereas leukemias and lymphomas are derived from blood forming cells and lymphoid tissue. Melanoma is derived from melanocytes and retinoblastoma, glioblastoma and neuroblastoma are derived from the stem cells of the retina, glia, and neurons, respectively. According to the American Cancer Society the lifetime risk for developing cancer in the United States is 1 in 3 for women and 1 in 2 for men; in 1999 about 1.2 million new cancer cases are expected to be diagnosed not including carcinoma in situ or basal or squamous cell skin cancer; and cancer is a leading cause of death in the United States and approximately 22% of all deaths are due to cancer.

15.2.1 Causes, Incidence, and Mortality Rates of Human Cancer

Cancer suceptibility is determined by complex interations between age, environment, and individual's genetic make up. It is estimated from epidemiological studies that 35–80% of all cancers are associated with the environment in which we live and work. The geographic migration of immigrant populations and differences in cancer incidence between communities has provided a great deal of information regarding the role of the environment and specific cancer incidences. It should be noted that the term *environment* is not restricted to exposure to man-made chemicals in the environment but applies to all aspects of our lifestyle including smoking, diet, cultural and sexual behavior, occupation, natural and medical radiation, and exposure to substances in air, water, and soil. The major factors associated with cancer and their estimated contribution to human cancer incidence are listed in Table 15.2. Only a small percentage of total cancer occurs in individuals with a hereditary mutation/hereditary syndrome. However, an individual's genetic background is the "stage" in which the cancer develops and susceptibility genes have

TABLE 15.2 Proportions of Cancer Deaths Attributed to Various Different Factors.

Major Factors	Best Estimate	Range of Acceptable Estimates
Diet	35%	10–70
Tobacco	30%	25–40
Infection	10%	1–?
Reproductive and sexual behavior	7%	1–13
Occupation	4%	2–8
Geophysical factors	3%	2–4
Alcohol	3%	2–4
Pollution	2%	<1–5
Food additives	1%	–5[a]–2
Medicines	1%	0.5–3
Industrial products	1%	<1–2
Unknown	?	?

[a] Allowing for possible protective effects of antioxidants and other preservatives.

Source: Adapted from R. Doll and R. Peto The Causes of Cancer 1981 Oxford Medical Publications.

been identified in humans. For example genetic polymorphisms in enzymes responsible for the activation of chemical carcinogens may represent a risk factor as is the case for polymorphisms in the N-acetyl-transferase gene and the risk of bladder cancer. These types of genetic risk factors are of low penetrance (low to moderate increased risk), however, increased risk is usually associated with environmental exposure (see Chapter 17). While the values presented in Table 15.2 are a best estimate, it is clear that smoking and diet constitute the major factors associated with human cancer incidence. If one considers all of the categories that pertain to man-made chemicals, it is estimated that their contribution to human cancer incidence is approximately 10%. The factors listed in Table 15.2 are not mutually exclusive since there is likely to be interaction between these factors in the multistep process of carcinogenesis.

Cancer cases and cancer deaths by sites and sex for the United States are shown in Figure 15.1. Breast, lung, and colon and rectum cancers are the major cancers in females, whereas prostate, lung, and colon and rectum are the major cancer sites in males. A comparison of cancer deaths versus incidence for a given site indicates that prognosis for lung cancer cases is poor whereas that for breast or prostate cancer cases is much better. Age-adjusted cancer mortality rates (1935–1995) for selected sites in males are shown in Figure 15.2 and for females are shown in Figure 15.3. The increase in the mortality rate associated with lung cancer in both females and males is striking and is due to cigarette smoking. It is estimated that 87% of lung

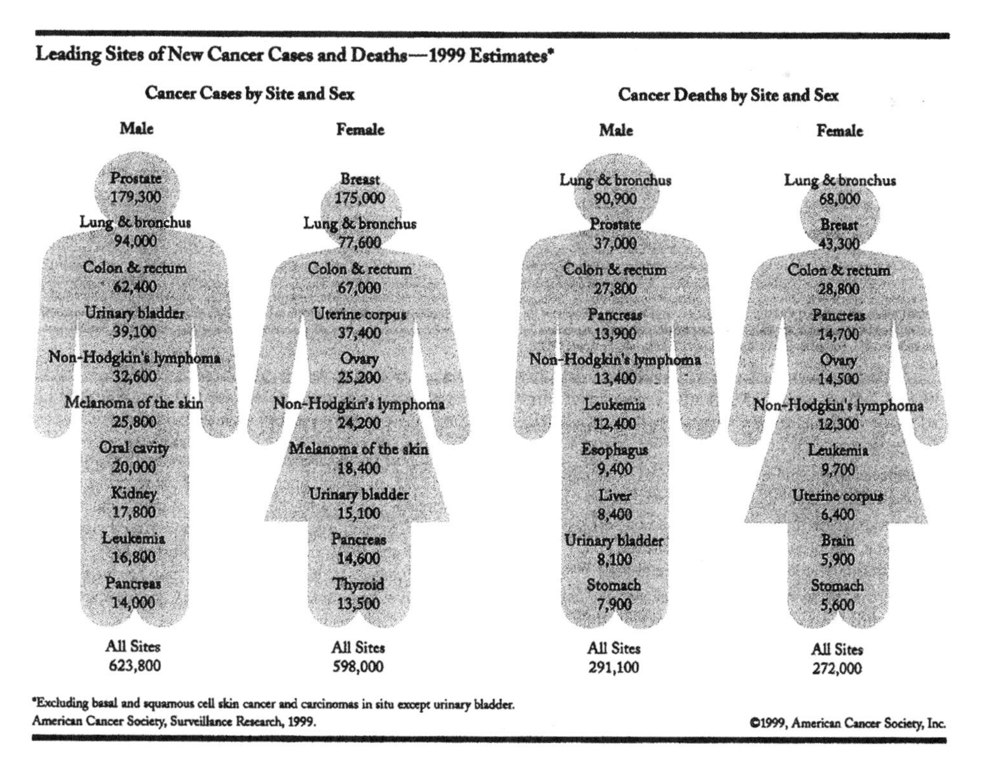

FIGURE 15.1 Cancer cases and cancer deaths by sites and sex 1999 estimates.
Source: Reprinted with the permission of the American Cancer Society, Inc.

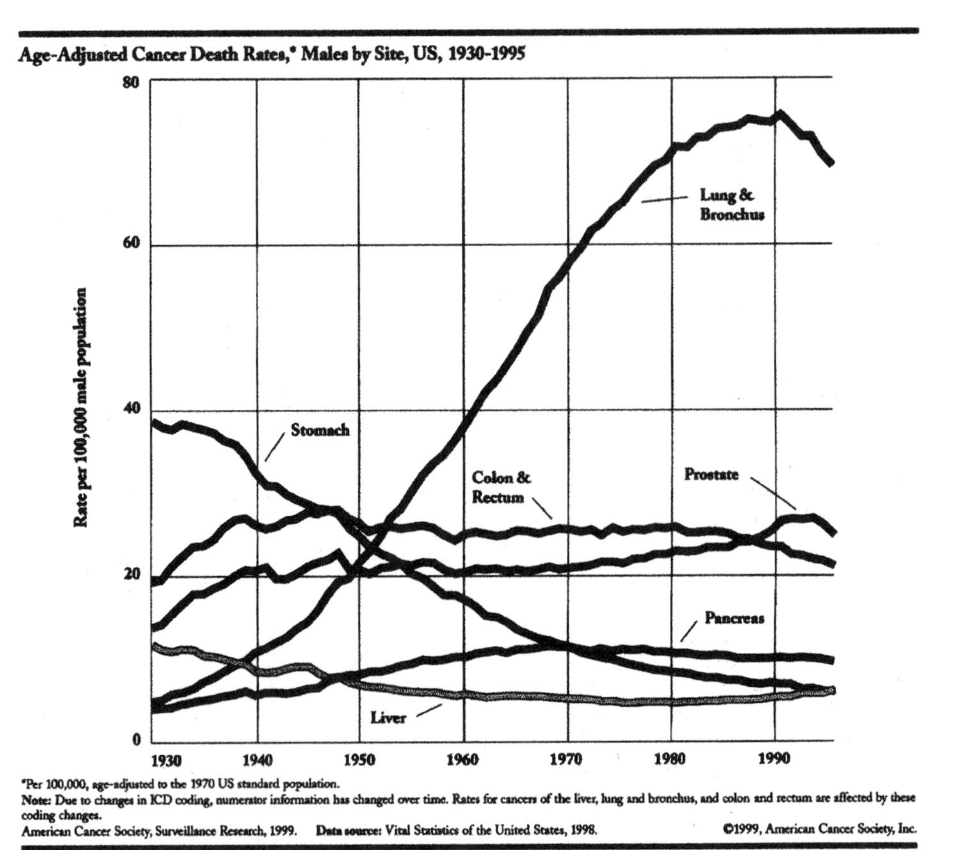

FIGURE 15.2 Age-adjusted mortality rates* (1935–1995) for selected sites in males.
Source: Reprinted with the permission of the American Cancer Society, Inc.

cancers are due to smoking. Lung cancer death rates in males and females began to increase in the mid-1930s and mid-1960s, respectively. These time differences are due to the fact that cigarette smoking among females did not become popular until the 1940s whereas smoking among males was popular in the early 1900s. Taking into account these differences along with a 20–25 year lag period for the cancer to develop explains the differences in the temporal increase in lung cancer death rates in males and females. Another disturbing statistic is that lung cancer, a theoretically preventable cancer, has surpassed breast cancer as the cancer responsible for the greatest number of cancer deaths in women. In addition to lung cancer, cigarette smoking also plays a significant role in cancer of the mouth, esophagus, pancreas, pharynx, larynx, bladder, kidney, and uterine cervix. Overall, the age-adjusted national total cancer death rate is increasing. In 1930, the number of cancer deaths per 100,000 people was 143. In 1940, 1950, 1970, 1984, and 1992 the rate had increased to 152, 158, 163, 170, and 172, respectively. According to the American Cancer Society, when lung cancer deaths due to smoking are excluded, the total age-adjusted cancer mortality rate has actually decreased by 16% between 1950 and 1993. However, it is important to realize that the death rates for some types of cancers are increasing whereas the rates for others are decreasing or remaining constant (Figs 15.2 and 15.3).

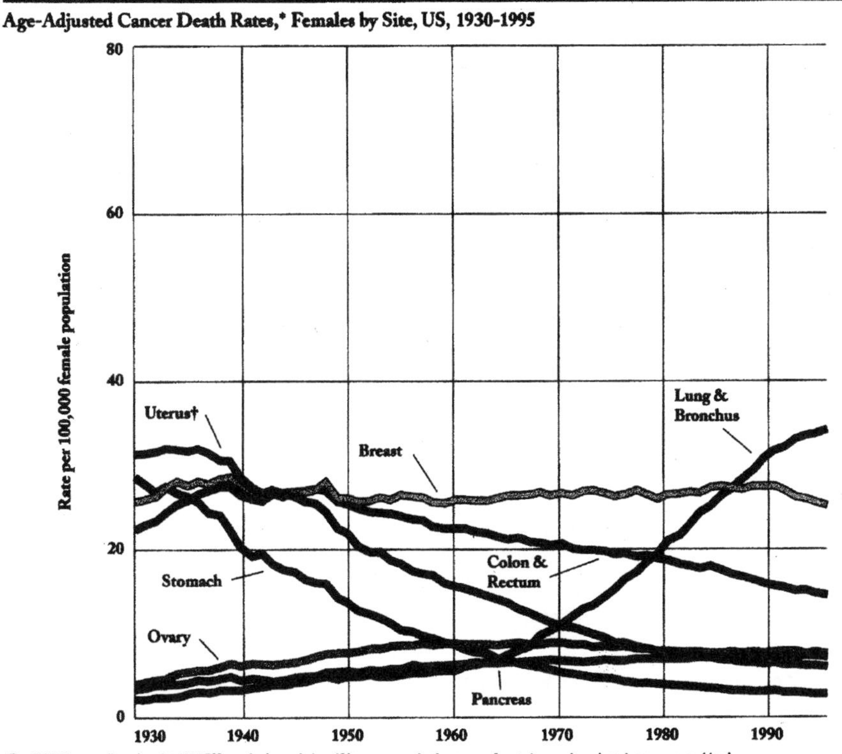

FIGURE 15.3 Age-adjusted Mortality Rates* (1935–1995) for selected sites in females. *Source*: Reprinted with the permission of the American Cancer Society, Inc.

15.2.2 Known Human Carcinogens

Since the time of Percival Pott's observation, epidemiological studies have provided sufficient evidence that exposure to a variety of chemicals, agents, or processes is associated with human cancer. For example, the following associations have emerged between exposure and the development of specific cancers: vinyl chloride and hepatic cancer, amine dyes and bladder cancer, benzene and leukemia, and cigarette smoking and lung cancer. Naturally occurring chemicals or agents such as asbestos, aflatoxin B_1, betel nut, nickel, and certain arsenic compounds are also associated with an increased incidence of human cancer. Both epidemiological studies and rodent carcinogenicity studies are important in the identification and classification of potential human carcinogens. The strongest evidence for establishing whether exposure to a given chemical is carcinogenic in humans comes from epidemiological studies. However, these studies are complicated by the fact that it often takes 20–30 years after carcinogen exposure for a clinically detectable cancer to develop. This delay is problematic and can result in inaccurate historical exposure information and additional complexity due to the interference of a large number of confounding variables. This lag period can prevent the timely identification of a putative carcinogen and result in unnecessary exposure. Therefore, an assay to identify potential human carcinogens is essential and would eliminate any human exposure. The most common method used to identify potential human carcinogens is the long-term rodent bioassay or the 2-year rodent carcinogenesis bioassay. It is clear that almost all if not all human carcinogens identified to date are rodent carcinogens, however, it is not known if all rodent carcinogens are human carcinogens. Indeed, identification of possible human carcinogens based on rodent carcinogenicity can be extremely complicated (see below). Table 15.3 contains the International Agency for Research on Cancer (IARC) list of the known human carcinogens. In addition, Table 15.3 also includes information on carcinogenic complex mixtures and occupations associated with increased cancer incidence. In vitro mutagenicity assays are also used to identify mutagenic agents that may have carcinogenic activity (see Chapter 16).

15.2.3 Classification of Human Carcinogens

Identification and classification of potential human carcinogens through the 2-year rodent carcinogenesis bioassay is complicated by species differences, use of high doses (MTD, maximum tolerated dose), the short life span of the rodents, sample size, and the need to extrapolate from high to low doses for human risk assessment. Although these problems are by no means trivial, the rodent 2-year bioassay remains the "gold standard" for the classification of potential human carcinogens. Criteria for the classification of carcinogens used by the National Toxicology Program, Eighth Report on Carcinogens 1998, are shown in Table 15.4; the criteria used by the Environmental Protection Agency (EPA) and the IARC are shown in Table 15.5. Carcinogens are classified by the weight of evidence for carcinogenicity referred to as sufficient, limited or inadequate based on both epidemiological studies and animal data. As of the year 2000, the EPA is scheduled to change their guidelines for carcinogen risk assessment and their carcinogen classification scheme. New guidelines will emphasize the incorporation of biological mechanistic data in the analysis, and will not rely solely on rodent tumor data. In addition, the six alphanumeric categories listed in Table 15.5 will be replaced by three descriptors for classifying human carcinogenic potential. Carcinogens will be classified by the EPA as: (i) known/likely to be a human carcinogen; (ii) cannot be determined to be a human carcinogen; and (iii) not likely to be a human carcinogen.

15.3 CLASSES OF AGENTS THAT ARE ASSOCIATED WITH CARCINOGENESIS

Chemical agents that influence cancer development can be divided into two major categories based on whether or not they are mutagenic in in vitro mutagenicity assay: *DNA damaging agents* (genotoxic), which are mutagenic in in vitro mutagenicity assays and are considered to produce permanent alterations in the genetic material of the host in vivo, and *epigenetic agents* (nongenotoxic), which are not mutagenic in in vitro assays and these agents are not believed to alter the primary sequence of DNA. However, this classification is complicated by the fact that some nongenotoxic agents may directly or indirectly induce mutagenicity

TABLE 15.3 IARC List of Agents, Substances, Mixtures, or Exposure Circumstances Known to Be Human Carcinogens.

Agents and Substances
Aflatoxins
4-Aminobiphenyl (4-aminodiphenyl)
Arsenic and certain arsenic compounds
Asbestos
Azathioprine
Benzene
Benzidine
Beryllium and certain beryllium compounds
N,N-Bis(2-chloroethyl)-2-naphthylamine (chlornaphazine)
Bis(chloromethyl) ether and technical-grade chloromethyl methyl ether
1,4-Butanediol dimethylsulfonate (Myleran®)
Cadmium and certain cadmium compounds
Chlorambucil
1-(2-Chloroethyl)-3-(4-methylcyclohexyl)-1-nitrosourea (MeCCNU)
Chromium and certain chromium compounds
Cyclophosphamide
Cyclosporin A (ciclosporin)
Diethylstilbestrol
Epstein-Barr virus
Erionite
Estrogen therapy
Estrogens nonsteroidal
Estrogens steroidal
Ethylene oxide

Helicobacter pylori
Hepatitis B virus (chronic infection)
Hepatitis C virus (chronic infection)
Human immunodeficiency virus type 1
Human papillomavirus type 16
Human papillomavirus type 18
Human T-cell lymphotropic virus type I
Melphalan
Methoxsalen with ultraviolet A therapy (PUVA)
MOPP and other combined chemotherapy including alkylating agents
Mustard gas
2-Naphthylamine
Nickel compounds

Opisthorchis viverrini
Oral contraceptives
Radon
Schistosoma haematobium
Silica
Solar radiation
Talc containing asbestiform fibers
Tamoxifen
2,3,7,8-Tetrachlorodibenzo-para-dioxin
Thiotepa [Tris(1-aziridinyl)phosphine sulfide]

TABLE 15.3 *Continued*

Thorium dioxide
Treosulfan
Vinyl chloride
Mixtures
Alcoholic beverages
Analgesic mixtures containing phenacetin
Betel quid with tobacco
Coal tar and coal pitches
Mineral oils
Salted fish
Shale-oils
Tobacco smoke and tobacco smokeless products
Wood dust
Exposure Circumstances
Aluminum production
Auramine, manufacture
Boot and shoe manufacture and repair
Coal gasification
Coke production
Furniture and cabinet making
Haematite mining with exposure to radon
Iron and steel founding
Isopropanol manufacture
Manufacture of magenta
Painter
Rubber industry
Strong inorganic acid mists containing sulfuric acid

in vivo, for example, through oxidative damage. Furthermore, many epigenetic/nongenotoxic agents cause the clonal expansion of cells containing DNA alterations to form tumors, however, in the absence of such DNA alterations these epigenetic agents have no effect on tumor formation. Thus, as we gain a better understanding of chemical carcinogenesis we find that there is functional and mechanistic overlap and interaction between these two major categories of chemical carcinogens.

15.3.1 DNA Damaging Agents

DNA damaging agents can be divided into four major categories. (1) *Direct-acting carcinogens* are intrinsically reactive compounds that do not require metabolic activation by cellular enzymes to covalently interact with DNA. Examples include N-methyl-N-nitrosourea and N-methyl-N′-nitro-N-nitrosoguanidine; the alkyl alkanesulfonates such as methyl methanesulfonate; the lactones such as beta propiolactone and the nitrogen and sulfur mustards. (2) *Indirect-acting carcinogens* require metabolic activation by cellular enzymes to form the ultimate carcinogenic species that covalently binds to DNA. Examples include dimethylnitrosamine, benzo[a]pyrene, 7,12-dimethylbenz[a]anthracene, aflatoxin B1 and

TABLE 15.4 Carcinogen Classification System of the National Toxicology Program.

Known to be a human carcinogen:
There is sufficient evidence of carcinogenicity from studies in humans that indicates a causal relationship between exposure to the agent, substance, or mixture and human cancer.

Reasonably anticipated to be a human carcinogen:
There is limited evidence of carcinogenicity from studies in humans, which indicates that causal interpretation is credible, but that alternative explanations, such as chance, bias, or confounding factors, could not adequately be excluded; *or*

There is sufficient evidence of carcinogenicity from studies in experimental animals that indicates there is an increased incidence of malignant and/or a combination of malignant and benign tumors: (1) in multiple species or at multiple tissue sites, or (2) by multiple routes of exposure, or (3) to an unusual degree with regard to incidence, site or type of tumor, or age at onset; *or*

There is less than sufficient evidence of carcinogenicity in humans or laboratory animals, however, the agent, substance, or mixture belongs to a well-defined, structurally related class of substances whose members are listed in a previous Report on Carcinogens as either a known to be human carcinogen or reasonably anticipated to be human carcinogen, or there is convincing relevant information that the agent acts through mechanisms indicating it would likely cause cancer in humans.

Conclusions regarding carcinogenicity in humans or experimental animals are based on scientific judgment, with consideration given to all relevant information. Relevant information includes, but is not limited to dose response, route of exposure, chemical structure, metabolism, pharmacokinetics, sensitive subpopulations, genetic effects, or other data relating to mechanism of action or factors that may be unique to a given substance. For example, there may be substances for which there is evidence of carcinogenicity in laboratory animals but there are compelling data indicating that the agent acts through mechanisms that do not operate in humans and would therefore not reasonably be anticipated to cause cancer in humans.

Source: From the Eighth Report on Carcinogens, U.S. Department of Health and Human Services, Public Health Service, National Toxicology Program.

2-acetylaminofluorene (Fig. 15.4). (3) *Radiation* and *oxidative DNA damage* can occur directly or indirectly. Ionizing radiation produces DNA damage through direct ionization of DNA to produce DNA strand breaks or indirectly via the ionization of water to reactive oxygen species that damage DNA bases. Ultraviolet radiation (UVR) from the sun is responsible for approximately 1 million new cases of human basal and squamous cell skin cancer that will be diagnosed each year. Reactive oxygen species can also be produced by various cellular process including respiration and lipid peroxidation. (4) *Inorganic agents* such as arsenic, chromium, and nickel are considered DNA damaging agents, although in many cases the definitive mechanism is unknown. DNA damaging agents can produce three general types of genetic alterations: (i) gene mutations that include point mutations involving single base pair substitutions that can result in amino acid substitutions in the encoded protein and frame shift mutations involving the loss or gain of one or two

TABLE 15.5 IARC and EPA Classification of Carcinogens.

IARC	EPA	Definition
1	Group A	Human carcinogens—sufficient evidence from epidemiological studies to support a causal association between exposure to the agents and cancer
2A	Group B	Probable human carcinogens
	Group B1	Limited epidemiological evidence that the agent causes cancer regardless of animal data
	Group B2	Inadequate epidemiological evidence or no human data on the carcinogenicity of the agent and sufficient evidence in animal studies that the agent is carcinogenic
2B	Group C	Possible human carcinogens—absence of human data with limited evidence of carcinogenicity in animals
3	Group D	Not classifiable as to human carcinogenicity—agents with inadequate human and animal evidence of carcinogenicity of for which no data are available
4	Group E	Evidence of non-carcinogenicity for humans— agents that show no evidence for carcinogenicity in at least two adequate animal tests in different species or in both adequate epidemiologic and animal studies

base pairs resulting in an altered reading frame and gross alterations in the encoded protein; (ii) chromosome aberrations including gross chromosomal rearrangement such as deletions, duplications, inversions, and translocations; and (iii) aneuploidy and polyploidy, which involve the gain or loss of one or more chromosomes.

15.3.2 Epigenetic Agents

Epigenetic agents that influence carcinogenesis are not thought to alter the primary sequence of DNA, but rather they are considered to alter the expression or repression of certain genes and/or produce perturbations in signal transduction pathways that influence cellular events related to proliferation, differentiation, or apoptosis. Many epigenetic agents favor the proliferation of cells with an altered genotype (cells containing a mutated oncogene(s) and/or tumor suppressor gene(s)) and allow the clonal expansion of these altered or "initiated" cells. Epigenetic agents can be divided into four major categories: (1) *hormones* such as conjugated estrogens and diethylstilbesterol, (2) *immunosuppressive xenobiotics* such as azathioprine and cyclosporin A, (3) *solid state agents*, which include plastic implants and asbestos, and (4) *tumor promoters* in rodent models including 12-O-tetradecanoylphorbol-13-acetate, peroxisome proliferators, TCDD, and phenobarbital (Fig. 15.5). In humans, diet (including caloric, fat, and protein intake), excess alcohol and late age of pregnancy are considered to function through a promotion mechanism. Although smoking and UVR have initiating activity, both are also considered to have tumor-promoting activity. By definition, tumor promoters are not

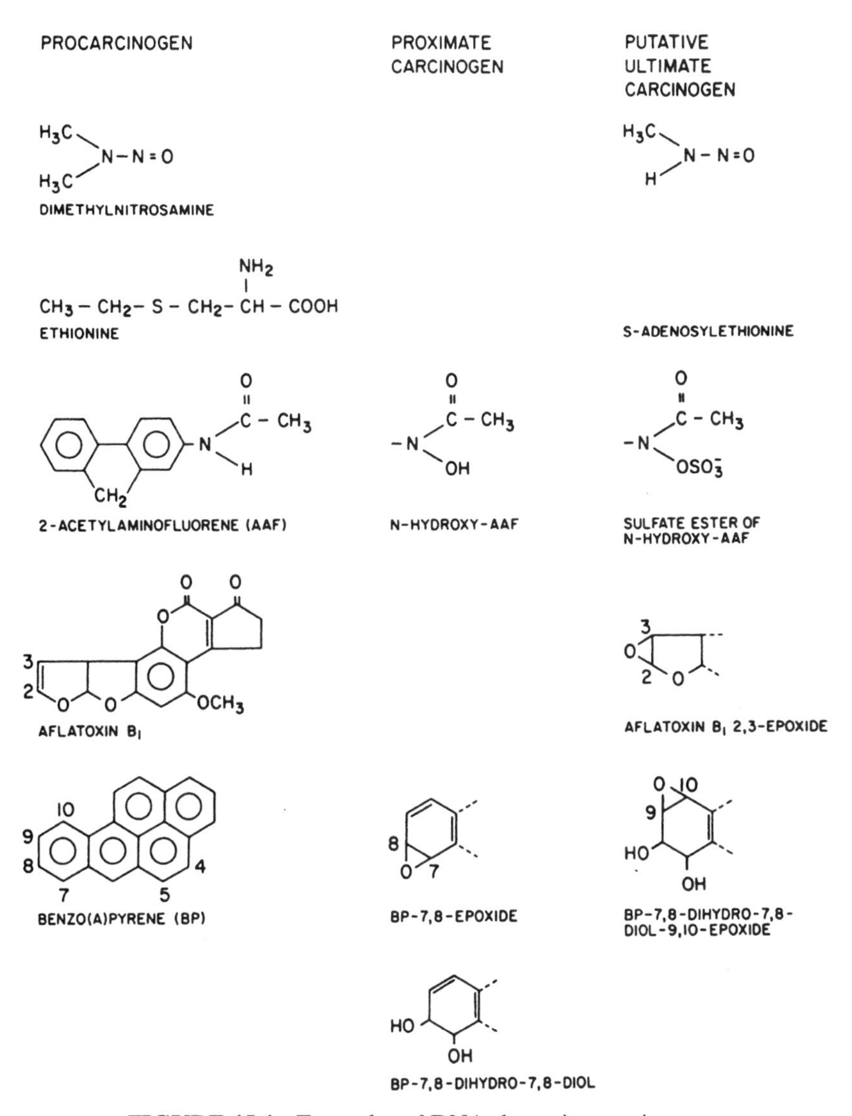

FIGURE 15.4 Examples of DNA-damaging carcinogens.

classified as carcinogens inasmuch as they are considered inactive in the absence of initiated cells. However, an altered genotype or an initiated cell can arise from spontaneous mutations resulting from imperfect DNA replication or oxidative DNA damage or can result from environmental carcinogens. Theoretically, in the presence of a tumor promoter these mutant cells would clonally expand to form a tumor. Therefore, the nomenclature becomes somewhat a matter of semantics as to whether the tumor promoter should or should not be classified as a carcinogen. Certain hormones and immunosuppressive agents are classified as human carcinogens, although it is generally considered that these agents are not carcinogenic in the absence of initiated cells, but rather like tumor promoters, may only allow for the clonal expansion of cells with an altered genotype.

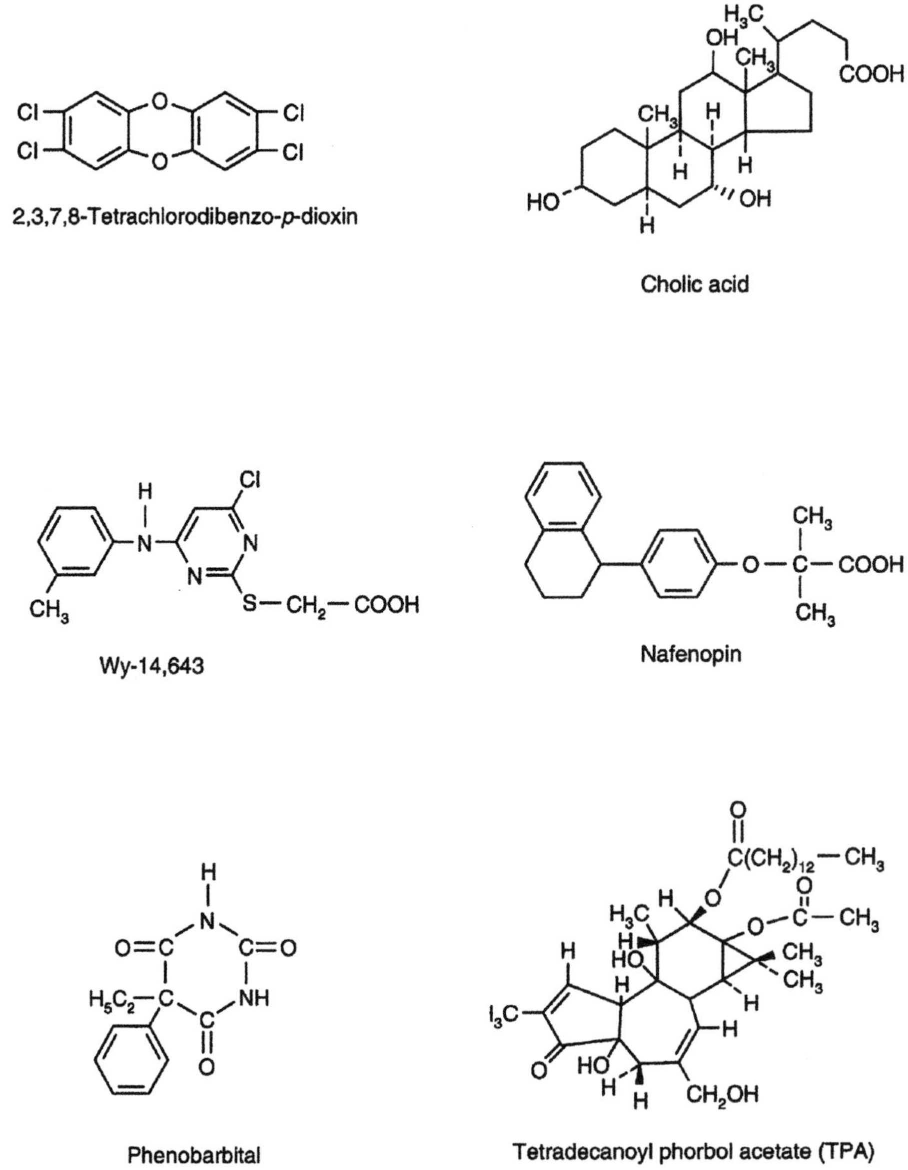

2,3,7,8-Tetrachlorodibenzo-*p*-dioxin

Cholic acid

Wy-14,643

Nafenopin

Phenobarbital

Tetradecanoyl phorbol acetate (TPA)

FIGURE 15.5 Examples of tumor promoters.

The distinction between epigenetic (nongenotoxic) and genetic carcinogens is further blurred as there is evidence that some nongenotoxic/epigenetic agents can induce oxidative DNA damage in vivo through the direct or indirect production reactive oxygen species. For example, certain estrogen hormones may possess this ability and such a characteristic may contribute to its carcinogenicity. As discussed above, it is important to keep in mind that there can be significant overlap in the mechanisms/classification/nomenclature scheme presented above.

TABLE 15.6 Major Evidence Supporting a Genetic Mechanism of Carcinogenesis.

1. Cancer is a heritable stable change.
2. Tumors are generally clonal in nature.
3. Many carcinogens are or can be metabolized to electrophilic intermediates that can covalently bind to DNA.
4. Many carcinogens are mutagens.
5. Autosomal dominant-inherited cancer syndromes provide direct evidence for a genetic component in the origin of cancer.
6. Autosomal recessive-inherited cancer syndromes associated with chromosome fragility or decreased DNA repair predisposes affected individuals to cancer.
7. Most, if not all, cancers display chromosomal abnormalities.
8. The transformed phenotype can be transferred from a tumor cell to a non-tumor cell by DNA transfection.
9. Cells can be transformed with oncogenes.
10. Proto-oncogenes are activated by mutation in cancer cells.
11. Tumor suppressor genes are inactivated by mutation in cancer cells.

15.4 SOMATIC MUTATION THEORY (GENETIC MECHANISM)

A great deal of evidence has accumulated in support of the *somatic mutation theory* of carcinogenesis, which simply states that alterations in DNA within somatic cells is necessary for neoplasia. Over the past three decades it has been demonstrated that many carcinogens produce a permanent change in DNA and that this change is involved in the carcinogenic process. The major pieces of evidence supporting the somatic mutation theory (genetic mechanism) of carcinogenesis are shown in Table 15.6.

15.4.1 Electrophilic Theory, Metabolic Activation, and DNA Adduct Formation

Elizabeth and James Miller observed that a diverse array of chemicals could produce cancer in animals. In an attempt to explain this, they hypothesized that many carcinogens are metabolically activated to electrophilic metabolites that are capable of interacting with nucleophilic sites in DNA, RNA, and protein. The Millers termed this the *electrophilic theory of chemical carcinogenesis*. From this concept of metabolic activation, the important terms *parent*, *proximate*, and *ultimate* carcinogen were developed. A *parent carcinogen* is a compound that must be metabolized in order to have carcinogenic activity; a *proximate carcinogen* is an intermediate metabolite requiring further metabolism resulting in the *ultimate carcinogen*, which is the actual metabolite that covalently binds to the DNA. Examples of chemical carcinogens that require metabolic activation by cellular enzymes include benzo[a]pyrene, 7,12-dimethylbenz[a]anthracene, aflatoxin B1 and 2-acetylaminofluorene. Some of the parent structures of these compounds as well as their proximate and ultimate carcinogenic species are shown in Figure 15.4. The electrophilic theory was further refined to suggest that if the DNA adduct is not repaired before the cell replicates, an error in the newly synthesized DNA could occur and a mutation would be permanently fixed in the DNA of the daughter cell. If this change has occurred in a critical gene, for example, in a proto-oncogene or tumor suppressor gene, it could represent an initiating event(s) in carcinogenesis. Therefore, error-prone repair or lack of repair followed by DNA replication is an important aspect of chemical carcinogenesis. Thus, the electrophilic theory of chemical carcinogensis is consistent with and supportive of the somatic mutation theory of carcinogenesis.

Metabolic activation of chemical carcinogens by cytochromes P-450 is well documented. In addition to the phase I enzymes, phase II enzymes, such as GSH transferases can also participate in metabolically activating or inactivating chemical carcinogens. Importantly human genetic polymorphisms have been characterized in some phase I and phase II genes (see Chapter 5). Such polymorphisms could alter an individual's response after exposure to certain chemical carcinogens; these genes are termed *metabolic susceptibility genes*. The metabolism of benzo[a]pyrene has been extensively studied and at least 15 major phase I metabolites have been identified. Many of these metabolites are further metabolized by phase II enzymes to produce numerous different metabolites. Extensive research has elucidated which of these metabolites and pathways are important in the carcinogenic process. As shown in Figure 15.7, benzo[a]pyrene is metabolized by cytochrome P-450 to benzo[a]pyrene-7,8 epoxide, which is then hydrated by epoxide hydrolase to form benzo[a]pyrene-7,8-diol. Benzo[a]pyrene-7,8-diol is considered the proximate carcinogen because it must be further metabolized by cytochrome P-450 to form the ultimate carcinogen, the bay region diol epoxide, (+)-benzo[a]pyrene-7,8-diol-9,10-epoxide-2. It is this reactive intermediate that binds covalently to DNA forming DNA adducts. (+)-Benzo[a]pyrene-7,8-diol-9,10-epoxide-2 binds preferentially to deoxyguanine residues, forming N-2 adduct (Fig. 15.6). (+)-Benzo[a]pyrene-7,8-diol-9,10-epoxide-2 is highly mutagenic in eukaroytic and prokaryotic cells and carcinogenic in rodents. It is important to note that not only is the chemical configuration of the metabolites of many polycyclic aromatic hydrocarbons important for their carcinogenic activity, but so is their chemical conformation/stereospecificity (Fig. 15.7). For example, four different stereoisomers of benzo[a]pyrene-7,8-diol-9,10 epoxide are formed, each one only differs with respect to whether the epoxide

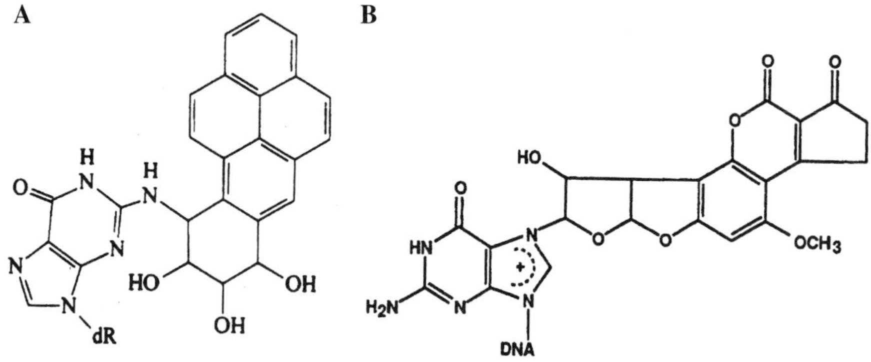

FIGURE 15.6 Examples of DNA adducts.
(A) (+)-Benzo[a]pyrene-7,8-diol-9,10-epoxide-2-N-2 guanine adduct; (B) aflatoxin epoxide-N7 guanine adduct.

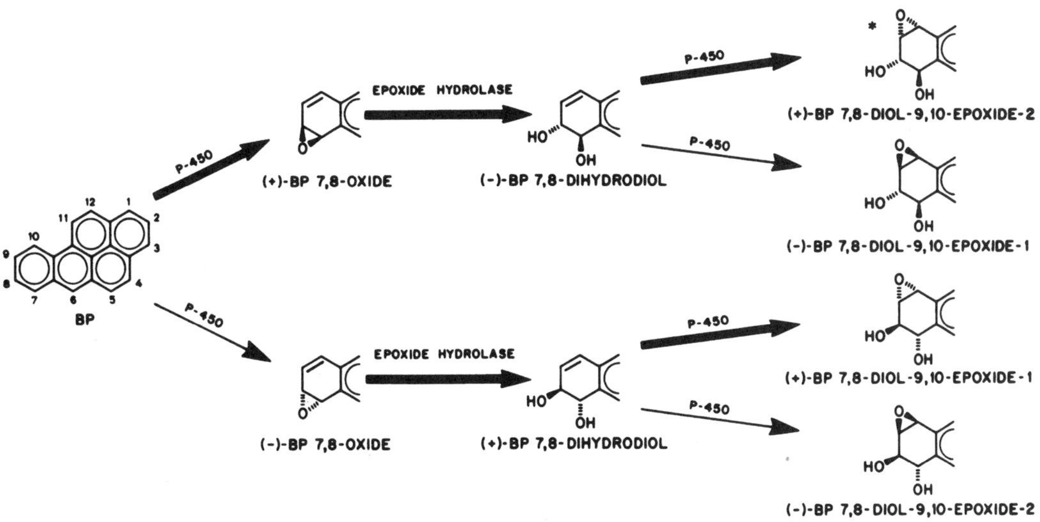

FIGURE 15.7 Benzo[a]pyrene metabolism to the ultimate carcinogenic species. Heavy arrows indicate major metabolic pathways, *represents ultimate carcinogenic species. *Source*: Adapted from Conney, A.H. *Cancer Res.* 42 (1982), 4875.

or hydroxyl groups are above or below the plane of the flat benzo[a]pyrene molecule, and yet only one, (+)-benzo[a]pyrene-7,8-diol-9,10-epoxide-2, has significant carcinogenic potential. Many polycyclic aromatic hydrocarbons are metabolized to bay region diol epoxides. The bay region theory suggests that the bay region diol epoxides are the ultimate carcinogenic metabolites of polycyclic aromatic hydrocarbons. Examples of bay region diol epoxides that are considered the ultimate carcinogenic species of a number of polycyclic aromatic hydrocarbons are shown in Figure 15.8.

DNA can be altered by oxidative damage, large bulky adducts, and alkylation. Carcinogens such as N-methyl-N′-nitro-N-nitrosoguanidine and methyl methanesulfonate alkylate DNA to produce N-alkylated and O-alkylated purines and pyrimidines. Ionizing radiation and reactive oxygen species commonly oxidize guanine to produce 8-oxoguanine. Formation of DNA adducts may involve any of the bases, although the N-7 position of guanine is one the most nucleophilic sites in DNA. Of importance is how long the adduct is retained in the DNA. As shown in Figure 15.6A, (+)-benzo[a]pyrene-7,8-diol-9,10-epoxide-2 forms adducts mainly at guanine N-2 whereas aflatoxin B1 epoxide, another well-studied rodent and human carcinogen, binds preferentially to the N7 position of guanine (Fig. 15.6B). For some carcinogens, there is a strong correlation between the formation of very specific DNA adducts and tumorigenicity. Quantitation and identification of specific carcinogen adducts may be useful as biomarkers of exposure. Importantly, the identification of specific DNA adducts has allowed for the prediction of specific point mutations that would likely occur in the daughter cell providing there was no repair of the DNA-adduct in the parent cell. As will be discussed in a later section, some of these expected mutations have been identified in specific oncogenes and tumor suppressor genes in chemically induced rodent tumors providing support that the covalent carcinogen binding produced the observed mutation. In several cases, spe-

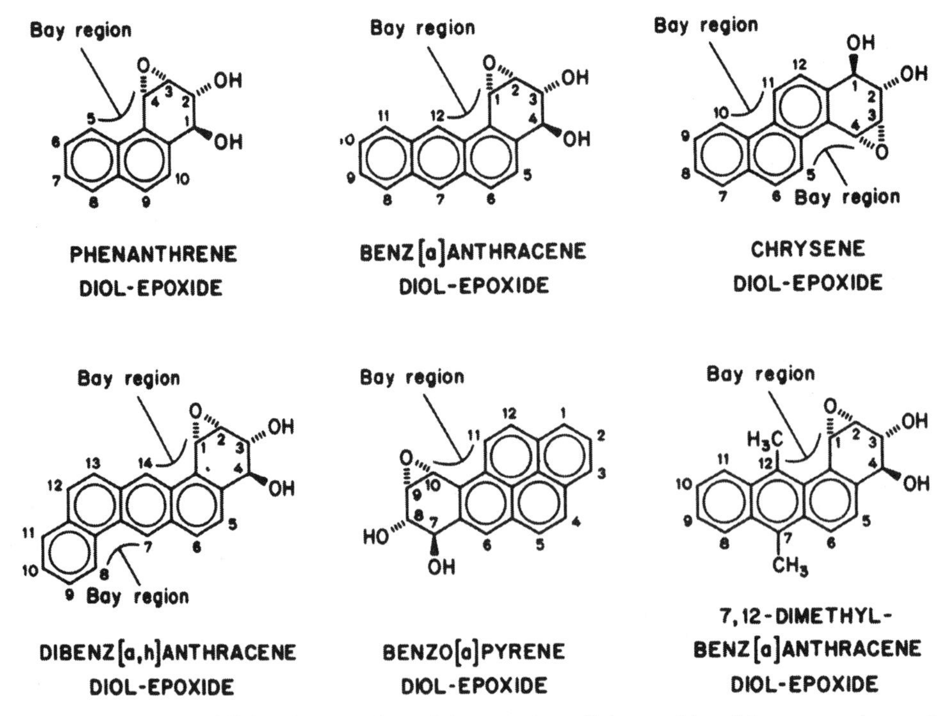

FIGURE 15.8 Additional examples of bay region diol epoxide ultimate carcinogenic species of other polycyclic aromatic hydrocarbons.

Source: Adapted from Conney, A.H. *Cancer Res.* 42 (1982), 4875.

cific base pair changes in *p53* tumor suppressor gene in human tumors are associated with a mutational spectrum that is consistent with exposure of the individual to a specific carcinogen. For example, the mutation spectrum identified in *p53* in human tumors thought to result from the exposure of the individual to UVR, aflatoxin, and benzo[a]pyrene (from cigarette smoke) are consistent with the observed specific mutational damage in *p53* induced by these agents in experimental cellular systems.

15.5 ONCOGENES AND CELLULAR TRANSFORMATION

Much evidence has accumulated for a role of covalent binding of reactive electrophilic carcinogens to DNA in chemical carcinogenesis. It is known that chemical mutagens and carcinogens can produce point mutations, frameshift mutations, and chromosome aberrations in mammalian cells. If the interaction of a chemical carcinogen with DNA, leading to a permanent alteration in the DNA, is a critical event in chemical carcinogenesis, then the identification of these altered genes and the function of their protein products is essential to our understanding of chemical carcinogenesis. While specific DNA-carcinogen adducts were isolated in the 1970s and 1980s, it was not until the early to mid-1980s that the identification of specific genes that were mutationally altered by chemical carcinogens became known. The ability to transfer isolated purified DNA from chemically induced and spontaneous tumors into immortalized cells in culture with subsequent transformation of the recipient cells was a critical advance that aided in the identification and understanding of cellular oncogenes. The general characteristics of transformed cells in culture are (1) anchorage independence (ability to grow in soft agar), (2) loss of contact inhibition (pile up and form foci), (3) immortality, (4) a reduced requirement for growth factors and/or serum, and (5) the ability to produce tumors when injected into immunodeficient mice (nude mice that lack a thymus and therefore do not have T-lymphocytes). Certain normal cellular genes, termed proto-oncogenes, can be mutated by chemical carcinogens providing a selective growth advantage to the cell. The mutational activation of proto-oncogenes is strongly associated with tumor formation, carcinogenesis, and cell transformation. Proto-oncogenes are highly conserved in evolution and their expression is tightly regulated. Their protein products function in the control of normal cellular proliferation and differentiation. However, when these genes are altered by a mutation, sequence deletion, virus integration, chromosome translocation, gene amplification, or promoter insertion, an abnormal protein product or an abnormal amount of product is produced. Under these circumstances these genes have the ability to transform cells in vitro and are termed *oncogenes*.

15.5.1 Acute Transforming Retroviruses Contain Cellular Genes

Oncogenes were first identified as genes contained in acute transforming retroviruses. With the exception of a rare human T-cell leukemia, retroviruses are not considered to play a role in human cancer. However, because these viruses were instrumental in the identification of cellular genes that are involved in cancer, it is instructive to understand how they transform cells and how they contributed to our current knowledge of the genetic basis of cancer and chemical carcinogenesis. Retroviruses encode their genetic material in RNA. The viral genome is composed of powerful retroviral promoters contained within the long terminal repeat (LTR) and three genes termed *gag* (derived from the historical designation *gr*oup *as*sociated *an*tigen), *pol*, and *env*. The *gag* gene encodes a polyprotein that is cleaved into three or four proteins that make the internal structural proteins of the virion. The *pol* gene also encodes a polyprotein that is cleaved to form RNA-dependent DNA polymerase or reverse transcriptase and integrase needed to insert the viral genome into the host genome. The *env* region encodes another polyprotein that produces the proteins of the viral envelope. Following infection, the virus RNA is transcribed into DNA by viral reverse transcriptase and double stranded viral DNA integrates into host DNA where the strong retroviral promoters drive the expression of viral genes to make mRNA and replicate the viral genome (Fig. 15.9). Retroviral mRNA is translated and proteins produced, the viral genome is then packaged and infectous retroviruses are released from the cell. Acute transforming retroviruses are a unique type of retrovirus as they have the ability to produce cancer in certain species (predominantly in rodents and avian species) in a relatively short time period (days to a few weeks). Examples of acute transforming retroviruses include the Rous sarcoma virus (RSV) (Fig. 15.10), which produces sarcomas in chickens and the Harvey murine sarcoma virus (Ha-MSV), which produces sarcomas and erythroleukemia in mice. Studies utilizing RSV provided the first evidence that onco-

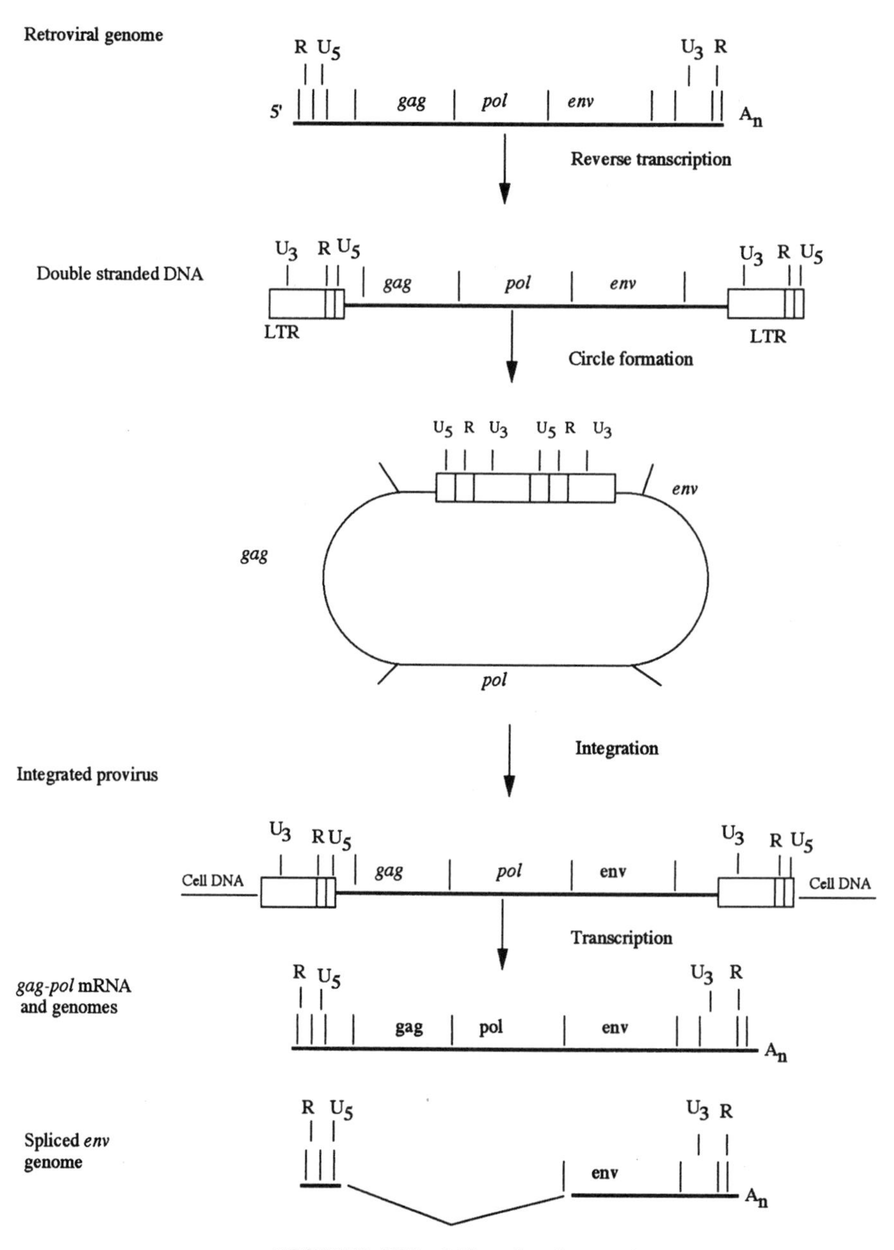

FIGURE 15.9 Lifecycle of retrovirus.

Source: Adapted from *Molecular Biology of the Gene* (4th Ed.), Watson JD., Hopkins NH., Roberts JW., Steitz JA. and Weiner AM (Eds).

genes exist in mammalian cells. It was known that RSV could quickly produce sarcomas in chickens in a few days as well as transform cells in culture. In the mid-1970s, a segment of the RSV genome was identified that was not required for viral replication but was essential for transformation. The transforming gene was cloned and called *src* (*src* for sarcoma), and a working hypothesis was advanced stating

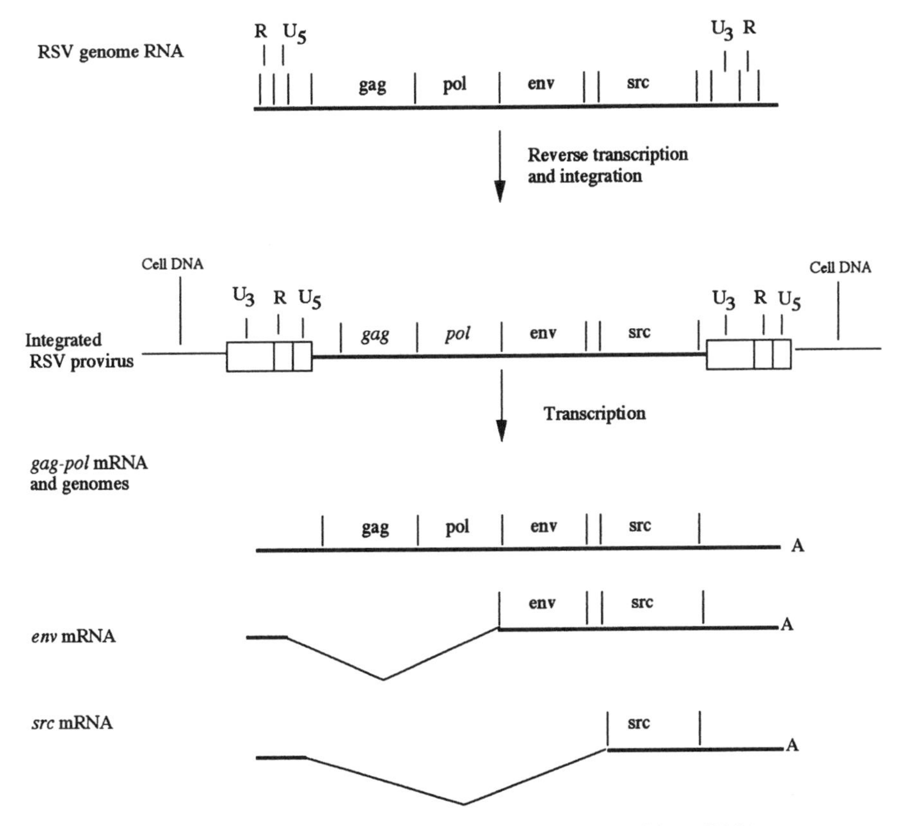

FIGURE 15.10 Rous sarcoma virus genome and its mRNAs.

Source: Adapted from *Molecular Biology of the Gene* (4th Ed.), Watson JD., Hopkins NH., Roberts JW., Steitz JA., and Weiner AM (Eds).

that, upon RSV infection, the retrovirus was incorporated into the genome, which was followed by *src* expression and transformation. However, Southern blot analysis revealed that the *src* sequence was present in uninfected chickens, in fact, *src* was detected in all vertebrate organisms as well as in flys, worms, and sponges, indicating that the *src* sequence was conserved in evolution for hundreds of millions of years! It was ultimately determined that a cellular gene, a proto-oncogene termed c-*src* (c for cellular) had been transduced (a rare recombination event occurred between retroviral genome and the *src* gene) into the transforming retrovirus. Thus, it is the cellular gene that had been assimilated from the host into the virus that is responsible for the oncogenic potential of RSV. Approximately 30 different cellular oncogenes have been discovered due to their transduction into the acute transforming retroviruses. Thus, the acute transforming retroviruses have been instrumental in the identification of cellular oncogenes as many of the genes transduced by retroviruses have also been found to be altered in human and rodent tumors. Retroviral transduction results in overexpression of the transduced gene and viral replication results in alterations in the primary sequence of the transduced cellular proto-oncogenes due to the high error rate of reverse transcriptase. These alterations include the acquisition of point mutations, deletions, or gene fusions within the coding region of the transduced proto-oncogene resulting in proteins with aberrant function. The cellular genes are labeled with the prefix c-, that is, c-*Ha-ras* and c-*src*, where c- indicates a cellular gene within the cellular genome and *Ha-ras* and *src* are the gene names. This designation distinguishes cellular genes from their viral counterparts where a v- prefix is used, for example, v-*src* where v- indicates that the gene has been transduced into the viral genome. If the v- or c- designation is not indicated, then it is understood that the reference is to the cellular gene. With exception of RSV, all the acute transforming retroviruses are replication deficient because a portion of their genome has been replaced with the transduced gene. Their replication is dependent upon co-infection with a helper retrovirus that provides the proteins encoded by the genes deleted in the acute transforming retrovirus.

In addition to the acute transforming retroviruses, there are the weak transforming retroviruses that produce cancer predominately in avian and rodent species over a period of months to a year. These weakly oncogenic retroviruses do not contain cellular genes, but disrupt normal regulatory elements of proto-oncogenes by insertion into the host DNA near a proto-oncogene where the strong retroviral promoters drive transcription of the host's proto-oncogene. Avian leukosis virus (ALV) produces B-cell lymphomas in birds and it was determined that the virus inserts itself adjacent to the c-*myc* genome in host genome and causes the 50- to 100-fold increase in the expression of c-*myc*. This event is termed *insertional mutagenesis* or *retroviral insertion* and has been implicated in tumorigenesis of a variety of weak transforming retroviruses. Retroviral insertion and subsequent activation of a proto-oncogene has led to the identification of approximately 50 oncogenes.

15.5.2 DNA Oncogenic Viruses Contain Viral Oncogenes

In contrast to acute transforming retroviruses that have transduced a cellular gene into their genome, the oncogenic DNA viruses do not contain cellular sequences. Oncogenic DNA viruses such as members of the papovavirus group (SV40, polyoma), adenoviruses (Ad12), and papilloma viruses have employed a different strategy for replication and transformation. Unlike the retroviruses, these viruses do not generally integrate into the host DNA to replicate their genome. Some of these DNA viruses do not possess all the genes necessary for DNA synthesis. Upon infection of nondividing cells, these viruses can stimulate these cells to initiate DNA synthesis to provide an adequate supply of precursors and enzymes needed for viral replication. Because cells that are in the G1 phase of the cell cycle do not synthesize the enzymes needed for DNA synthesis, DNA viruses have the ability to override the block in G1 and cause cellular proliferation. Soon after these viruses infect cells, the early viral genes are expressed that cause quiescent G1 cells to enter S-phase. S-phase entry is accompanied by an increase in cellular enzymes needed for DNA synthesis and these cellular enzymes are used in the synthesis of the precursors needed for viral replication. Soon after the induction of these enzymes, the virus and host DNA are replicated. Other viral genes referred to as late genes, responsible for synthesis of capsid proteins and the proteins involved in the lytic response, are subsequently expressed. Cells in which a virus multiplies and leads to cell lysis and release of infectious virus are called permissive cells; nonpermissive cells are those cells in which viral growth does not occur. In nonpermissive cells the virus can occasionally cause cell transformation. In these cells only the early viral genes that unlock the control of cellular proliferation are expressed; the late genes that are responsible for synthesis of capsid proteins and those involved in the lytic response are not expressed. It is this block in the expression in viral late genes in nonpermissive cells that allows the cells to become transformed. Transformation is also accompanied by integration of the DNA virus genome in the host's genome. Several of the viral early genes such as *SV40 large T-antigen*, *polyoma large and middle T-antigen*, *adenovirus E1A and E1B*, and *papillomas virus E6 and E7* have transforming ability. Several DNA viruses are important in the etiology of certain human cancers, for example, chronic infection with hepatitis B or C virus is associated with liver cancer, human papillomavirus type 16 and type 18 infection is associated with cervical cancer, and Epstein-Barr virus (EBV) infection is associated with Burkitt's lymphoma. As will be discussed in the tumor suppressor gene section, viral early gene protein products of oncogenic DNA viruses have the ability to bind to and inactivate growth regulatory proteins encoded by the tumor suppressor genes, *p53* and *Rb*.

15.5.3 Oncogenes and Signal Transduction

As mentioned above, certain normal cellular genes, termed proto-oncogenes, are highly conserved in evolution and their expression is tightly regulated. Their protein products function in the control of normal cellular proliferation, differentiation, and apoptosis. However, when these genes are altered by a mutation, chromosome translocation, gene amplification, or promoter insertion, they have the ability under certain conditions to transform cells and are termed oncogenes. Of the 200 oncogenes that have been identified, approximately 30 cellular oncogenes are linked to human cancer. Some oncogenes such as members of the Ras family are frequently mutated in a variety of cancers while some, such as the c-abl/bcr fusion protein, are associated with only two types of leukemia. Most oncogene protein products appear to function in one way or another in cellular signal transduction pathways that are involved in regulating cell growth, differentiation, or apoptosis. Signal transduction pathways are used by the cells to receive and process information to ultimately produce a biological cellular response. These pathways are the cellular circuitry conveying specific information from the outside of the cell to the nucleus. In the nucleus, specific genes are expressed, and their encoded proteins produce the evoked biological response. Much of the cellular circuitry involves the covalent modification of proteins through the phosphorylation of serine, threonine, or tyrosine residues by protein kinases. Phosphorylation results in conformational changes in the acceptor proteins; in the case of substrates such as protein kinases, phosphorylation often produces a catalytically active enzyme whereas phosphorylation of transcription factors often stimulates their transcription activity. Phosphorylation of tyrosine residues can also produce docking sites for other proteins that become activated when docked and communicate the signal downstream to other proteins in a signaling pathway. Signal transduction pathways are complex and multiple. In simplified generic terms, a growth factor receptor pathway can consist of an external signal such as a growth factor whose signal is received by a transmembrane growth factor receptor. The binding of the growth factor to the receptor causes the activation of the tyrosine kinase region of intracellular domain of the receptor, resulting in autophosphorylation of tyrosine residues of the intracellular domain. Docking proteins recognize the phosphorylated receptor regions and dock and subsequently communicate the signal downstream through protein-protein interactions. These interactions result in the activation of membrane-bound low-molecular-weight guanine nucleotide binding proteins. Activated guanine nucleotide binding proteins stimulate cytoplasmic serine/threonine kinases that transmit their signal through protein phosphorylation to activate a cascade of kinases, ultimately resulting in the phosphorylation and activation of DNA binding proteins/transcription factors. The activated transcription factors regulate the expression of specific genes whose protein products produce the evoked cellular response. Thus, the growth factor signal is communicated from the outside of the cell to the nucleus and results in the expression of specific genes. If a component of the circuit is altered, then the entire cellular circuit of which the component is a part is altered. It is not difficult to imagine how an alteration in a pathway that regulates cellular growth, differentiation, or apoptosis could have very profound effects on cellular homeostasis. Indeed, this is the molecular basis of how oncogenes contribute to the cancer process.

15.5.4 Activation of Proto-oncogenes by Gene Amplification, Chromosomal Translocation, and Point Mutation

Proto-oncogenes can be activated to oncogenes either qualitatively or quantitatively. Proto-oncogenes can be activated to oncogenes by a mutation, chromosome translocation, gene amplification, or promoter insertion (Fig. 15.11). Frequent qualitative changes include (i) point mutations in the coding region of a proto-onocogene, (ii) the generation of a hybrid or chimeric gene that results from chromosomal translocation, and (iii) sequence deletion. The best example of oncogenes that are activated by point mutation are the *Ras* family of genes. The location of these point mutations within specific regions of *ras* and the chemical carcinogens that produce such changes will be described later. An example of an oncogene that is activated by translocation resulting in a hybrid protein is c-*abl*. In 95% of the cases of human chronic myelogenous leukemia, an abnormal chromosome, termed the Philadelphia chromosome, is present. The Philadelphia chromosome is produced as a result of a translocation between chromosomes 9 and 22. The translocation juxtaposes a portion of the proto-oncogene *c-abl*, a tyrosine kinase, from chromosome 9 with the a portion of chromosome 22 termed *bcr*. *Bcr* encodes a serine/threonine kinase that also has a region with homology to guanine nucleotide exchange proteins. The fusion gene produces a bcr/c-abl hybrid mRNA resulting in a hybrid protein in which the amino terminus region is derived from the bcr protein and the carboxy terminus region from the c-abl protein. This fusion protein possesses both growth-promoting properties and elevated tyrosine kinase activity. Chromosome translocations resulting in fusion proteins are especially prominent in leukemias and lymphomas.

Quantitative changes include gene amplification (increase in the number of copies of a given gene within the chromosome) leading to overexpression, or chromosomal translocation of a proto-oncogene leading to aberrant high-level expression of an unaltered normal proto-oncogene. The best example of these processes involves the myc family. Amplification of *myc* in human or murine carcinomas,

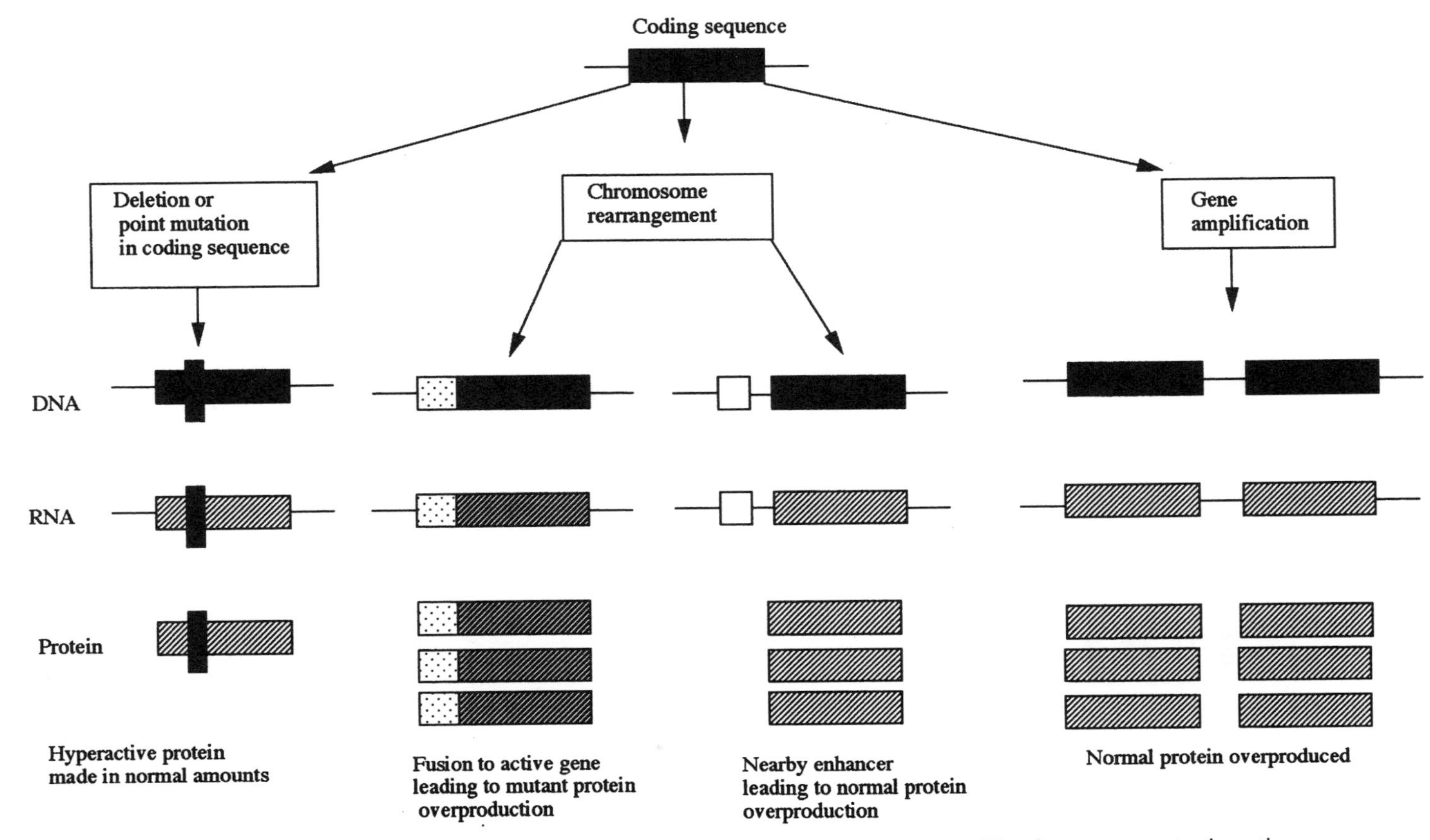

FIGURE 15.11 Activation of oncogenes by mutation, translocation, gene amplification or promoter insertion.

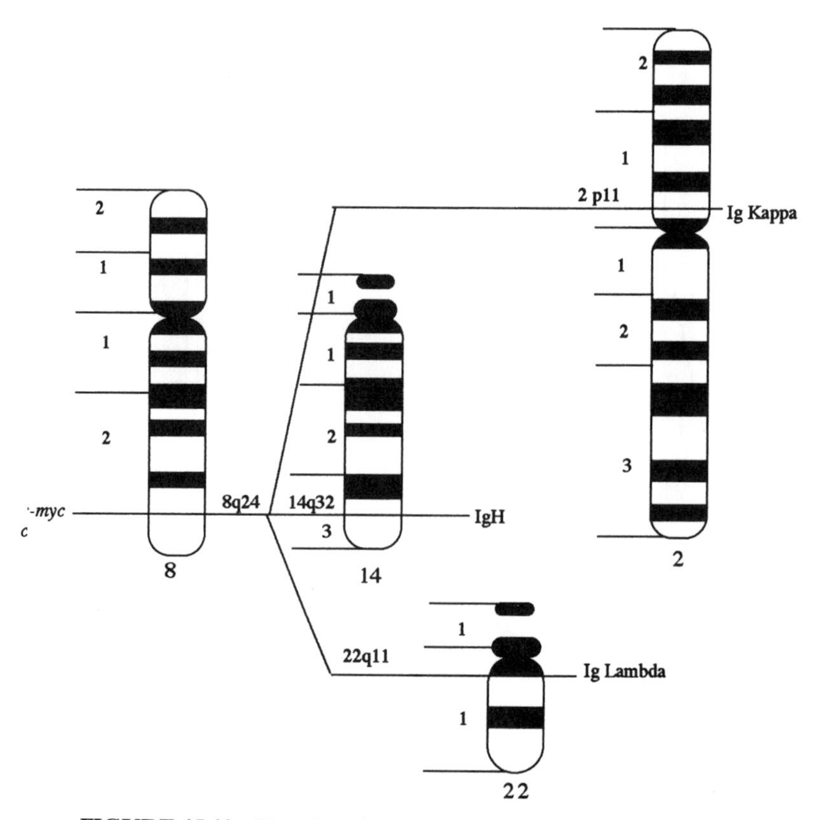

FIGURE 15.12 Translocation of c-myc in Burkitt's lymphoma.

leukemias or sarcomas is a late, progression related event. In human small cell lung cancer, c-*myc* and N-*myc* are amplified up to 50 times; in mammary carcinomas *erbB2/HER2* and c-*myc* are amplified up to 30–50 times. In Burkitt's lymphoma and mouse plasmacytoma, the c-*myc* gene is translocated and inserted near an immunoglobulin locus (Fig. 15.12). This juxtaposition allows for the deregulation of c-*myc* expression that is no longer under proper cellular regulation but rather under the control of the immunoglobulin promoter and is therefore constitutively expressed at high levels. Translocation of c-*myc* from chromosome 8 to an immunoglobulin region in either chromosome 14, 22, or 2 is a consistent feature found in the great majority of Burkitt's lymphoma tumor biopsies or cell lines derived from these tumors. In terms of proto-oncogene overexpression, hypomethylation of the promoter region of some proto-oncogenes can also contribute in proto-oncogene overexpression.

15.5.5 Oncogene Classification

Oncogenes are classified with respect to their biological function. As shown in Table 15.7, the protein products of the major families of oncogenes are involved in transduction of signals from the outside of the cell through the membrane and cytosol into the nucleus, where DNA binding proteins can regulate the expression of genes

TABLE 15.7 Oncogenes Classification.

Families	Genes
Growth factors	sis, hst-1, int-2, wnt-1
Growth factor receptor tyrosine kinases	erb B, fms, met/HGFR, neu/HER2, trk/NGFR
Nonreceptor tyrosine kinases	Abl, src, fgr, fes, yes, lck
Guanine nucleotide binding proteins	H-ras, K-ras, N-ras, TC21, Ga_{12}
Serine/threonine kinases	mos, raf, bcr, pim-1
DNA-binding proteins	myc, fos, myb, jun, E2F1, ets, rel

that contain specific enhancer sequences. A brief description of some oncogene protein products from each of the six classes follows.

15.5.5a Growth Factors

The v-*sis* oncogene of the simian sarcoma virus (SSV) is derived from the cellular gene encoding the B chain of platelet-derived growth factor (PDGF). The c-sis encoded protein is a 28kDa phosphoprotein, secreted by platelets and monocytes. Platelets release PDGF during clotting and within connective tissue cells, PDGF binds to the PDGF receptor, stimulates tyrosine kinase activity, and produces mitogenesis. In human connective tissue tumors, c-sis is expressed by the tumor cells (PDGF is not expressed in the normal tissue counterpart) resulting in an autocrine response in which the tumor cell-produced PDGF is released from the cell and binds to the PDGF receptor on the same tumor cell, evoking a mitogenic response. Transforming growth factor alpha (TGFα) is a growth factor that binds to the epidermal growth factor receptor (EGFR). TGF-α is overexpressed in many carcinomas that express EGFR. Thus, TGF-α also is thought to function in an autocrine manner by stimulating the growth of the tumor cells that produce it. TGF-α expression is increased in some chemically induced rodent liver and skin tumors.

15.5.5b Growth Factor Receptor Tyrosine Kinases

Many growth factors mediate their effects through receptors that possess intrinsic tyrosine kinase activity. These receptors consist of three general domains: an extracellular N-terminal domain that binds a specific growth factor, a transmembrane domain, and an intracellular C-terminal domain that contains intrinsic tyrosine kinase activity. Binding of the growth factor causes receptor dimerization and is accompanied by the activation of the tyrosine kinase activity of the internal region of the receptor resulting in autophosphorylation of tyrosine residues of the internal region of the receptor (Fig. 15.13). Docking proteins containing specialized regions, termed src-2 homology domains (SH2 domain) recognize the phosphorylated tyrosine residues of the receptor regions and dock, subsequently stimulate other proteins to communicate the signal downstream. Some examples of docking proteins that contain an SH2 domain include Shc and GRB2, which are adapter proteins that do not have catalytic activity but function to recruit additional proteins to the membrane/activated receptor. Additional examples include a variety of enymes such as phospholipase Cγ (PLCγ), phosphatidylinositol-3-kinase (PI3K), c-src, GTPase-activating protein (GAP p120), and the tyrosine phosphatase, SH-PTP2. In addition

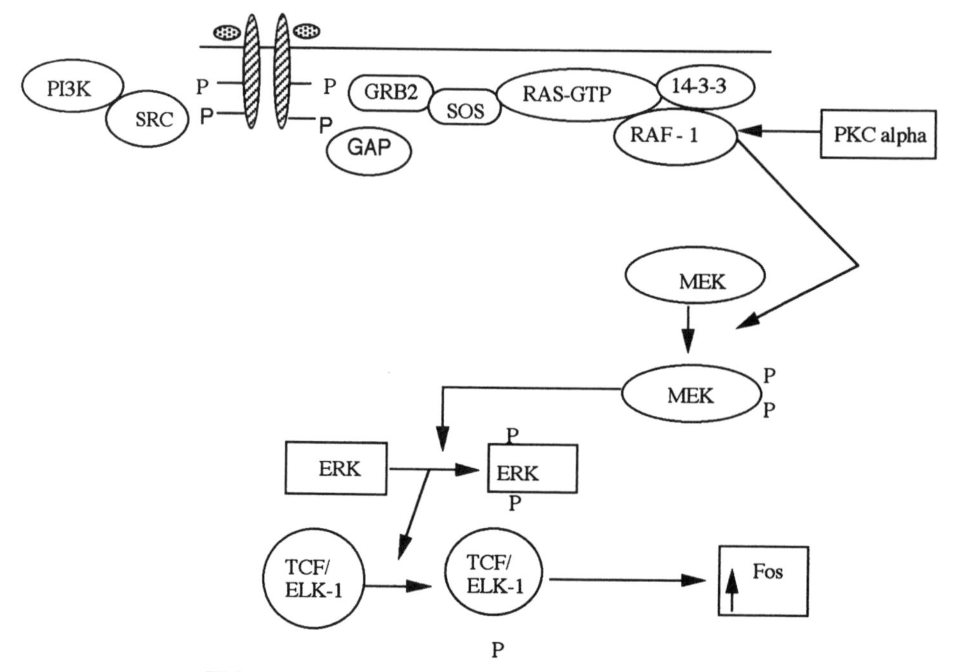

FIGURE 15.13 Receptor tyrosine kinase signaling.

to SH2 domains, Grb2 also contain an SH3 domain that recognizes proline rich motifs in signaling proteins. Grb2 is bound to the ras nucleotide exchange factor, Son of sevenless (SOS) protein, which allows SOS to interact with ras and catalyze the exchange GDP for GTP on the ras protein. When ras is in the GTP bound state it is activated and can stimulate downstream effector molecules such as the cytoplasmic serine/threonine kinase raf. Raf transmits its signal through protein phosphorylation to activate a cascade of kinases that ultimately result in the phosphorylation and activation of a DNA binding protein/transcription factor. The activated transcription factors regulate the expression of specific genes whose protein products produce the evoked cellular response. As mentioned, several proteins dock on the activated receptor. When PLCγ docks on the receptor it becomes activated and hydrolyzes membrane phospholipids such as phosphatidylinositol to produce two second messengers, sn-1,2-diacylglycerol (DAG) and inositol triphosphate. DAG binds to and activates the internal effector, in this case a family of enzymes called protein kinase C (PKC). The catalytically active PKC then phosphorylates specific serine/threonine residues of cellular target proteins that are involved directly or indirectly in producing the biological response. The other second messenger, inositol triphosphate can cause the release of intracellular calcium from the endoplasmic reticulum. Calcium can stimulate calcium-dependent enzymes and indirectly alter gene expression and/or cellular functions. Both GAPp120 and PI3K can bind to the ras protein and will be discussed below in the ras section.

Thus, growth factor receptor tyrosine kinase initiates a complex cascade of signaling events and several of these receptors can be activated to oncogenes by mutation, gene rearrangements/translocations and amplification. The v-*erb-B* encodes a truncated version of the epidermal growth factor receptor (EGFR). The v-*erb-B* oncogene is missing most of the amino terminal portion of the protein that is responsible for binding epidermal growth factor (EGF). In addition, the C-terminal portion of the protein that contains the tyrosine kinase activity has several mutations giving the protein the ability to constantly signal the cell to divide in the absence of EGF. In human glioblastomas and squamous cell carcinomas, the c-*erbB* or *EGFR* gene is frequently amplified and overexpressed. The neuroregulin receptor or the *Neu*/*HER2* oncogene is also a receptor tyrosine kinase and a single point mutation in the transmembrane domain is sufficient to produce a constitutively active receptor in the absence of ligand. Platelet-derived growth factor receptor (*PDGFR*), nerve growth factor receptor (*trk/NGFR*), hepatocyte growth factor receptor (*met/HGFR*), and the glial cell-dependent nerve growth factor receptor (*ret/GDNFR*) are additional examples of receptors that are activated by gene rearrangements/translocations to produce constitutively active oncogenic receptors. In addition, growth factor receptors such as *HGFR* and *HER2/Neu* are frequently amplified and overexpressed in stomach and breast cancer, respectively.

15.5.5c Nonreceptor Tyrosine Kinases

The src family of nonreceptor tyrosine kinases (src, fes, fgr, fps, lck and yes) contain both SH2 and SH3 domains and are associated with the inner surface of the plasma membrane. Many of these nonreceptor tyrosine kinases are rapidly activated by a variety of transmembrane signaling receptors. The src protein is myristylated on the amino terminus and is associated with the inner surface of the plasma membrane. Src kinase activity is elevated in some human colon cancers. As mentioned above the c-*abl* proto-oncogene is a cytoplasmic/nuclear tyrosine kinase. Following a chromosomal translocation c-abl forms a hybrid gene with the *bcr* gene to produce a protein with increased tyrosine kinase activity and growth promoting properties. This hybrid protein is present in 95% of the cases of human chronic myelogenous leukemia as previously mentioned.

15.5.5d Guanine Nucleotide Binding Proteins

Ras genes are frequently altered in chemically induced animal tumors and are the most frequently detected activated oncogenes in human tumors. Approximately 25–30% of all human tumors contain an activated *ras*. In humans, members of the Ras family are mutated in particularly high frequency in tumors of the pancreas (>90%), colon (~50%), and some types of lung tumors. A subset of acute nonlymphocytic leukemias and thyroid and skin cancers also contain activated ras. The *ras* genes are part of a larger superfamily of small molecular weight (20–25 kDa) guanine nucleotide binding proteins known as the Ras-related superfamily of proteins. The superfamily contains 60 members and can be divided into several subfamilies, which include Ras, Rho, Rab, Ran, Rad/Gem, and Arf. The Ras subfamily includes H-ras, K-ras (A and B generated from alternative splicing), N-ras, four Rap protein, three Ral proteins, three R-ras proteins, and the newly identified Rheb and M-ras proteins. Although several members of the subfamily have the ability to transform cells in culture, only H-*ras*, K-*ras*, and N-*ras* have been found to be frequently mutationally activated in numerous tumors from a large variety of species including humans. *TC21* is a member of the R-*ras* family and also has been found to be mutationally activated in some human tumors.

H-*ras*, K-*ras*, and N-*ras* code for highly related proteins of 188–189 amino acid residues, generically known as p21 (21 kDa). These *ras* genes are present in all eukaryotic organisms from yeast to humans. Mammalian *ras* can rescue ras deficient yeast cells and yeast RAS can transform NIH3T3 cells. The fact that *ras* genes are highly conserved in evolution suggests that the protein product they encode plays an essential role in normal cellular physiology. Mutations producing changes in specific amino acids as well as the overexpression of the normal protein alters normal cellular proliferation and differentiation and, in many cases, is associated with the neoplastic process.

Activated *ras* oncogenes have been detected in a large number of animal tumors induced by diverse agents including physical agents, such as radiation and a large number of chemical carcinogens (Table 15.8). Some chemical carcinogens bind covalently to DNA, forming specific adducts that upon DNA replication yield char-

TABLE 15.8 Activation of ras Oncogenes in Carcinogen-Induced Animal Tumors.

Species	Tumor	Carcinogen	Oncogene	Incidence (%)
Rat	Mammary carcinoma	NMU	H-ras	86
		DMBA	H-ras	23
	Kidney mesenchymal	NMU	K-ras	60
		DMN	K-ras	43
	Lung carcinoma	TNM	K-ras	74
	Hepatocellular carcinoma	AFB	K-ras	25
	Skin carcinoma	X-rays	K-ras	50
Mouse	Skin carcinoma	DMBA	H-ras	90
		DBACR	H-ras	80
	Mammary carcinoma	DMBA	H-ras	75
	Lymphoma	X-rays	K-ras, N-ras	57
		NMU	K-ras, N-ras	85
	Thymic lymphoma	MCA	K-ras	83
	Fibrosarcoma	MCA	K-ras	50
	Lung carcinoma	TNM	K-ras	100
	Hepatocellular carcinoma	HOAAF	H-ras	100
		EC	H-ras	100
		VC	H-ras	100
		HOAB	H-ras	100
		Vinthionine	H-ras	100
		Safrole	H-ras	100
		Estragol	H-ras	100
		Furfural	H-ras	85
		AFB	H-ras, K-ras	85
		B(a)P	K-ras, H-ras	80
Rabbit	Keratoacanthoma	DMBA	H-ras	60

NMU, nitrosomethylurea; DMBA, demethylbenz(a)anthracene; DMN, dimethylnitrosamine; TNM, tetranitromethane; AFB, aflatoxin B1; DBACR, dibenz(c,h)acridine; MCA, 3-methylcholanthrene; B(a)P, benzo(a)pyrene; HOAAF, N-hydroxy-2-acetylaminofluorene; HOAB, N-hydroxy-4-aminoazobenzene; VC, vinylcarbamate; EC, ethylcarbamate; MNNG, N-methyl-N′-nitro-N-nitrosoguanidine.

Source: Adapted from Klein, G. (Ed), *Cellular Oncogene Activation*. New York: Marcel Dekker Inc., 1988.

acteristic alterations in the primary sequence of the Ha-*ras* proto-oncogene. The study of the *ras* oncogene as a target for chemical carcinogens has revealed a correlation between specific carcinogen-DNA adducts and specific activating mutations of *ras* in chemically induced tumors (Table 15.9). For example 7,12-dimethylbenz[a]anthracene, a polycyclic aromatic hydrocarbon carcinogen, is metabolically activated to a bay region diol epoxide that binds preferentially to adenine residues in DNA. Skin tumors isolated from mice treated with DMBA contain an activated H-*ras* oncogene with an A to T transversion of the middle base in the 61st codon of H-*ras*. Therefore, the identified mutation in *ras* is consistent with the expected mutation based on the DMBA-DNA adducts that have been identified. Likewise, rat mammary carcinomas induced by nitrosomethylurea contain a G to A transition in the 12th codon of H-*ras* and this mutation is consistent with the modification of guanine residues by this carcinogen. Based on these events, the alteration of *ras* by specific chemical carcinogens appears to be an early event in carcinogenesis.

Ras proteins function as membrane associated molecular switches operating downstream of a variety of membrane receptors including receptor tyrosine kinases (PDGFR; EGFR) (Fig. 15.13), cytokine receptors without intrinsic tyrosine kinase activity (IL-12), T-cell receptors, and some receptors that are coupled to heterotrimeric G-proteins. Ras proteins are membrane-bound proteins and bind guanosine triphosphate (GTP) and guanosine diphosphate (GDP). Ras proteins are in the "off" position when they are bound to GDP and in the "on" position when they are bound to GTP. Ras protein also possess intrinsic GTPase activity. The non-stimulated inactive ras protein exists in the GDP-bound form; however, when stimulated by an upstream signal, GDP is then exchanged for GTP. Subsequently the ras protein undergoes a conformational change that allows it to interact with its downstream effector proteins (see below) in the cell. The effector proteins then directly or indirectly elicit the appropriate response. Once this information is trans-

TABLE 15.9 Specific Activating Mutations in ras in Chemically Induced Tumors.

Species	Tumor	Carcinogen	Expected Mutation	Activating Mutation	Incidence
Rat	Mammary carcinoma	NMU	G to A	G^{35} to A	61/61
		DMBA	A to N	A^{182}, A^{183} to N	5/5
Mouse	Skin papilloma or carcinoma	DMBA	A to N	A^{182} to T	33/37
		MNNG	G to A	G^{35} to A	4/4
Mouse	Hepatocarcinoma	None		C^{181} to A	15/25
				A^{182} to T	6/25
				A^{182} to G	3/25
		Ho-AAF	C to A	C^{181} to A	8/10
		HOAB	C to A	C^{182} to A	9/10
		VC, EC	?	A^{182} to T	13/16
		DMBA	A to N	A^{182} to T	10/10

For abbreviations, see Table 15.8.

Source: Adapted from Klein, G. (Ed), *Cellular Oncogene Activation*. New York: Marcel Dekker Inc., 1988.

duced, the ras protein hydrolyzes GTP to GDP through its intrinsic GTPase activity and with the aid of the GAPp120 protein (GTPase-activating protein) which binds to ras and increases the rate of GTP hydrolysis by ras by at least two to three orders of magnitude. The hydrolysis of GTP to GDP returns the protein to its inactive state or the "off" position. Ras is regulated by guanine (nucleotide) dissociation stimulators (GDS) also known as guanine (nucleotide) exchange factors (GEFs) such as the SOS protein, which promotes the formation of the active GTP bound state and GTPase-activating proteins (GAPs) such as GAPp120, which promote the formation of the inactive GDP bound state.

Generally, in vivo, point mutations in the 12th, 13th, 59th, 61st, or 63rd codon of *ras* produce cellular transformation and impair the intrinsic GTPase activity and also make GAP ineffective at increasing GTP hydrolysis. Mutations in the 15th, 16th, 116th, 119th, 144th, and 146th increase the rate of the exchange of GDP/GTP of the protein and therefore favor the active GTP form of the protein. By far the most common mutations in tumors are in the 12th, 13th, and 61st codon. The reduced GTPase activity or increased GDP-GTP exchange rate of the mutated ras allows for constitutively activated *ras* that is involved in cellular transformation. The net result being a ras protein that remains in the GTP "on" state. It is not understood if GAP's sole function is to increase the rate of GTP hydrolysis of ras or if it represents an effector molecule itself since it binds to the effector region of ras.

Downstream effectors of ras include several proteins that have been shown to bind to the effector domain (amino acid residues 26–35) of the ras GTP form including, raf, phosphatidylinositol-3-kinase (PI3K), GTPase-activating protein (GAPp120), neurofibromin (NF-1), Ral-GDS (guanine dissociation stimulator for ras related protein ral) and MEKK1. The best characterized of these effector pathways is the raf to MEK/Erk pathway (Fig. 15.13). As described in the growth factor receptor tyrosine kinase section, the binding of growth factor causes dimerization of the receptor, followed by the activation of the intrinsic tyrosine kinase activity and the autophosphorylation of tyrosine residues within the intracellular domain of the receptor. SH2-containing proteins such as the adapter protein Grb2 binds to certain phosphorylated tyrosine residues of the receptor via SH2-binding domains. Grb2 is associated via its SH3 domain with a proline rich motif of SOS protein. This association brings SOS in physical contact with membrane-bound ras and SOS catalyzes the release of GDP from ras and the binding of GTP. GTP-bound ras binds cytoplasmic raf-1, and translocates it to the membrane where it is activated by a poorly understood mechanism. Activated Raf-1, a serine threonine kinase phosphorylates MAP kinase-extracellular signal regulated kinase (MEK1), a dual-specificity kinase (phosphorylates threonine and tyrosine residues) which in turn phosphorylates threonine and tyrosine residues on extracellular regulated kinase Erk1 (p44) and Erk2 (p42). Erks are serine/threonine kinases that have numerous cellular substrates including cPLA2 and p90 ribosomal S6 kinase. Erk1 and Erk2 also translocate into the nucleus where they phosphorylate the transcription factor, ternary complex factor (TCF/Elk-1), which is associated with another transcription factor, serum response factor (SRF). This protein complex is bound to the serum response element (SRE) in the promoter region of genes such as fos. Phosphorylation of TCF/Elk-1 stimulates it transactivation functions and leads to the transcription of c-fos.

Although this ras/raf/MEK/Erk pathway is one of the best-characterized pathways of ras function and has been shown to be important in ras transformation and numerous ras dependent cell processes, it should be noted that other ras effectors (see above) and other ras downstream pathways are important in the cell. In addition, ras influences cell cycle control and has been shown to up-regulate cyclin A, D1, D3, E, and the E2F family of transcription factors. The induction of cyclin D1 by ras (likely mediated by raf/MEK pathway) is associated with the phosphorylation of retinoblastoma protein (Rb) by CDK4/6, and the shortening of the G1 phase of the cell cycle. The major effects of ras on the cell cycle are likely mediated through cyclin D1-mediated phosphorylation of Rb. Cyclin D is overexpressed in numerous tumors and under these circumstances is considered an oncogene.

15.5.5e Cytoplasmic Serine/Threonine Kinases

The v-*mos* gene is the gene that is responsible for the sarcomas induced by the Moloney MSV retrovirus. It is a cytosolic serine/threonine kinase. The expression of c-mos is normally confined to germ cells, however, the forced expression of c-mos in fibroblasts results in chromosome fragmentation. c-mos has been reported to phosphorylate MEKK. *v-Raf* is the gene responsible for the sarcomas induced by the murine sarcoma virus 3611. The v-raf protein has the first 250–300 amino acids of the c-raf-1 deleted, removing the negative regulatory domain of the protein resulting in a constitutively active enzyme. As described above raf-1 is an important effector of ras function and raf-1 can be activated by protein kinase Cα. Oncogenic *raf-1* has been detected by transfection of tumor DNA from several different types of sarcomas and carcinomas.

15.5.5f DNA-Binding Proteins

The *v-fos* gene encodes a protein that has DNA-binding characteristics. Fos is a phosphoprotein that forms a heterodimer with jun and together they form the transcription factor known as AP-1. AP-1 binds to the AP-1 consensus sequence (also referred to as the TPA responsive element (TRE)) and can increase the expression of a gene with such an enhancer sequence. Both *v-fos* and *v-jun* have mutations that result in the loss of the negative regulatory elements and produce a constitutively active protein. Fos is overexpressed in the majority of human osteosarcomas while jun is overexpressed in some lung cancers. Another transcription factor, v-rel, a member of the NF-kB transcription factor family, has lost its positive effector regions and functions in a dominant negative manner to suppress the expression of certain genes required for differentiation. In Burkitt's lymphoma the c-*myc* gene is translocated and inserted near an immunoglobulin locus where it is now under the regulation of the immunoglobulin promoter and constitutively expressed at high levels (Fig. 15.12). In addition and as described earlier, various members of the myc family are amplified in variety of human and rodent tumors.

15.5.6 Oncogene Cooperation

A single oncogene such as *H-ras* can transform immortalized cells such as NIH-3T3 fibroblasts, however, this is not generally the case for primary cells. Primary cells are not immortal, but undergo a limited number of cell divisions before they become senescent. It has been shown that transfection of a *myc* oncogene into rat primary embryo cells confers an immortal-like state; however, these cells are not tumorigenic when injected into nude mice. Similarly, transfection of a *ras* oncogene into rat primary embryo cells does not produce tumorigenic cells. If both *myc* and *ras* are introduced into these cells they become tumorigenic. Experiments such as these have demonstrated that certain oncogenes can cooperate with one another to transform primary cells. In fact, any of the following oncogenes—*H-ras*, *K-ras*, *N-ras*, *src*, and *polyoma middle T*—can cooperate with the any following oncogenes—*myc*, *N-myc*, *l-myc*, *adeno E1A*, *polyoma large T*, *SV40 large T*, and papillomavirus *E7*—to transform rat embryo fibroblasts. Additional studies have shown that other oncogenes such as *fos* and *ras* or *jun* and *ras* can cooperate to transform primary mouse keratinocytes. Oncogene cooperation has also been demonstrated in vivo in doubly transgenic mice. For example, mice doubly transgenic for *myc* and *ras* under the regulation of the mouse mammary tumor virus promoter display a synergistic increase in the incidence of mammary carcinomas compared with *ras* or *myc* single transgenic mice. Synergistic responses have also been demonstrated in oncogene transgenic mice treated with carcinogens suggesting oncogene cooperation between the carcinogen-mutated gene and oncogenic transgene. Although rodent cells are efficiently transformed by cooperating oncogenes, similar experiments with human cells have not demonstrated transformation. Recently, the ectopic expression of the catalytic subunit of telomerase in combination with the SV40 large-T antigen and mutant H-ras in both normal human epithelial cells and fibroblast cells resulted in transformation and tumorigenic conversion. Telomerase is expressed in germ cells and many different cancer cells. Telomerase prevents cellular senescence by not allowing shortening of the chromosome telomeres. Thus, human cells can be tranformed by three distinct genetic elements. The fact that it takes multiple oncogenes belonging to different classes to transform cells lends support to the multihit nature of carcinogenesis observed in vivo and to the notion that more than one cellular circuit must be altered for transformation.

15.5.7 Detection of Oncogenes

The ability to detect transforming oncogenes in tumors and tumorigenic cell lines is essential to our understanding of chemical carcinogenesis. Several methods are available and each has its limitations and usefulness. A brief discussion of some commonly used techniques follows. Several of these techniques are described in more detail in Chapter 2. If the reader is further interested, these techniques are described in detail in most laboratory manuals of molecular biology.

15.5.7a Northern, Southern, and Western Blot Analysis

Northern, Southern, and Western blot analysis can be used alone or in combination with a variety of techniques to detect quantitative and qualitative changes in oncogenes. Using Northern blot analysis and ^{32}P-labeled gene-specific probes it can be determined if a particular oncogene is overexpressed at the mRNA level. Using Southern blot analysis and ^{32}P-labeled specific probes, cellular DNA can be examined to determine whether a given oncogene is amplified. Using Western blot analysis and specific antibodies to oncogene protein products, the presence and level of these proteins can be determined. The activated or oncogenic ras p21 protein contains an amino acid substitution due to a specific mutation in the ras gene resulting in a different electrophoretic mobility in polyacrylamide gels relative to the normal protein. Using antibodies to p21 and Western blot analysis, the altered mobility of a mutant *ras* protein can be detected.

15.5.7b Restriction Fragment Length Polymorphism

A mutation in an oncogene can create a new restriction site in the gene. A restriction site is a specific palindromic sequence of nucleotides in the DNA that is recognized by a restriction endonuclease that cut the DNA at this site, releasing specific size fragments from the DNA. Using restriction fragment length polymorphism (RFLP) and oligonucleotide probing, it has been shown that 90% of DMBA-initiated, TPA-promoted skin tumors contain an A $\rightarrow$ T transversion in the 61st codon of H-*ras*. The A $\rightarrow$ T mutation results in the creation of a new restriction site for the restriction endonuclease XbaI. This polymorphism can be detected in XbaI-digested DNA using Southern blot analysis, which reveals two additional bands (4 and 8 kb) after hybridization with a ^{32}P-labeled H-*ras* probe in the DMBA-induced tumors but not the normal tissue.

15.5.7c Polymerase Chain Reaction and Oligonucleotide Hybridization

The polymerase chain reaction (PCR) can be used to amplify certain segments of a given gene. However, the sequence of the gene of interest must be known so that primers can be constructed that will provide starting and stopping points for DNA synthesis. Using PCR the segment of the gene of interest can be amplified thousands of times, greatly increasing the sensitivity to detect a mutation in the sample DNA. For example, using the appropriate primers, the exon containing the 61st codon of the H-ras oncogene could be amplified to determine if there is a mutation in the 61st codon in a sample of tumor DNA. Subsequently, this DNA could be slot blotted and probed with ^{32}P-labeled oligonucleotide probes (20 nucleotides in length) that contain the normal unmutated codon 61 or a mismatched oligomer that contains an A $\rightarrow$ T transversion in the 61st codon. If the mutation does exist in the amplified DNA from the tumor the mismatched oligonucleotide will hybridize to the tumor DNA but not to the normal DNA. Alternatively, the PCR product can be analyzed by direct DNA sequencing or by single-strand conformational polymorphism (SSCP) analysis (see Chapter 2). Mutant allele specific mutation amplification (MASA) is another technique to detect mutated oncogenes. MASA employs PCR primers that will only bind to the mutated region of the gene, selectively amplifying the mutant allele.

15.5.7d Gene Transfer—NIH 3T3 Transformation Assay

NIH 3T3 fibroblasts are an immortalized nontumorigenic mouse embryo fibroblast cell line. Under appropriate conditions, these cells will readily take up exogenous DNA. When NIH 3T3 cells take up high molecular weight DNA from tumors containing certain transforming oncogenes, the cells become morphologically transformed, pile-up, and form foci on the background of a normal monolayer of cells. Cells isolated from these foci can grow in soft agar (anchorage independent growth), which is a characteristic of transformed cells. When these transformed cells are injected into a nude mouse, they are tumorigenic. NIH 3T3 cells are efficiently transformed by ras genes that contain transforming mutations as well as by several other transforming oncogenes. If carcinogenesis is a multistep process, it may be asked how a single oncogene can transform these cells? It is generally thought that transformation of NIH 3T3 cells by a single oncogene is effectively accomplished because these cells are immortal prior to transformation and therefore not normal. While the NIH 3T3 assay is a powerful tool, it has some limitations in that many oncogenes are not detected because they do not cause NIH 3T3 cells to form foci.

15.5.7e Nude Mouse Assay

The nude mouse assay is used to determine the oncogenic potential of NIH 3T3 mouse fibroblasts that have been transfected with tumor DNA. This assay uses athymic, immunodeficient mice to detect transforming sequences of DNA. Tumor DNA is isolated and co-transfected with a selectable marker such as neomycin resistance into NIH 3T3 cells. Cells that have taken up the DNA are selected for on a medium containing neomycin. These cells will form colonies and after approximately 2 weeks the colonies are harvested and subcutaneously injected into nude mice. The tumors that develop are associated with dominant transforming genes and can be analyzed to characterize the oncogene and its activating mutation. The advantage of the nude mouse assay is that it does not rely on morphological transformation leading to focus formation and thus has the potential to identify oncogenes that are not detected in the NIH 3T3 focus assay.

15.6 TUMOR SUPPRESSOR GENES

Activation of oncogenes results in a gain of function whereas inactivation of tumor suppressor genes results in a loss of function. Tumor suppressor genes have also been termed anti-oncogenes, recessive oncogenes, and growth suppressor genes. Tumor suppressor genes encode proteins that generally function as negative regulators of cell growth or regulators of cell death. In addition, some tumor suppressor genes function in DNA repair and cell adhesion. The majority of tumor suppressor genes were first identified in rare familial cancer syndromes and some are frequently mutated in sporadic cancers through somatic mutation. Known tumor suppressor genes, their proposed function and the cancer syndrome they are associated with are shown in Table 15.10. When tumor suppressor genes are inactivated by allelic loss, point mutation, or chromosome deletion they are no longer capable of negatively regulating cellular growth leading to specific forms of cancer predisposition. Generally, if one copy or allele of the tumor suppressor gene is inactivated, the cell is normal and if both copies or alleles are inactivated, loss of growth control occurs. In some cases, a single mutant allele of certain tumor suppressor genes such as p53 can give rise to an altered intermediate phenotype, however, inactivation of both alleles is required for full loss of function and the transformed phenotype.

The concept that cancer involves loss of gene function and involves a recessive trait came from somatic cell fusion experiments conducted by H. Harris and colleagues in the late 1960s. When tumor cells were fused with normal cells some of the resulting hybrids were nontumorigenic. If these nontumorigenic hybrids were cultured for extended periods, some of hybrids reverted to the tumorigenic phenotype that was accompanied by chromosome losses. These experiments suggested that the normal cell was contributing genes to the tumor cell that imposed normal

TABLE 15.10 Tumor Suppressor Genes Involved in Human Cancers.

Cancer Syndrome	Gene	Locus	Location/Function	Neoplasms
Li-Fraumeni	p53	17p13.1	Nuclear/transcription factor	Most human cancers
Retinoblastoma	RB1	13q14	Nuclear/transcription modifier	Retinoblastoma, osteosarcoma carcinomas of breast, prostate, bladder, and lung
Familial adenomatous polyposis	APC	5q21	Cytoplasmic/may mediate cell cycle progression and adhesion, binds α and β catenin	Carcinoma of colon, stomach, and pancreas
Wilms' tumor	WT-1	11p13	Nuclear/transcription factor	Wilms' tumor
Neurofibromatosis type 1	NF-1	17q11	Cytoplasmic/p21ras-GTPase activating protein	Schwannomas
Neurofibromatosis type 2	NF-2	22q	Cytoplasmic/cytoskeletal membrane linkage	Schwannomas and meningiomas
Familial melanoma and pancreatic cancer	$P16^{INK4}$	9p21	Nuclear/cyclin dependent kinase inhibitor	Mesothelioma, pancreas, melanoma, glioblastoma
Familial breast cancer	BRCA1	17q21	Nuclear/unknown	Breast cancer
	BRCA2	13q12–13		Breast and ovarian carcinomas
Tuberous sclerosis	TSC2	16p13.3	?Golgi/p21rap-GTPase activator	Renal and brain tumors
Hereditary non-polyposis colon cancer	MSH2	2p22–21	Nuclear/DNA mismatch repair	Colon cancer
Hereditary non-polyposis colon cancer	MLH1	3p21.3–p23	Nuclear/DNA mismatch repair	Colon cancer
von Hippel-Lindau disease	VHL	3p25	Nuclear/regulates transcriptional elongation	Renal cell pheochromatomas, hemangiomas
Nevoid basal cell carcinoma	PTC	9q22.3	Transmembrane/transcription repressor	Basal cell carcinomas, fibromas of ovaries and heart, medulloblastomas, meningiomas
No known associated cancer syndrome				
	DPC4	18q21.1	Cytoplasm/TGF-β signaling	Pancreatic colon
	E-cadherin	16q22.1	Nuclear/adhesion, signaling	Gastric, breast, endometrial, ovarian
	α-catenin		Cytoplasm/cytoskeletal linker protein	Colorectal, brain, neuroblastoma
	DDC	18q21	Transmembrane/receptor?	Colorectal, brain, neuroblastoma
	TGFBR2	3p22	Transmembrane/receptor for TGF-β signaling	Colorectal, hepatomas, head and neck tumors

growth restraints on the latter and that cancer is a recessive trait. Subsequently, it was determined that these tumor cells contained specific chromosome deletions and such deletions were "corrected" in the normal-tumor cell hybrids because the previously deleted chromosome was now contributed by the normal cell. Furthermore, by using microcell-mediated chromosome transfer, which allows for the selective introduction of a specific chromosome or a portion of a chromosome into these tumor cells, it could be determined which chromosome or portion of a chromosome suppresses the tumorigenic phenotype and restores normal growth control. Because the transferred "genes" suppressed growth and/or tumorigenicity, these genes were collectively called tumor suppressor genes.

Additional early evidence for tumor suppressor genes in humans came from the studies of Knudson, who postulated that two mutational events or "hits" are necessary for the development of retinoblastoma, a rare tumor of the retina that occurs in 1 in 20,000 infants and children. Approximately 60% of the retinoblastomas are sporadic and 40% are familial (autosomal dominant inheritance). Knudson observed that the majority of familial cases developed bilaterally (both eyes were generally affected) and earlier than the sporadic form which developed unilaterally. To explain the familial and sporadic occurrence, Knudson proposed a model in which two hits were necessary for retinoblastoma to develop. He proposed in the familial cases one hit was inherited in the germline and was present in all somatic cells and the other hit occurred as a somatic mutation in the retinal lineage. For the sporadic form, he proposed that two somatic hits must occur within the same cell in the retinal lineage. Knudson's hypothesis was eventually substantiated by molecular and cytogenetic studies that led to identification of the retinoblastoma gene (Rb) and the discovery that both copies of the gene (two mutational events) are inactivated and/or deleted in retinoblastoma tumors. Thus, a child that carries an inherited mutant Rb allele on chromosome 13 in all somatic cells is normal except for increased risk of developing cancer. This child is heterozygous at the *Rb* locus (one mutant and one normal allele), however, if the normal allele becomes mutant, there is a loss of heterozygosity (LOH) and the cell in which this occurs is now homozygous for the mutant allele and predisposed to cancer development. LOH is an important molecular approach in the identification of tumor suppressor genes. LOH can result from (i) chromosome nondisjunction and reduplication of the chromosome containing the mutant allele (ii) mitotic recombination, and (iii) nondisjunction. Linkage analysis and cytogenetics are additional important approaches used to identify tumor suppressor genes.

15.6.1 Retinoblastoma Gene and Protein

The retinoblastoma (*RB1*) gene is a large gene of 190 kb. *RB1* gene has been shown to be inactivated by deletion and nonsense point mutations. This gene is inactivated in all retinoblastomas (see above) and is inactivated in a high proportion of small-cell lung carcinomas, 20–30% of non-small-cell lung and bladder carcinomas, and in a small proportion of mammary carcinomas. Individuals with an inherited *RB1* mutant allele are also susceptible to develop osteosarcomas and soft tissue sarcomas. The protein product of the *RB1* is a 105-kDa differentially phosphorylated nuclear protein (pRb) that binds nonspecifically to DNA in vitro. pRb via the "Rb pocket" binds to numerous cellular proteins (>20 proteins), many of which are tran-

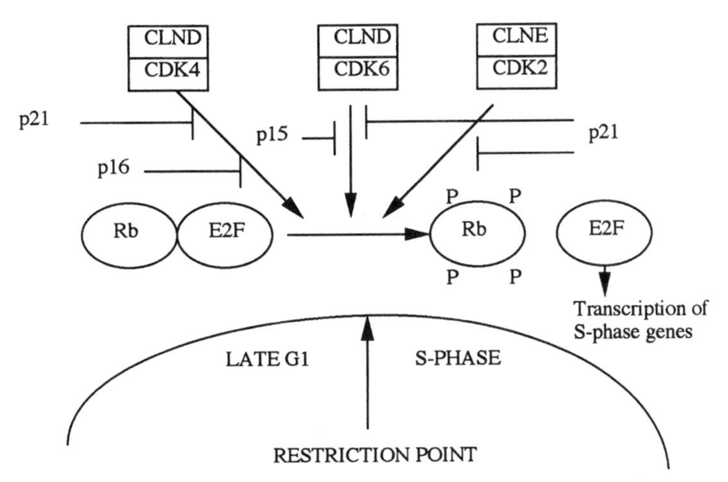

FIGURE 15.14 Retinoblastoma gene and the cell cycle.

scription factors. One of the better functionally characterized pRb binding proteins is the family of transcription factors collectively known as E2F. E2F is involved in the transcriptional regulation of S-phase genes, such as dihyrofolate reductase, thymidine kinase, EGFR, myc, myb, CDC2, and E2F1 itself. pRb is a regulator of the cell cycle clock through its modulation of transcription of genes important in cell proliferation. In G1/G0 of the cell cycle, pRb is hypophosphorylated and in S, G2, and M phases of the cell cycle pRb is hyperphosphorylated. A model has been proposed in which the hypophosphorylated form of pRb is bound to several proteins including E2F in G0/G1 phase of the cell cycle (Fig. 15.14). Under these conditions, hypophosphorylated pRb is thought to suppress cell growth and functions as a repressor of the trans-activation activity of E2F. Upon receiving mitogenic stimuli, the pRb protein is phosphorylated by cyclin D/cyclin dependent kinase 4 (CDK4), cyclin D/CDK6 or cyclin E/CDK2. Hyperphosphorylated pRb results in the inactivation of pRb and the cell progresses past the restriction point in G1 and into S-phase. Hyperphosphorylated pRb releases E2F allowing it to become transcriptionally active and critical genes necessary for S-phase are transcribed. pRb is dephosphorylated at the termination of M-phase and once again forms a complex with E2F. An alternative model for pRb cell cycle regulation proposes that pRb/E2F complex is a trans-repressor and that the phosphorylation of RB blocks the repressor function without releasing E2F. Regardless of the model, it is clear that pRb is a key protein in the regulation of the cell cycle. As mentioned, pRB binds numerous cellular proteins and although these proteins are less studied and understood than E2F, some of these may be important downstream effectors of pRb.

When both alleles of Rb are inactivated by mutation or deletion, pRb is unable to negatively regulate cell cycle progression. The cell is also compromised in its ability to respond to negative growth regulatory signals such as TGFβ and contact inhibition. The phosphorylation of Rb by cyclin D/CDK4, cyclin D/CDK6 can be blocked by the cyclin-dependent kinase inhibitors (CDI) p16 (INK4A), p15 (INK4B), or p21 (WAF1/Cip 1), which negatively regulate cell cycle progression.

p21 can also block the activity of cyclin E/CDK2. TGFβ, cell contact, and cAMP can induce certain of these CDIs and block pRB phosphorylation. p16 is a tumor suppressor gene that is mutated in many different cancers. As mentioned in Section 15.5.2, DNA viruses do not possess the genes that encode the enzymes necessary for DNA synthesis and are therefore dependent upon the host cell's DNA replicative enzymes for viral replication. Because cells that are in the G1 phase of the cell cycle do not synthesize the enzymes needed for DNA synthesis, DNA viruses have the ability to override the block in G1 and cause cellular proliferation. Transforming proteins encoded by DNA tumor viruses such as SV40, large T antigen, papilloma virus E7, and adenovirus type 5 E1A bind specifically to the hypophosphorylated form of pRb and block its growth-suppressive activity and allow for S-phase entry. The binding of these proteins to hypophosphorylated pRb is a critical event in the transforming activity of SV40, adenovirus, and papilloma virus. Thus the pRb circuit can be altered by mutation or deletion of *RB1*, inactivation of the CDI, p16, loss of TGFβ function, and by pRb's interaction with transforming proteins of DNA oncogenic viruses.

15.6.2 *p53* Gene and Protein

p53 is a 12.5 kb gene that encodes a 53 kDa protein. *p53* is mutated in 50% of all human cancer and is the most frequently known mutated gene in human cancer. The majority (~80%) of *p53* mutations are missense mutations and *p53* is mutated in approximately 70% of colon cancers, 50% of breast and lung cancers, and 97% of primary melanomas. In addition to point mutations, allelic loss, rearrangements, and deletions of *p53* occur in human tumors. *p53* is a transcription factor and participates in many cellular functions including cell cycle regulation, DNA repair, and apoptosis. The *p53* protein is composed of 393 amino acids and single missense mutations can inactivate the *p53*. Unlike ras genes, which have a few mutational codons that result in its activation, the *p53* protein can be inactivated by hundreds of different single point mutations in *p53*. Ninety-eight percent of *p53* mutations occur within codons 110–307, almost all of the mutations in *p53* occur in the DNA-binding domain (amino acids 100–293) with mutational hot spots occurring at codons 155, 245, 248, 249, 273, and 282. It has been proposed that the mutation spectrum of *p53* in human cancer can aid in the identification of the specific carcinogen that is responsible for the genetic damage, that is to say that different carcinogens cause different characteristic mutations in *p53*. Some of the mutations in *p53* reflect endogenous oxidative damage, whereas others such as the mutational spectrum in *p53* in hepatocellular carcinomas from individuals exposed to aflatoxin demonstrate a mutation spectrum characteristic aflatoxin. In sun-exposed areas where skin tumors develop, the mutations found in *p53* in these tumors are characteristic of UV light induced pyrimidine dimers and finally the mutation spectrum induced by (+)-benzo[a]pyrene-7,8-diol-9,10-epoxide-2 in cells in culture is similar to the mutational spectrum in *p53* in lung tumors from cigarette smokers. Thus, certain carcinogens produce a molecular signature that may provide important information in understanding the etiology of tumor development.

The *p53* protein functions as homotetramer and is involved in the transactivation of numerous genes including p21, mdm2, Gadd45, bax, and cyclin G1 (Fig. 15.15). *p53* also represses the expression of myc and bcl-2. In addition, *p53* also pos-

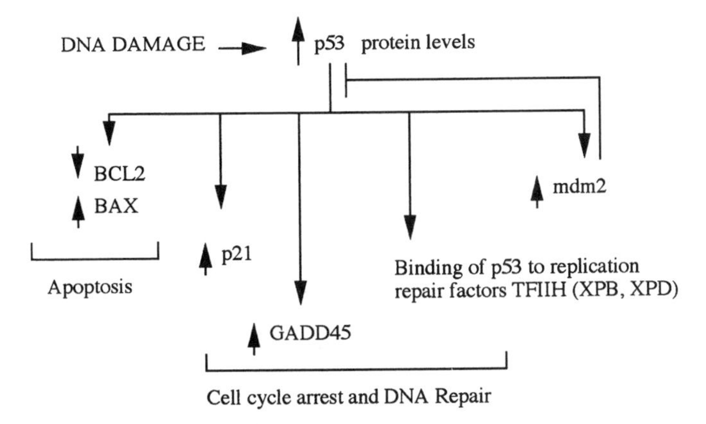

FIGURE 15.15 *p53* Function.

sesses 3′ to 5′ endonuclease activity and binds to variety of cellular proteins. DNA damage indirectly through an unknown mechanism, results in the phosphorylation of *p53*, which increases in short half-life of 15 minutes to 3–4 hours and activates its transcription function. Activated *p53* increases the transcription of numerous genes including p21, a CDI that inhibits CDK activity and blocks the phosphorylation of pRB, thus preventing the cell from entering S-phase. If a cell with DNA damage were to enter S-phase, mutations could result in the daughter cell, which could contribute to neoplastic transformation. In addition to p21, *p53* increases the expression of Gadd45 (growth arrest on DNA damage), which is thought to stimulate excision repair and form a complex with PCNA (proliferating cell nuclear antigen). *p53* also binds to XPB and XPD DNA helicase proteins in the TFIIH complex to modulate nucleotide excision repair. *p53* can interact with the human RAD51 protein, an important factor in homologous recombination and recombinatorial repair. Thus, *p53* appears to be involved in several DNA repair pathways. *p53* also increases the expression of mdm2 (murine double minute 2). Mdm2 binds to the N-terminal region of *p53* and inhibits *p53* trans-activation and also targets the *p53* protein for degradation through a ubiquitin-dependent pathway. Thus, mdm2 is thought to be a feedback regulator of the *p53*.

The accumulation and activation of *p53* inhibits the cell cycle until the DNA damaged is repaired, however, if the DNA damage is sustained and severe, *p53* can activate apoptosis. *p53* increases the expression of bax, a protein that is a promoter of cell death and decreases the expression of bcl-2, which is an inhibitor of cell death. In addition to *p53*'s role in cell cycle regulation and DNA repair, the *p53* protein is a key regulator of apoptosis. As discussed in the previous section phosphorylation of Rb and the release of the transcription factor E2F is an important event in G1 to S phase transition. Mutant ras can result in the overexpression of cyclin D and subsequent phosphorylation pRb and release of E2F, resulting in a proliferative response. If abnormally high levels of E2F are released, E2F can increase the expression of p19 (ARF), which inactivates mdm2 protein resulting in the accumulation of *p53* and the activation of an apoptotic response. Thus, the cell has a protective pathway to eliminate cells that possess abnormally elevated signaling through a ras/cyclin D/Rb pathway by integrating these signals from the pRb pathway to the *p53*. It is generally believed that circuitry involving both pRb and *p53* must be altered to produce a tumor. As discussed throughout this chapter, these alterations can occur at many different points within each circuit. Thus, *p53* has been termed the "guardian of genome" because it controls a G1 checkpoint and regulates DNA repair and apoptosis. Mutation of *p53* disrupts these functions leading to the accumulation of mutations (mutator phenotype; genetic instability) and further development of malignant clones.

DNA oncogenic viral proteins such as SV40 large T, adenovirus E1A and E1B, human papillomavirus 16 E6 bind to the wild-type *p53* but not mutant *p53* and inhibit the transactivation function of *p53*. Again we see a convergence between the action of DNA oncogenic viruses and tumor suppressor gene function inactivation.

15.6.3 Role of Tumor Suppressor Genes and Oncogenes in Human and Rodent Carcinogenesis

As mentioned earlier, epidemiology studies have revealed that the incidence of cancer increases exponentially with age. From these studies it has been suggested that three to seven mutations or "hits" are necessary for cancer development. Most cancers are monoclonal (derived from a single cell) in origin and do not arise from a single mutation but from the accumulation of sequential mutations in a single cell. Each additional mutation provides a further growth advantage for the developing tumor. These mutations can be the result of imperfect DNA replication, oxidative DNA damage or DNA damage caused by environmental carcinogens. As shown in Figure 15.16, the development of human colon cancer involves the mutational activation of oncogenes and the specific deletion and inactivation of tumor suppressor genes. The development and progression of a benign adenoma to a malignant carcinoma involves the inactivation of APC (*a*denomatosis *p*olyposis *c*oli), DCC (*d*eleted in *c*olorectal *c*ancers), and *p53* tumor suppressor genes and the activation of ras oncogene. The APC gene interacts with β-catenin and mutation in APC eliminates its ability to down regulate β-catenin, which in turn binds to and activates Tcf/Lef transcription factors that produce a mitogenic response. β-catenin itself has been found to be mutated in some tumors and these mutations make β-catenin resistant to the effects of APC, thus mutated β-catenin is thought to be an oncogene. In addition, the hypomethylation of DNA, presumably leading to alterations in gene expression, appears to play a role in colonic tumor development. Not all

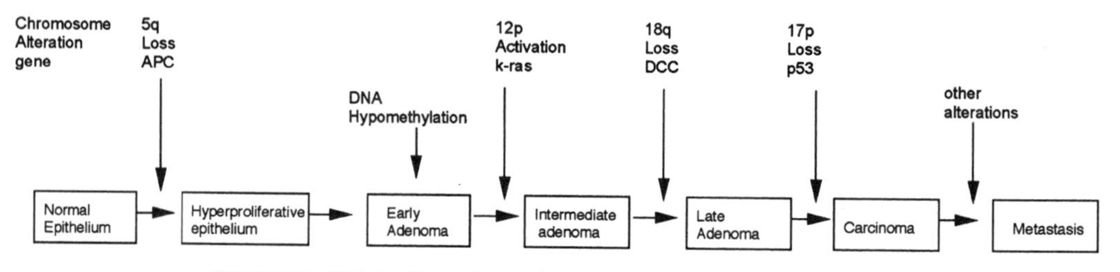

FIGURE 15.16 Genetic model for colorectal carcinogenesis.

Source: Adapted from Fearon ER. and Vogelstein B. A genetic model for colorectal tumorigenesis. *Cell* 61 (1990), 759–767.

tumors demonstrate all of the above changes and not all demonstrate the above sequence of events, however, what is critical is the accumulation of mutagenic events. In terms of experimental carcinogenesis, numerous molecular events involving inactivation of tumor suppressor genes and activation of oncogenes have been documented in neoplastic development in animal models of cancer including the mouse skin model of carcinogenesis (see Section 15.10.1a).

15.7 APOPTOSIS

Apoptosis or programmed cell death is the process by which cells actively undergo cell death characterized by loss of viability, chromosome condensation, and DNA fragmentation. Apoptosis is essential for normal development and for many years it has been defined based on its morphological characteristics. However, it is now appreciated that apoptosis is a genetically controlled response and numerous genes, belonging to the Bcl-2 family and other non Bcl-2 genes, are now known to regulate apoptosis (Table 15.11). Some of the gene products are activators of apoptosis whereas others are inhibitors of apoptosis. Altered regulation of these genes can produce an imbalance leading to either increased death potentially resulting in altered development and degenerative diseases or suppression of death that could contribute to cancer development and the accumulation of genetic lesions. As mentioned throughout this chapter, numerous oncogenes and tumor suppressor genes modulate apoptosis. For example, the oncogene bcl-2, an inhibitor of apoptosis, is

TABLE 15.11 Regulators of Apoptosis.

Gene Product	Function
Bcl-2 family	
Bcl-2	Inhibitor of apoptosis, binds Bax and Bak
Bcl-x	L form inhibits, S form accelerates apoptosis, binds Bax and Bak
Bcl-w	Inhibitor of apoptosis
Bax	Accelerator of apoptosis, binds Bcl-2, Bcl-x_L, EIB 19K
Bak	Accelerator of apoptosis, can also be inhibitor, binds Bcl-2, Bcl-x_L, EIB 19K
Mcl-1	Inhibitor of apoptosis
Bad	Accelerator of apoptosis, binds Bcl-2, Bcl-x_L
Nbk, Bik1	Accelerator of apoptosis, binds Bcl-2, Bcl-x_L, EIB 19K, and BHRF1
Non-Bcl-2-related	
TNF-R1	Cell surface receptor, ligand (TNFα) binding promotes apoptosis, in the absence of new protein synthesis, contains cytoplasmic death domain
Fas/Apol/CD95	Cell surface receptor of TNF receptor family, ligand (Fas) binding promotes apoptosis, in the absence of new protein synthesis, death induction requires cytoplasmic death domain
Caspase family	Family of cysteine proteases that induce apoptosis
Dad1	Inhibitor of apoptosis
Survival factors	Subset of growth factors and cytokines that act to promote cell survival

translocated in certain B-cell lymphomas resulting in the overexpression of bcl-2 and in a block in apoptosis. The tumor suppressor gene, *p53*, regulates apoptosis through the transcription activation of bax, a promoter of cell death and the repression of bcl-2, a protector or inhibitor of cell death. As will be discussed in Section 15.10.2, some hepatic tumor promoters suppress apoptosis. The regulation of apoptosis is extremely complex and involves a multitude of proteins. The current overall emerging theme is that inhibitors/suppressors and activators/promoters of cell death interact with each other and the outcome that prevails is dependent upon the ratio of death promoters to death suppressors. Cytochrome C release from the mitochondria plays an important role in caspase activation and the initiation of apoptosis.

15.8 DNA REPAIR/MUTATOR PHENOTYPE

Impaired DNA repair is known to be a critical event in cancer formation. The number of familial cancer syndromes that are associated with defects in DNA repair attest to the importance in cancer development (Table 15.12). Repair pathways for ionizing and nonionizing radiation repair are complex and involve multiple proteins. There are several types of DNA repair, for example, nucleotide excision repair (repair of UV-induced pyrimidine dimers and bulky lesions like the BPDE adduct), base excision repair (repair of deaminaton or base loss, oxygen free radical attack, and methylation of ring nitrogens), mismatch repair (repair mismatch bases that occur during DNA replication that result from recombination between divergent DNA sequences, or from modification of bases by chemicals or radiation), double-strand break repair (repair of double-stranded breaks induced by radiation). As previously discussed, the replication of carcinogen damaged DNA is a critical event in carcinogenesis and can result in permanent genetic alterations in the daughter cell. Defects in DNA repair pathways as well as cell cycle checkpoints can result in the accumulation of mutations in a single cell resulting in what is referred to a mutator phenotype (see Section 15.6.2).

TABLE 15.12 Human Disorders Affecting DNA Repair.

Disorder	Cancer	Proposed DNA repair defect
Ataxia-telangiectasia	Lymphoma	Kinase activity
Bloom syndrome	Various	DNA helicase
Fanconi anemia	Leukemia	Repair of instrand cross links
Hereditary non-polyposis colorectal cancer	Colon and other cancers	Defective DNA mismatch repair
Li-Fraumeni syndrome	Leukemia, Sarcoma, breast cancer	Via cell cycle control
Nijmegen breakage syndrome	Lymphoma	?
Retinoblastoma	Retinoblastoma and others	Via cell cycle control?
Werner syndrome	Various	DNA helicase
Xeroderma pigmentosum	Squamous cell carcinoma, basal cell carcinoma, and melanoma	Excision repair or daughter strand repair

TABLE 15.13 Evidence of Epigenetic Mechanisms of Carcinogenesis.

1. Cancer is associated with altered differentiation.
2. The cancerous state of tumors is sometimes reversible.
3. Carcinogenesis is induced by nonmutagenic agents.
4. Not all carcinogens are mutagens.
5. Carcinogenesis is associated with changes in DNA methylation.
6. Cell transformation can occur at very high frequencies in vitro.

15.9 EPIGENETIC MECHANISMS OF CARCINOGENESIS

As discussed throughout this chapter, there is ample evidence for a role of somatic mutation in carcinogenesis, however, there is also evidence for epigenetic mechanisms of carcinogenesis. Epigenetic agents do not alter the primary sequence of DNA; they alter the expression or repression of certain genes and/or produce perturbations in signal transduction that influence cellular events related to proliferation, differentiation, or apoptosis. Tumor promoters are generally considered to produce their effects through an epigenetic mechanism involving perturbation of signal transduction pathways. The major pieces of evidence in support of an epigenetic mechanism of carcinogenesis are shown in Table 15.13. As discussed in Section 15.4, epigenetic and genetic mechanisms can cooperate/interface to allow tumor development. The importance of epigenetic mechanisms in carcinogenesis is underscored by the fact that approximately 40% of all rodent carcinogens are not mutagenic in vitro. Some of these epigenetic carcinogens may effect the clonal expansion of cells that have spontaneous mutations and increase tumor incidence and multiplicity.

15.10 MULTISTAGE CHEMICAL CARCINOGENESIS

Carcinogenesis in humans and laboratory animals is a complex multistep process. This process involves epigenetic events such as the inappropriate expression of certain cellular genes and genetic events that include the mutational activation of oncogenes and the inactivation of tumor suppressor genes. A number of in vitro and in vivo models have been important in the identification of epigenetic and genetic events associated with carcinogenesis. In experimental models, carcinogenesis can be divided into at least three stages, termed *initiation*, *promotion*, and *progression* (Fig. 15.17). Initiation is a genotoxic or DNA-damaging event, one in which an alteration in the primary sequence of DNA is produced, whereas promotion is considered a nongenotoxic or epigenetic event. Initiation results in a permanent heritable change in the cell's genome; this cell is often referred to as an "initiated cell." It is generally considered that the initiated cell is likely a stem cell within the affected organ. Many factors can impinge upon the formation of the initiated cell. For example, the metabolism of the carcinogen to the ultimate carcinogen can be modified by the expression (induction) or polymorphisms in phase I and II enzymes. Once formed, the ultimate carcinogen may spontaneously decompose, bind to noncritical sites, or, in the case of carcinogenesis, bind to critical sites in the DNA. The adducted DNA can be repaired producing a normal cell. If there is error in the

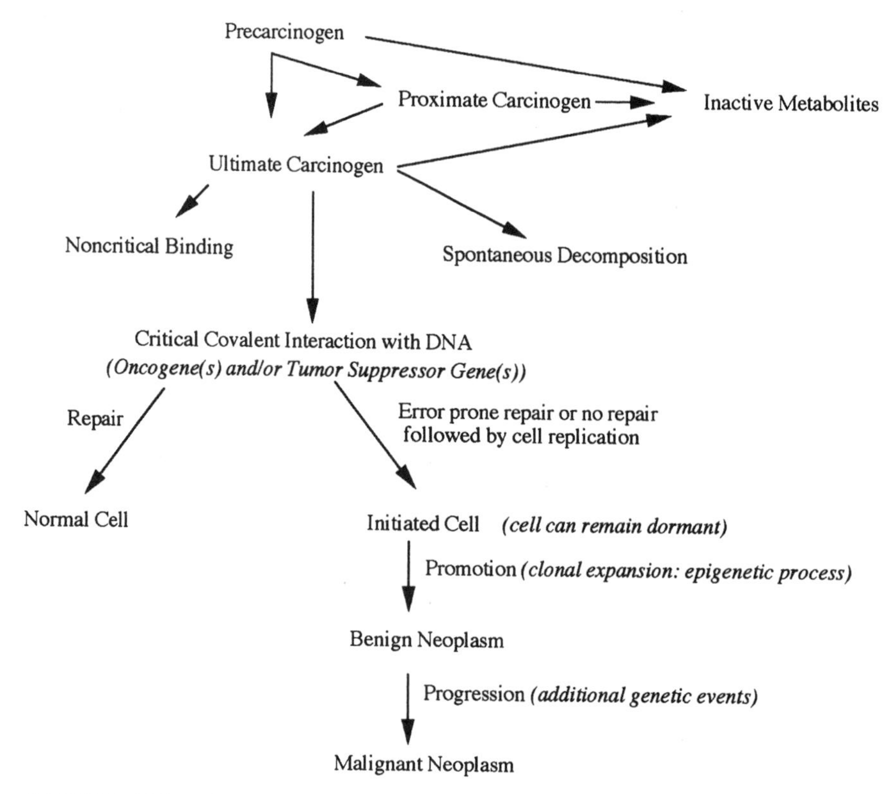

FIGURE 15.17 General aspects of multistage chemically induced carcinogenesis.

repair of the DNA or the DNA adduct is not repaired before the cell replicates, an error in the newly synthesized DNA could occur and a mutation would be permanently fixed in the DNA of the daughter cell. The "initiated cell" may remain dormant (not undergo clonal expansion) for the lifetime of the animal, however, if the animal is exposed to a tumor promoter, it will allow the "initiated cell" to clonally expand and eventually produce a benign tumor. This process is termed tumor promotion and is an epigenetic process favoring the growth of cells with an altered genotype. The development of a malignant tumor from a benign tumor encompasses a third step, termed *progression*, and involves additional genetic damage. In some models such as the mouse skin model, additional stages have been described, including initiation, promotion, premalignant progression, and malignant conversion. In this case, progression has been further subdivided into premalignant progression and malignant conversion.

15.10.1 Initiation-Promotion Model

Experimentally, the initiation-promotion process has been demonstrated in several tissues including skin, liver, lung, colon, mammary gland, prostate, and bladder as well as in several cell lines. Although tumor promoters have different mechanisms of action and many are organ specific, all have common operational features (Fig. 15.18). These features include (1) following a subthreshold dose of initiating car-

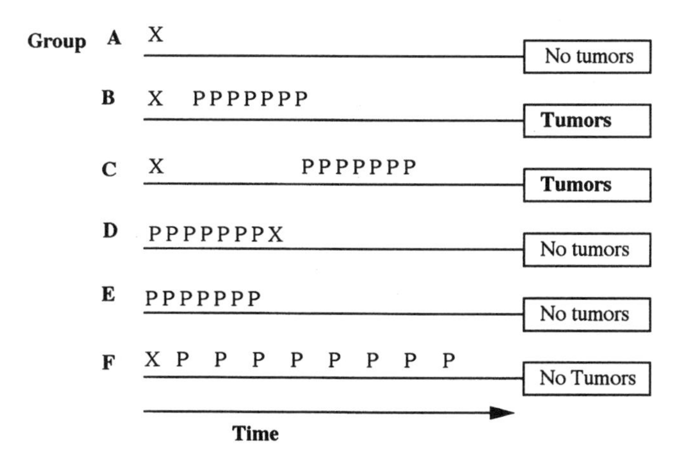

FIGURE 15.18 Initiation/promotion model.
X = Application of initiator; P = application of promoter.

cinogen, chronic treatment with a tumor promoter will produce many tumors; (2) initiation at a subthreshold dose alone will produce very few if any tumors; (3) chronic treatment with a tumor promoter in the absence of initiation will produce very few if any tumors; (4) the order of treatment is critical, that is, you must first initiate and then promote; (5) initiation produces an irreversible change; and (6) promotion is reversible in the early stages, for example, if an equal number of promoting doses are administered but the doses are spaced further apart in time, tumors would not develop or would be greatly diminished in number. Many tumor promoters are organ specific. For example, 12-O-tetradecanoylphorbol-13-acetate (TPA) also known as phorbol 12-myristate 13-acetate (PMA) belongs to a family of compounds known as phorbol esters. Phorbol esters are isolated from croton oil (derived from the seeds of the croton plant) and are almost exclusively active on mouse skin. Phenobarbital, DDT, chlordane, TCDD, and peroxisome proliferators Wy 24,643, clofibrate, and nafenopin are hepatic tumor promoters. TCDD is also a promoter in lung and skin. Some bile acids are colonic tumor promoters while various estrogens are tumor promoters in the mammary gland and liver. There are multiple mechanism of tumor promotion and this may explain the organ-specific nature of the many promoters. In addition, certain chemicals possess both a DNA-damaging and nongenotoxic tumor-promoting component depending on the dose. Under conditions in which the chemical produces tumors without tumor promoter treatment, the chemical agent is often referred to as a complete carcinogen.

15.10.1a Mouse Skin Model

In 1941, Berenblum demonstrated that croton oil, a potent tumor-promoting substance, could greatly increase the incidence of tumor formation on mouse skin if applied alternately with small doses of benzo[a]pyrene. Further experimentation by Mottram and co-workers demonstrated that mice treated with a single subeffective dose of benzo[a]pyrene (a dose that does not produce tumors in the lifetime of the animal) followed by repetitive croton oil treatment produced many tumors, whereas

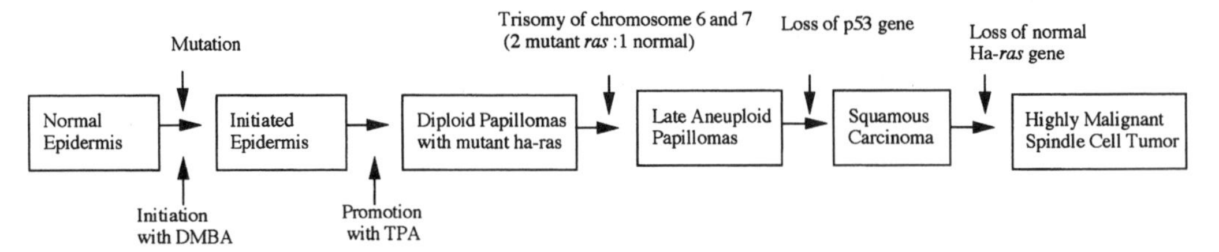

FIGURE 15.19 Molecular events in mouse skin carcinogenesis.

animals treated only with croton oil developed no or very few tumors. These were the earliest experiments that described a two-stage model. Currently the mouse skin model consists of initiation, promotion, and progression stage. The mouse skin carcinogenesis model is one of the best-defined in vivo models of experimental carcinogenesis and it has significantly contributed to our current understanding of the multistage nature of carcinogenesis. Many landmark findings in cancer research have come through studies of the mouse skin model including binding of carcinogens to DNA, monoclonal origin of benign and malignant tumors, powerful tumor-promoting phorbol esters, the modifying role of retinoids, steroids, certain dietary constituents, age, and caloric intake on tumorigenesis as well as the requirement for multiple genetic lesions in malignant conversion. A number of molecular events have been described in this model that are associated with the operational multistage terms *initiation*, *promotion*, *premalignant progression*, and *malignant conversion* (Fig. 15.19). For example, in 7,12-dimethylbenz[a]anthracene (DMBA)-initiated mouse skin, 12-O-tetradecanoylphorbol-13-acetate (TPA) treatment promotes diploid papillomas containing a mutant c-Ha-*ras* $A^{182} \rightarrow T$ allele (>90% contain this mutation). The transition to a later stage aneuploid papilloma is accompanied by nonrandom trisomies of chromosomes 6 and 7 resulting in an allelic imbalance in favor of mutant Ha-*ras*, which is located on chromosome 7. The late stage papilloma can convert to a squamous carcinoma with approximately 30% exhibiting loss of *p53* gene function, and the transition to a highly malignant spindle cell tumor is associated with the loss of the normal Ha-*ras* allele. The progression of papillomas to malignant tumors has also been characterized phenotypically by inappropriate expression of membrane receptor/adhesion molecules, keratins, growth factors, and cyclins/cyclin-dependent kinases. Recently the mouse skin model has been extended to a multihit, multistage model. Results from these experiments demonstrate that (i) promotion between the first hit (subthreshold dose of DMBA) and second hit (another subthreshold dose of DMBA) has a profound outcome on carcinogenesis, presumably by increasing the probability that a second hit will occur in a previously initiated cell; and (ii) continued promotion after the second hit is required for full expression of malignancy. Thus, tumor promoters not only effect the clonal expansion of initiated cells, but based on these experiments also appear to be influencing later stages of carcinogenesis.

15.10.1b Rat Liver Models

There are several rodent liver initiation/promotion models and three models are briefly described below. The liver models have been important in our understanding of multistage chemical carcinogenesis. The *sequential treatment protocol* involves feeding rats an initiating carcinogen such as 2-acetylaminofluorene (2-AAF) for a period of time (often 2–3 weeks). A few weeks after the cessation of initiator treatment, a promoter such as phenobarbital or DDT is fed to the rats for the remainder of the experiment. The experiment requires at least 250 days for the development of tumors. Some of the problems associated with this model are the 2–3 weeks of initiator treatment, which is not representative of a single treatment, and the extended period of time required for tumors to develop. Another model, the *selection model* postulates that the neoplastic process involves the acquisition of resistance to the "mitoinhibitory" effects of the carcinogenic stimuli. Under selective growth condition certain "preneoplastic" cells are thought to form foci or nodules. In this model, rats are exposed to a necrogenic dose of the initiating carcinogen, diethylnitrosamine (DEN). The selection environment is then produced by the generation of an intense proliferative stimulus such as a partial hepatectomy, while at the same time the rats are treated with a low level of 2-AAF, which is thought to be mitoinhibitory to normal cells. The regeneration of the bulk of the liver is blocked, and the hepatocytes rendered resistant by the DEN treatment proliferate and form visible nodules within one week and tumors within 8 weeks. Three general types of nodules form: (1) persistent nodules, those that do not progress to tumors; (2) nodules that regress; and (3) the persistent nodules that do progress to carcinomas. Some criticisms of this model include the following: too complicated, severe conditions, and high doses of DEN leading to multiple genetic changes. The *partial hepatectomy model* involves a single treatment with an initiating carcinogen during liver regeneration followed by promoter feeding. A non-necrotizing dose of DEN is administered 24 hours after a partial hepatectomy, which is then followed by prolonged exposure to a tumor promoter. The development of altered foci is monitored by histochemical techniques to determine ATPase, glucose 6-phosphatase, and gamma-glutamyltranspetidase activity in foci.

15.10.2 Possible Mechanisms of Tumor Promotion

It is generally accepted that tumor promoters allow for the clonal expansion of initiated cells by interfering with signal transduction pathways that are involved in cell growth, differentiation, and/or apoptosis (Table 15.14). Although the precise mechanisms of tumor promotion are not completely understood at the molecular/biochemical level, the role of growth factors, protein kinase C, cell cycle regulators, and TPA- and TCDD-PPAR-responsive elements in genes, expression, and/or repression of genes regulating growth, differentiation, and apoptosis; interaction of promoters with oncogenes and/or suppressor genes; and the production of reactive oxygen species are areas of current research that are providing new and promising mechanistic insights into how tumor promoters allow the selective growth of initiated cells.

In the mouse skin two-stage model of carcinogenesis, 12-O-tetradecanoylphorbol-13-acetate is a potent tumor promoter and appears to produce many of its effects through a specific phorbol ester cellular receptor. This receptor has been identified as protein kinase C (PKC). PKC is a multigene family of serine/threonine kinases that is involved in the transduction of extracellular signals conveyed by growth factors, neurotransmitters, hormones, and other biological molecules. To

TABLE 15.14 Some General Mechanisms of Tumor Promotion.

Selective proliferation of initiated cells
Increased responsiveness to and/or production of growth factors, hormone and other active molecules
Decreased responsiveness to inhibitory growth signals
Perturbation of intracellular signaling pathways
Altered differentiation
Inhibition of terminal differentiation of initiated cells
Acceleration of differentiation of uninitiated cells
Inhibition of apoptosis in initiated cells
Toxicity/compensatory hyperplasia
Resistance to toxicity by initiated cells

date, eleven PKC isoforms have been characterized. It has been demonstrated that PKCα, δ, ε, η, μ, and ξ are expressed in human and mouse epidermis in vivo. Among these six isoforms expressed in keratinocytes, each isoform is thought to play a specific role in keratinocyte function. As mentioned, PKCs are major cellular receptors for the potent mouse skin tumor promoter 12-O-tetradecanoylphorbol-13-acetate (TPA) and it is thought that TPA mediates many of its tumor promotional, inflammatory, and proliferative/differentiative effects through the activation of one or more PKC isoforms. When TPA binds to the regulatory domain of the enzyme, the enzyme becomes catalytically active and specific target proteins are phosphorylated and it has been proposed that these directly or indirectly regulate the expression or repression of specific genes or cellular events important in proliferation and tumor promotion. Okadaic acid, another skin tumor promoter, functions through its inhibition of serine/threonine protein phosphatases known as PP-1 and PP2A. Phosphatase inhibition provides further support for the importance of serine/threonine protein phosphorylation in skin tumor promotion. Treatment of mouse skin or cultured keratinocytes with TPA produces a pleotropic response. For example, TPA treatment increases the expression of proinflammatory mediators such as TNF-α, IL-1α, GM-CSF, and cyclooxygenase-2 (COX-2). TPA treatment also induces TGFα expression and hyperplasia, stimulates late markers of differentiation but decreases early differentiation proteins keratin 1 and keratin 10, and modulates the transactivation activity of the transcription factor AP-1. Diacylglyerols (DAGs) are intracellular second messengers that are produced by receptor-mediated hydrolysis of phospholipids and are the endogenous activator of PKC. When applied to skin, DAGs produce alterations in epidermal growth and are also effective tumor promoters. Collectively, these results support a function for PKC isoforms signaling pathways in a wide spectra of cutaneous biology, however, little is known about the function of individual PKC isoforms in proliferation, differentiation, inflammation, and tumor promotion. Transgenic mouse models in which different PKC isoforms are targeted to mouse skin with specific keratin promoters are beginning to shed some light on the role of individual PKC isoforms in inflammation, promotion, proliferation, and differentiation. Although the activation of PKC by TPA or diacylglycerol is considered to be important in tumor promotion, these promoters also produce the down-regulation of PKC, which may also be important in the mechanism of tumor promotion.

Some bile acids are colonic tumor promoters and have been shown to activate PKC and cause the release of endogenous diacylglycerols from membrane phospholipids. Such effects of bile acid are postulated to be important in colonic tumor promotion.

Mirex, an organochlorine pesticide, flame retardant, and persistent organic pollutant, is a potent mouse skin tumor promoter and appears to function through a novel mechanism involving estrogen. Female mice are four times more sensitive to mirex promotion than male mice and ovariectomized female mice are resistant to mirex promotion. 15β-estradiol implants in ovariectomized female mice restore the intact female response to mirex promotion. These results indicate that estrogen has a role in mirex tumor promotion. Benzoyl peroxide and chrysarobin are also skin tumor promoters that are thought to produce their effects through the generation of oxygen free radicals. Oxidative stress has been shown to activate the JNK pathway increasing the trans-activation activity of the c-jun transcription factor.

In sun-exposed areas of human skin, UVR can cause the mutational inactivation of the *p53* gene in keratinocytes, and the skin cancer that develops often contains mutated inactivated *p53*. Additional exposure to UVR causes additional DNA damage to both normal and *p53* mutant keratinocytes with the normal keratinocytes undergoing apoptosis via a *p53* mechanism. However, the *p53* mutant keratinocytes cannot undergo apoptosis and respond to the damage by proliferating to replace the normal cells that have undergone apoptosis. Thus, UVR is thought to function as an initiator and a promoter and can be termed a complete carcinogen. In addition, UVR can produce an immune suppression that may further enhance the development of skin cancers.

In the rodent liver initiation-promotion model, phenobarbital is a tumor promoter. Phenobarbital is a hepatic mitogen and recent evidence indicates that in preneoplastic foci and hepatic tumors phenobarbital can also suppress apoptosis and accelerate the growth of the foci or tumor. Withdrawal of phenobarbital treatment is accompanied by an increase in apoptosis. Similar results involving the inhibition of apoptosis have been reported for nafenopin, a peroxisome proliferator and hepatic tumor promoter. In isolated mouse hepatocytes, phenobarbital and nafenopin suppress both TGFβ and bleomycin-induced apoptosis, accompanied by alterations in the expression of bcl-2 family members. The inhibition of apoptosis may aid in clonal expansion of initiated cells and contribute to the further accumulation of additional genetic damage within a cell. In addition, isolated mouse hepatocytes respond to the bleomycin-induced DNA damage by arresting in G1 and G2 of the cell cycle. Phenobarbital can delay and attenuate the G1 checkpoint in *p53* dependent manner. Altered G1 checkpoint function represents an epigenetic mechanism by which phenobarbital may indirectly increase the frequency and the accumulation of genotoxic events.

Peroxisome proliferators are a diverse group of chemicals that include the fibrate class of hypolipidemic drugs (clofibrate, Wy 14,643, and nafenopin), plasticizers (di(ethyllhexyl)phthalate, DEHP), and herbicides (2,4-dichlorophenoxyacetic acid). Peroxisome proliferators cause an increase in the number and size of peroxisomes in the liver, kidney, and heart of rats and mice. Peroxisome proliferators are hepatic mitogens and tumor promoters. These agents function as hepatic tumor promoters through their binding as ligands to the peroxisome proliferator-activated receptor-α (PPARα), a member of the steroid/thyroid receptor superfamily. PPARα forms a heterodimer with the retinoid RXR receptor (ligand 9-cis retinoic acid) and binds to peroxisome proliferator. This complex binds to peroxisome proliferator response element (PPRE) in the promoter region of specific genes to increase transcription. It is thought that PPAR responsive genes are involved in producing alterations in proliferation and differentiation. As indicated above, peroxisome proliferators suppress apoptosis, presumably through the direct modulation of gene expression. Gene-knockout mice lacking PPARα are refractory to peroxisome proliferation and peroxisome proliferator-induced changes in gene expression and WY-14,643-induced hepatocarcinogenesis.

Halogenated hydrocarbons such as organochlorine pesticides, polybrominated biphenyls, and TCDD are representative of a large group of hepatic tumor promoters. TCDD is a potent promoter in rodent liver where it binds to and activates the TCDD receptor, a transcription factor. The activated receptor regulates gene expression by binding to the TCDD-responsive elements in the promoter region of certain genes thought to be important in tumor promotion. Many of these gene products and their importance in tumor promotion remain to be determined.

Many tumor promoters from different mechanistic classes, including the halogenated compounds mentioned above, inhibit gap junction intercellular communication (GJIC) or cell-to-cell communication. Cell communication can be mediated through gap junctions in which the connexon (formed from a six connexins) of one cell couples with a corresponding connexon in a contiguous cell, to form a pore between the cells and to join the cytoplasms. This serves to synchronize the functions of cells within a tissue. Most normal cells within solid tissues have functional GJIC. In contrast, cancer cells of solid tissues appear to have dysfunctional GJIC. The transition of a normal cells to cancer cell is often accompanied by a loss of GJIC. The role of inhibition in cellular communication has been postulated to play a role in the tumor promotion and progression stages of carcinogenesis. It is postulated that normal cells hold initiated cells in check through cell-cell communication. However, if this communication is blocked by a tumor promoter, the initiated cell population proliferates and clonally expands to ultimately form a tumor.

15.11 USE OF TRANSGENIC AND KNOCKOUT MICE

The manipulation of cells and animals through the introduction of genes into the host DNA is a powerful approach to determine the function of a specific genes in carcinogenesis. Of particular importance in toxicology is the ability to make transgenic animals and gene "knockout" animals where the function of a specific gene can be studied in vivo. The creation of transgenic and nullzygous mice provides powerful tools for cancer research and many such mice have provided and will continue to provide mechanistic information on the development of cancer. For example, mice deficient in certain forms of P450 genes have been created and are used to better understand the role of specific P450 genes in the metabolism of carcinogens. Mice that are heterozygous for mutant *p53* are being used as a sensitive model to detect carcinogens. Transgenic mice that overexpress Bcl-2, myc, fos, PKCα, TGFα, and ras to name just a few have provided a more in-depth understanding of the role of these proteins in carcinogenesis.

15.12 CONCLUSIONS

It is clear the environment, genetics, and age are the major determinants of susceptibility to cancer induction. A better understanding of the mechanisms through which chemicals cause cancer and the genes involved in tumor development will eventually provide for a rationale basis for regulation and the estimation of individual risk. Genomic technology and informatics along with transgenic and knockout animals will likely play a major role in this important endeavor.

SUGGESTED READING

Eighth Report on Carcinogens, U.S. Department of Health and Human Services, Public Health Service, National Toxicology Program 1998 (www.niehs.nih.gov).

Overall Evaluations of Carcinogenicity to Humans, International Agency for Research on Cancer (IARC) (www.iarc.fr).

Doll, R. and Peto, R. *The Causes of Cancer*. Oxford Medical Publications, Oxford England, 1981.

Vogelstein, B. and Kinzler, K.W. (Eds). *The Genetic Basis of Human Cancer*, New York: McGraw Hill Companies Inc., 1998.

Hesketh, R. (Ed). *The Oncogene and Tumor Suppressor Gene Facts Book*, 2nd ed. San Diego, CA, Academic Press, 1998.

CHAPTER SIXTEEN

Genetic Toxicology

R. JULIAN PRESTON

16.1 INTRODUCTION AND HISTORICAL PERSPECTIVE

Genetic toxicology is a branch of the field of toxicology that assesses the effects of chemical and physical agents on the hereditary material (DNA and RNA) of living cells. The field can be considered to have its roots in 1927 with the publication of the seminal paper by H.J. Muller on the induction of phenotypically described mutations in *Drosophila* by X-rays. In his studies he not only showed that radiation exposures could increase the overall frequencies of mutations but also that the types of mutations induced were exactly the same in effect, or phenotype, as those occurring in the absence of radiation exposure. This observation exemplifies that fact, that will be returned to later, that an induced mutagenic response must be assessed in relation to background mutations.

Karl Sax (1938) extended Muller's observations by showing that X-rays could induce chromosome alterations in plant pollen cells. Sax and his colleagues in an elegant series of studies showed that at least two critical lesions in a target were necessary for the production of exchanges within and among chromosomes. We know now that the "lesions" referred to by Sax are DNA double-strand breaks, base damage, or multiple damaged sites. The important conclusion that was derived from these studies is that the two lesions necessary for chromosome aberration formation are independently produced, leading to nonlinear dose-response curves. The relevance of this conclusion will become more apparent in subsequent discussions in this chapter.

Charlotte Auerbach and colleagues in 1946 described studies (actually conducted in 1941) that demonstrated that nitrogen mustards could induce mutations, and that these were similar at the whole-organism level in *Drosophila* to those induced by X-rays. Thus, research begun in the new area of chemical mutagenesis ran in parallel with radiation mutagenesis studies.

The next event of particular significance to the field of genetic toxicology was the development by William Russell in 1951 of a mouse mutagenesis assay that could be used to establish if previously described data on radiation-induced mutations in *Drosophila* could be replicated in a mammalian system. The mouse tester strain developed for the specific locus assay contained recessive mutations at seven loci coding for visible mutations such as coat color, eye color, and ear shape. This strain could be used to identify induced recessive mutations at wild-type loci in irradiated animals. Mutations were indeed detected following X-ray exposure of mouse germ cells at frequencies that were similar to those induced by X-rays in *Drosophila*. Subsequent studies showed that mutations could be induced at these seven loci by a range of chemical agents. Over the next 20 years, genetic toxicologists collected qualitative and quantitative information based on phenotype about induction of mutations, largely by radiation, in germ and somatic cells. The ability to culture mammalian cells in vitro and to assess chromosome alterations in mammalian cells, particularly in human lymphocytes following mitogen stimulation in culture, greatly facilitated progress. The potential utility of the mutagenicity data for genetic and cancer risk assessments was enhanced by using the parallelogram approach for estimating human responses from rodent data. This utility was enhanced by the knowledge that most tumors contained chromosome alterations, that the formation of tumors required mutations, and that cancer was a clonal process, with the addendum that tumors originated from single cells (Nowell, 1976). In the 1970s, two significant advancements allowed for a clearer understanding of the mechanisms of mutagenesis and the potential role of mutagens in carcinogenesis. Jim and Elizabeth Miller and their colleagues demonstrated that chemical carcinogens could react to form stable (covalent) derivatives with DNA, RNA, and proteins both in vitro and in vivo. They further demonstrated that for their formation these derivatives could require the metabolism of the chemical to primary and subsequent metabolites. Thus, the requirement for metabolism for some carcinogens was established. In vivo, such metabolism is endogenous, but for most in vitro cell lines metabolic capability has been lost. To overcome this, Heinrich Malling and colleagues developed an exogenous metabolizing system based upon a rodent liver homogenate (S9). This system has proven to be of utility, but with some drawbacks related to species and tissue specificity.

The second significant development occurring in the 1970s was that of Bruce Ames and colleagues, who designed an assay with the bacterium *Salmonella typhimurium* that can be used to detect chemically induced reverse mutations at the histidine locus. Not only was the assay able to play an important role in hazard identification, but also mutagens were predicted to be carcinogens on the basis that cancer required mutation induction. The next few years saw the development of about 200 short-term assays that were used to screen for potentially carcinogenic chemicals via the assessment of mutagenicity, DNA damage, or other genotoxic activity. Several international collaborative studies were conducted to establish the sensitivity and specificity of a set of short-term assays as well as to compare interlaboratory results (International Programme on Chemical Safety, 1988). The outcome was that most assays, to a greater or lesser degree, were able to detect carcinogens or noncarcinogens at only about 70% efficiency. Unfortunately, this mutagenicity/carcinogenicity relationship became somewhat of a hindrance to the field of genetic toxicology because it provided a study framework that was too rigid. The comparison of results for different assays in a standard battery by Tennant et al. (1987) emphasized this point by showing that mutagenicity only predicted a portion (~70%) of known carcinogens. This finding, in part, initiated a downturn in enthusiasm for short-term tests used solely for detecting carcinogens. Subsequently, the lack of a tight correlation between carcinogenicity and mutagenicity (and the converse noncarcinogenicity and nonmutagenicity) was determined to be due to the presence of chemicals that were not directly mutagenic but instead induced the genetic damage necessary for tumor development indirectly by, for example, clonally expanding preexisting mutant cells (i.e., tumor promotion). Such chemicals were categorized under the somewhat misleading term *nongenotoxic*. Although a chemical can be carcinogenic via a nondirectly mutagenic process, mutations have to be produced to move cells along the multistep pathway to tumor formation. With the appreciation that nondirectly mutagenic chemicals could be carcinogenic, an era arose of studies that were designed to identify cellular processes, other than directly induced mutagenicity, with the potential to be involved in carcinogenicity (e.g., cytotoxicity with regenerative cell proliferation, mitogenicity, or receptor-mediated processes). The types of assays that are most frequently used for identification of nondirectly mutagenic chemicals are described in the previous chapter (Chapter 15, Chemical Carcinogenesis).

In the past 15 years, the field of genetic toxicology has moved into the molecular era. The potential for advances in our understanding of basic cellular processes and how they can be perturbed has become enormous. The ability to manipulate and characterize DNA, RNA, and proteins has been at the root of this advance in knowledge. However, the development of sophisticated molecular biology techniques does not in itself imply a comparative advance in the utility of genetic toxicology and its application to risk assessment. Knowing the types of studies to conduct and knowing how to interpret the data developed remains as fundamental as always. Finer and finer detail of cellular processes and their alteration by mutation can perhaps help to complicate the utility of various mutagenicity assays. The following sections provide descriptions of standard genotoxicity assays and how the addition of molecular analysis can provide information pertinent to risk assessment.

The US Environmental Protection Agency's (EPA) *Proposed Guidelines for Carcinogen Risk Assessment* allow for the incorporation of mechanistic data into the risk assessment process. Thus, the use of genetic toxicology assays can provide data beyond hazard identification for incorporation into the cancer risk assessment process (National Research Council, 1994).

16.2 GENOTOXICITY ASSAYS

The aim of this section is not to provide detailed protocols of the range of prokaryotic and eukaryotic assays used to detect the presence or absence of a genotoxic response. Such descriptions can be found in the published literature, for example, in the series of reports of the US EPA's Gene-Tox committees. Rather, the principles to be followed for the assays and current approaches for obtaining underlying mechanistic data are presented.

16.2.1 DNA Damage and Repair Assays

Any one of a number of DNA damage and repair assays can be used to provide preliminary information on the potential mutagenicity of a chemical. Briefly, DNA damage can be quantitated by the assessment of specific DNA adducts using HPLC-based methods when information on the chemical is known, and for DNA adducts in general using ^{32}P-postlabeling techniques. DNA strand breakage can be measured by the DNA elution assay or by the relatively recently developed single-cell gel electrophoresis (Comet) assay. The advantage of the latter is that observations can be made at the level of the single cell. The Comet assay is based on the fact that DNA containing strand breaks migrates more rapidly in a gel when electrophoresed. The length of the DNA "tail" drawn by electrophoresis from the cell is a measure of the DNA damage present. The conditions under which the electrophoresis is conducted determines the nature of the DNA damage detected: double-strand breaks at neutral pH and single-strand breaks at high pH. One of the current concerns with the Comet assay is that it is overly sensitive and might lead to apparent DNA damage under nondamage-inducing exposure conditions.

DNA repair can be detected by the same assays (alkaline elution and Comet) that are used to detect DNA damage, but with various time intervals between the end of exposure and the time of sampling. An additional assay for assessing DNA repair has been utilized for many years, namely the unscheduled DNA synthesis (UDS) assay. The endpoint measured is the uptake of tritiated thymidine into DNA repair sites following exposure. The sensitivity of the assay is relatively low, especially in vivo, and this limits its predictive value for hazard identification.

The assessment of DNA repair per se provides rather little information on the quantitative assessment of the potential subsequent mutagenicity of a chemical because it is the frequency of misrepair that is of consequence to the cell with regard to an adverse outcome.

16.2.2 Gene Mutation Assay—*Salmonella typhimurium*

The *Salmonella* (Ames) assay utilizes a set of tester strains that have been constructed to select for different classes of reverse mutations at the histidine locus. The strains commonly used, together with the genetic alteration detected, are shown in Table 16.1. The resistance transfer factor (R factor) was incorporated, resulting in a greater sensitivity to mutation induction than with the parental strains.

The assay can be conducted with an exogenous metabolizing system (S9), allowing for the detection of mutagens that require metabolic activation to a DNA-reactive intermediate (e.g., polycyclic aromatic hydrocarbons). The *Salmonella* assay has been conducted with literally hundreds of chemicals, such that various calculations of specificity and sensitivity for detecting carcinogens and noncarcinogens can be performed for various chemical classes and modes of action for inducing DNA damage.

TABLE 16.1 Strains Commonly Used in the *Salmonella*/Histidine Reversion Assay.

Strain	Endpoint Detected
TA 1535	Base-pair substitution
TA 1537	Frameshift
TA 1538	Base-pair substitution
TA 98	TA 1538 with R factor
TA 100	TA 1535 with R factor
TA 102	Mutants from oxidative damage and crosslinks

More recently, it has been feasible to collect data on the molecular nature of the revertants induced and thereby better understand the mechanism of their induction and allow for interspecific extrapolation. The sequencing of the histidine revertants leads to the generation of mutation spectra for a chemical and its metabolites and for simple and complex mixtures. Of particular importance for assessment of effects at low exposure levels, it is possible in the majority of cases to distinguish between background and induced mutations, thereby enhancing the sensitivity of the assay. Comparisons of mutation spectra for a specific chemical across species from *Salmonella* to humans suggest that a similarity of formation of mutations holds across all species. Thus, data generated for prokaryotes have some application in predicting effects in humans. A similar assay to that developed for *Salmonella* is available for *Escherichia coli*, employing reversion at the tryptophan locus. The WP2 strains used can only detect base-pair substitutions given that the original trp mutation is a base-pair substitution. There are some obvious limitations to this approach given the difference in cellular and genomic complexity between prokaryotes and eukaryotes.

16.2.3 Yeast Assays

Forward and reverse mutation assays for a range of selectable loci can be conducted with *Saccharomyces cerevisiae* as well as with other yeast strains and fungal species (e.g., *Neurospora*, *Aspergillus*, and *Ustilago*) in much the same way as with bacteria. However, yeast have multiple chromosomes, haploid and diploid karyotypes, mitotic and meiotic cycles, and, of increasing significance, a considerable homology as regards cell cycle control and DNA repair capacities, for example, to those in mammalian species including humans. Thus, mutations in genes specific for pathways involved in cellular housekeeping processes can be assessed; such mutations can be considered as viable biomarkers of tumor responses. Mutational spectra can also be developed as an aid to interspecies extrapolation of response.

In general, *S. cerevisiae* exhibits a high degree of homologous, mitotic recombination. For example, transferred plasmids have a high probability of incorporation into targeted genes; much higher than for mammalian species. The advantage is that assay systems can be developed for assessing mitotic crossing-over and mitotic gene conversion as additional cellular responses to chemical exposure. As an example, Schiestl and colleagues have demonstrated that chemicals that do not interact with DNA (e.g., polychlorinated biphenyls and 2,3,7,8-tetrachlorodibenzo-p-dioxin) can induce deletions via intra chromosomal recombination in an assay established to detect such events (the DEL assay).

16.2.4 Mammalian Cell Assays (In Vitro)

16.2.4a Gene Mutation Assays

As noted above, mutational information can be obtained with prokaryotic and lower eukaryotic systems that can enhance the ability to predict effects in humans. However, because of differences in genome organization and genetic complexity, assays using mammalian cells have been developed. In vitro mammalian cell assays almost always utilize cells that are either immortalized spontaneously (e.g., Chinese hamster ovary cells, mouse lymphoma cells) or virally transformed, as in the case for human nontumor cell lines. The benefit of using such immortalized cells is that they can be cloned as single cells, a necessary requirement for mutation identification.

The loci that can typically be used as mutational sites are ones that can be selected for, although PCR methods under development will allow for analysis of any gene for which the DNA sequence is known. The selection procedure requires that a locus be heterozygous or hemizygous such that only a single mutation is required for the selectable phenotype to be assessed. Examples of such loci are the *hprt* (*h*ypoxanthine-guanine *p*hospho*r*ibosyl *t*ransferase) gene that is X-linked and is thus hemizygous in males, and effectively hemizygous in females because of X-chromosome inactivation; or the thymidine kinase locus that is heterozygous in a mouse lymphoma cell line (L5178Y) and a human lymphoblastoid cell line (TK6); and the *aprt* (*a*denine *p*hospho*r*ibosyl *t*ransferase) gene that is heterozygous in a selected CHO cell line. Because the mutations induced by the great majority of chemicals involve errors of DNA replication on a damaged template, at least part of the treatment period should be for cells in the DNA synthesis (S) phase.

The CHO/*HPRT* assay is presented in somewhat more detail as an example. After treatment with a chemical being assayed for its mutagenic potential, cells are subcultured for an expression period of about 6–8 days to deplete residual *HPRT* protein. Mutant clones can be identified as cells that grow in the selective agent 6-thioguanine. Wild-type cells are killed by the toxic metabolite, 6-thioguanine monophosphate, that they can form via the normal *hprt* gene; *hprt* mutant cells cannot perform this metabolic step. The analysis of the mutation spectra for a range of chemical and physical agents has shown that he assay can detect point mutations and small to mid-sized deletions. However, large deletions covering many megabases of DNA are not recovered, largely because viability genes are located in the proximity of the *hprt* gene. Because the *hprt* gene is X-linked, compensation by an active, homologous chromosome does not take place. Other assay systems employing autosomal loci are required if the detection of very large deletions is needed. An example of such an assay is that utilizing the heterozygous thymidine kinase locus (*tk*+/–) in the L5178Y mouse lymphoma cell line. The selection for mutants relies on the phenotype of trifluorothymidine resistance. A further utility of this assay has been proposed, namely that fast-growing mutant colonies are produced by cells containing point mutations whereas show-growing colonies arise from cells containing chromosome mutations, particularly deletions. The distinction is perhaps less clear-cut than this, and molecular characterization of mutants provides the definitive characterization of the induced mutations.

16.2.4b Cytogenetics Assays

Structural and Numerical Alterations. Chromosome aberration in vitro assays can be conducted with any higher eukaryotic cell type that divides or can be induced to divide in tissue culture. As mentioned above, the majority of cell lines routinely used are immortalized cells (e.g., CHO cells, human lymphoblastoid lines) although human peripheral lymphocytes can be stimulated to divide in vitro by the addition of a mitogen such as phytohemagglutinin. A diagram of a mitotic cell cycle is shown in Figure 16.1 to illustrate the various stages referred to in this section and Section 16.2.5.2 on in vivo cytogenetic assays.

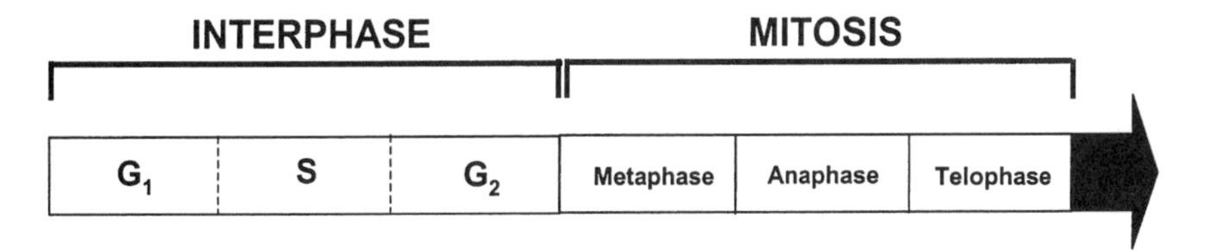

FIGURE 16.1 Mitotic cell cycle.

The endpoints utilized for assessing cytogenetic alterations are chromosome structural alterations for clastogenicity (chromosome breaking ability) and numerical alterations (most effectively gains of chromosomes). The classical method is to assess both structural and numerical aberrations in metaphase cells using a general stain such as Giemsa and light microscopy. As an alternative, chromosomal breakage and chromosome loss events can be assessed by the presence of micronuclei. Acentric chromosome fragments or whole chromosomes that fail to segregate at anaphase and are not incorporated into either daughter nucleus, become bounded by a membrane and remain as a micronucleus in the cytoplasm through one or more cell cycles. The frequencies of micronuclei are most accurately assessed in cells that have undergone one cell division, since they are not produced in the absence of a cell division and are reduced in frequency through loss (or extrusion) at the second or subsequent division after induction. Cytochalasin B can be used to inhibit cytokinesis and so micronuclei can be conveniently assessed in binucleate cells, namely ones that have undergone a single division (mononucleate, no division; tri and higher nuclear numbers, more than one cell division).

Another relatively recent modification of the assay is to stain micronuclei with a fluorescent antibody (CREST) to the centromeric region of chromosomes and view through a fluorescence microscope. A micronucleus that contains a whole chromosome will fluoresce whereas an acentric fragment (i.e., with no centromere) will not fluoresce (Tucker et al., 1997). It should be noted that the assessment of micronuclei only identifies subsets of cytogenetic alterations, acentric fragments for clastogenicity and chromosome loss for aneuploidy. Conventional metaphase analysis can be used for identifying all classes of cytogenetic alteration.

A broad range of fluorescence-based molecular techniques have been developed in the past 10 years or so, and these have considerably enhanced our ability to assess cytogenetic alterations. It is quite feasible to identify specific gene regions, repetitive DNA sequences, and whole chromosomes using fluorescent DNA probes for the appropriate chromosomal units. Thus, chromosomal structural alterations can be readily detected. For example, reciprocal translocations (exchanges of chromosomal material between two chromosomes) can be identified using whole chromosome painting techniques. This is of particular importance because this class of aberrations is generally transmissible from one cell generation to the next and many tumors are characterized by the presence of specific reciprocal translocations. With conventional metaphase analysis many of the aberration types that are recorded are cell lethal and are not per se a risk factor. An additional advantage of being able to assess transmissible cytogenetic alterations is that the accumulated effects of longer-term exposures can be measured, although this is of particular utility with in vivo exposures.

Additional, somewhat more complex assessment of genomes can be conducted by chromosome banding techniques, comparative genomic hybridization (CGH), and spectral karyotyping (SKY) (Preston, 1998). CGH is of particular utility when attempting to identify chromosomal differences between two genomes, such as those of normal and tumor cells, or between mutant clones and wild-type cells. The technique is basically not informative for standard genotoxicity assays. SKY uses computerized representations of karyotypes and allows for the identification of each pair of human chromosomes. Again, the greater utility of SKY is for identifying karyotypic changes in tumors, but there could be an application for identifying genomic instability that can develop at some time after a particular exposure. It has been shown that genomic instability is a hallmark of tumor formation. Adding its assessment to standard assays for chromosomal alterations or mutations would appear to have significant benefit to the risk assessment process.

As mentioned above, many aberration types are cell lethal, and so a requirement for measuring the induced frequency for total aberrations is that the sampling of cells has to be at the first mitosis after the exposure. In addition, the exposure needs to be of a duration of less than that of a cell cycle for the particular cell type being used. The great majority of chemicals produce aberrations via the conversion of DNA damage (e.g., DNA adducts, cross links) into an aberration as an error of replication on the damaged template. Thus, to maximize the sensitivity of a cytogenetic assay, cells should receive at least part of their exposure while they are in the DNA synthesis (S) phase of the cell cycle.

Sister Chromatid Exchange (SCE) Assay. It is possible to identify apparently reciprocal exchanges between the two sister chromatids of a chromosome by differential staining methods. The general approach is shown in Figure 16.2. Visualization of differentiation can be achieved by (a) treatment with Hoechst 33258 stain,

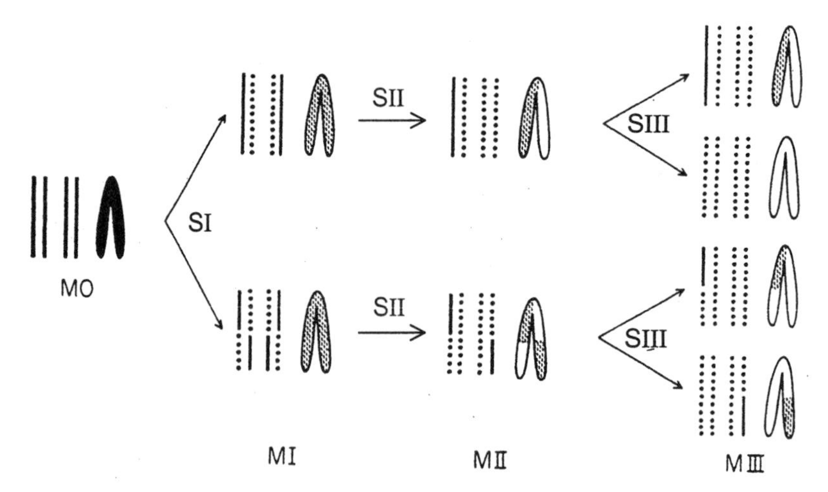

FIGURE 16.2 Sister chromatid exchange formation.
For differential staining of the sister chromatids, bromodeoxyuridine (BrdU) is incorporated into replicating DNA in place of thymidine (TdR). The *solid line* represents the original TdR-containing DNA strand and the *dotted line*, the DNA strand containing BrdU. An SCE is shown to occur at the first S phase (SI) but will be observed at the second mitosis (MII).

DNA elution with saline sodium citrate under black light, and staining with Giemsa or (b) use of an antibody to BrdU.

SCE appear to be produced largely as errors of DNA replication on a damaged template or by homologous recombination. The fact that they appear to be reciprocal in nature means that they do not represent a risk factor to cells. An increase in SCE can be via processes that do not involve directly induced DNA damage, making the interpretation of observed increases somewhat difficult. In review of test data for a range of chemicals, the SCE assay appears to provide rather equivocal information. For this reason, it is not currently recommended in a standard genotoxicity test battery.

16.2.5 Mammalian Cell Assays (In Vivo)

16.2.5a Gene Mutation Assays

As with the in vitro assays discussed above, the conduct of an in vivo gene mutation assay requires a selectable marker. The majority of human studies have been conducted on peripheral lymphocytes using the *hprt* gene. The same assay can be conducted with rodent lymphocytes but these are rather difficult to clone in vitro. Additional information can be obtained in human studies by assessing the T-cell antigen receptor status of mutant clones; the presence of an identical sequence in two or more clones indicates that these were derived by clonal expansion from a single mutant cell. Thus, an induced mutation frequency can be obtained by correcting the observed mutant frequency for clonal expansions. An earlier version of the human *hprt* assay relied on autoradiographic identification of mutant cells (i.e., those that could grow in 6-thioguanine) rather than cell cloning assays. The problem with such an assay is the inability to distinguish true mutations from phenocopies, leading to concerns with interpretation of the data in terms of significant increases in mutations.

Other genetic markers have been utilized for the assessment of gene mutations in vivo in rodents. Those include the *Dlb*-1 locus (binding site for Dolichosbiflorus agglutinin) and the *lacI* or *lacZ* transgene in mice or rats. There are a number of advantages to the use of the transgenic mouse assays, because mutagenicity can be assessed in any tissue, especially target tissues for carcinogenicity. There are drawbacks, however. First, that only point mutations and small deletions can be detected. Large deletions that remove the viral packaging sites are not recovered. Second, in vivo cell replication is required for the formation of the mutation in vivo, and third, the lac gene is not an endogenous gene, and is not transcribed. As with in vitro assays, PCR-based mutation analysis is reaching the levels of sensitivity at which any gene is potentially available for the assessment of mutations. Mutational spectra can be assessed, and for comparisons between rodents and humans, currently the *hprt* gene assay represents the most viable system.

16.2.5b Cytogenetic Assays

Structural and Numerical Alterations. The basic principles underlying in vivo cytogenetic assays are the same as those for in vitro assays discussed above (Section 16.2.4a). The cell types that are available for chromosome aberration analysis are those that are cycling cell populations, such as bone marrow and spermatogonial cells. For acute exposures, first division cells with S-phase treatment and analysis of all aberration types is appropriate for hazard evaluation. For chronic exposures, it is inappropriate to analyze unstable aberration types; the informative approach is to analyze stable types (e.g., reciprocal translocations and inversions) using fluorescence in situ hybridization (FISH).

For numerical chromosomal alterations, it is possible to analyze interphase cells using FISH for whole chromosome paints, or chromosomal specific probes. Thus, it is feasible to assess nondisjunctional events in any cell type, including target cells for tumorigenesis.

Other fluorescence-based methods described in the section on in vitro methods (Section 16.2.4a) can be used in vivo, but they provide rather limited value for basic mutagenicity assays.

A micronucleus assay can be conducted with bone marrow cells, peripheral lymphocytes and germ cells (spermatids) of mice. Rats can be used for assaying the bone marrow or spermatogenic cells for micronuclei, but the use of peripheral lymphocytes requires surgical removal of the spleen, since this is the predominant site of removal of micronucleated erythrocytes in rats.

Bone marrow cells, following exposure to the test chemical, are collected and stained with Wright-Giemsa or acridine orange, for example. Micronuclei are assayed in polychromatic erythrocytes (PCE), the most recently matured erythrocytes. Cytotoxicity is assessed as the ratio of PCE to normochromatic erythrocytes (NCE) in treated versus control animals. NCE are the more mature erythrocytes. A lower PCE/NCE ratio in the treated animals than in the control suggests cytotoxicity. There are specific criteria for establishing that a cytoplasmic inclusion is a micronucleus (Hayashi et al., 1994). As mentioned above, the micronucleus assay detects a subclass of cytogenetic alterations (deletions and chromosome loss) and is useful for hazard identification but not for identifying aberration spectra.

16.2.6 Germ Cell Assays

Several assays have been developed for assessing genetic alterations in germ cells. The information developed can be used as hazard identification for a cancer risk assessment or more appropriately in the development of a genetic risk assessment. The decision of when to conduct a germ cell assay is an important one because the assays are generally expensive and time-consuming. An exception is the use of *Drosophila* as a substitute for mammalian assays.

16.2.6a Drosophila Sex-Linked Recession Lethal Test (SLRL)

For the SLRL, the mutations being assessed are observed in the progeny of chemically treated males or females, and thus are heritable effects. The longer the time interval between treatment and mating, the further back in the spermatogenic cycle are the treated cells, thereby allowing for the assessment of effects in all germ cell stages. Another feature of the assay is that *Drosophila* have a xenobiotic metabolizing system that is quite similar to that in mammals. The particular genotypes of the flies used allow for the detection of recessive lethal mutations on the X chromosome by an absence of wild-type males in the F_2 generation (Lee et al., 1983).

16.2.6b Dominant Lethal Test (Rodent)

The dominant lethal assay is designed to detect induced mutations in germ cells that manifest themselves as embryonic lethals. The test system can be applied to mice and rats, and assessment can be made of effects upon male and female germ cells. Continuous exposure with matings for 1–2 weeks postexposure allows for the assessment of the cumulated effects of exposure for all germ cell stages. Serial matings over several weeks allows for assessment of effects on the different germ cell stages.

For determining dominant lethal effects in males, for example, the number of dead implants as a proportion of total implants is measured between days 13 and 19 of gestation. The test is not very sensitive for detecting mutagenic responses, and so is not particularly valuable for detecting potential carcinogens. It has a greater utility for detecting chemicals that *are* germ cell mutagens (Ashby et al., 1966).

16.2.6c Heritable Translocation Test (Rodent)

The heritable translocation test is based on the fact that carriers of reciprocal translocations have reduced litter sizes (semisterility) as a consequence of lethal (duplications and deficiencies) segregants of the translocations. Thus, the assay requires the observation of response in the F_2 of a treated animal. The reciprocal translocation is induced in a parental germ cell, leading to an F_1 animal carrying the translocation in every cell, and the F_2 showing semisterility. It is feasible to assess translocations by direct observation of diplotene-diakinesis spermatocytes. As with the dominant lethal test, the heritable translocation test is not useful as a screen for carcinogens because of its rather low sensitivity. It can be used for identifying chemicals that *are* germ cell mutagens.

16.2.6d Mouse Specific Locus Test

The mouse specific locus test was introduced in Section 16.1 because it represented the initial use of the mouse for assessing induced germ cell mutations. The tests relies upon the use of a tester stock that is homozygous recessive for seven loci or five loci depending on the particular version. The treated individuals are wild-type at the same loci. Thus, any mutation induced in any one of the wild-type genes will be observed as a phenotypic mutant in the offspring following mating to the tester stock. The mutants are saved for further testing. In fact, more recently, a number of the mutations have been characterized at the DNA level. The information generated can be used for comparison with data obtained in other assay systems, and in genetic risk assessment.

Clearly, the assay is labor intensive and very expensive. Its use is for specific instances where germ cell mutagenicity has to be assessed for genetic risk assessment purposes.

16.2.7 Summary

The preceding sections provide a brief overview of the conduct of prokaryotic and eukaryotic genotoxicity assays. The intent is to introduce the basic principles for each assay, the use of the data, and how current advances, particularly in molecular biology can enhance the predictive value of an assay. There is a long list of other assays that are available, the present one is for those assays that are most frequently used and consequently have the larger available data base. The question to be addressed at this point is how to best utilize a group or battery of tests to assess mutagenicity and predict potential carcinogenicity.

16.2.8 Test Batteries for Detecting Carcinogens

The development of the *Salmonella* mutagenicity assay for detecting mutagens, and by association carcinogens, was viewed initially as a one test approach for hazard identification. It quickly became apparent that this was more appropriate for positive results, but even in this case the predictivity for carcinogenicity in humans (or rodents) was far from 100%. Thus, the use of a battery of tests was developed to include mammalian assays and in vivo tests. Subsequently, as noted above, because a significant proportion of chemicals can be carcinogenic in the absence of direct mutagenicity, additional assays have been developed for assessing the potential for a chemical to be a "nongenotoxic" carcinogen. These assays are discussed in Chapter 15.

A variety of test batteries have been described, each with merit and potential drawbacks associated with predictivity for carcinogenicity. The particular scheme presented in Figure 16.3 is of utility because it allows for the detection of somatic and germ cell mutagens (as needed) and also provides a preliminary approach for identifying the mode of action for potential carcinogens. A more detailed description of the testing approach following this scheme can be found in Ashby et al. (1996). The basic components (or required tests) for initial hazard identification of this and the great majority of test batteries for genotoxicity are a microbial assay such as the Ames test, a mammalian gene mutation assay, and an in vitro assay for chromosomal effects. The aim of this group of assays is to screen products for mutagenicity in the decision-making process in product development. In order to receive approval for commercialization, regulatory agencies require additional tests to be conducted. Although these can vary according to the particular human exposure, in general an in vivo cytogenetics assay is required (chromosome aberrations or micronucleus). A negative outcome could require an evaluation of mutagenicity in a second tissue, depending upon the level of potential human exposure. The interpretation of the test data from such a battery can be complicated by conflicting outcomes, for example, a positive result for an in vitro cytogenetics assay and a negative result for an in vivo cytogenetics assay. A number of possible explanations are available. High concentrations of some chemicals in vitro can lead to chromosomal alterations by processes unrelated to direct interaction with DNA (Scott et al., 1991). Also, the frequency of false positives among the different genotoxicity assays is variable but generally between 10% and 20% (Shelby et al., 1993; Kirkland and Dean, 1994), with in vitro cytogenetic assays being at the upper end of this range. Despite the collection of enormous quantities of data with genotoxicity assays, caution is still required in their interpretation given that they are generally conducted at high exposure levels, frequently using transformed cells, and the assays have been developed to be sensitive for endpoint induction. However, for hazard identification it is prudent to use a conservative approach that is somewhat more likely to identify false positives than false negatives. Moving to the next steps of risk assessment requires a quantitative analysis that would be unreasonably compromised by the use of data from overly sensitive assay systems.

16.3 GENOTOXICITY DATA AND CANCER RISK ASSESSMENT

Section 16.2 described the use of genotoxicity assays for the identification of potential carcinogens. In addition, it has been appreciated for some time that data from

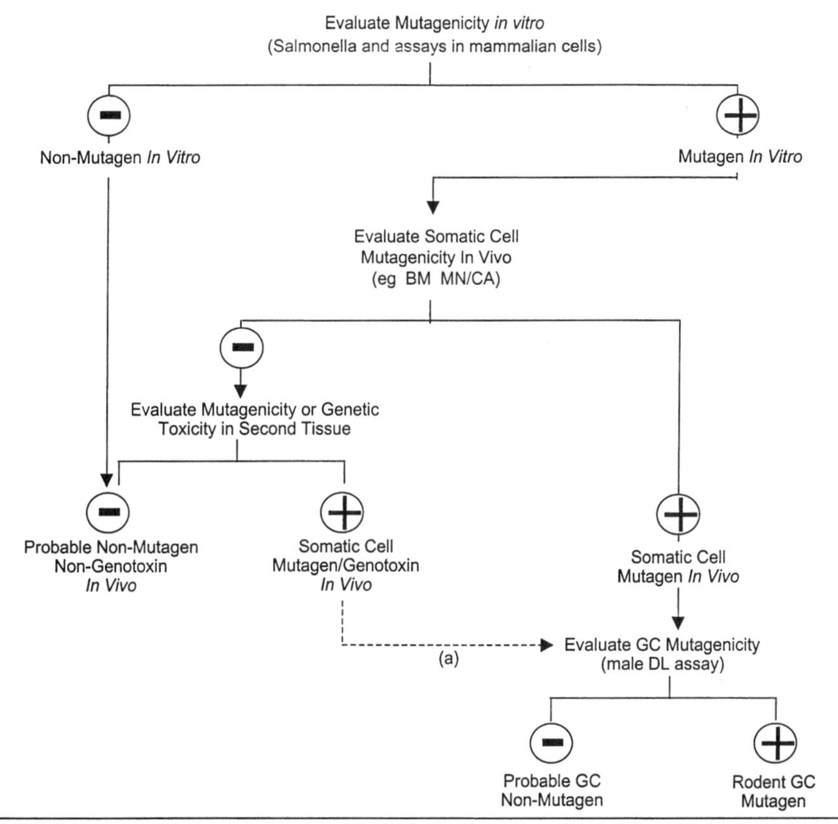

Agents found non-mutagenic in vitro may be evaluated in vivo as necessary, eg, in cases of high human exposure. Non-genotoxins/non-mutagens in vivo may be non-carcinogens or non-genotoxic carcinogens. Sites of possible non-genotoxic carcinogenesis for such compounds may be indicated by non-specific phenomena, eg, peroxisome proliferation, induced mitogenesis or inhibited apoptosis, etc. Mutagens/genotoxins in vivo are potential genotoxic carcinogens/germ cell mutagens. For such compounds, further information on potential genotoxic carcinogenesis may be provided by DNA adduction/damage/repair (UDS), or by TG mutation assays. All such studies are optimally conducted at the chronic MTD. Established rodent carcinogens may be studied using this approach to indicate their broad mechanism of carcinogenic action (genotoxic versus non-genotoxic). Agents active in the DL assay may be studied further, as individual circumstances dictate, using HT, SL or PB HGPRT assays. Such studies will confirm heritability and/or contribute to definition and quantification of possible human GC hazard. Agents inactive in the DL assay may be studied further as individual circumstances dictate, eg, using female DL, GC TG assays, etc.

FIGURE 16.3 Scheme for detecting mutations in vitro and in vivo in somatic and germ cells.

Scheme 1. Scheme for detecting mutagens in vitro, somatic and germ cell mutagens in vivo, rodent genotxins and potential rodent cartcinogens. Such data contribute to the recognition of germ cell mutagens and non-mutagens, genotoxic and non-genotoxic carcinogens, and non-carcinogens. TG = transgenic assay; GC = germ cell; BM = bone marrow; CA = chromosomal aberration assay; MN = micronucleus assay; HT = heritable translocation assay; SL = specific locus assay; MTD = maximum tolerated dose; PB HGPRT = peripheral blood HGPRT/HPRT assays in rodents/humans. (a) observation of genotoxicity in vivo for BM non-mutagens does not necessarily trigger the need for GC studies; the genetic event monitored, the magnitude of the induced effect and the dose level required, influence this decision.

Source: From Ashby et al., 1996 with permission.

assays measuring mutagenicity can be used as a component of quantitative cancer risk assessment. This stems, in part, from the knowledge that cancer is a genetic disease, whereby a series of mutations are required for passage of a cell from normal to transformed (i.e., multistage carcinogenesis). These required mutations are most frequently in proto-oncogenes or tumor suppressor genes leading to cell cycle perturbations or genomic instability, both of which are hallmarks of tumor cells. Thus, it can be argued that data on mutagenicity can be used as an effective surrogate for tumors, at least as far as shape of dose response is concerned, at exposure levels below those at which increases in tumors can be detected, but at which detection of increases in mutagenicity are practical. This forms the basis for the approach to cancer risk assessment presented in the US EPA's *Proposed Guidelines for Carcinogen Risk Assessment* (1996). In this approach it is proposed to use mechanistic data to support the assessment of cancer risk at low (environmental or occupational) exposure levels, absent human or laboratory animal tumor data in this exposure range. This represents a significant advance over previous practice whereby for mutagenic chemicals a default approach was used that required a linear extrapolation from the lowest observed tumor incidence for a particular chemical, without some regard to known mechanism of action. In some cases, a linear response is the appropriate one even when regard is paid to mechanism of action (e.g., for the production of point mutations involved in tumor production). On the other hand, nonlinear responses can be predicted and assessed for other mechanisms of action (e.g., when chromosomal alterations are involved in tumor production). For the majority of nondirectly mutagenic chemicals, a nonlinear or threshold response best describes the shape of the tumor dose response curve over the low exposure range. This is supported by the available data on mechanisms of action of this broad family of chemicals.

Suffice it to say that there is a clear need to enhance the database for the mechanism of action for tumorigenicity for the great majority of chemical (and physical) agents. However, the availability of new techniques for assessing effects at the cellular and molecular level should allow for the rapid expansion of this database. In parallel, there will be an improvement in the reliability of cancer risk assessment for human exposures.

16.4 NEW RESEARCH DIRECTIONS

The myriad of new research tools available to probe the consequences of the interaction of chemicals with cellular macromolecules have allowed for a new enthusiasm in addressing the question, "How do chemicals induce tumors?" Some of these recent advances are noted here for illustrative purposes.

In order to extrapolate from tumor data collected in animal studies to responses in humans, it is greatly advantageous to demonstrate that a similar etiology or underlying mechanism is involved. Three new approaches facilitate such an endeavor.

To ensure that analysis is being performed on tumor cells in the absence of contamination of adjacent normal cells, microdissection by laser capture microscopy can be used to select small areas of a tumor for subsequent molecular characterization. A very small number of cells is needed because reliable PCR amplification techniques are available to provide the amounts of DNA or RNA required for analysis.

As noted above, sophisticated techniques are available for the analysis of mutations at the gene and chromosomal level. These include ligation-based PCR techniques (Pfeifer et al., 1999) and fluorescence in situ hybridization that can be used in conjunction with clones that cover limited genomic regions to identify specific chromosomal breakpoints.

To provide global comparisons between the gene expression patterns of normal and tumor cells, for example, mRNA (cDNA) arrays have been developed. The approach is to hybridize expressed genes in the form of fluorescent cDNAs to an array of hundreds to thousands of fragments of known genes or expressed sequence tags (ESTs). The cDNAs can be derived from treated versus untreated cells or tumor versus nontumor cells. The resultant hybridization patterns can be compared for alterations in expression (Jelinsky and Samson, 1999). The analysis of the very large quantities of data generated by this techniques have proven to date to be quite difficult to conduct.

Another area of rapid advancement that will impact upon the interpretation of genotoxicity data is that of DNA repair. The complexity of DNA damage response pathways is beginning to be appreciated. For example, the apparently straightforward repair of ultraviolet light-induced pyrimidine dimers by the nucleotide excision pathway requires at least 17 enzymes for recognition and removal of the DNA damage and resynthesis at the site of excision. Repair of other types of DNA damage involve similar steps and an attendant complex of repair enzymes. The efficiency of DNA repair influences the responses to genotoxic chemicals and subsequently the probability of disease outcomes. In addition, there is a growing interest in the association of tumor suppressor genes with the DNA repair process.

An additional area of current research that impacts upon the magnitude of the mutagenic response is that of cell cycle control with the attendant cell cycle checkpoints and programmed cell death (apoptosis) (discussed in more detail in Chapter 15). A checkpoint halts the progression of a cell allowing additional time for repair. Abrogation of checkpoints can result in cells progressing into significant cell cycle phases (e.g., DNA replication and cell division) with unrepaired DNA damage. The result is an increase in mutation sensitivity. Apoptosis provides for the death of damaged cells. Failure to undergo apoptosis results in damaged cells being maintained in the proliferative pool, with a subsequent increase in mutation frequency. Understanding the mechanism of chemically induced apoptosis and also how chemicals can reduce apoptosis is currently a research challenge.

The examples noted above serve to highlight where advances in the understanding of basic cellular housekeeping processes and the consequences of their perturbation can have a very significant impact on determining the mechanism of adverse health outcomes and the subsequent assessment of risk. Thus, studies of mutagenicity play an important role in hazard identification, dose response assessment, and the characterization of risk.

16.5 CONCLUSIONS

The field of genetic toxicology has had an overall life of about 70 years, during which time it has served to address both practical and basic scientific issues. Much has been learned about the relative potency of chemicals and radiations to induce gene mutations and chromosome aberrations. To a great extent this information has been obtained from short-term assays using prokaryote and eukaryotic cells in vitro and insect and mammalian cells in vivo. The suggestion that carcinogens are mutagens led to the development of genotoxicity test batteries that were designed to detect potential carcinogens. The lack of a high level of predictivity was due to there being a large class of carcinogens that are not directly mutagenic. Assays for the detection of this class have been developed based upon the range of their different mechanisms of carcinogenicity. New molecular techniques have enhanced the potential utility of genotoxicity assays by allowing for a much clearer understanding of mechanisms of induction of gene mutations and chromosome alterations. The ability to sequence DNA, for example, allows for the detection of the specific nature of mutations. Fluorescence in situ hybridization techniques make the analysis of transmissible chromosome alterations feasible. The current view that cancer (and to some extent other diseases) is a genetic disease, requiring several mutations to convert a normal cell to a transformed one has allowed for the association between induced genetic alterations and tumors in laboratory animal models and humans to be considered. The way forward for the field of genetic toxicology would seem to be to provide data that can be used for the description of tumor dose-response curves (or other disease outcomes) below exposure levels at which tumor can be reliably observed. In addition, by understanding the mechanism of formation of tumors in different species, a greater reliability can be placed on extrapolating from tumor data collected in an animal model to predicted responses in humans. The expectation is that uncertainty in cancer (and perhaps noncancer) risk assessments will be reduced.

SUGGESTED READING

Ashby, J., Waters, M.D., Preston, J., Adler, I.D., Douglas, G.R., Fielder, R., Shelby, M.D., Anderson, D., Sofuni, T., Gopalan, H.N.B., Becking, G. and Sonich-Mullin, C. IPCS harmonization of methods for the prediction and quantification of human carcinogenic/mutagenic hazard, and for indicating the probable mechanism of action of carcinogens. *Mutat. Res.* 352 (1996), 153–157.

Hayashi, M., Tice, R.R., MacGregor, J.T., Anderson, D., Blakey, D.H., Kirsh-Volders, M., Oleson, F.B. Jr., Pacchierotti, F., Romangna, F., Shimada, H., Soutou, S. and Vannier, B. *In vivo* rodent erythrocyte micronucleus assay. *Mutat. Res.* 312 (1994), 293–304.

Ashby, J., DeSerres, F.J., Shelby, M.D., Margolin, B.H., Ishidate, M. and Becking, G. (Eds) International Programme on Chemical Safety, *Evaluation of Short-Term Tests for Carcinogens, Vols. I and II.* Cambridge, MA: University Press, 1988.

Jelinsky, S.A. and Samson, L.D. Global response of *Sacchraromyces cerevisiae* to an alkaylating agent. *Proc. Natl. Acad. Sci. U.S.A.* 96 (1999), 1486–1491.

Kirkland, D.J. and Dean, S.W. On the need for confirmation of negative genotoxicity results *in vitro* and on the usefulness of mammalian cell mutation tests in a core battery: experiences of a contract research laboratory. *Mutagenesis* 9 (1994), 491–501.

Lee, W.R., Abrahamson, S., Valencia, R., von Halle, E.S., Wurgler, F.E. and Zimmering, S. The sex-linked recessive lethal test for mutagenesis in *Drosophila melanogaster* (a report of the U.S. Environmental Protection Agency Gene-Tox Program). *Mutat. Res.* 123 (1983), 183–279.

National Research Council, Committee on Risk Assessment of Hazardous Air Pollutants. *Science and Judgment in Risk Assessment*. Washington, DC: National Academy Press, 1994.

Nowell, P.C. The clonal evolution of tumor cell populations. *Science* 66 (1976), 23–28.

Pfeifer, G.P., Chen, H.H., Komura, J. and Riggs, A.D. Chromatin structure analysis by ligation-mediated and terminal transferase-mediated polymerase chain reaction. *Methods Enzymol.* 34 (1999), 548–571.

Preston, R.J. New approaches in genetic toxicology and their possible applications to cancer risk assessment. *CIIT Activities* 18 (1998), 1–7.

Scott, D., Galloway, S.M., Marshall, R.R., Ishidate, M., Brusick, D., Ashby, J. and Myhr, B.C. Genotoxicity under extreme culture conditions. A report from ICPEMC Task Group—9. *Mutat. Res.* 257 (1991), 147–204.

Shelby, M.D., Erexson, G.L., Hook, G.J. and Tice, R.R. Evaluation of a three-exposure mouse bone marrow micronucleus protocol: Results with 49 chemicals. *Environ. Mol. Mutagen.* 21 (1993), 160–179.

Tennant, R.W., Margolin, B.H., Shelby, M.D., Zeiger, E., Haseman, J.K., Spalding, J., Caspary, W., Stasiewicz, S., Anderson, B. and Minor, R. Prediction of chemical carcinogenicity in rodents from *in vitro* genetic toxicity assays. *Science* 236 (1987), 933–941.

Tucker, J.D., Eastmond, D.A. and Littlefield, L.G. Cytogenetic end-points as biological dosimeters and predictors of risk in epidemiological studies. In: *IARC Scientific Publication 142*, Lyon, France: IARC, 1997, pp 185–200.

CHAPTER SEVENTEEN

Molecular Epidemiology

MARIANA C. STERN, P.E. NORBERT, P. KOPER, and JACK A. TAYLOR

17.1 INTRODUCTION

Molecular toxicology studies are primarily focused on the biochemical and toxicological mechanisms involved in the interaction between a potentially toxic compound and a living organism, and their possible effects in disease development. Similarly, epidemiological studies are also concerned with the continuous spectrum from exposure to disease. Epidemiology can be defined as the study of the distribution and determinants of health and diseases in populations. Molecular toxicology research is mainly of an experimental nature, characterized by relatively small scale and highly controlled interventions by the researcher. Although intervention studies are considered within the scope of epidemiology, most epidemiological studies are observational.

The most important challenge for epidemiologists is finding explanations of why a specific exposure results in disease. The development of new methods in the field of molecular biology has increased the possibilities for understanding the mechanisms of disease development. Epidemiological studies that apply such techniques along with other biological measurements are referred to as molecular epidemiological studies, emphasizing the fact that tools originating from molecular biology are applied. However, as discussed later, the molecular epidemiological approach not only consists of the incorporation of molecular biology techniques but, more importantly, the incorporation of a better understanding of the biologic events underlying the disease process. This includes determination of genetic susceptibility, assessment of individual exposure through molecular markers, and identification of critical alterations as early markers of disease. The implementation of new laboratory techniques and the increased use of biological specimens in epidemiological studies have important implications for the design of study protocols and data analysis. This calls for increased collaboration between epidemiologists and laboratory scientists such as molecular biologists, biochemists, toxicologists, and geneticists.

The objective of this chapter is to provide the reader with an introduction to the principles and methods used in this emerging field of research. We will briefly introduce standard epidemiological study designs, measures of assessment, and the concepts of bias, confounding, and interaction. In the second section we will focus more specifically on molecular epidemiological research.

17.2 THE EPIDEMIOLOGICAL APPROACH

17.2.1 Types of Epidemiological Study Designs

Epidemiological studies examine the relation between exposure (e.g., chemical compounds, diet, or environmental factors) and health outcomes. Some of the questions asked in epidemiological research are Who is at risk for a disease? Who gets a particular disease? How many people get the disease? Does a particular exposure increase the risk of the disease? There are several different approaches, or study designs, that are commonly utilized to answer these questions. Epidemiological studies are defined by:

- population studied
- criteria used for selection of subjects (based on disease or based on exposure)
- time of data collection with respect to disease development

The first step in epidemiological research, similarly to laboratory research, is to define the research hypothesis to be tested. This should include a clear definition of the type of exposure and health outcome to be measured. For instance, one might want to test the hypothesis that cigarette smoking (exposure) increases the risk for bladder cancer (health outcome). The next step is to choose the most appropriate study design for the particular hypothesis being tested.

In the next paragraphs we will introduce some of the more common designs and describe their strengths and limitations. Table 17.1 presents a summary of the different study designs described and how they differ according to the factors mentioned above. The most relevant aspects of data analysis and measures of effect are presented. Finally, we discuss the concepts of bias, confounding, and interaction.

17.2.1a Ecological Studies

Ecologic studies involve the observation of disease frequency in large population-based groups and the correlation of those frequencies with exposure information. The unit of observation is the group and not the individual. Groups can be defined according to different parameters such as geographic location, occupation, economic status, or nutritional status. For example, one might correlate newborn birth weight in different Western countries with the frequency of smoking for those same countries and find an association between lower birth weight and the frequency of smoking. Such studies frequently are done using existing data, such as government data on demographics and disease rates from registries, death certificates, or other sources. Therefore, in many instances, they are quick and inexpensive to carry out. However, they present some limitations. Because the unit of observation is a group and not the individual, the observed relations between an exposure and disease cannot be directly extrapolated to the individual level. This is due to the fact that the exposure levels represent average levels for the population group being analyzed, within the group not all individuals have the same level of exposure. For example, one might look in different population groups and correlate the average level of ovarian cancer mortality with the average family size and conclude that pregnancy protects against ovarian cancer. However, with this type of approach it is not known if the women who died of ovarian cancer in each population group

TABLE 17.1 Comparison of Different Epidemiological Study Designs According to Some of the More Relevant Characteristics.

Study Design	Ecologic	Cross-Sectional	Case-Control	Cohort	Experimental
Unit of observation	Population-based groups defined by specific factors. Examples: -residence -occupation -ethnicity	Individuals from a population-based group defined by specific factors. Examples: -exposures -geography	Individuals from a population-based group	Individuals from a population-based group	Individuals from a population-based group
Bases for subject selection	No individual subjects analyzed. Selection at group level	Individuals representative of target population. Usually selected by random sampling	Individuals selected by disease status	Individuals selected by exposure status	Random allocation of individuals to specific study groups
Time of *exposure* data collection with respect to *health outcome*	Simultaneous	Simultaneous	Simultaneous or retrospective	Prior to disease development	Assigned by researcher
Source of *exposure* data collection	Routine data. Examples: -government databases -cancer registries	Data collected by researcher. Example: -questionnaires -population registries	Data collected by researcher. Example: -questionnaire -occupational and medical records	Data collected by researcher. Example: -questionnaire -occupational and medical records	Experimental data generated by researcher
Outcome measure more frequently obtained	Correlations	Prevalence Odds ratios	Odds ratios	Risk ratio Rate ratio	Disease incidence relative risk

had children or not. This is known as the "ecologic fallacy." Consequently, although ecologic studies can be very useful for generating hypotheses of exposure-disease correlation, such associations must be tested using other study designs before they can be accepted.

17.2.1b Cross-Sectional Studies

Cross-sectional studies involve the analysis of a previously selected group of subjects from a defined population and the ascertainment of exposure factors and disease status at the same point in time. The unit of study is the individual. Individuals are selected, usually at random, and information on specific exposures and health outcomes is obtained. For example, the infection status for a particular virus of a random sample of inhabitants of a town is recorded and at the same time a questionnaire is administered to obtain information on socioeconomic status, nutritional habits, and lifestyle. Correlation of these data might reveal interesting information about risk factors for this particular viral infection. These data may also allow the estimation of the proportion of infected people in a certain population at a specific point in time. This measure of occurrence is called *prevalence* and is commonly calculated in cross-sectional studies.

This study design is certainly very simple and efficient in terms of necessary time and cost because it does not require the follow-up of individuals. It is good for analyzing genetic characteristics because these do not change over time. Moreover, because exposure and disease are recorded at the same time it is difficult to distinguish between exposures or conditions that cause disease and those that are simply associated with the disease process. That is, it is not good for investigating causal relationships or for exposures that change over time.

17.2.1c Case-Control Studies

In a case-control study a group of individuals diagnosed with a disease (cases) are compared with those without the disease (controls). The history of exposure to a particular factor or factors (e.g., occupation or exposure to some specific event), is obtained for each individual. A comparison of the proportion of individuals exposed to the specific factors among cases and controls is made. If differences are found, this comparison can lead to clues about potential causal factors for the specific health outcome being studied. The number of cases and controls enrolled in this type of study ranges from 50 cases and 50 controls to 1000 cases and 1000 controls.

This type of study design provides the advantage that several risk factors can be studied at the same time. It is also especially useful for rare diseases or diseases with long induction periods because it starts with the collection of people with the specific disease of interest. The main disadvantages of this study design are found in its logistics, how controls are selected, and how measures of past exposure are obtained. When this is improperly done it can lead to *bias*, a concept we discuss later in the chapter. Also, given the fact that exposure and health outcome are obtained simultaneously, it may be difficult to establish causal relationships between exposures and health outcomes.

17.2.1d Cohort Studies

Prospective cohort studies consist of the definition of a population-based group that is followed over time and monitored for the development of disease. Groups can be defined based on time (e.g., freshmen students at college) or based on specific criteria (e.g., factory workers, menopausal women). Individuals are recruited into a cohort and baseline information is collected on exposure(s) and lifestyle factors of interest, this information can be updated with subsequent data collection procedures. As time goes by some members of the cohort will develop disease whereas others will remain disease free. The range of individuals enrolled in cohort studies goes from 100 to 100,000. Cohort studies can also be *retrospective*, also called *historical cohort studies*. A group of people can be defined and their disease status and exposures assessed retrospectively using existing records. This is a useful approach for occupational epidemiology where usually occupational exposure and medical records are readily available. Cohort studies are usually considered the gold standard of epidemiological research because exposure is ascertained prior to disease. Using a cohort study it is possible to look at a variety of different disease outcomes or health effects. However, the major disadvantage of this type of study is that they require large numbers of people, can be very expensive, and take years to accumulate sufficient numbers of cases if a disease is not common.

In this context we should also mention the nested case-control study, which is a case-control study within a cohort study. Such studies start with the prospective follow up of a cohort. In time new cases of disease will occur, and one can compare individuals with disease to individuals without disease, all from within the cohort, using previously collected data. This way it is possible to study various disease outcomes using a case-control approach and still have the advantage of already collected exposure information. As will be discussed in a subsequent section, this design can be applied in molecular epidemiology, provided that relevant biological samples are available from the cohort.

17.2.1e Experimental Studies

The paradigm of scientific study design is the experiment. The epidemiological equivalent to the experiment is the randomized controlled trial. In this type of study, individuals are randomly allocated to different treatment. Large group sizes are usually needed to ensure that randomization equally divides the groups based on all factors except the treatment that is assigned to each group. Such experiments, done under stringently controlled conditions, are routine in the evaluation of efficacy and safety of new medications. At the end of the study it is possible to obtain information on the percentage of individuals that under a given treatment developed disease, a measure known as *disease incidence*. There are serious legal and ethical constraints that limit the possibilities for human experiments. Treatments are most often agents being tested for the prevention or cure of disease, because one cannot ethically expose a person to something that may lead to disease. In nonexperimental "observational" epidemiological study designs, the investigator attempts to approximate experimental conditions as much as possible.

17.2.2 Aspects of Data Analysis

The next sections will briefly discuss some aspects of data analysis done in epidemiological studies. In epidemiological studies, specific emphasis is placed on the statistical analysis of the data to control for other potential risk factors that might correlate with the exposure and health outcome of interest. We will briefly introduce the concepts of bias, confounding, and effect modification (interaction) that are important in epidemiological studies. The data used in these examples are not original, they have been modified to provide clearer illustrations of concepts.

17.2.2a Measures of Occurrence and Measures of Association

As we mentioned in Section 17.2.1b, one measure of occurrence that can be inferred from epidemiological studies is the *prevalence*. This measure refers to a single point in time and therefore might not always be enough to quantify disease risk factors. Another measure of occurrence that can be inferred from some epidemiological studies is *incidence*, that is the number of new cases of disease that develop in a population at risk during a specific time interval. There are three types of measures of incidence that can be inferred: *risk*, *odds of disease*, and *incidence rate*. These three measures have the same numerator in the equation, which is the number of new cases of disease that appeared during a specific time period. The denominator is what makes them different. In the case of *risk*, the denominator is the number of disease-free individuals at the beginning of the study, and, therefore, this measure indicates the proportion of individuals in a population that were initially free of disease but developed it at the end of the specified time period. In the case of *odds of disease*, the denominator is the number of disease-free individuals at the end of the specified time period, and therefore this measure is a ratio of the likelihood of developing disease to the likelihood of not developing disease during a given time period. In the case of *incidence rate*, the denominator is the "person-time at risk" during the specified time period. The "person-time at risk" is the sum of time each person remained under observation and at risk of developing disease.

In order to quantify the association between a given exposure and disease, the incidence of disease in a group of exposed individuals has to be compared with the incidence of disease in a group of nonexposed individuals. A ratio of incidence for the two groups can be calculated to estimate the likelihood of developing disease in exposed individuals relative to nonexposed. These are called relative measures of association or exposure effect. There are three types, depending on what type of incidence measure is used to obtain the ratio: *risk ratio*, *rate ratio*, and *odds ratio of disease*. These measures are usually referred to as *relative risk* (*RR*). In general, this is valid when the disease is relatively rare (e.g., cancer) but is not if the disease is common (e.g., the common cold).

The results of a cohort study are often expressed as *RR* of disease. This is illustrated by the 2 × 2 table shown in Example 1. The probability, or risk of disease in the exposed group (A/A + B), is divided by the risk of disease in the nonexposed group (C/C + D) to obtain an estimate of the RR of disease in exposed versus unexposed. A RR of 4 would imply a four times higher risk of developing the disease in exposed compared with nonexposed. Conversely, a RR estimate of 0.5 implies a halving of the disease risk in exposed compared with unexposed and thus a protective effect of the exposure.

The results of a case-control study can also be summarized using a 2 × 2 table (Example 1). However, one cannot directly calculate RR using data from a case-control study. Unlike a cohort study, the numbers of cases and controls are determined by the investigator so that the numbers are not representative of the proportion of cases and controls present in the population. Calculating a "relative risk" using these numbers would result in an erroneous risk estimate. However, the

	Disease (case)	No disease (control)	Total
Exposed	A	B	A+B
Non-exposed	C	D	C+D
Total	A+C	B+D	A+B+C+D=N

A. **Cohort Studies**

$$\text{Relative risk} = \frac{A/(A+B)}{C/(C+D)}$$

B. **Case-Control Studies**

$$\text{Odds Ratio} = \frac{A/C}{B/D} = \frac{AD}{BC}$$

EXAMPLE 1 Analysis of 2 × 2 tables.

data in a 2 × 2 table from a case control study can be used to calculate the odds of exposure among cases (A/C) can be compared with the odds of exposure among controls (B/D). The ratio of these odds is known as the *odds ratio of exposure* (OR) = (A/C)/(B/D) = AD/BC and is considered an approximation of the RR calculated in cohort studies.

For both the *relative risk* and the *odds ratio of exposure*, one can test whether the estimates are statistically different from the null value of 1 (where there is neither increased risk nor protective effect). This is done by calculating the chi-square statistic and reading the corresponding P value from a probability table. A more informative approach is to calculate a confidence interval around the risk estimate. Besides revealing statistical significance (if the confidence interval does not include 1) the width of the interval gives an indication of the statistical variance of the risk estimate. More detailed explanations of these procedures can be found in the epidemiology textbooks mentioned as suggested readings.

17.2.2b Bias

Bias can be defined as an effect that tends to skew results away from the true value. There are many sources of potential bias that can occur in epidemiological studies. Two types of bias can be introduced in the selection of the subjects (*selection bias*) or introduced in the measurement or classification of exposure (*information or measurement bias*). The results of these biases are that subjects in the study groups are not comparable or that there is a systematic difference in the information obtained between comparison groups. Consider, for example, a cross-sectional study designed to investigate the frequency of a specific disease among workers of a certain hazardous industry compared to a nonhazardous one at one point in time. Workers that already developed disease as a result of occupational exposure might have already quit, therefore only healthy people remain working. If a comparison were made between the frequency of a specific disease between people who work at the hazardous industry and people that do not, the selection of "healthy workers" at the hazardous workplace could result in the surprising conclusion that working in such environment prevents the development of disease. This is an example of selection bias. In another example, consider a case-control study designed to correlate the frequency of a specific neonatal malformation and certain occupational exposures. Women who have a baby with a malformation may be more likely to remember exposures they received during pregnancy than women with healthy babies, leading to a bias in recall of exposure. This is an example of information or recall bias.

17.2.2c Confounding

The concept of confounding is very important in epidemiology. In essence, confounding is the effect of a third variable that influences the exposure-disease relationship under study. The main characteristics of a *confounder* variable are that it is associated with the exposure being studied, it is also an independent risk factor for the health outcome under analysis and it must have a stronger effect than the exposure under study. To illustrate this let us consider a cohort study looking at the relation between alcohol consumption and the risk of lung cancer (Example 2). In this study a relative risk of 1.5 is observed. This would suggest that people that consume alcohol have a 50% higher risk of developing lung cancer. However, people who consume alcohol often also smoke. Smoking is a known risk factor for lung cancer. Therefore, the association between alcohol and lung cancer might be explained by concurrent smoking of the individuals who drink alcohol. To find the correct relative risk we could either restrict the study population to nonsmokers, or separate the smokers and nonsmokers. In the latter approach, the relative risk for lung cancer in relation to alcohol consumption is 1.1 for both nonsmokers and smokers (Example 2). The initially observed relative risk of 1.5 was confounded by the high frequency of smoking among people who drink alcohol. Because smoking is a known risk factor for lung cancer (here the relative risk is approximately 9, 12/1200 vs. 108/1200), the association between smoking and alcohol artificially increased the relative risk estimated for alcohol and lung cancer. A corrected estimate of the relative risk can be obtained by calculating a weighted average of the stratum-specific relative risks for smokers and nonsmokers. Statistical procedures such as logistic regression are used to adjust for confounding.

17.2.2d Interaction

Finally, there is the concept of interaction, also known as effect modification. In essence, interaction is similar to the biological/pharmacological phenomena of additivity, synergism, or antagonism. Interaction should be investigated considering the biological bases of the observed associations. An example of interaction is shown in Example 3. This illustrates a cohort study of risk factors for coronary heart disease. From the 2×2 table it can be calculated that individuals with high endogenous catecholamine levels have a relative risk of 2.4 of developing coronary heart disease. However, coronary heart disease is a disease with a complex multifactorial mecha-

Total Study Population:

Alcohol consumption	Lung cancer	No lung cancer	Total
Yes	72	1128	1200
No	48	1152	1200
Total	120	2280	2400

$$\text{Relative Risk (crude)} = \frac{72/1200}{48/1200} = 1.5$$

Non-smokers:

Alcohol consumption	Lung cancer	No lung cancer	Total
Yes	5	475	480
No	7	713	720
Total	12	1188	1200

$$\text{Relative Risk (non-smokers)} = \frac{5/480}{7/720} = 1.1$$

Smokers:

Alcohol consumption	Lung cancer	No lung cancer	Total
Yes	67	653	720
No	41	439	480
Total	108	1092	1200

$$\text{Relative Risk (smokers)} = \frac{67/720}{41/480} = 1.1$$

EXAMPLE 2 Example of confounding by smoking status in a cohort study looking at the association between consumption of alcohol and risk of developing lung cancer.
Although there is an apparent association between alcohol and lung cancer in the total study population, when this population is split into two groups or strata by smoking status, there is no evidence that alcohol increases lung cancer risk in either group. The apparent association between alcohol and lung cancer was produced by the confounding effect of smoking, a recognized risk factor for lung cancer and the association of smoking and alcohol consumption.

nism. Other factors, such as presence of an elevated serum cholesterol level, sex, and increased age, also have an impact on the disease risk and need to be included in the analysis. When the data are stratified according to cholesterol status the relative risk estimates differ considerably. In the low cholesterol stratum, the relative

Total Study Population:

High catecholamines	Coronary Heart Disease	No Coronary Heart Disease	Total
Yes	81	285	366
No	132	1329	1461
Total	213	1614	1827

$$\text{Relative Risk (crude)} = \frac{81/366}{132/1461} = 2.4$$

Low Serum Cholesterol:

High catecholamines	Coronary Heart Disease	No Coronary Heart Disease	Total
Yes	60	282	342
No	111	1059	1170
Total	171	1341	1512

$$\text{Relative Risk (low cholesterol)} = \frac{60/342}{111/1170} = 1.8$$

High Serum Cholesterol:

High catecholamines	Coronary Heart Disease	No Coronary Heart Disease	Total
Yes	21	3	24
No	21	270	291
Total	42	273	315

$$\text{Relative Risk (high cholesterol)} = \frac{21/24}{21/291} = 12.1$$

EXAMPLE 3 Example of interaction by serum cholesterol status in a cohort study looking at the association between endogenous catecholamines and risk of developing coronary heart disease.
Both cholesterol and catecholamines increase risk of coronary heart disease. When both are present, risks are more than the product of the two independent risk factors indicating a synergy or interaction between the two factors and risk of disease.

risk for coronary heart disease in relation to high endogenous catecholamine levels is 1.8. In contrast, in the high cholesterol stratum the relative risk is increased to 12.1. This suggests that interaction is present between these risk factors such that the combined risk of high cholesterol and high catecholamines is much more than the product of the two individual risks. When an interaction is observed one should try to elucidate the underlying biological mechanisms. Calculating a weighted average risk estimate would not make sense because it would ignore the biological clues that can be inferred from these observations.

The concept of interaction is important in molecular epidemiological studies that consider the interplay between genes (e.g., genetic polymorphisms resulting in different enzyme activity) and environmental exposures.

17.3 MOLECULAR EPIDEMIOLOGY

Molecular epidemiology involves the incorporation of biologic events at the physiologic, cellular, and molecular levels into epidemiological research. This application of molecular tools to population-based studies may help to better dissect and understand the development of disease. A key contribution of molecular epidemiology is that it helps to define better the events between exposure and disease. This not only contributes to our knowledge of the different pathways and mechanisms of disease, but it may provide useful markers for prevention, detection, treatment, and prognostic purposes.

An important issue in epidemiological research is the availability of accurate measures of exposure. The molecular epidemiological approach can provide a better assessment and reconstruction of past exposures, a more sensitive measure of current exposure and how people respond to exposure.

The combination of better exposure assessment and better understanding of disease pathways helps reduce misclassification of these two important variables. Another challenge to epidemiological research has been to explain the observation that, given the same exposure, only a fraction of individuals develop disease. The concept of genetic susceptibility, and the incorporation of genetic analyses into recent epidemiological research, may help explain part of this variability and has generated widespread interest in molecular epidemiology. We will focus on this aspect in more detail later in this chapter.

17.3.1 The Use of Biological Markers in Epidemiological Research

The development of molecular epidemiological research has been characterized by the search for biological markers that influence or predict the incidence or outcome of a particular disease. Biological markers, commonly referred to as biomarkers, are any substance, structure, or process that can be measured in the human body or its products. They can consist of biochemical, molecular, genetic, immunologic, or physiologic signals of events in biological systems. Biomarkers have been classified in different categories according to where they occur in the progression from exposure to disease. Given the biological nature of such markers, these classifications are imperfect and reflect our limited understanding of the complex process from exposure to disease.

The issue of what constitutes a good biological marker for epidemiological research is complex. Does the marker measure what it is supposed to? Are the conclusions obtained applicable to other populations besides the one under analysis? Is a putative marker for an exposure specific for that exposure? As with all epidemiological studies, one must be cautious in extrapolating the results of one study to different populations or between exposure doses or different target tissues.

The characteristics of any biologic measure must be well described before using them in large-scale epidemiological studies. The first step involves basic research on specific disease models, usually using a variety of animal, human, in vitro, and in vivo systems. The bridge between the laboratory research and epidemiology consists of the application of biomarkers in small population-based studies in what has been referred to as *transitional studies*. These studies deal mainly with such issues as sample processing, assay accuracy and precision, and the analysis of possible confounders. It is important to determine whether a particular biomarker assay is reliable and accurate and which are the optimal conditions for collection, processing, and storage of samples. Finally, the biomarker may be applied in larger studies.

17.3.2 Types of Molecular Epidemiological Studies

We discuss below four common types of molecular epidemiological studies that focus on different aspects of the exposure disease continuum. We introduce as well several types of biological markers that have been studied in these different types of molecular epidemiological studies.

17.3.2a Exposure Dose Studies

Biological markers can be used in epidemiological studies to assess doses of exposure and their effects in the outcome of disease. In traditional epidemiological studies, exposures are frequently assessed by questionnaire information, exposure records, or medical histories. The use of biological markers that measure how much of an agent is present in the body may provide a more direct and accurate measure of exposure, and helps eliminate some of the problems of questionnaire-based exposure information.

An important issue is the half-life of the agents being measured. The excretion half-life of biological markers in blood and urine ranges widely from less than 1 hour, as in the case of styrene, to more than 10 years, as in the case of cadmium. Different tissues might provide different information as in the case of lead. In blood, the half-life of lead is approximately 1 month, which limits the use of this tissue for estimating past exposures. However, in bone tissue the half-life of lead ranges from 10 to 40 years, making such measurements a good marker for assessing past exposures.

Very few markers are stable for long periods of time and thus they only reflect recent exposure. This limits their application in the study of chronic disease where past exposure, often years prior to disease, is most important. For example, cotinine, a compound present in tobacco, has been used in many studies to estimate levels of exposure to tobacco smoke. However, this compound is rapidly excreted and therefore it can only be used to assess recent exposure and not lifetime exposure to cigarette smoke.

Biological markers are sought that measure how much of the exposure agent has actually interacted with the organism. Because individuals process exposures differently, the use of such markers may provide a more accurate measurement of relevant exposure or biological dose. An example of this type of marker is the use of protein and/or DNA-adducts, usually measured in blood, serum, urine, or exfoliated cells. Again, such measures often reflect only recent exposure since most adducts are short-lived species and may have limited utility in the study of chronic disease, besides being quite expensive to measure.

17.3.2b Physiologic Studies

Another variation of a molecular epidemiological approach is the analysis of physiologic alterations and their associations with disease. These studies may provide insights into disease mechanisms. They also provide measures that can be used in diagnosis or prevention.

For example, there are recent reports of an association between plasma levels of insulin-like growth factor 1 (IGF-1) and colorectal cancer. IGF-1 is a hormone that has been shown to be involved in promoting postnatal somatic growth, maintaining lean tissue mass and inhibition of apoptosis. Animal experiments have shown that dietary restrictions lower serum levels of IGF-1 as well as the incidence of cancer development. The action of IGF-1 can be regulated by the IGF-1 binding protein (BP3). This protein has been shown to inhibit cell proliferation and induce apoptosis by preventing the binding of IGF-1 to its receptor (IGF-1R). A cohort study was carried out in which plasma samples of each individual were collected and stored. After years of follow-up, a number of these individuals developed colorectal cancer. A nested case-control study was set up within this cohort to analyze a possible association between colorectal cancer development and plasma levels of IGF-1 and BP3 in the samples obtained prior to disease. Those individuals that developed the disease had higher levels of IGF-1 and lower levels of BP3 at enrollment when compared with those individuals who did not develop colorectal cancer. Because the plasma samples had been collected prior to the development of disease, the levels were unlikely to have been affected by the disease process. Instead, they are believed to reflect presumably inherited differences between individuals prior to disease, and suggest that these two factors are important determinants of colon cancer risk.

This example illustrates how biologic measures can be used in epidemiological studies to identify causes of disease.

17.3.2c Analysis of Gene-Environment Interactions

One of the more recent areas of molecular epidemiology has been the study of *genetic susceptibility*, that is the contribution of the inherited genetic background in the development of disease, and *gene-environment interactions*, the interplay between the inherited genetic background and environmental factors in the development of disease. In most human diseases, there is a wide interindividual variability in occurrence, prognosis, and outcome. In a limited number of situations, the incidence of disease can be explained by pure environmental or genetic etiologies. For example, a few viral or bacterial infections where almost everyone who is exposed develops disease or specific chromosomal aberrations such as Down's syndrome, where everyone with the aberration is affected, albeit to different degrees. More commonly, a complex combination of genetics and exposure, so-called "nature and nurture" appears to control who develops disease. The risk from exposure to a particular agent may only be evident in a genetically susceptible subgroup of the population. Conversely, the risk of carrying a particular genotype may only be evident in an exposed population. It is important to keep in mind that when we refer to "environmental" exposures we include all that is not genetic: lifestyle exposures, diet, occupational, and environmental agents among others. The analysis of gene-environment interactions is an emerging issue in modern epidemiological research and the molecular epidemiological approach provides a framework with which to carry out such studies. We discuss two aspects of gene-environmental interactions: the analysis of genetic susceptibility and the analysis of critical target gene alterations.

Analysis of Genetic Susceptibility. When a germline variation is present in the population at a frequency of at least 1% it is referred to as a *polymorphism*. There are three types of DNA variations: deletions, insertions, and base substitutions. Such variation may have functional consequences in terms of transcription, translation, or in the activity of the resulting protein, or may be "silent" and not have a functional consequence. For susceptibility studies functional polymorphisms are the most relevant. The underlying hypothesis is that common inherited variations in genes can increase the risk of developing disease. The usual study design for susceptibility gene analysis is case-control studies.

Until now the lack of candidate genes for analysis has probably been the limiting factor in this type of study. Even though we may have DNA sequence information for a number of genes, the presence of allelic variants is only known for a small proportion of them. Fortunately, the search for common polymorphisms in a variety of genes is an area of increased research. This is facilitated by faster and more accurate DNA sequencing techniques that allow the screening of large numbers of individuals, and make efforts of allele discovery feasible.

To date, most studies of gene-environment interactions have been focused on the analysis of metabolism genes involved in the activation (phase I) and/or detoxification (phase II) of different carcinogens in a variety of cancers. Polymorphisms for a number of phase I and phase II enzymes have been studied and related to disease risk. For example, N-acetyl-transferase 1 (NAT1) and N-acetyl-transferase 2 (NAT2), two enzymes involved in the O-acetylation and N-acetylation of arylamines, have been linked to risk of bladder cancer (Table 17.2). Arylamines are car-

TABLE 17.2 Data from a Case-control Study Looking at the Association Between Bladder Cancer, Smoking, and Polymorphisms in the NAT1 and NAT2 Genes.

Smoking (Environment)	Risk Polymorphisms for NAT1 and NAT2 (Genotype)	Cases	Controls
−	Both absent	6	13
−	Only one present	24	47
−	Both present	6	10
Total nonsmokers		*36*	*70*
+	Both absent	42	32
+	Only one present	102	77
+	Both present	35	12
Total smokers		*179*	*121*
Total individuals		*215*	*191*

Source: Data derived from Taylor et al. Cancer Research 58: 3603–3610, 1999.

cinogenic compounds present in tobacco smoke and in a variety of industrial and occupational exposures. Functional polymorphisms have been found for the genes encoding these enzymes. In the case of the NAT2 gene, variant alleles produce proteins that have lower activity for the detoxification of arylamines by N-acetylation. In the case of the NAT1 gene, a variant allele produces a protein that activates arylamines to a more reactive compound. Cigarette smoking is a risk factor for developing bladder cancer. A bladder cancer case-control study of polymorphisms in these two genes and exposure to cigarette smoke demonstrated a statistically significant three-way interaction between these two genes and exposure. This study is illustrated in Example 4 as an example of one of many types of possible gene-environment interactions. Smokers, regardless of their genotype, had an increased risk of less than three for developing bladder cancer. Those individuals who had the two variants, regardless of smoking status, had an increased risk of less than 2 for developing bladder cancer. However, the combination or interaction, between the presence of these two variants with the exposure to cigarette smoke increased the risk of developing bladder cancer up to sixfold over nonsmokers without the variants. If an individual exposed to arylamine compounds in cigarette smoke has the variant NAT2 allele, the arylamine pro-carcinogens will not be efficiently detoxified. The excess arylamine compounds may be activated by NAT1 to active carcinogens. Another type of gene-environment interaction is illustrated in Example 5. In this case, neither the environmental exposure, smoking, nor the genetic background, TGFα polymorphism, increases the risk for developing a specific birth defect in newborn babies. However, when both are present the risk increases fivefold.

Analysis of Critical Target Gene Alterations. The main goal in these studies is to identify patterns of acquired (rather than inherited) DNA mutation in genes that are critical to the development of disease. This type of study provides insights into the mechanisms by which an exposure leads to disease and may identify pathways for diagnosis, prevention, and treatment. This type of analysis is particularly relevant in the study of cancer given the genetic nature of carcinogenesis.

Common gene alterations that can be analyzed are point mutations, deletions, amplifications, inversions, and changes in DNA methylation patterns. In the case of carcinogenesis, critical target genes include tumor suppressor genes, oncogenes, genes involved in cell cycle control, angiogenesis, metastasis, and DNA repair. To date, the tumor suppressor gene p53 has been the most intensively studied critical target gene. This gene is involved in cell cycle control, DNA repair, and apoptosis. It is the most commonly mutated gene in human cancers and mutations throughout the gene can disrupt its function. One example of exposure-specific critical target gene mutation is in skin cancer. Studies were carried out where DNA was extracted from tumor tissue, the p53 gene was amplified by PCR, and screened for mutations. Skin tumors from people with extensive sunlight exposure showed frequent C to T mutations usually at the sites of pyrimidine dimers. In a few cases there were double mutations: CC to TT, a mutation that is almost unique for UV light exposure. These types of alterations are infrequently caused by other agents and have been rarely observed in other malignancies where UV light exposure is not an etiological factor.

The most common study design to analyze critical target gene alterations is a case-case design. In cancer studies, tumors from individuals exposed to a particular

Effect of Smoking (environment):

Smoking	Individuals with Bladder Cancer (cases)	Individuals without Bladder Cancer (controls)	Total
Yes	179	121	300
No	36	70	106
Total	215	191	406

Odds Ratio of exposure (environment) $= \dfrac{179/36}{121/70} = \dfrac{179*70}{36*121} = 2.8$

Effect of NAT1 and NAT2 at risk genotypes (gene):

Both risk polymorphisms present	Individuals with Bladder Cancer (cases)	Individuals without Bladder Cancer (controls)	Total
Yes	41	22	63
No	48	45	93
Total	89	67	156

Odds Ratio of exposure (genotype) $= \dfrac{41/48}{22/45} = \dfrac{41*45}{48*22} = 1.8$

Combined effect (Gene-Environment interaction):

	Individuals with Bladder Cancer (cases)	Individuals without Bladder Cancer (controls)	Total
Smoking and both Risks polymorphisms	35	12	47
No smoking and no risk polymorphisms	6	13	19
Total	41	25	66

Odds Ratio of exposure $= \dfrac{35/6}{12/13} = \dfrac{35*13}{6*12} = 6.3$

EXAMPLE 4 Example of gene-environment interaction (GEI) in a case-control study looking at the association between bladder cancer, smoking and polymorphisms in the NAT1 and NAT2 genes.

Please refer to Table 17.2 to follow this example. In this particular type of GEI the environmental exposure alone increases the risk considerably, the genotype alone has some contribution to the development of bladder cancer, but not as strong as the environment. However, when both are considered together the increase in risk is higher than what would be expected from the product of both factors alone. There is a synergistic effect or interaction between the two risk factors, genotype and environment, in the development of bladder cancer. In this example for simplicity we consider the "at risk" genotype as the combination of the two functional polymorphisms for each of the two genes analyzed, NAT1 and NAT2.

Source: Data derived from Taylor et al. *Cancer Res.* 58 (1999), 3603–3610.

Effect of Smoking (environment):

Smoking	Children with cleft palate (cases)	Children without cleft palate (controls)	Total
Yes	13	69	82
No	36	167	203
Total	49	236	285

$$\text{Odds Ratio (smoking)} = \frac{13/36}{69/167} == \frac{13*167}{36*69} = 0.9$$

Effect of TGFα polymorphism (gene):

Risk polymorphism	Children with cleft palate (cases)	Children without cleft palate (controls)	Total
Yes	7	34	41
No	36	167	203
Total	43	201	244

$$\text{Odds Ratio (gene)} = \frac{7/36}{34/167} = \frac{7*167}{36*34} = 1.0$$

Combined effect (Gene-Environment interaction):

	Children with cleft palate (cases)	Children without cleft palate (controls)	Total
Smoking and Risk polymorphism	13	11	24
No smoking and No risk polymorphism	36	167	203
Total	49	178	227

$$\text{Odds Ratio (Gene-Environment)} = \frac{13/36}{11/167} = \frac{13*167}{36*11} = 5.5$$

EXAMPLE 5 Example of gene-environment interaction (GEI) in a case-control study looking at the association between maternal smoking, genetic polymorphisms in the transforming growth factor alpha (TGFα) gene and the risk of cleft palate in children. In this case neither the environmental exposure alone (smoking) nor the presence of an "at risk" genotype alone (TGFα polymorphism) increases the risk of developing this particular facial malformation in newborns known as cleft palate. However, when both risk factors are considered together there is a 5-fold increase in the risk of developing this particular malformation.

Source: Data derived from Yang et al., *Epidemiol. Rev.* 19 (1997), 33–43.

agent are compared with tumors from individuals unexposed to that agent. The most important factors in the success of this type of study are the availability of good exposure information, good tissue samples, and the knowledge of critical genes involved in the development of disease.

17.3.3 Aspects of Field Work in Molecular Epidemiology

As in traditional epidemiology, the next step after the careful design of most studies that will test a specific hypothesis regarding an exposure and a health outcome, is the *field work*. During this part of the study all the exposure data will be collected. In the case of molecular epidemiological studies, biological samples will also be collected.

17.3.3a Informed Consent and Confidentiality

In many cases every individual that participates in a study must sign an informed consent. An informed consent document provides a summary of the research, gives permission to collect data and samples, specifies what will be done with the samples obtained, and informs the participants about risks and confidentiality. Such documents usually include the following points:

- scientific reasons for the study
- the logistics of the procedure and the specific contribution of the participant (e.g., filling out a questionnaire and having blood sample taken by a nurse)
- risks or benefits of participation
- precautions taken to maintain confidentiality of data and test results,
- the possibilities to be notified about individual test results (now and/or in the future)
- freedom to decline participation without any consequence

If a subject agrees to participate an informed consent statement is signed. Usually informed consents are written to anticipate for future use of the samples and questionnaire information. If someone declines participation the researcher usually records the reason for nonparticipation in order to assess possible selection bias.

17.3.3b Sample Collection and Storage

The availability of appropriately collected and correctly processed tissue is a central concern in molecular epidemiological studies. The required sample has to be relatively easy to obtain and in sufficient amounts without unreasonable burden for the subject. For epidemiological studies that may include hundreds or thousands of subjects, it is important to make sample collection as complete, simple, and inexpensive as possible. The organization of large-scale data and biological sample collection takes considerable effort and money.

After informed consent has been obtained biological samples and questionnaire data are collected. A unique identification number is assigned to each participant, which makes it possible to store biological samples and questionnaire data separated from personal identifiers. This secures each subject's privacy and at the same time blinds the researchers. A master list is usually kept under limited access to provide optimum privacy of the participants.

There are many types of specimens that can be a source of material suited for long-term storage in large epidemiological studies. Studies have been reported using hair, toenails, semen, maternal milk, saliva, sweat, exfoliated skin, and bladder cells. Pathology tissue collections are also useful to this end. However, blood and urine are by far the most commonly used sources of material for large-scale prospective studies. The methods and materials used for obtaining, processing, and storing of samples must be carefully chosen. Therefore, collaboration between epidemiologists and experienced molecular biologists and analytical chemists is needed in a very early phase of the study design. Sample collection must be done in a way that offers the best possibilities and flexibility for storage and future laboratory analysis. It is beyond the scope of this text to fully expand on all steps involved. More details can be found in the suggested reading at the end of this chapter.

17.3.3c Quality Assurance Program

One of the biggest challenges for large molecular epidemiological studies in which biological samples are collected, stored, and analyzed is the establishment of a quality assurance system. This should cover the specimen from collection through the reporting of the results. Control samples are useful for quality control. Storage of several aliquots of a control sample at different places and/or temperatures gives the opportunity to assess the influence of storage duration, storage location, freeze/thaw history, equipment failures, and possible contamination. This is important because it is sometimes unclear what the exact effect of storage conditions and duration can be for certain specimens and assays. Quality control during analysis in the laboratory is also important, such as routine experiments to check precision and reproducibility of an assay.

Epidemiological studies may stretch over a considerable time period because both sample collection and processing of large numbers of samples takes time. Therefore, it is critical to implement longitudinal control of laboratory performance. This can be a challenge because of sample instability over time and subtle changes in laboratory procedures may be difficult to distinguish. Any changes in assay procedure over time should be carefully documented and evaluated. Because of the large scale of epidemiological studies any procedure should be standardized, simplified, and automated as much as possible.

17.4 SUMMARY AND CONCLUDING REMARKS

Molecular epidemiological research is a field that directly benefits from the implementation of laboratory techniques. Given the large size of most epidemiological studies there is a need for development and use of high-throughput devices in order to apply molecular techniques at a large scale with optimal quality and speed. The continuous development of new DNA analysis techniques makes its application in molecular epidemiological studies very promising. This also calls for close collaboration and exchange of knowledge between epidemiologists and laboratory scientists in all phases of the study. Such collaborations best extend throughout the study inasmuch as careful attention should be paid to the sample collection methods, storage, data analysis, and interpretation. The power of this field is in the combination of strengths of epidemiology and biology, providing a better understanding of disease events at the physiologic, cellular, and molecular levels.

SUGGESTED READING

Recommended Text on Epidemiology

Dos Santos Silva. *Cancer Epidemiology: Principles and Methods*. Lyon, France: IARC Scientific Publications, 1999.

Hennekens, C.H. and Buring, J.E. *Epidemiology in Medicine*. Boston, MA: Little Brown, 1987.

Rothman, K.J. and Greenland, S. *Modern Epidemiology*. Philadelphia, PA: Lippincott-Raven, 1998.

General Texts on Molecular Epidemiology

Hulka, B.S., Wilxosky, T.C. and Griffith, J.D. Biological Markers in Epidemiology. New York: Oxford University Press, 1990.

Schulte, P.A. and Perera, F.P. *Molecular Epidemiology*. San Diego, CA: Academic Press, 1993.

Toniolo, P., Boffetta, P., Shuker, D.E.G., Rothman, N., Hulka, B. and Pearce, N. *Application of Biomarkers in Cancer Epidemiology*. Lyon, France: IARC Scientific Publications, 1997.

Biomarkers

Rothman, N., Stewart, W.F. and Schulte, P.A. Incorporating biomarkers into cancer epidemiology: a matrix of biomarker and study design categories. *Cancer Epidemiol. Biomarkers Prevention* 4 (1995), 301–311.

Sample Collection and Storage

Austin, M.A., Ordovas, J.M., Eckfeldt, J.H. et al. Guidelines of the National Heart, Lung, and Blood Institute Working Group on Blood Drawing, Processing, and Storage for Genetic Studies. *Am. J. Epidemiol.* 144 (1996), 437–441.

Yates, J.R.W., Malcolm, S. and Read, A.P. Guidelines for DNA Banking: Report of the Clinical Genetics Society Working Party on DNA Banking. *J. Med. Genet.* 26 (1989), 245–250.

Genetic Epidemiology

Khoury, M.J., Beaty, T.H. and Cohen, B.H. *Fundamentals of Genetic Epidemiology*. New York: Oxford University Press, 1993.

Journal Focused on Epidemiology and Molecular Epidemiology

American Journal of Epidemiology

Cancer Epidemiology, Biomarkers and Prevention

Epidemiology

CHAPTER EIGHTEEN

Biochemical Toxicology of the Peripheral Nervous System

PIERRE MORELL, JEFFRY F. GOODRUM, and THOMAS W. BOULDIN

18.1 INTRODUCTION

All parts of the nervous system are susceptible to toxic injury. When toxic injury alters the function of the peripheral nervous system (PNS), the disease process is referred to as *toxic neuropathy*. Neuropathy is a common manifestation of toxic injury to the nervous system and may lead to serious and sometimes permanent disability. Recognition of the mechanisms responsible for toxic injury of the PNS is critically important if we are to better predict which of the new chemicals constantly being introduced into our pharmacopoeia, workplace, and environment have the potential for causing neuropathy. A partial list of neurotoxic agents causing neuropathy is provided in Table 18.1. Only a few of these toxicant-induced neuropathies have been studied in detail, and thus our understanding of the biochemical toxicology of these neuropathies is very limited. This chapter reviews primarily the biochemical mechanisms implicated in axonal degeneration and segmental demyelination, the two common noncarcinogenic responses of the PNS to toxic injury, and discusses possible molecular markers of peripheral neuropathy that may be useful in evaluating the potential toxicity of novel compounds.

The PNS is defined as that part of the nervous system external to the brain and spinal cord (Fig. 18.1). As such, the PNS includes the cranial nerves, dorsal and ventral spinal roots, spinal nerves and their branches, and ganglia. The primary function of the PNS is to convey sensory and motor information, including information from the autonomic nervous system, between the central nervous system (CNS, brain, and spinal cord) and the rest of the body. Nerve fibers convey the information between the periphery and the CNS. The nerve fibers are bundled together to form individual peripheral nerves (Fig. 18.2).

Each *nerve fiber* is composed of an axon and its supporting Schwann cells. The *axon* is a long, highly specialized cytoplasmic process of the neuron and conducts the nerve impulse. Axons measure 0.2–22 μm in diameter and may be over 1 m meter in length (Fig. 18.3). The neuronal cell bodies giving rise to these peripheral nerve axons may be located in the CNS (e.g., motor neurons innervating skeletal muscle) or PNS (e.g., sensory neurons innervating skin and other organs). The sensory

TABLE 18.1 Some Examples of Chemicals (Neurotoxic Agents) Causing Neuropathies in Mammals.

Drugs	Environmental Agents
Amiodarone	Acrylamide
Chloroquine	Allyl chloride
Cisplatin	Arsenic
Colchicine	Buckthorn toxin
Dapsone	Carbon disulfide
Disulfiram	Chlordecone
Glutethimide	Dimethylaminopropionitrile
Gold compounds	Diphtheria toxin
Hydralazine	Ethylene oxide
Isoniazid	n-Hexane
Lithum carbonate	Methyl n-butyl ketone
Misonidazole	Lead
Nitrous oxide	Mercury
Nitrofuratoin	Methyl bromide
Perhexiline maleate	Polychlorinated biphenyls
Phenytoin	Tellurium
Pyridoxine (vitamin B6)	Tetrachlorobiphenyl (TCB)
Sodium cyanate	Thallium salts
Suramin	Trichloroethylene
Thalidomide	Vacor
Vincristine	Zinc pyridinethione

neurons of the dorsal root ganglia have an axon that branches, with one axonal process traveling to the CNS via a spinal nerve root and another axonal process traveling to the periphery via a peripheral nerve.

Schwann cells, which are supporting cells analogous to the glial cells of the CNS, envelop all axons of the PNS. The Schwann cells covering the larger diameter axons show further specialization by covering the axon with a *myelin sheath*, which is an elaboration of each enveloping Schwann cell's plasmalemma (Fig. 18.4). The myelin sheath permits much more rapid transmission of a nerve impulse by the nerve fiber. The differentiation of the Schwann cell into a myelinating Schwann cell is dictated by the axon.

The biochemistry of neurons and Schwann cells is discussed below. It is, however, important to realize that there are certain other structural features of the PNS that greatly influence the effects of systemically administered toxicants. Of great importance is the fact that the PNS has a *blood-nerve barrier* (BNB), which is analogous to the blood-brain barrier (BBB) of the CNS and regulates the chemical composition of the extracellular (interstitial) fluid surrounding the individual nerve fibers within a peripheral nerve. The BNB is located at the walls of the capillaries within peripheral nerves (endoneurial capillaries), and at the perineurium, which is a multilayered cellular sheath that surrounds each peripheral nerve. The endothelial cells lining the endoneurial capillaries form a "nonporous" barrier between the blood

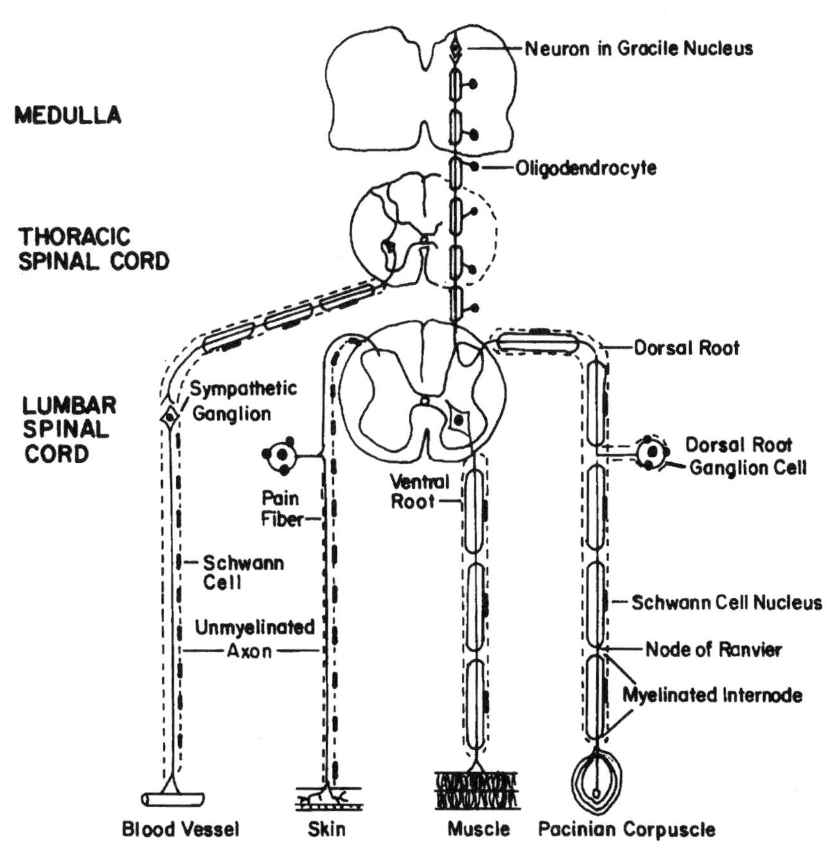

FIGURE 18.1 Peripheral nervous system. The major components of the PNS are illustrated in this diagram.

Source: Adapted from Schaumburg et al., *Disorders of Peripheral Nerves*, Philadelphia PA: F.A. Davis Co., 1983.

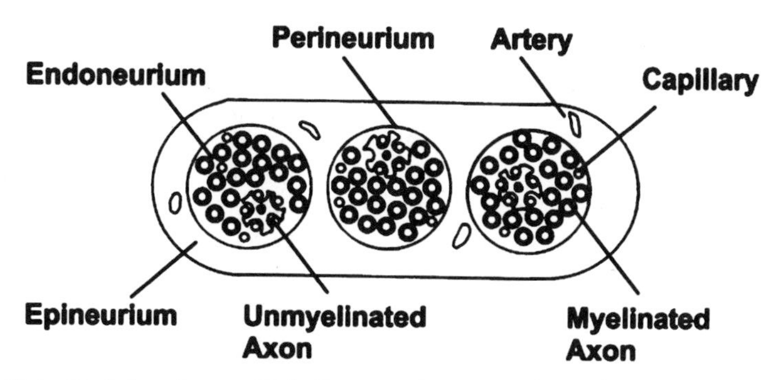

FIGURE 18.2 Peripheral nerve. This diagram shows a peripheral nerve in cross section. The nerve contains three bundles (fascicles), with each fascicle containing a mixture of myelinated and unmyelinated axons.

Source: Adapted from Schaumburg et al., *Disorders of Peripheral Nerves*, Philadelphia PA: F.A. Davis Co., 1983.

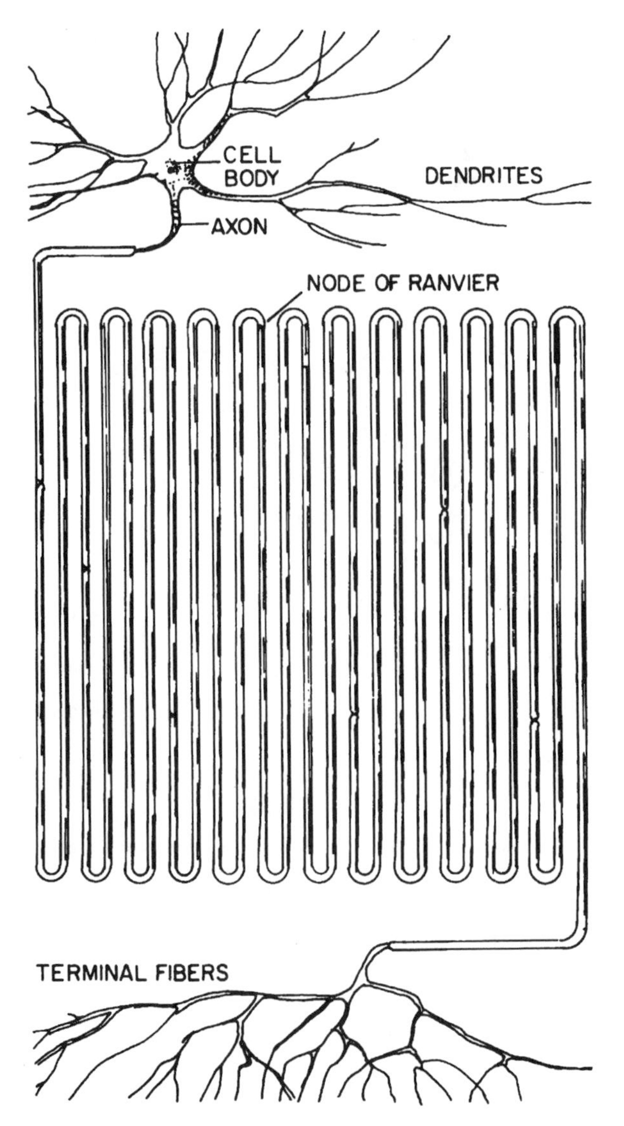

FIGURE 18.3 Neuron. This diagram is drawn to scale for a neuron with a 1-cm axon (the axon is folded for diagrammatic purposes). Many PNS axons may be more than a meter in length.

Source: Adapted from Stevens C.F. The neuron. *Sci. Am.* 241 (1979), 54.

and the interstitial fluid within the nerve, whereas the perineurial cells form a nonporous barrier between the interstitial fluid outside and inside the nerve (see Fig. 18.2). These cellular barriers greatly restrict the movement of macromolecules, ions, and water-soluble nonelectrolytes, but not of lipid-soluble molecules, into the peripheral nerve. In addition to these nonporous structural barriers, the endothelial cells of the BNB have carrier-mediated transport systems for selectively moving

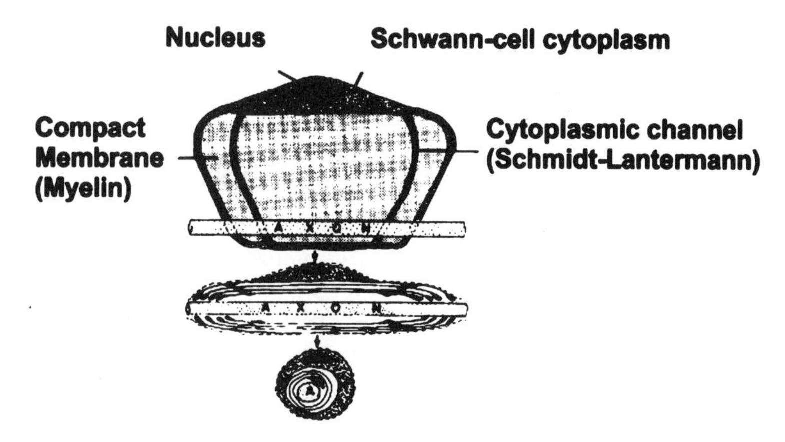

FIGURE 18.4 Myelinating Schwann cell of PNS. The same Schwann cell is shown unwrapped (*top*), in longitudinal section (*middle*), and in cross-section (*bottom*). Note the channels of cytoplasm (Schmidt-Lantermann clefts) and the large expanses of compacted cell membranes (myelin). (These drawings are not to scale.)

Source: Adapted from Raine C.S. Morphology of myelin and myelination. In: Morell, P. (Ed) *Myelin*, 2nd Ed. New York: Plenum Press, 1984.

certain required molecules (e.g., glucose) across the barrier into the nerve. The combination of these structural barriers and carrier-mediated transport systems permits the BNB to closely regulate the composition of the interstitial fluid surrounding the individual nerve fibers within each peripheral nerve.

The importance of the BNB in protecting the PNS from toxic injury is illustrated by the toxic neuropathy associated with diphtheria. Diphtheria is a human disease caused by an infection of a wound or the upper respiratory tract by the bacterium *Corynebacterium diphtheriae*. The polypeptide toxin released by some strains of this bacterium is capable of entering the myelin-forming cells (Schwann cells of PNS; oligodendrocytes of CNS), inhibiting myelin synthesis, and producing breakdown of the myelin sheath (demyelination). It has long been recognized that the demyelination associated with diphtheria is remarkably localized to the ganglia of the PNS and completely spares the CNS. Experimental studies in animal models have revealed that the BBB, which effectively excludes the toxin from the CNS, is responsible for the absence of demyelination in the CNS. Likewise, the BNB excludes the toxin from much of the PNS. However, the ganglia of the PNS do not have a BNB and are exposed to the blood-borne diphtheria toxin. It is this absence of a BNB in the PNS ganglia that explains the localization of the demyelination to the PNS ganglia in diphtheria neuropathy.

18.2 SPECIALIZED ASPECTS OF NEURONAL METABOLISM

Discussion of vulnerability of the axon to toxicants must be based on an appreciation of specialized aspects of the metabolism of the neuron. This cell type has been discussed in the chapter dealing with the CNS. The concepts of chemically mediated neurotransmission, second messenger systems, specialized ion channels, and conduction of waves of depolarization with consequent high demand for energy to restore Na^+ and K^+ ion gradients, are applicable to the PNS as well as the CNS. Of particular relevance is the fact that the neuromuscular junction is cholinergic and there is a wide range of toxic agents which act at this site, for example, many insecticides such as organophosphorous compounds inhibit cholinesterase.

Another aspect of the PNS that renders its neurons particularly sensitive to certain types of toxic insult is the sheer size of many of the motor and sensory neurons. Neurons have a highly asymmetric distribution of cytoplasm and plasma membrane; in the larger PNS neurons the volume of the axon and its surface area exceeds that of the cell body by several orders of magnitude (Fig. 18.3). The cell body, however, contains most of the machinery for synthesis and degradation of macromolecules. Thus, maintenance of the vast expanse of the axon and of the distant and metabolically active nerve terminals is absolutely dependent on the continued delivery (*axonal transport*) of materials to and from the cell body. The processes involved in axonal transport presumably involve mechanisms similar to those utilized by all eukaryotic cells to transport materials from their site of synthesis to their site of utilization and, eventually, to their site of degradation. However, the unique structure of neurons makes the significance of these transport processes, relative to other metabolic processes, much greater than in most other cell types. It is considered likely that many toxicant-induced metabolic insults may directly or indirectly affect axonal transport. Thus, this process is a focus of many studies exploring the mechanisms by which certain toxicants bring about axonal degeneration.

18.2.1 Anterograde Axonal Transport

Anterograde axonal transport (movement from the cell body to the nerve ending) involves movement of macromolecules at one of several relatively discrete velocities. The extremes of this spectrum are referred to respectively as rapid and slow axonal transport. *Rapid transport* involves the movement of cytoplasmic vesicles at rates from 200 to 400 mm/day. Included are vesicles involved in neurotransmission (and containing, among other components, various enzymes of neurotransmitter metabolism), vesicles presumed to be carrying various surface membrane components, as well as agranular endoplasmic reticulum. It has long been suspected that microtubules form the framework on which such vesicles move and, in recent years, it has been clearly demonstrated that a microtubule-stimulated kinase, kinesin, is involved in conversion of the energy of ATP to anterograde movement of vesicles along the microtubules. There is evidence that mitochondria, and possibly certain other subcellular organelles, are transported at slower rates than are the rapidly transported vesicles. This may be explained without presumption of another anterograde transport system if it is assumed that the time of association with the transport system may vary among different classes of particles. For example, vesicles carrying certain enzymes involved in neurotransmitter synthesis might remain strongly associated with the transport vector and thus move rapidly while mitochondria might spend less time attached to the vector and therefore have a slower net transport rate.

Slow axonal transport involves primarily components of the cytoskeleton and its protein precursors. Slow transport can be resolved into two rates, one of less than 1 mm/day and one of several millimeters per day. Components traveling at the slower rate include tubulin, and its associated proteins, and a group of three other structural proteins known as the neurofilament triplet. Components traveling at the faster rate include some tubulin, as well as actin and various soluble enzymes. It has been noted that the rate of rapid transport is constant in an animal over a range of axon types and is relatively constant among various species if correction is made for temperature differences; in contrast, slow transport has some dependence on the length or diameter of the axon. The mechanism of slow axonal transport is not well understood.

The slow transport of tubulin and neurofilament proteins has been interpreted as the movement of the microtubules and neurofilaments, which are the major components of the axonal cytoskeleton. Intact microtubules are thought to move at the slower of the two "slow" rates of transport, whereas the soluble dimers or oligomers of tubulin subunits move at the faster of the two slow rates. There is an equilibrium between the two phases; tubulin subunits are constantly being added or removed from microtubules. The three neurofilament peptides are associated almost exclusively in insoluble polymers. Part of this polymer moves down the axon at the slower rate of slow transport; the rest is in a stationary component. The equilibrium between the two pools of neurofilament proteins may be governed by phosphorylation and dephosphorylation, with the most phosphorylated forms being preferentially represented in the stationary component.

18.2.2 Retrograde Axonal Transport

Retrograde axonal transport (from nerve ending back to cell body) has been studied in a number of systems and consists of a single phase of transport of vesicles at a rate slightly more than half that of rapid anterograde transport. Another microtubule-stimulated kinase, dynein (distinct from kinesin), is specifically involved in using energy from ATP to bring about movement of vesicles from the nerve terminal back to the cell body. The retrogradely traveling vesicles are different from those moving anterogradely and appear related to those formed in connection with endocytic and degradative pathways. Both anterograde and retrograde transport are dependent on local oxidative metabolism within the axon for a continued supply of ATP and have certain requirements for calcium ions.

18.3 SPECIALIZED ASPECTS OF SCHWANN CELL METABOLISM

Schwann cells in contact with axons may undergo further specialization and make myelin. Myelin is a greatly extended and modified plasma membrane wrapped around the axon in a spiral fashion (Fig. 18.4). Myelin accounts for almost half of the protein and an even larger proportion of the lipid of the sciatic nerve of mammals (because of its size and accessibility this nerve is the object of most studies of PNS myelination). As is the case in the CNS, much of the myelin in the PNS is deposited in a restricted period of development; for example, axons in the sciatic nerve of the rat go from being unmyelinated at birth to being well myelinated at 2 weeks of age. It should be noted that in many cases rapid synthesis of myelin continues well beyond the time when, by morphological criteria, much of the deposition of myelin appears to be nearing completion. This is because body growth continues long after peripheral nerves are myelinated and functional. As limbs grow, so do the peripheral nerves, and the Schwann cells must produce new myelin to accommodate the elongating peripheral nerves.

Schwann cells share with neurons the characteristic of having a cell body that must support an enormous amount of peripherally located cell components: myelin in the case of the Schwann cell, and axon in the case of the neuron (Fig. 18.3 and 18.4). Thus, as in the case of axonal transport in neurons, it is assumed that highly specialized metabolic processes are involved in supporting the topologically distant myelin. In addition, myelin itself has a unique composition and structure (see following section), suggesting other possible sites of action of toxicants.

18.3.1 Myelin Composition and Metabolism

Myelin is modified plasma membrane. Myelin of the PNS resembles that of the CNS with respect to lipid composition. There is an enrichment in such specialized lipids as cerebroside and ethanolamine plasmalogen, and the high content of cholesterol plays an important role in control of membrane fluidity. The protein composition of PNS myelin is, however, distinct from that of CNS myelin. A single protein, P_0, accounts for half of all protein of PNS myelin. Of the other proteins present, most are expressed in the CNS as well as the PNS but in quantitatively different amounts. Prominent among these proteins are myelin basic proteins and myelin-associated glycoprotein.

Both the lipid and protein components of the myelin sheath are involved in metabolic turnover. Although this metabolism is less vigorous than that of the plasma membrane of other cell types, myelin is far from being metabolically inert. It is now known that there are certain components of myelin that turn over extremely rapidly; prominent in the PNS are the phosphate groups of P_0 protein and of the polyphosphoinositides. It should also be noted that the tight apposition of the cytoplasmic faces of the membranes (viewed in the electron microscope as the "dense" line of the repeat structure) is often split by a cytoplasmic channel (Schmidt-Lantermann cleft), which may be a metabolically active compartment. The possibility that this region of cytoplasmic contact with myelin is active metabolically is a relatively new concept, because for many years it was thought that myelin was, metabolically and structurally, a highly stable structure. Regions of metabolic activity in the membrane may also be suggestive of pores or channels, and this hypothesis is reinforced by data indicating the presence in compact myelin of carbonic anhydrase and other enzymes that might be related to transport. The observation, detailed below, that certain compounds may induce vacuolation (edema) in myelin raises the possibility that these compounds may act by interfering with some mechanism of water and/or ion transport involved in the structural stability of myelin membrane.

18.4 TOXIC NEUROPATHIES

18.4.1 Selective Vulnerability

The neuron and the Schwann cell are the principal cell types in the PNS. There are great morphological, biochemical, and functional differences between neurons and Schwann cells, and this is reflected in the considerable variation in their vulnerability to toxic injury. Some toxic neuropathies are characterized primarily by injury of the neuron or its axon, as evidenced by the presence of axonal degeneration in peripheral nerve, whereas other toxic neuropathies are characterized primarily by Schwann cell injury, as evidenced by the presence of demyelination. Those neuropathies characterized by axonal injury are often categorized as "axonal neuropathies," whereas those characterized by demyelination are categorized as "demyelinating neuropathies."

18.4.2 Characteristics of Axonal Neuropathies

The vast majority of neurotoxic agents that affect the PNS preferentially cause axonal injury rather than Schwann cell injury. Axonal injury is usually manifested as *axonal degeneration*, a pathologic process characterized by complete dissolution of the axon. If the degeneration involves a myelinated axon, then the myelin sheath enveloping the degenerating axon also breaks down. This myelin breakdown, which occurs in the context of axonal degeneration, is not considered demyelination, since "demyelination" refers to loss of the myelin sheath from an intact axon.

Detailed studies of the PNS in animal models of toxic neuropathy reveal that in most instances the axonal degeneration initially involves only the distal end of the axon, with the more proximal axon and neuronal cell body remaining intact (Fig. 18.5, A and B). In addition, these studies reveal that usually this distal axonal degeneration is initially limited to the longest and largest-diameter axons, which are those innervating the distal extremities. This distal axonal degeneration is often referred to as a 'dying-back" type of axonal degeneration because continued exposure to the toxic agent results in progression of the degeneration to more proximal portions of the axon and may eventually lead to degeneration of the entire axon and the neuronal cell body. Fortunately, if exposure to the toxic agent is ended before the axonal degeneration extends centripetally to involve the proximal axon and neuronal cell body, the intact proximal axon may regenerate and extend distally to reestablish contact with the periphery and permit return of nerve function (Fig. 18.5C). This potential for axonal regeneration and return of function is unique to the PNS; there is no significant regeneration of axons within the CNS.

Some of the postulated mechanisms for this increased vulnerability of the distal ends of the largest and longest axons to toxicant-induced axonal degeneration are discussed in following sections. Of considerable interest is whether the dying-back axonal degeneration is due to direct toxic injury of the neuronal cell body or its axon. Because the neuronal cell body is responsible for maintaining both itself and its axon, some hypotheses suggest that the toxic agent initially compromises the metabolism of the cell body, such that the cell body can no longer adequately maintain its entire axon. It is further reasoned that the distal axon, being farthest

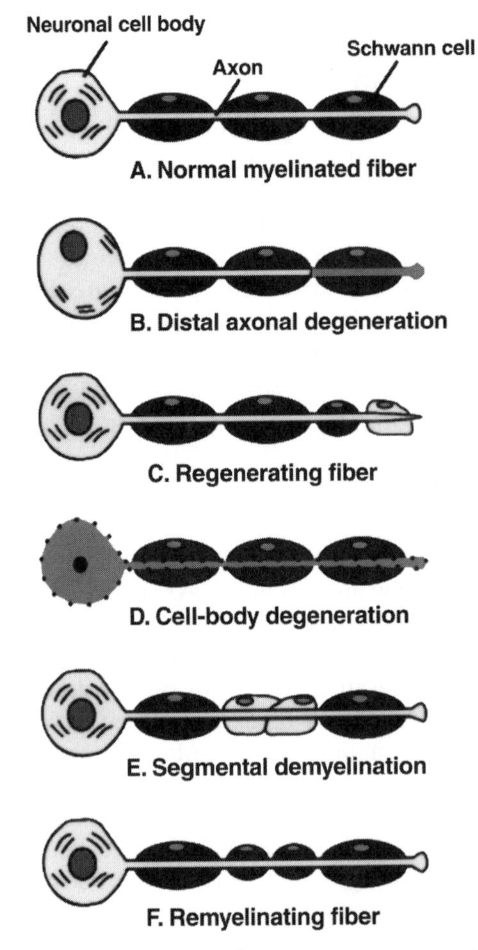

FIGURE 18.5 Responses of the peripheral nervous system to toxicant-induced injury (A) Normal myelinated axon in PNS. (B) Distal axonal degeneration (dying-back neuropathy). The distal end of the axon has degenerated. The myelin covering the degenerated axon has also broken down as a consequence of the axonal degeneration. The proximal axon, its overlying myelin sheath, and the neuron's cell body remain intact. (C) Regenerating myelinated axon. Subsequent to distal axonal degeneration, there is potential for regeneration of the axon and the return of nerve function. Note that Schwann cells remyelinate the regenerated axon. (D) Cell-body degeneration (neuronopathy). The neuron's cell body and its entire axon have degenerated. There is no potential for regeneration when the cell body degenerates. Note that the myelin also breaks down as a consequence of the degeneration of the underlying axon. (E) Demyelination. Several myelin internodes (segments), each representing the myelin of one Schwann cell, have undergone degeneration (segmental demyelination). Note that the underlying axon remains intact despite breakdown of the myelin internodes. (F) Remyelinated axon. Subsequent to demyelination there is remyelination of the segments of demyelinated axon. The remyelinated segments characteristically have shorter internodal distances than the original myelin internodes. Remyelination permits return of function to the affected axon.

removed from the neuronal cell body, would be most severely affected by the metabolic compromise and would be the first part of the neuron to degenerate.

Competing hypotheses suggest that the distal axonal degeneration is due to the toxic agent having a direct effect on the axon. The vulnerable region might be at the distal axon, thus directly accounting for the localization of the degeneration to the distal axon. It is also possible that the deleterious effect of the toxicant (e.g., destruction of an axolemmal component) is uniformly expressed along the axon but that the limited capacity of axonal transport to replace the damaged component results in the more distal axon being less likely to be repaired.

A few toxic agents, such as mercury, initially produce degeneration of the neuronal cell body rather than the distal axon (Fig. 18.5D). These toxic neuropathies are sometimes referred to as "neuronopathies" to emphasize that the cell body of the neuron is the initial site of degeneration. Degeneration of the axon also occurs in these toxic neuronopathies, because the axon is totally dependent on the neuronal cell body for survival. In contrast to dying back neuropathies, neuronopathies have no potential for axonal regeneration and recovery of function, because an intact cell body is a prerequisite for axonal regeneration.

A variety of morphologic abnormalities in the axon or neuronal cell body may precede axonal degeneration in toxic neuropathies. These associated morphologic abnormalities are often very characteristic of a particular toxic agent or group of agents and are of great interest because they may give clues as to the basic mechanisms by which a particular toxic agent produces axonal injury and degeneration. Among the best studied of these morphologic abnormalities associated with toxic neuropathies are the large masses of neurofilaments that accumulate locally within axons during intoxication with 2,5-hexanedione (2,5-HD) and related γ-diketones, with acrylamide, β, β-iminodipropionitrile (IDPN), and carbon disulfide.

18.4.3 Biochemical Mechanisms of Axonal Degeneration

The following sections discuss the major categories into which PNS toxicants have traditionally been grouped and the major hypothetical mechanisms of action. These categories are not mutually exclusive; a toxicant may manifest characteristics of more than one type, and have more than one mechanism of action. The discussion in the following sections is meant more to give insight as to how relevant hypotheses are framed rather than to imply that the mechanisms of action of toxicants inducing neuropathies are well understood.

18.4.3a Inhibitors of Energy Production

Most of the published biochemical investigations of toxicant-induced axonopathies are interpreted in terms of the specialized aspects of neuronal metabolism. The preferential vulnerability of neurons with very long axons is of special relevance to this focus. These neurons have a great metabolic demand to maintain the ion gradients necessary for nerve conduction. The other main energy demand is for synthesis and axonal transport of the macromolecules needed for maintenance of the axon and nerve endings. It is well known that the process of rapid axonal transport is absolutely dependent on production of ATP, which in turn is dependent on glycolysis, oxidative phosphorylation, and oxygen.

Based on these current concepts of energy requirements in the axon, a hypothesis has been proposed to account for the preferential vulnerability of the distal regions of long axons to certain toxicants. It has been suggested that the supply of enzymes for glycolysis may be a rate-limiting step in energy production inasmuch as the glycolytic enzymes must be supplied to the axon by slow axonal transport. If a particular toxicant has partially inactivated an enzyme, the cell body may be unable to respond with an adequate increase in production and axonal transport of the needed enzyme, causing the distal region of the axon to become energy deficient. This hypothesis assumes that the demand for production of energy is relatively constant along the axon, and that successively fewer enzyme molecules are delivered to progressively distal regions. Although postulated to be the mechanism of action of several neurotoxicants causing distal axonal degeneration, this hypothesis has not been adequately tested for any particular set of compounds. In this connection, a promising model utilizes the rodenticide Vacor (N-3-pyridylmethyl-N′-*p*-nitrophenylurea, a structural analog of nicotinamide). Ingestion by humans brings on the onset of neuropathy within hours. Studies with rats have shown that, although axonal transport does not appear affected in the proximal region of the sciatic nerve, there is a considerable reduction of material reaching the distal nerve regions. It is assumed that the block in axonal transport is connected with the development of neuropathy, and that it could well be causal. The defect in axonal transport is not known, but the possibility that this nicotinamide analog might perturb energy metabolism by inhibiting NAD-NADH utilizing enzymes is of interest.

18.4.3b Inhibitors of Protein Synthesis

As mentioned above, some neurotoxicants initially produce degeneration of the neuronal cell body rather than the distal axon. These neuropathies are referred to as "neuronopathies." In some cases the toxic agent appears to act by inhibition of protein synthesis. Examples include the antimitotic, doxorubicin, and the plant lectin, ricin. Doxorubicin binds to DNA and blocks RNA transcription, whereas ricin binds to the 60s ribosomal subunit and blocks RNA translation. In both cases, the resulting block in protein synthesis leads to neuronal degeneration. Doxorubicin cannot cross the BNB, so that the neuropathy is restricted to those regions of the PNS where there is no barrier (see Section 18.1). Ricin is unique in that the toxin gains access to the neuronal soma by being taken up by the axon and carried by retrograde axonal transport to the cell body. Ricin neuropathy is thus restricted to those neurons whose axons have taken up ricin.

Methylmercury and trimethyltin also produce neuronopathies. Methylmercury inhibits protein synthesis, but experimental evidence suggests that additional or alternative mechanisms must exist to account for the neuronal degeneration. The mechanism of action of trimethyltin is unknown.

18.4.3c Structural Components of the Axon as Sites of Vulnerability

The structural components intrinsic to the axonal transport system are also a point of vulnerability. A prominent example of this is the sensorimotor neuropathy that may complicate treatment of leukemia or lymphoma with the antineoplastic agent vincristine. This alkaloid functions as a mitotic spindle inhibitor by binding to tubulin. This same action of binding to tubulin leads to inhibition of axonal transport in experimental systems and could account for the neuropathy in humans. Two other major structural components of the axon that are potential targets of neurotoxicants are discussed below.

Neurofilaments. Neurofilaments are presumed to be the target site for several neurotoxicants of environmental significance. Exposure to certain toxicants induces focal accumulations of neurofilaments within neurons and their processes. Accumulation may be within the cell body, or in proximal, middle, or distal regions of the axon. Within myelinated axons, these accumulations of neurofilaments generally occur at multiple, paranodal sites and produce axonal swellings. This common feature of multifocal accumulations of neurofilaments suggests that these toxicants share a common mechanism of action.

Based on the observation that many of the toxicants causing axonal accumulations of neurofilaments react covalently with proteins, including neurofilament proteins, a "unified" hypothesis has been advanced to explain the pathogenesis of this class of neuropathies. Briefly, it is suggested that certain toxicants cause a specific covalent modification of neurofilament proteins. These modifications are postulated to destabilize the cytoskeletal framework. As the neurofilaments become increasingly modified they eventually become nontransportable and accumulate. The exact position along the nerve at which transport fails would depend on the reactivity, concentration, and time course of administration of the toxicant. The somewhat different ultrastructural pathology seen with each of these agents is presumed to be a result of the specific covalent modification of neurofilaments that each toxicant makes.

This hypothesis is best supported in the case of the hexacarbons. Hexacarbons can react both in vitro and in vivo with lysine ε-amino moieties of proteins to yield pyrrole adducts, which permit secondary auto-oxidative crosslinking. It has been postulated that pyrrole derivitization and secondary crosslinking of neurofilaments are the initiating events in hexacarbon neuropathy. Although this mechanism has been challenged, many studies support this mechanism for the hexacarbons. However, this hypothesis offers no explanation for the early effects on retrograde transport observed in hexacarbon neuropathy.

Axonal Smooth Membranes. Several neurotoxicants, including *p*-bromophenylacetylurea, zinc pyridinethione, acrylamide, and certain organophosphorous compounds cause distal axonal swellings composed partly or entirely of accumulations of vesiculotubular membranous structures derived from the endoplasmic reticulum. These neuropathies manifest a distal "dying-back" degeneration of nerve fibers. Although a range of axonal transport abnormalities have been reported in these neuropathies, it is unknown whether these accumulations of membrane reflect a direct effect of the toxicant on axonal membranes, a direct effect on the transport mechanism, or merely a nonspecific reactive response of the axon.

18.4.4 Characteristics of Demyelinating Neuropathies

In contrast with the large number of toxic agents that produce axonal degeneration in the PNS, only a few agents selectively cause Schwann cell injury. Toxic injury to Schwann cells is restricted, with rare exceptions, to Schwann cells ensheathing myelinated axons and is manifested morphologically as *demyelination*, that is, loss of myelin from an intact axon (Fig. 18.5E). Less commonly, the toxic injury results in fluid accumulation within the myelin sheath (intramyelinic edema). The degenerating myelin is catabolized by Schwann cells and by macrophages that are attracted to the degenerating myelin. Although the demyelinated axons retain their connections with the periphery, the loss of continuity of the myelin sheath results in loss of the axon's ability to convey nerve impulses. Shortly after the onset of demyelination, Schwann cells begin proliferating and cover the demyelinated segment of axon. Within a few days these Schwann cells begin producing a new myelin sheath to cover the demyelinated segments. If there is cessation of exposure to the toxic agent, the process of demyelination ends and the reparative process of remyelination is able to restore function to the demyelinated nerve fibers (Fig. 18.5F).

18.4.5 Biochemical Mechanisms of Demyelination

Several aspects of Schwann cells metabolism emerge as potential points of vulnerability to toxicants. The Schwann cell perikaryon (cell body) supports an enormous peripheral structure, the myelin sheath, which if unwrapped would dwarf the body of the Schwann cell (see Fig. 18.4). Thus, as in the case of axonal transport in neurons, there may be specialized processes involved in supporting the topologically distant myelin. Furthermore, myelin has a specialized lipid and protein composition and a relatively rigid and ordered structure as compared with other membranes. Metabolic perturbations that potentially cause alterations in the composition of lipids and proteins assembled to form this membrane may cause destabilization and collapse of the myelin membrane. In this context, myelin might be much more vulnerable than plasma membranes of other cells.

The toxic agents that selectively damage myelin can usefully be separated into those that bring about myelin alterations without apparent injury to the myelinating cell, and those that injure both the myelinating cell and its myelin. Triethyltin and hexachlorophene are examples of the former. These neurotoxicants characteristically produce a reversible vacuolation (edema) of the myelin sheath through splitting of the myelin lamellae at the intraperiod line. The intramyelinic edema is much more prominent in CNS myelin than in PNS myelin. The specificity of these myelinotoxic agents suggests that these toxicants are located primarily in myelin (this would also be consistent with the known lipid solubility of triethyltin and hexachlorophene). It has been suggested that triethyltin produces edema by a direct action on myelin membranes, analogous to its action in mitochondria where it acts as an anion exchanger and collapses the proton gradient across the mitochondrial membrane. The specific mechanisms, however, by which these myelinotoxic agents produce intramyelinic edema are not yet established.

Diphtheria toxin, inorganic lead, and tellurium are considered toxicants that cause demyelination through injury to the myelinating cell. The mechanism of action of diphtheria toxin was discussed in Section 18.1. The mechanism by which inorganic lead causes Schwann cell injury and demyelination is not well understood, but may be related to uncoupling or inhibition of oxidative phosphorylation secondary to interference of lead with some aspect of ion transport across the mitochondrial membrane.

The study of tellurium neuropathy has provided some insights as to how a toxic agent may cause demyelination. Inclusion of elemental tellurium in the diet of 20-day-old rats results in a rapid demyelination of peripheral nerves that is maximal at about 5 days after starting the diet (the CNS does not show demyelination). Available evidence suggests that the active agent is tellurite (Te^{+4}), and that its primary metabolic action is inhibition of the conversion of squalene to squalene epoxide, thereby causing marked accumulation of squalene and almost complete inhibition of cholesterol synthesis. The inhibition of the squalene epoxidase reaction occurs in all tissues including brain, raising the question as to why the demyelination is limited to the PNS. An explanation may lie in the fact that, in young animals, the rate of membrane deposition and of cholesterol synthesis is normally much higher in peripheral nerve than in brain and other tissues. Thus, the rate of accretion of membranes deficient in cholesterol is more rapid in PNS myelin than in brain myelin. Furthermore, plasma membrane systems of other cell are presumably not as vulnerable to a deficiency in cholesterol because they are not as highly ordered and cholesterol-rich as myelin.

18.4.6 Molecular Markers for Toxicant-Induced Neuropathy

A question of great practical importance is how to determine whether exposure of an animal to a putatively neurotoxic agent results in damage to the peripheral nervous system. This can be tested by study of behavioral parameters such as limb strength or sensitivity to pain, or electrophysiological parameters such as conduction velocity, or morphological parameters such as ultrastructural evidence or nerve degeneration. An obvious question is whether some parameter at the level of biochemistry or molecular biology would be a useful addition to this armamentarium. Considerable strides have been made toward this goal in recent years. It is possible to measure, quantitatively and with high specificity, levels of mRNA coding for proteins relatively specific to a particular cell type. Thus, if an animal is treated with a particular toxicant that may preferentially act on Schwann cells to cause demyelination, an expectation is that mRNA coding for P_0 protein would be down-regulated prior to actual demyelination and morphologically observable damage. This is indeed the case, but it does presuppose a hypothesis as to what cell is preferentially affected by the toxicant. Of particular utility would be a more general marker that might be perturbed in a variety of pathological conditions that may cause damage to the PNS. Levels of mRNA coding for nerve growth factor receptor (NGFR) is indeed such a marker. Among other functions, NGFR is presumed to play a role in the process of development that involves interaction between the growing axon and the Schwann cells that envelop this axon. The message levels for this protein are therefore high during development and low in the mature nerve. If, however, the relationship between axon and Schwann cell is disturbed, a presumably compensatory upregulation of this protein and the mRNA coding for it takes place in Schwann cells. This is irrespective of whether the disturbance is due to initiation of primary demyelination—by, for example, tellurium—or whether it is related to an axonopathic agent such as acrylamide or isoniazid. Thus, elevation of mRNA for NGFR within the sciatic nerve has been shown to be a sensitive marker of an early stage in peripheral neurotoxicity—possibly more sensitive than all except the most high-resolution, and therefore difficult and labor intensive, morphological studies. Whether this or some other molecular marker of PNS neurotoxicity should be established as a screening technique is a matter of debate.

18.5 CONCLUSION

The intent of this chapter is to provide an appreciation for the unique features of the PNS that may make it vulnerable to toxicant injury, and to discuss possible ways in which that injury may come about. The reader should remember that the categories and conceptualizations used in this chapter (e.g., axonal neuropathy versus demyelinating neuropathy; toxicants affecting energy metabolism versus structural components) are not mutually exclusive. Few toxicants have such specific effects that they fall exclusively in one category. Placing a toxicant in one category or another is necessary and useful for framing testable hypotheses, but should not prevent the consideration of alternative possibilities. We would also reemphasize that in very few cases have the specific mechanisms of action been well established for PNS toxicants. Furthermore, in those instances where a specific biochemical alteration has been demonstrated (e.g., crosslinking of proteins by hexacarbons, or inhibition of cholesterol synthesis by tellurium), a causal link to the neuropathy is yet to be firmly established. The techniques of molecular biology should be helpful both in gaining understanding of mechanisms and in formulating approaches to screen for compounds or environmental conditions that may be neurotoxic. The references cited below are intended to provide the reader with an introduction to the literature on neuronal (axonal) and Schwann cell (myelin) metabolism and on PNS toxicants.

SUGGESTED READING

Chang, L.W. (Ed). *Principles of Neurotoxicology*. New York: Marcel Dekker, Inc., 1994.

Dyck, P.J., Thomas, P.K., Griffin, J.W., Low, P.A. and Poduslo, J.F. (Eds), *Peripheral Neuropathy, 3rd Ed.* Philadelphia: W.B. Saunders Co., 1993.

Goodrum, J.F. The role of axonal transport in toxicant-induced peripheral neuropathy. In: Tilson, H.A. and Harry, G.J. (Eds), *Neurotoxicology, 2nd Ed.* Philadelphia: Taylor and Francis, 1999, pp 81–98.

Graham, D.G., Amarnath, V., Valentine, W.M., Pyle, S.J. and Anthony, D.C. Pathogenetic studies of hexane and carbon disulfide neurotoxicity. *Crit. Rev. Toxicol.* 25 (1995), 91–112.

Hirokawa, N. Axonal transport and the cytoskeleton. *Curr. Opin. Neurobiol.* 3 (1993), 724–731.

Morell, P. and Quarles R. Myelin formation, structure and biochemistry. In: Seigel, G., Agranoff, B., Albers, R.W. Fisher, S.K. and Uhler M.D. (Eds), *Basic Neurochemistry, 6th Ed.* Philadelphia: Lippincott-Raven, 1999, pp 69–93.

Morell P. and Toews A.D. Schwann cells as targets for neurotoxicants. *Neurotoxicology* 17 (1996), 685–696.

Olsson, Y. Microenvironment of the peripheral nervous system under normal and pathological conditions. *Crit. Rev. Neurobiol.* 5 (1990), 265–311.

Sayre, L.M., Autilio-Gambetti, L. and Gambetti, P. Pathogenesis of experimental giant neurofilamentous axonopathies, a unified hypothesis based on chemical modification of neurofilaments. *Brain Res. Rev.* 10 (1985), 69–83.

Spencer, P.S. and Schaumburg, H.H. (Eds). *Experimental and Clinical Neurotoxicology, 2nd Ed.* New York: Oxford University Press, 2000.

Stenoien, D.L. and Brady, S.T. Axonal transport. In: Seigel, G., Agranoff, B., Albers, R.W., Fisher, S.K. and Uhler M.D. (Eds), *Basic Neurochemistry, 6th Ed.* Philadelphia: Lippincott-Raven, 1999, pp 565-588.

Toews, A.D. and Morell, P. Molecular biological approaches in neurotoxicology. In: Tilson, H.A. and Harry G.J. (Eds), *Neurotoxicology, 2nd Ed.* Philadelphia: Taylor and Francis, 1999, pp 1–36.

CHAPTER NINETEEN

Biochemical Toxicology of the Central Nervous System

BONNIE L. BLAKE, CINDY P. LAWLER, and RICHARD B. MAILMAN

19.1 CNS FUNCTION

The central nervous system (CNS) is the origin of emotions, thoughts, control of movement, and the regulation of neuroendocrine, immune, and circulatory function. The CNS receives information from peripheral nerves and the endocrine and immune systems. It integrates this information, and modulates the very systems providing the input. Because of this important role in coordinating physiological and cognitive functions, damage to the CNS ultimately impairs the entire organism. Although a great deal of research has been dedicated to understanding the mechanisms and sequelae of neurotoxicity, xenobiotic insult to the CNS is often poorly understood. One reason for this lack of understanding, as we shall see later in this chapter, is the presence of a structural and metabolic status peculiar to neurons. For example, very few cells in the body are so dependent on aerobic respiratory mechanisms as are those in the brain and spinal cord. Another is the unusual cellular complexity of the brain, which often results in region- or cell-selective toxic effects. Finally, an especially frustrating manifestation for clinicians and neurotoxicologists is that CNS toxicity often induces subtle lesions that do not cause visible signs or pathological changes. In some cases, toxicological changes may be evident only when a challenge (either environmental or chemical) is present, inasmuch as the insult may alter the ability of the organism to respond to perturbation. The purpose of this chapter is to provide an overview of CNS function, a review of some known mechanisms of neurotoxicity, and, most importantly, an awareness of the close association of molecular and biochemical neurotoxicology with the study of neurodegenerative diseases and basic neurosciences.

19.1.1 Organization of the CNS

The nervous system is divided into two major parts: the central nervous system, formed by the brain and spinal cord, and the peripheral nervous system, formed by the cranial and spinal nerves that enter (afferent fibers) or originate from (efferent fibers) the brain and spinal cord (see Chapter 18). For the purposes of this chapter, two types of cells in the CNS are most important—neurons (nerve cells) and neuroglia (supporting cells). The importance of neurons in the reception and transmission of electrochemical signals is apparent in their specialized structure and physiology. Neurons vary widely in size and shape (see Fig. 19.1). They generally are composed of a cell body (soma or perikaryon), a long process extending from the soma (axon), and shorter processes (dendrites) that extend from the cell body and receive local information. The cell body is essential for survival of the axon and dendrites.

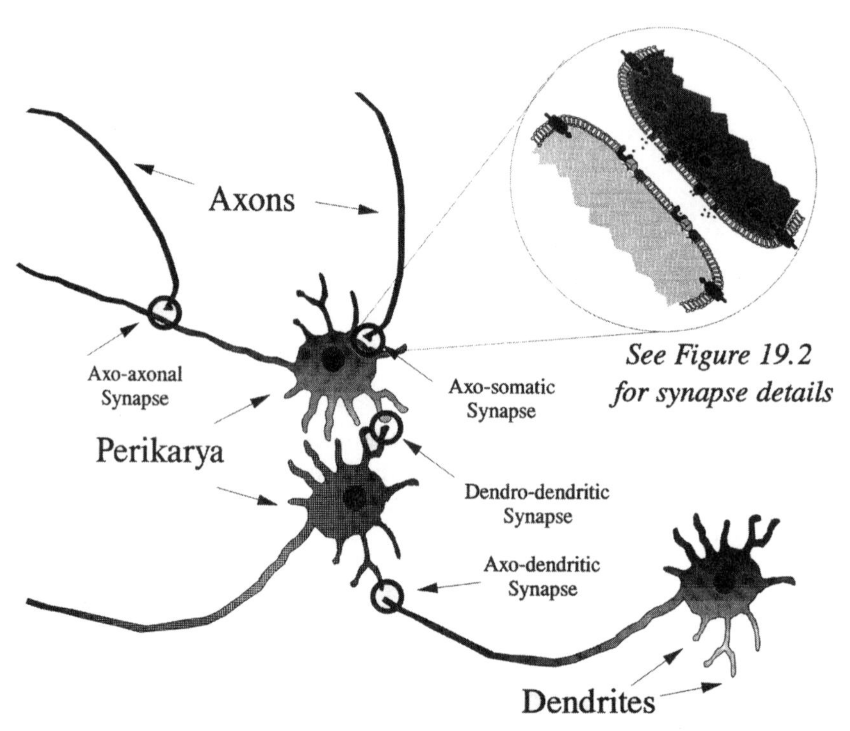

FIGURE 19.1 Schematic of neuron illustrating major structural features including dendrites, axon, soma, and terminals, and some possible synaptic connections.
Neurons are specialized cells that receive and conduct information in the form of bioelectric signals. Despite a tremendous morphological diversity, several general features of neurons can be described. The cell body or soma contains the nucleus and the associated structures for protein production. Dendrites are specialized extensions of the cell body that provide a large network for receipt of incoming signals from other neurons. Neurons typically possess a single axon, which conducts information in the form of action potentials away from the cell body and toward other target neurons. Impulses are transmitted between cells at specialized appositions called synapses, which occur between an axon terminal and a dendrite or cell body (axo-dendritic and axo-somatic synapse) or between two axons (axo-axonic synapse) or two dendrites (dendro-dendritic synapse). The presynaptic membrane contains vesicles and enzymes for production and storage of neurotransmitters. The postsynaptic membrane contains specific receptors and associated effector mechanisms (e.g., G-proteins, adenylate cyclase). The ultimate effect of synaptic transmission is either an inhibition or excitation of the postsynaptic cell.

Impulses generally travel unidirectionally from the dendrites to the axon, and the axons terminate in a series of specialized structures called synapses. Although experiments using electrical stimulation have demonstrated that information, in the form of action potentials, may travel bidirectionally in axons, the synapse (discussed below) acts as a rectifier to ensure unidirectional flow under physiological conditions in the intact organism. Other types of signals, for example, those that convey regulatory information or support functions, are exchanged interactively among cells in the CNS. This type of interchange is especially important for CNS activities involving long-term or permanent alterations in neuronal structure or function such as occurs during neurodevelopment.

Although neurons (of which there may be as many as 100 billion in humans) are clearly of great importance, neuroglial cells constitute the large majority (ca. 90%) of the cells in the brain. There are several types of neuroglia. Oligodendrocytes, like the Schwann cells of the peripheral nervous system (see Chapter 18), produce myelin, a complex of lipid and protein that is critical to the function of many neurons. Microglia are a second major group of specialized, non-neuronal cells in the CNS. These cells resemble small macrophages both functionally and morphologically. A third major type of glial cell is the astrocyte, a small-bodied cell with numerous serpentine processes thought to provide structural support as well as isolation of the neurons from direct contact with blood vessels. The latter function serves an important role in maintenance of the blood-brain barrier (see Section 19.1.4). Astrocytes appear to respond to neuronal activity, because they manufacture and secrete some neurotransmitter and express some neurotransmitter receptors. In addition, astrocytes biotransform xenobiotics, and probably help to regulate the ionic microenvironment around neurons. Following toxicant or traumatic injury to the CNS, astrocytes promote glial "scarring" by proliferation and hypertrophy in the injured area. This proliferation, or gliosis, is accompanied by the increased expression of glial fibrillary acidic protein (GFAP). The finding of increased tissue GFAP represents a hallmark sign of response to injury in the central nervous system. Examples given later in this chapter will show how glia can play other critical roles in toxicant action.

19.1.2 Neurotransmission

19.1.2a Electrochemical Neurotransmission

An essential characteristic of neurons is the ability to integrate diverse inputs (from many afferent neurons) and then transmit information to efferent neurons. Following a stimulus in the dendritic region of a neuron, an impulse that is sufficiently large to overcome the threshold for electrical propagation is transferred down the axon in a single direction. The ability to perform this feat lies partially in the fact that neurons maintain a separation of positive and negative charges across the cell membrane. At rest, the interior of the neuron is more negatively charged than the exterior; in other words, the resting neuronal membrane is polarized. Voltage-gated sodium channels spanning the membrane are opened upon the arrival of a nerve impulse or action potential. Through these channels, sodium ions flood inward and create a relative positive potential on the inside of the neuron. In response to this depolarization, voltage-gated potassium channels open rapidly to allow potassium ions to exit and restore the original polarization. Restoration of the ionic gradients of sodium and potassium is maintained by the constant activity of the ATP-dependent sodium-potassium pump (Na^+/K^+ATPase). Ion channels thus mediate the electrical excitability of neurons in the CNS, and as discussed in Section 19.2.3, they are often a target for neurotoxicants.

Electrical information may also be shared between cells in the CNS through interconnective electrical synapses, known as gap junctions. These channels are formed between cells by the protein connexin, and they serve in the transfer of ions, nutrients, and metabolites, and allow electrical currents to be exchanged between cells. Clusters of gap junctions facilitate communication between neighboring glial cells and between glial cells and neurons. In this respect, gap junctions may represent a transport route from the blood to neurons, one that circumvents the blood-brain barrier, allowing access to nutrients as well as toxicants. The blood-brain barrier is discussed further in Section 19.1.4.

Although some information transfer between neurons is electrical, the huge majority is chemical in nature. This communication involves a specialized structure called the synapse where small amounts of molecules called neurotransmitters are released by the neuron. These chemical messengers are stored in synaptic vesicles and released when the action potential reaches the synapse. Among the endogenous chemicals used for neurotransmission are the traditional neurotransmitters acetylcholine and serotonin, the catecholamines (norepinephrine, dopamine, and epinephrine), and the amino acids glutamate, GABA, glycine, and aspartate. Recent advances in the neurosciences, however, have increased the complexity of this picture. The use of molecular methods has led to the discovery of other molecules believed to subserve the function of information transfer in the nervous system. Table 19.1 provides some examples of the hundreds of molecules now accepted as serving such roles in chemical message transmission.

TABLE 19.1 Examples of Chemical Messengers (Neurotransmitters/Modulators).

Monoamines	***Neuropeptides***
Acetylcholine	Opioid peptides (endorphins, enkephalins, dynorphins)
Dopamine	Tachykinins: substance P, neurokinin B
Norepinephrine	Somatostatins
Serotonin	Gastrin and cholecystokinin
Histamine	Vasoactive intestinal polypeptide (VIP)
	Oxytocin, vasopressin, neurophysins
Amino acids	Neuropeptide Y
γ-Aminobutyric acid (GABA)	Neurotensin
Glutamate	Bombesin
Glycine	Angiotensin
Aspartate	Calcitonin gene-related peptide (CGRP)
Purine-based	***Small molecules***
Adenosine	Nitric oxide
Adenosine monophosphate (AMP), diphosphate (ADP), and triphosphate (ATP)	

In the title of Table 19.1, we have used the terminology "neurotransmitters/neuromodulators." The term *neuromodulator* was coined because many of these molecules, often peptides or small molecules such as nitric oxide (NO), do not meet the formal criteria of neurotransmitters. The distinction between neuromodulators and neurotransmitters, however, is becoming increasingly blurred. Many of these molecules may work more slowly than neurotransmitters and over greater distances (rather than at a single synapse). On the other hand, some neuromodulators are likely to be constitutively released or released only at times of high or low metabolic activity. Several neuromodulators have little or no direct action in the CNS except to enhance or reduce the action of neurotransmitters. Often, a classical neurotransmitter (e.g., dopamine) may be released simultaneously with a neuromodulator (e.g., substance P). Several studies have demonstrated that such cotransmission provides an exquisite mechanism for regulation of cellular function. In the remainder of this chapter, we use the term *neurotransmitter* (or transmitter) generically, describing all molecules released by neurons to convey information to adjacent cells.

19.1.2b Synapses

The biochemical events occurring at the synapse are important to toxicologists for two reasons. First, toxicants may exert their effects by acting directly on these loci. Second, after many types of toxic insults, the nervous system may compensate by altering synaptic function, such as by increasing the activity of synapses or by changing the number of receptors. From this perspective, the synapse can be both a site of toxicant action and a mirror of toxicity, and therefore measurements of synaptic function are often used in the assessment of neurotoxicity. Figure 19.2 provides a schematic representation of a synapse. The biochemical mechanisms illustrated in Figure 19.2 are important potential loci of toxicant action.

Normal synaptic transmission depends on the orchestrated interaction of multiple biochemical and electrochemical events. Provided the conditions are appropriate, the arriving action potential causes the release of neurotransmitter from the presynaptic cell (the one sending the chemical message). Although the molecular mechanisms of neurotransmitter release are only beginning to be understood, it is known that the presence of extracellular calcium is of absolute necessity. Upon arrival of the action potential, voltage-dependent calcium channels near the sites of release are opened. Calcium, more highly concentrated on the outside of the presynaptic membrane than inside, rushes in through the open channels to form localized areas of high calcium concentration near the channels. These calcium "clouds" then activate calcium-sensing proteins that have assembled with neurotransmitter vesicles at the inner presynaptic membrane surface. The activation of these proteins results in the creation of a fusion pore between the vesicle and the membrane. The contents of the vesicles are then extruded into the synapse where the neurotransmitter(s) can interact with receptors. Receptors are proteins that neurotransmitters bind with high specificity, in order to convey specific chemical information into the cell. This information then may be reconverted into electrical signals or modulate other neurochemical events. The ability of some neurotransmitters to bind and signal receptors is terminated by enzymatic breakdown in the extracellular synapse. Other neurotransmitters (e.g., catecholamines) are transported back into the axon terminal, where they are either repackaged into synaptic vesicles or destroyed by

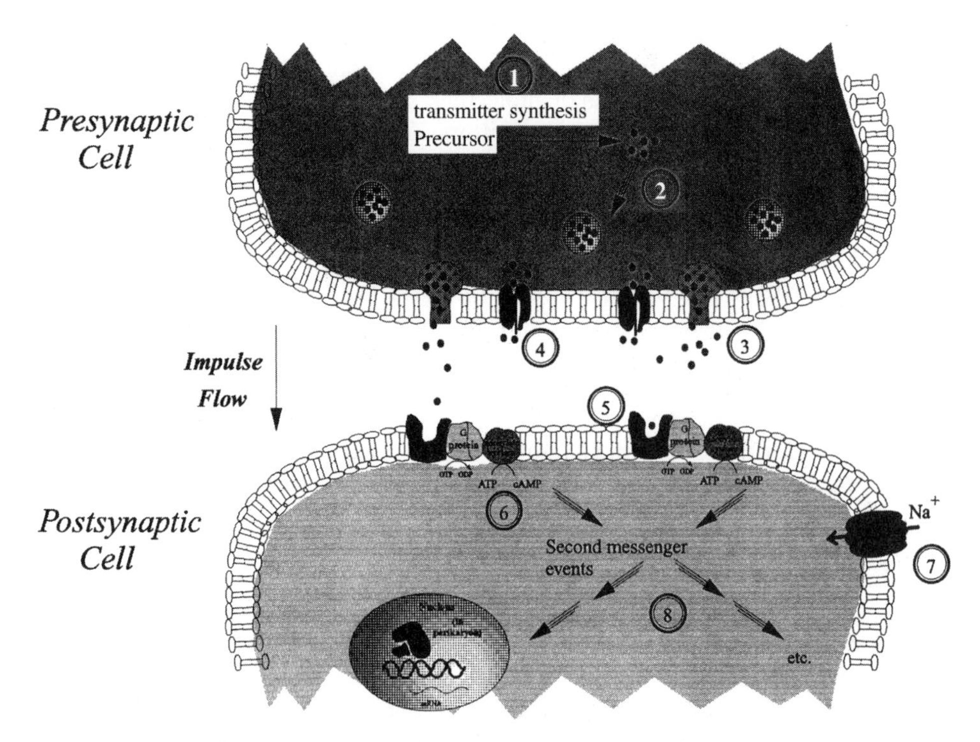

FIGURE 19.2 Schematic representation of synapse.
(1) Toxicants can affect the synthesis of neurotransmitter (represented by small solid dots) from its precursor by affecting the synthetic enzymes or precursor availability (e.g., ethanol). (2) Toxicants can affect the storage of neurotransmitter in vesicles (e.g., reserpine). (3) Toxicants can affect the release of neurotransmitter (e.g., methyl mercury; ephedrine). (4) Toxicants can affect the inactivation of neurotransmitter [uptake (e.g., cocaine), degradation (not shown in figure, e.g., organophosphates)]. (5) Toxicants can interact with receptors, or affect the interaction of the endogenous transmitter with the receptors (e.g., atropine; α-bungarotoxin). (6) Toxicants can affect synthesis or degradation of second messengers, or affect transduction system (e.g., aluminum; caffeine). (7) Toxicants can affect the pumping or transport of ions like Na^+, K^+, or Ca^{2+} (e.g., tetrodotoxin; pyrethrins). (8) Toxicants can affect a variety of cellular functions occurring after second messenger synthesis, including nucleic acid and protein synthesis. *See text for further details and examples.*

monoamine oxidase (MAO). These events are subject to disruption by the selective activity of some neurotoxicants, as shown in Section 19.2.

19.1.2c Receptors and Signal Transduction

CNS transmitter receptors can be divided into two types that differ in their transduction mechanisms: metabotropic (e.g., G protein-coupled receptors) and ionotropic (e.g., ligand-gated ion channels, and allosteric sites on voltage-gated channels). (Chapter 12 provides a more detailed description of receptors and their superfamilies.) Both ionotropic receptors (which act directly) and metabotropic receptors (which act indirectly) modulate the function of postsynaptic neurons by altering the activity of their ion channels. Importantly, the stimulation of these receptors initiates several events including the opening or closing of electrochemical gates, ultimately resulting in the activation of excitatory postsynaptic potentials (EPSPs) or inhibitory postsynaptic potentials (IPSP) in the neuron. Provided the excitatory threshold is exceeded, the summation of these potentials may then propagate the signal as an action potential.

The signaling events mediated by stimulation of neurotransmitter receptors are not limited to those that result in excitation or inhibition of postsynaptic potentials. Other receptor-mediated signals are important for the future signaling capability of the neuron, and even for neuronal survival. These signaling pathways are activated primarily by metabotropic receptors that interact with intracellular proteins. Most metabotropic neurotransmitter receptors are coupled to a signaling molecule known as a G protein (or guanine nucleotide-binding protein). Through this interaction, G protein-coupled receptors send signals to receptor-specific effectors, which may be ion channels, or enzymes such as adenylate cyclases, phospholipases, and cGMP phosphodiesterases. Dopamine receptors provide a useful example of such a sequence of cellular events. The binding of dopamine to the D_1 dopamine receptor is coupled with activation of adenylate cyclase via the action of a specific G protein. In turn, adenylate cyclase increases the synthesis of the second messenger cyclic AMP (cAMP), which activates an important signaling molecule, protein kinase A (PKA). PKA phosphorylates a number of proteins within the cell, thus modulating their activity. Among the proteins modulated by PKA are receptors, phospholipases, potassium channels, Na^+/K^+ATPase, and calcium channels; all participate in other signaling processes occurring coincidentally within the neuron. One important PKA substrate is the cAMP response element binding protein, which, upon phosphorylation, binds DNA and activates transcription. In this way, D_1 receptors may activate long-term or permanent signaling changes. These changes are not only important in future molecular functions such as increases or decreases in subsequent neuronal signaling, but also in future brain functions such as the development of memory. It should be self-evident, then, that signals transmitted between and within neurons are tightly controlled, significant, and potentially vulnerable events. Each of the signaling steps we have described is a possible target for biochemical attack by toxicants resulting in marked alterations in CNS transmission.

19.1.3 Susceptibility of the Central Nervous System to Toxicants

It is clear from the description above that neurons possess many specialized features that are adapted particularly for rapid signal transmission. These same features, however, also predispose neuronal tissue to potential xenobiotic insult. For example, the high metabolic rate and electrical excitability of neurons are critically dependent on the maintenance of aerobic metabolism and membrane integrity. The extended length of many axons poses the logistical problem of tightly regulating transport of materials and information, thus representing a highly sensitive component of the CNS. In addition, the chemical constituents of many cells may cause them to be particularly vulnerable to injury. For example, there is evidence to show that the selective loss of dopamine neurons in Parkinson's disease (discussed further in Section 19.3.2) may be related to the oxidative stress imposed as a result of the oxidative metabolism of dopamine in the substantia nigra. Regional differences in neurochemical architecture of the nervous system are often an important issue in neurotoxicological studies. Finally, unlike glial cells, which can reproduce, neurons are unable to replace dead or dying cells by dividing. An exception to this dogma may be the granule cells of the hippocampus, reportedly able to maintain the ability to divide and proliferate even into adulthood. Nevertheless, in order to compensate for the enhanced vulnerability of the CNS to toxicants, nature has developed the blood-brain barrier.

19.1.4 The Blood-Brain Barrier

Most tissues have a single membrane (the plasmalemma) separating the intra- from the extracellular compartment. As noted in Chapter 18, the peripheral nervous system is partially isolated from contact with the circulation by the blood-nerve barrier. In the brain, the blood-brain barrier and the choroid plexus serve a similar function. The blood-brain barrier limits the influx of circulating substances from capillaries into the immediate brain interstitial space, whereas the choroid plexus separates the blood from the cerebrospinal fluid. Direct central neurotoxicity depends upon sufficient concentrations of the toxicant reaching sensitive sites in the brain or spinal cord. Compounds, therefore, must pass the blood-brain barrier, or there must be physical breaches in the integrity of the barrier (traumatic or inflammatory).

The blood-brain barrier exists because of specific features in the brain microvasculature. The two most important ones are the "tight junctions" between adjacent capillary endothelial cells, and the presence of processes from adjacent cells (i.e., "astrocytic end feet") in these areas. This makes the healthy brain capillary relatively impervious to passage of macromolecules, ions, and water-soluble nonelectrolytes. The use of marker proteins in conjunction with electron microscopic techniques has demonstrated that the barrier is localized to the endothelium. For example, a microperoxidase (1800 Da) that readily transverses capillaries in other tissues will not pass through capillaries in the brain. Some tracers, however, can pass readily through the open clefts between glial foot processes, indicating that these processes do not play an active part in the operation of the barrier. Instead, the glial end feet play a more long-term role in inducing and maintaining the structure of the epithelial barrier. It is of teleological interest that those physiological systems essential to survival have these protective features (in the adult the gut membranes, in the developing organism the placental membranes, and in the most sensitive tissue, the brain, the blood-brain barrier).

Since the brain is dependent on polar molecules (amino acids, sugars, inorganic ions, etc.), mechanisms exist that permit the selective entry of these materials into the brain. For example, the endothelial cells of the blood-brain barrier have carrier-mediated transport systems for selectively moving certain required molecules (e.g., glucose and amino acids) across the barrier into the brain. The combination of these structural barriers and carrier-mediated transport systems permits the composition of the interstitial fluid to be closely regulated.

19.1.4a Breaching the Barrier

In order to understand the biochemical toxicology of the CNS it is necessary to know whether a toxicant can pass through these protective barriers or can alter or damage them, permitting secondary events to occur. Several generalizations may be made about these events.

First, large molecules, such as larger peptides or proteins, do not usually pass through or their passage is severely retarded. Consequently, ingestion of a proteinaceous toxin will result in only peripheral toxicity until a sufficiently large concentration "forces" the toxin past the blood-brain barrier. It is this fact that often makes central effects of the toxicant moot, since the peripheral actions may be lethal.

Second, polar molecules are generally physically excluded from the brain, or at least physically hindered from contacting sensitive sites, because the blood-brain barrier is a highly lipophilic membrane, as are most biological membranes. Nonpolar, lipid-soluble molecules, however, penetrate into the brain more easily. Thus, alkylated mercury derivatives (e.g., dimethyl mercury) are absorbed several orders of magnitude more readily than comparable inorganic ions from both the gastrointestinal tract into the blood, and from the blood into the brain. This was an essential toxicokinetic factor in Minamata disease, the extreme CNS dysfunction (including permanent neurological changes or death) that resulted from biomethylation of mercury and consumption of fish containing these methylated derivatives.

Third, the specific transport systems cited earlier may facilitate the passage of toxicants into the brain. Although these systems are usually highly specific in nature, potential toxicants with structures analogous to physiological substrates may be transported into the brain, thus bypassing the blood-brain barrier. Again, mercury serves as an excellent example. Elemental mercury forms complexes with cysteine, which are recognized by amino acid transporters as methionine. Mercury thus may enter the brain by both energy-dependent (active transport) and energy-independent (diffusion) mechanisms.

A fourth factor to consider is the presence of extra blood-brain barrier structures in the brain (known as supraependymal structures). These structures are bathed in cerebrospinal fluid and can be exposed to molecules that will not penetrate into deeper brain tissue. A possible role for supraependymal structures is to monitor neuroendocrine function. Presumably, the function of the openings (fenestrations) in the barrier is to permit the access of specific endogenous molecules. Other chemicals, however, including putative CNS toxicants, may penetrate through these fenestrations, although they may pass through other portions of the blood-brain barrier only with great difficulty. These supraependymal sites may thus be particularly vulnerable to toxicants. Finally, many factors can influence blood-brain permeability. The most important of these is the age of the target animal. It has been shown that for embryonic or neonatal organisms, xenobiotics have higher accessibility to the CNS.

Alterations in blood-brain barrier function may occur also as a direct effect of exposure to many types of toxicants. For example, substances that alter membrane function directly, such as the bile salt sodium desoxycholate or high concentrations of various organic solvents (including alcohols) disrupt the blood-brain barrier. Increases in blood-brain barrier permeability are also seen after exposure to cobra venom, presumably because the phospholipases in the venom hydrolyze membrane lipids. Manganese is thought to alter the permeability of blood-barrier to other metals such as iron and aluminum. It should be noted that changes caused by these and other agents may be reversible or irreversible and may also profoundly influence the CNS toxicity of subsequent exposure to other materials.

19.2 CNS TARGETS FOR TOXICANT ACTION

As we have seen, the complexity of the CNS endows this system with multiple targets for neurotoxic action that may occur through a variety of mechanisms. In this section, we will discuss important toxicants that are grouped loosely by their mechanism of action. It should be remembered while reading this section, however, that many neurotoxicants have multiple loci of action, and that damage to one aspect of the CNS, such as membrane permeability or oxidative phosphorylation, has consequences for other functions as well.

19.2.1 Agents That Increase Neurotransmission

Molecules that pass the blood-brain barrier may modulate neurotransmission in several ways. First, toxicants may interact with neurotransmitter receptors to mimic the in situ neurotransmitter and cause agonist-like actions. On the other hand, some compounds activate the release of neurotransmitter into the synapse, resulting in nonphysiologically high levels of signaling. Others may inhibit reuptake or destruction of the neurotransmitter, prolonging its effects. In any case, if the modulation of neurotransmission by the toxicant is sufficiently strong, or of sufficient duration, the result may be a substantial alteration in homeostatic, compensatory mechanisms that ultimately may result in a variety of secondary changes.

Neuronal function is tightly regulated in the CNS. Many agents, however, are capable of enhancing normal neurotransmitter function despite the systems that control the level of neurotransmitter in the synapse. One of the best-known examples of this is the action of organophosphates, inhibitors of acetylcholinesterase enzymes. As might be expected, poisoning by these insecticides and nerve gases (e.g., soman or sarin), as well as carbamates, results in increased synaptic concentrations of acetylcholine. Another agent that increases synaptic acetylcholine is scorpamine, a type of scorpion venom. Scorpamine exerts its toxic effects primarily on motor neurons and postganglionic autonomic neurons. Neurotransmitter release also is affected by certain invertebrate venoms. Black widow venom (latrotoxin), for example, is one of the most powerful stimulators of neurotransmitter release in vertebrates. This neurotoxin causes explosive, nonspecific release of vesicle-bound neurotransmitters followed by destruction of prejunctional nerve endings. Other chemical compounds that increase synaptic levels of neurotransmitters, without directly stimulating receptors themselves, include methamphetamine, amphetamine, ephedrine, and other amphetamine derivatives. These agents increase the release of catecholamine transmitters. Methyl mercury (MeHg) neurotoxicity also involves increased spontaneous release of neurotransmitters, however, this effect is thought to be secondary to alterations in calcium homeostasis. On the other hand, cocaine and its analogs increase neurotransmitter levels by inhibiting the synaptic reuptake of dopamine and other monoamines. Finally, ethanol, responsible for multiple CNS effects, also affects the release, uptake, and metabolism of catecholamines and stimulates GABAergic activity. While the latter effect may be the result of direct stimulation of GABA receptors, chronic ethanol ingestion has been correlated with low MAO activity, and consequently elevated dopamine in the blood of alcoholics during remission. Consequently, it has been hypothesized that elevated levels of dopamine in the brain are related to alcohol craving, abstinence delirium, and physical dependence.

Many of the direct-acting chemicals that affect neurotransmission have been exploited by society as pharmaceuticals, food additives, drugs of abuse, and so forth, often because of the potent and selective effects of these compounds. For example, the methylxanthenes like caffeine (found in coffee and cola) and theobromine (found in chocolate) act principally via a direct receptor mode; their target is believed to be purine receptors. Nicotine binds directly to nicotinic cholinergic neurons in both the CNS and PNS. Most severe in the CNS are the effects of developmental exposure, which have been implicated in decreased birth weights and attention deficit hyperactivity disorder (ADHD). Another common direct-acting agent is monosodium glutamate (MSG), the sodium salt of the amino acid glutamate. Glutamate is an endogenous neurotransmitter and can be actively transported into the brain. MSG is added to many processed foods to enhance flavor, and virtually all foods have some glutamate. For the general population, MSG is thought to pose no health risk, although some individuals are hypersensitive to its effects. Large doses of MSG given in utero or to neonatal rodents produce hypothalamic and retinal degeneration, however, effects on human development have not been reported. Many other neurotoxins with direct receptor effects are constituents of various fungi (some of which make mushroom picking an artful and nonroutine pastime). Mescaline, a derivative of the peyote cactus, has powerful actions on the brain that are believed to be due to activity at central serotonin receptors. Ergot fungus alkaloids like ergotamine and other LSD-like derivatives can act as agonists or antagonists at central serotonin and dopamine receptors. The hallucinogenic effects of these toxins are not unexpected in view of the importance of both serotonin and dopamine systems for emotion and sensory processing. In fact, poisoning caused by ingestion of ergot-contaminated grain has been hypothesized to be responsible for medieval European cities becoming "possessed by the devil"; a similar involvement has been suggested for the Salem (Massachusetts) witch trials.

19.2.2 Agents That Selectively Inhibit Neurotransmission

Molecules that bind to receptors without causing activation work by blocking access of the endogenous ligand to the receptor, thus inhibiting neurotransmission. Another mechanism of inhibition is the disruption of the storage and/or release of transmitters. An example of the latter is the mechanism of botulism. Botulinum toxins, produced by the anaerobic bacteria *Clostridium botulinum*, are the sole cause of botulism, a toxicity that results from ingestion of food in which this ubiquitous spore-forming anaerobic bacterium grew (e.g., improperly canned vegetables). Although they act primarily at neuromuscular junctions, the botulinum toxins are sufficiently potent such that microgram doses may be lethal to adult humans. The toxins block release of the neurotransmitter acetylcholine at the neuromuscular junction, inducing flaccid paralysis. Another interesting bacterial toxin is tetanospasmin, a toxin produced by the bacterium *Clostridium tetani*. *C. tetani* infection occurs in deep wounds under anaerobic conditions, and results in muscle stiffness that progresses to rigidity and convulsive spasms that are often fatal. The toxin blocks the release of amino acid neurotransmitters from inhibitory interneurons, particularly in the spinal cord. Another example is the *Rauwolfia* alkaloid reserpine. Reserpine can enter certain monoamine neurons (e.g., those using serotonin or catecholamines as transmitters) and disrupt storage mechanisms, thereby causing a massive release of neurotransmitter, followed by a long-term depletion. Some venoms also can affect transmitter release. β-Bungarotoxin is one of the neurotoxins isolated from the venom of elapid snakes like the Indian Krait (*Bungarus multicinctus*). β-Bungarotoxin acts presynaptically to block neuromuscular transmission through a specific reduction in transmitter release, thus making the motor endplate unresponsive to nerve stimulation. Ultrastructurally, synaptic vesicles disappear after exposure to β-bungarotoxin.

On the other hand, α-bungarotoxin, another of the neurotoxins isolated from the venom of elapid snakes, acts by receptor inhibition. This toxin is a highly specific blocker of the postsynaptic nicotinic acetylcholine receptor. Another receptor blocker is the historically important agent strychnine. A product of seeds from the *Strychnos nux vomica* tree, this alkaloid was once used as an aphrodisiac and an appetite suppressant. Toxic symptoms, including apprehension, nausea, spasms, convulsions, and coma occur rapidly upon ingestion. Strychnine blocks glycine receptors in the spinal cord preferentially, however, higher doses can also affect these receptors in the brain. Today, strychnine is used to study the pharmacology and function of glycine receptors. Finally, atropine is an extremely important antidote for muscarinic acetylcholinergic overactivity, however, in high doses it can also cause both central and peripheral toxicity. This drug is derived from a group of alkaloids from the "deadly nightshade" (*Belladonna* sp.), which share specificity for the muscarinic receptor.

19.2.3 Agents That Disrupt Ion Channels and Calcium Homeostasis

Increases or decreases in neurotransmission may occur through direct neurotoxicant effects on a neurotransmitter or its receptor, or through activation or inactivation of ion channels. For obvious reasons, compounds that affect such mechanisms usually have a high degree of toxicity. Many agents specifically affect the activity of sodium channels. Tetrodotoxin is a compound isolated from the ovary and liver of the puffer (or globe) fish, so named because it can engorge air and blow itself up to ferocious dimensions. It blocks the increase in conductance of sodium channels, and in so doing, disrupts the generation of action potentials. Conversely, a group of plant alkaloids, the grayanotoxins (found in the Ericaceae family) have been demonstrated to cause increases in resting sodium permeability. Human exposure to these agents has been reported most commonly after ingestion of honey made from the nectar of these plants. Another interesting and potent neurotoxin is batrachotoxin, one of the toxic principles contained in the secretions from the skin of the Columbian arrow poison frog, *Phyllobates aurotaenia*. It is a potent steroid-based molecule with actions similar to those of grayanotoxins. Veratridine is a steroidal alkaloid (found in *Veratrum* and *Zygadenus* species) that also depolarizes nerve membranes. Saxitoxin and related agents (like the gonyautoxins) are found in dinoflaggelate phytoplankton like *Alexandrium catenella*. These heat stable compounds, sometimes found in shellfish that have eaten the phytoplankton, block sodium channels in nerve membranes much as tetrodotoxin does. High doses of pyrethroid pesticides prolong the open time of sodium channels in mammalian neurons, resulting in repetitive discharges and excessive neuronal activity. These effects are not neuron-specific and so they produce generalized signs such as hyperactivity, ataxia, convulsions, and, eventually, paralysis. Similar effects are seen with chlorinated hydrocarbons; however, even low doses of DDT can produce hyperactivity in animals. DDT has also been shown to inhibit Na^+/K^+ATPase, as have a number of other neurotoxins, including ethanol, copper, and lead. In some cases, inhibition of this pump is an epiphenomenon. Ouabain, however, a glycoside obtained from several plant species, directly inhibits Na^+/K^+ATPase. Radioreceptor studies (see Chapter 12) have shown that ouabain acts by binding with high affinity to a site on the enzyme.

Another energy-dependent transporter, the Ca^{2+}/Mg^{2+}ATPase, is a second target of pyrethrins, as is the calcium-binding protein calmodulin. Inhibition of these proteins results in elevated intracellular levels of calcium, a critical component of intracellular signaling pathways. Excessive levels of intracellular calcium have also been associated with heavy metals such as mercury, lead, and aluminum. These metals, particularly lead, probably compete with calcium for binding sites on all types of calcium binding proteins. For example, lead blocks calcium channels on neuronal and mitochondrial membranes. Blockade of calcium channels on the plasmalemma not only prevents calcium from being pumped out of the cell, but also interferes with neurotransmitter release. Blockade of mitochondrial calcium channels prevents the uptake of cytosolic calcium by mitochondria, helping to drive the rising concentration of calcium within the cell. As we shall see below, perturbations of calcium flux have multiple additional effects on neurons, including the activation of apoptosis.

19.2.4 Agents That Alter Intracellular Signaling Molecules

Of all the neurotoxicants discussed in this chapter, probably the most difficult to categorize is mercury. Exposure to mercury leads to a diffuse distribution of neuronal injury and thus to diffuse encephalopathy. Mercury targets multiple cellular processes, and its primary site of action is unknown. One proposed mechanism is the ability of mercury to couple with cysteine and other thiol-containing groups, promoting its binding to many different proteins. For example, glutathione conjugates mercury and other metals, and consequently those neurons with relatively little glutathione, such as small cerebellar and cortical cells, are particularly vulnerable. Microtubules are also specifically disrupted prior to irreversible cell injury by mercury. Some studies have identified abnormally high levels of tubulin subunits suggesting decreased synthesis of tubulin. Mercury also interferes with protein synthesis. Although methyl mercury has greater affinity for the brain, it is unclear whether the ultimate toxicant in these insults is methyl mercury or liberated ion.

The ability of aluminum to perturb intracellular signaling pathways is also related to its chemical nature. First, the coordination chemistry of aluminum is similar to iron, and so not surprisingly, competes with iron for cellular uptake pathways. Secondly, although not a transition metal, it appears that aluminum participates in redox cycling and oxygen radical formation. For example, the lipid peroxidation produced by iron is enhanced in the presence of aluminum. Third, aluminum is remarkably adept at promoting the aggregation and precipitation of certain proteins, synthetic neurofilaments, and DNA. Proteins that can be precipitated by aluminum include beta amyloid and tau, which are found in the senile plaques and neurofibrillary tangles, respectively, in the brains of patients with Alzheimer's disease (AD). Because of this relationship, as well as other evidence, aluminum has been linked with the pathogenesis of AD. Recently, however, contrasts have been drawn between the neurofibrillary tangles induced by aluminum toxicity (by animal dosing or dialysis encephalopathy), and those seen in the brains of AD patients. Although still controversial, many researchers now believe that aluminum may ultimately contribute to the formation of tangles, but is not a direct cause of AD. Their skepticism is in part due to the fact that patients suffering from aluminum intoxication and aluminum factory workers do not have high incidence of AD. Furthermore, there is evidence to suggest that aluminum preferentially accumulates in compromised neurons, and thus contributes to the lesions as a secondary event. Interestingly, aluminum interacts with a number of other intracellular signaling molecules including G proteins and calmodulin.

Cytoskeletal proteins are intricately involved in intraneuronal signaling pathways. As described in Chapter 18, agents that bind to cytoskeletal proteins may cause distal neuropathy or "dying back" in both the peripheral and central nervous systems. Animals affected with this type of neuropathy typically exhibit splaying of the hind limbs, and this response is quantified as a measure of neuromuscular dysfunction in toxicology testing. One compound that causes these types of effects is acrylamide. Acrylamide is thought to disrupt axonal transport, but some researchers believe that the transport disruption may be secondary to other axonal lesions, such as specific inhibition of glycolytic enzymes or to disruption of the repair mechanisms of the axon. Acrylamide has been shown to bind specifically to cytoskeletal proteins, an important component of axon transport systems. A few studies have attempted to determine the effects these perturbations may have on neurotransmission. Radioligand binding studies (see Chapter 12) examined the receptor sites of several neurotransmitter systems after high- and low-dose exposure to acrylamide. At the lower doses, only striatal dopamine receptors were affected, being increased reversibly. Other neurotransmitter receptor systems were affected at the highest doses. Interestingly, the ultimate toxicant in this lesion may have been a metabolite of acrylamide, because the cytochrome P450 inhibitor SKF525A abolished the increases in dopamine receptor number. In related behavioral studies, low doses of acrylamide that caused no outward changes in rodent behavior did change the animals' response to dopaminergic agents, suggesting that homeostatic adaptations in the CNS were sufficient to prevent overt signs of dopaminergic perturbation.

Another compound that interacts with cytoskeletal proteins is the alkaloid colchicine, which has been used clinically for the treatment of gout and is used experimentally as a mitotic poison. Colchicine is a CNS toxin because it prevents tubulin formation, on which axonal transport mechanisms are dependent. The inherent peripheral toxicity of the compound coupled with the ability of the blood-brain barrier to prevent central influx of the alkaloid minimizes its central effects.

Ethanol produces a broad spectrum of toxic effects in the CNS, as well as other organs. Space does not permit an exhaustive review of ethanol neurotoxicity. For this, we recommend the reader refer to the Suggested Reading at the end of this chapter, and to other, more focused dialogues on this subject. Although the behavioral effects of ethanol intoxication are familiar to many of us, the mechanisms of ethanol activity in the CNS remain enigmatic. From a toxicological standpoint, the most devastating consequences of ethanol exposure in the CNS are clearly those imposed on the developing fetus. The consumption of excessive amounts of ethanol during pregnancy is one of the leading causes of mental retardation in the Western world. The pattern of neurodevelopmental defects caused by ethanol is so familiar as to have been distinguished as "fetal alcohol syndrome" (FAS) characterized by small stature, microcephaly, poor muscular coordination and mental deficiency. Ethanol interferes with all stages of brain development and the severity of the lesions is correlated directly with the level of exposure. Cell division, proliferation, growth, differentiation, and migration have all been identified as targets for interference by ethanol. Many if not all neurotransmitter systems are altered directly or indirectly by prenatal alcohol exposure. The neocortex, hippocampus, and cerebellum are particularly affected by developmental ethanol exposure.

The subcellular foci of ethanol neurotoxicity remain under investigation. Mechanistically, the accumulated evidence suggests a role for GABA-A receptors in mediating many of ethanol's effects. Ethanol is also thought to block NMDA receptors, thus altering calcium homeostasis, NO synthesis, and production of cGMP. In neuronal cultures, the ability of ethanol to induce cell death can be lessened by the application of NMDA or GABA, but the mechanisms of cell rescue are unknown. Other recent studies have suggested an important role for growth factors in mediating the proliferative, survival and death responses of neurons to ethanol. The inhibition of cellular proliferation caused by ethanol in some cellular systems may be due to interference with signaling pathways mediated by mitogenic growth factors (e.g., bFGF, PDGF, EGF and IGF-1). Furthermore, NGF has been shown to interact both positively and negatively with ethanol in mediating neuronotoxic effects by altering calcium homeostasis. These studies and other evidence suggest that growth factors and/or their receptors are among the primary targets for ethanol-induced neurotoxicity.

Lithium is a common divalent metal used in industry and medicine. It is used in the manufacture of electronic tubes, alkaline storage batteries, alloys, ceramics, and lubricants, and serves as a catalytic agent in chemical synthetic reactions. Pharmacologically, lithium is used in treatment of manic depression. Although the mechanism of pharmacological activity by lithium is by no means clear, recent studies are suggestive of its involvement in modulating receptor-G protein coupling and/or G-protein expression. The body distribution of lithium is similar to that of sodium, and it may compete with sodium at some sites, such as kidney tubule reabsorption. Lithium also competes with other cellular cations such as potassium, and this property may be responsible for the effects exerted on ion exchange and intracellular metabolic disturbances. Cardiovascular and nervous system changes are the most prevalent, and the distribution of lithium to the brain is slow but cumulative. Blood levels in excess of 1.5 mEq/dl may cause CNS responses such as blurred vision, increased thirst, muscular weakness and tremor, drowsiness, dizziness, and confusion. Even higher blood levels are associated with more severe effects including blackouts, seizures, slurred speech, myoclonic movements (muscular contractions), choreoathetoid movements (writhing, turning, and twisting), and coma.

The methylxanthines caffeine and theophylline are alkaloids found in coffee and tea. While the principal mode of action of these compounds is via purine receptors, at higher concentrations they also affect cyclic nucleotide systems by inhibiting the phosphodiesterase enzymes that degrade cAMP or cGMP. Medically, caffeine is widely used to treat apnea in newborns; however, high or repeated doses increase oxygen consumption in neurons resulting in cell death. The methylxanthenes produce convulsions at high doses and may cause neurotoxicity by promoting the release of NO from astrocytes. Both organotins (e.g., trimethyltin) and ortho-substituted PCBs have been shown to promote calcium accumulation and activate protein kinase C (PKC). Mushrooms of the *Amanita* sp. produce a series of cyclic octapeptides, the amanitines. These compounds cause a delayed cerebrotoxicity that may be related to their specific inhibition of nuclear RNA polymerases. Lectins of various sorts (like ricin and volkensin) can be taken up by nerve terminals, transported great distances by retrograde transport, and ultimately interfere with protein or nucleic acid synthesis, thereby causing neurotoxicity.

19.2.5 Agents That Cause Hypoxia

Any agent capable of depriving the CNS of oxygen is neurotoxic. Neuronal subpopulations in the hippocampus and neocortex are particularly sensitive to hypoxic insult because their metabolic activity is very high. The cellular events associated with sequelae to neuronal hypoxia are similar to those of excitotoxicity (see Section 19.4), probably because glutamate is released under hypoxic conditions by an unknown mechanism. Hypoxia may be anoxic, ischemic, or cytotoxic, depending on the insult. Anoxic hypoxia occurs when the oxygen supply to the brain is compromised, despite adequate blood flow. Examples of this type of hypoxia include agents that cause respiratory insufficiency, such as petroleum distillates and derivatives (e.g., gasoline, methane), and agents that interfere with the oxygen-carrying capacity of the blood, such as carbon monoxide. Ischemic hypoxia is produced by agents that block the blood supply to the brain. These include any agent causing cardiovascular failure such as digitalis glycosides, catecholamines, and chlorinated hydrocarbons. Cytotoxic hypoxia is caused by agents that interfere with cellular respiration such as cyanide, azide, and hydrogen sulfide. It should be remembered that hypoxia affects not only neurons, but also oligodendrocytes. Prolonged exposure to hypoxic agents, for example, may result in the loss of ability by oligodendrocytes to maintain myelin.

19.2.6 Agents That Affect Membranes

Among its many effects, organic and elemental lead damages membranes, probably by disrupting ion channels. This results in ultrastructural damage to mitochondria, breakdown of active transport systems, and damage to myelin-containing membranes. An increased susceptibility to infections may also be associated with membrane compromise by lead and other heavy metals. Copper intoxication has been associated with the incidence of Alzheimer's disease. Copper is capable of participating in the formation of reactive oxygen species (ROS) and lipid peroxidation, thought to be the cause of copper-induced vacuolation of astrocytes.

Solvents and vapors, including ethanol, are lipid soluble and thus easily pass within and through membranes. These agents are capable of changing the lipid composition of membrane, thus altering membrane fluidity. These types of changes further alter the protein-lipid and protein-protein interactions that occur within and near the membrane. For example, increases in membrane resistance and ion fluxes resulting in alterations in action potentials have been measured in neurons after exposure to ethanol. These effects have been hypothesized to underlie the actions of ethanol on neurotransmitter metabolism and release.

19.2.7 Agents That Interfere with Oxidative Phosphorylation

The process of mitochondrial oxidative phosphorylation is critical to survival in all cells, but the CNS is particularly sensitive to uncoupling agents due to its high need for energy. The maintenance of ionic gradients in excitable cells is thought to be the primary energy-drawing process. The classical inhibitors of oxidative phosphorylation (e.g., dinitrophenol, cyanide, and hydrogen sulfide) are lethal largely because they disrupt neuronal function resulting in respiratory arrest. In previous sections, we have seen that lead, mercury, and other heavy metals affect mitochondrial function in neurons. Because of mitochondrial insult, oxidative phosphorylation is compromised. Many agents disrupt oxidative phosphorylation because of their effects on membrane integrity or their disruption of ion channel function. Section 19.3.3 describes the mechanism of action of MPTP, a direct inhibitor of the oxidative phosphorylation pathway.

19.2.8 Agents That Damage Myelin

Finally, the function of nerve cells depends on the integrity of myelin. As was discussed in Chapter 18 (Peripheral Nervous System Toxicity), axons of many neurons are surrounded by sheaths of concentric layers of the surface membrane of specialized glial cells (in the CNS called oligodendrocytes). Myelin sheaths, like neuronal cell surface membranes, are subject to attack by toxicants. Many of the same agents that act peripherally, such as hexachorophene, isoniazid, tellurium, and organotins, may also cause demyelination in the CNS. Agents that damage membranes (see Section 19.2.6) such as lead and organic solvents (e.g., hexane) are known to demyelinate neurons in the CNS.

19.2.9 Biotransformation in the CNS

The capacity of the CNS to metabolize xenobiotics is thought to be much lower than that of other major organ systems. Although some biotransformation processes can and do occur in the brain, the relative ability of brain cells (i.e., neurons, glial cells, and endothelial cells) to metabolize exogenous compounds is miniscule compared to that of hepatocytes. Several reasons for the relative deficiency of brain metabolism may be speculated. First, the brain is protected from the entry of many xenobiotics by the blood-brain barrier; therefore, there is less need for the large-scale expression of transforming enzymes. Second, the brain is downstream of the route of entry for many toxicants, the gut, and thus spared by first-pass metabolism. Third, the high metabolic activity of neurons would suggest that most of the cellular energy expended for protein synthesis goes into fulfilling critical neuronal functions, rather than synthesis of macromolecules of biotransformation. Fourth, the biotransformation needs of the CNS are different from other tissues. As discussed in other sections of this chapter, there is a very high potential for free-radical formation in the brain, while the risk of exposure to molecules like aminopyrene, for example, is much lower. Finally, the diversity of cellular subpopulations in the CNS indicates that the needs of each region are likely to be different as well. Specific types of biotransformation would thus need to occur in situ, localized to the region at risk, and might be different from those in a neighboring brain region. Because of localized expression, enzymes involved in xenobiotic metabolism would probably not be easily detectable. The remainder of this discussion will focus on the detoxication mechanisms that have been identified in the brain.

Protection from oxidative damage is probably the most important role of detoxication in the CNS. The brain has the highest rate of oxidative activity of any organ, and endogenous oxygen-derived free radicals are thought to be important in the pathogenesis of many neurodegenerative diseases (see Section 19.3). Both neurons and glia contain protective mechanisms against oxidative stress. Often, the neurons benefit from secreted enzymes manufactured in the glia. For example, glutathione peroxidase and superoxide dismutases have been identified mainly in astrocytes, and catalase is found predominately in oligodendrocytes. Glutathione is a critical element in the antioxidant armamentarium of the CNS. Not surprisingly, this thiol is distributed ubiquitously in the brain. Glutathione chelates transition metals and prevents their redox cycling, while also buffering other redox cycling events in the cell. In situations where reduced glutathione is depleted, lipid peroxidation ensues, targeting mitochondrial membranes. Loss of mitochondrial inner membrane integrity leads to uncoupling of oxidative phosphorylation and the further production of radicals.

Monooxygenases, NADPH cytochrome P450 reductase, and epoxide hydrolase have been identified in astrocytes, neurons, and endothelial cells. Many of the cytochrome P450 isoforms present in the brain are those that endogenously metabolize fatty acids (e.g., CYP2C11) and their role in xenobiotic metabolism in this organ is unknown. On the other hand, MAO-B, as we shall see later in this chapter, is an essential component of the bioactivation of MPTP (see Section 19.3.3). The cytochrome P450 isoforms of the 2D and 2E subfamilies have been localized to neurons. Interestingly, both of these subfamilies have been linked to the pathogenesis of Parkinson's disease (PD). CYP2E1 is found in neurons of the substantia nigra that are known to be affected in PD (see Section 19.3.2). Because this enzyme is a potent generator of free radicals and associated with the formation of endogenous toxins, a contribution to the nigral pathology of PD has been proposed. A longer association with PD has occurred with CYP2D isozymes. A genetic polymorphism resulting in altered expression of CYP2D6 was thought to increase susceptibility to PD. Evidence in favor of this hypothesis pointed to the ability of the unaffected CYP2D6 enzyme to metabolize MPTP to nontoxic metabolites, to expression of the enzyme in dopaminergic cell areas, and to various studies suggesting that the polymorphism may be overrepresented, at least in some populations of PD patients. Recent genetic studies, however, have failed to support a direct link between the CYP2D6 polymorphism and PD.

19.3 NEURODEGENERATIVE DISEASES AND NEUROTOXIC MECHANISMS

19.3.1 Introduction

There is a diverse group of chronic progressive neurodegenerative disorders that are among the most devastating illnesses in medicine and whose etiology has been particularly challenging. Recently, mutations in specific genes have been found to cause at least some of the cases in these diseases, including Alzheimer's disease, amyotrophic lateral sclerosis (ALS), and Parkinson's disease. Yet most of the cases of these disorders are sporadic in nature (i.e., without a clear genetic cause), requiring that both genetic and environmental factors be considered in elucidating their molecular and cellular basis. Environmental chemicals have been suggested as being causative in those diseases without a clear genetic basis, and it may be that some fraction of neurodegenerative diseases have such a component. Understanding the mechanism involved in these diseases is informative about the ways in which chemical or physical perturbations of the CNS initiate neurotoxicity.

Neurodegenerative diseases, including Alzheimer's, Parkinson's, and ALS, are age-related disorders associated with a variety of risk factors, including the presence of mutations in specific genes. Probably the most common neurodegenerative condition is Alzheimer's disease, the majority of which are sporadic cases with no family history. Several molecules and their genes appear to be linked to the neurodegeneration in AD, including β-amyloid, the amyloid precursor protein, ubiquitin, and presenilins, although a clear mechanism is not yet established. As stated above, because aluminum is itself a toxic metal, and is found in high concentrations in the plaques in AD brains, some have suggested an etiologic role for this metal. Most investigators believe that the accumulation of aluminum is a consequence of, rather than a cause, of neurodegeneration. The motor neuron disease, ALS, is typified by muscular atrophy, spasticity, and paralysis. Approximately 10% of adult cases of ALS are familial, and 20% of these are linked to mutations in the SOD1 gene encoding the free-radical scavenger Cu^{2+}/Zn^{2+} superoxide dismutase (SOD). The scavenging activity of Cu^{2+}/Zn^{2+} SOD protects molecules and organelles from damage by free radicals. As reductions in Cu^{2+}/Zn^{2+} SOD activity have been demonstrated to accelerate cell death, it was initially hypothesized that the reduced enzymatic activity of Cu^{2+}/Zn^{2+} SOD resulting from mutation of the SOD1 gene resulted in ALS. Recent studies, however, indicate that mutant forms of the SOD1 gene may result in increased cellular toxicity through other mechanisms, possibly by dysregulation of copper availability.

Huntington's disease (HD) is caused by the loss of specific populations of neurons, resulting in progressive motor, psychiatric, and cognitive impairments. HD is a dominantly inherited defect caused by an expanded stretch of CAG trinucleotide repeats that encode polyglutamine segments in a protein named huntingtin.

The clinical manifestations of HD gradually advance to an incapacitating "chorea" resulting from selective neuronal loss in the cerebral cortex and striatum. The normal function of huntingtin has been difficult to elucidate, although its intracellular distribution in neurons suggests that it could play a role in the anchoring of vesicles and organelles within the cytosol, and that it could be involved in vesicle tracking.

19.3.2 Etiology of Parkinson's Disease

Parkinson's disease is an idiopathic condition affecting about upwards of one million Americans. It is known that the disease is due to the death of neurons in the brain region called the pars compacta of the substantia nigra. A primary characteristic of these neurons is their dark color, for which they are named, due to the presence of neuromelanin. The substantia nigra neurons use dopamine as their neurotransmitter, releasing it at terminals in the striatum and globus pallidus. These terminal fields are critical for the initiation and control of movement, hence the signs of this disease include motor deficits expressed as tremor, rigidity, bradykinesia, and postural instability. There has been significant progress in our knowledge of the cause, the pathogenesis, and the nature of the mechanism of cell death in Parkinson's disease (PD). While there are rare familial cases that have been linked to a specific mutation of a specific gene (e.g., the synaptic protein α-synnuclein), current evidence suggests that the huge majority of PD does not involve a strong genetic component. For this reason, many investigators have attempted to find evidence linking environmental factors to the etiology of PD.

19.3.3 MPTP Provides Support for a Toxic Etiology of Neurological Disorders

One of the compelling scientific findings that spurred this hypothesis was two clusters of parkinsonism described in the late 1970s and early 1980s. A few young, otherwise healthy, individuals presented in emergency rooms with parkinson-like symptoms. Their treating physicians were initially baffled until they realized that all of these patients were drug (specifically opioid) users. Some detective work revealed that, in each case, the individuals had bought a synthetic heroin derivative that was made by a home chemist who had inadvertently synthesized not 1-methyl, 4-propion-oxypiperidine (MPPP) as intended, but 1-methyl-4-phenyl-1,2,3,6-tetrahydropyridine (MPTP). After these drug addicts injected themselves with MPTP, they developed within hours a condition that looked like Parkinson's disease. We now know that MPTP itself is not the cause of idiopathic PD, but the similarities between MPTP-induced and idiopathic parkinsonism strongly suggest that naturally occurring neurotoxicants may selectively damage the nigrostriatal system, thus causing or contributing to neurodegeneration in PD. Finally, when given to non-human primates, MPTP is accepted as providing a very predictive model of human Parkinson's disease.

MPTP easily traverses the blood-brain barrier. The compound enters specific neuroglial cells in the brain, and is there metabolized by the enzyme monoamine oxidase B (MAO-B) to 1-methyl-4-phenylpyridinium (MPP+; see Fig. 19.3). MPP^+ is released from glia and is then taken into catecholamine neurons via the active

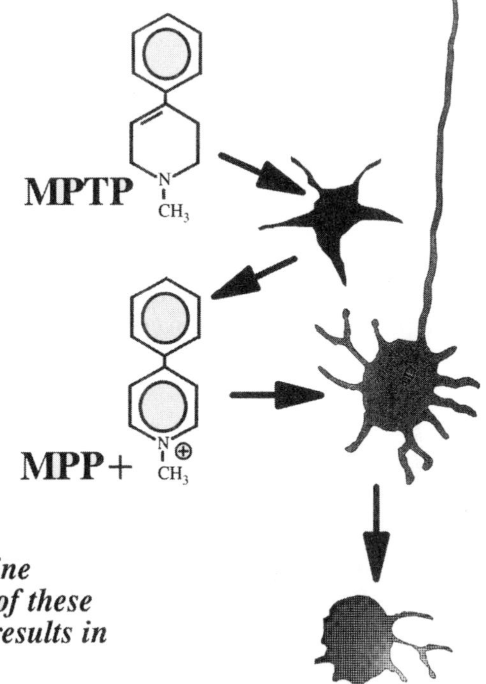

FIGURE 19.3 Schematic of biotransformation and toxicity of MPTP.
MPTP is a neurotoxicant with significant selectivity for dopamine neurons in the primate substantia nigra. Peripherally administered MPTP is a lipophilic compound which easily crosses the blood-brain barrier. The parent compound MPTP is relatively nontoxic; however, in brain it is taken up by glial cells and undergoes oxidation by an isozyme of monoamine oxidase (MAOB). One of the oxidation products, MPP+, is extremely toxic. This toxic metabolite is transported to the dopamine cell interior via the dopamine uptake system. The mechanism of subsequent cell death by MPP+ is unclear, although the current prevailing hypotheses involve effects on mitochondrial respiration. Compensatory processes in surviving dopamine cells prevent gross behavioral symptoms until the huge majority of dopamine neurons are damaged (>90%), at which time features of Parkinson's disease result.

uptake system, which normally scavenges the endogenous transmitter (see Fig. 19.2). Although not completely understood, MPP^+ may destroy dopamine neurons by several processes that work alone or in concert. One target of this compound is mitochondrial Complex I, which is inhibited by high levels of MPP^+. Second, MPP^+ may induce free-radical formation within the neurons. Direct infusion of MPP^+ into dopaminergic nerve terminals results in hydroxyl radical formation and lipid peroxidation, and these responses can be inhibited with antioxidants. Interestingly, the MAO inhibitor selegiline not only prevents the formation of MPP^+ in the glia by inhibition of MAO-B, but also suppresses the generation of hydroxyl radicals in neurons by an antioxidant-like effect.

19.3.4 The Role of Endogenous Factors in Parkinsonism

A potential mechanism related to the direct formation of free radicals involves the sustained efflux of dopamine from substantia nigra nerve terminals caused by MPP^+ and related analogs. As pointed out above, MPP^+ is taken up by dopamine transporters on the nerve terminals. At the same time, MPP^+ competes with endogenous dopamine for sites on these transporters, resulting in the decreased uptake of dopamine. The excess synaptic dopamine may itself represent a source of oxidative stress. Excessive synaptic dopamine is metabolized by oxidation. There are two mechanisms for the oxidative breakdown of dopamine. The enzymatic oxidation of dopamine (DA) by MAO-A or MAO-B results in the formation of the deaminated product, 3,4 dihydroxyphenyl acetate (DHPA), ammonia, and hydrogen peroxide.

$$DA + O_2 + H_2O \rightarrow DHPA + NH_3 + H_2O_2 \quad (19.1)$$

Hydrogen peroxide, it will be remembered, is then converted by ferrous iron to hydroxyl radical by the Fenton reaction.

$$H_2O_2 + Fe^{2+} \rightarrow {}^{\bullet}OH + OH^- + Fe^{3+} \quad (19.2)$$

The nonenzymatic breakdown of dopamine is by auto-oxidation, a two-step reaction of dopamine and divalent oxygen to a semiquinone, hydrogen ion_2 and a superoxide radical. The radical may then interact with another molecule of dopamine to form a second semiquinone, as well as two more hydrogen ions and H_2O_2.

$$DA + O_2 \rightarrow SQ^{\bullet} + {}^{\bullet}O_2^- + H^+ \quad (19.3)$$

$$DA + {}^{\bullet}O_2^- + 2H^+ \rightarrow SQ^{\bullet}\ H_2O_2 \quad (19.4)$$

The metabolism of dopamine thus results in the formation of highly reactive free radicals and other oxidizing species. The semiquinones are thought to contribute to the synthesis of neuromelanin. Neuromelanin granules in the substantia nigra of PD patients accumulate iron, which may further promote the generation of ROS.

Dopamine is toxic to neurons in culture. A causative role for free-radical formation and reactivity in dopamine-mediated neuronal death is supported by the bulk of evidence, which includes the presence of markers for lipid peroxidation and nuclear changes consistent with apoptosis. In fact, dopamine and L-dopa, the dopamine precursor and therapeutic agent for PD, are equipotent in their ability to promote cytotoxicity. Consequently, many scientists now worry that L-dopa administration may accelerate the progression of PD by increasing the rate of nigrostriatal degeneration. In vivo studies have failed to show a causative relationship, however, between L-dopa therapy and progression of PD.

There is evidence that some patients with idiopathic Parkinson's disease have a mutation in mitochondrial DNA (mtDNA) that results in lowered Complex I activity of the electron transport chain. Deficiencies in mitochondrial electron transport have been reported in parkinsonian patients that bear striking similarity to those caused by MPP^+. It is not known, however, whether these mitochondrial defects represent a primary etiology of Parkinson's disease, or perhaps are secondary to neurotoxic events that occur in the pathogenesis of the disease.

19.3.5 The Role of Environmental Agents in Parkinsonism

Epidemiological studies have found an association between PD and possible exposure of environmental factors, including agricultural chemicals and well water contaminants. Evidence in favor of an environmental etiology of PD has prompted the search for agents causally related to the disease and, through epidemiological and toxicological studies, several types of candidate toxins may now be identified. First, a number of population-based studies have found an association between PD and at least one of the following environmental risk factors: rural residence, farming, well water drinking, and exposure to agricultural chemicals. This association supports the hypothesis that toxicants present in the rural environment, such agrichemicals as pesticides, herbicides, and other chemical groups with the capability of destroying undesirable organisms may cause damage to the nigrostriatal system. Interestingly, the suspect compounds are structurally diverse, and share no apparent common mode of action.

Some of these candidates have been studied in the laboratory. The bipyridyl herbicide paraquat is similar structurally to MPP^+, the active metabolite of MPTP. Although direct intracerebral injection of paraquat into the rat brain can cause a dose-dependent depletion of dopamine and a concomitant loss of neurons in the substantia nigra, access of this pesticide into the brain would be expected to be limited (see Section 19.1.4.) because of its formal positive charge. In fact, after systemic administration, effects in the brain are seen only after repeated systemic administration, if at all. Another class of compounds suggested as etiological agents is the chlorophenoxy pesticides such as 2,4-dichlorophenoxyacetic acid (2,4-D) and 2,4,5-trichlorophenoxyacetic acid (2,4,5-T). These two compounds were the major herbicides used in military operations in Vietnam. Compounded in a 50:50 mixture, 2,4-D and 2,4,5-T are together known as Agent Orange. Alleged health problems (including neurologic effects) have been attributed to exposure to Agent Orange, but often are assumed to be due to 2,4,7,8-tetrachlorodibenzo-p-dioxin (TCDD), a contaminant in Agent Orange. TCDD has recently been reported to induce oxidative stress in neural tissue, consistent with the idea sometimes proposed that dopamine neurons are particularly vulnerable to oxidative injury. In addition to bipyridyl and chlorophenoxy derivatives, other pesticides have been implicated. Thus, some workers have reported that organochlorine pesticides such as dieldrin are found in higher concentrations in subjects with PD but not in controls. Another organochlorine heptachlor has been reported to affect the function of dopamine neurons (i.e., altering expression of the dopamine transporter and increasing dopamine release) in ways that might increase the vulnerability of dopaminergic neurons to neurodegeneration. Yet despite the epidemiological and laboratory evidence, the case for environmental chemical involvement is not overwhelming and further research is required to ascertain the true role of such agents.

19.3.5a Chemical Interactions in Parkinson's Disease

Humans are likely to be exposed to complex mixures of chemicals, and it has been suggested that the pathogenesis of nigrostriatal degeneration is neuronal damage arising from the combined actions of several chemicals. For example, several investigators have reported that the thiocarbamate fungicide diethyldithiocarbamate (DDC), although not neurotoxic itself, markedly enhances nigrostriatal damage when coadminstered with MPTP. The mechanism of this synergism is unknown, and may involve the copper chelation caused by DDC, actions at specific enzymes (e.g., superoxide dismutase and aldehyde dehydrogenase), excitotoxicity, or affects on CNS disposition of MPTP or its toxic metabolite MPP^+. In fact, if environmental compounds play a role, it is likely that they interact either with other dietary constituents, or with a genetic factor that alone is inadequate to cause the disease.

19.3.5b Other Candidate Compounds for Parkinson's Disease

The awareness of the neurotoxicity of MPTP/MPP^+ led to the search for other compounds that might be potential neurotoxicants selective for the nigrostriatal system. Two classes of compounds that have been of interest include tetrahydroisoquinolines (TIQs) and β-carbolines. Members of both families can be bioactivated to quinolinium and carbolinium metabolites that may have neurotoxic potential, for example, to impair mitochondrial respiratory chain activity much like MPP^+. TIQs may be formed endogenously from nonenzymatic condensation of the amine group of dopamine or related amines with the reactive carbonyl groups of aldehydes and α-ketoacids. Both TIQs and β-carbolines have been detected in the human brain, including in the cerebrospinal fluid of PD patients. TIQs and β-carbolines are naturally occurring alkaloids present in a variety of foods, but can also be generated within the brain by reactions involving biogenic amines. Nonetheless, both TIQs and β-carbolines are much less toxic than MPTP, and by themselves cannot cause acute damage of the type seen with MPTP.

19.3.6 A Movement Disorder Caused by Manganese

A xenobiotic that produces Parkinson-like symptoms bears mention at this point. Manganese is an essential element that participates in multiple cellular functions, but it is also neurotoxic at high doses. Persons occupationally exposed to manganese such as miners, iron and steelworkers, and welders, as well as some individuals with severe liver disease, suffer from deficits in movement similar to those of Parkinson's disease. Until recently, this disorder was also though to arise from a destruction of dopamine neurons, however, recent findings suggest an alternative explanation. The pathogenesis of manganese toxicity differs somewhat from Parkinson's disease. First, the selective location of manganese-mediated destruction is in the globus pallidus, and a separate area of the substantia nigra, the pars reticulata. In contrast to the substantia nigra pars compacta, these brain areas receive dopamine innervation, but do not contain dopaminergic cell bodies. Second, magnetic resonance imaging (MRI) and positron emission tomography (PET) studies have suggested that manganese intoxication damages dopamine output pathways downstream of the nigrostriatal pathway. Perhaps because of this, patients with manganese toxicity respond poorly to the antiparkinsonian drug L-dopa. Earlier work suggested that manganese toxicity in dopaminergic neurons might result in part from its ability to oxidize dopamine. Indeed, manganese is a transition metal, and as such is capable of participating in redox reactions. It appears, however, that other mechanisms of oxidative stress may be operative, such as mitochondrial compromise and entry into the brain of other heavy metals due to blood-brain barier damage.

19.4 EXCITOTOXICITY IN THE CNS

Although a role for excitotoxicity in the etiology or progression of several human neurodegenerative conditions or disease has been studied intensively, no direct evidence of excitotoxicity as a causative agent of these disorders has been provided. Even so, this area is widely regarded as being critical in understanding central neurotoxicity and neurodegeneration. There is an ongoing controversy of the contributions of apoptosis (see chapters 15 and 21) and necrosis in neurodegeneration. In fact, it may be that neurotoxic events can involve both processes along a continuum, rather than being one or the other. Thus, excitotoxic injury may well result in neurodegeneration that involves both apoptosis and necrosis, with the relative degree of insult related to the degree of neuronal activation, and the brain region and neuronal cell type being targeted.

19.4.1 Mechanisms of Excitotoxicity

Excitotoxicity is a general neurotoxic mechanism that has received increasing attention during the past few years. Basic neurobiological studies had demonstrated that some drugs that bound to receptors for the neurotransmitter glutamate (glutamic acid) could cause the death of the neurons on which these receptors resided. Later it was shown that glutamate-producing neurons could be similarly affected when electrically overstimulated. Glutamate is an excitatory neurotransmitter, meaning that its postsynaptic binding results in a depolarizing or excitatory effect, rather than a hyperpolarizing or inhibitory effect. The neurotoxicity resulting from glutamate neurotransmission is thus referred to as excitotoxicity. Glutamate is the most widespread excitatory transmitter in the vertebrate CNS, producing both acute and long-lasting changes in neuronal excitability, synaptic structure and function, neuronal migration during development, and neuronal viability. Two types of CNS neurotransmitter receptors are represented by glutamate receptors: the ionotropic glutamate receptors including the kainic acid receptor, the N-methyl-D-aspartate (NMDA) receptor, and the AMPA receptor; and the metabotropic glutamate receptors (formerly the called the quisqualate receptor). Excitotoxicity appears to be meditated primarily by the NMDA receptors. The role of metabotropic glutamate receptors is controversial; disparate reports link these receptors to causative and to neuroprotective roles in excitotoxicity. The remaining discussion will be limited to neurotoxicity mediated by NMDA.

NMDA receptors are part of a calcium ion channel complex that has binding sites for glutamate, NMDA, and some other compounds; these binding sites can modulate the activity of the channel. The release of endogenous glutamate, the failure of synaptic uptake mechanisms, or the application of NMDA or glutamate toxins may result in hyperstimulation of NMDA receptors. In response, large amounts of calcium enter the neuron through these channels, overwhelming the ability of homeostatic mechanisms to restore ion balance. This causes many secondary events that determine the ultimate fate of the cell. First, in the event that balance can be restored, the cell may recover from injury as described below. A second possibility is that the ionic imbalance leads to a sudden influx of sodium and chloride, creating a hyperosmotic state and finally necrotic cell death. Yet another possibility is that the calcium excess initiates a series of events leading eventually to apoptotic cell death.

One of the interesting features of excitotoxicity is the response to injury that occurs if the neuron is allowed to recover. Evidence suggests that localized areas within the neuron containing elevated calcium may be selectively damaged. Neuronal NMDA receptors are clustered on dendrites; virtually none are found on the axons. Consequently, localized areas of high calcium, as might occur early in excitatory stimulation, are confined to areas where NMDA receptors are found. The secondary events induced by calcium thus also occur locally, resulting in the selective death of dendrites, a phenomenon known as dendritic "pruning." Axons and perikaryon, spared of these neurotoxic changes in the initial stages of excitotoxicity, remain functional, but the character and intensity of signals transmitted by these neurons is altered by the pruning.

Continued hyperstimulation of NMDA receptors renders the neuron incapable of overcoming ionic imbalances. Excessive calcium accumulation occurs within mitochondria in response to the cytosolic calcium overload. Oxidative phosphorylation is uncoupled by this response, leading to decreased synthesis of ATP and the release of oxygen radicals from the mitochondria. At the same time, elevated cytosolic calcium affects a number of calcium-dependent signaling proteins, including proteases and nitric oxide synthase (NOS). The activated NOS then elevates levels of NO inside the neuron, which interact with superoxide radicals to form highly reactive peroxynitrites. Peroxynitrites modify proteins by interacting with tyrosine residues, and certain of these proteins respond by consuming NAD and/or ATP. As ATP is further depleted, permeability transition pores open in the mitochondria causing the continued collapse of mitochondrial membrane potential, and the release of cytochrome c. Cytochrome c is though to be an important trigger for the initiation of apoptotic signaling, initiating the activation of caspases, the proteolytic effectors of cytoskeletal and nuclear degradation.

It should be clear from this discussion that mitochondria and reactive oxygen species are critical elements of excitotoxicity, as they are for the pathogenesis of many neurotoxic agents. Some neurotoxicants that utilize excitotoxicity are discussed below.

19.4.2 Examples of Excitotoxicity

19.4.2a Cycad Poisoning

One of the most complex and fascinating epidemics of neurotoxicity is the Western Pacific Amyotrophic Lateral Sclerosis/Parkinsonism-Dementia (ALS/PD) complex found on the Island of Guam, West New Guinea, and the Kii peninsula in Japan (for a very readable history of this epidemic, and the scientific arguments that have ensued, see Monmaney, 1990). ALS/PD was first identified by military physicians stationed on Guam during World War II, who noted an unusually high prevalence of the syndrome (referred to locally as Lytico-Bodig) among one particular ethnic group, the Chamorros. The clinical symptoms of the syndrome are those of either ALS or PD, with both sets of symptoms often occurring in the same individual.

These include bradykinesia, rigidity, and tremor, accompanied by progressive deterioration of memory, comprehension, and orientation in time and space. Neuropathological symptoms include generalized cerebral atrophy, neuronal loss, and Alzheimer's type of neurofibrillary tangles in the several brain regions, depigmentation in the substantia nigra, and widespread astrocytic gliosis. ALS/PD patients differ somewhat from patients showing the typical PD or ALS, however, in the pattern of dopaminergic depletion in the basal ganglia and the brain amino acid contents. Reduced striatal dopamine uptake is found in patients showing either the ALS or the PD symptoms, and reduced dopaminergic uptake has been found in some Chamorros who do not exhibit the clinical symptoms of either ALS or PD.

In the 1950s, the ALS prevalence ratios and death rates for Chamorro residents of Guam were 50 to 100 times greater than those of the sporadic form that occurs throughout the world, and that accounts for about 1 out of every 1000 adult deaths. Symptoms of PD also occurred at a far greater rate than usual, with both ALS and PD symptoms often found in the same individual. ALS/PD has declined on Guam since 1955, and the age of onset has steadily increased from 25 to 30 in the late 1950s to over 50 today. No cases have been reported among Chamarros who were born on Guam after 1935. Although ALS/PD clinically is similar to late onset postencephalitic parkinsonism and ALS, the absence of any evidence suggesting a hereditary or viral cause led to a search for environmental agents.

One hypothesis was that the disease was related to an unusual mineral imbalance, due to chronic calcium deficiency, coupled with an excess of aluminum. Evidence for this was based upon observations that the disease was much less prevalent in the northern part of the island, where the water and soil contain much more calcium, and upon the discovery of unusually high concentrations of aluminum in the brains of ALS/PD victims. According to this theory, chronic deficiency of calcium would result in a greater absorption of other minerals such as aluminum, which may be toxic to cells. Epidemiological data, however, have shown no relationship between calcium concentrations in soil or water and the incidence rates of ALS/PD in various districts on Guam.

A second hypothesis, which has gained more experimental support, is that the syndrome is caused by toxins in the false sago palm, a cycad plant which is native to Guam, and which was used as a traditional source of food and medicine until the westernization of Guam following World War II. Flour prepared from the cycad seed became a major food source during the Japanese occupation of Guam (1941–1944), when rice was hard to obtain. Natives of Guam had long recognized the toxicity of the plant, and traditional preparation of flour from the seed involved days or weeks of repeated washing and soaking to remove the toxins. Laboratory analysis of the cycad revealed the presence of two toxins, cycasin (which is also a potent carcinogen), and α-amino-β-methylaminopropionic acid (L-BMAA), a compound similar to a neurotoxin found in the chickpea, which has been linked to a central motor system disorder (see 19.4.2c Other Excitotoxins). Initial research with cycasin and BMAA showed that while they were both clearly neurotoxins, they did not produce any experimental disorder in rodents that was similar to ALS/PD. When BMAA was fed chronically to cynomolgus monkeys, however, they developed corticomotorneuronal dysfunction, parkinsonian features, and degeneration of motor neurons in the cerebral cortex and spinal cord.

Various data have shown that BMAA acts as an indirect agonist at the NMDA type of glutamate receptor, although its precise mechanism of action is unknown. The idea of a "slow acting" neurotoxin is intriguing, and much of the criticism of the BMAA hypothesis of ALS/PD is based upon observations that there are no toxic symptoms that immediately follow exposure to cycad, and that the latencies between exposure to cycad and the onset of symptoms range from years to decades. If normal aging processes in the central nervous system are mediated, to some extent, by glutamate receptors, however, it may be that the clinical symptoms do not appear until after several decades of age-related or other neuronal compromise. For example, extensive depletion of catecholamines is required before behavioral symptoms can be easily observed in animal models of Parkinson's disease. Furthermore, neurofibrillary tangles and decrements in dopamine uptake have been found in Chamarros who do not show the clinical symptoms of ALS/PD, suggesting that subclinical neuronal insult in this population may be widespread.

19.4.2b Domoic Acid

Interest in the mechanism of domoic acid intoxication was stimulated by an outbreak of food poisoning in Canada in late 1987, caused by ingestion of blue mussels from the Prince Edward Island region of Newfoundland. The acute symptoms included gastrointestinal distress within 24 hours, followed by neurologic symptoms (convulsions, agitation, memory impairment, coma) within 48 hours. Four fatalities were reported among the 107 documented cases, and autopsies on these individuals revealed brain lesions in several forebrain areas including the hippocampus, amygdala, thalamus, and cerebral cortex. Chemical analysis of the mussels showed that they contained high concentrations of domoic acid, and the source of the domoic acid was traced to a species of plankton that synthesizes the acid in large quantities while in bloom. This plankton serves as a food source for filter-feeding shellfish such as mussels.

Although a purified form of domoic acid was not available until 1989, earlier work with an unpurified form showed that domoic acid binds specifically to the kainic acid receptor. More recent work with the purified form has shown that domoic acid activates responses in cultured hippocampal cells that are indistinguishable from kainic acid, and is several times more potent than kainic acid in destroying neurons in chick embryo retina. When given to rats, domoic acid (in doses of 2.5–3.0 mg/kg s.c.) elicits hypoactivity for approximately 60 minutes, followed by repetitive head scratching behavior, wet dog shakes, and clonic seizure activity that may progress to status epilepticus, ending in recovery or death, depending on dose. Monkeys given domoic acid (5.0–10.0 mg/kg p.o.) exhibit a different pattern of behavior, dominated by salivation, retching, vomiting, lip smacking, and mastication. Ultrastructural studies have shown that the toxic action of domoic acid is localized to the postsynaptic dendrosomal membranes where glutamate receptors are located. Domoic acid, like kainic acid, is thought to bind to postsynaptic dendrites near the synapse, where it either induces the release of excessive amounts of glutamic acid or prevents its absorption by the presynaptic axon ending and the resident astrocyte process.

A major issue in understanding the mechanism of domoic acid-induced brain lesions is the separation of direct damage due to domoic acid from the indirect damage mediated through seizure activity. Evidence suggests that the lesions are due to seizures resulting from excessive release of glutamate, because the cytopathological changes observed following domoic acid and kainic acid administration are indistinguishable from those following seizures elicited by a variety of other methods. It has been proposed that domoic acid and kainic acid trigger persistent seizure activity even at doses that do not cause a direct toxic action, and that this seizure activity is sufficient to cause brain damage by an excessive release of glutamate, which exerts excitotoxic actions at NMDA receptors. If this hypothesis is true, then NMDA antagonists may be useful as antidotes to domoic acid poisoning.

One observation made following the domoic acid outbreak in Canada was that the severity of the symptoms was related to the age of the individual, with the most severe neuropsychological deficits occurring in victims over 60 years of age. If this observation can be verified experimentally, it suggests that at least one, and possibly other types of glutamate receptors may mediate excitotoxic events as the human central nervous system ages.

19.4.2c Other Excitotoxins

Zinc colocalizes with glutamate receptors within excitatory synapses and may function as a neuromodulator under normal conditions. The activation of NMDA receptors by glutamate is attenuated by zinc, whereas activation of AMPA-type glutamate receptors is enhanced. Furthermore, in cortical neuronal cultures zinc is neuroprotective against neurotoxicity induced by either glutamate or NMDA. In contrast, activation of AMPA receptors in the presence of zinc produced a synergistic neurotoxic effect. β-N-oxalylamino-L-alanine (BOAA), which is similar to BMAA described above, is found in the chickpea, (also known as the chickling pea or the grass-pea, *Lathyrus sativas*), which is grown as a crop in Ethiopia, India, and Bangladesh. Heavy consumption of this crop produces Lathyrism, a disorder of the central motor system, which is characterized initially by a spastic gait, and exaggerated deep tendon reflexes, followed by the appearance of the Babinski reflex, and eventual loss of leg use. Lathyrism has been known since the time of Hippocrates (460–377 BC), and the relationship between Lathyrism and the consumption of chickpea was established in the 1960s. It continues to be grown as a cash crop however, because it grows well in a wide range of soil conditions, is easy to cultivate, and tolerates both drought and flooding. The excitotoxic actions of BOAA appear to be mediated by direct interaction with subclasses of glutamate receptors, whereas BMAA is not a direct acting glutamate agonist.

19.4.3 Neuronal Interactions

Although most of the examples provided in this chapter have emphasized insults targeted toward a single neurotransmitter system, the complexity of the nervous system raises the obvious possibility that two compounds that alone might have modest effects, could cause significant effects when present together. There have been some interesting examples of this that can be generally illustrative.

In addition to glutamate-induced excitotoxicity, it also is known that antagonists of NMDA glutamate receptor (e.g., both noncompetitive channel blockers like phencyclidine or ketamine, as well as competitive antagonists) can produce pathological changes. The neurotoxicity of the NMDA antagonists can be blocked by atropine (a muscarinic cholinergic antagonist), yet can be increased by the coadministration of pilocarpine (a muscarinic agonist). This raises an obvious toxicological situation, as both agricultural workers and people in the military may be exposed to cholinesterase inhibitors. The anticholinesterase insecticides of both the organophosphate and carbamate classes are used environmentally. Pyridostigmine, a cholinesterase inhibitor, is administered as a military prophylactic agent when there is a danger of chemical attack with nerve gas agents that are cholinesterase inhibitors. Although pyridostigmine does not normally cross the blood-brain barrier, it may do so under stressful conditions when the blood-brain barrier is compromised. In addition, there are several cholinesterase inhibitors used as therapy in individuals with moderate dementia. Thus, there may be specific concerns if people with otherwise low levels of cholinesterase inhibition also are exposed to NMDA antagonists. In fact, memantine, a drug that is proposed to have some utility in Parkinson's disease, has NMDA antagonist properties, as does felbamate, a drug that has been used as an anticonvulsant.

Another interaction that has received attention is that of glutamate and dopamine. As the chemoarchitecture of dopamine in the basal ganglia (including the striatum) has been elucidated, is has become clear that one role for dopamine is neuromodulation of glutamatergic inputs. For example, dopamine can, via activation of D_1 dopamine receptors, potentiate responses mediated by activation of NMDA glutamate receptors. Conversely, dopamine-mediated activation of D_2 receptors attenuates responses mediated by activation of non-NMDA receptors. The functional implications of such interactions are important in a fundamental neurobiological context, but are also of direct neurotoxicologal importance. For example, there is some data suggesting that the neurotoxicity of agents as diverse as methamphetamine, carbon monoxide, and MPTP may have a link to glutamate-mediated excitotoxicity and subsequent degeneration of dopamine cells.

19.4.4 Future Directions for Neurodegeneration and Neuroprotection

During the past decade, there has been explosive growth in our understanding of neural development, maintenance, and death. For example, it has been postulated that neurons of the same type in developing tissues compete for a limited supply of target-derived neurotrophic factor; neurons that first innervate their targets survive, while neurons that reach the same target later die. This idea leads to the hypothesis that neurons retrogradely transport vesicles containing signaling molecules (i.e., neurotrophins) to the cell body and nucleus, where they may modulate gene transcription. Studies of such interactions and resultant functional changes are likely to have relevance to how environmental chemicals may participate in the etiology of these neurodegenerative diseases. As the molecular events that are involved in these processes are elucidated, it should be possible to understand how specific chemicals cause neurotoxicity. Yet, although neurotrophic factors can mediate neuronal repair and viability in vitro, delivery of these agents (proteins or polypeptides) cannot be done orally or parenterally. Using these agents therapeutically will require advances in gene or protein delivery.

19.5 THE SCIENCE AND POLITICS OF CNS NEUROTOXICOLOGY

19.5.1 Detecting CNS Toxicity

The complexity of the CNS makes it difficult to pinpoint when and if an agent has caused toxicity. It is sometimes assumed that the activities of man are responsible for toxicity resulting from environmental agents. In fact, as can be noted from the many examples listed earlier in this chapter, natural products have been a rich source of neurotoxicants. Often such toxic products (or agents derived from them) have become clinically useful drugs. The traditional use of plant, animal, and microbial products in folk medicine is a reflection of the toxicopharmacological actions of these materials. This is important in two contexts. First, "natural" or "organic" does not mean that something may not have potential toxic consequences for an organism. Secondly, a neurotoxic response seen when a clear etiology is unknown may not always be due to an "environmental pollutant." Evaluation and detection of neurotoxicity thus becomes an important public health issue.

19.5.2 Approaches to Neurotoxicological Questions

Historically, pathological studies provided the primary, often sole way, to demonstrate the effects of toxicants. Clever use of differential staining, coupled with biological intuition, has often provided elegant hypotheses about neurotoxicants. In the past, such approaches relied upon traditional preparation of tissue (e.g., Golgi and other metallic impregnations, basophilic Nissl stains, Weigert's myelin method). Later, the availability of the electron microscope provided a new dimension for such studies. By the beginning of the last decade, the armamentarium of tools that was available had started to grow geometrically, and forced modern neurotoxicologists to chose carefully techniques for given purposes.

While modern immunohistochemical, autoradiographic, and cytochemical methods have permitted the localization and quantification of small molecules, proteins, and nucleic acids in cells, new techniques have also permitted ever more sensitive functional studies, even in the intact organism. A useful example is the type of approaches that might be used to study one of the problems discussed earlier, such as MPTP induced neurotoxicity. Classically, one would have waited for the availability of post-mortem tissue from an MPTP intoxicated individual. Such studies would have ranged from the neurochemical (is there a decrease in the amount of dopamine in the striatum, the terminal field of the substantia nigra), the biochemical (what are the relative activities of enzymes associated with dopamine neurons), and the neuropathological (is there a loss of melanin-containing neurons in the substantia nigra). Today, one could ask even more elegant questions from such tissue, such as what cell types are affected and in what way (are there losses of cell specific markers; are there compensatory changes in surviving cells; etc.) Yet, it is the ability to query the living tissue that has changed most dramatically. In vitro, one can construct cells of specific phenotypes to elucidate neurotoxicological mechanisms, using electrophysiological, imaging, and molecular end points. These methods can often be extended in vivo, with advances in chemical sampling (e.g., in vivo microdialysis and electro- and flurochemistry) and noninvasive imaging (PET, SPECT, and MRI). Thus, it has been demonstrated that the density of dopamine transporter protein (the site of uptake of dopamine from the synapse) is proportional to the density of dopamine neurons. Thus, using a PET or SPECT radioligand specific for the dopamine transporter, it is possible to estimate the degree of damage to substantia nigra dopamine neurons in an individual with Parkinson's disease, MPTP intoxication, etc.

Complementing such mechanistically oriented approaches are the use of a broad battery of behavioral endpoints as a screen for neurotoxicity. To meet guidelines under the Toxic Substances Control Act (TSCA), two approaches are used in which behavior is envisioned as an apical measure of nervous system function, that is, one reflecting the integrated output of many neuronal systems. The two strategies are to use motor activity and a battery of observational tests called the Functional Observational Battery (FOB). Changes in these end points are hypothesized to be useful to assess gross, integrated changes in behavior caused by toxicants. There are more than two dozen parameters that constitute the FOB, including horizontal, vertical, and total activity, convulsions, tremors, stereotyped (repetitive) behavior, respiratory pattern, gait, urination, startle response, piloerection, pupil size and light response, salivation, excessive vocalization, lacrimation, grip strength, muscle tone, etc. Although this is an appealing approach for screening for toxicity, there are inherent problems, including dose selection and the relatively small numbers of animals that are used.

Approaches such as the FOB often can provide important clues as to the mechanism of action of neurotoxicants. If a toxicant primarily affects one neuronal system (i.e., neurons using a particular transmitter), it may cause changes of a predictable pattern, often resembling known therapeutic agents. Probably the best known examples are organophosphate insecticides, which cause signs typical of cholinergic stimulation (mydriasis, salivation, urination, etc.). This is known to be due to the increase in synaptic concentrations of acetylcholine due to inhibition/inactivation of the enzyme acetylcholinesterase. Another example is the fungicide triadimefon, which increases activity and causes stereotyped (patterned) behavior much like that seen with amphetamine and cocaine. The actions of amphetamine and cocaine are a result of increased synaptic concentrations of the neurotransmitter dopamine. This suggested that tridimefon might act similarly, and recent studies have shown that it inhibits dopamine uptake, also leading to increased synaptic concentrations of dopamine.

19.5.3 Human Disease: When Is Neurotoxicity Involved?

No situation causes a greater interest than when a direct human health risk has been demonstrated or suspected. One research arena that has involved neurotoxicological hypotheses has been the area of human developmental disorders. It is well documented that insult to the neonatal nervous system (via drugs like ethanol, other toxicants, febrile illness, anoxia, trauma, or infection) can result in profound, often devastating developmental disorders (including severe and profound mental retardation). Such developmental problems are a major research arena. Yet, more subtle problems have garnered even wider public attention. For example, there is a childhood ailment now known diagnostically as attention deficit hyperactivity disorder (ADHD). This condition (previously named "minimal brain dysfunction," "hyperactivity," or "hyperkinesis") afflicts many children (estimates ranging from 0.5 to 3%). Usually, there is no etiological factor that can account for these behavioral problems, and the inability of the scientific and medical establishment to either prevent or cure this condition has led to the proffering of many hypotheses.

Several years ago, there was a specific neurotoxicological controversy regarding ADHD that provides an excellent example of misuse of current techniques. In the 1970s and 1980s, there was a widely held belief in the lay public that diet was the primary cause of most ADHD. The most widely known hypothesis was proposed in 1975 by Feingold, who argued that diets having no artificial flavors, colors, or certain natural ingredients would dramatically alleviate symptoms in one-third or more of ADHD children. Although large numbers of parents gave testimony to the Feingold diet, appropriate double-blind and challenge trials found no such effect. The Feingold diet received a boost when it was reported in 1979 that one of the eight food colors that had been used in the clincical studies had an important neurochemical action. This color, Red #3 (erythrosine), inhibited the uptake of dopamine in rat brain membranes, causing an increase in synaptic concentrations of dopamine much like cocaine or amphetamine. Such a result was thought to explain "hyperactivity," and provided the first testable biochemical hypothesis for the dietary theory. Another group of scientists expanded on the initial idea by suggesting that Red #3 inhibited neurotransmitter uptake by binding to the "ouabain receptor" (see Section 19.3.3), an ATPase that translocates sodium, a process necessary for dopamine uptake.

Although these data were highly publicized in both the popular and medical press, this elegant picture soon disintegrated. A literature search suggested that Red #3, a compound of known lipid solubility, affected several biochemical mechanisms unrelated to these neural systems. Based on this, our laboratory was able to show that Red #3 had effects not only on neurotransmitter uptake, but also nonspecifically on a host of biochemical mechanisms. All of this could be explained by the lipid solubility of this dye, rather than by a neural-specific mechanism. From this, we hypothesized that Red #3 was specifically unlikely to be a neurotoxicant, an idea that was later confirmed in both rodent and human studies. There is a human tendency to find relevance for one's research, yet overextrapolation of basic molecular and biochemical data can be at some cost—in the present example it led to a significant number of laboratory and clinical studies that were never warranted. With Red #3, one might argue that a thorough literature search at the outset would have led to appropriate experimental controls that would have prevented significant scientific detours, as well as false hope for parents of children with ADHD. The final footnote to this story is that erythrosine was later removed from the market—because it was a suspected carcinogen.

We provide this example because of the great interest in the role of xenobiotics in, for example, the various neurodegenerative disorders discussed earlier (see Section 19.3). Despite the increasingly powerful techniques available to explore such questions, a well-reasoned, hypotheses-oriented, conservative approach is essential. The value of adherence to fundamental principles (e.g., dose-response and dose-time relationships; conservative interpretations of correlative data; differentiation of primary toxicant mechanism vs. toxic sequelae; etc.), and the coordinated use of multidisciplinary approaches, cannot be overstated. As we have noted, many neurotoxicological questions can affect issues of public policy. Inappropriate use of the best scientific tools can result in spurious conclusions that are of great expense, both scientifically and socially (or in the words of a well-known neuropathologist, "A fool with a tool is still a fool."). Providing a synopsis of CNS toxicology is a daunting task for a whole book, an impossible one for a single chapter. We hope that this chapter has provided a basic understanding of the CNS, and spurred an interest in how toxicants may affect brain and spinal cord function.

SUGGESTED READING

The successful investigation of the biochemical toxicology of the CNS is dependent on first understanding fundamental aspects of neurobiology. The references below provide a starting point for further reading.

Abou-Donia, M.B. *Neurotoxicology*. Boca Raton: CRC Press, 1992, 621 pp. A good reference book, with emphasis on mechanistic aspects of neurotoxicology.

Chang, L.W. and Slikker, W.N. *Neurotoxicology: Approaches and Methods*. San Diego: Academic Press, 1995, 851 pp. Comprehensive overview of neurotoxicological methods.

Conner, J.R. *Metals and Oxidative Damage in Neurological Disorders*. New York: Plenum Press, 1997, 374 pp. Timely synthesis of scientific and clinical aspects of neurological disorders of toxicological importance.

Hutchins, J.B. and Barger, S.W. Why neurons die: cell death in the nervous system. *Anat. Rec. (New Anat.)* 253 (1998), 79–90.

Klaassen, C.D. *Casarett and Doull's Toxicology: The Basic Science of Poisons, 5th Ed*. New York: Pergamon Press, 1996, 1033 pp.

Monmaney, T. This obscure malady. *The New Yorker*, October 29, 1990, 85–113.

Nicholls, J.G., Martin, A.R. and Wallace, B.G. *From Neuron to Brain: A Cellular and Molecular Approach to the Function of the Nervous System, 3rd Ed*, Sunderland, MA: Sinauer Associates, 1992, 807 pp. An excellent, readable text that emphasizes molecular aspects of neurobiology.

CHAPTER TWENTY

Hepatotoxicity

SHARON A. MEYER and ARUN P. KULKARNI

20.1 INTRODUCTION

Liver is a frequent target tissue of toxicity from specific members of all classes of toxicants and natural toxins. Experimental hepatotoxicity of the rat from the solvent CCl_4 is a prototype model system for study of toxicodynamics. In humans, toxicity to liver has often limited usage of new pharmaceuticals such as troglitazone, an important drug for adult-onset diabetes mellitus that has recently been restricted to nonsusceptible individuals as determined from serum enzyme measurements during therapy. Liver damage from repeated exposure to toxic doses of ethanol is a major human health problem. The risk of chemically induced hepatotoxicity is increased by multiple, simultaneous exposures that can occur with chemical mixtures in the environment or from multiple, concurrent therapies. The rising prevalence of viral liver disease is another interacting factor that can be expected to aggravate the hepatotoxicity of chemicals. Understanding the molecular and cellular mechanisms mediating these toxic effects within the context of the function and structure of the liver is important knowledge necessary for predicting avoidable toxicity and adverse interactions.

Several factors consequential to the physiological function of the liver paradoxically predispose it to chemical toxicity. Hepatocytes are exposed to orally administered xenobiotics without systemic modification or dilution because they directly flow to the liver by the portal venous blood that delivers absorbed nutrients from the gastrointestinal tract. Coupled with an opposing movement of bile from its site of synthesis at the hepatocyte canalicular membrane to the duodenum, this routing establishes a loop—the enterohepatic circulation—which can slow toxicant clearance and facilitate hepatocyte reexposure. Routing of portal venous blood to the afferent vasculature also has the adverse effect of decreasing pressure and oxygenation on the arterial side of the hepatic sinusoids compared with capillary beds of other tissues. The predominance of hepatocyte oxidative metabolism of hydrophobic xenobiotics, a necessity for their renal and biliary excretion into aqueous compartments, can generate locally reactive metabolites if uncoupled from protective conjugating reactions. Also, the highly active, anabolic activities that occur within hepatocytes make this cell type especially susceptible to adverse effects of toxicants that act as antimetabolites or compromise mitochondrial energy production (see Chapter 13). The importance of these diverse liver functions for metabolite availability to critical organs such as brain causes severe effects of liver toxicants to be threatening to the viability of the organism.

20.2 LIVER ORGANIZATION AND CELLULAR HETEROGENEITY

The liver is a heterogeneous tissue composed primarily of parenchymal hepatocytes plus several additional nonparenchymal cell types. In addition, hepatocytes are functionally diverse depending upon their positional relationship to the afferent blood supply of the sinusoids. The liver parenchyma is comprised of a mesh of plates, each being the thickness of one hepatocyte (Fig. 20.1). Liver plates are separated by sinusoids, an arrangement ensuring that two surfaces of each hepatocyte are in close proximity to the blood supply. In two-dimensional microscopic sections, this organization appears as rows of hepatocyte cords separated by nearly parallel sinusoids. A specialized region of the plasma membrane, the canalicular membrane, runs as a band around each hepatocyte midway between the sinusoidal surfaces. Canalicular regions from neighboring hepatocytes align and are joined by tight intercellular junctions to form bile canaliculi.

Several organizational models of liver structure have been proposed. The most frequently used are the morphologically defined lobule and functionally defined acinus. The hepatocyte lobule is visualized in two dimensions with light microscopy as a hexagon centered on the terminal branch of the hepatic vein. The first few hepatocytes of cords radiating from this terminal hepatic venule (THV) or central vein are classified as perivenous or centrilobular. At the vertices of the imagined hexagon are the portal triads, cross-sectioned connective tissue tracts in which are embedded a bile duct (BD) with its distinctive cuboidal epithelial cells and preterminal branches from the hepatic artery (HA) and portal vein (PV). Hepatocytes located around the portal triads and following the circumference of the hexagonal lobule are classified as periportal. Midzonal hepatocytes intervene between centrilobular and periportal hepatocytes. The acinus is centered on the terminal distributing vessels that branch from the preterminal hepatic arterial and portal venous branches of the portal triads that form the sides of the lobular hexagon. Acinar hepatocytes are grouped into three zones that are layered concentrically around the distributing vessels. The similarity of intrazonal hepatocytes is determined by their exposure to a nearly equivalent blood composition, whereas interzonal hepatocyte differences reflect gradients in blood constituents known to exist across the sinusoids. Although conceptually different, there are approximate correspondences between lobular periportal and acinar zone 1 hepatocytes, centrilobular and zone 3 hepatocytes and midzonal and zone 2 hepatocytes.

20.2.1 Zonal Heterogeneity of Hepatocytes

Gradients in hepatocyte function occur over the periportal to perivenous distance. Periportal (zone 1) hepatocytes are exposed to the highest concentrations of oxygen, hormones, nutrients, and xenobiotics coming from the intestine or the systemic circulation. The partial pressure of oxygen (pO_2) is about 65 mm Hg

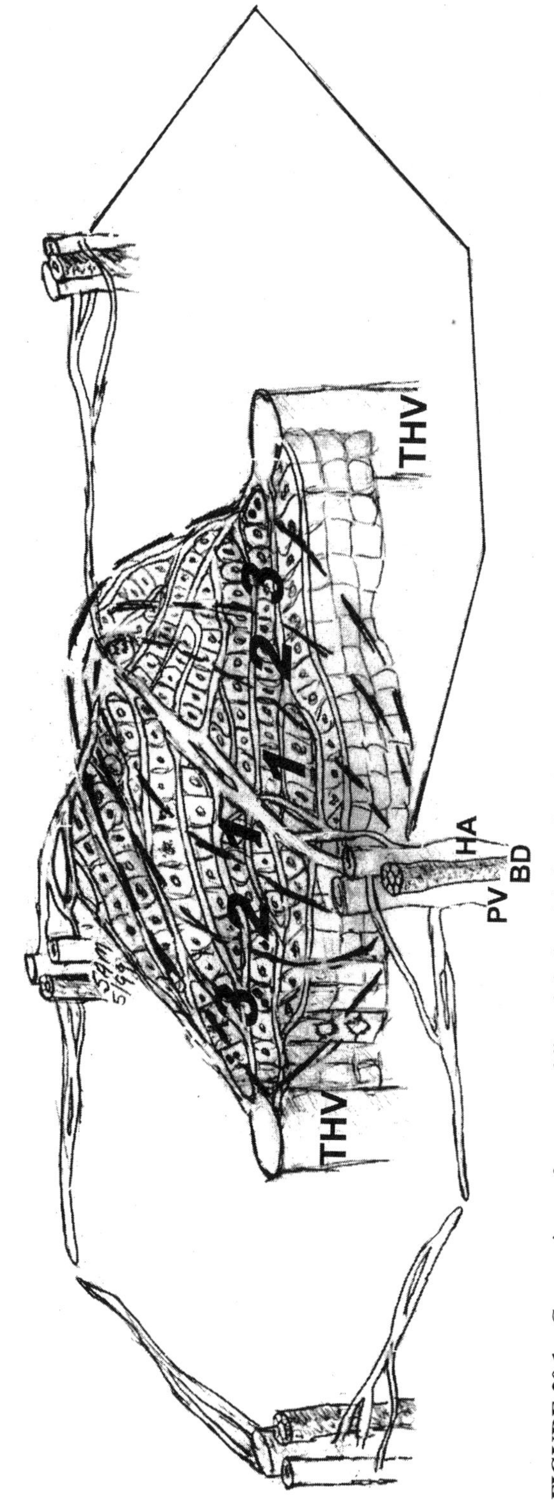

FIGURE 20.1 Comparison of structural liver lobule with functional liver acinar regions. The liver lobule is centered on the terminal hepatic venule (THV), also called central vein, and assumes a roughly hexagonal shape with its vertices at the portal triads, which contain the portal vein (PV), hepatic artery (HA) and bile duct (BD). The liver acinus is centered upon the tract of blood vessels that branch from the hepatic artery and portal vein of the portal triads. Hepatocytes within the acinus are grouped within functional zones 1, 2 and 3 located at increasing distances from the vascular tracts that interconnect adjacent portal triads.

(corresponding to about 9–13% O_2) in this zone. Cells are rich in mitochondria with more and longer cristae, and oxidative energy metabolism predominates. Glucose for systemic delivery to extrahepatic sites is primarily synthesized in periportal hepatocytes due to their enrichment in gluconeogenic enzymes. Co-localization of amino acid transaminases in periportal hepatocytes provides one source of gluconeogenic substrates. By contrast, perivenous (zone 3) hepatocytes are exposed to blood with reduced oxygen content and increased concentration of metabolic products. Perivenous pO_2 is about 30–40 mm Hg (4–5% O_2) and mitochondria are matrix-rich and cristae-poor. Enzymes of anaerobic glycolysis and lipogenesis and preferentially located in these cells.

Metabolic zonation of xenobiotic metabolizing enzymes determines the zone-specific localization of toxicant damage. Many hepatotoxicants elicit perivenous damage because cytochrome P450 enzymes are preferentially localized in zone 3. This coupled with a reversed distribution for conjugating glutathione (GSH) makes perivenous hepatocytes especially susceptible to reactive electrophiles. Differential distributions also exist for glutathione S-transferases, UDP-glucuronosyltransferases, epoxide hydrolase and alcohol dehydrogenase, which have higher perivenous activities, while sulfotransferase is predominant in periportal hepatocytes. Ammonia detoxication to urea occurs periportally, but is of limited capacity. Residual ammonia reaching zone 3 hepatocytes is incorporated into glutamine by perivenous glutamine synthetase.

Mechanisms maintaining metabolic zonation differ depending upon the process. Zonation of carbohydrate metabolism can be reversed by changing the direction of blood flow and thus appears dependent upon substrate availability and post-transcriptional signaling events. Repression of periportal transcription by hormonal influences, especially thyroid and growth hormones, appears to be responsible for cytochrome P450 distribution. However, plasticity exists for these systems as changes in zonal size occur to accommodate functional needs. In contrast, glutamine synthetase expression is a stable phenotype of perivenous cells.

20.2.1a Liver Stem Cells

Unlike most mature cells, differentiated hepatocytes retain the ability to proliferate. The basal rate of hepatocyte cell proliferation is very low, but under conditions necessitating an increase in liver capacity, hepatocytes will act as facultative stem cells. For example, hepatocytes will enter the cell cycle almost synchronously after surgical resection until liver functional capacity is restored. Also, chemicals like phenobarbital will cause an increase in liver mass, in part, through hyperplasia. During continuous exposure to these chemicals, hepatocytes transiently undergo cell proliferation, but thereafter return to a basal rate while augmented liver mass is maintained through chemically mediated inhibition of cell loss. In contrast, high doses of hepatotoxicants inhibit hepatocyte proliferation. If a growth stimulus is received during exposure to hepatotoxicants, another cell type will proliferate. These so-called "oval" cells are named after their distinctive oval nuclei and are derived from the portal areas. They are bipotential in that they can differentiate into hepatocytes or bile duct epithelial cells; hence, they are believed to be a true liver stem cell. A well-defined model for oval cell proliferation is surgical resection of liver of rats during treatment with the potent hepatotoxicant, 2-acetylaminofluorene (2-AAF).

20.2.2 Nonparenchymal Liver Cells

Approximately 30% of liver cells are nonparenchymal cells. Cell types located in the interlobular portal tracts include bile duct epithelial cells and connective tissue fibroblasts. Those located within the lobule interior are the rare fixed, natural killer lymphocytes (pit cells) and the more abundant endothelial cells lining the sinusoids, Kupffer cells, and fat-storing Ito cells. The recent availability of isolation and cell culture techniques for these cell types has allowed characterization of toxicant effects on nonparenchymal cells that are consistent with the tissue pathology. An example is the susceptibility of sinusoidal endothelial cells to toxicity of chemicals that cause veno-occlusive disease such as monocrotaline.

Endothelial cells of the sinusoids are highly fenestrated. This porosity plus the absence of highly structured connective tissue, as found in underlying basement membrane of true capillaries, determines the high permeability of hepatic sinusoids. Plasma proteins and small particles pass into the space of Disse, the interval between hepatocytes and sinusoids, but particles larger than 150 nm (e.g., chylomicrons, erythrocytes, viruses) are impeded. Sinusoidal endothelial cells participate in the liver inflammatory response through secretion of cytokines and expression of adhesion molecules for neutrophils. Both sinusoidal endothelial cells and Kupffer cells are active in endocytosis, including that for clearance of modified plasma proteins such as oxidized lipoproteins. Kupffer cells are the fixed macrophages of the liver located within the sinusoidal lumen where they are active in phagocytosis of particulate matter. They are important for clearance of systemic endotoxin and initiation of the liver inflammatory response through secretion of tumor necrosis factor-α (TNFα). Gadolinium chloride ($GdCl_3$) and dextran sulfate, inhibitors of Kupffer cell activation, have become important experimental tools for determination of the role of inflammation in chemically induced hepatotoxicity.

Quiescent, perisinusoidal fat-storing cells (Ito cells) accumulate vitamin A in large cytoplasmic droplets. However, these cells also express the muscle protein desmin and become activated to myofibroblasts upon exposure to hepatotoxicants such as CCl_4. Activated Ito cells synthesize extracellular matrix proteins, especially when stimulated by transforming growth factor-β (TGFβ). These cells, induced by Kupffer cell-derived TGFβ, are thought to be the primary source of collagen in liver fibrosis.

20.3 TYPES OF CHEMICALLY INDUCED LESIONS

Classification of chemically induced hepatotoxicity is primarily based upon pattern of incidence and histopathological morphology. *Intrinsic* hepatotoxic drugs demonstrate a broad incidence, dose-response relationship and will usually give similar results in humans and experimental animals. The incidence of liver damage from *idiosyncratic* hepatotoxicants is limited to susceptible individuals and results from hypersensitivity reactions or unusual metabolic conversions that can occur due to polymorphisms in drug metabolism genes (see Chapter 5).

Morphological liver damage differs qualitatively depending upon duration of exposure. The hallmarks of acute exposure are impaired parenchymal cell function and viability that are manifested histopathologically as *steatosis*, *cholestasis*, and *necrosis*. Many toxicants induce steatosis, that is, hepatocellular lipid accumulation to greater than 5% of weight. This response is generally considered to be reversible, however, when severe it is associated with more progressive injury as, for example, with the anticonvulsant valproic acid. Cholestasis occurs when impaired hepatocyte function results in inhibition of bile formation and secretion. A consequence of cholestasis is impaired clearance of bilirubin, a yellow degradation product of heme whose accumulation leads to jaundice. Damage to bile duct epithelium can also result in acute cholestasis. A notable example of human cholestasis resulted from accidental exposure to 4,4′-diaminodiphenylmethane, an epoxy resin hardener, which was a bread contaminant causal to Epping jaundice.

Loss of viability or necrosis is the most severe form of acute hepatoxicity. Necrosis can be focal or diffuse throughout the acinus, massive to include many acini, or zone-specific with potential bridging across equivalent zones of adjacent acini depending upon severity. Zonation of heptotoxicant damage is due to acinar compartmentalization of mediators of toxicity, such as higher oxygen concentration in zone 1 and localization of cytochromes P450 in zone 3. Acute, drug-induced hepatitis follows necrosis as activated neutrophils infiltrate and, if damage is sublethal, regenerative cell proliferation of residual hepatocytes ensues. A more limited form of lethality is apoptosis or programmed cell death. Apoptosis occurs in isolated hepatocytes and without an inflammatory reaction. Acute exposure to lower doses of necrogenic toxicants is associated with apoptosis, whereas drugs that increase liver size, such as phenobarbital, inhibit apoptosis.

The pathology following chronic exposure to hepatotoxicants includes a broad variety of lesions. Steatosis and cholestasis can result from chronic, as well as acute, exposures. For example, both acute and chronic exposures to the phenothiazine chlorpromazine induce cholestasis. Immune-mediated hepatotoxicity in susceptible individuals is also observed upon repeated exposure to some chemicals. The most well-studied example is hepatitis from reexposure to the general anesthetic halothane that presumably results from immunization against liver proteins adducted to halothane metabolite formed upon initial exposure. *Fibrosis* often accompanies repeated hepatocellular damage and can result in extracellular matrix deposition in the perisinusoidal space, around the terminal hepatic venule and in portal tracts. *Cirrhosis* results when exaggerated deposition of collagen and associated matrix proteins occurs in interconnected bands of fibrous scar tissue. Cirrhosis impairs regeneration, is not reversible, and results in severe liver dysfunction. Fibrosis is also associated with damage to biliary epithelium and periductal cirrhosis can result from prolonged exposure.

Vascular lesions. Chemical damage to the hepatic efferent vasculature results in various disorders causing impairment of sinusoidal perfusion. Obstruction of sinusoidal drainage through terminal hepatic venules related to endothelial cell damage and fibrosis occurs in veno-occlusive disease. Blockage of flow through larger hepatic veins from thrombosis occurs in Budd-Chiari syndrome. Both of these vascular disorders, as well as sinusoidal dilation, have been associated with chronic exposure to oral contraceptives.

Neoplasia. Approximately one-fourth of the chemicals tested in the National Toxicology Program rodent bioassay are liver carcinogens, however, the significance of these high-dose exposures to human risk has been questioned. The livers of the C3B6F1 tester strain of male mice are very susceptible to chemical induction of tumors. In humans, the mycotoxin aflatoxin B_1 is highly associated with hepatocellular carcinoma, especially in combination with hepatitis B viral infection. Alcoholic beverage consumption is to be listed as a known human carcinogen in the 9th Report on Carcinogens based, in part, upon its epidemiological association with liver cancer, presumably due to chronic exposure to ethanol. Tumors of the vasculature endothelia (angiosarcomas) have occurred with industrial exposure to vinyl chloride and administration of the radioactive contrast agent, thorium dioxide. Further discussion of chemical carcinogenesis can be found in Chapter 15.

20.4 MECHANISMS OF CHEMICALLY INDUCED HEPATOTOXICITY

At the cellular and molecular level, the general mechanisms of cellular injury outlined in Chapter 10 are applicable here. The most direct mechanism of toxicity is through specific interaction of a chemical with a key cellular component and consequential modulation of its function. More common mechanisms, however, involve secondary effects of toxicant interaction. These include depletion of cellular molecules, such as ATP and GSH; free radical and oxidant damage, in particular to membrane lipids; covalent binding of reactive metabolites to critical cellular molecules; and collapse of regulatory ion gradients. The following discussion will highlight how these cellular and molecular mechanisms contribute to specific types of chemically induced hepatic lesions.

20.4.1 Fatty Liver

Lipid accumulates in hepatocytes primarily as triglyceride and this accumulation occurs when there is an imbalance between uptake of extrahepatic triglyceride and precursors compared with hepatic secretion of triglyceride-containing lipoprotein and fatty acid catabolism. The primary source of extrahepatic lipid to the liver is nonesterified fatty acids that have been mobilized from adipose tissue and bound to serum albumin. A secondary source is residual triglyceride of chylomicron remnants that originate from intestinal processing of dietary lipid. Within the hepatocyte, fatty acids are synthesized *de novo* and esterified when acetyl CoA and glycerol are available from glycolysis. Intrahepatic catabolism of fatty acids occurs primarily through mitochondrial β-oxidation with a smaller contribution from peroxisomal oxidation. Fatty acid β-oxidation is limited by translocation of fatty acylcarnitine derivatives across the inner mitochondrial membrane. Hepatocyte triglycerides are combined with phospholipids, cholesterol, cholesterol esters and glycosylated apolipoproteins, then secreted as very low density lipoproteins (VLDL). Further processing of VLDL occurs in the systemic circulation to yield other lipoprotein classes of lesser triglyceride content that become a source for apoprotein recycling by the liver.

This multicomponent process, summarized in Figure 20.2, provides for many sites of chemical interaction. Excessive free fatty acid (FFA) delivery to the liver seems to be one component of the mechanism in steatosis induced by CCl_4, ethionine, and phosphorus. These agents trigger lipolysis in the adipose tissue and cause a dramatic increase in circulating FFA. Under these conditions, triglycerides accumulated in

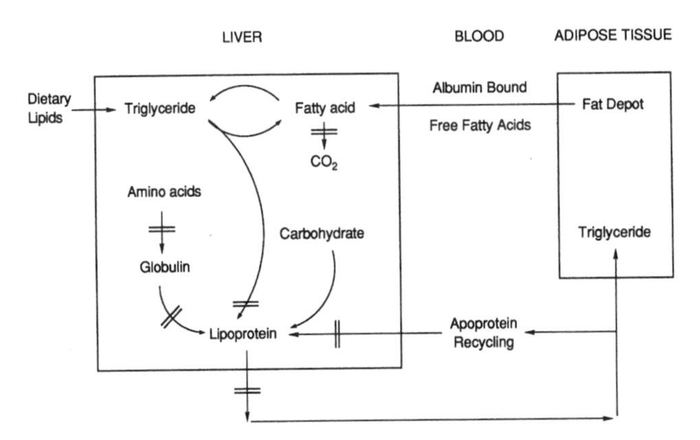

FIGURE 20.2 Triglyceride cycle in the pathogenesis of fatty liver.
The scheme shows the transport of free fatty acids from fat depot to liver and the return of surplus as triglyceride. Also shown are possible metabolic blocks (=) in hepatotoxicity.

the liver due to saturation of the rate-limiting step of synthesis of VLDL. Mobilization of depot far from adipose tissue is under the control of the pituitary-adrenal axis, the stimulation of which triggers massive release of catecholamines leading to activation of hormone-sensitive lipase in adipose tissue and mobilization of FFA. Prolonged systemic stress by chemicals such as DDT, nicotine, and hydrazine also stimulate this system. Apparently, a similar mechanism causes fatty liver in starved animals treated with a large dose of ethanol.

Intrahepatic events contributing to increased fatty acid availability include enhanced *de novo* synthesis and impaired β-oxidation, whereas increased esterification would lead to triglyceride accumulation. Each of these processes is sensitive to the cellular redox state and thus thought susceptible to ethanol through elevation of NADH. Increased diglyceride made available through ethanol induction of phosphatidic acid phosphatase may also enhance esterification. Valproic acid steatosis is related to inhibition of β-oxidation. The synthesis of apoprotein by hepatocytes is a highly endergonic process and depletion of ATP, such as that which occurs with ethionine, an antimetabolite of methionine that competes for S-adenylation, is causal to fatty liver. Triglyceride accumulation would also result from impaired assembly and secretion of VLDL. Acetaldehyde produced from ethanol oxidation has been shown to adduct tubulin and inhibit microtubule polymerization. Microtubules are required for trafficking of VLDL-containing vesicles to the plasma membrane for exocytosis. Similarly, components of the hepatocyte secretory process are involved in fatty liver by orotic acid and CCl_4.

20.4.2 Cholestasis

Hepatocyte formation of bile depends upon energy-dependent transport of bile acids into the canalicular lumen, a multicomponent process involving vectoral movement across the cell (Fig. 20.3). The osmotic pressure generated by canalicular bile acids is the driving force for the fluid component of bile that helps propel bile flow. Bile acids are taken up from the blood against a concentration gradient

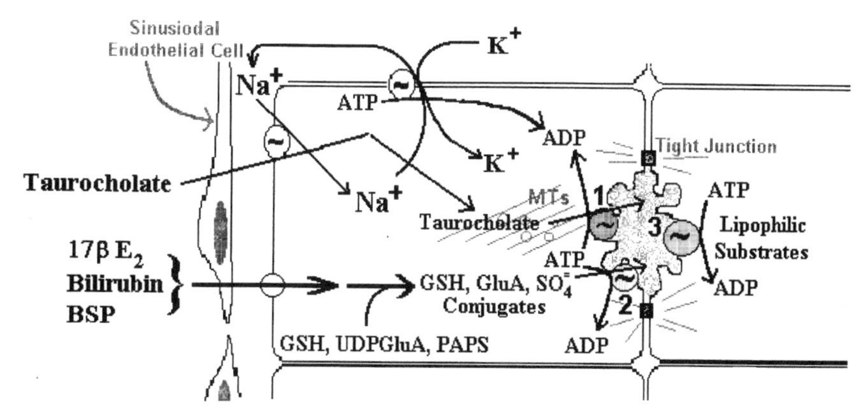

FIGURE 20.3 Transport of bile salts and other constitutents across the hepatocyte. The Na^+-dependent bile salt (taurocholate) transporter is shown on the sinusoidal membrane that utilizes the Na^+ gradient maintained by the Na^+K^+-pump, shown here on the lateral aspect of the plasmalemma. Bile salt transcellular transport involves microtubules (MTs), which then deliver substrate to the canalicular bile salt transporter (1). Bilary excretion of GSH, glucuronate (GluA) and sulfate conjugates of compounds such as 17β-estradiol (17βE2), bilirubin, and bromosulfothalein (BSP) is catalyzed by the multispecific organic anion transporter (MOAT; 2). Both 1 and 2 are members of the ABC family of ATP-dependent transporters that also includes P-glycoprotein (3), another canalicular transporter catalyzing excretion of lipophilic compounds such as the chemotherapeutic drug, daunorubicin.

by transport proteins located on the hepatocyte sinusoidal membrane. The best characterized of these is the Na^+/taurocholate cotransporter whose energy is derived from the downhill gradient of Na^+ that is maintained by the plasma membrane Na^+K^+-ATPase. Intracellular binding proteins coordinate transcellular transport of bile acids and cytoskeletal elements mediate trafficking within vesicles to the canalicular membrane. An ATP-dependent bile salt transporter localized in the canalicular membrane catalyzes the rate-limiting, concentrative movement of taurocholine into the canalicular space.

Another sinusoidal transporter catalyzes Na^+-independent uptake of organic anions and is instrumental for biliary clearance of glucuronidated and sulfated steroids, the diagnostic chemical bromosulfophthalein (BSP) and possibly bilirubin. Canalicular transport of glucuronidate and GSH conjugates is coupled to ATP hydrolysis catalyzed by the multispecific organic anion transporter (MOAT), a multidrug resistance protein related to the mdr2 gene product. MOAT catalyzed excretion of BSP-GSH and bilirubin glucuronidate occurs after intracellular conjugation. Another member of the multidrug resistance protein family, the P-glycoprotein product of the mdr1 gene, has also been localized to the canalicular membrane and catalyzes ATP-dependent efflux of a wide variety of hydrophobic polycyclic compounds.

Cholestatic hepatotoxicants that target these molecular and cellular events are well known and some have deleterious effects on multiple components. Compounds with global effects on membrane permeability, such as chlorpromazine through its cationic detergent action, can inhibit Na^+K^+-ATPase and collapse the Na^+ gradient necessary for sinusoidal bile acid uptake. A detergent effect of lithocholate on canalicular membranes is also evident from ultrastructural morphology. Cholestasis from the 17β-glucuronide conjugate of estradiol has been related to its activity as a substrate of both MOAT and P-glycoprotein. The cholestatic effects of cholchicine, a microtubule poison, as well as phalloidin and cytochalasin B, inhibitors of microfilament function, emphasize the importance of cytoskeletal elements in bile formation.

Toxicity to bile duct epithelial cells also causes cholestasis. Necrosis of these cells can result in sloughing of an obstructive cast into the duct lumen. The experimental chemical α-naphylisothiocyanate (ANIT) is known to target bile duct epithelial cells in addition to hepatocytes. Bile duct epithelial toxicity is GSH-dependent and is thought to depend upon concentrative accumulation within bile ducts subsequent to hepatocanalicular transport of the ANIT-GSH conjugate.

20.4.3 Necrosis

The ultimate event in cell death is irreversible loss of plasma membrane integrity. Previous mechanistic studies have emphasized that the inability to maintain transmembrane calcium gradients is the early determinant, initiating an irreversible sequence of events leading to cell death. However, new microscopic techniques have enabled observation of temporal relationships of multiple events at the level of a single toxicant-exposed cell. These elegant studies have demonstrated that loss of mitochondrial function with an associated depletion of ATP can precede elevation of cytosolic Ca^{2+}. Thus, ATP depletion and disruption of Ca^{2+} homeostasis are both important determinants of lethality and, when extreme, each may act independently. Alternately, sublethal responses of one may lead to and synergize with the other such that the temporal order is inconsequential. A variety of different types of primary toxic effects would then be expected to converge upon mitochondrial function and/or maintenance of low cytosolic Ca^{2+} levels by the ATP-dependent Ca^{2+} pumps of the plasma membrane and endoplasmic reticulum (ER).

Phosphorylation of ADP to ATP by mitochondria is driven by an electrochemical proton gradient established across the inner mitochondrial membrane as a consequence of vectoral transport of protons from NADH and succinate during oxidation by the respiratory chain (see Chapter 13). Hence, lipophilic weak acids or bases such as 2,4-dinitrophenol that can shuttle protons across membranes will dissipate the proton gradient and uncouple oxidation from ADP phosphorylation. Intramitochondrial ADP can be rate-limiting as demonstrated by inhibition of the mitochondrial adenosine nucleotide carrier by atractyloside. Inhibition of ATP synthesis also is a consequence of the mitochondrial permeability transition (MPT). Exposure to several different types of toxicant effectors opens a mitochondrial high conductivity pore that nonspecifically allows passage of molecules with molecular weights <1.5 kDa to result in the MPT. The involvement of such a pore in necrosis is suggested by the protection against lethality from several toxicants by cyclosporin A, an inhibitor of the MPT.

Induction of the MPT requires Ca^{2+}; conversely, plasma membrane and ER Ca^{2+} pumps require ATP. One of the earliest effects of CCl_4 is metabolism-dependent inhibition of the ER Ca^{2+} pump through oxidation of a critical – SH residue. A similar oxidant sensitive – SH residue is present in the plasma membrane Ca^{2+} pump. Thus, the reciprocal requirements of Ca^{2+} for the MPT and ATP for cellular Ca^{2+} pumps provide a mechanism for linkage of ATP depletion and disruption of Ca^{2+} homeostasis as mediators of cell death. Most proximal toxic effects that lead to hepatocellular mitochondrial and Ca^{2+} dysfunction can be generalized into two classes, those involving production of oxidative stress or formation of reactive metabolites.

20.4.3a Oxidative Stress

Cellular metabolism of dioxygen O_2 is tightly regulated to yield primarily 4 e^- reduction to $2O^{-2}$. However, some leakage of the 1 e^- partial reaction products, $O_2^{\bullet -}$ superoxide radical and hydrogen peroxide (H_2O_2), occurs from processes such as mitochondrial respiration and cytochrome P450-catalyzed oxidation. Peroxisomal oxidative metabolism of fatty acids is also a significant source of endogenous cellular H_2O_2. Metabolism of superoxide to H_2O_2 by superoxide dismutase and reduction of H_2O_2 via GSH peroxidase and catalase complete the reduction to H_2O. Under normal conditions, these pathways complemented by additional hepatocellular antioxidants such as GSH, vitamin E, and ascorbate have sufficient capacity to prevent deleterious oxidation of other cellular constituents (see Chapter 9).

Exposure to certain hepatotoxicants can enhance the rate of production of reactive oxygen intermediates such that they exceed the capacity of these protective pathways. A structurally diverse group of compounds cause proliferation of peroxisomes within rodent hepatocytes and hence elevate endogenous levels of H_2O_2 leaked from this organelle. Peroxisome proliferators include the hypolipidemic agent clofibrate, plasticizer diethylhexyl phthalate (DEHP) and herbicides 2,4-dichloro- and 2,4,5-trichloro-phenoxyacetic acid (2,4-D and 2,4,5-T). The liver is particularly susceptible to injury from iron overload since it is a major site of iron storage. Iron is usually stored in hepatocytes associated with proteins ferritin and hemosiderin; however, catalytically active iron can be released from storage by superoxide. Ferritin iron is also reduced and mobilized by NADH, which may play a role in ethanol-induced oxidative stress in liver. Released iron can react with reducing agents and O_2 to form superoxide. Superoxide and H_2O_2 react with iron to form highly reactive hydroxyl radicals ($^{\bullet}OH$). Copper ions can also catalyze the Fenton reaction to produce $^{\bullet}OH$. Superoxide is also produced by single electron transfer to O_2 by nonoxidant free radicals. Formation of the trichloromethane radical $CCl_3^{\bullet}$ from cytochrome P450-dependent metabolism of CCl_4 contributes to hepatocyte oxidative damage through this mechanism.

Certain phenolic derivatives are conducive to 1 e^- reduction, especially as catalyzed by microsomal NADPH-cytochrome P450 reductase, with the formation of oxygen-centered free radical anion. The 1 e^--reduction products of these can then transfer an electron to O_2 resulting in the formation of superoxide with reoxidation to the parent compound (Fig. 20.4). This type of reaction is particularly destructive because the parent compound undergoes redox cycling through repeated rounds of 1 e^- reactions. Superoxide, continuously produced during redox cycling, will be detoxified through H_2O_2 and GSH peroxidase until NADPH becomes limiting for GSSG reduction. GSSG is then effluxed from the hepatocyte and total glutathione levels fall. Compounds that undergo redox cycling include the *p*-quinones,

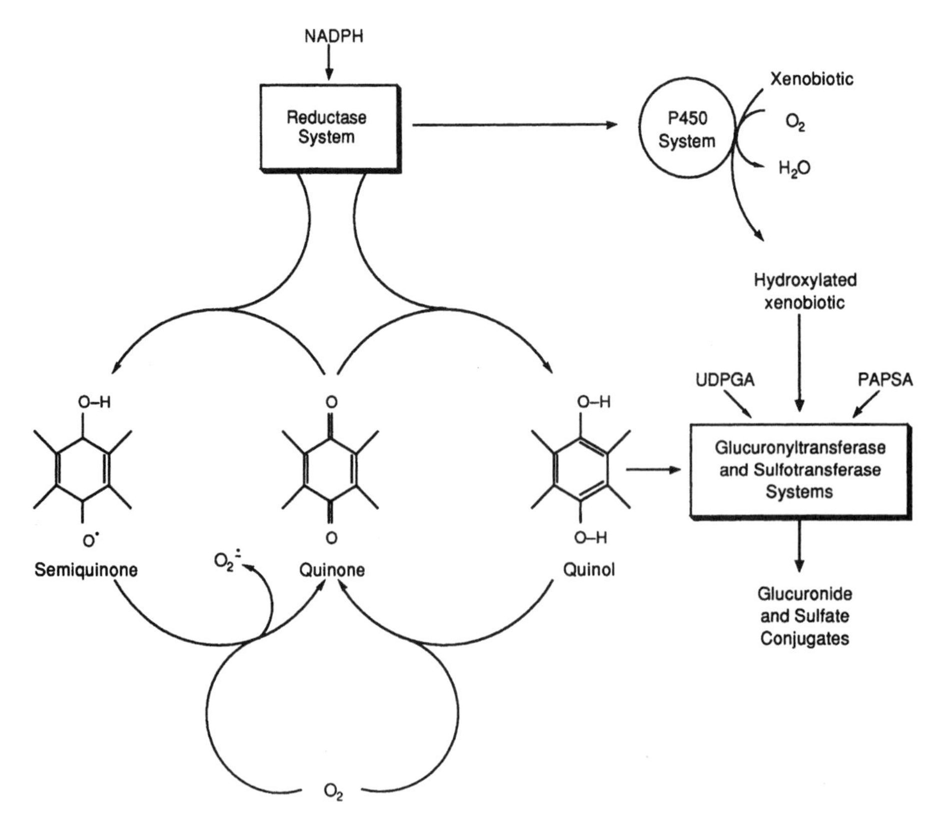

FIGURE 20.4 Mechanism of quinone drug redox cycling by liver microsomal NADPH:cytochrome P450 reductase.
One-electron pathway competes with two-electron process (catalyzed by DT-diaphorase) and is coupled with the monoxygenation and glucuronidation and sulfation systems.

o-quinones (catachols), quinoneimines, and quinone methides and include such hepatotoxicants as metabolites of estrogen, acetaminophen, eugenol, and bromobenzene. The antimicrobial agent nitrofurantoin and the bipyridyl herbicides, diaquat and paraquat, also cause hepatotoxicity through redox cycling. A protective mechanism against quinone redox cycling is afforded by NAD(P)H:quinone oxidoreductase (DT diaphorase), which competes for the quinone substrate and catalyzes 2 e^- reduction to the hydroquinone.

Extrahepatocellular factors also contribute to oxidative toxicity in liver. Xanthine oxidase is primarily located in sinusoidal endothelial cells and catalyzes 1 e^- reduction of O_2. Kupffer cells produce reactive oxygen species as part of their phagocytic function. Both activated neutrophils and macrophages recruited to liver upon chemically induced damage produce a respiratory burst of superoxide through the activity of membrane-bound NADPH oxidase. Also, much of the damage associated with reperfusion of liver after ischemia is associated with production of reactive oxygen species.

Several adverse consequences result from exposure of the cell to oxidative stress. Cellular proteins with critical – SH residues and membrane polyunsaturated fatty acids appear to be especially susceptible to oxidative damage that can lead to necrosis. Reactive oxygen species can directly oxidize protein – SH groups to form disulfides, an effect that is exaggerated by impaired repair due to depletion of GSH consequential to its consumption by GSH peroxidase. Two critical – SH containing proteins whose functions are impaired by oxidation are the plasmalemma and ER Ca^{2+} pumps. Such damage to these activities would provide a mechanism for oxidant-induced derangement of cellular Ca^{2+} homeostasis. Mitochondrial ATP synthesis may also be directly affected since sulfhydryl oxidation is an inducer of the MPT.

Oxidant radicals, $O_2^{\bullet -}$ and $^{\bullet}OH$, as well as other free radicals such as semiquinone and $CCl_3^{\bullet}$, are very active in initiating peroxidation of polyunsaturated lipids. Lipid peroxidation is a multiphasic process and involves initiation, propagation, and termination (Fig. 20.5). The hydrogen atoms on methylene carbons of unsaturated fatty acids are highly susceptible to free radical attack. The abstraction of methylene hydrogen atom yields lipid free radical. Propagation continues as lipid free radical abstracts methylene hydrogen from a neighboring lipid to generate a second free radical until termination reactions yield nonradical products. These lipid radicals are unstable and undergo a series of transformations, including rearrangement of double bonds to give conjugated dienes. Lipid free radicals react rapidly with O_2 to form lipid peroxyl radicals, which decompose to aldehydes, the most abundant being malondialdehyde and 4-hydroxy-2,3-nonenal.

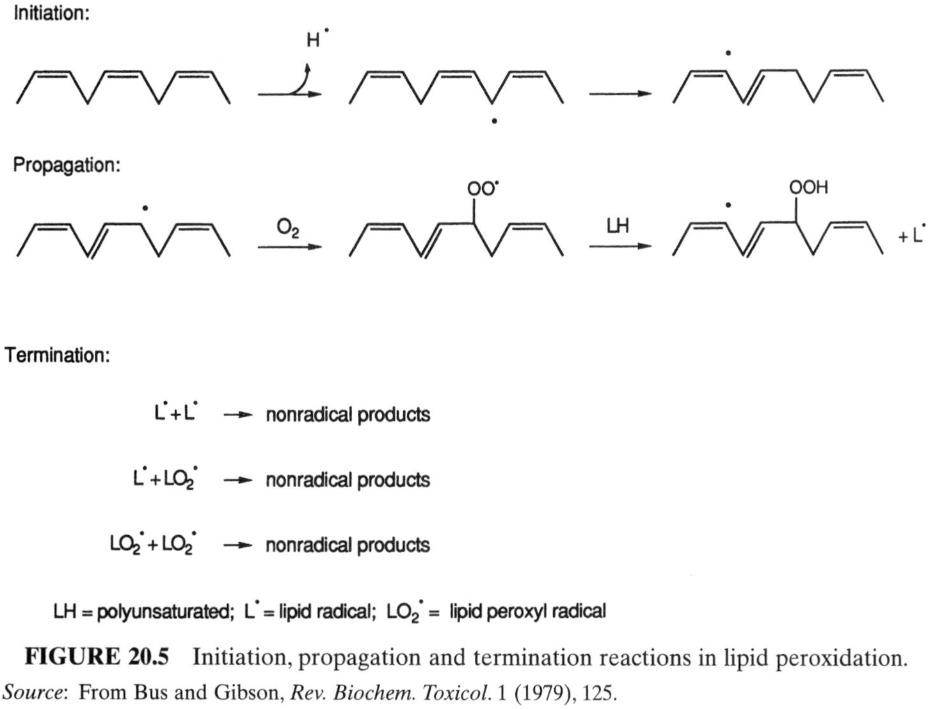

Initiation:

$H^{\bullet}$

Propagation:

O_2 $OO^{\bullet}$ LH   OOH $+ L^{\bullet}$

Termination:

$L^{\bullet} + L^{\bullet} \rightarrow$ nonradical products

$L^{\bullet} + LO_2^{\bullet} \rightarrow$ nonradical products

$LO_2^{\bullet} + LO_2^{\bullet} \rightarrow$ nonradical products

LH = polyunsaturated; $L^{\bullet}$ = lipid radical; $LO_2^{\bullet}$ = lipid peroxyl radical

FIGURE 20.5 Initiation, propagation and termination reactions in lipid peroxidation. *Source*: From Bus and Gibson, *Rev. Biochem. Toxicol.* 1 (1979), 125.

Subcellular membranes are rich in unsaturated fatty acids esterified to phospholipid and are thus obvious targets of lipid peroxidation. Membrane lipid peroxidation will result in loss of both structural integrity and function of the affected organelles. Also, lipid modification in the annula surrounding membrane transport proteins is known to modulate their activity. Thus, lipid peroxidation can significantly impair membrane maintenance of ion gradients, such as the mitochondrial proton gradient necessary for ATP synthesis and plasma membrane and ER Ca^{2+} gradients.

20.4.3b Production of Reactive Electrophiles

Deleterious effects of reactive electrophiles are generally attributed to two general mechanisms, through binding to nucleophilic sites of cellular macromolecules and depletion of GSH. Toxicity from macromolecule adduction is best understood for binding to nucleophilic sites of DNA and proteins. Although electrophilic adduction of nucleotides can contribute to cytolethality through formation of DNA single strand breaks, it is of greatest toxicological interest with respect to mutagenesis and carcinogenesis as discussed in Chapter 15.

Metabolism-dependent binding of toxicants to cellular protein is a common observation. However, determination of the functional consequences of binding and relating these to toxic effects is challenging. In those cases where general binding of reactive metabolites to protein creates neoantigens, toxicity is mediated by sensitization and the subsequent inflammatory response. The anesthetic halothane is a prototypical hepatotoxicant with this type of mechanism. Another situation in which clear effects result from binding is that resulting from suicide inhibition of cytochrome P450 isozymes by irreversibly bound reaction intermediates at the active site. Diallyl sulfide, a component of garlic, is a suicide inhibitor of cytochrome P450 2E1, and 4-hydroxy-2,3-nonenal, a byproduct of lipid peroxidation, has recently been shown to inhibit a several cytochrome P450 isozymes through adduction of the prosthetic heme group. The functional consequence of cytochrome P450 inhibition through adduction will depend on whether activation or detoxication reactions are affected.

Acetaminophen provides the most thoroughly studied example of hepatotoxicity associated with protein adduction. Although acetaminophen primarily is conjugated with glucuronide and sulfate and excreted, metabolism to a reactive electrophile, N-acetyl-p-benzoquinoneimine (NAPQI) by cytochrome P450 becomes significant when the conjugation pathways are saturated (see Chapter 10). Through the use of antibodies that recognize protein-bound NAPQI, several specific adducted proteins have been identified. More recently, application of two-dimensional gel electrophoresis and matrix-assisted laser desorption ionization (MALDI) mass spectrometry has expanded this list to include 23 adducted mouse liver proteins, most of known function. However, even with knowledge of the biochemical function of these proteins, the molecular mechanism involving a critical target protein(s) causal to acetaminophen hepatotoxicity remains to be determined.

Numerous hepatotoxicants are converted to reactive toxic electrophiles by cytochrome P450. Cytochrome P450 metabolism of chloroform to phosgene and bromobenzene to its 3,4-epoxide are classic examples. In addition, reactive electrophiles are generated upon activation of hepatotoxicants by other metabolic enzymes. Alcohol dehydrogenase catalyzes formation of acetaldehyde from ethanol and acrolein from allyl alcohol. Thioacetamide hepatotoxicity results from flavin monooxygenase-catalyzed activation through S-oxidation and aflatoxin B_1 epoxidation can be catalyzed by lipoxygenase. Glucuronidation of the carboxylic acid of diclofenac, a nonsteroidal anti-inflammatory drug, facilitates its adduction of protein through acylation.

Electrophiles not only complex with macromolecular nucleophilic sites, but also with those of small molecules. The most important of these is the –SH group of GSH. Conjugation of electrophiles with GSH is an important detoxication reaction catalyzed by GSH transferases (see Chapter 6). Although GSH concentration in hepatocytes is high relative to other tissues (4–8 mM), depletion of GSH can occur upon acute exposure to protoxicants whose electrophilic metabolites conjugate GSH. GSH lost to conjugation is stoichiometric and thus occurs at high doses and with a threshold. When GSH levels are depleted to ~25% or less, significant risk of oxidative damage ensues as GSH becomes limiting for GSH peroxidase activity. Because GSH conjugates are actively effluxed from the hepatocyte, recovery of GSH requires *de novo* synthesis through the sequential activities of γ-glutamylcysteine synthetase and GSH synthetase. GSH synthesis also requires an adequate supply of cysteine. These properties enable the use of several experimental tools to indicate GSH depletion as an effector of chemically induced hepatotoxicity. Potentiation will be observed with diethyl maleate, which depletes GSH through direct conjugation, and buthionine sulfoximine, an inhibitor of γ-glutamylcysteine synthetase. Also, overnight food withdrawal, as is frequently used in acute oral exposure studies, will deplete GSH and enhance toxicity. Conversely, N-acetylcysteine serves as a source of cysteine and is protective.

20.5 INTERACTIONS

20.5.1 Metabolism-Dependent

Exposure to two or more chemical agents can result in altered expression of hepatotoxicity. However, qualitatively different interactions may be achieved depending upon the relative timing of exposures. Simultaneous exposure to competing substrates of a specific isozyme of cytochrome P450 will often slow metabolism and can be protective against reactive metabolite-mediated hepatotoxicity. Alternately, pretreatment with one agent may induce metabolic enzymes that either protect against or potentiate the hepatotoxicity of a second chemical depending upon whether induction results in increased detoxication or activation. Acetaminophen and ethanol are two hepatotoxicants that illustrate these relationships. If ethanol precedes an acute dose of acetaminophen, but is not present simultaneously, potentiation of acetaminophen hepatotoxicity results. Prior exposure to ethanol induces cytochrome P450 2E1, an isoform that activates acetaminophen to NAPQI. However, when acetaminophen and ethanol are administered concurrently, ethanol protects against hepatotoxicity because both are cytochrome P450 2E1 substrates. The cytochrome P450 2E1 suicide inhibitor, diallyl sulfide, also protects against acetaminophen toxicity. Prior exposure to ethanol and other inducers of cytochrome P450 2E1 such as isopropanol, acetone, and the antimicrobial agent isoniazid also potentiate the hepatotoxicity of CCl_4. Several organochlorine pesticides are also potent inducers of cytochrome P450 isozymes and consequently modify hepatotoxicant potency. Examples are potentiation of hepatotoxicity of acetaminophen by mirex and CCl_4 by DDT. Hepatotoxicity of CCl_4 is also enhanced by chlordecone pretreatment, but in this situation the mechanism is related to suppression of tissue repair. For pharmaceuticals, interactions involving hepatic cytochrome P450 isozyme 3A are of concern because a large number of drugs are metabolized by this enzyme. Such interactions can alter the pharmacokinetics of a drug to effect a change in therapeutic efficacy. Well-studied examples are interactions with the 3A substrates erythromycin and ketaconozole, and the inducer rifampin.

20.5.2 Protective Priming

Preexposure to a sublethal dose of CCl_4 has been shown to protect against lethality from a second, independently lethal dose, a phenomenon called "autoprotection." Comparable histopathological damage to liver is seen independent of the early priming dose, which discounts a lessening of reactive metabolite formation by a previously inhibited cytochrome P450 as the protective mechanism. The kinetics of autoprotection correlate with regenerative cell proliferation induced by the priming dose. This observation has led to the hypothesis that newly born hepatocytes are more resistent to the necrogenic effects of CCl_4. Importantly, protective priming also occurs with heterologous pairs of chemicals as shown by protection against acetaminophen hepatotoxicity by pretreatment with a sublethal dose of thioacetamide. Although the molecular basis of protective priming is unknown, it represents another instance in which chemicals interact to modulate heptotoxicity.

20.6 DETECTION AND PREDICTION OF HEPATOTOXICITY

20.6.1 Clinical

Necrosis of hepatocytes leads to discharge of intracellular contents into the serum. Some of these provide noninvasive parameters of hepatotoxicity that can be used in the clinical setting. The most important and specific serum marker of hepatocellular damage is the enzyme alanine aminotransferase (ALT; also called serum glutamate pyruvate transaminase, SGPT). Another commonly used, but less specific, indicator is aspartate aminotransferase (AST; also called serum glutamate oxaloacetate transaminase, SGOT). In milder forms of hepatocellular injury, serum levels of these enzyme may be elevated to as much as three times normal values. Similar values are also characteristic of more serious cholestatic injury, but this can be distinguished by coincident elevation in the levels of alkaline phosphatase. With severe hepatocellular necrosis, serum aminotransferases may be elevated as much as 50 times normal. Because albumin and prothrombin are synthesized by hepatocytes, serum albumin concentration and prothrombin clotting time are useful parameters of hepatocellular toxicity. Elevated serum levels of bilirubin and bile acids are indicative of either hepatocellular or bile duct epithelial damage. More specific tests for liver function include rate of biliary clearance of injected BSP and determination of urinary metabolites of probes for hepatocyte metabolic enzymes, such as caffeine for cytochrome 450 1A2.

20.6.2 Experimental

The serum enzymes ALT, AST, and alkaline phosphatase are also important indicators of chemically induced heptotoxicity in experimental animal studies. An additional heptocyte enzyme, sorbitol dehydrogenase (SDH), is also used frequently as a more sensitive complement to the aminotransferases. In experimental studies, histopathology is the most definitive indicator of injury and highly informative about mechanistic events.

Additional information about mechanisms of chemically induced toxicity to liver cells is obtained in experiment studies through use of *ex vivo* systems. Isolated perfused liver is especially useful for assessment of cellular metabolite fluxes and bile flow. Isolated hepatocytes are readily obtained by collagenase perfusion and can be used in suspension or in monolayer cultures (see Chapter 4). Although isolated hepatocytes lose activity and inducibility of certain cytochrome P450s with time, they offer a cellular context in which to evaluate mechanism of toxic effects over the short term (<24 hr). The availability of hepatocyte couplets with intact canalicular lumens has provided a valuable system for mechanistic studies of bile formation. Nonparenchymal cells can also be obtained by collagenase perfusion and isolated cell types can be isolated with additional separation procedures. Cocultures can then be constructed to determine cellular interactions in mediating chemical toxicity. This approach has been especially fruitful in understanding mechanisms involved in hepatotoxicity initiated through toxicant effects on inflammatory cells.

Isolated subcellular fractions, especially microsomes, are most frequently used to obtain metabolite profiles of drugs and identify metabolizing enzymes. Production of a metabolite with structures of known high reactivity may indicate potential toxicity through a previously known mechanism. Preparations made from cells engineered to express specific drug metabolizing enzymes are commercially available and demonstrate the potential for metabolism in vivo by that enzyme. Liver slices are also frequently used in metabolite profiling now that newer methods for preparation and maintenance have improved their viability, although loss of cytochrome P450 activity limits their use to short-term studies. These precision-cut liver slices offer the simplicity of an in vitro system, but with retention of cellular context, normal tissue architecture and cell-cell interactions. Mechanistic knowledge gained from these in vitro systems has been instrumental for prediction of parent drug toxicity across species and in susceptible individuals and of interactions with other drugs.

20.7 COMPOUNDS CAUSING LIVER INJURY

20.7.1 Drugs

Several therapeutic agents used as tranquilizers, antidepressants, anticonvulsants, antibiotics, and anti-inflammatory drugs, as well as some prescribed for cardiovascular, endocrine, rheumatic, and neoplastic diseases have been found to cause hepatotoxicity. Recent episodes of liver toxicity have prevented development of fialuridine, for treatment of hepatitis B, and the nonsteroidal anti-inflammatory drugs bromfenac and benoxaprofen. Hepatotoxicity has limited the usage of troglitazone for treatment of diabetes and tacrine for Alzheimer's disease. One of the most extensively studied and widely available hepatotoxic drugs is acetaminophen. Although therapeutic doses are well below toxic levels, alcohol potentiation of acetaminophen hepatotoxicity is a concern. Protein adduction by reactive electrophilic metabolite NAPQI, depletion of GSH, and redox cycling are all contributors to acetaminophen hepatotoxic effects and have been discussed in detail above.

The antipsychotic drug chlorpromazine is a prototype heptotoxicant for production of cholestasis. Pleiotropic effects of chlorpromazine on membrane permeability and associated ion gradients and microfilament-mediated canalicular contraction have been attributed to detergent effects. Valproic acid, an anticonvulsant, is associated with microvesicular steatosis. Inhibition of mitochondial fatty acid β-oxidation is an important component of this toxic effect and is apparently related to carnitine availability as evidenced by the protection afforded by L-carnitine supplements. The hypolipidemic drugs, clofibrate, fenofibrate and gemfibrozil, are peroxisome proliferators in rodent liver, but not in humans. Isoniazid, an antibiotic used to treat tuberculosis, exhibits an approximately 1% incidence of hepatotoxicity. Although toxicity is known to be metabolism-dependent and protein adduction has been well correlated with toxicity, the identity of the reactive metabolite and target protein(s) is unknown.

20.7.2 Ethanol

Alcoholic liver disease is a progressive, multifaceted disorder. Thus, it is not surprising that a large number of molecular targets are affected by ethanol. Several points of ethanol interference with hepatic lipid metabolism have been identified above (Section 20.4.1). The role of Kupffer cells, through their secretion of TNF-α has been clearly demonstrated in ethanol-induced necrosis and inflammation and may be related to ethanol effects of their responsiveness to endotoxin. Kupffer cells isolated from ethanol-treated rats also have an increased capability to produce superoxide. Considerable evidence suggests that protein adduction and oxidative stress are causal to ethanol hepatotoxicity. Microtubule adduction by acetaldehyde, the product of ethanol oxidation by alcohol dehydrogenase, has a role in impaired VLDL secretion. Pro-oxidants are primarily from partial reactions of the ethanol-inducible cytochrome P450 2E1 with secondary contributions from mobilization of active iron and free radical reactions originating with formation of 1-hydroxyethyl radical and xanthine oxidase metabolism of acetaldehyde. In addition, ethanol inhibits signaling by the receptor for the hepatomitogen, epidermal growth factor, and thus may have detrimental effects on tissue repair.

20.7.3 Halogenated Aliphatic Hydrocarbons

Exposures to hepatotoxic levels of halogenated aliphatic hydrocarbons have been reduced considerably with improved industrial hygiene and use restriction. Important exposure scenarios that remain involve the use of fluorinated general anesthetics, replacement of ozone-depleting refrigerants with hydrochlorofluorocarbons (HCFCs), and storage of high volumes of certain haloalkanes in EPA Superfund sites. Hepatotoxicity from fluorinated anesthetics is thought to result from hypersensitivity to metabolism-based adduction of liver proteins. The brominated, fluorinated anesthetic halothane has had the highest incidence of toxicity and has largely been replaced by safer, nonbrominated analogs. The reactive metabolite of halothane, trifluoroacetyl chloride, is formed by cytochrome P450-catalyzed oxidative debromination. A similar mechanism has been suggested for a recent occurrence of hepatotoxicity in industrial workers repeatedly exposed to HCFCs. HCFCs have also been shown to be peroxisome proliferators in rodent liver. Of the high volume chemicals at Superfund sites, those with hepatotoxicity include CCl_4, chloroform, trichloroethylene, and tetrachloroethylene. The human hepatocarcinogen vinyl chloride is also a Superfund site contaminant.

20.7.4 Pesticides

Liver failure from acute exposures to pesticides is rare as lethality usually results from toxicity to other organ systems. Fatty liver, zone 3 necrosis, and cholestasis with destruction of bile duct epithelia have resulted from exposure to the herbicide paraquat and elevated serum enzymes have been detected in workers employed in the production of chlordecone. The fumigant ethylene dibromide causes experimental hepatotoxicity similar to CCl_4. The herbicides 2,4-D, 2,4,5-T, dicamba, lactofen, and tridiphane are peroxisome proliferators in rodent liver. The most consistent effect of pesticides in liver is induction of cytochrome P450 isozymes, especially by the organochlorine insecticides. Thus, although hepatotoxicity as an independent toxic effect of pesticides is uncommon, the potential for toxic interactions is significant.

20.7.5 Natural Toxins

The plant species yielding natural hepatotoxins include *Lantana camara*, *Sassafras albium*, and more than 200 species of *Crotalaria*. A large number of species of the genera *Senecio*, *Heliotropium*, *Cyanoglossum*, and *Symphytum* are sources of pyrrolizidine alkaloids, etiological agents of veno-occlusive disease noted with excessive consumption of some herbal teas. Medicinal plants containing the diterpenoid glycoside, atractyloside, have caused fatalities associated with zone 3 hepatic necrosis. Atractyloside is a specific inhibitor of the mitochondrial adenosine nucleotide carrier and an inducer of the MPT. Unripe akee fruit of the *Blighia surpida* tree of Jamaica leads to acute steatosis from hypoglycin A, an inhibitor of fatty acid oxidation. Hepatotoxic effects of tannins at high doses and of phalloidin and α-amanitin from poisonous mushrooms are also known. Several metabolic products of bacteria and lower fungi cause liver injury. An important endogenous hepatotoxicant is endotoxin lipopolysaccharide (LPS) from enteric Gram-negative bacteria that is normally present in the systemic circulation at low levels, but whose concentration can increase when gastrointestinal integrity is compromised. The mycotoxin aflatoxin B_1 from *Aspergillus* is acutely hepatotoxic as well as carcinogenic. Molecular epidemiology has associated a site-specific mutation of the p53 tumor suppressor gene characteristic of aflatoxin B_1 exposure with incidence of human hepatocarcinogenesis in the People's Republic of China.

20.8 CONCLUSION

Much is known about the biochemical toxicology of hepatotoxicants, yet much remains to be learned. Hepatotoxicity resulting in either cell necrosis or fatty infiltration is known to be a widespread phenomenon, potentially of importance to human health. It is caused by numerous drugs and environmental agents and its incidence is expected to increase as confounding viral liver disease becomes more prevalent. Much is known about mechanisms based upon comprehensive studies with a few prototypical chemicals—namely, CCl_4, ethanol and acetaminophen—which support a convergence of varied primary effects on the ultimate failure of mitochondrial function and Ca^{2+} homeostasis. The extensive metabolic activity of the liver exposes its cells to a continuous flux of pro-oxidants. The importance of metabolic activation for the production of reactive metabolites is well recognized. The high specific activity of these enzymes in the liver insures that the potential for metabolism-dependent toxic effects is extensive.

The recognition of the roles of the various liver cell types in mediation of toxicity is a model demonstrating the value of cellular toxicology in mechanistic research. The delineation of key metabolic pathways has been made possible by the availability of systems engineered to express individual molecular pathway components. This knowledge has been readily applied to drug discovery and development. Perhaps in no other tissue has the knowledge gained from fundamental research in biochemical toxicology had such obvious implications for protecting human health.

SUGGESTED READING

Berkowitz, C.M., Guibert, E., Bilir, B. and Gumucio, J.J. Functional compartmentation of hepatocytes. In: LeBouton, A.V. (Ed), *Molecular and Cell Biology of the Liver*. Boca Raton: CRC Press, 1993, p 347.

Cohen, S.D., Hoivik, D.J. and Khairallah, E.A. Acetaminophen-induced hepatotoxicity. In: Plaa, G.L. and Hewitt, W.R. (Eds), *Toxicology of the Liver, 2nd Ed.*, Washington, DC: Taylor & Francis, 1998, p 159.

Day, C.P. and Yeaman, S.J. The biochemistry of alcohol-induced fatty liver. *Biochim. Biophys. Acta* 1215 (1994), 33.

Hodgson, E. and Meyer, S.A. Pesticides. In: Sipes, I.G., McQueen, C.A. and Gandolfi, A.J. (Eds), *Comprehensive Toxicology, Vol. 9, Hepatic and Gastrointestinal Toxicology*. New York: Elsevier Science, 1997, p 369.

Ishak, K.G. and Zimmerman, H.J. Morphological spectrum of drug-induced hepatic disease. *Gastroenterol. Clin. North Am.* 24 (1995), 759.

Jaeschke, H., Smith, C.W., Clemens, M.G., Ganey, P.E. and Roth, R.A. Mechanisms of inflammatory liver injury: adhesion molecules and cytotoxicity of neutrophils. *Toxicol. Appl. Pharmacol.* 139 (1996), 213.

Kourie, J.I. Interaction of reactive oxygen species with ion transport mechanisms. *Am. J. Physiol.* 275 (1998), C1.

Lemasters, J.J. Necrapoptosis and the mitochondrial permeability transition: shared pathways to necrosis and apoptosis. *Am. J. Physiol.* 276 (1999), G1.

Mehendale, H.M., Roth, R.A., Gandolfi, A.J., Klaunig, J.E., Lemaster, J.J. and Curtis, L.R. Novel mechanisms in chemically induced hepatotoxicity. *FASEB J.* 8 (1994), 1285.

Oinonen, T. and Lindros, K.O. Zonation of hepatic cytochrome P-450 expression and regulation. *Biochem. J.* 329 Pt.1 (1998), 17.

Qiu, Y., Benet, L.Z. and Burlingame, A.L. Identification of the hepatic protein targets of reactive metabolites of acetaminophen in vivo in mice using two-dimensional gel electrophoresis and mass spectrometry. *J. Biol. Chem.* 273 (1998), 17940.

Zimmerman, H.J. and Ishak, K.G. General aspects of drug-induced liver disease. *Gastroenterol. Clin. North Am.* 24 (1995), 739.

CHAPTER TWENTY-ONE

Dermatotoxicology

NANCY A. MONTEIRO-RIVIERE

21.1 INTRODUCTION

One of the principal portals through which environmental toxicants can enter the body is the skin. Skin has emerged as an organ of interest, in part because it is a highly visible organ that serves as an interface with the environment. This makes skin vulnerable to the damaging effects of toxicants. The paradigm of a structure-to-function relationship is central to modern biology and infers, based on our current knowledge, that the structural and functional relationships of the skin are likely to be complex. Anatomical factors may affect the barrier function causing an altered rate of absorption. Therefore, the anatomical complexity of this organ must be fully understood before percutaneous penetration, metabolism, and cutaneous responses to specific toxicants can be investigated.

21.2 FUNCTIONS OF SKIN

Skin, the largest organ of the body, is considered to be the barrier between the well-regulated "milieu interieur" and the outside environment. Skin is a very heterogeneous, yet integrated, dynamic organ that has a myriad of biological functions that go far beyond its role as a barrier to the external environment. One of its primary roles is to act as a communicator between the interior and exterior environments. It acts as an environmental barrier by protecting major internal organs, as a diffusion barrier that helps to minimize insensible water loss that could result in dehydration, and as a metabolic barrier that can metabolize a compound to more easily eliminate products after absorption has occurred. The skin or integument participates in thermoregulation where blood vessels constrict to retain heat and dilate to dissipate heat. In addition, hair in humans and the fur of lower mammals act as insulation devices, while sweating facilitates heat loss by evaporation. Skin can serve as an immunological affector axis by having Langerhans cells process antigens and as an effector axis by setting up an inflammatory response to a foreign insult. Skin has a well-developed stroma, which supports all other organs. In addition, the skin has receptors that sense the modalities of touch, pain, and heat. The skin can function as an endocrine organ. It participates in the synthesis of vitamin D, and is a target for androgens, which regulate sebum production, and for insulin, which regulates carbohydrate and lipid metabolism. The skin has numerous sebaceous glands that can secrete sebum, a complex mixture of lipids that function as antibacterial agents or as a water-repellent shield in some animals. In addition, the skin contains both apocrine and eccrine sweat glands that produce a secretion that contains scent that functions in territorial demarcation. The integument also plays a role in the biosynthesis of keratin, collagen, melanin, lipids, and carbohydrates, as well as in respiration and biotransformation of xenobiotics.

In order to have a basic understanding and appreciation of how chemicals may interact with skin, its anatomy, physiology, and chemical composition must be fully grasped. All of the aforementioned biological functions and structural adaptations have a substantial impact on the skin's barrier properties and the rate and extent of percutaneous absorption. When a compound or toxicant is applied topically, it must penetrate through several cell layers of the skin in order to be picked up by the capillaries for systemic distribution. Alternatively, it may have a direct effect on the keratinocytes themselves. Skin can be anatomically divided into two principal components, the outermost epidermis and the underlying dermis.

21.3 EPIDERMIS

The epidermis consists of keratinized stratified squamous epithelium derived from ectoderm and forms the outermost layer of the skin. Two primary cell types based on origin, the keratinocytes and nonkeratinocytes, comprise this layer. The classification of epidermal layers from the outer or external surface is as follows: stratum corneum (horny layer), stratum lucidum (clear layer), stratum granulosum (granular layer), stratum spinosum (spinous or prickle layer), and stratum basale (basal layer). The nonkeratinocytes include the melanocytes, Merkel cells, and Langerhans cells that reside within the epidermis but do not participate in the process of keratinization. The epidermis is avascular and undergoes an orderly pattern of proliferation, differentiation, and keratinization that as yet is not completely understood. Various skin appendages, such as hair, sweat and sebaceous glands, digital organs (hoof, claw, nail), feathers, horn, and glandular structures are all specializations of the epidermis. Beneath the epidermis is the dermis, or corium, which is of mesodermal origin and consists of dense irregular connective tissue. A thin basement membrane separates the epidermis from the dermis. Beneath the dermis is a layer of loose connective tissue, the hypodermis (subcutis), which is superficial fascia that contains elastic fibers and aids in binding the skin to the underlying fascia and skeletal muscle.

In general, the basic architecture of the integument is similar in all mammals. However, differences exist in the thickness of the epidermis and dermis, the number of cell layers, and the blood flow patterns between species. Additionally, differences exist within the same species at different body sites. Skin is thickest over the dorsal and lateral surfaces and thinnest on the ventral and medial surfaces of the body. The stratum corneum is thickest in glabrous skin regions such as the palmar and plantar surfaces where considerable abrasive action occurs. The epidermis is thin in areas where there is a heavy protective coat of hair or fur. Understanding these variations in the skin is important in studies involving biopharmaceutics, dermatological formulations, cutaneous pharmacology, and dermatotoxicology.

21.3.1 Epidermal Keratinocytes

21.3.1a Stratum Corneum

The stratum corneum is the outermost layer of the epidermis and consists of several layers of completely keratinized flattened cells without nuclei or cytoplasmic organelles (Fig. 21.1). The most superficial layers of the stratum corneum that undergo constant desquamation are called the stratum disjunctum. The stratum corneum cell layers vary in thickness between body sites and species. These individual cells are highly organized and stacked upon one another to form vertical interlocking columns and have a flattened 14-sided polygonal (tetrakaidecahedron) shape. This structure provides a minimum surface-to-volume ratio, which allows for space to be filled by packing without interstices that yield a tight, water-impermeable barrier. Between the stratum corneum cells, the intercellular substance derived from the lamellar granules forms the intercellular lipid component of a complex stratum corneum barrier, which prevents both the penetration of substances from the external environment and the loss of body fluids. A plasma membrane and a thick submembranous layer that contains involucrin encompass these cells. This protein is synthesized in the lower stratum spinosum layers and cross-

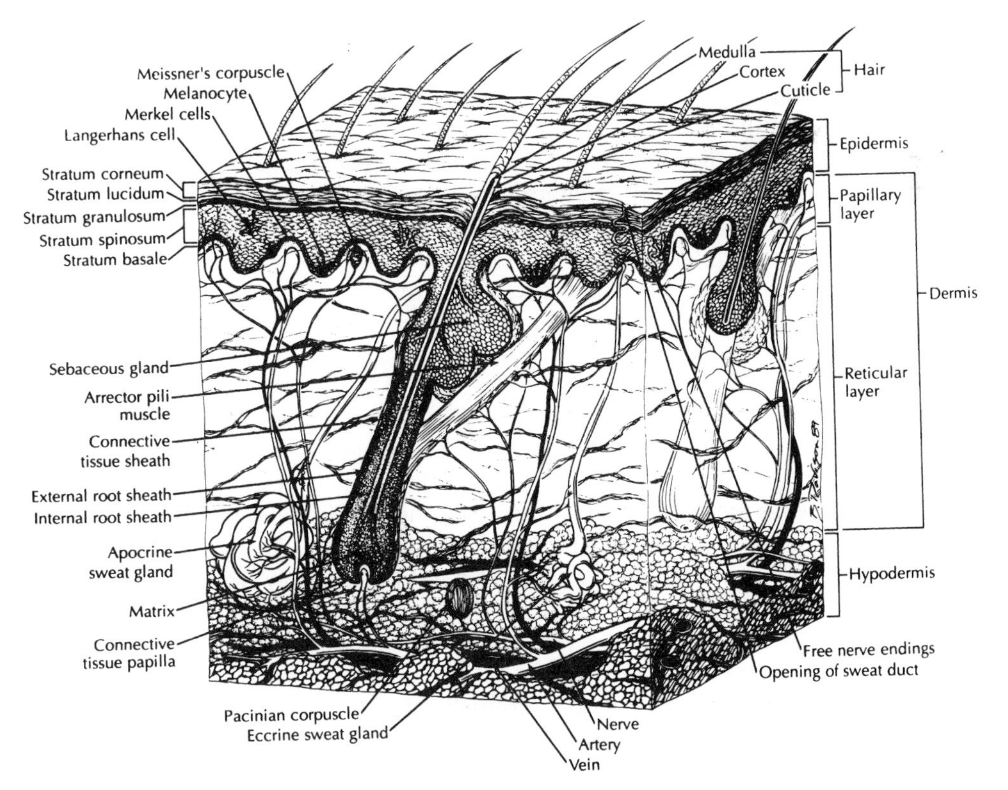

FIGURE 21.1 Schematic diagram of the integument depicting skin in various regions of the body.

Source: Reprinted from Monteiro-Riviere, 1991.

linked in the stratum granulosum by an enzyme that makes it highly stable. Involucrin provides structural support to the cell, thereby allowing the cell to resist invasion by microorganisms and destruction by environmental agents, but does not appear to regulate permeability.

21.3.1b Stratum Lucidum

The stratum lucidum consists of a thin, translucent, homogeneous line between the stratum granulosum and stratum corneum layers (Fig. 21.1). It is only found in specific areas of the body where the skin is exceptionally thick and lacks hair (e.g., plantar and palmar surfaces). This stratum consists of several layers of fully keratinized, closely compacted, dense cells devoid of nuclei and cytoplasmic organelles. Their cytoplasm contains protein-bound phospholipids and eleidin, a protein that is similar to keratin but has a different histologic staining affinity.

21.3.1c Stratum Granulosum

The next layer down is the stratum granulosum, which consists of several layers of flattened cells lying parallel to the epidermal-dermal junction (Fig. 21.1) containing irregularly shaped, nonmembrane-bounded, electron-dense keratohyalin granules. These granules contain profilaggrin, a structural protein and a precursor of filaggrin, and are thought to play a role in keratinization and barrier function. An archetypal feature of this layer is the presence of membrane-bound lamellar granules also known as Odland bodies, lamellated bodies, or membrane-coating granules. These granules increase in number and size as they move toward the cell membrane, where they fuse and release their lipid contents by exocytosis into the intercellular space between the stratum granulosum and stratum corneum. These lipids are responsible for coating the cell membrane of the stratum corneum cells and are the primary component of the barrier to chemical absorption across the skin. These lipids include the ceramides, cholesterol, fatty acids, and small amounts of cholesteryl esters, and hydrolytic enzymes such as acid phosphatases, proteases, lipases, and glycosidases are also present. The content and mixture of lipids can vary between species.

21.3.1d Stratum Spinosum

The next deeper layer of the epidermis is the stratum spinosum, or "prickle cell layer," (Fig. 21.1) that consists of several layers of irregular polyhedral-shaped cells in which the uppermost layers contain small lamellar granules. Desmosomes connect these cells to the adjacent stratum spinosum cells and to the stratum basale cells below. The most notable characteristic feature of this layer is the numerous tonofilaments, which differentiates this layer morphologically from the other cell layers.

21.3.1e Stratum Basale

The next layer, the stratum basale, consists of a single layer of cuboidal or columnar-shaped cells that are attached to the underlying irregular basement membrane by hemidesmosomes, and laterally to each other and to the overlying stratum spinosum cells by desmosomes (Fig. 21.1). The basal cells continuously undergo mitosis, which causes the daughter cells to be distally displaced, keeping the epidermis replenished as the stratum corneum cells are sloughed from the surface epidermis. Depending on the region of the body, age, disease states, and other modulating factors, cell turnover and self-replacement in normal human skin is thought to take approximately 1 month. The mitotic rate increases after mechanical (tape stripping, incisions) or chemical induced injuries.

21.3.2 Epidermal Nonkeratinocytes

21.3.2a Melanocytes

Melanocytes are located within the basal layer of the epidermis, external root sheath and hair matrix of hair follicles, the sweat gland ducts, and sebaceous glands (Fig. 21.1). Melanocytes possess dendritic processes that extend between adjacent keratinocytes and clear cytoplasm except for pigment-containing ovoid granules commonly referred to as melanosomes. Following melanogenesis, the melanosomes emigrate to the tips of the dendritic processes; the tips pinch off and are phagocytized by the adjacent keratinocytes. They are randomly distributed within the cytoplasm of the keratinocytes and sometimes localized over the nucleus, forming a caplike structure that protects the nucleus from ultraviolet radiation. Skin color is determined by several factors, such as the number, size, distribution, and degree of melanization.

21.3.2b Merkel Cells

These cells are located in the basal region of the epidermis in both hairless and hairy skin with their long axis running parallel to the skin surface (Fig. 21.1). Ultrastructurally, Merkel cells possess a lobulated and irregular nucleus with a clear cytoplasm, lack tonofilaments, and are connected to adjacent keratinocytes by desmosomes. Their cytoplasm is vacuolated on their dermal side and contain spherical electron-dense granules. Merkel cells are associated with axonal endings, and as the axon approaches the epidermis, it loses its myelin sheath and terminates as a flat meniscus on the basal aspect of the stratum basale cell. Merkel cells act as slow-adapting mechanoreceptors for touch.

21.3.2c Langerhans Cells

Langerhans cells are found in the upper stratum spinosum layers and have long dendritic processes that traverse the intercellular space to the granular cell layer (Fig. 21.1). They have a clear cytoplasm containing organelles and an indented nucleus, but lack tonofilaments and desmosomes. They are very apparent in toluidine blue-stained sections embedded in epoxy and appear as dendritic clear cells in the suprabasal layers of the epidermis. A unique characteristic of this cell is a distinctive rod- or racket-shaped granules within the cytoplasm called Langerhans (Birbeck) cell granules, which may function in antigen processing. Langerhans cells are derived from bone marrow and are functionally and immunologically related to the monocyte-macrophage series. They play a major role in the skin immune response because they are capable of presenting antigen to lymphocytes and transporting them to the lymph node for activation. They are considered to be the initial receptors for cutaneous immune responses such as delayed-type hypersensitivity and to contact allergens, and can play an initiating role in some forms of immune-mediated dermatologic reactions.

21.3.3 Keratinization

The process in which the epidermal cells differentiate and migrate upward to the surface epithelium is referred to as keratinization. This is designed to provide a constantly renewed protective surface. As the cells proceed through the terminal differentiation stage, many cellular degradation processes occur. The spinosum and granular layers have lost their proliferative potential and thus undergo a process of intracellular remodeling. The cytoplasmic volume increases, and tonofilaments, keratohyalin granules, and lamellar granules also become abundant. Keratin is the structural protein abundantly synthesized by the keratinocytes that consists of many different molecular types. A loose network of keratins K5 and K14 are located within the basal cells. The active stratum spinosum cells secrete K1 and K10 and contain coarser filaments than those in the stratum basale. As the cells become flatter, their cellular contents increase, the nuclei disintegrate, and the lamellar granules discharge their contents into the intercellular space coating the cells. The nucleus and other organelles disintegrate and the flattened cells become filled by filaments and keratohyalin. This envelope consists of the precursor protein involucrin and the putative precursor protein cornifin-α/SPRR1. The final product of this epidermal differentiation and keratinization process can be thought of as a stratum corneum envelope consisting of interlinked, protein-rich cells containing a network of keratin filaments surrounded by a thicker plasma membrane coated by multilamellar lipid sheets. This forms the typical "brick and mortar" structure in which the lipid matrix acts as the mortar between the cellular bricks. This intercellular lipid mortar constitutes the primary barrier and paradoxically the pathway for penetration of topical drugs through skin.

21.3.4 Basement Membrane

The basement membrane zone or epidermal-dermal junction is a thin extracellular matrix that separates the epidermis from the dermis. Ultrastructurally, it can be divided into four component layers: (1) the cell membrane of the basal epithelial cell, which includes the hemidesmosomes; (2) the lamina lucida (lamina rara); (3) the lamina densa (basal lamina); and (4) the sub-basal lamina (reticular lamina), which contains a variety of fibrous structures (anchoring fibrils, dermal microfibril bundles, microthread-like filaments). In addition, it has a complex molecular architecture comprised of many large macromolecules that play a key role in adhesion of the epidermis to the dermis. The molecular components of the epidermal basement membrane, common to all basement membranes, include type IV collagen, laminin, entactin/nidogen, and heparan sulfate proteoglycans. Constituents limited to the skin include bullous pemphigoid antigen (BPA), epidermolysis bullosa acquisita (EBA), fibronectin, GB3, L3d, and 19DEJ-1. The basement membrane has several functions: it helps to maintain epidermal-dermal adhesion, serves as a selective barrier between the epidermis and dermis by restricting some molecules and permitting the passage of others, influences cell behavior and wound healing, and serves as a target for both immunologic and nonimmunologic injury. Pertinent to toxicology, the basement membrane is the target for specific vesicating agents such as bis (2-chloroethyl) sulfide and dichloro (2-chlorovinyl) arsine, which causes blisters on the skin after topical exposure.

21.3.5 Dermis

The dermis lies directly under the basement membrane and consists of dense irregular connective tissue with a feltwork of collagen, elastic, and reticular fibers embedded in an amorphous ground substance of mucopolysaccharides. The predominant cells of the dermis are fibroblasts, mast cells, macrophages, plasma cells, chromatophores, fat cells, and extravasated leukocytes often found in association with blood vessels, nerves, and lymphatics. Sweat glands, sebaceous glands, hair follicles, and arrector pili muscles are present within the dermis. Arbitrarily, the dermis can be divided into a superficial papillary layer that blends into a deep reticular layer. This layer is thin and consists of loose connective tissue, which is in contact with the epidermis and conforms to the contour of the basal epithelial ridges and grooves. When it protrudes into the epidermis, it gives rise to the dermal papilla. When the epidermis invaginates into the dermis, epidermal pegs are formed. The thicker reticular layer is made up of irregular dense connective tissue with fewer cells and more fibers.

One of the primary components of the dermis is the extensive network of capillaries that help to regulate body temperature. Blood flow through skin can vary 100-fold, depending on environmental conditions, making it one of the most highly perfused organs in the body. The dermis participates in several functions: mechanical support, exchange of metabolites between blood and tissues, fat storage, protection against infection, and tissue repair.

Deep to the reticular layer of the dermis is the hypodermis (subcutis) consisting of very loose connective tissue with adipose cells. This layer helps to anchor the dermis to the underlying muscle or bone. This thermal barrier and mechanical cushion is sometimes considered a site that acts as a depot or reservoir for certain toxic compounds.

21.3.6 Appendageal Structures

Appendageal structures commonly found within the skin are the hairs, hair follicles, associated sebaceous glands, apocrine and eccrine sweat glands, and arrector pili muscles. Hairs are formed by epidermal invaginations. These keratinized structures traverse the dermis and may extend into the hypodermis. The free part of the hair above the surface of the skin is the hair shaft and the part deep within the dermis is the hair root, which forms an expanded knoblike structure called the hair bulb. This is composed of a matrix of epithelial cells in different stages of differentiation. Hair is composed of three concentric epithelial cell layers, the outermost thin cuticle, a densely packed keratinized cortex, and a central medulla of cuboidal cells. The hair follicle consists of four major components: (1) internal root sheath (internal root sheath cuticle, granular layer, pale epithelial layer); (2) external root sheath (several layers similar to the epidermis); (3) dermal papilla (connective tissue); and (4) hair matrix (comparable to the stratum basale of the epidermis).

The process of keratinization is continuous in the surface epidermis, while in the hair follicle the matrix cells undergo periods of quiescence during which no mitotic activity occurs. This cyclic activity of the hair bulb allows for the seasonal change in the hair coat of domestic animals. The length of time to regrow new hairs depends on the growth stage of the hair follicle. The period of the hair cycle where cells are mitotically active and grow is called anagen. After this growth phase, catagen occurs where metabolic activity slows down and the hair bulb atrophies. The final stage called telogen occurs when the hair follicle enters a resting or quiescent phase and growth stops. In this stage, the base of the bulb is located at the level of the sebaceous canal. As the new hair grows beneath the telogen follicle, it gradually pushes the old follicle upward toward the surface, where it is eventually shed. This intermittent mitotic activity and keratinization of the hair matrix cells constitute the hair cycle. It is controlled by several factors, including length of daily periods of light, ambient temperature, nutrition, and hormones, particularly estrogen, testosterone, adrenal steroids, and thyroid hormone. If a chemical's mechanism of action requires interaction with an active metabolic process, then toxicity may be exerted only during the anagen growth phase. Exposure at other times may not elicit any response. The regulation of hair growth by endocrine factors may provide another mechanism for chemical interactions.

Bundles of smooth muscle fibers commonly seen attached to the connective tissue sheath of the hair follicle toward the papillary layer of the dermis are referred to as arrector pili muscles (Fig. 21.1). The hair sits at an obtuse angle to the skin surface and the arrector pili is situated at the lower portion of the follicle. Contraction of this muscle in cold weather will elevate the hairs, forming "goose pimples" on the skin. These muscles are well-developed in humans and are supplied by postganglionic adrenergic sympathetic nerve fibers. Also, contraction of this muscle plays a role in emptying the sebaceous glands.

Many cytotoxic chemicals (e.g., cancer chemotherapeutic drugs and immunosuppressants like cyclophosphamide) whose mechanism of action is to kill dividing cells will produce hair loss (alopecia) as an unwanted side effect of nonselective activity (e.g., thallium). Damage may occur with the use of alkylating agents such as antimetabolites or colchicine, which effect the matrix cells. Alkali and oxidizing agents such as peroxides cause keratolytic damage.

Inorganic constituents of human hair are receiving attention because of the potential in diagnostic medicine. Trace elements present in hair may come from the water supply, which provides calcium and magnesium. Transition metals such as iron and manganese can deposit on the hair from the general water supply, and copper from swimming pools. Metals can also originate from sweat deposits, diet, air pollution, and metabolic problems. In addition, metal contamination in hair can come from hair products like dandruff shampoos that can add zinc or selenium and lead from lead acetate hair dyes. Accumulations of heavy metals in hair are usually at low levels, however, if the concentrations are well above the norm then it can be utilized as a diagnostic tool. There has been a good correlation of cadmium levels in hair with other target organs. Learning disorders such as dyslexia in children can be detected early by cadmium analysis of hair due to the high cadmium levels that exist in dyslexic children. In cases of arsenic poisoning, hair has served as a good source for the localization of arsenic. Also, mercury can be detected in hair when subjects have been exposed to a diet containing high levels of mercury. Iron deficiencies can be detected by analyzing the iron content of hair. Hair analysis is a useful tool for toxicology for diagnosing other illnesses or conditions of exposure.

21.3.7 Sebaceous Glands

These glands are usually found all over the body and are associated with hair follicles (Fig. 21.1). They are evaginations of the epithelial lining and histologically are simple, branched, or compound alveolar glands containing a mass of epidermal cells enclosed by a connective tissue sheath. They produce an oily secretion called sebum, which is derived from disintegrating glandular cells and contain a mixture of lipids that can vary between species. These lipids act as antimicrobial agents and, in hairy mammals, as a waterproofing agent. In humans, the principal lipids are squalene, cholesterol, cholesterol esters, wax esters, and triglycerides. In addition, the density of these glands can vary between anatomical site and between individuals. These cells move inward through mitotic activity and accumulate lipid droplets to release their secretory product, sebum, by the holocrine mode of secretion. During early adolescence, hormonal activity increases human sebum production, which results in acne vulgaris. Sebum production is thus involved in the screening of anti-acne drug candidates.

Some toxicants that interact with sebaceous gland function can induce comedons, an acne-like response, to cause a condition termed chloracne. Several chloracnegens have been considered a hazard in the occupational setting. They include chloronaphthalenes, polychlorinated biphenyls, tetrachloroazoxybenzene, tetrachloroazobenzene, polychlorinated dibenzodioxins, polychlorinated dibenzofurans, and polybrominated biphenyls. Many of these chloracnegens induce cytochrome P-450-mediated microsomal monooxygenase activity.

21.3.8 Apocrine Sweat Glands

The apocrine sweat glands are much larger than the eccrine sweat glands and are located in specific body areas such as the axillary, pubic, areolae, and perianal regions (Fig. 21.1). Microscopically, they consist of simple sacular or tubular structures with a coiled secretory portion located in the deep dermis and a straight duct that runs parallel to the hair follicle and penetrates the follicular epidermis to open alongside the follicle at the surface. Apocrine sweat glands release a milky oily fluid that contains a mixture of lipids, proteins, lipoprotein, and saccharides. When surface bacteria metabolize this secretion, they produce a body scent in humans and other mammals that may be related to communications between species. It may also act as a sex attractant and in domestic species, such as the dog and cat, can serve as a territorial marker (pheromone-secreting role).

21.3.9 Eccrine Sweat Glands

Eccrine (merocrine) sweat glands are simple tubular glands that open directly onto the skin surface (Fig. 21.1). In humans, they are found over the entire body surface except for the lips, external ear canal, clitoris, and labia minora. Myoepithelial cells located in the secretory portion of these glands are specialized smooth muscle cells that, upon contraction, aid in moving the secretions toward the duct. The eccrine sweat gland duct is comprised of two layers of cuboidal epithelium resting on the basal lamina that opens in a straight path onto the epidermal surface. Some investigators postulate that the duct of these glands provide an alternate pathway for polar molecules, normally excluded by the stratum corneum, to be absorbed through skin.

The exocrine gland, whose principal function is thermoregulation, is one of the major cutaneous appendages and is functionally very active in humans. Sweating in humans refers to a distinct physiological function of excreting body fluids to the surface of the skin. This is necessary for fluid and electrolyte homeostasis. Physiologically stressed individuals can excrete 2 liters/hr to support evaporative heat loss. Only the higher primates have a built-in mechanism that can accommodate this large volume loss without circulatory collapse. The secretory portion secretes isotonic fluid that is low in protein and similar in ionic composition and osmolarity to plasma. On passage down the duct portion, it becomes hypotonic and reabsorption of sodium chloride, bicarbonate, lactate, and small amounts of water occur. Abnormality in this fluid and electrolyte transport system leads to cystic fibrosis. In fact, analysis of this secretion is a prime diagnostic tool for this disease.

21.4 ANATOMICAL FACTORS TO CONSIDER IN MODEL SELECTION

One of the most important questions in dermatotoxicology is what model system should be used. What is the most appropriate model? Of course, humans would be ideal and the most logical choice, however, the study of potent or toxic compounds would be unethical. Toxicologists thus often utilize animal models and extrapolate the data to humans. What are the key factors important in selecting animal models?

21.4.1 Species Differences

In general, the basic architecture of the integument is similar in all mammals, although differences exist in the number of stratum corneum cell layers and in the thickness of the epidermis (Table 21.1). The number of epidermal cell layers varies among species and body sites. The mouse, rat, and rabbit have one to two viable cell

TABLE 21.1 Comparative Stratum Corneum Thickness, Epidermal Thickness, and the Number of Epidermal Cell Layers in the Ventral Abdomen (VAB) and at the Thoracolumbar Junction (TLJ) of Several Species. (H&E, Paraffin Sections).

Species	Area	Cell Layers Mean ± SE	Epidermis (μm) Mean ± SE	Corneum (μm) Mean ± SE
Monkey	TLJ	2.67 ± 0.24	26.87 ± 3.14	12.05 ± 2.30
	VAB	2.08 ± 0.08	17.14 ± 2.22	5.33 ± 0.40
Pig	TLJ	3.94 ± 0.13	51.89 ± 1.49	12.28 ± 0.72
	VAB	4.47 ± 0.37	46.76 ± 2.01	14.90 ± 1.91
Rabbit	TLJ	1.22 ± 0.11	10.85 ± 1.00	6.56 ± 0.37
	VAB	1.50 ± 0.11	15.14 ± 1.42	4.86 ± 0.79
Rat	TLJ	1.83 ± 0.17	21.66 ± 2.23	5.00 ± 0.85
	VAB	1.44 ± 0.19	11.58 ± 1.02	4.56 ± 0.61
Mouse	TLJ	1.75 ± 0.08	13.32 ± 1.19	2.90 ± 0.12
	VAB	1.75 ± 0.25	9.73 ± 2.28	3.01 ± 0.30

Source: Modified from Monteiro-Riviere et al., *J. Invest. Dermatol*, 1990.

layers and the pig has four to five viable epidermal cell layers. In addition, the types and concentration of lipids within the stratum corneum vary between different species and body sites. In man, 50% of the total lipid mass consists of ceramides, 20–27% cholesterol, 10% cholesterol esters, and 26% fatty acid content. Alternatively, pig epidermis contains only 1–2% cholesterol esters.

21.4.2 Body Sites

There are variations in the thickness of the epidermis and dermis within species in different regions of the body (Table 21.1). Skin is the thickest over the dorsal and lateral surfaces of limbs, and thinner on the ventral and medial surfaces of limbs. The back (thoracolumbar lumbar junction) is usually thicker than the abdomen. In areas possessing high hair density, the epidermis is thin; whereas in glabrous areas such as mucocutaneous junctions, the epidermis is thicker. The palmar and plantar surfaces consist of extremely thick stratum corneum due to the fact that it is an area where abrasive action occurs.

21.4.3 Hair Follicles

Along with body site differences, there are differences in the density and arrangement of hair follicles (hf) between species. Hair density in pig, like man is sparse with 11 hf/cm^2, whereas the Fisher-344 rat has 289 hf/cm^2, mouse has 658 hf/cm^2, and the nude mouse has 75 hf/cm^2. Not only is the density different in different species, but the arrangement of hair follicles can vary. In young pigs, hf occurs in groups of three, whereas the horse and cow have individual follicles that are evenly distributed. Dogs have compound follicles consisting of a single primary hair and a group of smaller secondary hairs. The cat possesses a single, large primary guard hair surrounded by clusters of two to five compound hair follicles.

21.4.4 Blood Flow

It is important to be aware of the anatomical and physiological differences in blood flow between species and within different body sites of the same species. Cutaneous blood flow varies at different body sites. The ventral abdomen has a higher blood flow than the back (thoracolumbar junction) (Table 21.2). Laser Doppler velocimetry (LDV) or laser Doppler perfusion imaging (LDPI) are noninvasive techniques used to assess cutaneous blood flow. These techniques can evaluate the vascular response during acute inflammation, ultraviolet light, or drug therapy such as nitroglycerin and corticosteroid. In addition, the LDPI provides a two-dimensional mapping of peripheral blood perfusion that allows visualization of spatial tissue blood flow distribution in wound healing studies, irritant or allergenic patch testing, and in vasoactive drug evaluation studies.

A comprehensive study comparing the epidermal histologic thickness and the cutaneous blood flow as assessed by LDV was conducted in nine species (mouse, rat, rabbit, cat dog, pig, cow, horse, and monkey) at five cutaneous sites (buttocks, abdomen, skin over the humeroscapular joint, skin over the thoracolumbar junction and ear). Blood flow did not correlate to skin thickness across species and body sites

TABLE 21.2 Comparative Blood Flow Measurements at the Ventral Abdomen (VAB) and at the Thoracolumbar Junction (TLJ) in Five Species.

Species	TLJ	VAB
Monkey	2.40 ± 0.82	3.58 ± 0.41
Pig	2.97 ± 0.56	10.68 ± 2.14
Rabbit	5.46 ± 0.94	17.34 ± 6.31
Rat	9.56 ± 2.17	11.35 ± 5.53
Mouse	20.56 ± 4.69	36.85 ± 8.14

Source: Modified from Monteiro-Riviere et al., *J. Invest. Dermatol*, 1990.

but rather were independent variables, suggesting that they must be evaluated separately in pharmacology, dermatology, and toxicology studies.

21.4.5 Age

The cutaneous barrier function may also be affected with age. The ability to differentiate actinically damaged skin from chronologically aged or environmental influenced such as chronic sun exposure, wind, cold, low humidity, chemical exposure, or physical trauma is difficult. Skin of the elderly have been characterized as dry and wrinkly. Aged skin is microscopically different and is associated with vascular thickening and a decrease in lipid content. All of these changes may affect the clearance and absorption of topically dosed chemicals. Age-related differences have been observed in the absorption of 14 different pesticides, where the absorption of some compounds increased whereas in others it decreased. This indicates that the physiological properties of the compound are extremely important in assessing the percutaneous absorption of a compound.

21.4.6 Disease State

Skin diseases including essential fatty acid deficiency and ichthyosis may also effect the transdermal delivery of a compound. Studies have shown that the epidermal barrier function is altered by abnormal lipid composition in noneczematous atopic dry skin. Numerous other dermatologic conditions affect the anatomical structure and function of skin which may impact on the nature of the toxic responses seen.

21.4.7 Metabolism

Skin plays a role in drug and chemical metabolism and therefore is an active site for xenobiotic metabolizing enzymes. The metabolites generated by a compound in skin can diffuse through the skin to become absorbed by the cutaneous vasculature and exert its influence systemically. This results in a first-pass metabolic effect for some topically applied chemicals. There are multiple isozymes of cytochrome P450 present in skin. The precise location of these drug metabolizing enzymes is controversial. Studies have demonstrated that the epidermis contains the highest specific activity of monooxygenases. P450 dependent monooxygenase activity in the skin has been shown with a variety of substrates including benzo(a)pyrene, ethoxy coumarin, ethoxyresorufin, aldrin, and parathion. Human skin contains 5α steroid reductase activity that metabolizes testosterone. Other studies have demonstrated cortisone conversion to cortisol.

The skin has a well-documented role in vitamin D metabolism. 7-Dehydrocholesterol (provitamin D_3) is activated by exposure to ultraviolet radiation in the skin to previtamin D_3, which isomerizes to vitamin D_3. Recently, further metabolism of 24, 25 dihydroxyvitamin D to biologically active 1, 25 dihydroxyvitamin D, has also been demonstrated in skin, a conversion previously assumed to occur only in the kidney.

Subcellular localization studies have identified P450 dependent monooxygenase activity in adult hairless mice sebaceous glands. Phase II conjugation pathways have also been identified in skin. Extracellular enzymes including esterases are present in skin, which has been utilized to formulate lipid soluble ester prodrugs that penetrate the stratum corneum and then are cleaved to release active drug into the systemic circulation. Finally, co-administration of enzyme inducers and inhibitors modulate cutaneous biotransformation and thus alter the systemic toxicity profile. These metabolic interactions that occur in skin have attracted a great deal of research attention and clearly illustrate that skin is more than a passive barrier to toxin absorption.

21.4.8 Summary

Once information is gathered concerning the physical and physiological differences between species, one then has a basis to select the correct species for study. In order to comprehend how a toxicant or drug can permeate the skin, the vehicle and its components, disease state, hair follicle density, age, thickness, body site, blood flow, and metabolism must be considered because they can modulate chemical absorption and affect the skin barrier. Toxicants, "inert" components of chemical mixtures, and vehicles as well as products generated by the skin such as sweat, sebum, dead cells, and metabolic byproducts can alter the physicochemical properties of skin that can influence the rate of a toxicant's penetration or toxicological activity. Many studies do not consider these factors. For example, hair follicle growth cycle and body site were significant factors in hydrocortisone absorption. The challenge for investigators is to determine which key factors have the greatest impact on a specific chemical's transdermal passage or toxicity before it can exert a toxic effect.

21.5 PERCUTANEOUS ABSORPTION AND PENETRATION

The tendency for a toxicant to traverse the skin is a primary determinant of its dermatotoxic potential. That is, a chemical must penetrate the stratum corneum in order to exert toxicity in lower cell layers. The quantitative prediction of the rate and extent of percutaneous penetration (into skin) and absorption (through skin) of topically applied toxicants is complicated by the biological complexity discussed above.

The skin is generally considered to be an efficient barrier preventing absorption (and thus systemic exposure) of most topically administered toxicants. It is relatively impermeable to most ions and aqueous solutions. It is, however, permeable to a large number of lipophilic solids, liquids, and gases, suggesting that the term barrier is inappropriate for these substances. A number of agricultural workers have experienced acute dermal poisoning from direct exposure to pesticides such as parathion during application, or from secondary exposure by contact with vegetation previously treated with such insecticides. In fact, percutaneous absorption of nicotine in tobacco workers led to the concept of the modern transdermal nicotine "patch" delivery systems.

As discussed above, compared with most routes of drug absorption, the skin is by far the most diverse across species (e.g., sheep versus pig) and body sites (e.g., human forearm compared with scalp). The stratum corneum affords the greatest deterrent to absorption. Although highly water retarding, the dead, keratinized cells are also highly water absorbent (hydrophilic), a property that keeps the skin supple and soft as they absorb water on its way to being evaporated from the surface. This is the mechanism of action of most cosmetic moisturizers. Additionally, sebum appears to augment the water-holding capacity of the epidermis, but has no appreciable role in retarding the penetration of xenobiotics.

A number of studies have demonstrated that disruption of the stratum corneum removes all but a superficial deterrent to penetration. This is supported by "tape-stripping" experiments, in which an adhesive (cellophane tape) is placed on the skin repeatedly to remove progressive layers of the corneum. The skin ultimately loses its ability to retard penetration and compound flux increases greatly. This can be noninvasively assessed by measuring the skin's ability to prevent insensible evaporative water loss from the body to the environment by utilizing water as a marker of molecular transport across the epidermal barrier. This is performed by measuring transepidermal water loss (TEWL) with an instrument called an evaporimeter. TEWL increases greatly when the stratum corneum is either stripped, removed by extracting the intercellular barrier lipids with solvents such as acetone, or damaged in response to cutaneous toxicants.

The stratum corneum has been estimated to contribute 1000 times the diffusional resistance to chemical penetration as the layers beneath it, except for extremely lipid soluble compounds with tissue/water partition coefficients greater than 400. As in most other epithelial tissues, the two other layers of the skin (dermis and subcutaneous tissue) offer little resistance to penetration. Once a substance has penetrated the outer epithelium, these tissues are rapidly traversed. This may not be true for highly lipid soluble compounds, because the dermis may function as an additional aqueous barrier preventing a chemical that has penetrated the epidermis from being absorbed into the blood.

21.5.1 Dermatopharmacokinetics

The rate of diffusion of a topically applied toxicant across the rate-limiting stratum corneum is directly proportional to the concentration gradient across the membrane, the lipid/water partition coefficient of the drug, and the diffusion coefficient for the compound being studied. This can be summarized by Fick's law of diffusion in the equation:

$$\text{Rate of diffusion} = \frac{D\,P}{h}(\Delta C)$$

where D is the diffusivity or diffusion coefficient for the specific penetrant in the membrane being studied, P is the partition coefficient for the penetrant between the membrane and the external medium, h is the thickness or actual length of the path by which the drug diffuses through the membrane, and ΔC is the concentration gradient across the membrane. The diffusivity is a function of the molecular size, molecular conformation, and solubility in the membrane milieu, as well as the degree of ionization. It should be noted that if the compound is dosed in an organic vehicle, the vehicle itself may penetrate into the intercellular lipids of the stratum corneum and change the estimated diffusional coefficient. The partition coefficient reflects the ability of the penetrant to gain access to the lipid membrane. Depending on the membrane, there is a functional molecular size/weight cutoff that prevents very large molecules from being passively absorbed across any membrane. The total flux of drug across a membrane is dependent upon the area of membrane exposed and thus is usually expressed in terms of square centimeters. This relationship, which works well in an in vitro experiment, is only an approximation in vivo because penetration may be slow and a long period of time is required to achieve steady state.

If the lipid/water partition coefficient is too great, the toxicant may be retained in the membrane rather than traverse it and thus some fraction of compound will actually not be available for diffusion through the system. However, passage through the skin generally correlates with experimentally determined lipid/water partition coefficients in octanol/water and olive oil/water. In some cases where the specific lipid composition of the membrane is known (e.g., the stratum corneum from a specific species), the slurry of the actual lipids may be employed. This is becoming more sophisticated with the advent of advanced organ culture techniques where, for example in skin, lipid membranes very similar in composition, structure and function to those in vivo can be prepared in culture and used to study drug transport.

In dermatotoxicology, the amount of toxicant per area of skin (e.g., mg/cm^2), rather than the amount of toxicant per unit of body weight (e.g., mg/kg) used in oral and parenteral studies is the primary determinant of dose. This explains why infants, with a relatively large ratio of skin surface area to body mass, are particularly prone to systemic toxicity from topical poisons when large areas of skin are exposed. This is further potentiated in neonates, who do not have a fully developed cutaneous barrier.

Occlusion of the skin, seen with application of water-impermeable drug vehicles or patches, alters the rate and extent of toxicant absorption. As the skin hydrates, a threshold is reached where transdermal flux dramatically increases (approximately 80% relative humidity). When the skin becomes fully hydrated under occlusive conditions, flux can be dramatically increased. This occlusive effect must be accounted for when extrapolating toxicology studies conducted under occlusive conditions to field scenarios where the ambient environmental conditions are present. Hydration may also markedly affect the pH of the skin, which varies between 4.2 and 7.3. Therefore, dose alone is often not a sufficient metric to describe topical doses when the method of application and surface area become controlling factors. Dose must be expressed as mg/cm^2 of exposed skin.

The dermis is a highly vascular area, providing direct access for systemic absorption once the epithelial barrier has been passed. The blood supply in the dermis is under complex, interacting neural and local humoral control whose temperature-regulating function can have an effect on toxicant distribution and absorption by altering blood supply to this area. The absorption of a chemical possessing vasoactive properties would be affected through its action on the dermal vasculature; vasoconstriction would retard absorption and increase the size of a dermal depot, whereas vasodilation may enhance absorption and minimize any local dermal depot formation. For a systemic toxicant, vasodilation would potentiate activity whereas a vasoconstriction might blunt the response. However, if the chemical is directly toxic to the skin, the reverse occurs with vasoconstriction preventing removal of drug to the systemic circulation away from cutaneous toxic sites, thereby potentiating local effects.

21.5.2 Routes of Absorption and Penetration

The appendages of the skin that extend to the outer surface may play a role in the penetration of certain compounds and also be selective targets after topical exposure. Anatomically, percutaneous absorption might occur through different routes. The majority of nonionized, lipid-soluble toxicants appear to move through the intercellular lipid pathway between the cells of the stratum corneum (Fig. 21.2). Movement across keratinocytes does not generally occur. Some absorption may occur through the appendages such as hair follicles or sweat ducts. Very small and/or polar molecules appear to have more favorable penetration through appendages or other diffusion shunts, but only a small fraction of drugs are represented by these molecules. However, the epidermal surface area is 100 to 1000 times the surface area of the skin appendages, depending on species. The only exception to this rule is par-

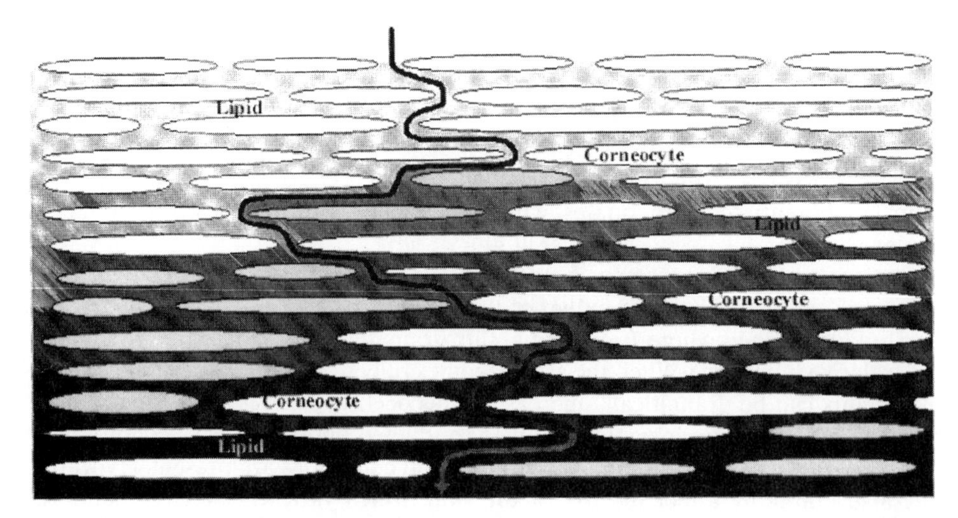

FIGURE 21.2 Schematic representation of the barrier property of skin composed of proteinaceous keratinocytes (corneocytes) embedded in an extracellular nonhomogenous matrix of lipid. *Arrow* depicts intercellular lipid pathway.

ticulate exposures (microspheres, liposomes), which may lodge in the opening to hair follicles and provide a unique access to the dermal circulation.

In addition to movement through shunts, polar substances may diffuse through the outer surface of the protein filaments of the hydrated stratum corneum, whereas nonpolar molecules dissolve in and diffuse through the nonaqueous lipid matrix between the protein filaments. The rate of percutaneous absorption through this intercellular lipid pathway is correlated to the partition coefficient of the penetrant, as presented above in Fick's Law.

Penetration of drugs through different body regions varies due to the anatomical factors discussed previously. In humans, it is generally accepted that for most nonionized toxicants, the rate of penetration is in the following order: scrotal > forehead > axilla = scalp > back = abdomen > palm and plantar. The palmar and plantar regions are highly cornified and their much greater thickness (100 to 400 times that of other regions) introduces an overall lag time in diffusion. In addition to thickness, differences in stratum corneum cell size and hair follicle density may also effect absorption of some molecules. The scalp should thus be considered in a different light than the rest of the body. Differences in cutaneous blood flow in different body regions may be an additional variable to consider in predicting the rate of percutaneous absorption.

21.5.3 Factors and Chemicals That Affect Percutaneous Absorption

The most damaging substances routinely applied to skin include soaps and detergents. Whereas organic solvents must be applied in high concentrations to damage skin, 1% aqueous solutions of detergents and surfactants (e.g., sodium lauryl sulfate) increase the penetration of toxicants through human epidermis.

Organic solvents can be divided into damaging and nondamaging categories relative to their effects on the barrier properties of skin. Methanol, acetone, ether, hexane, and mixed solvents such as chloroform: methanol or ether: ethanol are in the damaging category because they are able to extract epidermal lipids and proteolipids that alter permeability. Another mechanism for this solvent effect is that the solvents themselves partition into the intercellular lipid pathway, changing its lipophilicity and barrier property that result in an increased diffusion coefficient. Use of more polar or amphoteric solvents may enhance the penetration of polar molecules, in some cases, by forming "ion pairs" that have a greater ability to penetrate the lipid domain. In contrast, solvents such as higher alcohols, esters, olive oil, etc. do not appear to damage skin appreciably. However, the penetration rate of solutes dissolved in them is often reduced. This is best explained by partitioning and retention of the penetrant into the nonabsorbed solvent, preventing release of the toxicant into the stratum corneum. Thus, one can appreciate that for a specific chemical, the rate of penetration can be drastically modified by the solvent system used.

These phenomenon bring into question the wisdom of using organic solvents to decontaminate skin after exposure to lipophilic toxicants, because they will be easily absorbed into the skin and may enhance toxicant absorption through their interaction with intercellular lipids. This practice should be strongly discouraged. Not surprisingly, it has been found that lipid-soluble toxicants may be markedly resistant to washing within a short time after application due to subsequent depot formation.

For example, 15 minutes after application, a substantial portion of parathion cannot be removed from exposed skin by soap and water.

In environmental exposures, the chemical may come into contact with the skin as a mixture or in contaminated soil. In mixtures, other components may function as solvents and modulate the rate of absorption. This may be a determining factor in the toxicity of a complex chemical mixture. Our laboratory has extended this concept to classify chemical mixtures based on how components (mixtures of defined solvents, surfactants, reducing agents, vasoactive compounds) may modulate the absorption or direct cutaneous toxicity of suspected toxicants using a classification paradigm termed Mechanistically Defined Chemical Mixtures (MDCM). This approach would allow complex environmental mixtures to be assessed for the presence of such modulating compounds to triage them according to potential toxic potential. In soil, a large fraction of the toxicant may remain bound to soil constituents thereby reducing the fraction absorbed.

21.5.4 Experimental Techniques Used to Assess Absorption

Although generalizations are difficult, human skin appears less permeable than the skin of the cat, dog, rat, mouse, or guinea pig. The skin of pigs and some primates serves as useful approximations to human skin, but only after a comparison has been made for each specific substance. The major determinants of species differences are keratinocyte thickness, cell size, hair density, lipid composition, and cutaneous blood flow.

Whole animal studies assess the percent of the applied dose absorbed into the body using classic techniques of bioavailability, where absorbed chemical is measured in the blood, urine, feces, and tissues with mass balance techniques. Recently, methods have been developed to assess absorption by measuring the amount of chemical in the stratum corneum inasmuch as it is the driving force for diffusion. Cellophane tape strips are collected 30 minutes after chemical exposure and the amount of drug assayed in these tape strips correlates to the amount systemically absorbed. If the focus of the research is to determine the amount of chemical that has penetrated into skin, core biopsies may be collected, serially sectioned, and a profile of the chemical as a function of skin depth may be obtained.

In vitro approaches are often used to assess topical penetration. Most employ diffusion cell systems that sandwich skin sections between a donor and a receiver reservoir. The chemical is placed in the donor compartment (epidermis) and compound flux into the receiver (dermal) solution is monitored over time. This system can use a variety of "skin" sources, ranging from full-thickness specimens (epidermis and dermis) to epidermis alone to various "artificial" membranes such as lipid layers. In most skin studies, the donor compartment is open to the ambient air. A diffusion cell in which the receiver solution is a fixed volume is termed a static cell. A cell in which the receiver solution flows through the dermal reservoir is termed a "flow-through" system. This mimics the in vivo setting where blood continuously removes absorbed compound. Various cell and organ culture approaches have also been developed that assess absorption across cultured epidermal and/or dermal membranes. The rate of steady state flux can directly be used to calculate a permeability constant for the chemical under study using Fick's law above.

In vitro studies should be conducted at 30° to 37°C. Debate exists as to the choice of receptor fluids to use. In pharmaceutical studies involving relatively hydrophilic drugs, saline is often the receptor of choice. In contrast, toxicological investigations generally involve the assessment of lipophilic compounds, which require a receptor fluid in which the penetrant is soluble (e.g., albumin based buffers, solvent/saline mixtures). These systems have also been used to assess cutaneous metabolism, although maintenance of viability using oxygenated perfusate and glucose-containing receptor fluid is then required.

The next level of in vitro systems employed is the use of isolated perfused skin flap preparations, which are surgically prepared vascularized skin flaps harvested from pigs and then transferred to an isolated organ perfusion chamber. This model allows absorption to be assessed in skin that is viable, anatomically intact, and has a functional microcirculation. Studies conducted to assess the percutaneous absorption of drugs and pesticides in this model compared with humans show a high correlation.

21.6 DERMATOTOXICITY

If a cytotoxic chemical is capable of being absorbed across the stratum corneum barrier, it has the potential to cause toxicity to the skin. Chemical-induced damage to skin can be assessed by determining which key anatomical structures or physiological processes are perturbed after exposure to topical compounds.

A large number of cutaneous irritants specifically damage the barrier properties of skin, which results in an irritation response. These include the organic solvents discussed above that extract the intercellular lipid and perturb the skin's barrier. Some chemicals digest or destroy the stratum corneum and underlying epidermis. Those that cause chemical burns are properly termed corrosives and include strong acids, alkalis, and phenolics. They essentially attack the epidermal barrier and chemically destroy the underlying viable cell layers. The best treatment in these cases is to dilute and remove the offending agents by flushing with water or aqueous solvents. The exception is CaO (quicklime), which violently reacts with water and generates heat, which causes further thermal damage, and metallic (e.g., tin, titanium) tetrachloride compounds, which hydrolize to hydrochloric acid causing further damage. These types of reactions are easy to assess using in vitro models such as the Corrositex® system, which detects macromolecular damage to a collagen matrix resulting in a chemical color change in an associated detector system.

In other cases, epidermal cells are affected, which may then initiate other sequelae. If an absorbed chemical is capable of interacting with the immune system, the resulting manifestations will be dependent upon the type of immunologic response elicited (e.g., cellular versus humoral, acute hypersensitivity, etc.). It should be stressed that immune cells (e.g., Langerhans cells, lymphocytes, mast cells) may modulate the reaction or the keratinocytes themselves may initiate the response. This is in contrast to the previous role of epidermal keratinocytes in the skin, which was believed to be related only to maintaining the biophysical integrity of the skin by producing keratins and lipids for the formation of an intact stratum corneum barrier. Keratinocytes may thus act as the key immunocyte in the pathophysiology of allergic contact and irritant contact dermatitis. Figure 21.3 illustrates the path-

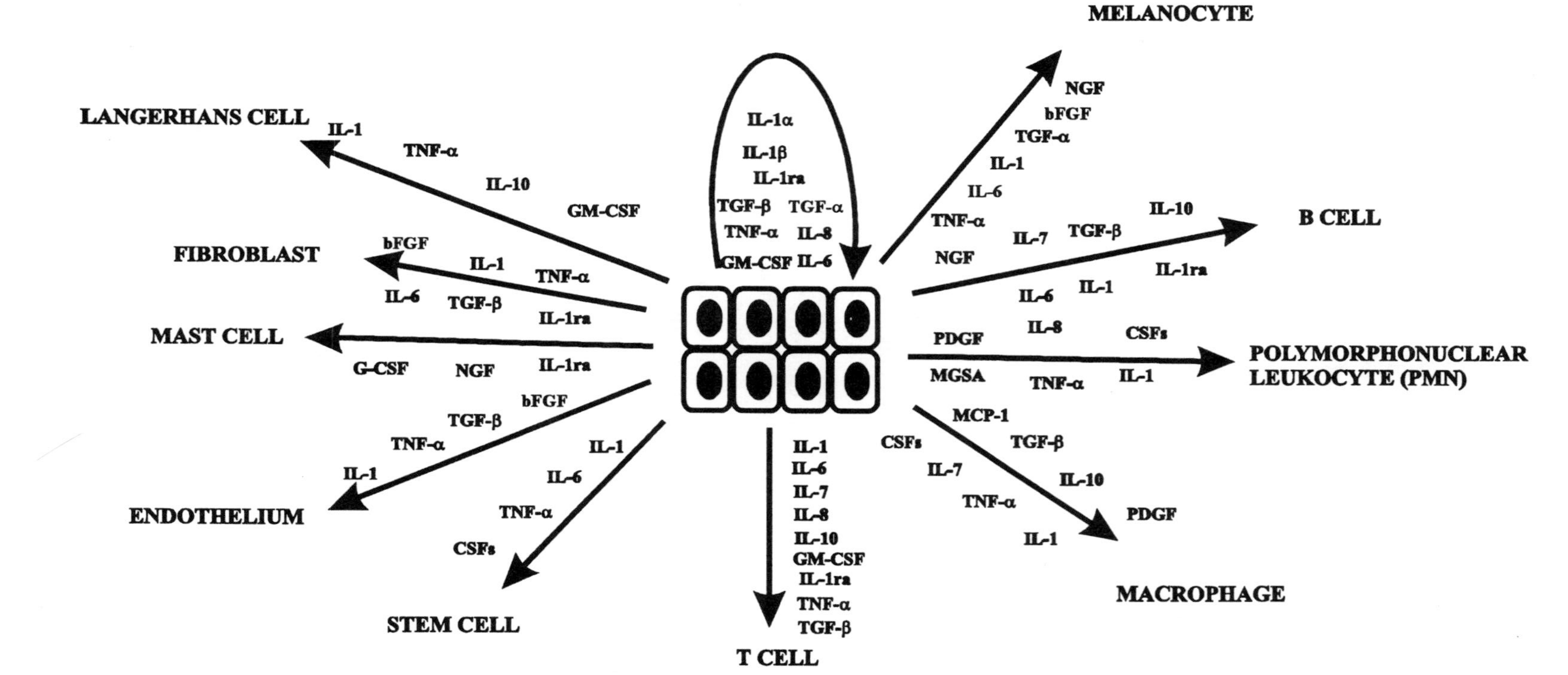

FIGURE 21.3 Schematic illustrating the role keratinocytes as modulators of cutaneous irritation.

ways that may trigger the production of proinflammatory cytokines after topical exposure to chemicals. Direct chemical-induced irritation of keratinocytes may also initiate this cytokine cascade without involvement of the immune system, blurring the distinction between direct and indirect cutaneous irritants.

21.6.1 Mechanisms of Keratinocyte Mediation of Skin Irritation/Inflammation

Evidence over the past decade has shown that keratinocytes are dynamic contributors that can produce growth factors, chemotactic factors, and adhesion molecules and thus act as initiators of inflammation. Environmental cues such as contact allergens (urushiol, nickel sulfate), toxic chemicals (croton oil, TPA, SLS), or physical stimuli (ultraviolet radiation) cause perturbations of the skin barrier that can induce a complex sequence of events known as the acute inflammatory response (Fig. 21.3). This biochemical cascade is responsible for localizing tissue damage to the site of injury and coordinating the wound healing mechanism of the body in an attempt to repair damage. The classic signs of this phenomenon include erythema, edema, heat, and pain. These symptoms are indicative of a process whereby resident cells release mediators that increase blood flow and capillary permeability to the injured area. Langerhans cells were once thought to be the only active instigator of this response. Keratinocytes, which make up 95% of the epidermal cell population, were found to act as signal transducers capable of the initiation and amplification of the acute inflammatory response by producing cytokines that may affect the surrounding tissue and incite migrating immune cells to differentiate. Tumor necrosis factor-alpha (TNF-α) and interleukin-1 (IL-1) are considered to be the two primary proinflammatory modulators of this response.

During the initiation phase of cutaneous inflammation, external stimuli, as mentioned above, can trigger a cutaneous inflammatory response by directly inducing epidermal keratinocytes to produce specific proinflammatory cytokines and adhesion molecules (Fig. 21.3). It is known that keratinocytes in conjunction with Langerhans cells are responsible for the release of many acute inflammatory cytokines such as TNF-α, and IL-1 and expression of the surface protein intercellular adhesion molecule-1 (ICAM-1). IL-1 and TNF-α secretion leads to activation of the dermal endothelial cells' production of the surface leukocyte adhesion ligands ICAM-1, endothelial cell adhesion molecule-1 (ELAM-1), vascular cell adhesion molecule-1 (VCAM-1) and consequent recruitment and sequestration of mononuclear cells from the circulation. Concomitant release of IL-8 and IL-1 by the keratinocytes promotes T-lymphocyte migration along a chemotactic concentration gradient to the epidermis.

Further interaction of recruited T-lymphocytes with keratinocytes leads to amplification of cytokine production by epidermal cells as well as promoting T-cell proliferation. Other immune-mediated processes can occur, resulting in the production of antigen-specific memory T-cells. If memory T-cells recognize this same antigen in the skin, then this reaction amplifies to a fulminating immune-mediated inflammatory response. If immune recognition does not occur, the response is limited. Although there are many cytokines that play a significant role in the evolution and perpetuation of an immuno-inflammatory response, TNF-α is considered to be the primary instigator.

The main sources of TNF-α in the skin are mast cells, resident keratinocytes, and Langerhans cells. Mature mast cells contain significant quantities of preformed TNF-α in their granules which are released in an IgE-dependent manner. Investigators studying the inflammatory response in mouse skin revealed by Western blot analysis that exposure to acetone and tape stripping increased TNF-α expression by 72% at 2.5 hours after treatment. The mRNA levels were highest at 1 hour after acetone, and then decreased to control levels by 8 hours. These studies indicate an immediate response of TNF-α to irritants that perturb the epidermal barrier.

It must be stressed that the primary mechanism of many topical irritants (e.g., organic solvents, corrosives) is the impairment to the stratum corneum barrier properties discussed above, and reflected by an increase in TEWL. If the stratum corneum barrier is perturbed, the feedback response mediated by cytokines (especially TNF-α) may be initiated whereby regeneration of the barrier occurs. However, additional responses to these inflammatory mediators may in themselves launch an irritation response mediated by the keratinocytes or lead to an immune reaction if the antigen is recognized. Regardless of the initiating mechanism, the sequelae to many irritants is the same, epidermal cell death.

21.6.2 Cell Death: Apoptosis or Necrosis

There are two primary mechanisms of cell death that occur in response to an irritant response, apoptosis or necrosis (Fig. 21.4). Apoptosis is a form of programmed cell death that occurs when the offending toxin interferes with the cell's metabolism without immediately rupturing cell membranes or mitochondria. Similarly, a toxicant that damages DNA or results in slower damage to a cell's metabolic function may also lead to apoptotic cell death. Finally, some cells undergo a genetically programmed cell death, such as that involved in the normal turnover of basal keratinocytes, via apoptosis. Apoptotic cell death minimizes any further inflammatory response because cellular contents are not released. This is also a protective response to some toxicants, for example, in the case of carcinogens, cells with damaged DNA will undergo a programmed death rather than proliferate and transmit the damaged genome. In contrast, if the offending toxic agent compromises the cell's membrane or mitochondrial integrity and causes ischemia, the cell may lyse, which leads to necrosis, now termed oncosis. This response releases cellular contents into the surrounding tissue, which may illicit an inflammatory response as described earlier.

The primary, and most widely accepted, method to differentiate apoptosis from necrosis is histology or supplementary transmission electron microscopy. There are no generally accepted specific histochemical markers for apoptosis. Apoptotic cells are characterized by condensation of chromatin, nuclear fragmentation or karyorrhexis, nuclear membrane dissolution, cell membrane budding, and cell shrinkage (Fig. 21.4). Necrosis is characterized by cell swelling, mitochondrial damage, cell membrane blebbing, karyolysis, or pyknosis. Apoptosis tends to occur in single cells whereas necrosis tends to involve multiple cells. Finally, apoptotic cells tend to be phagocytized by macrophages or parenchymal cells (e.g., liver). Necrotic cells induce inflammation and are phagocytized by granulocytes.

As can be appreciated from the above discussion, cutaneous irritants may induce a wide range of responses in the skin, ranging from mild irritation to complete

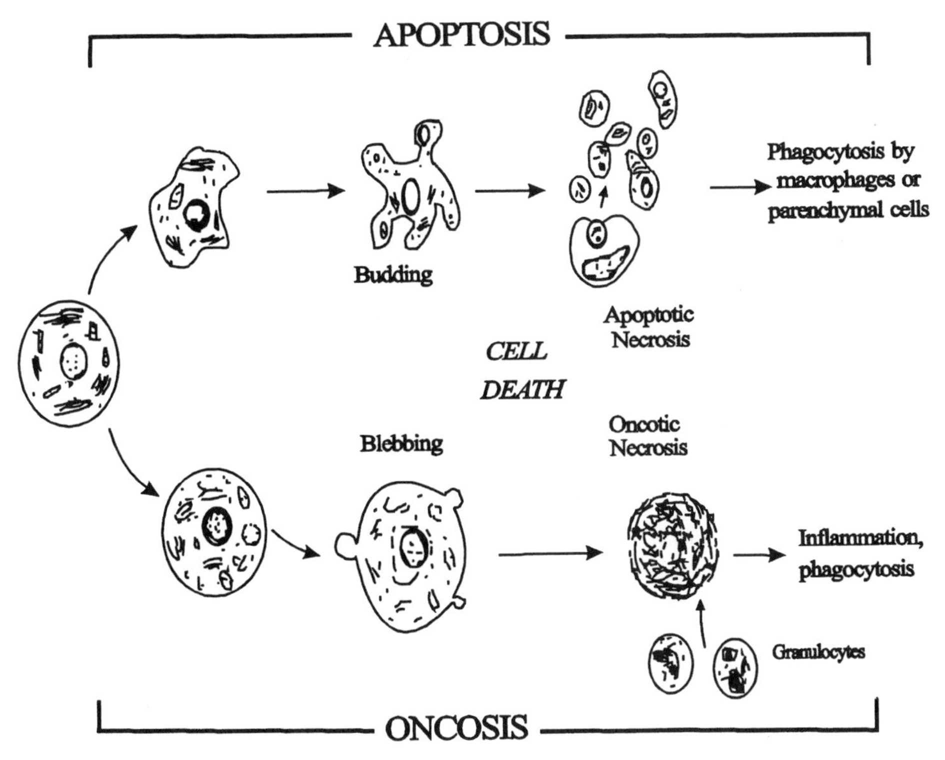

FIGURE 21.4 Schematic of cell death.

Source: Modified from Majno and Joris, 1995.

destruction with widespread necrosis. The degree of damage seen is dependent upon the chemical type of irritant involved (e.g., induce release of cytokines from keratinocytes, damage a cell's metabolic processes, corrosion and immediate cell lysis, and necrosis), the dose, the ability to penetrate the stratum corneum barrier, the duration of exposure, and its tendency to induce an immune response. Chemicals that are carcinogens can also transform basal keratinocytes into squamous cell carcinomas, a response fully discussed in another chapter of this text. Because of this widespread range of responses possible, toxicologic tests designed to detect cutaneous toxicity are varied and not always appropriate for all types of toxins.

21.6.3 Irritancy Testing Protocols

Direct irritation may thus be defined as an adverse effect of chemicals directly applied to the skin that does not involve prior sensitization and thus initiation by an immune mechanism. Irritation is usually assessed by a local inflammatory response characterized by erythema (redness) and/or edema (swelling). Other responses may be present that do not elicit inflammation such as an increase in skin thickness. Irritant reactions may be classified as acute, cumulative, traumatic, or pustular. However, two classifications are generally used by toxicologists. Acute irritation is a local response of the skin usually caused by a single agent that induces a reversible inflammatory response. Cumulative irritation occurs after repeated exposures to the same compound and is the most common type of irritant dermatitis.

There are several types of irritancy testing protocols that are used to comply with federal and international safety regulations. The classic Draize test was developed in 1944 to measure acute primary irritation. The test compound is applied in an occluded fashion to a clipped area of abraded and intact skin of at least six albino rabbits and evaluated 24 hours and 72 hours after patch removal. The degree of erythema and edema, ranging from one to four, is recorded to reflect severity of the irritation. Since these tests are occluded, irritancy is potentiated due to hydration, which reduces the skin barrier. The Draize test may be modified to assess sensitization by preexposing animals to a sensitizing dose of the study chemical and then rechallenging the animals at a later date to illicit the immune-mediated response.

Studies have correlated the Draize scores with biophysical estimates of cutaneous barrier function and erythema. Erythema may be assessed using a variety of different color measuring instruments. These systems are based on reflectance principles and operate by irradiating skin with specific light wavelengths and measuring the color of the reflected light. These systems can also detect toxin reactions that alter the melanization process and produces altered skin pigmentation. Erythema, which results from increased blood flow, may be directly assessed using noninvasive laser Doppler velocimetry discussed earlier.

A focus of recent research has been to develop humane alternatives to the Draize test. The majority of these approaches use a skin organ culture that attempts to provide the stratum corneum barrier and viable epidermal cells that can react to penetrated compounds. Because these are essentially organ culture systems, cell viability may be assessed by sampling the bathing culture medium. Glucose utilization is easily assayed. Neutral red (3-amino-7-dimethyl amino-2-methylphenazine hydrochloride) can be used to probe lysosomal integrity and thiazol blue (MTT) is a mitochondrial enzyme substrate that can assess mitochondrial function. Leakage of enzymes such as lactate dehydrogenase (LDH) can be assayed to detect cell membrane damage. Reactive products (cytokines, prostaglandins) produced by keratinocytes from an irritant response may be detected in the culture medium. Alternatively, histochemistry can be performed on tissue samples and specific enzymes examined that more closely reflect the mechanism of the offending chemical. The isolated perfused skin model described above has also been a useful model for the assessment of direct chemical toxicity caused by numerous chemicals.

21.6.4 Phototoxicity

There are additional environmental factors that lead to cutaneous irritation other than chemical toxicants. The most prevalent is ultraviolet (UV) radiation from sunlight. UV radiation encompasses short-wave UVC (200–290 nm), mid-wave UVB (290–320 nm), and long-wave UVA (320–400 nm). This topic is becoming increasingly important with the recent identified increase in UVB exposure in certain geographical regions apparently secondary to atmospheric ozone depletion. Although the ozone layer surrounding the earth filters out all UVC radiation, enough UVB and UVA reach the earth's surface to pose chronic and acute health hazards. Most known biological effects, including sunburn and skin cancer, may be attributed to UVB exposure. The most frequent and familiar phototoxic reaction is sunburn, caused by excessive cutaneous exposure to UVB radiation. Phototoxicity is a nonimmunological UV induced response. Sunburn is characterized by erythema, edema, vesication, increased skin temperature, and pruritis followed by hyperpigmentation.

The most characteristic feature of sunburn is the presence of "sunburn cells" (SBC) in the stratum basale layer of the epidermis (Fig. 21.5). These cells are dyskeratotic with a bright eosinophilic cytoplasm and a pyknotic nucleus in hematoxylin and eosin (H&E) stained tissue sections. This is an excellent example of cell death by apoptosis (Fig. 21.5), in this case induced by UV exposure rather than chemical toxicants. Ultrastructurally, these cells possess cytoplasmic vacuoles and condensed filament masses mixed with remnants of other cytoplasmic organelles (Fig. 21.6). In vitro models used to study phototoxicity (UVA and UVB) include *Candida albicans*, photohemolysis of red blood cells, tissue culture, and isolated human fibroblasts. Such simple models lack the complexity of living skin systems and are often unreliable. In vivo models include guinea pigs, rabbits, mice, and opossum. However, the skin of most animal models is not anatomically comparable to human skin. Studies have shown that UVB-induced erythema and SBC expression in pigs are comparable to humans, making the pig an accepted animal model for studying UV induced phototoxicity.

The optimal in vitro cutaneous model should possess viable cells and structures similar to intact skin as well as a functional vasculature. This would allow all manifestations of the pathogenesis of UVB exposure to be investigated. Our laboratory has utilized the previously described isolated perfused skin model to study UV

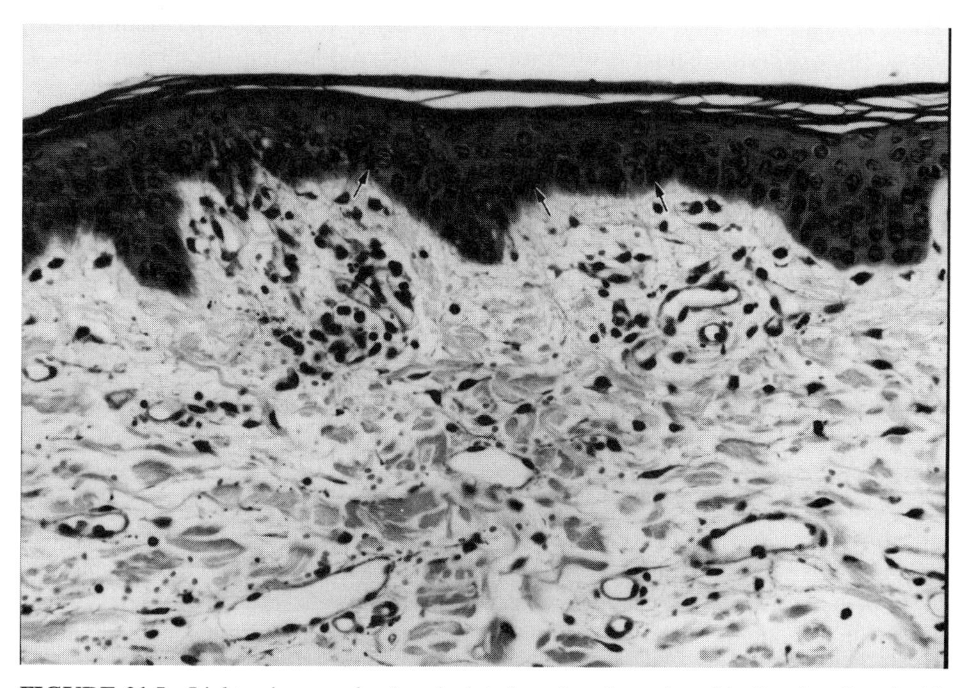

FIGURE 21.5 Light micrograph of an isolated perfused porcine skin flap irradiated with 1260 mJ/cm^2 depicting sunburn cells in the stratum basale layer (*arrows*) (400×).

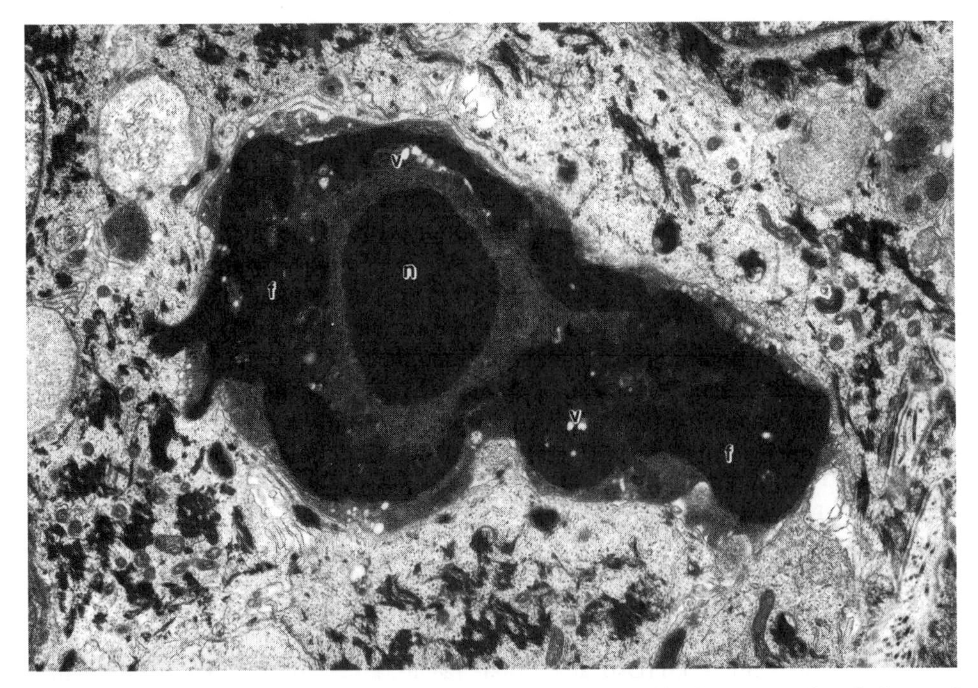

FIGURE 21.6 Electron micrograph of a sunburn cell located in the stratum basale layer of pig skin exposed 8 hours after being irradiated with 1260 mJ/cm^2. Note pyknotic nucleus (*n*), cytoplasmic vacuoles (*v*), and condensed filaments (*f*) (11,000×).

phototoxicity. In addition to morphologically assessing the dose dependent formation of SBCs, decreased glucose utilization was observed accompanied by increased vascular resistance and decreased cell proliferation by assaying proliferating cell nuclear antigen (PCNA). Levels of prostaglandin E_2 in the perfusate from UVB exposed skin flaps also increased in a dose dependent manner reflecting the cutaneous irritation response.

In addition to the direct effects of UV light on skin, some chemicals may be photoactivated to toxic intermediates which then cause skin toxicity. Exogenous phototoxic chemicals can be found in therapeutic, cosmetic, industrial, or agricultural formulations. These include, among others, tetracyclines, furosemide, chloroquin, organic dyes, furocoumarins, polycyclic aromatic hydrocarbons, and some nonsteroidal anti-inflammatory agents. Phototoxicity may also be aggravated by natural plant products and endogenous substances resulting from disorders of intermediates of heme biosynthesis such as porphyrins produced by inherited or acquired enzymatic defects.

Several types of phototoxic events may occur, those that are oxygen dependent and those that are nonoxygen dependent. When a reaction involves oxygen, the molecules can absorb the photons and transfer the energy to oxygen molecules thereby generating singlet oxygen, superoxide anions, and hydroxyl radicals that cause damage to the skin. This is what happens to protoporphyrins when they are irradiated. A photochemical reaction involves absorbance of photons by the chemical resulting in an excited state that will react with the target molecules to form photoproducts. A good example of this type of reaction is 8-methoxypsoralen (xanthotoxin), which reacts with specific sites on the DNA by forming covalent bonds between the pyrimidine base and the furocoumarin. This phototoxic reaction, which may be restricted to areas of skin exposed to UV light, has been used as a therapeutic technique to treat accessible skin tumors and psoriasis. With other compounds like chlorpromazine and protriptyline, the molecules absorb photons to form stable photoproducts that then induce cutaneous toxicity. The sequelae to these photoxic reactions is similar to that described earlier for cutaneous irritation.

21.6.5 Vesication

The final type of chemical toxicity that will be presented are the vesicants, chemicals that cause blisters on the skin. There are two classes of blisters that implicate different mechanisms of vesication. Intraepidermal blisters are usually formed due to the loss of intercellular attachment caused by cytotoxicity or cell death. The second class occurs within the epidermal-dermal junction (EDJ) due to chemical-induced defects in the basement membrane components. The classic chemical associated with EDJ blisters is the chemical warfare agent sulfur mustard (bis-2-chloroethyl sulfide; HD). HD is a bifunctional alkylating agent that is highly reactive with many biological macromolecules, especially those containing nucleophilic groups such as DNA and proteins.

The HD-induced dermal lesion is characterized by vesication and slow wound healing. Our laboratory has shown that the epidermal-dermal separation associated with vesication occurs in the upper lamina lucida of the epidermal basement membrane after HD-induced dermal injury in the perfused skin model. It was previously believed that alkylation of DNA with subsequent DNA cross-links or breaks was the primary and initial event responsible for HD cutaneous toxicity. Thus, a hypothesis regarding DNA alkylation, metabolic disruption, and proteolytic activity has been proposed. In this scenario, DNA repair processes are induced, including the activation of poly (ADP-ribose) polymerase, which consumes NAD^+ as a substrate. As repair continues, NAD^+ is depleted, which decreases epidermal glycolysis and stimulates activation of the $NADP^+$ dependent hexose monophosphate shunt, resulting in protease release. Extracellular proteases digest dermal tissue causing cell death, inflammation, and blister formation. Recently, results from several laboratories failed to support this hypothesis. Inhibitors of NAD^+ synthesis did not cause vesication. Additionally, when nicotinamide was used to increase NAD^+ levels, there were normal levels of glycolysis, although cell death and microblister formation still occurred.

Our laboratory has shown that gross blisters and microvesicles were present 5 hours after HD exposure. This suggests that a basement membrane component is the target of HD-alkylation. Other DNA alkylating agents do not cause vesication and the other pathological changes characterized by HD. Therefore, cell death may not directly cause HD-induced vesication, but may only be associated with blister formation due to direct toxicity to epidermal cells. In fact, HD may have unique protein targets in the basement membrane zone, which leads to diminished stability of the epidermal-dermal junction leading to vesication. Alkylation of laminin would delay wound healing requiring therapeutic strategies to remove this damaged laminin scaffolding before normal skin regeneration can occur.

21.7 CONCLUSION

Skin is a complex organ that is a focus of toxicology research. Because it is a primary portal for entry of numerous toxicants, its barrier properties have been extensively studied. Similarly, the ease by which chemicals can be exposed to skin increases the chances that direct dermatotoxicity may occur. Its role as an immune organ further increases its susceptibility to damage when allergens are encountered. In order to fully understand toxicologic effects in skin, one must take into account the numerous anatomical and physiological factors that differentiate species and body regions and are important in proper experimental design.

SUGGESTED READING

Barry, B.W. *Dermatological Formulations: Percutaneous Absorption*. Marcel Dekker, Inc., New York, 1993.

Barker, J.N.W.N., Mittra, R.S., Griffiths C.E.M., Dixit V.M. and Nickoloff B.J. Keratinocytes as initiators of inflammation. *Lancet* 377 (1991), 211–214.

Bos, J.D. *Skin Immune System (SIS), 2nd Ed.*, CRC Press, Boca Raton, FL, 1997.

Drill, V.A. and Lazar, P. *Cutaneous Toxicity*. Raven Press, New York, 1984.

Elias, P.M. Epidermal lipids, barrier function, and desquamation. *J. Invest. Dermatol.* 80 (1983), 44–49.

Feldman, R.J. and Maibach, H.I. Regional variations in percutaneous penetration of ^{14}C cortisol in man. *J. Invest. Dermatol.* 48 (1967), 181–183.

Levin, S., Bucci T.J., Cohen S.M., Fix, A.S., Hardisty J.F., LeGrand, E.K., Maronpot R.P. and Trump, B.F. The nomenclature of cell death: recommendations of an ad hoc committee of the society of toxicologic pathologists. *Toxicol. Pathol.* 27 (1999), 484–490.

Majno, G. and Joris, I. Apoptosis, oncosis, and necrosis; an overview of cell death. *Am. J. Pathol.* 146 (1995), 3–14.

Marzulli, F.N. and Maibach, H.I. *Dermatotoxicology, 5th Ed.*, Taylor & Francis, Washington, DC, 1996.

Monteiro-Riviere, N.A., Bristol, D.G., Manning, T.O. and Riviere, J.E. Interspecies and interegional analysis of the comparative histological thickness and laser Doppler blood flow measurements at five cutaneous sites in nine species. *J. Invest. Dermatol.* 95 (1990), 582–586.

Monteiro-Riviere, N.A. Comparative anatomy, physiology, and biochemistry of mammalian skin. In: Hobson, D.W. (Ed), *Dermal and Ocular Toxicology: Fundamentals and Methods*. CRC Press, New York, 1991, pp 3–71.

Monteiro-Riviere, N.A. Anatomical factors affecting barrier function. In: Marzulli, F.N. and Maibach, H.I. (Eds), *Dermatotoxicology*. Taylor & Francis, Washington, DC, 1996, pp 3–17.

Monteiro-Riviere, N.A. Integument. In: Dellmann, H.D. and Eurell, J. (Eds), *Textbook of Veterinary Histology, 5th Ed.* Williams & Wilkins, Baltimore, 1998, pp 303–332.

Mukhtar, H. *Pharmacology of the Skin*, CRC Press, Boca Raton, 1992.

Nickoloff, B.J. The cytokine network in psoriasis. *Arch. Dermatol.* 127 (1991), 871–883.

Rietschel, R.L. and Spencer, T.S. *Methods for Cutaneous Investigation*, Marcel Dekker, Inc., New York, 1990.

Webster, R.C. and Maibach H.I. Percutaneous absorption in diseased skin. In: Surber, C. and Maibach, H.I. (Eds), *Topical Corticosteroids*, Karger Publishing, Basel, Switzerland, 1992, pp 128–141.

CHAPTER TWENTY-TWO

REPRODUCTIVE AND DEVELOPMENTAL TOXICITY

STACY BRANCH

22.1 INTRODUCTION

Reproductive toxicity refers to any adverse effect on any aspect of male or female sexual structure, function, or lactation including effects on the reproductive potential and viability of the offspring. This concept may also include aspects of xenobiotically induced teratogenicity (developmental toxicity), the ability to cause dymorphogenesis, and altered function in the developing fetus and neonate. Developmental toxicity may be more specifically defined as any xenobiotically induced alteration in the normal growth, maturation, and function of the developing offspring. This chapter will be divided as follows: (1) reproductive toxicity relating to the effects on the male and female, and (2) adverse effects on the developing conceptus.

22.2 PHYSIOLOGY OF THE MALE REPRODUCTIVE SYSTEM

22.2.1 Hypothalamic-Pituitary-Testicular Axis

Under the influence of gonadotrophic hormone releasing hormone secreted by the hypothalamus, the anterior pituitary is stimulated to release the gonadotrophic hormones, leutenizing hormone (LH), and follicle stimulating hormone (FSH). Additionally, prolactin is released by the anterior pituitary. However, the release of prolactin is inhibited by dopamine. The cells of the testis are the targets of action of LH and FSH. LH is believed to stimulate steroidogenesis whereas FSH effects the sertoli cells. Secreted prolactin modulates the effects of LH in the gonadal tissue. Under control of the nervous system (with some involvement of endocrine factors) are libido, potency, and ejaculatory function. Stimulation of the central and autonomic nervous system affect sexual behavior. Modification of the nervous system can lead to altered endocrine function thus affecting reproductive behavior. Different control points within the hypothalamic-pituitary-testicular axis may be adversely affected by xenobiotics leading to various manifestations of male reproductive toxicity (Fig. 22.1).

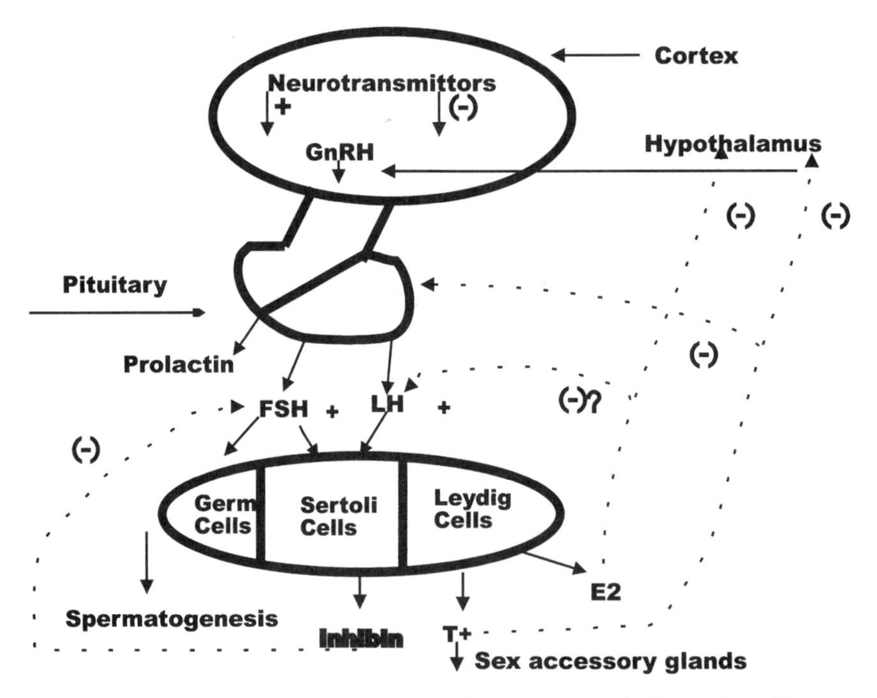

FIGURE 22.1 Diagram of the hypothalamic-pituitary-testicular axis with secreted hormones.

22.2.2 The Testis and Accessory Sex Organs

Two principal components of the testicular tissue are the seminiferous tubules (where spermatogenesis takes place) and the interstitial compartment (contains Leydig cells, which produce testosterone under the influence of LH). The seminiferous tubules are further broken down into two major cell types, the spermatogonial cells and the sertoli cells (Fig. 22.2). Spermatogonia undergo several mitotic divisions within the seminiferous tubules (Fig. 22.2). These cells progress through spermatocyte and spermatid stages to spermatozoa. Androgen stimulation leads to the maturation of the spermatozoa within the epididymis. This process of spermatogenesis, the growth of and activity of the accessory sex glands, masculinization, male behavior, and a variety of metabolic functions are controlled by androgens. Secretion of androgenic steroids by the developing fetal testis is essential for differentiation of the gonads. This differentiation is permitted via the regression of the Müllerian ducts with subsequent development of Wolffian ducts.

Sertoli cells have a number of functions. These cells provide nutritional and structural support for the developing sperm cells. They also synthesize androgen binding proteins, which bind both testosterone and estrogens. These complexes are carried to the seminiferous tubules for involvement in sperm maturation. Inhibin (involved in the regulation of LH and FSH and the maintenance of the blood-testis barrier) is also released by the Sertoli cells. Inhibin has an inhibitory effect on the anterior pituitary gland thus preventing oversecretion of FSH. As previously mentioned, gonadal differentiation occurs via the regression of the Müllerian ducts. The Sertoli

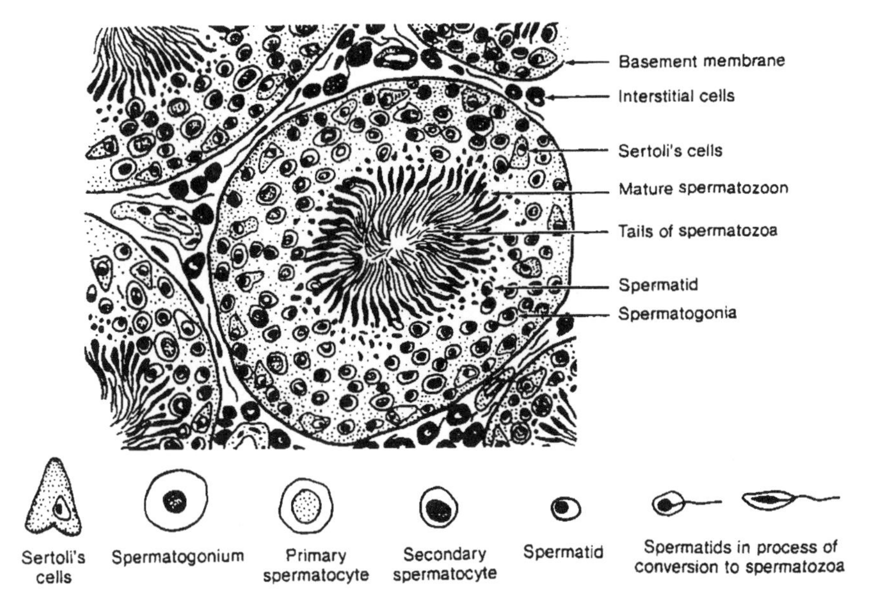

FIGURE 22.2 Sertoli's cells and the progression of spermatogenesis from spermatogonia to spermatozoa within the seminiferous tubules.

Source: Adapted from J.A. Thomas, *Toxic Responses of the Reproductive System*. In: Klassen, C.D. *Casarett and Doull's Toxicology, The Basic Science of Poisons*, 5th Ed. New York, McGraw-Hill, 1996.

cells secrete mullerian inhibitory substance (MIS) during fetal development to inhibit the formation of fallopian tubes from the Müllerian ducts. Further, expression of the male phenotype requires expression of the SRY gene on the Y chromosome. Expression of the Sry protein initiates a cascade of genetic events leading to testicular differentiation. Possible gene targets are P450 aromatase (inhibited) and MIS (activated). Estradiol (a metabolite of testosterone) is also secreted by the Sertoli cells and is a stimilatory factor in spermatogenesis. This functional role of estrogen in male reproduction is evidenced by studies using estrogen receptor knockout mice that are infertile and have abnormal testicular cellular morphology.

The leydig cells of the interstitial compartment of the testis secretes testosterone under the influence of LH. Testosterone is converted into the active hormone 5α-dihydrotestosterone in target tissue. The Steroidogenic Acute Regulatory (StAR) protein facilitates the transport of cholesterol into Leydig cell mitochondria. Pregnenolone is then formed with the eventual production of testosterone, 5α-dihydrotestosterone (DHT), and estradiol (Fig. 22.3). DHT binds to the androgen receptor with higher affinity than testosterone and is the principal androgen bound to the androgen receptor in vivo. DHT is required for the normal development of the external genitalia and prostate.

The male reproductive secretory organs produce seminal fluid. These organs include the prostate, bulbourethral and urethral glands, and the seminal vesicles. These organs are under the influence of androgens. The prostate gland secretes a milky, alkaline fluid that adds bulk to the seminal fluid. Prostatic fluid contains citric acid, calcium, acid phosphatase, and a clotting enzyme (profibrinolysin). In most mammals (excluding dogs and cats), the seminal vesicle produces more than 50%

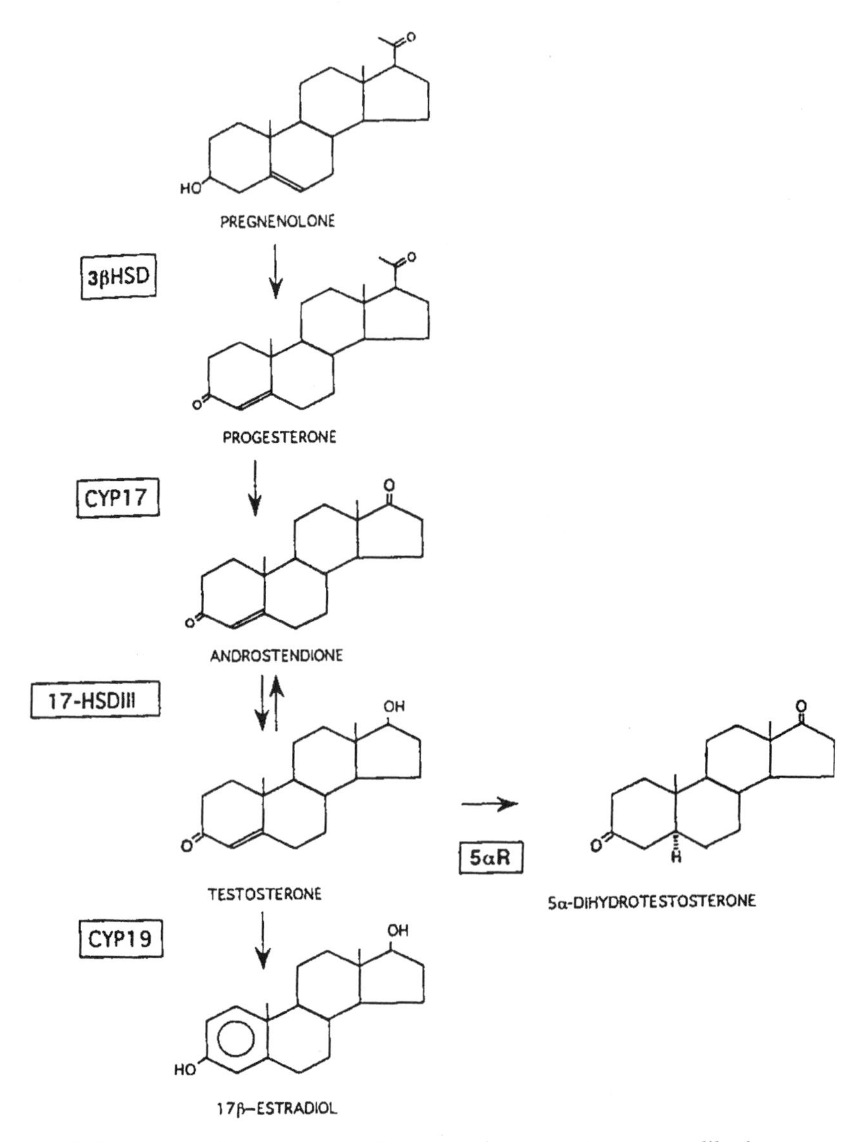

FIGURE 22.3 Enzymatic conversion of pregnenolone to testosterone, dihydrotestosterone and estradiol (with related intermediates).

Source: Adapted from McPhaul, M.J. *The biology of the male reproductive tract.* In: Korach, K.S. (Ed) *Reproductive and Developmental Toxicology*, New York, Marcel Dekker, 1998.

of the seminal fluid. As with the prostate, the seminal vesicular fluid contains proteins, enzymes, and polysaccharides.

22.3 MECHANISMS AND TARGETS OF MALE REPRODUCTIVE TOXICANTS

Many mechanisms of agents known to cause effects have not been fully elucidated. These mechanisms may involve multiple sites of action. Certain xenobiotics can interfere with endocrine function at key control points within the hypothamic-pituitary-gonadal axis (Fig. 22.1). Other compounds may primarily affect the CNS, which can subsequently lead to altered secretion of gonadotropins and releasing factors. The testes may be directly affected thus affecting the endocrine system and spermatogenesis. The cells of the testis have differential sensitivity (possibly due to differences in compound metabolizing abilities) to reproductive toxicant: spermatogonia > Sertoli cells > Leydig cells (most → least sensitive). A given agent may affect any or a combination of different sites of action making definitive elucidation of mechanisms more complex.

Cytotoxic agents are capable of altering spermatogenesis and viability of spermatozoa. The cells supporting the spermatogenis and sperm viability (leydig and sertoli cells) may also be directly adversely affected by cytotoxic compounds. Accessory sex gland function can be affected leading to abnormal sperm function, or altered seminal characteristics. Sperm may then have altered motility or inability to penetrate the ovum or even to survive in the female reproductive tract. Toxicants (endocrine disruptors) may mimic endogenous compounds such as hormones, thus acting as their agonists (mimicking sex hormones) or antagonists (blocking hormonal function or receptor binding). Xenobiotics or physical processes can alter normal spermatogenesis by disrupting chromosome structure or altering the normal chromosome number. These abnormal processes may also induce gene mutations in the germ cells. Resulting abnormal sperm may be unable to fertilize an oocyte. If fertilization does take place, the conceptus may fail to survive or develop with congenital defects. Adverse effects on the nervous system may alter sexual behavior leading to reduced or inhibited mating behavior, thus modifying reproductive function. Key enzymes involved in androgenic steroid synthesis may be inhibited at the synthesis or receptor level.

The ability of the male reproductive system to respond to toxic injury depends on a number of factors. There is some resistance offered by the blood-testis barrier. With the exception of the placental barrier, a similar barrier is absent in the female tract. The blood-testis barrier is located between the interstitial capillary lumen and the seminiferous tubule lumen. Biotransformation (systemically or within the male tract) of xenobiotics can determine the toxic injury response. The interstitial tissue of the testis contains primarily monooxygenase activity whereas the seminiferous tubles contains primarily phase two enzymes. Although local biotransformation processes are less abundant than that found within the liver, compounds penetrating the blood-testis barrier may be activated or detoxified locally.

22.3.1 Specific Examples

22.3.1a Effects on Spermatozoa

Antimetabolites used in cancer treatment (i.e., cyclophosphamide) are spermatotoxic and affect semen quality. Gossypol, a natural pigment found in the cottonseed, inhibits enzymes important for the synthesis of steroidal hormones within leydig cells. Abnormalities of the spermatozoa results in functional deficits leading to infertility. The fungicide benomyl interferes with normal meiosis during spermatogenesis by affecting spindle formation. Physical processes such as heat and ionizing radiation are spermatotoxic. Prolonged heating of the scrotum adversely affects early spermatogenic stages. This effect may be associated with a decrease in the binding of testosterone to androgen binding proteins. Spermatogenesis is inhibited by ionizing radiation with the spermatogonial cells being the most sensitive.

22.3.1b Endocrine Disruption

Reproductive toxins may act as agonists or antagonists of endogenous hormones thus affecting normal reproductive endocrine function. This class of toxicants is referred to as endocrine disruptors (ED). The components of the hypothalamic-pituitary axis (HPA) regulate the growth, development, and functions of various tissues including that of the reproductive system. Endocrine disruptors are xenobiotics that affect the normal hormonal homeostasis and functions. A given ED may affect estrogen and/or androgen function directly or by affecting various critical points within the HPA. EDs can bind to hormone receptor sites eliciting a response completely different from that of the endogenous hormone. Besides modulating hormone-receptor interactions directly, EDs can alter the synthesis and breakdown of the endogenous hormones or their receptors.

A given ED may bind to receptors that are normally the targets of endogenous hormones. This may result in a mimic or block of androgen binding sites thus impairing normal cell activity due to androgenic antagonism. These antiandrogens cause a failure in the activation of the androgen receptor subsequently affecting the cascade of events that follow (Fig. 22.4). For example, the fungicide vinclozolin is biotransformed into metabolites that are androgen receptor antagonists. Once in the nucleus, these metabolites fail to initiate DNA transcription. The historical ED diethylstilbestrol (DES) was administered to pregnant woman primarily during the 1950s to assist in the maintenance of pregnancy. DES affects the developing male fetus by antagonizing fetal testosterone activity resulting in testicular hypoplasia, abnormal semen parameters, and infertility. Cimetidine (used to treat peptic ulcers) competes with dihydrotestosterone for androgen receptors in the testis and accessory sex glands. The results are low sperm count and gynecomastia. Occupational exposure to oral contraceptives is also associated with the induction of gynecomastia in exposed males.

Ketoconazole can enter the seminal fluid where it can immobilize sperm. It also affects Leydig cell steroidogenesis by directly inhibiting P450 catalyzed cholesterol cleavage. This effect is reversible as only LH-stimulated testosterone synthesis is inhibited and not Leydig cell viability. Polychlorobiphenyls (PCBs) and polychlorodibenzo-p-dioxins (i.e., 2,3,7,8-TCDD) have been associated with the induction of reproductive defects in utero. Experimantal rats orally exposed to DDE, PCBs, vinclozolin, and 2,3,7,8-TCDD exhibited genital defects, reduced testicular weights, and low sperm counts. Epidemiological data indicates the possibility of similar effects in humans. The aryl hydrocarbon (Ah) TCDD activates the Ah receptor. TCDD exposure impairs testosterone biosynthesis. The resultant low testosterone level is not accompanied by an increase in LH. This is due to an alteration in the feedback mechanism controlling LH. Dietary phytoestrogens can influence human fertility. These compounds can act as estrogen agonists and antagonists. As a component of alcoholic beverages, phytoestrogens may be involved in the induction of altered reproductive function in chronic alcoholics. Chronic alcohol consumption is associated with hypoandrogenism resulting in testicular atrophy, hypoplastic spermatogonia, and impotence.

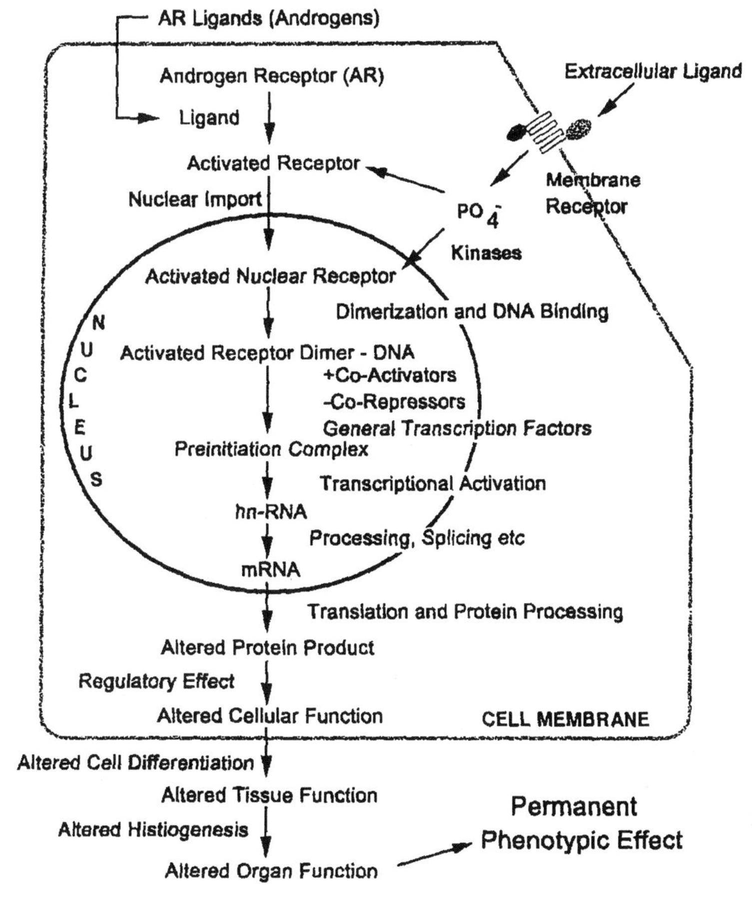

FIGURE 22.4 Cascade of cellular responses to androgens.

Some EDs bioaccumulate and become incorporated into the tissues and fat of animals and humans. A developing embryo of an exposed animal or human may exhibit developmental defects subsequent to exposure to EDs released by the tissues over time. Combinations of two or more endocrine disruptors may act synergistically to elicit abnormal reproductive development and function.

The normal pattern of endogenous hormone production and maintenance can be altered, resulting in altered homeostasis of hormones. Some endocrine disrupting contaminants induce monooxygenases. Monooxygenases are involved in the production of enzymes involved in the biosynthesis of sex steroids as well as influence the degradation of hormones. Xenobiotics capable of modulating monooxygenase activity alter the level of sex hormones in the body. Polycyclic chlorinated hydrocarbons are capable of inducing monooxygenases, thus increasing the clearance of endogenous steroids. Microsomal activation of n-hexane to 2,5-hexanedione is associated with the induction of testicular atrophy. 2,5-hexane disrupts sertoli cell microtubules and seminiferous tubule secretion.

The primary effect of endocrine disruption in males is a reduction in sperm production and also in the sperm's ability to fertilize an egg. The mechanism of ED-induced sperm count reduction is not fully understood. Alterations in hormone activity in the developing male embryo or neonate can affect Sertoli cells. Studies of men occupationally exposed to dioxin revealed a decreased level of circulating testosterone, an effect also seen in experimental rats exposed to dioxin. The reduced testosterone levels observed are caused by decreased secretion from testes and increased metabolism due to induction of the monooxygenases. Rat dams exposed to minute levels of dioxin (64 ng/kg), give birth to male offspring exhibiting decreased sperm count, altered sexual behavior, and shortened penises at puberty. Prostate enlargement in older human males is associated with exposure to estrogenic xenobiotics. However, the exact mechanism of the prostate enlargement is not fully elucidated.

22.3.1c Cellular and Genetic Effects

Toxicants may interfere with chromosomal rearrangement, meiotic activity, and DNA synthesis and replication. Such toxicants are particularly toxic to rapidly dividing cells. The antibiotics netropsin and distamycin A form complexes with DNA causing an inhibition of DNA and RNA synthesis. Actinomycin D disrupts the structure of DNA by intercalcating with DNA thus altering DNA function. The potent mutagen ethylnitrosourea induces reversible sterility in the mouse. Cytotoxic heavy metals (Pb, Hg, Cd) bind to phosphate and sulphydryl groups within proteins thereby disrupting the activity of various enzymes including DNA polymerase. These metals also interfere with the activity of monooxygenases.

22.3.1d Alterations in Sexual Behavior

Drugs acting on the central (CNS) or peripheral nervous system (PNS) have been shown to affect libido, potency, and ejaculatory function. Tricyclic antidepressants are anticholinergic causing impotence. Benzodiazipines decrease libido and impair ejaculation at the central level as well as by inhibiting peripheral autonomic nerve transmission. After prolonged use, CNS acting drugs of abuse (i.e., marijuana, alcohol, LSD) and anabolic steroids affect libido and potency in men. Industrial compounds such as boric acid, kepone, and carbon disulphide also affect libido and/or potency.

22.4 PHYSIOLOGY OF FEMALE REPRODUCTIVE TRACT

22.4.1 Hypothalamic-Pituitary-Ovarian Axis

As described above for the male, the female hormonal system consists of three primary levels: release of gonadotropin-releasing hormones from the hypothalamus, release of FSH and LH from the anterior pituitary, and release of estrogen and progesterone from the ovaries. Ovaries that are not stimulated by the pituitary hormones remain inactive as during early childhood. At puberty, the pituitary begins to secrete more gonadotropic hormones initiating the monthly sexual cycles.

As a follicle grows, it secretes increasing levels of estrogen. Estrogen has a negative feedback effect on the hypothalamus. This negative feedback results in a decrease in the release of FSH from the anterior pituitary gland. The principal site of ovarian hormone action is the uterus. The hormones affect changes in the uterus enabling conception and fetal development. The oocytes are formed before birth

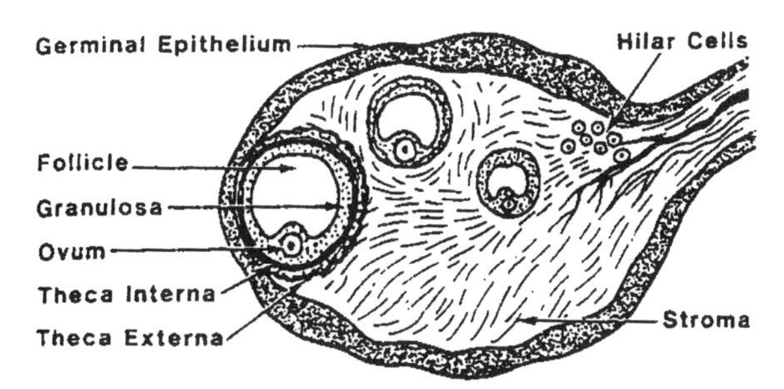

FIGURE 22.5 Development of the Graafian follicle.

Source: Adapted from Thomas, J.A. *Toxic responses of the reproductive system*. In: Klassen, C.D. (Ed) *Casarett and Doull's Toxicology, The Basic Science of Poisons*, 5th Ed. New York, McGraw-Hill, 1996.

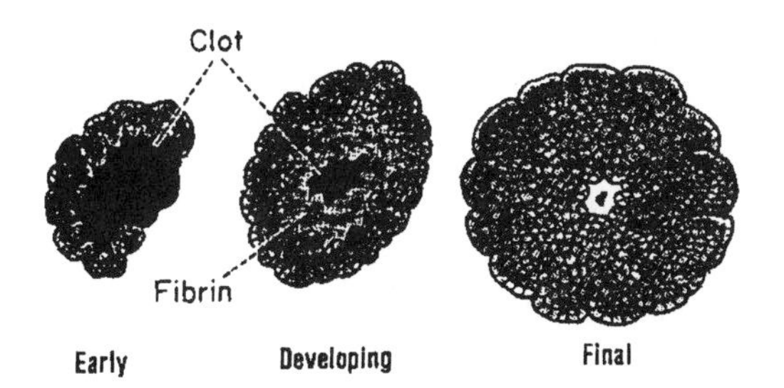

FIGURE 22.6 Formation of the corpus luteum.

Source: Adapted from "*Female physiology before pregnancy, and the female hormones*." In: Guyton, A.C. *Textbook of Medical Physiology*, 7th Ed. Philadelphia, W.B. Saunders, 1986.

and develop to the primary oocyte stage (after meiosis). At onset of puberty, gonadotropin release stimulates the oocytes to the typical mammalian graafian follicle (Fig. 22.5)

22.4.2 Effects of Hormonal Action on Reproductive Tissues

Estrus is the period when the female is most receptive to the male (coinciding with high levels of circulating estrogen). Rodents are considered polyestrus since they have a succession of estrus cycles. Dogs are monestrus; they exhibit one long, sustained estrus once or twice a year. Humans and higher primates experience cycles at monthly intervals. Most mammals ovulate spontaneously, whereas others must undergo induced ovulation (stimulated by mating).

Release of gonadotropin-releasing hormones by the hypothalamus is controlled by hormones (especially estrogen) secreted by the ovaries. Via a negative feedback loop, released estrogen causes a decrease in the release of FSH from the anterior pituitary. Follicular growth is stimulated by tonic release of FSH. As estrogen release increases, a reflex discharge of FSH and LH release occurs. This results in a release of a preovulatory LH surge. The increased LH concentration facilitates oocyte maturation and preovulatory progesterone. The dominant follicle ruptures and releases the oocyte (ovulation). After ovulation, the dominant follicle collapses leading to granulosa cell proliferation. This results in formation of the corpus luteum, the matured ovarian follicle that has discharged its ovum (Fig. 22.6). The primary hormone released by the corpus luteum is progesterone. If fertilization of the ovary occurs, breakdown of corpus luteum is inhibited. However, if fertilization does not occur, luteolysis of the corpus luteum occurs halting its endocrine function (involution). Estrogen increases uterine blood flow and stimulates endometrial development. Progesterone promotes secretory changes in the uterine endometrium in preparation for implantation of a fertilized ovum. Estrogen and progesterone act synergistically to prepare the uterus for implantation of the ovum. This is characterized by increased uterine vascularity and permeability and discharge of secretory products into the endometrial lumen of the glands allowing implantation.

The primary function of estrogen is to affect cellular proliferation and growth of reproductive tissues. The estrogens cause the fallopian tubes, uterus, and vagina to increase in size. The vaginal epithelium changes from cuboidal to stratified. The stratified form is more resistant to trauma and infection. Estrogens cause the glandular tissue of the fallopian tubes to proliferate and increase the number of ciliated epithelial cells lining the tubes. Estrogens affect the breasts by facilitating the development of the stromal tissues, ductile system growth, and fat deposition in the breasts. Progesterone and prolactin cause the growth of the lobules and alveoli of the breast tissue. During the female's reproductive years, the growth rate becomes rapid for several years. This is due to the effect of estrogens on increased osteoblastic activity. In addition, estrogens cause the early uniting of the epiphyses of the long bone shafts. Furthermore, estrogens cause deposition of fat in the breasts and subcutaneous tissues. One of the chief differences between the protein anabolic effect of estrogens and that of testosterone is that estrogen effects specific organs whereas testosterone's effects are generalized throughout the body.

As in the male, stimulation of the central and autonomic nervous system affect sexual behavior. The females' erectile tissue is controlled by the parasympathetic system. This parasympathetic action causes secretion of mucus by the vaginal epithelium.

22.5 MECHANISMS AND TARGETS OF FEMALE REPRODUCTIVE TOXICANTS

Reproductive toxicants alter normal functions of the cells/organs of the reproductive system. These xenobiotics may prevent ovulation or impair ovum transport or its fertilization and implantation. Toxicants may act by mimicking endogenous hormones of the hypothalamic pituitary ovarian axis. Other compounds may be directly cytotoxic. Compounds such as PCBs induce monooxygenases due to their hormone agonistic properties. Oral contraceptives mimic endogenous ovarian hormones thus suppressing gonadotrophin release. The ovaries have microsomal monooxygenases, and transferases capable of metabolizing xenobiotics.

22.5.1 Specific Examples

22.5.1a Effects on the Ovaries and Uterus

Alkylating agents (cyclophosphamide, vincristine) and radiation can lead to gonadal dysfunction. This dysfunction may be characterized by premature menopause and sexual dysfunction. The cytotoxic cyclophosphamide induces ovarian dysfunction or failure leading to amenorrhea or abnormal hormone levels. This compound has an affinity for resting oocytes. The age-dependent effects of cyclophosphamide are due to the age-related differences in the detoxification pathways for the compound. Busulphan treatment is associated with ovarian atrophy. This effect also appears to be age-related. Premature ovarian failure can be induced in offspring exposed in utero by active metabolites such as 6-mercaptopurine.

Polycyclic aryl hydrocarbons (PAHs) are converted into active metabolites that adversely affect oocytes directly or by affecting granulosa cells that support the ovary located within the follicle. Exposure to the conceptus in utero may destroy the oocytes of the developing female. PAHs in cigarette smoke may destroy oocytes at an earlier than normal age leading to premature onset of menopause in women who smoke. Di-(2-ethylhexyl) phthalate (DEHP) affects preovulatory follicles and granulosa cell differentiation. This results in hypoestogenism, hypoprogestinism, and anovulatory cycles in experimental rats. Via multiple modes of action, ovarian toxicants are responsible for endocrine disruption, early ovarian failure and infertility.

TCDD exposure may be linked to endometriosis, a painful disease currently affecting 10% of reproductive-age women. Endometriosis causes fragments of uterine lining to migrate to other pelvic organs and can cause infertility, internal bleeding, and other serious reproductive abnormalities. Certain PCBs (i.e., arochlor), kepone, and DDT can induce uterine growth in immature rats. The antiestrogens tamoxifen and clomiphene can inhibit uterine decidual induction in pseudopregnant rats. Antiandrogens can also affect uterine function. Hydroxyflutamide has been shown to suppress decidualization in pseudopregnant rats. However, hydroxyflutamide can delay implantation, fetal development, and parturition in pregnant rats.

22.5.1b Endocrine Disruption

As described previously for the male, reproductive toxins may act as agonists or antagonists of endogenous hormones thus affecting normal reproductive endocrine function. Endocrine disruptors (EDs) consist of synthetic and naturally occurring compounds that can adversely affect the normal hormonal homeostasis and functions in animals. Female species may be more prone to higher levels of accumulation of pollutants than males because they generally have a higher percentage of body fat. Pregnant and lactating women are more likely to mobilize contaminants stored in fatty tissues. This can then result in substantial fetal and neonatal exposure. One of the first recognized synthetic EDs was diethylstilbestrol (DES), a pharmaceutical product administered to pregnant women in the 1950s to prevent miscarriages. It induced vaginal carcinomas in female offspring (DES will be discussed in Section 22.7).

Because women are normally exposed to estrogen, the effects of EDs on females are more difficult to track due to the existence of an estrus cycle and the resulting differences in circulating hormone concentrations at different stages of the cycle. Estrogenic substances produce accelerated sexual maturation in experimental mice, irregular estrous cycles and prolonged estrous. Female mice also displayed masculine behavior in conjunction with the other effects. The presence of estrogen mimicking compounds in adult women can impair reproductive function by altering natural hormone cycles. This would adversely affect the females' reproductive potential.

Breast cancer may be linked to the estrogenic contaminants. Fat and fiber alter the intestinal resorption of estrogen. High-fat diets increase circulating estrogen levels whereas high-fiber diets are known to decrease estrogen levels. Exposure to estrogen-mimicking compounds can subsequently increase the potential for increased estrogen levels. Certain environmental EDs may function as promoters or inducers of carcinogenesis. Among estrogenic xenobiotics, PCBs and DDT are persistent in the environment. Serum DDE (a DDT metabolite) levels have been found to correlate with breast cancer incidence. However, work is ongoing concerning these findings. Experimental animal models as well as epidemiological studies in humans have revealed that estrogens induce the formation of DNA adducts that can lead to the cause of DNA mutations.

TCDD, an endocrine disruptor, has been used as a model to study the toxicity of halogenated aromatic hydrocarbons. TCDD induces a number of phase I and phase II enzymes in addition to altering the expression of various genes including TGFα, protooncogenes (*c-fos* and *c-jun*), and estrogen receptor gene. TCDD binds the Ah receptor leading to release of heat shock protein 90 from the receptor. The AhR-TCDD complex inters the nucleus and dimerizes with AhR nuclear translocator (ARNT). This resulting complex is a transcription factor that affects gene transcription of xenobiotic metabolizing enzymes via binding to dioxin response elements. The induced enzymes may metabolize endogenous compounds necessary for normal cellular function.

22.5.1c Cellular and Genetic Effects

Similar to effects on the male reproductive system, toxicants may affect normal chromosomal structure, meiosis, and DNA synthesis and replication. Mutations induced by radiation and mutagenic compounds can be inherited via the gametes. The resultant abnormalities are dependent on the specific mutation as well as the extent of chromosomal abnormalities. The sensitivity to genetic effects are dependent on the age of the animal exposed and the stage of germ cell maturation.

22.5.1d Alterations in Sexual Behavior

In most mammals, sexual activity is linked to ovulation. The use of oral contraceptives and psychoactive drugs (methadone, cannabis, alcohol) induces changes in female libido. Xenobiotic-induced ovarian failure in women has been associated with a decrease in libido. The antiandrogen cyproterone acetate (therapeutic agent for hirsutism), is also associated with ablation of libido in women. Fenfluramine, used to combat obesity, produces loss of libido. Certain xenoestrogens (i.e., kepone and methoxychlor) can masculinize female rats. These masculinized rats do not ovulate, lack stimulation of the LH surge, and exhibit male sexual behavior. Some narcotics cause a decrease LH levels. This subsequently impairs oocyte maturation, ovulation, and inhibits the FSH-LH feedback loop. Chronic marijuana use is associated with impaired reproductive function. Tetrahydrocannabinol may inhibit FSH and LH secretion causing secondary effects on the ovary. Among other reproductive effects, decrease in libido has been associated with chronic TCH exposure.

22.6 ELUCIDATING OF MECHANISMS OF REPRODUCTIVE TOXICITY

The general approach involves defining the target organ/cell in vivo and performing metabolism and pharmacokinetic studies, followed by the study of the biochemical and molecular mechanisms at the target cell level. Due to the natural complexity of the female reproductive physiology, this process is more difficult to perform for the female than the male.

22.6.1 Selected Current Approaches

22.6.1a Transillumination-Phase-Contrast Microscopic Techniques (Evaluation of Male Germ Cell Toxicity and Mutagenicity)

As previously described, spermatozoal development occurs in three major phases, production of spermatogonia, spermatocyte formation, and spermatid maturation. These stages occur in a wavelike fashion within the seminiferous tubules. This wave can be realized by transilluminating freshly isolated unstained rat seminiferous tubules. A microdissection technique was developed using transillumination. This approach is useful in detecting cyclic hormonal events in seminiferous tubules. Combining this technique with phase-contrast microscopy can allow the study of mutagenic affects on spermatogenesis. A rat testis is placed on a petri dish followed by dispersing (with forceps) of the tubules under a transilluminating stereomicroscope. Different spermatogenic stages have characteristic absorption levels.

22.6.1b Computer-Assisted Sperm Analysis (CASA) of Rodent Epididymal Sperm Motility

Sperm motion data can provide information regarding the adverse effects on spermatogenesis. The CASA technique allows the evaluation of individual sperm tracks in a rapid and repeatable fashion. A motility analyzer (e.g., Hamilton-Thorn-2000 Motility Analyzer) is used for this technique. Sperm samples are prepared and placed in the sample chamber of the motility analyzer. The sample is videotaped and viewed at a later session. This approach allows the measurement of sperm motility as a means of assessing reproductive toxicity.

22.6.1c Assessment of Ovarian Toxicity via Follicle Quantitation and Morphometrics

As stated earlier, the assessment of reproductive toxicants on the ovary can be difficult due to the complex endocrine control associated with the normal ovarian cycle. Determination of follicle numbers and ovarian morphometry can serve as indicators of normal and altered ovarian responses. A video camera transmits the analog image of the ovarian section to a computer. The image is digitalized and displayed on the video monitor. Follicle counts can be made and differential counts stored. The area and volume of sections can be calculated.

22.6.1d Isolation and Culture of Mouse Uterine, Epithelial, and Stromal Cells

The mechanisms by which reproductive hormones regulate uterine cell function is not fully elucidated. Purified uterine cell cultures that are hormonally responsive are valuable for the study of hormone action. Uterine epithelial cells cultured on basement components undergo a process of morphological differentiation (development of gland-like structures and polarity). This system more closely resembles the in vivo situation, and may exhibit similar responses to xenobiotic and endogenous steroids. One can then apply immunohistochemical, genetic, and gene expression techniques to study the effects of steroid hormones on proliferation and differentiation.

22.7 MECHANISMS OF DEVELOPMENTAL TOXICITY

Developmental toxicology refers to any morphological or functional alteration caused by chemical or physical insult that interferes with normal growth, homeostasis, development, differentiation, and/or behavior. Teratology is a specialized area of embryology comprising the study of the etiology of abnormal development or the study of birth defects. Teratogens are xenobiotics and other factors (stress, food and water deprivation, nutritional deficiencies that cause malformations in offspring. The proximal teratogen is often known, but knowledge of how the biochemical effects result in malformations is lacking.

22.7.1 Overview of Embryonic Development

Elucidating mechanisms of developmental toxicity is facilitated by understanding current available information regarding normal development. The fertilized oocyte (zygote) implants into the endometrium. The zygote divides into a solid mass of cells called the morula. This then develops into a blastocyst stage. This blastocyst consists of a layer of cells surrounding a fluid-filled cavity. A mass of cells appears in one region of this blastocyst and is referred to as the inner cell mass (which gives rise to the embryo proper). This is followed by a process called gastrulation. Gastrulation is characterized by the formation of three germ layers (ectoderm, mesoderm, and endoderm). Cells proliferate and migrate along the primitive streak and progressively to the endoderm. This process of cell movement is involution. Organ structures begin to develop during the period of organogensis. Continued development leads to the formation of the early fetus which continues to grow and differentiation prior to birth.

22.7.2 Principles of Teratology

In 1959, James Wilson proposed six principles of teratolgoy:

1. *Susceptibility to teratogenesis depends on the genotype of the conceptus and the manner in which this interacts with adverse environmental factors.* A given teratogen can cause species-specific effects. As seen with thalidomide, humans and rabbits are affected similarly (discussed later), whereas rodents exhibit a completely different pattern of defects and sensitivity.
2. *Susceptibility to teratogenesis varies with the developmental stage at the time of exposure to an adverse influence.* Gestational stage of exposure is a determinant of fetal outcome. This is referred to as critical periods of exposure. There are three major periods of development in mammals: preimplantation (resistant or susceptible to embryotoxicity), organogenesis (high susceptibility to

TABLE 22.1 Major Fetal Outcomes Depend on Stage of Pregnancy Affected.

Stage of Exposure	Outcome(s)
Preimplantation	Embryonic lethality
Implantation → time of organogenesis	Morphological defects
Fetal → neonatal stage	Functional disorders, growth retardation, carcinogenesis

dysmorphogenesis), and the fetal period (minimal susceptibility to defect induction) (Table 22.1). At the early pre-implantation stage, these embryos may be capable of repair of toxicological insult (esp. at low doses) due to totipotency retained at this stage. Larger doses of a given compound can lead to severe growth retardation, embryonic death, or failure of implantation. The period of rapid cell division and differentiation (organogenesis) is most susceptible to teratogenic insult. Toxicant exposure during the fetal period is primarily characterized by growth retardation, functional impairment, and fetal death. At this stage most of the organs are completely formed with continued growth as the primary activity.

3. *Teratogenic agents act in specific ways (mechanisms) on developing cells and tissues to initiate sequences of abnormal developmental events (pathogenesis).* Developmental toxicants can elicit effects via a number of mechanisms:

   - Altered gene expression
   - Alterations in programmed cell death
   - Oxidative stress
   - Altered signal transduction
   - Altered cell membrane function and structure
   - Impaired enzymatic processes
   - Alterations in DNA and RNA synthesis
   - Chromosomal aberrations
   - Mitotic impairments
   - Mutations
   - Epigenetic effects (i.e., altered DNA methylation)

   Various compounds such as retinoic acid, purine, and pyrimidine analogs may alter normal gene expression patterns. Cells with altered gene expression may be removed via programmed cell death (apoptosis).

4. *The access of adverse influences to developing tissues depends on the nature of the influence (agent).* As previously stated, a number of different processes/agents may be teratogenic: xenobiotics and other factors (stress, food and water deprivation, nutritional deficiencies). Some agents are cytotoxic (e.g., cyclophosphamide) causing early embryonic death or diffuse anatomical anomalies. Restraint stress has been shown to induce axial skeletal malformations in mice. Hyperthermic exposures of rat enbryos in vitro results in morphological alterations and growth impairment. Limb and neural tube defects are associated with folic acid deficiency. Other maternal factors

(metabolic disturbances) can also lead to dysmorphogenesis and functional defects.

5. *The four manifestations of deviant development are death, malformation, growth retardation, and functional deficit.* A given exposed litter may manifest more than one possible outcome. Most teratogens induce a combination of outcomes in a test group. Subsequent to toxicant exposure during organogenesis, a particular litter may contain dead and malformed (living) embryos. Any of three possible outcomes may occur: lethality, growth/functional retardation, and malformations. This pattern is primarily elicited by compounds such as cytotoxic agents (affecting processes at the DNA level), and chemotherapeutic agents (affecting processes involving rapid cell division). At later stages (i.e., the fetal stage), functional deficits and growth retardation predominates. This pattern is primarily elicited by compounds not considered teratogenic. They are considered fetotoxic or lethal. These compounds affect cellular functions such as glycolysis, cellular respiration and cell membrane integrity.
6. *Manifestations of deviant development increase in frequency and degree as dosage increases from no effect to the totally lethal level.* The dose of a teratogen may influence pregnancy outcome (i.e., embryotoxic vs. teratogenic). Generally, as the dosage of a teratogen is increased, the severity and frequency of malformations increase (as well as potential for embryonic and fetal death).

22.7.3 Teratogens and Their Sequelae (Historical Examples)

22.7.3a Thalidomide

Thalidomide was previously used as a sedative in Europe. It is being studied in the United States as a treatment for AIDS and is currently used in South Africa as a treatment for leprosy. Thalidomide exposure causes phocomelia (severe shortening of the limbs) in humans and rabbits. These effects are not seen in commonly used laboratory rodents. Although the mechanism has not been definitively elucidated, much progress has been made in determining the primary biochemical process involved in the teratogenicity of this compound:

1. Significant down-regulation of surface adhesion receptors on early limb bud cells during early organogensis (primate embryos). There is still the need to determine why this effect is specific for limb buds. Further, structure activity studies indicate that an intact phthalimidine group is needed for teratogenicity.
2. Possibly targets the cells of the mesoderm or specific genes associated with mesodermal differentiation because limb defects are usually accompanied by kidney and heart defects.
3. Elimination of the mesonephros (primative kidney) results in limb reduction defects. This suggests that the mesonephros is involved in the regulation of initial limb bud proliferation. Further, thalidomide appears to bind to the human mesonephros.
4. Fetuses affected by thalidomide exhibit reductions in dorsal root ganglia and their neurons. If these neurons are necessary for maintaining limb development, thalidomide may interfere with or destroy the neurons.

5. Thalidomide teratogenicity is ameliorated by acetylsalicylic acid. This protective mechanism suggests that thalidomide may be bioactivated by prostaglandin H synthase (prostoglandin endoperoxide synthase).
6. Co-treatment of rabbits with thalidomide and a P450 inhibitor results in an increase in skeletal malformations. However, soft tissue defects were reduced (more evidence of the role of bioactivation).

22.7.3b Valproic Acid (2-Propylpentanoic Acid)

Valproic acid (VA) is an anti-epileptic drug legalized in 1978. This drug alters the synthesis of the neurotransmitter gamma-amino butyric acid (GABA). VA prevents complete closure of the neural tube. In humans this results in the development of fetuses with spina bifida (absence of vertebral arches with protrusion of spinal cord tissues). In the milder form (s. bifida occulta) the neural tube does not protrude. The opening is covered by hair or is seen as a dimple. In mice, exencephaly is the usual result. The mechanism has not been fully elucidated, but studies to date indicate that the yolk sacs of exposed animals in vitro fail to vascularize properly. Most recent studies have shown that VA restricts proliferation in the mid-G1 phase of the cell cycle and alters the prevalence and/or glycosylation state of cell surface glycoproteins, which have the potential to mediate cell-cell and cell-matrix interactions critical to development.

22.7.3c Cyclophosphamide (CP)

First used as a defleecer for sheep, cyclophosphamide (CP) is now used as a treatment for lymphocytic leukemia and in the prevention of transplanted organ rejection. Children born to mothers receiving the drug during the first trimester exhibit ectrodactyly (absence of all or parts of digits), cardiac defects, and decreased fertility. In rats, effects include chondroplasia and cleft palate. Via the use of embryo culture systems, it has been shown that hepatic fractions and cofactors are needed to elicit abnormal development of CP. CP must be metabolically activated to be teratogenic. Metyrapone and carbon monoxide inhibit CP activation. This indicated the involvement of P-450 monooxygenases. The resultant toxic metabolites are acrolein (toxic) and phosphoramide mustard (highly toxic). Treatment of rats on gestation day 10 causes an S-phase cell cycle block and widespread cell death in the embryo. This cell death is observed in areas of rapid cell proliferation.

22.7.3d Retinoic Acid (RA)

Also called isotretinoin or accutane, these are vitamin A derivatives and are used to treat acne. Therapeutic doses are teratogenic. There have been two cited cases of retinoid teratogenesis in humans: renal anomalies including hydronephrosis. In rodents, the compound causes spina bifida, cleft palate, and a variety of other skeletal defects. RA has been thought of as a morphogen because it causes limb duplications when applied to chick embryonic limb buds. Up-regulation of the patterning gene *sonic hedgehog* mimics this effect. Search for the actual mechanism is ongoing. Other studies have shown that down-regulation of retinoic acid receptors accompany RA-induced teratogenesis. The decrease in receptors may allow increased amount of free RA available to cellular or molecular targets, thus leading to the dymorphogenesis. RA has been shown to alter *Hox* (homeobox) gene expression. Mutations in *Hox* genes lead to the development of a tissue/organ into the likeness

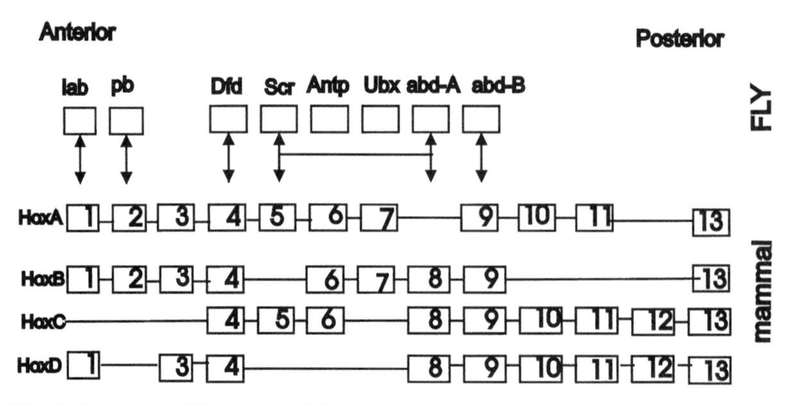

FIGURE 22.7 The fly (*Drosophila*) HOM complex and the mammalian Hox complexes. *Source*: Adapted from "*Cellular mechanisms of development*." In: Alberts, B., Bray, D., Lewis, J., Raff, M., Roberts, K., Watson, J.D. *Molecular Biology of the Cell*, 3rd Ed. New York, Garland Publishing, Inc. 1994.

of another. *Hox* genes are chromosomal clusters that are homologous to the Antennapedia and bithorax complexes of *Drosophila* (Fig. 22.7) and are involved in embryonic pattern formation.

22.7.3e Ethanol

Fetal alcohol syndrome (FAS) is characterized by varying degrees of growth and mental retardation. The syndrome is accompanied by subtle facial abnormalities: shortened palpebral fissure, epicanthal folds, broadened nasal bridge, thinning of upper lip, and upturned nose. These facial effects must be present for a diagnosis of FAS. More severe defects may occasionally be observed (i.e., microcephaly and hydrocephaly). Associated with facial abnormalities are decreased brain size, reduction or agenesis of the corpus callosum, and decrease in the size of the basal ganglia. A diagnosis of fetal alcohol effect (FAE) is made for the more subtle manifestation of embryonic ethanol toxicity. Hypothalamic-pituitary pathways may also be affected in ethanol exposure. However, the effects may be direct or maternally mediated.

A possible mechanism involves cell death by ethanol-induced superoxide radicals. It has been shown that the superoxide inhibitor, superoxide dismutase, prevents alcohol-induced cell death. Another possible mechanism involved the prevention of proper function of the cell adhesion molecule (L1). Ethanol blocks (in vitro) the adhesive functions of L1 proteins. Also, human L1 gene mutations can cause mental retardation, syndromes and malformations similar to that seen in severe cases of fetal alcohol syndrome.

22.7.3f Diethylstilbestrol (DES)

DES is a synthetic estrogen used from 1940 to 1970 for pregnancy maintenance. Effects were not seen in offspring until they reached puberty. In females, exposure induced vaginal adenocarcinoma (transplacental carcinogenesis) and adenosis and cervical erosion. In males, hypotrophic testes, hypospadias, and poor semen volume and quality resulted. Recent studies indicate that females exposed to DES in utero can transmit neoplastic phenotypes to normal animal blastocysts. Developmental exposure to DES impairs normal differentiation of the Müllerian duct and normal regression of the Wolffian duct. Neonatal exposure to DES causes demethylation of an estrogen-responsive gene in the mouse uterus.

22.7.3g Male-Mediated Developmental Toxicity

There have been reports of male-mediated genetic effects on fetuses. The genetic make-up of the paternal gametes is a determinant of the viability of offspring. Epigenetic changes are also of importance. Genetic imprinting (a process by which certain genes are expressed differentially based on whether they are maternal or paternal) influence fetal outcome. Xenobiotics and other factors that alter normal imprinting in male gametes may impair normal embryonic development. Further, toxicants may be transmitted to the oocyte via the semen or exposure to the maternal environment. Vinyl chloride and Pb-exposure in men has been associated with spontaneous abortions among the wives of exposed men. Occupational exposure of men to ionizing radiation has been linked to leukemia in their offspring. This is due to the transmittance of altered and aberrant spermatozoal genetic material.

22.8 ELUCIDATING MECHANISMS OF TERATOGENESIS

There is abundant data at the gross and cellular level that characterizes teratogenic mechanisms at these levels. However, molecular data remains scarce.

22.8.1 Selected Current Approaches

22.8.1a Subtractive Hybridization and Differential Display

Subtraction hybridization (SH) is a strategy that allows enrichment of specific nucleotide sequences prior to DNA cloning. It allows the detection of expression differences in one cell or tissue as compared to another.

Differential display has the same goal as SH but all cDNA of a given tissue is compared with another. The objective is to locate the DNA band(s) that are unique to one of the tissues. The band(s) that differ in intensity (decrease or increase in constitutive expression) must also be identified.

22.8.1b In Situ Hybridization (Histological and Whole Mount Techniques)

In situ hybridization involves the specific annealing of a labeled nucleic acid probe to complementary sequences retained "in situ" in a fixed tissue. The location of the probe is then visualized. The approach allows the monitoring of qualitative changes in gene expression (anatomic and temporal). With the histological technique, tissues are fixed, embedded in paraffin, and sectioned. These are then mounted on slides. With the whole mount technique, these fetuses are not sectioned or embedded in paraffin. This gives a three-dimensional view of the expression pattern.

Messenger RNA transcripts in embryonic tissues can be detected. One can monitor alterations in embryonic gene expression in response to chemical reagents in an attempt to understand mechanisms which lead to birth defects. For example, retinoic acid receptors play an important role during pattern formation in embryonic development. Pregnant mice can be treated with teratogenic doses of retinoic acid on the susceptible day, then embryos collected at different stages following the treatment. Untreated embryos of the same developmental stages are also collected. Ribonucleic acid probes (riboprobes) representing the above genes can be synthesized and radiolabeled. (Nonradioactive labels (biotin, digoxigenin) can also be used especially for whole mount techniques). The sections or whole embryos are incubated with the labeled probe. Via microscopy, one can determine if there are changes in the expression patterns of the genes of interest by comparing the treated and untreated specimen.

22.8.1c In Situ Transcription

With this approach, tissue is fixed as explained earlier for in situ hybridization. An oligo-dT primer is hybridized to the tissue. This primer will bind to poly-A sequences (mRNA). The mRNA is reverse transcribed to cDNA. The mRNA-cDNA hybrids are eluted via alkaline hydrolysis. This product can be amplified using uracil instead of thymidine to obtain RNA. One can use this to construct a library for further analysis.

22.8.1d Antisense Technology

This is a technique being investigated for its use in studying gene function during organogenesis. The antisense probes are DNA oligonucleotides that are complimentary to mRNA. This is done in cell, organ, or whole embryo cultures. The following are possible results of antisense treatment: triplex formation with DNA (no transcription); mRNA prevented from attaching to ribosomes (no translation); or activation of RNase H (destruction of transcripts).

With the use of antibodies, protein can be localized by immunohistochemistry, mRNA can be detected (RNase protection assays, reverse-transcription PCR, Northern analysis), mRNA can be localized by in situ hybridization, and early malformations may be evaluated. This approach can yield results in less time than the use of gene targeting and transgenics, however, more than one gene can be disabled. Also, only a limited number of embryonic stages can be studied because embryo cultures can not be maintained for more than approximately 3 days. Further, one will not be able to define the eventual phenotypic manifestations as in transgenics.

SUGGESTED READING

Ballantyne, B., Marrs, T. and Turner, P. *General and Applied Toxicology: College Edition*. Macmillan, New York, 1995.

Chapin, R.E. and Heindel, J.J. *Methods in Toxicology: Male Reproductive Toxicology*. Academic Press, San Diego, CA, 1993.

Heindel, J.J. and Chapin, R.E. *Methods in Toxicology: Female Reproductive Toxicology*. Academic Press, San Diego, CA, 1993.

Hood, R.D. *Handbook of Developmental Toxicology*. CRC Press, Boca Raton, FL, 1997.

Klaasen, C.D. *Cassarett and Doull's Toxicology: The Basic Science of Poisons,* 5th Ed., McGraw-Hill, New York, 1996.

Korach, K.S. *Reproductive and Developmental Toxicology*. Marcel Dekker, Inc., New York, 1998.

Naz, R.K. *Endocrine Disruptors*. CRC Press, Boca, Raton, FL, 1997.

Wassarman, P.M. and DePamphilis, M.L. *Methods in Enzymology: Guide to Techniques in Mouse Development*. Academic Press, San Diego, CA, 1993.

CHAPTER TWENTY-THREE

Immunotoxicity

MARYJANE K. SELGRADE, DORI R. GERMOLEC, ROBERT W. LUEBKE, RALPH J. SMIALOWICZ, MARSHA D. WARD, and DENISE M. SAILSTAD*

23.1 INTRODUCTION

The immune system defends the body against infectious agents (bacteria, viruses, fungi, and parasites), which are ubiquitous in our environment. It also has a role in defending the body against certain tumor cells that may arise spontaneously or as the result of environmental insults (viral, radiation, and chemical). A crucial part of this function is the ability to distinguish endogenous components ("self") from potentially harmful exogenous components ("non-self"). The immune system is composed of several tissues and cell types. Immune cells include a variety of leukocytes, which are derived from bone marrow, circulate in blood and lymph, and are widely distributed in the body in lymphoid as well as other tissues (Fig. 23.1).

A properly functioning immune system is essential to good health. In some individuals, the immune system is compromised by primary immune deficiencies resulting from genetic defects or secondary immune deficiencies resulting from diseases (e.g., AIDS or leukemia) or drug therapies. These individuals are more susceptible to infectious diseases and certain types of cancer, the consequences of which can be life-threatening. On the other hand, the immune system mediates certain types of disease. It may react to foreign substances that would otherwise be relatively innocuous, such as certain chemicals, pollens, or house dust. The resulting allergic reactions can produce an array of pathologies ranging from skin rashes and rhinitis to more life-threatening asthmatic and anaphylactic reactions. In addition, although the immune system can generally distinguish between what is "self" and what is "non-self," in some circumstances it may mistakenly react against "self" components resulting in another type of immune-mediated health effect, autoimmune disease.

Table 23.1 describes the major consequences that may result from interactions between toxic agents and the immune system. There is evidence that xenobiotic

* This chapter has been reviewed by the National Health and Environmental Effects Research Laboratory, U.S. Environmental Protection Agency, and approved for publication. Approval does not signify that the contents necessarily reflect the views and policies of the agency, nor does mention of trade names or commercial products constitute endorsement or recommendation for use.

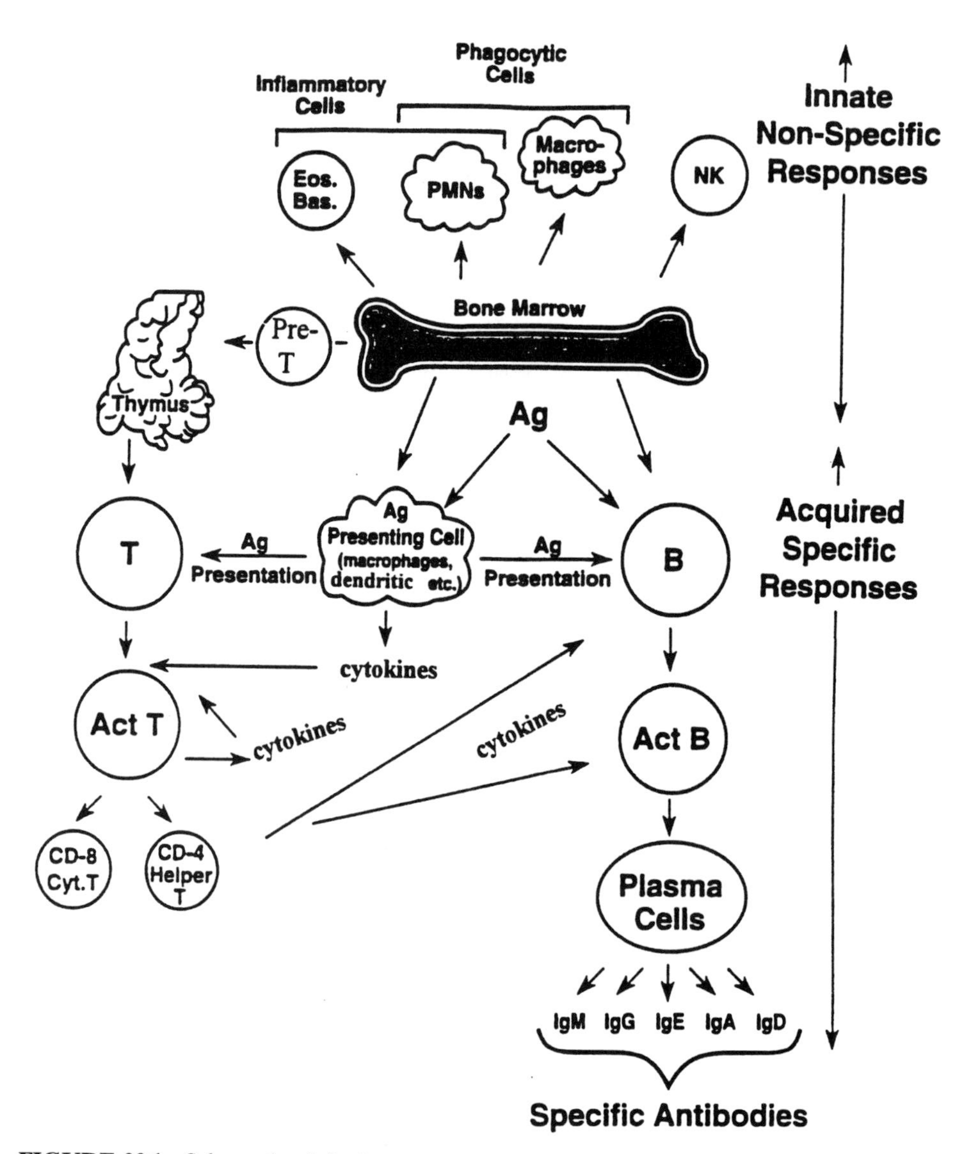

FIGURE 23.1 Schematic of the immune system.
The immune system is composed of primary lymphoid organs (bone marrow and thymus), secondary lymphoid organs (not shown) and several cell types. In addition a number of mediators including cytokines, antibodies, and complement regulate and/or are produced by the immune system. Abbreviations: Eos, eosinophil; Bas, basophil; PMN, polymorphonuclear leukocyte (neutrophil); NK, natural killer cell; Ag, antigen.

compounds, as well as other environmental stresses, such as ultraviolet and ionizing radiation, can alter the immune system either as immunosuppressants, allergens, or potentiators of allergic disease. There is less evidence about, but growing concern over the possible impact of xenobiotic exposures on autoimmune disease. A number of resources, noted under "Suggested Reading" at the end of this chapter provide information on the physiology, biochemistry, and toxicology of the immune system.

TABLE 23.1 Potential Consequences of Interactions Between the Immune System and Toxic Chemicals.

Nature of Interaction	Disease Enhanced
Suppression	Infectious Neoplastic
Stimulation	Allergic Autoimmune

In this chapter we will provide a brief overview of the immune system and immunotoxicology and then describe a variety of biochemical mechanisms underlying the toxicity of representative immunotoxicants.

23.2 ORGANIZATION OF THE IMMUNE SYSTEM

23.2.1 Cells, Tissues, and Mediators

Cells of the immune system include several different types of leukocytes (white blood cells): neutrophils (also called polymorphonuclear leukocytes or PMNs), eosinophils, and basophils/mast cells* (also collectively known as granulocytes), monocytes/macrophages,* natural killer (NK) cells (which are large granular lymphocytes), T and B lymphocytes, and plasma cells (B cells that produce antibodies (Fig. 23.1). Primary lymphoid tissues include the bone marrow, from which immune cells are derived, and the thymus, which has a major role in the differentiation of T lymphocytes. Secondary lymphoid tissues include the spleen, lymph nodes (scattered throughout the body), tonsils, and adenoids. There are also lymphoid aggregates at the three major portals of entry for environmental agents: lung, gut, and skin. These localized aggregates are referred to, respectively, as bronchus-, gut- and skin-associated lymphoid tissue (BALT, GALT, and SALT). From the preceding description it is clear that the immune system provides a diffuse target for toxic insult.

The maturation, differentiation, and mobilization of immune cells are controlled by cytokines (e.g., interleukins, interferons, and chemokines), which are soluble mediators produced by immune cells and/or by cells outside the immune system (e.g., epithelial cells and cells of the nervous system). Other soluble (humoral) mediators produced by immune cells include antibodies (immunoglobulins) and complement proteins (also produced by hepatocytes). Mediators are important in the implementation and regulation of immune responses.

23.2.2 Innate Responses

Immune responses are divided into innate responses directed nonspecifically against foreign substances and acquired responses directed against specific antigens (see Table 23.2 for important definitions). Innate immunity is generally viewed as the provider of rapid, although usually incomplete, antimicrobial host defense, whereas

* Different designations for similar cells found in blood (to left of slash) versus tissue (to right of slash).

TABLE 23.2 Important Definitions.

Term	Definition
Antigen	Molecules (generally proteins or carbohydrates) that evoke specific immune responses; usually foreign to the host
Antibody	Soluble proteins collectively known as immunoglobulins; these molecules circulate freely and react specifically with invoking antigen; subclasses include IgM, IgG, IgE, IgD, and IgA
CD + a number	CD stands for cluster of differentiation; followed by a number that designates a particular cell surface molecule
Complement	A series of nonimmunoglobulin plasma proteins that are sequentially activated by antigen-antibody complexes. They damage target cell membranes and are active in host defense.
Cytokines	Soluble substances (lymphokines and interleukins) that are secreted by cells and have a variety of effects on other cells
Hapten	Molecule that can react specifically with an antibody, but is too small to elicit an antibody response unless coupled to a protein
Hypersensitivity	Excessive humoral or cellular response to an antigen leading to tissue damage
Major Histocompatibility complex (MHC) class I and class II molecules	Cell surface molecules critical to antigen presentation

acquired immunity is a slower but more definitive response. There is considerable interaction between these two types of immunity (see Fearon and Locksley in Suggested Reading).

Macrophages and neutrophils (also called PMNs) are phagocytic cells that, as part of the innate response, engulf and in many cases destroy infectious agents and other foreign particles. This process can be nonspecifically enhanced by certain complement proteins or enhanced by antibodies specific for a particular microbe (a process called opsonization). Macrophages are also activated by certain cytokines, and once activated can kill certain tumor and virus-infected cells. Eosinophils, like neutrophils, contain lysozymes and other mediators and seem to be particularly important as a defense against helminthic parasites, which are not easily phagocytized due to size. Eosinophils are also typically present in certain types of allergic responses. In general, leukocytes have a significant role in inflammation, a major component of the body's defense mechanism. In addition to an influx of these cells, inflammation is characterized by activation of clotting mechanisms, increased blood flow, and increased capillary permeability. These responses facilitate mobilization of immune cells to the site of injury and result in the swelling and reddening associated with inflammation.

Other components of the innate response include natural killer (NK) cells and a number of cytokines. NK cells lyse cells bearing tumor or viral antigens and hence are thought to play an important role in host resistance to both neoplastic and viral disease. Type I interferons (interferon α and interferon β) are produced by a number of different cell types and appear very rapidly after viral infection. Type I interferons inhibit viral replication, inhibit cell proliferation, and increase the lytic potential of NK cells and therefore play a role in controlling viral and neoplastic disease. Several cytokines are important in the initiation of inflammatory responses. Those that have received the most attention include tumor necrosis factor α (TNF-α), interleukin (IL)-1, and IL-6. There are also a number of chemotactic cytokines (including IL-8), called chemokines, which help to mobilize immune cells to the site of injury.

23.2.3 Acquired Immune Responses

In addition to their phagocytic function, macrophages and related cells found in selected tissues (Langerhans cells in the skin, Kupffer cells in the liver, interstitial dendritic cells in the lung, and other tissues) also have an important role in the development of specific immune responses to pathogens in that they process and present antigens to T lymphocytes. The T-cell antigen receptor (TCR) recognizes and binds proteolytically processed short peptide fragments (antigens) bound to self major histocompatibility complex (MHC) molecules on the surface of an antigen-presenting cell. There are two major classes of MHC molecules (class I and class II) that present antigen to different types of T cells (CD8 and CD4, respectively, described below). In addition to the TCR, other molecules contribute to activation of T cells either by functioning as coreceptors, by increasing the avidity of the interaction with the antigen presenting cell, or by inducing separate signal transduction events that influence the cellular response. Protein-tyrosine phosphorylation is important in the initiation of cellular responses that follow TCR recognition of the MHC-antigen complex. The subsequent signal transduction events that lead to T cell activation, and the role of numerous coreceptors on the surface of both T cells and antigen-presenting cells have been the subject of intense study. Signal transduction events are described in more detail in Section 23.4.1 in conjunction with the mechanisms associated with cyclosporin A immune suppression. (Also see review by Robey and Allison, 1995.) The net result of T-cell activation is clonal expansion (proliferation) and activation of T cells specific for a particular antigen.

There are two major divisions of T cells that are distinguished by expression of different cell surface markers, CD4 and CD8. Pre-T cells migrate from the bone marrow to the thymus, which plays a key role in T-cell differentiation. As relatively immature cells, T cells express CD3 associated with the T-cell receptor (TcR) and both CD4 and CD8 molecules. As maturation progresses, these cells undergo both positive and negative selection. During positive selection, T cells are "screened" by the MHC molecules on the cortical epithelial cells of the thymus. Only cells that bind to MHC with a certain affinity survive. Cells that bind with higher or lower affinity undergo apoptosis (programmed cell death). As a result of this process, T cells become MHC restricted, that is, they will only respond to antigen presented in association with MHC. Cells that survive positive selection are potentially able to respond to self proteins. However, as the cells move from the thymic cortex to the medulla, it is believed that they undergo negative selection, during which self-reactive cells are removed or functionally inactivated. During the course of positive and negative selection, CD4+ CD8+ cells down-regulate the expression of one of these coreceptor molecules such that mature T cells express only CD4 or CD8. All continue to express CD3. Mature T cells leave the thymus and populate secondary lymphoid organs.

CD4 and CD8 T cells have different functions when activated. Cytotoxic T lymphocytes (CD8) lyse cells expressing specific viral or tumor antigens. CD8 cells have also been shown to down-regulate (or suppress) other immune responses under some circumstances, probably by the production of certain cytokines. Hence, they are sometimes referred to as suppressor T cells. CD4 helper T cells consist of two major subpopulations, illustrated in Figure 23.2. These subpopulations appear to regulate different sets of immune responses. T helper (Th)1 cells produce interleukin-2 (IL-2) and interferon-γ (IFN-γ). These cells are involved in delayed-type hypersensitivity (DTH) responses, facilitate production of certain antibody subtypes including complement-fixing antibodies, activate macrophages, and are particularly important for recognizing antigens expressed on cell surfaces, such as viral and tumor antigens. Th2 cells produce a different array of cytokines, including IL-4 and IL-5, which promote IgG1, IgA, and IgE responses, and enhance eosinophil

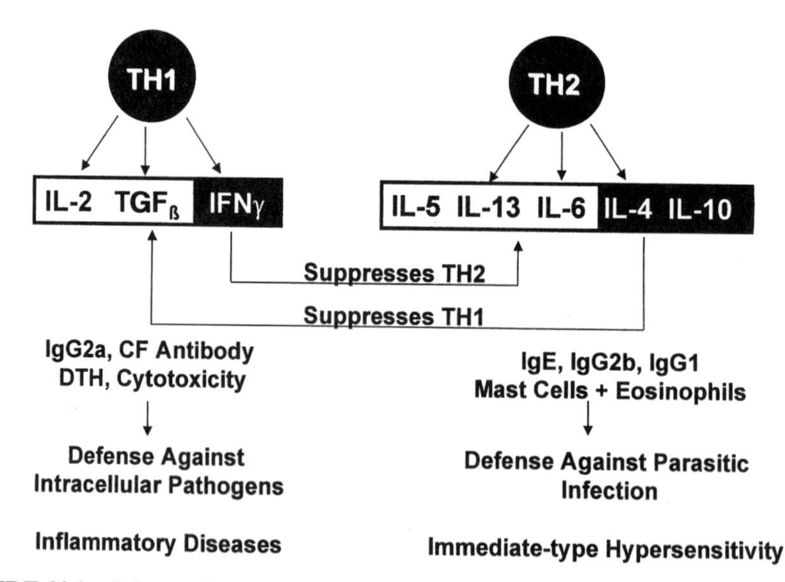

FIGURE 23.2 Sub-populations of T helper (Th) lymphocytes characterized by production of different cytokines.
The two populations are mutually antagonistic and have different roles in immune defense and pathogenesis. Abbreviations: IL, interleukin; INF, interferon; TGF, transforming growth factor; DTH, delayed-type hypersensitivity.

differentiation. Hence, Th2 cells may be particularly important in responding to certain parasitic infections, and also play an important role in immediate-type hypersensitivity, including reactions to common allergens such as pollen and dust mites. Th1 and Th2 cells are mutually antagonistic in that cytokines produced by each subtype tend to down-regulate the other. Recently, it has been suggested that there are also similar subpopulations of CD8 T cells distinguished by different cytokine profiles. (For more information see Kimber and Selgrade.)

The B-cell receptor (membrane-bound immunoglobulin) recognizes native or denatured forms of proteins or carbohydrates in soluble, particulate, or cell-bound form. B cells are particularly important in defending the host against extracellular pathogens and their toxins. Although B and T cells recognize distinct forms of antigen using very different receptors, the signal transduction events that result from the interaction of their antigen receptors with antigen are quite similar. B cells specific for a given antigen also expand clonally in response to that specific antigen, differentiate into plasma cells, and produce antibodies (immunoglobulins) that specifically bind to the eliciting antigen. There are two types of antigens that stimulate B cells: T-independent antigens, which have highly repeating antigenic determinants, cross-link immunoglobulin (Ig) receptors on the B cell surface, and activate the cell without T cell help, and T-dependent antigens, which do not have repeating determinants; i.e., T-cell help (via the secretion of important cytokines such as IL-2) is required in order to activate B cells. Most antigens belong to this latter category.

The structure of the immunoglobulin molecule is shown in Figure 23.3. All immunoglobulins consist of a basic unit of four polypeptide chains, two identical light chains and two identical heavy chains, held together by a number of disulfide bonds. The antigen binding region is located at the N-terminus. Papain digestion of the molecule results in two fragments at the antigen binding end (referred to as Fab) and a fragment (called Fc) at the C-terminus. The biologic functions of the antibody molecule derive from the properties of a constant region, which is identical for antibodies of all specificities within a particular class (isotype defined below). The constant region occupies about three quarters of the molecule starting at the C terminus. The specificity of antibodies for a particular antigen is determined by the amino acid sequence in the variable region, which differs from antibody to antibody. Phagocytic cells such as macrophages have Fc receptors facilitating phagocytosis of antigens bound to antibody. In addition to facilitating phagocytosis, antibodies may, in the presence of complement, specifically lyse bacteria, or cells bearing tumor or viral antigens on the surface, or neutralize viruses. Antibodies may also facilitate macrophage and NK cell mediated cytotoxicity.

There are several classes (called isotypes) of immunoglobulin molecules based on the structure of their heavy chains—IgM, IgG, IgA, and IgE. IgM is the predominant antibody in the primary immune response (following initial exposure to antigen). IgG appears later following a primary infection but is the predominant antibody in the secondary response (following subsequent exposure to the same antigen). IgG has been further characterized into four subclasses: IgG_1, IgG_2, IgG_3, and IgG_4. IgE acts as a mediator of allergy and parasitic immunity. IgA is found in secretions such as mucous, tears, saliva, and milk, as well as serum, and acts locally to block entrance of pathogens through mucous membranes. A fifth class, IgD, is mainly membrane bound on B cells. Its function is unknown. Whereas a given B cell forms antibody of just one single antigen specificity, during the lifetime of this cell,

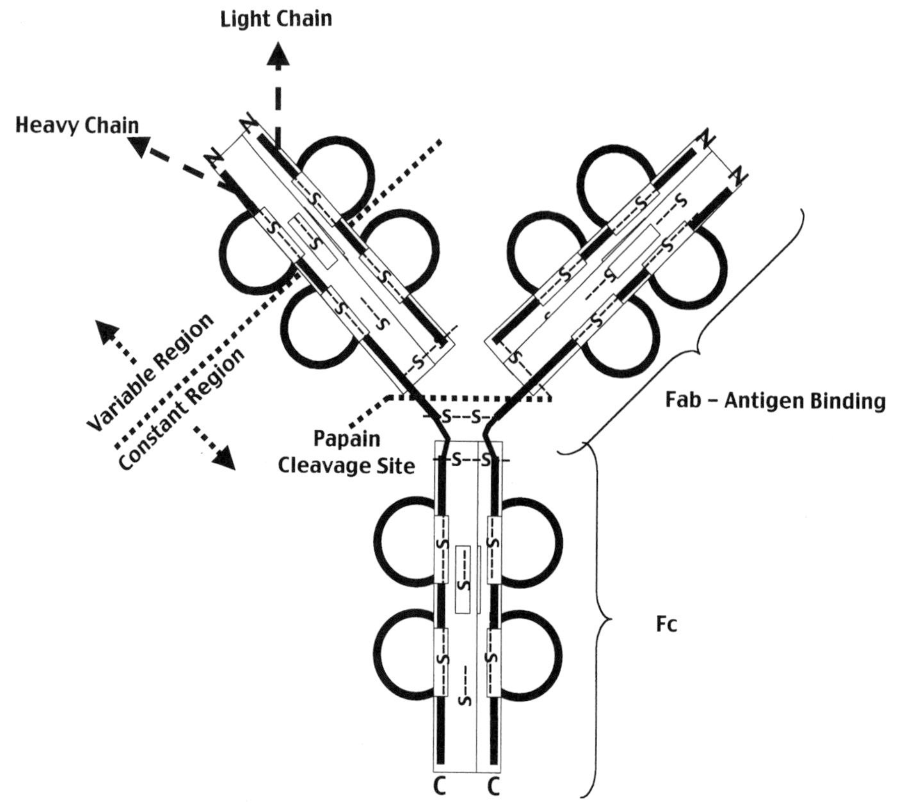

FIGURE 23.3 The structure of the antibody molecule.
All immunoglobulins consist of a basic subunit made up of four polypeptide chains (two light and two heavy) bound together by disulfide bonds.

it can switch to make a different class of antibody. For example, in allergic reactions, class switching within a given B cell clone involves progression from synthesis of IgM to synthesis of IgG1 to synthesis of IgE.

Memory is built into the immune system such that specific immune T- and B-cell responses to most antigens are activated more rapidly in cases where the immune system has encountered the antigen previously. Memory is mediated by long-lived T and B cells. It is this rapid recall that is responsible for the success of vaccination in preventing subsequent infection. T-independent antigens are an exception in that they give rise to predominantly IgM responses and relatively poor, if any, memory is generated by repeated exposure.

23.3 IMMUNOTOXICOLOGY

23.3.1 Identifying Immune Suppressants

There is ample evidence that a number of chemicals, certain microbial products (e.g., mycotoxins), and ionizing and ultraviolet radiation (UVR) can suppress various

TABLE 23.3 Selected Examples of Chemicals and Radiation That Are Immunosuppressive.

Chemical Class	Example[a]
Polyhalogenated aromatic hydrocarbons	TCDD, PCB, PBB, hexachlorobenzene
Aromatic hydrocarbons	Organic solvents
Polycyclic aromatic hydrocarbons	DMBA,[b] benzo[a]pyrene[b]
Aromatic amines	Benzindines
Heavy metals	Lead, cadmium, methylmercury
Oxidant gases	NO_2, O_3, SO_2, Phosgene[b]
Organotins	DOTC,[b] DBTC[b]
Radiation	Ionizing, ultraviolet
Mycotoxins	Aflatoxic,[c] ochratoxin A,[c] trichothecenes T-2 toxin[c]
Other	Asbestos, diethylstilbestrol (DES), dimethylnitrosamine

[a] *Abbreviations*: TCDD (2,3,7,8-tetrachlorodibenzo-p-dioxin); PCB (polychlorinated biphenyls); PBB (polybrominated biphenyls); DMBA (7,12-di-methylbenz[a]anthracene); DOTC (di-n-octyltin dichloride); DBTC (di-n-butyltin dichloride).

[b] Effects in humans are unknown; for all other compunds without superscripts, changes have been demonstrated in both rodents and humans.

[c] Effects in humans unknown, but veterinary clinicians have noted immunosuppression in livestock ingesting mycotoxins at levels below those that cause overt toxicity.

components of the immune system and enhance susceptibility to both infectious and neoplastic disease in laboratory animals. Chemical agents that have been shown in experimental studies to suppress immune function include a diverse array of environmental contaminants such as metals, pesticides, polycyclic hydrocarbons, halogenated aromatic hydrocarbons (HAHs), organic solvents, mycotoxins, and particulate and gaseous air pollutants (Table 23.3). In addition, a limited number of clinical and epidemiologic studies have reported suppression of immune function and/or increased frequency of infectious and/or neoplastic disease following exposure of humans to some of these agents.

As described above, exposure to a foreign antigen triggers a series of complex but highly integrated responses. One of the challenges in devising tests to assess chemicals for immunosuppressive effects has been to determine the most appropriate endpoints for adequate analysis of the integrity of the immune system. Because of the complexity of the immune system, tiered approaches to testing have frequently been employed. In some cases the first level of the tier relies solely on structural endpoints including changes in the weight of thymus and other lymphoid organs, histopathology of these organs, or differential blood cell counts. However, although these nonfunctional endpoints may be effective in identifying immunotoxic effects following exposure to high doses of chemicals, these endpoints by themselves are not very accurate in predicting changes in immune function or alterations in susceptibility to challenge with infectious agents or tumor cells at lower chemical doses (see Luster et al. for more information). Hence, the first tier of tests often includes a limited number of functional assays designed to assess (1) antibody-mediated responses, (2) T-cell-mediated responses, and (3) NK cell activity. Occasionally, a test for macrophage phagocytosis is included. The most commonly used immune function assay in laboratory animals assesses the ability of a mouse or rat to respond to challenge with sheep red blood cells. The response is assessed by determining the number of antigen-specific antibody-forming cells in the spleen (Jerne assay) or by assessing antigen-specific antibodies in serum using an enzyme-linked immunosorbent assay (ELISA). Because the sheep red blood cell is a T-dependent antigen, T and B cells, as well as antigen presenting cells, must be functional to have a successful immunization. In addition, T-cell function may be determined by assessing the proliferation response of spleen or lymph nodes cells. These cells may be obtained from a previously immunized and treated animal and challenged with specific antigen in vitro, or they may be taken from mice that have not been immunized and treated with nonspecific mitogens in vitro. Proliferation is usually measured via incorporation of tritiated thymidine. NK activity is usually assessed by the cytotoxic action of lymphocytes (usually spleen cells) against target cells that are particularly sensitive to NK activity.

When a more in-depth evaluation of an immunotoxicant is desired (usually in a research setting), a second tier of more sophisticated tests may be conducted. This tier may include an evaluation of animal resistance to challenge with infectious agents or transplantable tumor cells. More sophisticated immune system tests might involve flow cytometric analysis of cell surface markers to assess the proportion of T cells (total and CD4 and CD8 subsets) and B cells or to assess other cell surface markers associated with lymphocyte activation. Tier II could also include assessment of cytokine profiles by measuring messenger RNA expression, the actual cytokine proteins (by ELISA) in cell supernatants or body fluids, or cytokine function using indicator cell lines to test for cytokine activity in these fluids. For some immunotoxicants very sophisticated studies have been done to assess the impact of immunotoxicants on the cell signaling that leads to lymphocyte activation. There are several bacterial, parasitic, viral, and tumor host resistance models that have been used to determine whether immune suppression translates into enhanced disease. Both the use of host resistance models and interpretation of the data are complicated and require thorough understanding of the model being used, the role played by various components of the immune system in defending against that particular pathogen, and the degree to which it mimics human disease. Quite different results can be obtained simply by changing the challenge dose or strain of mice used in a study. Selection of appropriate tier II tests to be performed is usually based on results of tier I.

Similar to animal studies, information obtained from routine hematology (differential cell counts) and clinical chemistry (serum immunoglobulin levels) may provide general information on the status of the immune system in humans. However, as with the animal studies, these may not be as sensitive or as informative as assays that target specific components of the immune system and/or assess function. Increased availability of monoclonal antibodies against specific surface molecules on leukocytes, and improved flow cytometry technology have made enumeration of lymphocyte subsets in peripheral blood possible. These assays are relatively accurate and reproducible although, as with most immune tests, considerable variability exists within the human population. The assessment of certain lymphocyte surface antigens has been successfully used in the clinic to detect and monitor the progression or regression of leukemias, lymphomas, and HIV infections, all diseases associated with severe immunosuppression. However, the clinical significance of slight to moderate quantitative changes in the numbers of immune cell populations has not been established. Hence, unlike their counterpart in animal testing, the usefulness of these tests for detecting more subtle forms of immune suppression in humans remains to be demonstrated. There is consensus within the immunotoxicology community that tests that measure the response to an actual antigen challenge are likely to be more reliable predictors of immunotoxicity than flow cytometric assays for cell surface antigens because the latter generally only assesses the state of the immune system at rest. For ethical reasons it is difficult to conduct controlled studies of human immune responses to antigen challenge following exposure to toxic chemicals. One approach under consideration is that of assessing responses to vaccines in chemically exposed populations.

23.3.2 Identifying Chemicals That Cause or Exacerbate Allergic Disease

To date, toxicologists have primarily been concerned with two types of allergic effects: immediate hypersensitivity reactions in the lung that can lead to rhinitis and/or allergen-triggered asthma and DTH reactions in the skin, referred to as contact hypersensitivity or allergic contact dermatitis (e.g., poison ivy reactions). In each case, the initial exposure primes the immune response (induction or sensitization). Reactions are manifested following subsequent exposures (challenge or elicitation). Chemicals may be involved in allergic reactions in the lung either directly as the allergen or indirectly by enhancing allergic sensitization to common allergens (e.g., dust mites, cockroaches, and pollen) or by exacerbating the symptoms associated with subsequent exposure. Agents that enhance sensitization are sometimes referred to as adjuvants. Recently, food allergy has become an issue as a result of genetically engineered crops. However, research to understand the mechanisms associated with food allergies and approaches to routine testing for food allergens are in their infancy and will not be discussed further in this chapter.

Proteins and certain low molecular weight compounds (molecular weight <3000, equivalent weight <1000) have the potential to cause allergic sensitization. Responses to subsequent inhalation exposure include pulmonary inflammation, mucus secretion, specific bronchial hyperreactivity to the offending allergen, and nonspecific bronchial hyperresponsiveness to challenge with an agonist such as methacholine. These symptoms are generally considered to be the hallmarks of allergic asthma and represent a significant health hazard. In both animal models and in humans, genetic predisposition is an important determinant in the development of allergy and asthma. Also, although allergens are important triggers for asthma, the majority of individuals who have allergies do not develop asthma.

All proteins are not equally allergenic. In general, the potential to induce this type of response and the relative potency of different protein or chemical agents has been assessed in guinea pigs sensitized by the respiratory route and monitored for the development of cytophilic antibody (IgG1 in guinea pigs; IgE in mice and humans) as well as increased respiratory rate following pulmonary challenge. These approaches are expensive and not amenable to routine testing. As more is learned about the mechanisms associated with allergic asthma (described in more detail in Section 23.5), it is hoped that better test methods will be developed. Because low molecular weight chemicals are haptens and must react with a host protein in order to be allergenic, efforts have also been made to identify chemicals with potential for allergenicity based on the probability that a given chemical structure will react with protein. Two classes of compounds, diisocyanates and acid anhydrides, fit this description, and some but not all chemicals in these two groups have been identified as respiratory allergens following occupational exposure of humans and/or experimental exposure of guinea pigs.

In humans, similar endpoints are used to assess individuals for allergic responses. The skin prick test, in which different proteins are injected under the skin, tests for the presence of cytophilic antibodies and helps to identify which proteins are causing a response in an individual. Under very controlled situations, patients may be exposed via the respiratory route to potential allergens (broncho-provocation test) and respiratory function monitored to pinpoint the offending allergen. For example, this approach has been used to determine the etiology of occupational asthma.

In laboratory rodent studies, air pollutants, such as diesel exhaust and residual oil fly ash, have been shown to behave as adjuvants (i.e., they enhanced allergic sensitization to common allergens). These studies have been undertaken in a research settings using a variety of strategies. There is no standard approach to assessing chemicals for such effects. Likewise, epidemiology has shown an association between episodes of high air pollution and exacerbation of asthmatic symptoms requiring hospital emergency room visits. Again, however, there is no standard approach to testing chemicals for the capacity to exacerbate allergic symptoms in sensitized individuals.

In contrast to respiratory hypersensitivity, methods to assess low molecular weight chemicals (including drugs, pesticides, dyes, cosmetics, and household products) for potential to induce contact sensitivity (dermatitis) are well established and several protocols using guinea pigs have been in use since the 1930s. These protocols assess the actual disease endpoint, skin irritation, following sensitization and challenge with the test agent. Two tests that are commonly used are the guinea pig maximization test and the Buehler occluded patch test. The endpoint that is assessed, erythema and edema, is somewhat subjective. A chemical is considered to be a sensitizer if 30% (maximization) or 15% (Buehler) of the animals respond. (See Klecak for detailed descriptions of the guinea pig tests.)

Recently, a more economical, less subjective, test for contact hypersensitivity has been developed using mice. This test, the local lymph node assay (LLNA), assesses the proliferative response of lymphocytes in the draining lymph node following application of the agent to the ear and is based on our understanding of the immunologic mechanisms responsible for sensitization (see Section 23.5.6). The LLNA has recently been approved as a stand-alone alternative to the guinea pig tests by the Interagency Coordinating Committee on the Validation of Alternative Methods (ICCVAM).

23.4 MECHANISMS OF IMMUNE SUPPRESSION

From the preceding brief description of the immune system it is clear that toxic substances may potentially disrupt this system at many levels. For example, disruption of cell proliferation and/or differentiation, or interference with cell surface receptors, signaling pathways, or the metabolic activities needed for the production and function of assorted cytokines and antibodies would be expected to alter immune responses. This section provides a more detailed description of the different mechanisms that have been shown to account for immunotoxicity of representative compounds.

23.4.1 Cyclosporin A

Cyclosporin A (CsA) is an immunosuppressive drug used in humans to control rejection of organ transplants and to treat certain autoimmune diseases. CsA is one of approximately 25 cyclosporins isolated from the fungus *Tolypocladium inflatum*. CsA and tacrolimus (a metabolite of the fungus *Sacromyces*, also used as an immunosuppressive drug and formerly referred to as FK506) share a common mode of action in that both bind to similar intracellular receptors and inhibit the early phase of T-cell activation. They are both directly immunosuppressive, that is, metabolic activation is not required. CsA has been used extensively by immunologists as a tool to evaluate the mechanisms underlying T-cell functions and is sometimes used by immunotoxicologists as a positive control. As a result, the mechanisms of immunosuppression have been more clearly elucidated for CsA than for most other immunotoxicants.

Unlike cytotoxic drugs, CsA prevents activation of lymphocytes rather than affecting cell viability. CsA is very hydrophobic and readily permeates cell membranes; the molecule enters cells by diffusion. One of the best-characterized effects of CsA (and tacrolimus) is inhibition of T-cell receptor-mediated signal transduction pathways, resulting in decreased cytokine (IL-2, IL-3, IL-4, TNFα, and GM-CSF) production. Effects on IL-2 production have received the most research attention to date (Fig. 23.4). In normal T cells, antigen binding to the T-cell receptor (TCR) leads to the phosphorylation and activation of phospholipase Cγ (PLCγ) by the enzyme ZAP 70. PLCγ then catalyzes the metabolism of phosphatidylinositol bisphosphate to diacylglycerol (DAG) and inositol triphosphate (IP3). IP3 stimulates the release of Ca^{++} from intracellular stores, thus activating a serine/threonine phosphatase enzyme, calcineurin. Calcineurin then dephosphorylates a constitutively expressed, T-cell-specific cytosolic protein, nuclear factor of activated T cells ($NF\text{-}AT_c$). Dephosphorylation enables $NF\text{-}AT_c$ to enter the nucleus where it combines with nuclear $NF\text{-}AT_n$ to form the active dimer NF-AT, a potent transcription factor that binds to the promoter region of the IL-2 gene. Other more widely distributed transcription factors are also produced following T-cell activation. These involve protein kinase C (PKC) pathways that are activated by ligation of the TCR, leading to increased intracellular calcium levels via PKC-α activity and to production of several transcription factors via a MAP kinase cascade, including Fos. Also, DAG stimulates PKC-β, culminating in the production of other proteins, including Jun. In the nucleus, Jun combines with Fos to produce the transcription factor AP-1. These events are represented schematically in Fig. 23.4.

CsA has been shown to antagonize IL-2 gene transcription by interrupting at least two pathways: inhibition of $NF\text{-}AT_c$ dephosphorylation by calcineurin, preventing translocation into the nucleus, and blockade of PKC-β activity. Although CsA-mediated suppression of IL-2 synthesis requires binding of CsA to cyclophilin, it is the binding of the CsA/cyclophilin complex to calcineurin, and the ensuing

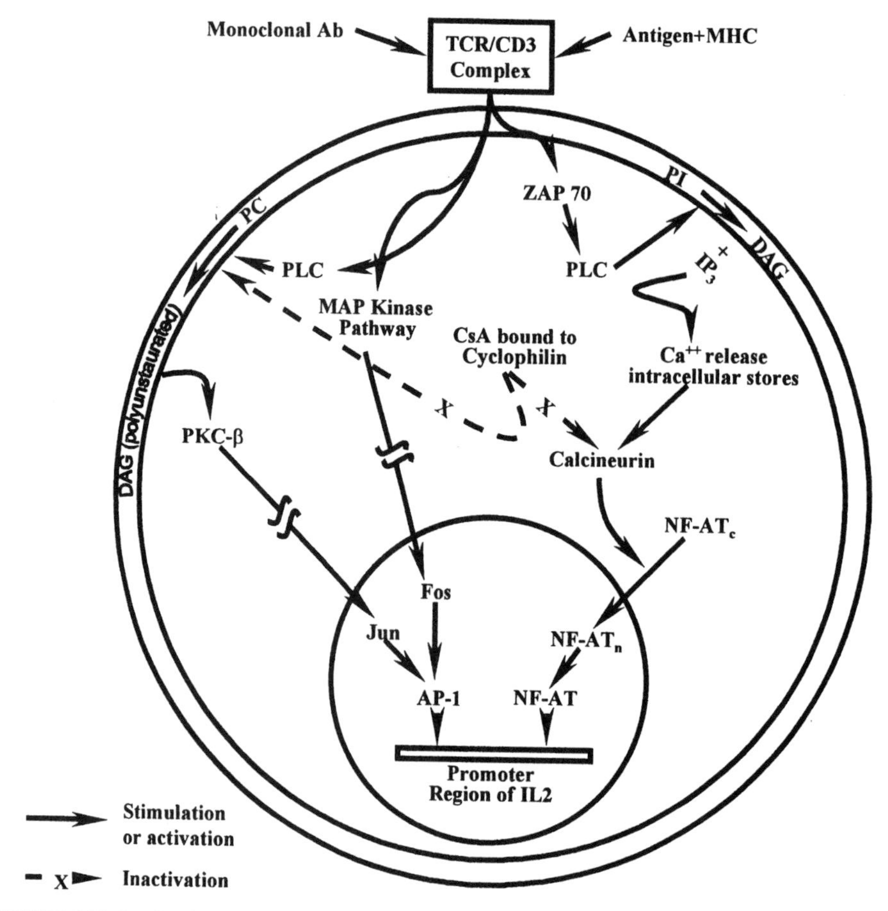

FIGURE 23.4 Cyclosporin A (CsA) disruption of signal transduction pathways leading to IL-2 production.
CsA binds to cyclophilin in the cytoplasm. The complex disrupts at least two signalling pathways that lead to activation of genes involved in cytokine production. See text for detailed explanation. Addreviations: TCR, t cell receptor; PLC, phopholipase C; IP, inositol triphosphate; PKC, protein kinase C; DAG, diacylglycerol; NF-AT, nuclear factor of activation.

inactivation of calcineurin, that prevents dephosphorylation and entry of NF-AT$_c$ into the nucleus. Likewise, blockade of the PKC-β pathway of transcription factor production is not a direct effect of CsA on the enzymatic activity of PKC. Instead, activation of T-cell fatty acid metabolism is blocked by CsA/cyclophilin, preventing the normal switch from predominantly saturated fatty acids in the resting T-cell membrane to higher levels of polyunsaturated fatty acids in activated cells. For diacylglycerols to be effective stimulators of PKC activity, they must contain polyunsaturated fatty acids. The low content of unsaturated fatty acids in DAG of CsA-exposed T cells is thought to provide a poor stimulus for PKC-β activity. The ultimate effect is decreased availability of the protein Jun and hence decreased AP-1. CsA has also been shown to antagonize the binding of other transcription factors to enhancer elements in the IL-2 promoter region, including Oct/OAP and

NFκB, probably via the same cyclophilin/calcineurin mechanism that inhibits NF-AT activity.

In summary, CsA binds to a cytoplasmic receptor, cyclophilin. This drug/receptor complex subsequently disrupts at least two signal transduction pathways that begin with the T-cell receptor on the cell surface and lead to activation of genes and the production of cytokines that are critical in the early activation of T cells.

Although CsA is a potent immunosuppressant, under certain conditions it may increase some responses, particularly those mediated by Th2-associated cytokines. Although the mechanism(s) controlling up-regulation are not yet known, disruption of the Th1 : Th2 balance, secondary to suppression of IL-2 and IFNγ production, may be involved.

23.4.2 Halogenated Aromatic Hydrocarbons

2,3,7,8-Tetrachlorodibenzo-p-dioxin (TCDD) is the most toxic of the group of structurally related compounds known as halogenated aromatic hydrocarbons (HAHs). These HAHs include the polychlorinated dibenzodioxins (PCDDs), of which TCDD is a member, polychlorinated dibenzofurans (PCDFs), and polychlorinated biphenols (PCBs). Based on the structure of these HAHs (see Fig. 23.5), chlorine substitutions can exist in any or all of 8 or 10 positions in the compound. As a result, there are over 400 chlorinated HAH congeners possible. PCDDs and PCDFs are "true" environmental contaminants because they are primarily formed as by-products during the production of chemicals derived from chlorinated phenols. On the other hand, because of their physical and chemical properties, PCBs have been used in a number of commercial applications. Approximately 30 of the chlorinated HAHs (i.e., 7 PCDDs, 10 PCDFs, and 13 PCBs) display toxic properties similar to that of TCDD, albeit having lower potencies. Of the myriad toxicities associated with TCDD exposure of experimental animals, thymic and lymphoid tissue atrophy and immune system suppression are most pertinent to this chapter.

TCDD exposure of experimental animals results in thymic atrophy and alterations in T-cell-mediated immune function, such as delayed-type hypersensitivity (DTH), cytotoxic T lymphocyte (CTL) activity, and T-cell-dependent antibody responses. Because both cell-mediated and humoral immunity are suppressed by TCDD and related HAHs, it is not surprising that administration of these compounds to experimental animals results in increased susceptibility to challenge with viral, bacterial, or parasitic diseases, as well as syngeneic tumors.

Suppression of T-cell-mediated immunity (i.e., DTH response) occurs at lower doses of TCDD when rodents are exposed perinatally compared to exposure as adults. TCDD exposure during immune system development also results in a persistent suppression of immune function, as demonstrated by suppression of the DTH response in 19-month-old rats exposed perinatally. Such exposure to TCDD results in alterations in fetal and neonatal CD4 and CD8 thymocyte precursor populations. This suggests that the mechanism underlying immunotoxicity in the case of perinatal exposure to TCDD may be interference with maturation of T cells in the thymus, including effects on positive or negative selection.

In adult mice, the primary antibody response to sheep red blood cells (SRBC) is one of the most sensitive and reproducible immune end points that is suppressed

2,3,7,8-TETRACHLORODIBENZO-p-DIOXIN

(TCDD)

POLYCHLORINATED DIBENZODIOXINS

(PCDDs)

POLYCHLORINATED DIBENZOFURANS

(PCDFs)

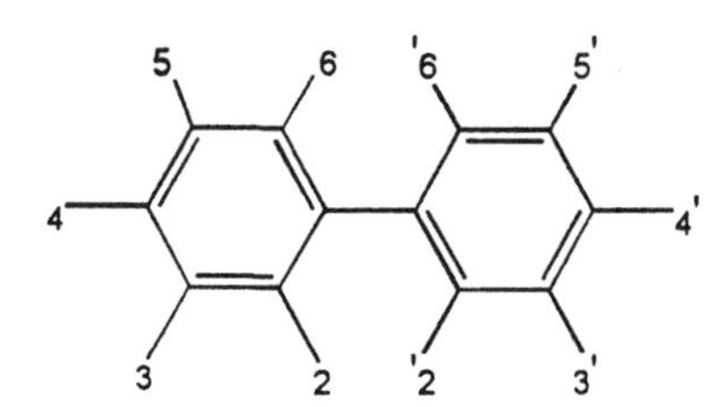

POLYCHLORINATED BIPHENYLS

(PCBs)

FIGURE 23.5 Molecular structures of chlorinated aromatic hydrocarbons.

by TCDD and related HAHs. For example, the TCDD dose that suppresses this response by 50% ($ED_{50\%}$) in mice is approximately 1.0 μg TCDD/kg. The antibody response to SRBCs is dependent on the collaborative interaction of antigen presenting cells (i.e., macrophages, dendritic cells), regulatory T lymphocytes (i.e., CD4 and CD8 cells), and antibody producing cells (i.e., plasma cells). A significant effort has been made to determine the cellular target(s) for TCDD-induced suppression of the anti-SRBC response, however, no clear picture has emerged. Evidence from in vivo work suggests that the T cell and/or macrophage components are primarily involved. On the other hand, in vitro experiments suggest that terminal differentiation of B cells is affected by TCDD. In these studies TCDD caused elevation of intracellular calcium levels and increased protein phosphorylation in resting B cells. These changes may result in inappropriate activation, making the B cell less responsive to specific antigen stimulation.

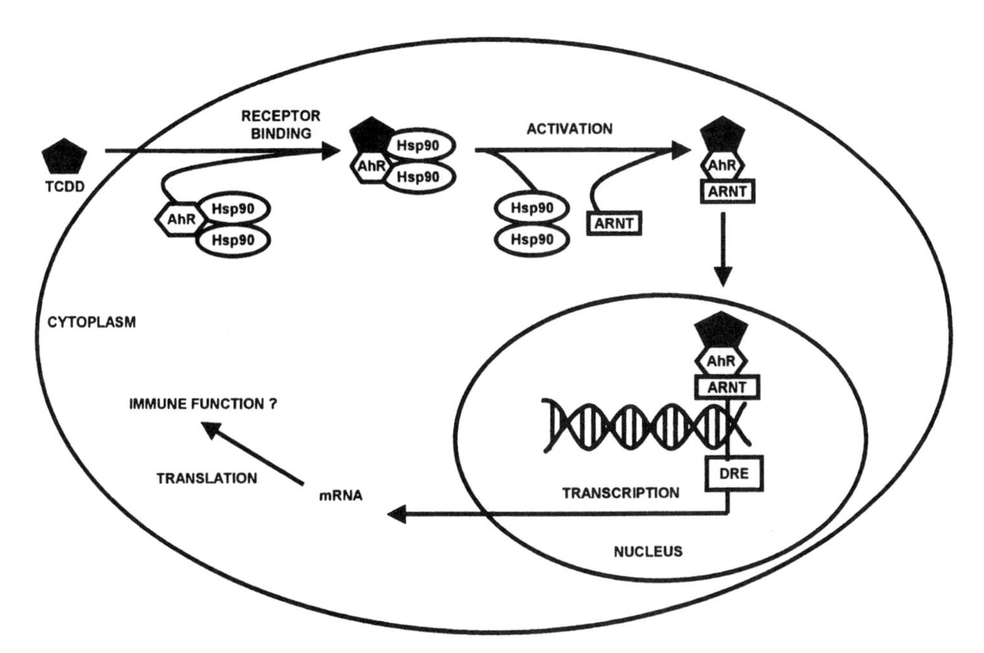

FIGURE 23.6 Model of AhR-mediated mechanism of action of TCDD and related HAHs. TCDD binds to the AhR in the cytooplasm. This complex interacts with another cytosolic protein (ARNT) and is translocated into the nucleus where it binds to specific DNA enhancer sequences upstream of TCDD responsive genes. Exactly which genes are important in the resulting immune suppression is unclear. Abbreviations: AhR, Aryl hydrocarbon receptor; ARNT, AhR nuclear-transporter; DRE, dioxin responsive elements.

Research on the mechanism(s) by which TCDD and related HAHs induce the biochemical and toxic effects associated with exposure to these chemicals has primarily focused on the aryl hydrocarbon receptor (AhR) (Fig. 23.6). The AhR functions as a ligand-activated transcription factor, which works in a fashion similar to that of steroid receptor-mediated responses (see Chapter 12). First, there is stereospecific binding of the chemical and AhR within the cytoplasm of the cell followed by the interaction of the AhR-ligand complex with another cytosolic protein called the AhR-nuclear-transporter (ARNT). This ligand-receptor complex is then translocated into the nucleus, where it binds with high affinity to specific DNA enhancer sequences called dioxin-responsive elements (DRE). This binding to the DRE occurs upstream of TCDD responsive genes resulting in the transcription of structural genes such as CYP1A1, which encode mRNA for production of the enzyme cytochrome P4501A1, as well as other gene products that regulate differentiation and proliferation of cells. Exactly which gene products are important in the resulting immune suppression remain unclear.

Evidence of a central role for the AhR in TCDD-induced immunosuppression comes from two areas of investigation. First, susceptibility to TCDD-induced immunosuppression in inbred strains of mice has been demonstrated to be associated with allelic variation at the Ah locus. Second, structure-activity relationships (SAR) for HAH-induced immunosuppression have been found to be associated with the AhR binding affinity of TCDD relative to that of TCDD-like HAHs.

Allelic variations at the Ah locus occurs in mice resulting in AhRs that differ in their ability to bind TCDD. For example, C57B1/6 (Ah^{bb}) mice are sensitive (i.e., highly responsive) to TCDD-induced immune suppression, whereas DBA/2 (Ah^{dd}) mice are much less sensitive (i.e., low responders). DBA/2 mice require a 10-fold higher dose of TCDD, compared with C57B1/6 mice, to achieve comparable suppression of the anti-SRBC response. This difference in strain sensitivity also has been demonstrated for other immune endpoints (e.g., CTL response). F_1 and backcross mice also provide evidence for the role played by the AhR in TCDD-induced immunosuppression. C57B1/6 mice congenic at the Ah locus (i.e., Ah^{dd}) are less sensitive to TCDD-induced suppression of the anti-SRBC response compared with wild-type mice (i.e., Ah^{bb}).

TCDD, compared with related HAHs, has the greatest binding affinity for the AhR. Studies of structure activity relationships have demonstrated that the immunosuppressive potency of HAHs is associated with the binding affinity of these congeners with the AhR in Ah^{bb} mice. PCDDs, PCDFs, and PCBs that bind the AhR to varying degrees, have been found to variably suppress the anti-SRBC response, whereas those that do not bind the AhR fail to suppress this response. Recent in vitro work with B cell lines, which differ in their expression of AhR, has further demonstrated an association between AhR expression and sensitivity to TCDD. Inhibition of antibody secretion and Ig heavy chain transcription, in a B cell line that displays marked expression of AhR protein, was found to follow a rank order potency of HAHs that was related to the congeners' binding affinity. This rank order potency of HAHs for immunosuppression correlated with the induction of CYP1A1 expression in these cells.

There is also in vivo and in vitro work suggesting that suppression of the anti-SRBC response may not be entirely Ah-mediated, and research with nonimmune cells suggests that other pathways may exist for the alteration of gene expression that may not be dependent on the AhR. Whether non-AhR pathways contribute to the immunosuppressive effects of TCDD remains to be determined.

Taken together, it is apparent that TCDD-induced immunotoxicity involves primarily AhR-mediated mechanisms and possibly non-AhR-mediated mechanisms. Therefore, it may be that TCDD-induced modulation of signal transduction and gene expression, which leads to immunotoxicity, may involve two pathways. The first and most plausible, based on the extensive existing database, involves a direct interaction of the AhR with the gene-associated DRE and consequent transcription and translation of gene products that regulate growth and differentiation of lymphoid cells. The second pathway would involve the initiation of a phosphorylation/dephosphorylation cascade leading to the modulation of other nuclear transcription factors and inappropriate activation of B cells, resulting in altered immune cell function, particularly affecting terminal differentiation of B cells.

23.4.3 Ultraviolet Radiation

Ultraviolet radiation (UVR) is one of the non-ionizing radiations in the electromagnetic spectrum and lies within the range of wavelengths 100–400 nm. The wavelengths of concern with respect to immune suppression are in the mid-range,

280–315 nm (UVB). The sun is the principal source of UVR exposure for people, although sun lamps used in medical therapy and cosmetic tanning also represent an exposure source of concern. Solar UVR undergoes significant absorption by the atmosphere. Exposure to UVR is expected to increase with the depletion of stratospheric ozone by anthropogenic chemicals such as chloroflorocarbons and halons. In addition to the immune system, the skin and eye are major targets for UVR.

Mice exposed to UVR have suppressed resistance to UVR-induced skin cancers, decreased responses to contact sensitizers, suppressed DTH responses to antigens injected in the footpad, and decreased resistance to a variety of intracellular infectious agents including the causative agents of Lymes disease, leprosy, tuberculosis, herpesvirus infections, and leishmaniasis. IgM and IgG antibody responses are not affected. Contact hypersensitivity (CHS) responses (described in more detail in Section 23.5) are suppressed when the sensitizer is applied to the UV-irradiated skin (local suppression) or when the sensitizer is applied at a site distant from that of UV radiation (systemic suppression). UVR-induced immune suppression appears to involve the induction of antigen-specific "suppressor" T cells. Hence, when lymphocytes from UVR-treated and immunized mice are adoptively transferred into untreated syngeneic mice that are subsequently challenged with the antigen, immune suppression similar to that observed in the donor is also observed in the recipient. When exposed to UVR at the time of initial immunization, mice become tolerant (immunologically unresponsive) to the immunogen, such that it is subsequently difficult to immunize them with the same antigen even in the absence of UVR. Again, this suggests the presence of suppressor cells. Suppression of contact sensitivity responses and the development of tolerance have also been demonstrated in humans following local exposure, and UVR-induced immune suppression is thought to play a role in the development of non-melanoma skin cancers in humans.

Because UVR does not penetrate below the skin, systemic effects must occur indirectly via mediators produced in the skin. Target cells in the skin include Langerhans cells (dendritic cells in the skin) and keratinocytes (Fig. 23.7). As indicated in Section 23.5, exposure to contact sensitizers causes increased expression of MHC II molecules on Langerhans cells and increased expression of intracellular adhesion molecules (ICAM), both of which facilitate antigen presentation. In UVR-exposed mice however, the expression of MHC II and ICAM are suppressed. Antigen presentation (both locally via Langerhans cells and in the spleen) is altered in such a way that activation of Th1, but not Th2, cells is suppressed. (The "suppressor" cells that mediate UVR-induced immune suppression are probably Th2 cells.) In response to UVR, keratinocytes produce a number of mediators, including TNFα, IL-10, cis-urocanic acid, and prostaglandin E_2 (PGE_2), which have immunosuppressive potential. It is postulated that these affect immune sensitization locally and also spill into the circulation to cause systemic suppression. Several studies support this theory. When supernatants from keratinocyte cultures exposed to UVR were injected into mice, suppression of the DTH response occurred in a manner similar to that observed in UVR exposed mice. When IL-10 activity in the supernatant was blocked with antibody, immune suppression did not occur, indicating that keratinocyte derived IL-10 could be responsible for the systemic suppression observed following UVR exposure. Likewise, suppression of DTH in UVR-irradiated mice was reversed by injecting mice with antibodies to IL-10. Studies also showed that

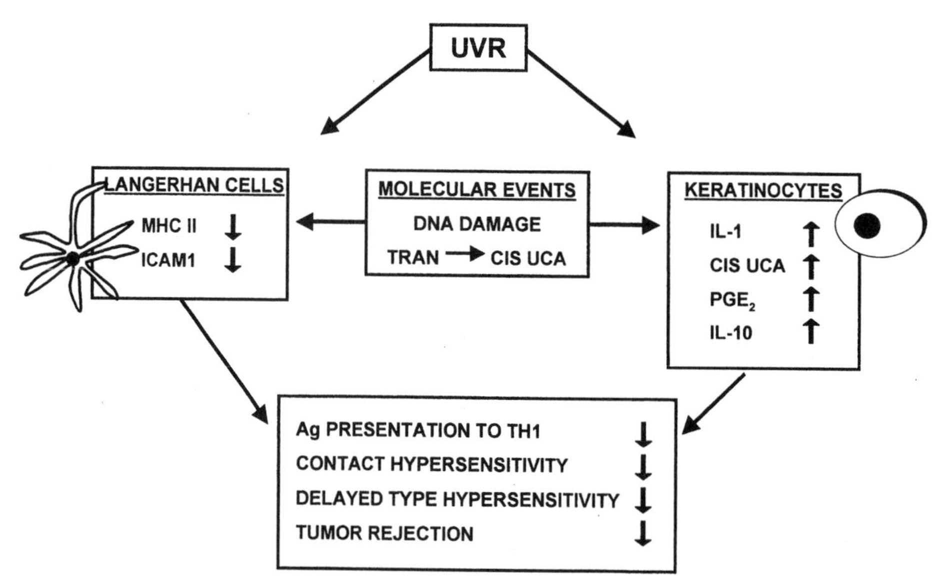

FIGURE 23.7 Mechanism underlying UVR induced immune suppression. Langerhans cells and keratinocytes are the cellular targets. Uroconic acid and DNA are molecular targets. Mediators produced by skin cells have both local and systemic effects. Abbreviations: MHC, major histocompatibility complex, ICAM, intracellular adhesion molecule; UCA, urocanic acid; IL, interleukin; PGE, prostaglandin E.

injection of IL-12, which drives Th1 responses and counteracts IL-10, overcame UVR-induced immune suppression.

Urocanic acid is another mediator that has been studied extensively with respect to UVR-induced immune suppression. UVR exposure causes isomerization of urocanic acid in the skin from trans-urocanic acid to cis-urocanic acid. When cis-urocanic acid was painted on the skin or injected subcutaneously or intravenously, immune suppression occurred that mimicked that observed with UVR. DTH responses were completely restored in mice treated with antibody to cis-urocanic acid. Hence, there are at least two mediators produced in the skin following UVR exposure that have the capacity to suppress DTH responses. Interestingly, neither cis-urocanic acid nor IL-10 have been definitively shown to effect suppression of contact sensitivity responses. Although both may be involved, neither appears to be sufficient.

TNFα is thought to be an important mediator in the suppression of CHS. This hypothesis is supported by the finding that injection of antibodies to TNFα reverses the suppression of CHS that usually occurs following UVR exposure. Conversely, intracutaneous injection of recombiant TNFα at the site of sensitization suppresses the development of a contact sensitivity response. Also, the genetic difference between UVR-resistant and UVR-susceptible strains of mice appears to involve a polymorphism in a regulatory region of the TNFα gene.

There is great debate concerning the identity of the molecular target(s) that triggers UVR induction of immune suppression. Urocanic acid, as mentioned above, is one candidate. The other candidate is DNA. UVR exposure is known to induce DNA damage in the form of pyrimidine dimers. In studies that introduced a bacteriophage derived pyrimidine dimer repair enzyme into keratinocytes (via liposomes), UVR-induced suppression of both DTH and CHS did not occur. Also, induction of nonspecific DNA damage by introducing into the skin (again via liposomes) the restriction enzyme HindIII, which causes double-stranded DNA breaks, resulted in suppression of CHS both locally and systemically and induced the production of IL-10. These data strongly suggest that DNA damage plays an important role in UVR-induced immune suppression.

The mechanisms thought to account for UVR-induced immune suppression are summarized in Figure 23.7. Langerhans cells and keratinocytes appear to be the cellular targets. The resulting damage to DNA and isomerization of urocanic acid, results in release of mediators including IL-10, cis-urocanic acid, and TNFα, which have both local and systemic effects. Antigen presentation by Langerhans cells and dendritic cells at other sites is altered such that Th1 but not Th2 cell activation is impaired. This results in suppression of Th1-associated responses including DTH and CHS and enhances susceptibility to skin cancer and certain infectious agents. UVR is frequently referred to as a complete carcinogen because it has both initiation and promotion properties. Immune suppression may be responsible for promotion. It should be noted that many chemical carcinogens are also immunosuppressive. Hence, the observation that DNA damage induces immune suppression has ramifications that extend beyond UVR-induced immune suppression. Furthermore, epithelial cells at other sites (the lung for instance) also produce cytokines in response to exposures to various chemicals suggesting that the mechanisms described here are probably not unique to UVR.

23.4.4 The Role of Enzymatic Transformation in Immunotoxicity

In addition to those chemicals that cause direct damage to immune cells and tissues, there are many compounds, including organic solvents such as benzene, cytoreductive drugs, pesticides, mycotoxins, and polycyclic aromatic hydrocarbons (PAHs), that induce immune alterations only after undergoing enzyme-mediated reactions within various tissues (Table 23.4). The biochemical pathways that have evolved to metabolize exogenous chemicals generally make them less toxic (deactivation reactions) and more water-soluble, thus facilitating their elimination from the body. However, in a number of cases exogenous compounds may be transformed to active or more toxic metabolites resulting in activation or bioactivation reactions (see chapters 5 and 6). Bioactivation reactions that result in immunotoxicity are frequently due to the formation of reactive intermediates, including epoxides, free radicals, or N-hydroxyl derivatives.

The P450s are the principal phase I enzymes involved in metabolic activation of foreign compounds. These enzymes are widespread in nature, tend to be concentrated in the portals of entry to the body (skin, respiratory system, and digestive system), and are thought to function as a first line of defense by detoxifying exogenous compounds before toxic insult can occur. The regulation of P450s has been studied extensively, and a number of factors may affect P450 expression, including gender, age, nutritional status, disease, genetic predeterminants, environmental pollutants, and stress. Of particular importance with respect to immunotoxicity is the

TABLE 23.4 Examples of Immunotoxic Compounds Requiring Metabolic Activation.

Class	Compounds
Mycotoxins	Aflotoxin Ochratoxin A Wortmannin
Solvents	Benzene Carbon tetrachloride Ethanol N-hexane 2-Methoxyethanol
PAHs	Benzo(a)pyrene Dimethylbenzanthracene 3-Methylcholanthrene
Pesticides	Chlordane Malathion Parathion
Other	Cyclophosphamide Dimethylnitrosamine

fact that both very young and very old organisms are deficient in many of the constitutive P450 enzymes, although certain P450 isozymes can be induced in these groups by exposure to xenobiotics. Thus, compounds can be more or less toxic as a result of age or nutritional status. Tissue concentrations of various P450s can also be influenced by dietary components as well as by any of a large number of lipophilic xenobiotics.

The elucidation of the metabolic pathways for enzymatic biotransformation of xenobiotics and immunotoxicity studies are areas that do not frequently overlap. However, as can be seen from the examples described below, knowledge of metabolism and identification of the ultimate toxic species are critical in understanding the mechanisms of immunotoxicity and the target cell populations for a wide variety of xenobiotics.

23.4.5 Cyclophosphamide

The drug cyclophosphamide (Cy) is a cytostatic phosphamide derivative of bis-(β-chloroethyl)-amine (nitrogen mustard). Cy has been used extensively in humans, alone and in combination with other drugs, to treat a variety of neoplastic, lymphoproliferative, and autoimmune diseases and as an immunosuppressant to prevent rejection of foreign tissue grafts. Although used therapeutically, Cy is also a human carcinogen.

Cy causes profound immunosuppression in rodents. Doses in the range of 30–100 mg/kg/day have often been used as a positive control for immunosuppression in immunotoxicology studies. Antibody production, delayed hypersensitivity, cytotoxic T-cell activity, mixed lymphocyte responses, natural killer cell activity, and resistance to viral, parasitic bacterial, or tumor cell challenge are all suppressed by Cy. Antibody-mediated immunity is more sensitive to suppression by Cy than cellular responses. However, if immunity is established prior to Cy exposure, the effect of daily Cy administration on antibody production is slight and is quickly reversed once the drug is withdrawn. Although Cy is a potent immunotoxicant, certain immune responses can be increased by Cy, under specific conditions. Many investigators have attributed elevated responses to shifts in lymphocyte cell populations because of the greater sensitivity of CD8 T suppressor cells (compared with CD4) to Cy-mediated cytotoxicity.

Cy immunosuppression is the result of cytotoxicity. Metabolites of Cy are alkylating agents that convalently cross-link DNA leading to inhibition of DNA synthesis, cell cycle arrest, and, ultimately, apoptosis, if DNA repair is unsuccessful. The ultimate result of these cytotoxic events is blockade of cell proliferation and clonal expansion, central components of the immune response.

Cy itself has no alkylating or cytotoxic activity; rather, hepatic oxidation via microsomal P450 enzymes (specifically, the CYP2B isoform), followed by other enzymatic (primarily aldehyde dehydrogenase) oxidative steps, is required to produce reactive metabolites (Fig. 23.8). Induction of CYP2B activity by pretreatment with barbiturate increases the rate of Cy metabolism, but not the total amount of metabolites. Cy also induces its own metabolism, which causes more rapid clearance at high doses. The initial P450 oxidation product is 4-hydroxycyclophosphamide (4-OHCy), which has little immunosuppressive activity; 4-OHCy and its tautomeric form, aldophosphamide, are present in equilibrium. The 4-OHCy metabolite enters cells easily by diffusion, and may represent a transport form of Cy. Acrolein, a toxic by-product, and phosphoramide mustard, the metabolite believed to have the greatest biological activity, are formed by the spontaneous decomposition of aldophosphamide. Both of these metabolites are immunosuppressive. 4-Ketocyclophosphamide and carboxyphosphamide are formed by oxidation of 4-OHCy and aldophosphamide, respectively. These metabolites are the main decomposition products, have little immunosuppressive activity, and represent the major metabolic products excreted in the urine. Carboxyphosphamide, the major metabolic product of Cy, is further metabolized to the immunotoxic metabolite nornitrogen mustard, also present in the urine. Rates of metabolite formation and elimination are quite variable in humans, and have been associated with aldehyde

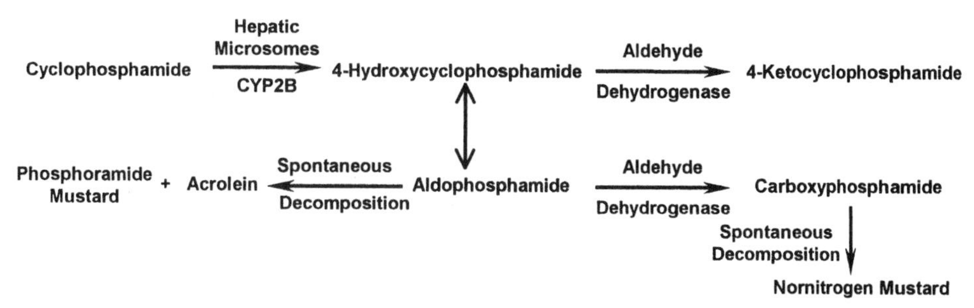

FIGURE 23.8 Metabolic activation pathway for cyclophosphamide. Acrolein, phosphoramide mustard, and nornitrogen mustard are the immunotoxic metabolites.

dehydrogenase gene variants that determine levels of ADH activity. Mouse strain differences exist as well in sensitivity to both the toxic and immunosuppressive effects of Cy.

In summary, highly reactive metabolites of Cy disrupt DNA synthesis, thus inhibiting cell proliferation. As a result, production of lymphocytes and accessory cells is suppressed, and host immunocompetence is compromised.

23.4.6 Polycyclic Aromatic Hydrocarbons (PAHs)

PAHs such as benzo(a)pyrene (B(a)P) are metabolized to reactive electrophilic intermediates primarily in the liver, and reactive metabolites are transported by serum proteins to other tissues. These reactive metabolites, primarily diol-epoxides, alter normal cellular function by binding covalently to RNA, DNA, and proteins. A number of other mechanisms have also been proposed to explain PAH-induced immunosuppression, including membrane perturbation resulting in altered signal transduction, calcium mobilization, gene expression, and/or cytokine production. In laboratory animals studies there is a significant correlation between suppression of antibody-forming cell responses and the carcinogenic activity of PAHs, suggesting that there may be a common mechanism.

PAHs have been shown to increase intracellular calcium levels in rodent and human lymphocytes. In T lymphocytes, PAH-induced activation of protein tryosine kinases Fyn and Lck and increased tyrosine phosphorylation of PLCγ results in depletion of intracellular calcium stores. Activation of the T cell following antigen binding to TCR requires calcium mobilization (see Section 23.4.1, Fig. 23.4). PAH apparently interferes with this process by depleting intracellular calcium stores, causing premature signaling, leading to tolerance. In B lymphocytes, PAH also elevates intracellular calcium levels. P450 metabolism appears to play an important role in this altered cell signaling, as PAH metabolites are significantly more effective in elevating intracellular calcium levels than their parent compounds, and treatment with P450 inhibitors has been shown to prevent calcium increases. Prolonged elevation of intracellular calcium precedes apoptosis, and at higher concentrations PAHs induce apoptosis in lymphoid precursors in the bone marrow, circulating B and T lymphocytes, and lymphocytes in the thymus, spleen, and lymph nodes.

In vivo exposure to B(a)P has been shown to target both primary and secondary immune tissues, and to significantly alter lymphoid cell numbers and cell surface antigen expression in the spleen, thymus, and bone marrow. In rodents, in vivo exposure to B(a)P inhibits both antibody- and cell-mediated immunity, as well as some aspects of innate immunity including macrophage phagocytosis and interferon production. Although DNA adduct formation is similar in immune and nonimmune tissues after in vivo exposure, in vitro cultures of murine splenocytes have little ability to generate B(a)P/DNA adducts, suggesting that hepatic bioactivation is an important mediator of B(a)P-induced immunotoxicity. In vitro studies using a murine splenocyte-rat hepatocyte coculture system demonstrated that biotransformation of B(a)P by phase I enzymes is required in the suppression of T-dependent antibody responses against sheep red blood cells. Direct addition of reactive metabolites, such as the 4,5-epoxide and 7,8 diol, to in vitro cultures results in suppression of immune responses similar to that seen with B(a)P in the presence of metabolizing enzymes.

In mice, there is some evidence that splenic macrophages can activate B(a)P to the highly reactive 7,8-dihydroxy-9,10-epoxy-7,8,9,10-tetrahydro-B(a)P in vitro. Freshly isolated splenic macrophages have demonstrated CYP1A1 activity and the ability to produce reactive metabolites after exposure to PAHs, whereas freshly isolated lymphocytes and thymocytes do not demonstrate significant metabolic activity. Splenic and alveolar macrophages also showed inducible ethoxyresorufin O-deethylase (EROD) activity, an enzymatic measure of CYP1A1 function, after in vivo exposure to TCDD, however, purified populations of lymphocytes and thymocytes did not demonstrate significant enzyme activity. The relative contribution of splenic versus hepatic bioactivation has yet to be determined.

23.4.7 Organic Solvents

Organic solvents that induce CYP2E1 are comprised of a few broad chemical classes, including hydrocarbons such as benzene and toluene, halogenated aliphatic compounds such as carbon tetrachloride and dichloroethane, aliphatic clacohols such as ethanol, and hydroxyethers such as 2-methoxyethanol. Industrial solvents are frequently mixtures of several compounds. Whereas there is some exposure to the general population via contaminated groundwater or vaporization of commercial solvents and gasolines, the most frequent solvent-associated toxicity occurs from occupational exposure. A number of organic solvents have been examined for their effects on the immune system, and the requirement for their bioactivation to produce immunotoxicity has been well established.

Of the organic solvents that induce CYP2E1, benzene and its metabolites have been most extensively studied with respect to immunotoxicity because they have long been associated with hematologic and immunologic disorders including leukemia in humans. A number of experimental studies suggest that benzene metabolites, including hydroquinone, catechol, and phenol, are responsible for its hematotoxicity. Benzene is metabolized by hepatic CYP2E1 primarily to phenol and in turn to hydroquinone and/or catechol. The phenolic metabolites preferentially accumulate in the bone marrow and lymphoid tissues of rodents. In the bone marrow, enzymatic conversion of both phenol and hydroquinone to more reactive binding species such as the semiquinone radical may be P450-independent and involve myeloperoxidases and prostaglandin synthetases. The semiquinone radical binds covalently to cellular proteins and forms DNA adducts, disrupting normal cellular functions such as cell division and mitochondrial RNA synthesis. Rapidly proliferating cells, such as lymphoid and myeloid progenitor cells in the bone marrow or clonally expanding lymphocyte subpopulations, are highly sensitive targets for such effects.

The earliest manifestation of benzene toxicity in exposed workers is a decrease in lymphocyte counts. A variety of blood disorders, including leukopenia, thrombocytopenia, granulocytopenia, and aplastic anemia, have been associated with benzene exposure. Studies in laboratory animals have demonstrated that treatment with benzene or its metabolites induces myelo- and immunosuppression. Benzene and its metabolites appear to be particularly cytotoxic to progenitor cells within the bone marrow, targeting the lymphocyte, monocyte, granulocyte, and erythrocyte lineages. There is also considerable evidence that benzene metabolites alter the stromal cell population in the bone marrow, which supports the differentiation and maturation of progenitor cells. In vivo exposure to benzene and/or its metabolites inhibits T-dependent antibody responses, B- and T-cell lymphoproliferative responses and cytotoxic T-lymphocyte-mediated tumor cell killing, and increases susceptibility to challenge with infectious agents.

Structure activity studies suggest that the polyhydroxy metabolites of benzene have the most immunosuppressive activity, and that benzene and phenol are significantly less toxic to both lymphoid and myeloid cells. Co-administration of thiol-reactive agents blocks hydroquinone-induced suppression of mitogen-stimulated lymphoproliferation and agglutination in rat splenocytes, indicating that oxidation to thiol-reactive quinones may be a critical step in bioactivation. Induction of CYP2E1 in rats by treatment with ethanol enhances the myelotoxicity of benzene and inhibition of CYP2E1 activity by administration of propylene glycol partially prevents benzene-induced immuno- and myelotoxicity.

23.4.8 Summary of Immunosuppressive Mechanisms

The various mechanisms associated with immunotoxicity for the example chemicals described above are summarized in Table 23.5. Several generalizations can be made. At the molecular level, damage to DNA of immune cells can have profound effects on proliferation needed for clonal expansion. DNA damage of nonimmune cells has also been shown to modulate cytokine responses in a manner that suppresses the Th1 response (and conversely may enhance Th2 responses). Alternately, toxic

TABLE 23.5 Some Mechanisms Associated with Immune Suppression.

Mechanism	Example
Disruption of T cell maturation in the thymus	In utero/perinatal TCDD?
Inactivation of calcineurin disrupting signals needed for IL-2 gene transcription	Cyclosporin A
Covalent binding to DNA disrupting replication, cell proliferation and function leading to cytotoxicity	Cyclophoshamide, polyaromatic hydrocarbons, organic solvents
Same as above targeting bone marrow progenitor cells	Benzene
Interaction with intracellular receptor which acts as a ligand-activated transcription factor the product of which affect immune response by unknown mechanisms	TCDD, HAHs
Blockade of fatty acid metabolism in the membrane, thus impairing diacyglycerocapacity to stimulate protein kinase C activity	Cyclosporin A
Cytokine-mediated alterations in antigen presentation resulting in suppression of Th1 activation and skewing towards a Th2 response	Ultraviolet radiation
Depletion of intracellular calcium stores, premature signaling leading to tolerance	Polyaromatic hydrocarbons

chemicals may interact with intracellular receptors and in so doing impact signal transduction pathways. In some cases, induction of transcription factors occurs; in other cases, the signals needed to activate transcription factors are interrupted. The net result is modulation of cytokine, cell surface molecule, or receptor expression leading to alterations in the course of the immune response. Depletion of intracellular calcium stores may also interfere with signal transduction needed for lymphocyte activation. There are a number of cellular targets. Alteration of T cell responses (frequently as a result of changes in cytokine responses), changes in the function of antigen-presenting cells, toxicity to stem cells in the bone marrow, and change in the production of mediators by nonimmune cells all provide mechanisms that can lead to immune suppression. At the developmental level, depletion of bone marrow stem cells or interference with maturation and selection of cells in the thymus can lead to permanent immune suppression. Finally, the induction of suppressor cells may dampen immune responses to specific antigens. The modulation of immune responses may result in suppression of some components of the immune system without affecting or even enhancing other components of the immune response. This will become more evident as some of the mechanisms associated with immune stimulation are described below.

23.5 MECHANISMS ASSOCIATED WITH HYPERSENSITIVITY

23.5.1 Classification of Immune-Mediated Injury Based on Mechanisms

Under certain circumstances, immune responses, rather than providing protection (and sometimes in addition to providing protection), produce damage in response to antigens that might otherwise be innocuous. These deleterious reactions are collectively known as hypersensitivity or allergy. Classically, hypersensitivity reactions have been divided into four types originally proposed by Gell and Coombs (Table 23.6). Although there is some overlap between the classifications (i.e., mechanistically different types of hypersensitivity may produce similar effects), the classification scheme provides a useful means of understanding the different mechanisms associated with immune-mediated injury. Inflammation plays a role in all of these responses.

Type I hypersensitivity (sometimes referred to as atopy) is mediated by antigen-specific IgE, which binds via the Fc receptor to mast cells and basophils. On subsequent exposure, the allergen binds to these cell-bound antibodies and cross-links

TABLE 23.6 Gell and Coombs Classification of Hypersensitivity Reactions.

Type I	Cytophilic antibody (IgE) binds to mast cells, antigen binds to antibody and cross-links receptors causing mediator (histamine) release; immediate hypersensitivity
Type II	IgM, IgG mediated cytolysis of cells
Type III	IgM, IgG complexed with antigen (immune complexes) leading to inflammation
Type IV	T-cell mediated; delayed-type hypersensitivity

IgE molecules, causing the release of mediators such as histamine and slow-reacting substance of anaphylaxis (SRS-A). These mediators cause vasodilation and leakage of fluid into the tissues, plus sensory nerve stimulation (leading to itching, sneezing, and cough). Type I is frequently referred to as immediate-type hypersensitivity because reactions occur within minutes after exposure of a previously sensitized individual to the offending antigen. Examples of type I reactions include allergic asthma, allergic rhinitis (hay fever), atopic dermatitis (eczema), and acute urticaria (hives). The most severe form is systemic anaphylaxis (e.g., in response to a bee sting), which results in anaphylactic shock, and potentially death.

Type II hypersensitivity is the result of antibody-mediated cytotoxicity that occurs when antibodies respond to cell surface antigens. Frequently, blood cells are the targets, as in the case of an incompatible blood transfusion or Rh blood incompatibility between mother and child. Antibodies or antigen-antibody complexes bind to the cell and activate the complement system, which leads to enhanced phagocytosis or lysis of the target cell. Autoimmune diseases, such as immune hemolytic anemia and thrombocytopenia, can result from drug treatments with penicillin, quinidine, quinine, or acetaminophen. Presumably these drugs interact with the cell membrane such that the immune system detects "foreign" antigens on the cell surface. Hemolytic disease may also have unknown etiologies although the mechanism is a type II response. This form of hypersensitivity has been encountered by toxicologists primarily in the form of drug-induced autoimmune disease.

Type III reactions are the result of antigen-antibody (IgG) complexes that accumulate in tissues or the circulation, activate macrophages and the complement system, and trigger the influx of PMNs, eosinophils, and lymphocytes (inflammation). This is sometimes referred to as the Arthrus reaction and includes postinfection sequella such as rheumatic heart disease. Farmer's lung, a pneumonitis caused by molds, has been attributed to both type III and type IV, and some of the late phase response (4–6 hours after exposure) in asthmatics may be the result of Arthrus-type reactions.

Unlike the preceding three types, type IV or delayed-type hypersensitivity (DTH), involves T cells and macrophages, not antibodies. Activated T cells release cytokines that cause accumulation and activation of macrophages, which in turn cause local damage. This type of reaction is very important in defense against intracellular infections such as tuberculosis, but is also responsible for contact hypersensitivity responses (allergic contact dermatitis) typified by the poison ivy response. Inhalation of beryllium can result in a range of pathologies including acute pneumonitis, tracheobronchitis, and chronic beryllium disease, all of which are associated with type IV reactions. As the name indicates, the expression of type IV responses following challenge is delayed, occurring 24 to 48 hours after exposure.

23.5.2 Features Common to Respiratory and Dermal Allergies

The remainder of this section will focus in more detail on type I (immediate) and type IV (delayed) hypersensitivity with emphasis on the lung and skin as target organs. In both cases the adverse effects of hypersensitivity develop in two stages: (1) Induction (sensitization) requires a sufficient or cumulative exposure dose of the sensitizing agent to induce immune responses that cause no obvious symptoms. (2) Elicitation occurs in sensitized individuals upon subsequent exposure to the allergen (immunogen) and results in adverse antigen specific responses that include inflammation. Because inflammatory responses also occur without an immunological basis, a distinction must be made between an irritant and a sensitizer. An irritant is an agent that causes local inflammatory effects but induces no immunological memory. Therefore, on subsequent exposures local inflammation will again result, but there is no enhancement of the magnitude of the response and no change in the dose required to induce the response. In immunologically mediated inflammation, there may be no response to a sensitizer during the induction stage, but responses to subsequent exposures are exacerbated. The dose required for elicitation is usually less than that required to achieve sensitization. It should be noted that some sensitizers produce irritation, particularly at higher doses.

The immunogens involved in hypersensitivity may be proteins or haptens. Although most proteins stimulate an immune response, only some proteins cause allergic responses. However, no amino acid sequence motifs specific for allergenicity have been identified thus far. Protein sensitizers are generally hydrophilic, heat and digestion stable, and may have enzymatic activity. Examples include house dust mite, cat, and cockroach allergens. A hapten is a small molecule ($\leq$3000 molecular weight) that is not itself immunogenic, but after conjugation to a carrier protein becomes immunogenic and induces hapten-specific antibody, which can then bind the hapten in the absence of the carrier. Therefore, the hapten or its metabolite must have a reactive site for conjugation with a protein. This conjugate complex is processed and presented to lymphocytes. When conjugated with proteins residing in the respiratory system or in the skin, haptens become sensitizing agents. Haptens associated with allergy include low molecular weight chemicals that are generally man-made substances (e.g., trimellitic anhydride and toluene diisocyanate) and metals.

The events that initiate hypersensitivity responses in the respiratory tract and skin have homologous features. As with any adaptive immune response, the antigen must be presented to the immune system and ultimately to T lymphocytes. Bone marrow-derived dendritic cells (DC) have been identified as highly efficient antigen presenting cells (APC) in both the epidermis (where they are called Langerhans cells) and the respiratory tract. These cells function to process the antigen and migrate to the draining lymph node where the antigen is presented in the context of the MHC to lymphocytes. The majority of the resident tissue DC are functionally immature DC that are very efficient at antigen uptake and processing, including the association of the antigenic peptides with MHC molecules and transport of the complex to the cell surface. However, in the respiratory tract both APC activity and T-cell proliferation in response to stimulatory signals are usually inhibited by signals from resident macrophages and other adjacent (primarily epithelial) cells. As the DC mature and migrate to the draining lymph nodes, they become more effective at antigen presentation and less efficient at antigen uptake and processing. The DC are quite efficient at stimulating primary immune responses. The environment in which initial interactions with T cells occur appears to play a role in immune outcome, specifically whether a Th1 (delayed-type) or Th2 (immediate type) response ensues. As described in Section 23.4.3, cytokines produced in the local milieu can influence antigen presentation to favor either Th1 or Th2 responses. Immunological outcomes will be further addressed in the discussion of type I hypersensitivity in the respiratory tract and type IV hypersensitivity in the skin.

Toxicologists are concerned not only with chemicals that may themselves be allergenic, but also with chemicals that act as adjuvants modulating immune responses in a manner that favors allergic sensitization. In addition, chemicals may exacerbate responses following challenge of a sensitized individual. Most of the work in this area has focused on air pollutants and respiratory allergy and on conditions that favor Th2 responses. Hence, these issues will be addressed in the following sections on respiratory hypersensitivity and modulation of Th cell responses.

23.5.3 Respiratory Allergy/Allergic Asthma

A major determinant in occupationally and environmentally induced respiratory diseases is immediate (type I) bronchial hypersensitivity to specific allergens often referred to as atopy. Atopy, which occurs in genetically predisposed individuals, is the presence of cytotropic, antigen-specific immunoglobulin IgE (or IgG1 in guinea pigs) that can lead to allergic rhinitis. A significant number of asthmatic individuals are atopic and have respiratory allergies that may act as stimuli for asthma attacks. In these individuals, antigen challenge causes an immediate response as the result of cross-linking of IgE molecules on mast cells and release of mediators responsible for inflammation and bronchoconstriction. Between 2 and 8 hour after this event, a more severe and prolonged (late phase) reaction occurs, which is characterized by mucous secretion, bronchoconstriction, airway hyperresponsiveness to a variety of nonspecific stimuli (e.g., histamine, methacholine), and airway inflammation characterized by eosinophils. Late phase responses may last up to 12 hours and do not appear to be mediated by IgE. The mechanisms for the late phase response are not fully understood. However, Th2 cells and associated cytokines (particularly IL-5 and IL-13) and eosinophils are thought to play a significant role.

Protein allergens include common indoor air contaminants such as house dust mite, cat, and cockroach antigens and a variety of microbial antigens, such as alcalase, a serine protease produced by *Bacillus subtilis* Carlsberg and used as a detergent enzyme. These proteins have been shown to produce sensitization and respiratory allergic responses (both early and late phase) in humans, guinea pigs, and mice. Additionally, they have been strongly associated with asthma morbidity. Of the low molecular weight chemicals that have been associated with allergic asthma, certain diisocyanates and acid anhydrides have received the most attention. Whereas protein allergens characteristically result in antigen-specific IgE, the presence of allergen-specific IgE in low molecular weight chemical respiratory allergy has not been universally demonstrated. In addition, the early phase response does not always occur in individuals with occupational asthma triggered by these low molecular weight compounds. The mechanisms underlying low molecular weight chemical respiratory allergy are still under investigation.

Mechanisms underlying the IgE-mediated responses have been studied extensively. There are two primary requirements for B-cell Ig isotype class switching from IgM to IgE: (1) The presence of the Th2-associated cytokine IL-4 (or another Th2 cytokine IL-13) and (2) direct cell-to-cell interaction via CD40 (B cell) and CD40 ligand (CD40L), expressed on T cells, basophils, and mast cells. IL-4 exhibits autocrine activity in Th2 cell differentiation and promotes mast cell development. The majority of IgE is found associated with mast cells via the $F_{c\epsilon}RI$ (high-affinity IgE receptor). Bone marrow derived immature mast cell precursors localize under

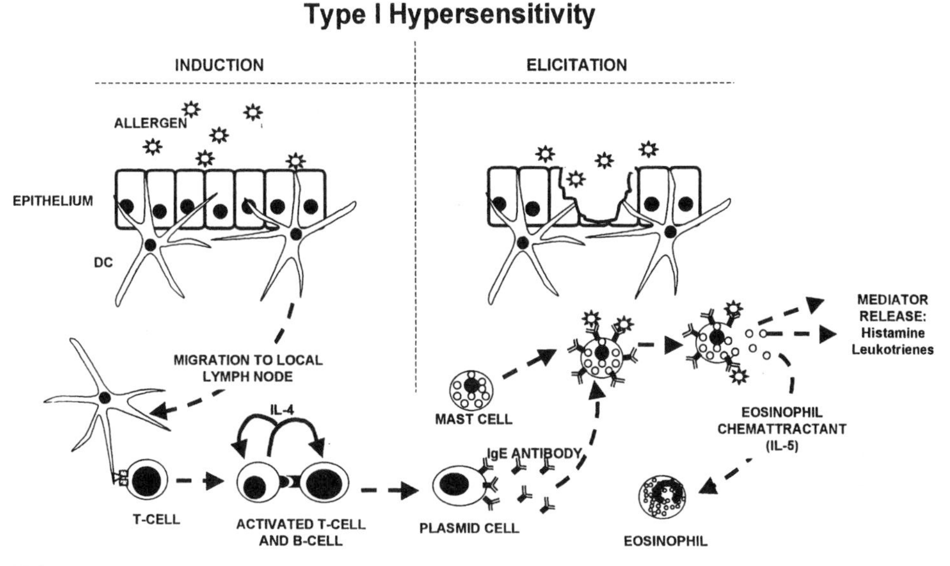

FIGURE 23.9 Schematic representation of type I hypersensitivity.
Induction: Resident respiratory tract dendritic cells (DC) take and process antigen, mature, migrate to the draining lymph nodes, and present antigen to lymphocytes. Activated T-lymphocytes in turn activate B-cell differentiation into antibody producing plasma cells. IL-4 promotes Ig isotype class switching from IgM to IgE and promotes mast cell development. IgE is associates with mast cells. *Elicitation*: allergen crosslinks the mast cell bound IgE, performed mediators and cytokines are released. (See Table 23.7). Inflammation and bronchoconstriction occur.

epithelium of mucosal areas (respiratory tract and gut) and the skin where tissue-specific maturation and expansion occur. These mast cells contain performed mediators and are capable of producing other effector molecules including IL-4 and IL-5. With these IgE, armed mast cells in the respiratory tract the stage is set for elicitation of antigen-specific allergic events (Fig. 23.9). Upon subsequent exposure, the specific allergen cross-links the mast cell-bound IgE resulting in the release of the preformed mediators and newly synthesized substances (summarized in Table 23.5). Upon subsequent exposure, the specific allergen crosslinks the mast cell-bound IgE resulting in the release of the preformed mediators and newly synthesized substances (summarized in Table 23.7). Chief among the preformed mediators is histamine, which acts through two receptors (H_1 and H_2). Ligation to either receptor causes increased vascular permeability. Whereas ligation to the H_1 receptor also causes smooth muscle contraction, vascular constriction, and nasal mucus production, ligation to the H_2 receptor results in mucus production in the lower airway. In addition to histamine, there are a variety of performed cellular chemotactic factors and enzymes. Products from the activation of two metabolic pathways of membrane-derived arachidonic acid, the lipoxygenase and cyclooxygenase pathways, provide other important mediators of allergic inflammation. The Th2-associated cytokine IL-5 secreted by mast cells has been shown to be essential in proliferation and maturation of eosinophil precursors as well as viability and

TABLE 23.7 Mediators of Allergic Inflammation.

Source	Mediator	Action
Mast cells	Histamine	Smooth muscle contraction, increased vascular permeability, vascular constriction, mucus production
	Enzymes: tryptase, lysosomal hydrolases, proteoglycans	Tissue damage
	Arachidonic acid metabolites	
	Lipoxygenase pathway: 5-hydroxyeicosatetraenoic acid (5HETE), leukotriene B_4 (LTB_4)	Smooth muscle contraction, increased vascular permeability, mucus production, chemotactic activity
	Cyclooxygenase pathway: prostaglandins (PGE_1, PGE_2, PGF_{2a}, PGD_2)	Increased vascular permeability, smooth muscle contraction
	Prostacyclins	Vasodilation
	Thromboxanes (thromboxane A_2)	Bronchoconstriction
Eosinophils	Major basic protein (MBP), eosinophil peroxidase (EPO), eosinophil cationic protein (ECP)	Microbial killing, tissue damage
	EPO/H_2O_2/halide	Mast cell degranulation

eosinophil granule protein release, and chemotaxis for eosinophils. Eosinophils that are recruited into the lung release their own set of mediators, which are thought to be important in late-phase responses. Furthermore, cytokines (IL-5, IL-3, and GM-CSF) secreted by activated eosinophils found at the site of allergic inflammation may result in a positive-feedback loop.

23.5.4 Modulation of the Balance Between Th1 and Th2 Responses

A number of both intrinsic and extrinsic factors influence the balance between Th1 and Th2 responses. The intrinsic factors include genetics, age, and gender (hormonal status). Extrinsic factors include route of antigen exposure, respiratory infection, stress, ultraviolet radiation, (described in Section 23.4.3) and exposure to pollutants. Many of these extrinsic factors affect conditions in the local antigen presenting milieu, which sets the stage for Th1 and Th2 responses.

Genetics and age are two of the most influential intrinsic factors contributing to allergic sensitization. Atopy and allergic asthma have genetic links and tend to run in families. Likewise certain strains of mice and rats are better models for allergic disease than others, and guinea pigs have traditionally been used to study both allergy and asthma because in addition to being prone to respond immunologically to allergens, guinea pig airways are particularly hyperreactive. With respect to age, there is evidence that fetal immune responses are of the Th2 type; this is thought to protect the placental tissue from the toxic effects of IFN-γ and to prevent rejection of the fetus. Furthermore, there is a growing body of evidence suggesting that this Th1/Th2 balance is reversed in the neonate as the adaptive immune response matures and is driven toward Th1 type responses by early infections. Impediments to this Th1 maturation may lead to a higher incidence of Th2 memory. There has been speculation that vaccinations, which are designed in many cases to produce antibody (Th2) responses, may interfere with this process. Exposure early in life to any agent that might prime Th2 cells or suppress Th1 cells could have lifelong consequences. On the other end of the age spectrum, Th1 responses appear to decline more readily with age.

Of most interest to toxicologists is the possibility of air pollutants (e.g., diesel exhaust, ozone, nitrogen dioxide, sulfur dioxide, particulate matter) enhancing both the induction and expression of allergic lung disease. A number of epidemiological studies have found that the incidence and severity of asthma attacks increase as levels of air pollution increase. Although many of these effects are thought to be a result of direct irritation of an already diseased tissue (i.e., enhanced expression of disease in individuals who are already asthmatic), a number of animal studies have demonstrated that air pollutants can also act as adjuvants to enhance allergic sensitization. These studies have demonstrated not only pollutant-enhanced IgE production and bronchial reactivity, but also enhanced production of Th2 cytokines. Most of these air pollutants cause nonspecific inflammation, which may establish a local environment in the lung that favors Th2 sensitization. In response to air pollution exposure, epithelial cells lining the lung and alveolar macrophages produce several mediators that suppress Th1, but not Th2 responses including PGE_2, IL-6, and NO. Hence, exposure to toxicants could possibly exacerbate allergic responses and allergic asthma in two ways, by acting as an adjuvant to enhance sensitization or by aggravating inflammatory responses that occur at the time of elicitation.

23.5.5 Immediate-Type Hypersensitivity Skin Reactions

The respiratory tract is not the only site of atopic responsiveness. Atopic dermatitis/immunologic contact urticaria is an IgE mediated inflammatory skin disease and is mechanistically generated in much the same manner as IgE-mediated respiratory allergy described above. The symptoms (redness, pruritis, and skin lesions) occur within minutes to an hour or so following contact by a sensitized individual. With the exception of the kinetics of onset, clinically atopic dermatitis/immunologic contact urticaria is similar to contact hypersensitivity (described below). However, mechanistically they are quite different.

23.5.6 Contact Hypersensitivity and Delayed-Type Hypersensitivity

Contact hypersensitivity (CH) results from the repeated encounter of a contact allergen with the skin of a sensitized individual and is a common occupational health problem resulting from the exposure to haptens. CH was traditionally considered the primary example of a T-cell-mediated delayed-type hypersensitivity (DTH) reaction. More recently, differences between CH and DTH have been identified. Whereas DTH antigen presentation results in activation of MHC Class II restricted CD4 T cells, CH antigen presentation involves predominately CD8 T cells and may include antigen presentation by both MHC Class II and Class I. These differences may be due to the different routes of exposure that lead to DTH and CH. Typically, DTH occurs following subcutaneous or mucosal antigen exposure thus bypassing the epidermis. Whereas the macrophage provides the cytokine environment for DTH, it is the epidermal Langerhans cells and keratinocytes that produce the cytokine environment to drive the CH response.

The mechanisms involved in CH sensitization (induction) originate upon epicutaneous application of a hapten that couples to carrier proteins on dermal and epidermal cells to become fully immunogenic. The Langerhans cells take up and process the antigen and migrate to the regional draining lymph node where they present the antigen to lymphocytes, as described in Section 23.5.2. Once antigen is effectively presented to cells within the lymph node, activation and rapid proliferation of cells occur. Cellular proliferation is considered a hallmark of the induction phase of CH.

The application of hapten to the skin induces cytokines that activates a cascade of events (Fig. 23.10). Epidermal keratinocytes secrete inflammatory cytokines that facilitate the maturation of Langerhans cells at the site of antigen application. Cytokines initially produced include interleukin (IL)-1β, IL-6, and IL-12. Langerhans cells also produce cytokines in an autocrine self-regulatory fashion. IL-1β appears to have a selective effect for CH responses and is not observed with non-specific irritation. IL-1β, tumor necrosis factor α (TNF-α) and granulocyte-macrophage colony-stimulating factor (GM-CSF) all contribute to Langerhans cell activation and migration to the draining lymph node.

During Langerhans cell maturation, expression of cell-surface molecules, including MHC class I and II, and adhesion and costimulatory molecules, is enhanced, thus facilitating antigen presentation and subsequent signal transduction leading to cytokine production. Although T cells are thought to be the key effector cells in the development of CH, both T and B lymphocytes proliferate in response to contact sensitizers. Enhanced proliferation can be measured and used as a tool to test chemicals for their potential to induce contact sensitivity (the murine Local Lymph Node Assay, see Section 23.3.2).

Events associated with the elicitation of contact sensitivity are not as well characterized as those associated with induction. It is evident from histopathology that mast cell degranulation, vasodilation, and the influx of neutrophils and mononuclear and T cells occur (classic type IV hypersensitivity responses described in Section 25.5.1). Yet, clinical manifestation, histopathology and immunohistology of allergic and irritant contact dermatitis are virtually indistinguishable and, in fact, most haptens exhibit dose-dependent cutaneous toxicity such that they act as irritants at higher doses. One difference is that the erythema, edema, and pruritus that is characteristic of the elicitation phase occurs 24 to 72 hours after epicutaneous application of the allergen whereas irritant responses occurs much sooner.

Exposure of a sensitized individual to the relevant hapten triggers the same cytokine responses that occur following induction (Fig. 23.10). Two theories have been advanced to described subsequent events in the elicitation phase. One theory suggests that Ag-specific primed (memory) T cells, which carry skin homing receptors, are constantly patrolling throughout the skin. As with induction, Langerhans take up and process antigen, however, unlike induction, antigen is presented to these primed T cells in the skin (rather than the lymph node) triggering the local responses

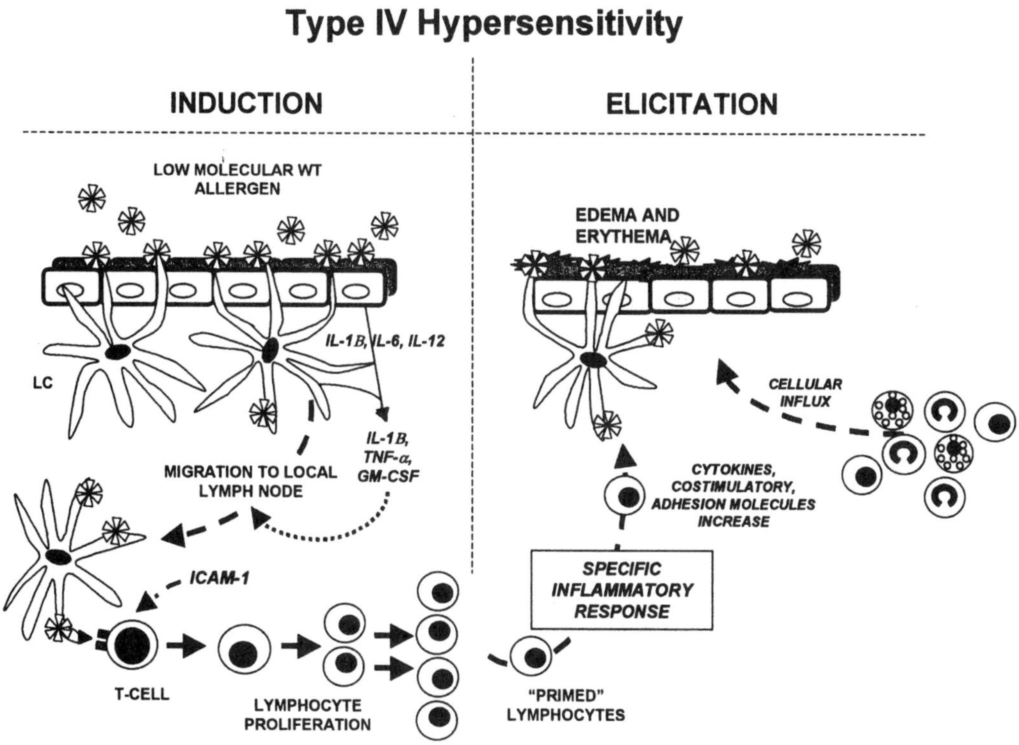

FIGURE 23.10 Schematic representation of type IV contact hypersensitivity. *Induction*: Immature dendritic Langerhans Cells (LC) take up and process the low molecular weight allergen. Simultaneously, keratinocytes and LCs release cytokines that promote LCs migration to the draining lymph node and maturation into effective antigen presenting cells. LC to T-cell interactions result in lymphocyte proliferation of "primed" effector lymphocytes with allergen-specific memory. *Elicitation*: Allergens cause release of nonspecific inflammatory mediators, responsible for at least some of the cellular influx into the site of allergen challenge. "Primed" lymphocytes home to the area in a very specific response. Cytokines, co-stimulatory factors and adhesion molecules act to produce erythema and edema.

classically associated with type IV hypersensitivity. Several pieces of evidence argue against this theory. Because there are few T cells present within the epidermis, it seems unlikely that sufficient numbers of Ag-specific T cells are present at any give time point in any part of the skin to elicit CH responses. Also, depletion of resident epidermal Langerhans cells actually enhances rather than diminishes the expression of CH, thus suggesting that Langerhans cells are not necessarily required in the elicitation phase of CH. It is very possible that other cell types present Ag to T cells during this phase. These cells may include keratinocytes, dermal mast cells, and endothelial cells or perhaps cells that home to the area early during the response such as macrophages.

A more recent theory suggests that the initial response to elicitation is a nonspecific inflammatory (irritant-like) response triggered by the production of proinflammatory cytokines and other mediators such as histamine, serotinin, and prostaglandins produced by the keratinocytes. These mediators activate local endothelial cells, attract leukocytes, including antigen specific lymphocytes, and, as during induction, up-regulate surface expression of MHC, costimulatory, and adhesion molecules on resident cells. It is also believed that IL-1 and TNF-α play a stimulatory role in this process and that monocyte chemotactic protein 1 (MCP-1), interferon-inducible protein 10 (IP-10) and macrophage inflammatory protein 1α (MIP-1α) are expressed during this phase of CH.

Adhesion molecules play an essential role in the elicitation phase of CH by activation of vascular endothelium and the cascade of leukocyte rolling, activation, adhesion, and transmigration through vessel wall. Specific molecules that seem to be of particular importance to this process include ICAM-1, CD54 and its receptor, leukocyte function-associated antigen 1 (LFA-1;CD11a/CD18), and selectins such as E-, L-, and P-selectin. Certain adhesion molecules may be more prominently involved at different times during the CH response and certain adhesion molecules may influence the type of response that develops. For instance, there is recent evidence that P- and E-selectin may be involved in preferential recruitment of Th1 versus Th2 cells, whereas L-selectin may play a role in sensitization but not in elicitation of CH.

As mentioned, CD8 T cells play an important role in CH. Subpopulations of CD8 cells (Tc1 and Tc2) similar to those described for CD4 (Th1 and Th2) have been distinguished based on cytokine profiles that are similar to their CD4 counterparts. The traditional concept of CH has been that of a Th1, Tc1 cytokine profile. Recently, however it has been observed that T-cell cytokines from both Th1 (Tc1) and Th2 (Tc2) profiles are found at sites of CH responses. Whereas the Th1(Tc1) cytokines stimulate the elicitation response, the Th2 (Tc2) cytokines may down-regulate or control the elicitation response. Production of mRNA for IL-4 occurs only late in the course of an ongoing CH response, suggesting that IL-4 may actually terminate the CH response.

In summary, CH like other hypersensitivity responses is characterized by an induction and an elicitation phase. Keratinocytes, Langerhans cells, and CD8 T cells are important cellular components of CH. A number of cytokines, chemokines, and adhesion molecules are involved in the CH response. The exact role that each of these components plays in the CH response has not been fully elucidated and is currently the subject of intense investigation.

23.5.7 Hypersensitivity Pneumonitis

DTH responses are not limited to the skin. Hypersensitivity pneumonitis (HP), an allergic lung syndrome considered to be a mix of type III and type IV responses, is caused by occupational inhalation exposure to a wide variety of organic dusts. These dusts contain antigenic substances including fungal/bacterial components, serum proteins, and some chemicals. Clinical diagnosis refers to HP as allergic alveolitis. However, there are a variety of additional HP syndromes named by occupation or antigen association. Farmer's lung is associated with exposure to moldy hay/grain; bagassosis with moldy sugarcane; bird-breeder's lung with bird droppings or other avian proteins; and maple bark stripper's disease with mold spores. HP is a diffuse disease that is difficult to characterize. It is associated with the predominance of mononuclear inflammation of the lung interstitium, terminal bronchioles, and alveoli. This inflammation is often associated with the most significant clinical feature, granuloma formation, possibly progressing to lung fibrosis (hardening of the lungs). These diseases have not been studied as extensively as disease associated with immediate-type allergic responses.

23.6 SUMMARY

In summary, there are analogous events in the various hypersensitivity profiles such as antigen processing and presentation to the T cells of the draining regional lymph nodes by antigen presenting cells, which are often the dendritic cells of the exposed or challenged organ. Likewise, inflammation is a common component in the expression of hypersensitivity responses. However, kinetics of events and symptoms differ with the type of hypersensitivity induced. Additionally, a complex network of cytokines, chemokines, and adhesion and costimulatory molecules, the nature of which is not entirely understood, are involved in the various hypersensitivity responses. In all cases, T cells appear to have an important role, although different types of hypersensitivity are characterized by different types of T-cell responses. The net result is that the immune system can be stimulated to induce tissue damage, and toxic chemicals may act either directly as antigens in this process or may exacerbate induction or expression of allergic responses induced by other antigens.

SUGGESTED READING

Benjamini, E. Sunshine, G. and Leskowitz, S. *Immunology: A Short Course*. Wiley-Liss, New York, 1991. A well-illustrated, user friendly text.

Dean, J.H., Luster, M.I., Munson, A.E. and Kimber, I. *Immunotoxicology and Immunopharmacology*, 2nd edition, Raven, New York, 1994.

Fearon, D.T. and Locksley, R.M. The instructive role of innate immunity in the acquired immune response. *Science* 272 (1996), 50.

Grabbe, S. and Schwarz, T. Immunoregulatory mechanisms involved in elicitation of allergic contact hypersensitivity. *Immunol. Today* 19(1) (1998), 37–44.

Holsapple, M.P. Immunotoxicity of halogenated aromatic hydrocarbons. In: Smialowicz, R.J. and Holsapple, M.P. (Eds), *Experimental Immunotoxicology*. CRC Press, Boca Raton, FL, 1996, pp 265–305.

Janeway, C.A., Jr., Travers, P., Walport, M. and Capra, J.D. *Immunobiology: The Immune System in Health and Disease*, 4th Ed., Current Biology/Garland Publishing, Longdon, UK, 1999.

Kerkvliet, N.I. Immunotoxicology of dioxins and related chemicals. In: Schecter, A. (Ed), *Dioxins and Health*. Plenum Press, New York, 1994, pp 199–225.

Kimber, I. and Selgrade, M.J.K. *T Lymphocyte Subpopulations in Immunotoxicology*. John Wiley & Sons, Chicester, England, 1998.

Kimber, I. and Maurer, T. *Toxicology of Contact Hypersensitivity*. Taylor and Francis, London, UK, 1996.

Klecak, G. Identification of contact allergens: predictive tests in animals. In: Marzulli, F.N. and Maibach, H.I. (Eds), *Dermatotoxicology*. Hemisphere, New York, 1987, pp 305–339.

Luster, M.I. et al., Risk assessment in immunotoxicology I. Sensitivity and predictability of immune tests. *Fundam. Appl. Toxicol.* 18 (1992), 200; and II. Relationships between immune and host resistance tests. *Fundam. Appl. Toxicol.* 21 (1993), 71.

Robey, E. and Allison, J.P. T-cell activation: integration of signals from the antigen receptor and costimulatroy molecules. *Immunol. Today* 16 (1995), 307.

CHAPTER TWENTY-FOUR

Respiratory Tract Toxicity

ROGENE F. HENDERSON and KRISTEN J. NIKULA

24.1 INTRODUCTION

The respiratory tract is the portal of entry for the air we breathe and, as such, is the first point of contact of airborne toxins within the body. In this chapter the normal anatomy, function, and biochemical capacity of the respiratory tract are described followed by a discussion of the potential acute and chronic toxic effects of airborne pollutants and certain systemic toxins on the respiratory tract.

24.2 ANATOMY OF THE RESPIRATORY TRACT

The main functions of the respiratory tract are olfaction and gas exchange. These functions require inhalation of large volumes of air over a lifetime, contact between molecules in the air and cells of the respiratory tract, and circulation of the entire cardiac output through the lung. These functional requirements make the respiratory tract vulnerable to irritants, oxidants, allergens, carcinogens, and other toxicants in the inhaled air as well as systemic toxicants delivered in the pulmonary circulation. Several defense mechanisms protect the lung, but the lung can be injured at sites where the local dose overwhelms the defenses or where cells are particularly susceptible because of their specialized function. The purpose of this section is to describe the functional anatomy of the respiratory tract.

The respiratory tract consists of the conducting portion that warms, moistens, and filters inspired air and the pulmonary parenchyma where gas is exchanged. The conducting portion consists of the nose, mouth, pharynx, larynx, trachea, bronchi, and bronchioles. The parenchyma consists of respiratory bronchioles, alveolar ducts, and alveoli.

24.2.1 Nose, Mouth, Pharynx, and Larynx

The mouth, nose, and pharyngeal region are important for collection of particles, humidification and temperature adjustment of inspired air, olfaction, and detection

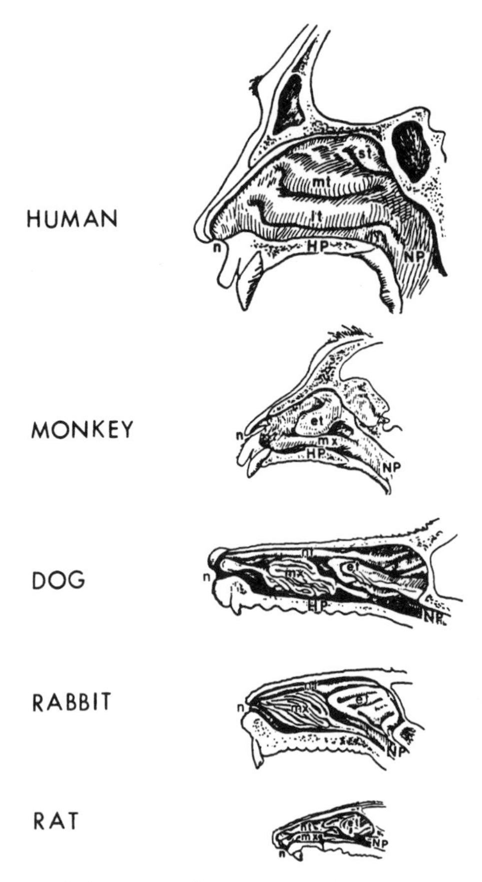

FIGURE 24.1 Diagrams of the exposed mucosal surfaces of the nasal lateral wall of the human, monkey, dog, rabbit, and rat.
HP = hard palate; N = naris; NP = nasopharynx; et = ethmoturbinate; nt = nasoturbinate; mx = maxilloturbinate; mt = middle turbinate; it = interior turbinate; st = superior turbinate.

Source: From Harkema, J.R. et al., *The Airway Epithelium: Physiology, Pathophysiology, and Pharmacology*, Vol. 55, Farmer, F.G. and Hay, D.W.T., eds., pp. 3–39, 1991. Reprinted with permission.

of inhaled irritants. The mouth or oral cavity is a variably sized space that can be a site of entry of inhaled materials and impaction of particles in species such as humans that can breathe through their mouths. The nose is the major site of entry for inhaled materials. Some species, such as rodents, are obligate nasal breathers, so the nose is the only site of entry into the respiratory system for air and inhaled materials.

The nose extends from the nostrils to the pharynx. It is divided longitudinally by a septum into two nasal cavities. In most species, for example, rodents, rabbits, cats, and dogs, each nasal cavity is divided into a dorsal, ventral, and middle (lateral) meatus by two turbinate bones, the nasoturbinate and maxilloturbinate. These turbinates project from the dorsolateral and ventrolateral wall of the cavity, respectively. In the posterior portion of the nose, the ethmoid recess contains the ethmoturbinate. The turbinate structure is less complex in primates (Fig. 24.1). The nasal cavity is lined by a mucosa that is very vascular and consists of four distinct epithelia. In rodents, these epithelia are (1) the stratified squamous epithelium that lines the nasal vestibule and floor of the ventral meatus in the anterior portion of the nose; (2) the nonciliated, pseudostratified, transitional epithelium that lies between the squamous epithelium and the respiratory epithelium and lines the lateral meatus; (3) the ciliated, respiratory epithelium that lines the remainder of the nasal cavity anterior and ventral to the olfactory epithelium; and (4) the olfactory epithelium (neuroepithelium) that lines the dorsal meatus and ethmoturbinates in the caudal portion of the nose. These same epithelia line the nasal cavities of other mammals, including humans, but their relative abundance and exact locations differ among species.

The squamous epithelium of the nasal vestibule contains hairs that trap large particles. The respiratory and olfactory epithelia are covered by mucus, which is produced by goblet cells and glands. The mucus moves posteriorly due to ciliary beating and is swallowed. The different cell types described in the respiratory epithelium are the ciliated cells, mucus (goblet) cells, nonciliated columnar cells, cuboidal cells, basal cells, brush cells, and small mucus granule cells. Not all of these cell types have been observed in all species. The olfactory epithelium is composed of basal, neuronal (olfactory), and sustentacular (support) cells. The respiratory and olfactory epithelia of rats, other animals, and humans contain many xenobiotic-metabolizing enzymes such as carboxylesterases, aldehyde and formaldehyde dehydrogenases, cytochromes P450, epoxide hydrolase, rhodanese, UDP-glucuronyl transferase, and glutathione (GSH) S-transferases. These enzymes are distributed heterogeneously among the cell types comprising the respiratory and olfactory epithelia. Nasal metabolism, which can result in detoxification or activation of inhalants, may render particular cell types or regions more susceptible or more resistant to chemically induced injury.

Posterior to the nose and oral cavity, the nasopharynx and oropharynx join to form the pharynx. These pharyngeal cavities are lined by a mucociliary epithelium and are coated with mucus. Large particles tend to impact the walls of the pharynx. The larynx (voice box) controls the flow rate and volume of air passing into the trachea. It is lined by transitional and mucociliary respiratory epithelium. Mucus in the larynx moves upward toward the pharynx and is swallowed. The larynx of rodents is one of the predilection sites for damage by a variety of inhaled materials, including cigarette smoke, glycerol, and copper sulfate heptahydrate.

24.2.2 Tracheobronchial Region

The trachea, bronchi, and bronchioles conduct air to the pulmonary parenchyma. The trachea extends from the larynx distally where it divides to form the two main bronchi, which enter the right and left lungs. The bronchi continue to divide and eventually they decrease in diameter to form bronchioles, which continue to divide for several generations. The most distal, conducting segment of the tracheobronchial tree is called the terminal bronchiole (Fig. 24.2).

The walls of the human trachea and main bronchi contain cartilaginous rings that change to cartilaginous plates in smaller bronchi. In the rat, cartilaginous rings are found in the trachea and extrapulmonary bronchi, but cartilage is not found along intrapulmonary airways. The cartilage in the walls, along with smooth muscle, helps

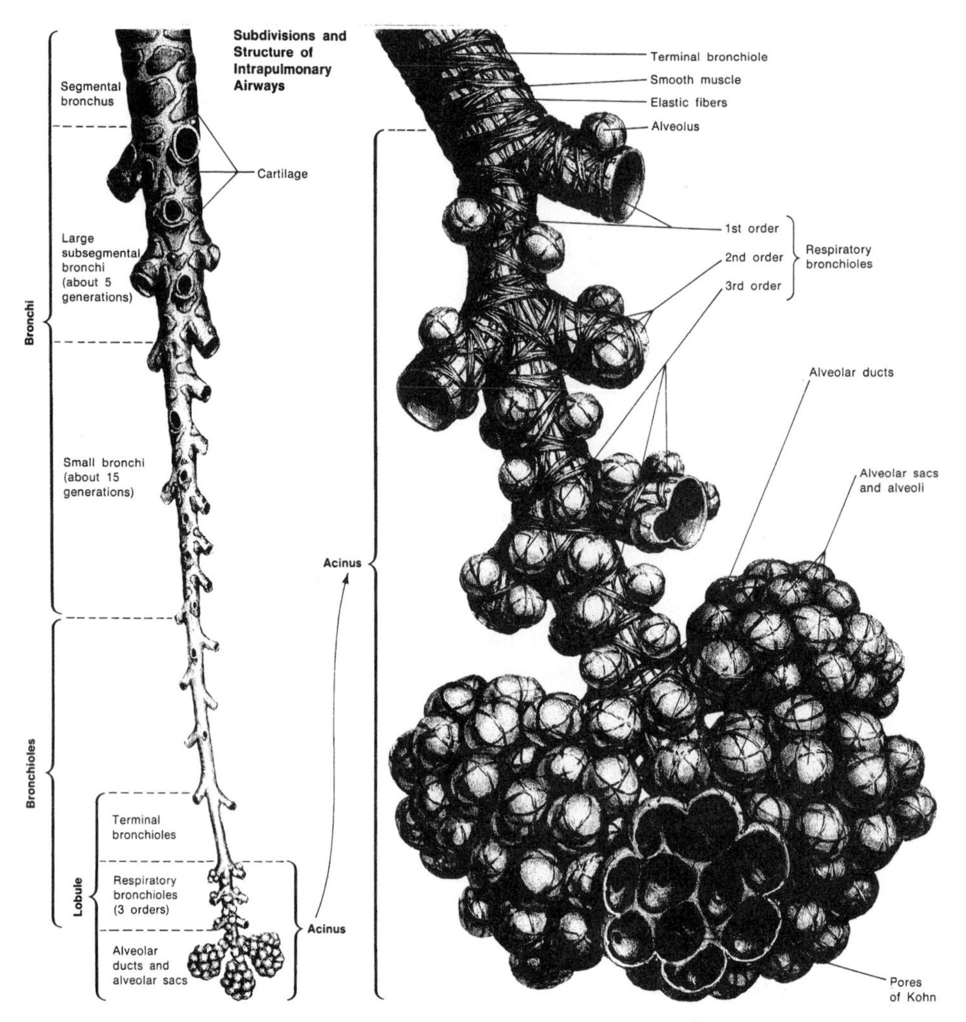

FIGURE 24.2 Structure of intrapulmonary bronchi, bronchioles, and acinar units in humans.
Note that cartilage is located in the walls of bronchi, but not distally. Smooth muscle is found in the walls of bronchi, and bronchioles, and a few smooth muscle fibers are found in alveolar ducts, but not in the alveolar walls.

Source: From the CIBA collection of medical illustrations, Vol. 7, *Respiratory System*, Frank H. Netter, p. 24, 1979. Reprinted with permission.

prevent collapse of the trachea and bronchi. The walls of bronchioles are devoid of cartilage, but contain smooth muscle and elastic fibers. The trachea and bronchi are lined by a pseudostratified, columnar, ciliated respiratory epithelium similar to that lining the nasopharynx (Fig. 24.3). Secretory cells in this epithelium may be predominately mucus cells or serous cells depending on species and previous exposure to inhaled materials or organisms. Also depending on species, serous and mucus glands are present in the submucosa with variable frequency. For example, submucosal glands are present in human bronchi, and they may enlarge and secrete more

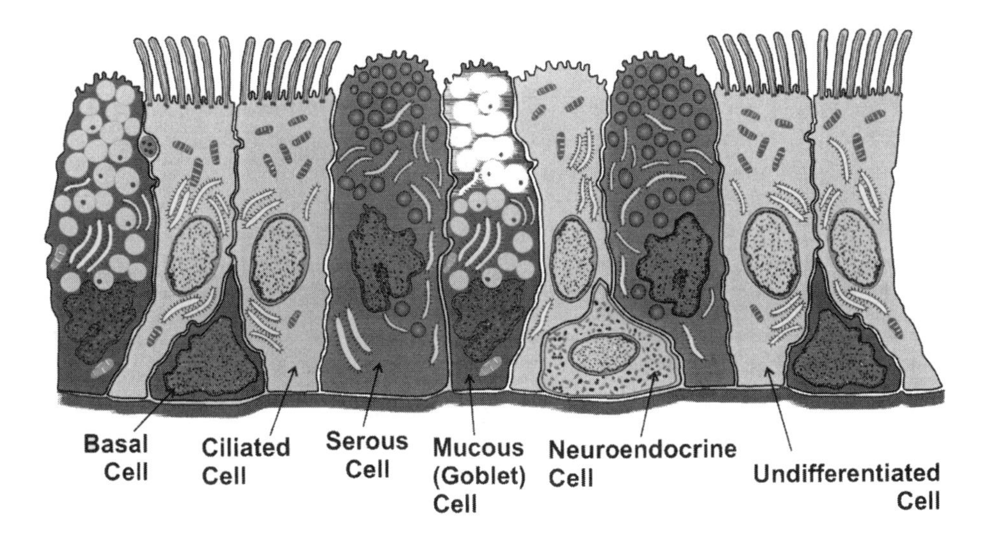

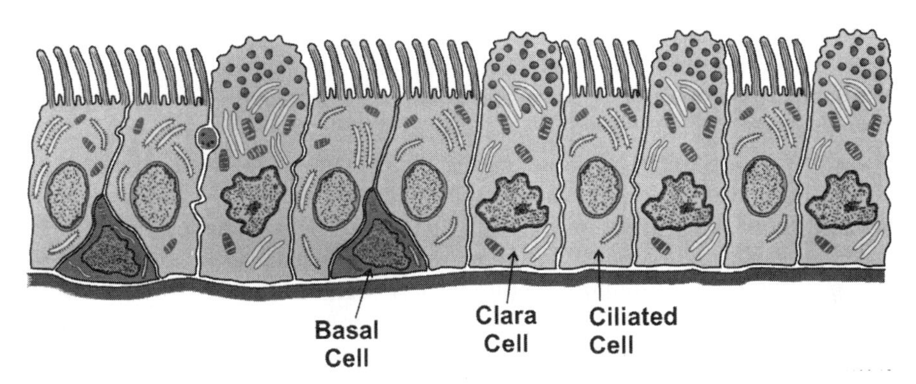

FIGURE 24.3 (A) Illustration of the typical mucocilliary epithelium lining the trachea and bronchi. In humans, most secretory cells are mucus-secreting goblet cells. In some other species, e.g., rats, serous cells are the predominant secretory cell type. Ciliated cells are common in the mucociliary epithelia of all species and basal cells are also numerous. Brush cells, undifferentiated cells, and Kulchitsky (neuroendocrine) cells are rare. (B) Illustration of the typical epithelium lining the bronchi. Ciliated cells and Clara cells predominate. Basal cells are found in proximal but not distal bronchioles.

Source: Adapted with permission from the CIBA collection of medical illustrations, Vol. 7, *Respiratory System*, Frank H. Netter, p. 26, 1979.

abundant mucus with chronic irritation such as cigarette smoking. Submucosal glands are absent in the bronchi of rats and mice, as well as in bronchioles, which are lined by a simple columnar ciliated epithelium. In the small bronchioles, mucus and serous cells gradually disappear and are replaced by nonciliated cuboidal (Clara) cells (Fig. 24.3). These cells have a high concentration of P450 enzymes and may be selectively damaged by toxicants that require metabolic activation by cytochrome P450s. Throughout the tracheobronchial region, mucociliary clearance, that is, the active propulsion of mucus and any foreign material embedded in it toward the pharynx, is an important clearance and defense mechanism.

The dimensions (diameters and lengths), numbers of generations, and geometry (branching patterns and angles) of the tracheobronchial airways vary considerably by species and largely determine deposition of particles in this region versus penetration to the parenchymal region. Two general branching patterns are observed in mammals. Most mammals exhibit markedly asymmetric monopodial branching, characterized by long, tapering main airways and numerous small branches, each typically making a 60° angle with the main airway. Dogs, rabbits, guinea pigs, hamsters, rats, and mice exhibit this type of asymmetric, monopodial branching. Humans exhibit the other pattern, called regular dichotomous branching. This branching is nearly symmetric and consists of two daughter airways at each branch point that are nearly equal in diameter and branching angle.

24.2.3 Pulmonary Parenchyma

The pulmonary parenchyma is the region where gas is exchanged. The main structures are the respiratory bronchioles, alveolar ducts, alveolar sacs, and alveoli (Fig. 24.2). Respiratory bronchioles are the bronchioles distal to the terminal bronchioles. They are lined by a simple cuboidal epithelium consisting of ciliated and Clara cells and have alveoli opening into their lumina. Because these alveoli participate in gas exchange, the respiratory bronchioles do not serve solely to conduct air. Humans, monkeys, dogs, cats, and ferrets have well-developed respiratory bronchioles. Rodents, cattle, sheep, and pigs have rudimentary to no respiratory bronchioles. Respiratory bronchioles end in alveolar ducts. In those species without respiratory bronchioles, the terminal bronchioles end in alveolar ducts.

Alveolar ducts are tubular structures whose walls are covered by alveoli. Alveolar ducts may branch to form two daughter alveolar ducts, or they can branch to form two blind-ended alveolar sacs. The alveoli resemble incomplete polyhedra. The open portion of the polyhedron exchanges gas with a respiratory bronchiole, alveolar duct, or alveolar sac. The wall of the alveolus is composed of the alveolar epithelium, a small amount of collagenous and elastic connective tissue, and network of capillaries lined by endothelial cells. CO_2 and O_2 are exchanged between air and blood by diffusion across the thin alveolar epithelium and adjacent endothelium and their fused basement membranes (Fig. 24.4). This distance, known as the blood-air barrier, is only about 0.4 μm thick. Tight junctions exist between alveolar epithelial cells, whereas zonulae occludens-type junctions are less tight than the junctions between epithelial cells and exist between the capillary endothelial cells. Therefore, the main permeability barrier to the leakage of fluid into the alveolar lumen is the alveolar epithelium.

The major cells in the alveoli are type I and type II epithelial cells, capillary endothelial cells, interstitial cells, and macrophages. Type I cells constitute 8–11% of the structural cells found in the alveolar region, and they cover 90–95% of the alveolar surface. Their main function is to prevent fluid leakage across the alveolar wall into the lumen, while allowing gases to equilibrate across the air-blood barrier. They are thin cells with a smooth surface and are vulnerable to damage from a variety of inhaled materials because of their large surface area. Their repair capacity is limited because they have few organelles associated with energy production and macromolecular synthesis.

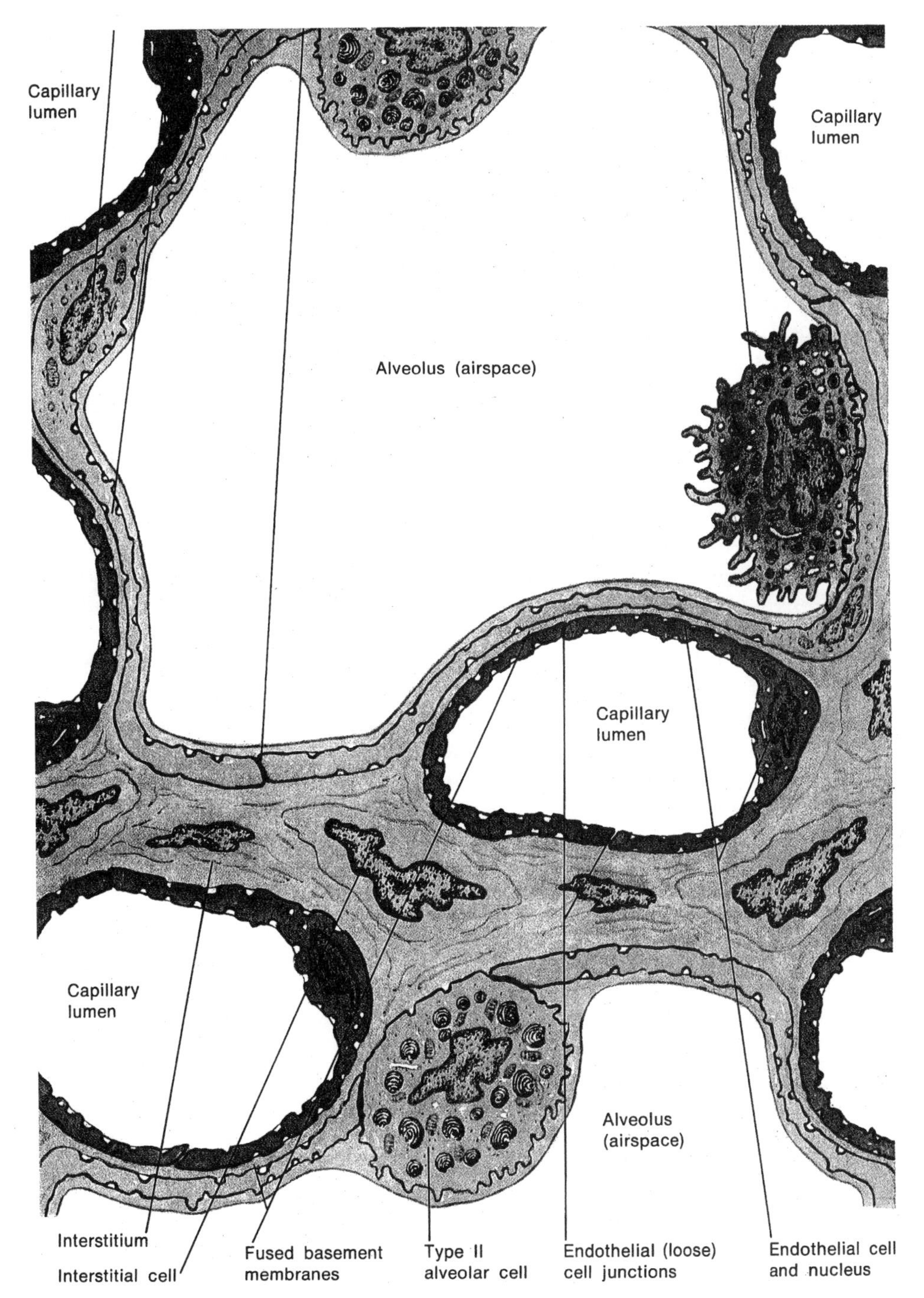

FIGURE 24.4 Ultrastructure of pulmonary alveoli and capillaries.
Note how the capillary network weaves through the interstitium (alveolar septa), so "thin" areas and "thick" areas alternate. Gas is diffused across the thin area where endothelial and epithelial basement membranes are fused, and there are no intervening interstitial cells. The thick areas contain interstitial cells and a small amount of connective tissue. Edema fluid first collects here when the endothelium is damaged, and this is where inflammatory cells first infiltrate.

Source: From the CIBA collection of medical illustrations, Vol. 7, *Respiratory System*, Frank H. Netter, p. 29, 1979. Reprinted with permission.

Type II cells constitute 12–16% of the structural cells in the alveolar region, but they cover only about 7% of the alveolar surface. They are cuboidal cells with a microvillus surface and characteristic organelles called lamellar bodies, which store surfactant. Their main function is to secrete surfactant that lowers the surface tension in the alveoli, thus reducing the tendency for the alveoli to collapse. Type II cells also serve as the progenitor cells for type I cells, which cannot replicate; thus, they repair the alveolar epithelium after injury. Type II cells can metabolize some xenobiotics.

A type III alveolar epithelial cell known as a brush cell has been described in some species. It comprises only a small percentage of the epithelial cells, and its function is unknown.

Capillary endothelial cells comprise 30–42% of cells in the alveolar region. The endothelium forms a continuous, attenuated cell layer that transports respiratory gases, water, and solutes. It forms a barrier to the leakage of excess water and macromolecules into the interstitium. Endothelial cells also metabolize vasoactive amines, prostaglandins, adenine nucleotides, peptides, lipids, hormones, and drugs. Thus, the endothelial cells help regulate the quantities of these substances in circulation and may influence activities of other organs. Pulmonary endothelial cells are susceptible to injury from inhaled substances and substances in the systemic circulation. Injury to endothelial cells may result in pulmonary interstitial edema.

The interstitium consists of resident interstitial cells, mostly fibroblasts, and scant connective and elastic tissue. Although the amount of collagen and elastin in the pulmonary parenchyma is small, these structural proteins are key to normal pulmonary mechanics. Increases or decreases in these proteins lead to impairment as in pulmonary fibrosis or emphysema. The resident interstitial cells comprise approximately 35% of the structural cells in the alveolar region. Interstitial macrophages, lymphocytes, plasma cells, and mast cells are also present in the interstitium. The numbers of these cells, as well as neutrophils, may increase with inflammation, which may also cause the interstitial space to thicken due to edema.

24.2.4 Macrophages

Three populations of macrophages have been described in the lung. The best studied are the alveolar macrophages, which are recognized as large mononuclear cells with numerous vacuoles located in alveolar lumens. The vacuoles correspond to lysosomes, and the size of the macrophages and the number and size of lysosomes generally increase with activation of the macrophages. Macrophages can move along the surface of the alveolus. These cells have important roles in lung defense as well as inflammatory and immune responses. Their main defense roles are phagocytosis, killing, and clearance of microorganisms, such as bacteria, and phagocytosis and clearance of inhaled particulate material, such as dusts. Once phagocytosis occurs, clearance can be by dissolution or mechanical clearance via the mucociliary escalator or the interstitium and lymphatics. Phagocytosis also triggers release of cytokines and chemokines, growth factors, oxygen radicals, and enzymes that mediate inflammatory responses. These mediators recruit and activate other cells and can eventually lead to resolution of the inflammation or structural alterations such as fibrosis or emphysema. Macrophages also participate in specific immune responses by presenting antigen to lymphocytes and releasing cytokines that are important in modulating the immune response.

Another population of pulmonary macrophages is derived from the bone marrow and resides in the interstitium. These cells can phagocytose particulate material that crosses the alveolar epithelium into the interstitium, or the cells can move into alveolar lumens to become alveolar macrophages. Both alveolar macrophages and interstitial macrophages are capable of cell division, so their numbers can increase in response to particulate material via replication in situ and via recruitment of mononuclear cells from bone marrow to the interstitium and the alveolar lumens.

The third population of pulmonary macrophages, the intravascular macrophage, is less well studied and is not present in all species. Intravascular macrophages are present in humans, ruminants, pigs, and cats. They are not present in rats, mice, rabbits, or dogs. These cells appear to be similar to Kupffer cells in the liver, and they remove particulate material such as cell debris, damaged red blood cells, and bacteria from the pulmonary capillary bed.

24.2.5 Circulatory System

Two separate arterial systems supply blood to the lung. The bronchial system is a high-pressure arterial system. Oxygenated blood from the left ventricle travels through the aorta to the bronchial arteries that supply the large conducting airways, the pleura, and the large pulmonary vessels. Bronchial arteries end at the level of the terminal bronchioles. Their capillaries merge with pulmonary capillaries and with bronchial venules. Bronchial veins drain blood from the bronchial tree and return this blood to the right heart.

The pulmonary system is a low-pressure arterial system. Oxygen-poor blood from the right ventricle travels through pulmonary arteries to supply capillary beds of the respiratory bronchioles, alveolar ducts, and alveoli where gas is exchanged. Pulmonary venules in the parenchyma merge into pulmonary veins that run through the intralobular septa to the hilus. This well-oxygenated venous blood is returned to the left heart.

24.2.6 Lymphatic System

The pulmonary lymphatic system is often divided into a superficial and deep portion, but these portions are connected. The superficial portion is located in the connective tissue of the pleura lining the lung. The deep portion is in the connective tissue surrounding the bronchovascular tree. The two portions connect in the interlobular septa. The lymphatic vessels are structurally similar to thin-walled veins. The presence of valves in the lymphatic vessels and the movement of the lung during respiration promote flow of lymph from the periphery and pleura toward the hilus. Afferent lymphatics from the lung drain into the tracheobronchial lymph nodes. Species differ in the number and location of lymph nodes in the hilar and tracheal area. Lymph from these nodes drains into the thoracic, right, and left lymphatic ducts and from these ducts into the systemic venous system.

The major function of the pulmonary lymphatic system is to remove excess fluid from the connective tissue spaces. The lymphatic system is also important in clearing particulate material from the lung to the lymph nodes.

24.2.7 Nerves

The respiratory tract has sensory (afferent) and motor (efferent) innervation. Both the parasympathetic and sympathetic portions of the autonomic nervous system provide the motor innervation. Preganglionic parasympathetic fibers descend in the vagus nerves to ganglia located around airways and blood vessels. The postganglionic fibers innervate the smooth muscle of airways and blood vessels, bronchial glands, and epithelial mucous cells. In general, the same structures are also innervated by postganglionic fibers from the sympathetic ganglia. Vagal stimulation causes airway constriction, dilation of the pulmonary circulation, and increased glandular secretion. Conversely, sympathetic nerve stimulation causes bronchial relaxation, constriction of pulmonary blood vessels, and inhibition of glandular secretion.

Sensory receptors that respond to irritants or mechanical deformation are located throughout the respiratory tract. Stimulation of these receptors leads to reflex responses, for example, stimulation of nasal receptors may cause sneezing. Three main vagal sensory reflexes and their corresponding receptors are known as, (1) bronchopulmonary stretch receptors, (2) irritant receptors, and (3) C-fiber receptors. Stretch receptors are associated with smooth muscle of the trachea and bronchi. They are stimulated by lung inflation and normally function to terminate inspiration. Rapidly adapting irritant receptors are located in the epithelium of extrapulmonary and to a lesser extent intrapulmonary bronchi. They respond to a variety of stimuli, including inhalation of gaseous irritants, mechanical stimulation of the larynx and airways, mechanical stimulation by large inflations and deflations, hyperpnea, and anaphylactic reactions in the lung. The reflex responses that follow stimulation of irritant receptors include bronchoconstriction, hyperpnea, constriction of the larynx during expiration or cough, and increased mucous secretion. Pulmonary C-fiber receptors are located in the pulmonary parenchyma and include Paintal's juxtacapillary (or J) receptors that respond to stimuli in and around capillaries in alveolar walls. It has been suggested that pulmonary C-fibers may contribute to the sensation of dyspnea that accompanies pulmonary edema, pneumonia, and inhalation of noxious gases. Bronchial C-fibers are located along the conducting airways, and they respond to stimuli in and around the bronchial arterial system. When stimulated, they cause airway constriction and may contribute the sensations experienced during asthma and anaphylaxis. Stimulation of both pulmonary and bronchial C-fibers causes a reflex increase in airway secretion.

24.3 PULMONARY FUNCTION

The major function performed by the respiratory tract is the transfer of the life-giving gas oxygen from the inhaled air to the blood for transport throughout the body and the transfer of the metabolic gas CO_2 from the blood to the exhaled air. Toxicity to the respiratory tract can interfere with this gas-exchange function by (1) altering the tone of the airways resulting in altered airflow; (2) damaging the delicate alveolar/capillary barrier of the deep lung resulting in impaired transfer of gases; or (3) causing tissue damage that leads to long-lasting structural changes in the lung, resulting in altered lung volumes (volumes of air that can be inhaled or exhaled) or lung mechanics (degree of elasticity of the lung) or alterations in the distribution of gas within the lung.

Pulmonary function tests are used to evaluate the net effect of inhaled toxicants on the ability of the respiratory tract to perform its major function; such tests can be performed on living animals or humans. Although pulmonary function tests can detect the functional impacts of structural changes, the tests cannot determine specific structural changes or substitute for histopathologic evaluations of the respiratory tract tissues. The tests that are commonly used to evaluate pulmonary function will be discussed in this section.

24.3.1 Evaluation of Breathing Patterns

Breathing patterns are a common means of evaluating pulmonary function and are usually documented by measures of respiratory frequency, tidal volume, and minute volume. Figure 24.5 indicates the standard subdivisions of lung volume. Tidal volume is the volume of air inhaled in a single breath during normal breathing. The minute volume is the tidal volume times the respiration rate per minute and the volume of air inhaled per minute. Tidal volume is usually measured by some type of pneumotachograph or plethysmograph, which record respiration rate and tidal volume. Such measurements are readily influenced by factors such as anesthetics, excitement, exercise, and body temperature, factors that must be considered in the interpretation of the values. The breathing patterns of exposed mice are sometimes used in toxicology to evaluate the irritancy of an airborne chemical or mixture toward the respiratory tract. Breathing patterns are also used in toxicology studies to record the total amount of an airborne pollutant to which an animal is exposed by inhalation.

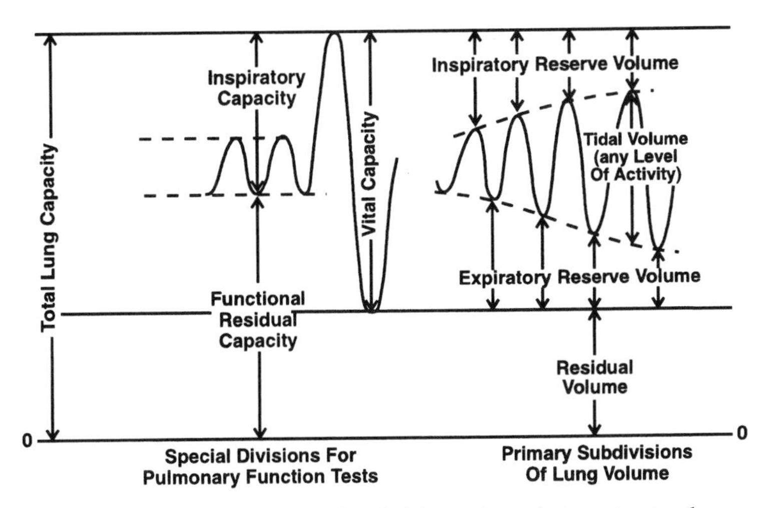

FIGURE 24.5 Illustration of the divisions of respiratory tract volumes.

24.3.2 Evaluation of Lung Volumes

An evaluation of lung volumes can provide information on the total capacity of the lung and the extent to which that capacity has been altered by a toxicant. As shown in Figure 24.5, the total lung capacity is the total volume of the lung at maximum inspiration. The vital capacity is the maximum volume that can be exhaled starting at the total lung capacity. In humans these measurements can be made by voluntary effort and recorded by plethysmography. Techniques have been developed in laboratory animals to make similar measurements by applying positive airway pressure or negative pressure surrounding the body of the animal while it is in a whole-body plethysmograph. Total lung capacity and vital capacity are reduced in fibrotic lung disease in which the lung becomes smaller and stiffer. This type of change in the lung is called restrictive lung disease and is marked by smaller lung volumes and little change in airflow. In an emphysematous lung, on the other hand, the total lung capacity may increase as a result of the breakdown of alveolar walls and loss of elastin fibers that allow the lung to deflate on exhalation, but vital capacity is often reduced due to airway collapse during exhalation. This type of change in the lung is called obstructive lung disease and is marked by reduced airflow. Measures of lung volumes can be used in toxicology to determine if the degree of structural alteration (fibrotic or emphysematous) is sufficient to change the lung function.

24.3.3 Evaluation of Lung Mechanics

Measures of the ability of the lung to expand to fill with air during inspiration and to deflate during exhalation are termed as lung mechanics. These properties depend on the elasticity of the lung and the caliber of the airways. Lung mechanics are commonly reported as compliance and resistance. Compliance is the volume change per unit pressure change. Resistance is the pressure difference per change in airflow. Compliance and resistance are commonly measured during steady tidal breathing by recording transpulmonary pressure, tidal volume, and airflow rate. Because the direct measurement of transpulmonary pressure requires an esophageal balloon or intrapleural catheter, indirect measures approximating compliance and resistance are typically applied to humans using plethysmography.

The most common test of lung function in humans is the forced exhalation test, which evaluates both lung volume and flow performance. The person inhales to his/her total lung capacity, then exhales as rapidly and deeply as possible to his/her residual volume. The volumes of air exhaled per unit time and the airflow during exhalation are recorded. The test provides information on both lung volumes and lung mechanics. The forced expiratory volume in 1 second (FEV_1) as well as the FEV_1 as a percentage of the forced vital capacity are typically reported. The shape of the tracings of the flow/volume curve also provides useful information. Reductions in peak flow and flow at high lung volumes are generally indicative of large airway obstruction. Reductions of flows at lower lung volumes are generally indicative of obstruction in smaller airways. The equivalent of the forced exhalation tests as performed in humans can be conducted in animals using negative airway driving pressures.

24.3.4 Evaluation of Gas Distribution

The uniformity of the distribution of gases within the lung can provide information on the uniformity of pathologic lesions within the lung. Multiple-breath or single-breath gas washout curves are used to assess this characteristic. A subject is switched from breathing air to breathing another gas, and the rapidity with which the new gas washes out the original gas in the lung is recorded. A common approach is to switch the person from breathing air to breathing oxygen, then to measure the breath-by-breath decrease in nitrogen in exhaled air. If the gases are distributed uniformly throughout the lung, the amount of nitrogen in the exhaled air decreases quickly following a multiexponential decay function. There is an initial rapid loss of nitrogen from the conducting airways, followed by a less rapid release of nitrogen from the major portion of the lung volume, and finally, a slower release of nitrogen from poorly ventilated areas of the lung or from release of nitrogen from the blood. These tests can be performed in both humans and animals.

24.3.5 Evaluation of Alveolar-Capillary Gas Exchange

A key measure of pulmonary function is the partial pressure of arterial blood gases, as measured by electrode techniques. Normal partial pressures of oxygen in arteries are in the range of 85–100 mm Hg, depending on the altitude at the place of measurement. Arterial pH is closely linked to the partial pressure of arterial CO_2. Based on the following reaction, increases in CO_2 in the blood will push the reaction to the right leading to a more acidic condition (more hydrogen ions), and less CO_2 will lead to a more alkaline condition (less hydrogen ion):

$$CO_2 + H_2O \Leftrightarrow H_2CO_3 \Leftrightarrow H^+ + HCO_3^-$$

If the elevated CO_2 is caused by hypoventilation, the acid condition is called "respiratory acidosis." If hyperventilation leads to reduced arterial CO_2, the condition is known as "respiratory alkalosis."

A more sensitive indicator of impairment in alveolar-capillary gas exchange is the alveolar-arterial difference in O_2 and CO_2 partial pressures. An increased pressure difference can reflect a diffusion barrier in subjects still able to maintain normal blood gases by increasing ventilation. Alveolar gas pressures are estimated by analyzing the last portion of an exhaled breath, and blood gases are measured as described above.

Another measure of the gas exchange characteristics of the alveoli is the carbon monoxide (CO) diffusing capacity. This measurement is one of the most sensitive measures of impaired gas transport across the alveolar-capillary barrier. A common method for determining the diffusing capacity of the lung for CO is to inflate the subject's lung with a large, single breath of gas containing CO and an inert gas such as neon or helium. The gas is allowed to remain in the lung for a measured length of time (5–10 seconds). The dilution of the inert gas allows calculation of the gas volume in the lung. The CO difference in the inhaled gas versus the alveolar gas (the last gas in an exhalation from the lung) indicates the fractional uptake of CO. CO diffusing capacity is calculated as the rate of CO uptake into the blood per unit of alveolar CO pressure.

24.4 BIOCHEMISTRY OF THE RESPIRATORY TRACT

24.4.1 Energy Sources

The major sources of energy for the lung are fatty acids and glucose. The lung has a high rate of oxidation of fatty acids, which enter the lung readily via hydrolysis of blood triglycerides by lipoprotein lipases in the endothelial cells of the lung. The lung uses the fatty acids both for energy and phospholipid synthesis (see Section 24.4.3 on surfactant synthesis). The lung has a high level of glycolysis and resultant lactate production, despite the aerobic nature of the organ, which is constantly exposed to oxygen in the air. The lung also has a relatively high rate of oxidation of glucose via the hexose monophosphate shunt, a pathway that leads to production of reduced nicotamide adenine dinucleatide phosphate (NADPH). This production is essential for the lung because it is required for fatty acid synthesis (needed to repair the membranes in the lung), cytochrome P450 oxidation of xenobiotics (needed to protect the body from pollutants that enter the lung via inhalation), and for formation of reduced GSH (a tripeptide-reducing agent needed to protect the lung against oxidant damage).

24.4.2 Oxidative and Antioxidative Processes in the Lung

The lung is constantly exposed to an atmosphere of 20% oxygen and on occasion to oxidant pollutant gases such as ozone and nitrogen dioxide. In addition, phagocytic cells in the alveoli generate reactive oxygen species to kill infectious agents in the lung. These species, also generated upon phagocytosis of dusts, are the basis for some of the lung damage produced by those particles (see Section 24.4.5 on biochemical activity in the alveolar space). The lung must therefore have a means of protecting against oxidant stress.

24.4.2a Oxidants in the Respiratory Tract

Life-giving oxygen is also the source of superoxide anion ($O_2^{\bullet-}$, the product of a one-electron reduction of diatomic oxygen), an oxidizing agent. $O_2^{\bullet-}$ is formed by mitochondrial metabolism, molybdenum hydroxylase reactions (xanthine, sulfite, and aldehyde oxidases), arachidonic acid metabolism, and NADPH oxidase-dependent processes in phagocytic cells. Hydrogen peroxide (H_2O_2) is formed by the further oxidation of $O_2^{\bullet-}$ as catalyzed by superoxide dismutase (SOD):

$$2O_2^{\bullet-} + 2H^+ \rightarrow H_2O_2 + O_2$$

Hydrogen peroxide can be formed directly from diatomic oxygen via a two electron reduction by enzymes such as D-amino acid oxidase, glycolate oxidase, or urate oxidase. H_2O_2 can serve as a source of the highly toxic hydroxyl radical (•OH) via the iron-catalyzed Fenton reaction:

$$Fe(II) + H_2O_2 \rightarrow Fe(III) + \bullet OH + OH^-$$

The $O_2^{\bullet-}$, in the presence of H_2O_2 and a divalent metal can also produce •OH via the iron-catalyzed Haber-Weiss reaction:

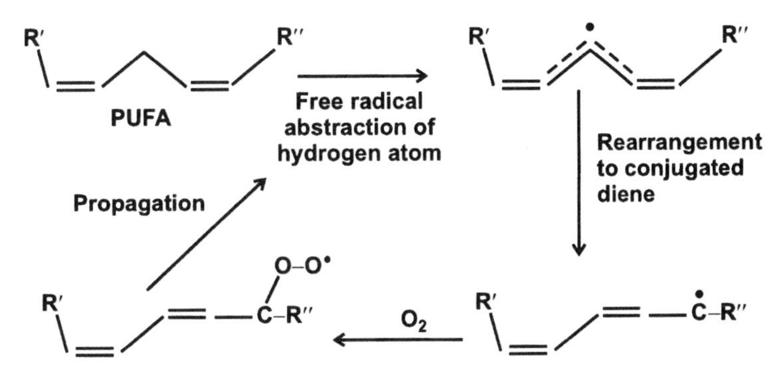

FIGURE 24.6 Peroxidation of polyunsaturated fatty acids (PUFA) by free radicals. The free radical abstracts a hydrogen atom from the methylene group leaving a fatty acid free radical, which rearranges to a conjugated diene radical. This radical can interact with oxygen to form a peroxy radical, which can abstract another hydrogen from a PUFA molecule and propagate the reaction. R′ and R″ are the two ends of the fatty acid of unspecified chain link.

$$Fe(II) + H_2O_2 \rightarrow Fe(III) + \bullet OH + OH^-$$

$$Fe(III) + O_2^{\bar{\cdot}} \rightarrow Fe(II) + O_2$$

$$O_2^{\bar{\cdot}} + H_2O_2 \rightarrow \bullet OH + OH^-$$

The •OH is a more potent oxidizing agent and thus potentially more damaging to cells than is $O_2^{\bar{\cdot}}$.

The function of such reactive molecules is thought to be the killing of infectious organisms in the airways. However, if these reactive oxygen species are not balanced by antioxidant molecules (see below), and particularly if oxidative stress is added from exposure to pollutants, such as ozone and nitrogen dioxide, the cells of the lung will be damaged by lipid peroxidation of the polyunsaturated fatty acids (PUFAs) in the cell membranes (Fig. 24.6). The peroxidation begins with the abstraction of a hydrogen from a methylene group of the PUFA by a radical species, such as $O_2^{\bar{\cdot}}$. The superoxide is what is known as a free radical and has an unpaired electron in its outer shell. Such a radical is highly reactive because it requires another electron to drop to a lower, more stable energy state. A free radical tends to abstract hydrogen atoms (which contain one electron) from other molecules, resulting in the formation of another radical species. Such reactions tend to expand and propagate more free radicals unless the reaction is quenched by reducing agents (hydrogen donors) such as vitamin E. In the case of PUFA, a fatty acid radical is formed that rearranges to form a peroxy radical. This radical can either abstract a hydrogen atom from another fatty acid or, if the PUFA peroxy radical contains three or more olefinic bonds, can undergo internal (within the molecule) cyclization to form monocyclic or bicyclic endoperoxides, as takes place in the formation of prostaglandins.

In addition to reactive oxygen species, the lung is a source of reactive nitrogen species. Nitric oxide (NO, sometimes written as NO• to emphasize that the compound has an unpaired electron in its outer shell) is a free radical and a vaso- and bronchodilator serving to regulate airway tone. It is produced by a constitutive form of the NADPH-requiring enzyme, NO synthase, in endothelial cells and in neutrophils. Macrophages and airway epithelial cells have an inducible form of this enzyme and release NO upon induction by bacterial products or inflammatory cytokines. NO has been implicated in cell injury and repair in association with inflammation and with the induction of pulmonary edema. Excess NO is produced in ischemia-reperfusion injury and in the inflammatory response to hyperoxia, ozone, and bacterial products (endotoxin). Under the same conditions there is excess production of the $O_2^{\bar{\cdot}}$ and H_2O_2. $O_2^{\bar{\cdot}}$ reacts with NO to form peroxynitrite, a cytotoxic oxidant that has been implicated in lipid peroxidation, DNA damage, damage to surfactant, pulmonary edema, and acute lung injury.

24.4.2b Antioxidants in the Respiratory Tract

The respiratory track is protected against oxidant stress by molecules such as vitamin E (a-tocopherol), GSH, vitamin C (ascorbic acid), and uric acid, as well as enzymes such as SOD, catalase, GSH peroxidase, GSH synthetase, GSH reductase, GSH transferase, and semidehydroascorbate reductase. GSH and ascorbic acid are both water-soluble antioxidants (also called reducing agents or hydrogen donors) and are both present in lung tissue at concentrations of 1–2 μmol/g wet weight.

GSH is a tripeptide (L-γ-glutamyl-L-cysteinyl glycine) that is present in high concentrations intracellularly, but is also present extracellularly. GSH is synthesized within cells via the enzymes of the γ-glutamyl cycle, with γ-glutamylcysteine synthetase being the enzyme that catalyzes the initial and rate-limiting step. Exposure of rats and mice to oxidant gases such as ozone induces high levels of GSH synthesis in the lung. Inhibition of this enzyme with buthionine sulfoximine leads to GSH deficiency. This deficiency caused by inhibition of GSH synthesis is not alleviated by administration of GSH, because it is not readily transported into cells. Esters of GSH, on the other hand, readily enter cells and are hydrolyzed to form GSH; and these esters show promise as therapeutic agents to relieve GHS deficiency. N-acetylcysteine, a substrate for γ-glutamylcysteine synthetase, is used to treat oxidative stress in the lung for conditions such as chronic obstructive pulmonary disease (COPD), idiopathic pulmonary fibrosis, or oxidative damage to endothelial cells.

The lipid-soluble antioxidant vitamin E is present at less than a tenth of the concentration of the water-soluble antioxidants. Vitamin E is integrated into cell membranes where it can protect against peroxidation of the membrane lipids. The water-soluble antioxidants GSH and ascorbic acid are thought to be the reducing agents that regenerate the oxidized form of vitamin E. GSH, ascorbic acid, and uric acid are present in the epithelial lining layer of the respiratory tract and can be sampled by bronchoalveolar lavage. The concentration of GSH in the lining fluid (~400 μm) is approximately 100-fold higher in plasma. Ascorbic acid and uric acid are present in the lining fluid at about half the concentration of GSH. Lavage and lung tissue levels of GSH and ascorbic acid are increased by exposures to oxidant gases or cigarette smoke. However, in cigarette smokers with chronic bronchitis, decreased lavage fluid GSH correlates with decreased lung function. Thus, the lung appears to have defense mechanisms to compensate for some oxidant stress, but

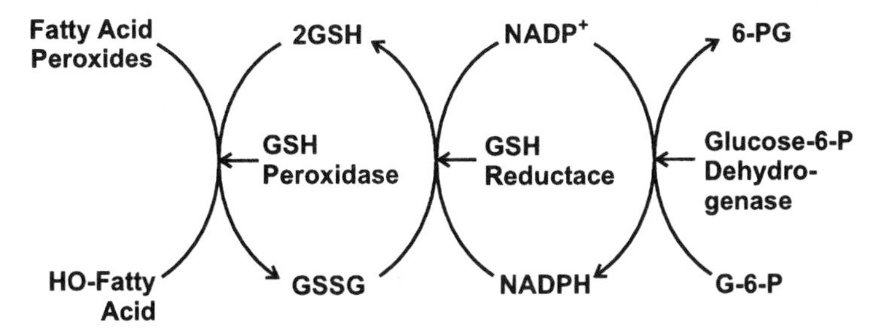

FIGURE 24.7 Antioxidant interactions in the lung.
GSH = reduced glutathione; GSSG = oxidized glutathione; $NADP^+$ = oxidized nicotinamide adenine dinucleotide phosphate; NADPH = reduced nicotinamide adenine dinucleotide phosphate; G-6-P = glucose-6-phosphate; 6-PG = 6-phosphogluconic acid.

once those defenses are overwhelmed, the lung is injured and lung function decreases.

The enzyme, SOD, as mentioned above, converts the superoxide free radical to H_2O_2. This is considered to be a protective reaction because the H_2O_2 is less reactive than the superoxide radical. SOD is found in all oxygen-consuming organisms; copper/zinc SOD is found in the cytosol, and manganese SOD is found in the mitochondrial matrix of eukaryotic cells.

Both catalase and the selenium-dependent GSH peroxidase convert H_2O_2 to water. Catalase is located in peroxisomes, which are associated with the smooth endoplasmic reticulum of cells. In the lung the enzyme is present mainly in type II cells, Clara cells, and in macrophages. GSH peroxidase exists in a selenium- and a nonselenium-dependent form, both of which are found in the lung. The nonselenium-dependent form catalyzes the breakdown of organic, but not inorganic, peroxides. The selenium-dependent form catalyzes the breakdown of both H_2O_2 and organic hydroperoxides.

Another important antioxidant molecule is NADPH, the cofactor necessary for the regeneration of GSH from its oxidized form, GSSG. The NADPH, in turn, is regenerated by the enzyme, glucose-6-phosphate dehydrogenase, as part of the hexose monophosphate shunt pathway of energy metabolism, a pathway that has high activity in the lung. Figure 24.7 illustrates the interaction of some oxidant/antioxidant activities.

24.4.2c Indicators of Oxidative Stress

Oxidative stress can lead to exhalation of H_2O_2 or hydrocarbon gases such as pentane or ethane. The latter come from the reaction of lipid hydroperoxides with antioxidants to form short-chain hydrocarbons or unsaturated aldehydes (malondialdehyde). DNA damage in the form of hydroxylated bases, such as 8-hydroxy-2′-deoxyguanosine, is also used as an indicator of oxidative stress. Oxidized plasma proteins containing carbonyl groups are elevated in the bronchoalveolar lavage fluid and the plasma from cigarette smokers. Bioactive prostaglandin F_2-like compounds that are made by free-radical catalyzed peroxidation of arachidonic acid (F_2-isoprostanes) are elevated in the plasma of smokers versus nonsmokers. All of these indicators of oxidant stress are decreased after antioxidant treatment.

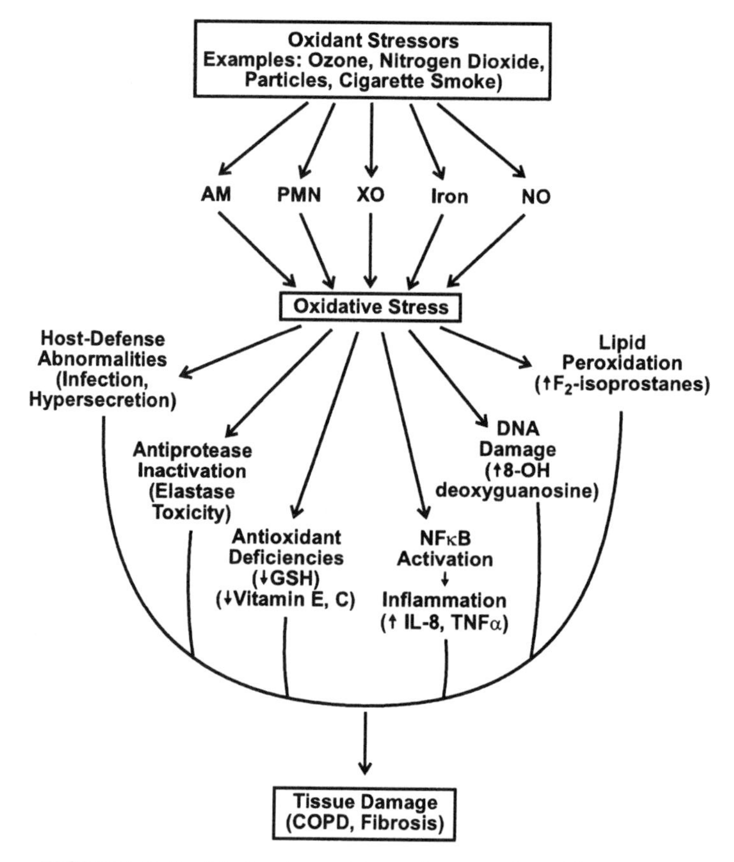

FIGURE 24.8 Mechanisms of lung damage via oxidant stress.
Source: Modified and used by permission from Repine et al., *Am. J. Respir. Crit. Care Med.* 156; 341–357, 1997.

Oxidants not only damage DNA, lipids, and proteins, but also induce a variety of processes that relate to lung disease. Oxidants increase mucus production by epithelial cells and cause mucous cell hyperplasia, induce apoptosis, reduce surfactant activity, impair cilia function, and increase epithelial cell permeability. Figure 24.8 summarizes some of the various ways in which oxidant stress can cause lung injury.

24.4.3 Synthetic Processes in the Lung

A major synthetic activity of the lung is synthesis of a surfactant that is a part of the alveolar epithelial lining fluid. The small alveolar sacs where gas is exchanged would not remain patent during respiration if not for the presence of a surfactant material that lines the alveoli and lowers the surface tension of the sacs during exhalation. Pulmonary surfactant is synthesized mainly in the alveolar type II cell, where it is stored prior to secretion in lamellar bodies that are distinctive morphological markers for this cell. The Clara cell also secretes glycoproteins and lipids and may contribute to the surfactant lining layer. The marker protein for Clara cell secretions is known as Clara cell secretory protein (CCSP). The molecular function of the CCSP is not known, but in knockout mice that do not have this protein, there is an increase in susceptibility to infections.

The surfactant of the lung contains several proteins (referred to as SP-A, -B, -C, or -D) and a special type of phospholipid. Two of the proteins (SP-A and SP-D) are water soluble and are collagen-like proteins called C-lectins. SP-A is tightly bound to the surfactant phospholipids and influences the structure and properties of the pulmonary surfactant. Two of the proteins (SP-B and SP-C) are hydrophobic and are small cationic proteins that promote the adsorption of the surfactant phospholipids at the air-liquid interface. SP-A and SP-D are members of the collectin family of host defense molecules functioning as opsonins that facilitate phagocytosis of various bacterial and viral pathogens.

The phospholipids of the surfactant differ from phospholipids that are a part of membranes, in that both long-chained fatty acids that are attached to the glycerol backbone of the phospholipid are fully saturated. Membrane phospholipids have one saturated fatty acid in the alpha position of the glycerol and an unsaturated, cis fatty acid in the beta position. The crook in the cis fatty acid chain accommodates a molecule of cholesterol in the membrane. The presence of two saturated fatty acids in the surfactant phospholipids allows them to be pushed together very tightly (there is no crook in the saturated fatty acid chains), producing a large outward force and lowering the surface tension. The surfactant is necessary for life; if the surfactant is not present, due to immaturity in newborns or due to removal or inactivation by lung injury, the alveoli collapse and can no longer perform their function of gas exchange. Surfactant proteins, and consequently surfactant function, can be damaged by the elastase released from neutrophils drawn into the alveolar space in an inflammatory response in the lung. The surfactant also modulates immune and stress responses in the alveoli.

Other synthetic processes in the lung include the synthesis of proteins such as collagen, elastin, and mucous glycoproteins. Collagen synthesis is increased in pulmonary fibroblasts under conditions leading to fibrosis (see Section 24.7.1 on chronic disease). Mucus synthesis and secretion are increased in the goblet cells of the upper airways in response to inhaled irritants and oxidants. Exposure to toxicant gases such as ozone leads to mucous cell metaplasia (i.e., an increase in the number of mucous cells in the epithelial lining cells), an apparent defense mechanism to help protect the epithelia lining the airways.

Enzyme synthesis can also be induced upon provocation by inhaled toxicants. For example, exposure to high levels oxygen induces increased activity of SOD and glucose-6-phosphate dehydrogenase (an enzyme in the hexose monophosphate shunt necessary for the production of NADPH), both enzymes needed to protect against oxidant toxicity. Enzymes responsible for the synthesis of the antioxidant tripeptide, GSH, are also induced during oxidant stress. Exposures to xenobiotic vapors or gases often induce higher activities of xenobiotic-metabolizing enzymes.

24.4.4 Catabolic Processes in the Lung

The lung is unique among the organs of the body in that the total cardiac output passes through the lung. Thus, it is a major site for clearance of vasoactive compound from the blood. Lung endothelial cells are the site of inactivation of substances such as serotonin, bradykinin, norepinepherine, adenine nucleotides and various prostaglandins. Angiotensin I is activated to angiotensin II by the angiotensin-converting enzyme via cleavage of two amino acids in caveolae and plasma membranes of the pulmonary endothelium.

Proteases in the lung are active in connective tissue breakdown and remodeling in lung injury. Antiproteases present in pulmonary macrophages and airway epithelia and released into the epithelial lining fluid inhibit the protease activity and prevent destruction of lung tissue (see Section 24.7.2 on chronic obstructive pulmonary disease [COPD]).

24.4.5 Biochemical Activity in the Alveolar Space

The alveolar space contains the epithelial lining fluid and cells free in the alveolar space. The major free cell in the healthy lung is the macrophage, a highly metabolically active cell that removes particulate matter and bacteria from the alveoli via phagocytosis and in the case of infectious agents, cell killing. The macrophage contains numerous acid hydrolases in intracellular bodies known as lysosomes; when foreign matter is phagocytized by the macrophages, the material is fused with a lysosome to form a phagolysosome, where the foreign matter is broken down. The macrophages may travel up the mucociliary escalator in the respiratory tract and be swallowed into the gut. If the phagocytized material is toxic, the macrophage may die and release the material to be phagocytized by another macrophage. Pulmonary toxins activate the macrophages, which then release a variety of cytokines, growth factors, proteases, antiproteases, and reactive oxygen and nitrogen species into the alveolar milieu, which in turn induce the release of other factors from the surrounding epithelial cells. All of these various factors influence the resolution of the disease process. Activated macrophages produce $O_2^{\bar{\cdot}}$ via membrane-associated NADPH oxidases, and this radical dismutates to form H_2O_2. If divalent cations are present, the H_2O_2 and $O_2^{\bar{\cdot}}$ react to form •OH, which contributes to cell injury during the inflammatory process. The reactive oxygen species can lead to peroxidative damage to the delicate membranes of the alveolar region of the lung and to formation of vasoactive lipids such as prostaglandins, thromboxanes, and leukotrienes.

Macrophages and epithelial cells have both been found to release NO, a potent vasodilator that plays an important role in regulating airway tone. The mediator is produced in cells via NADPH and the *l*-arginine-dependent enzyme, NO synthase, and in macrophages only after activation by toxins or cytokines.

Stimulated macrophages also release chemotactic factors for neutrophils, leading to an influx of these inflammatory cells. If an allergic component is present, eosinophils will also be attracted to the alveolar space. In large animals and in humans, an influx of lymphocytes accompanies some responses to toxic substances. Each of these inflammatory cells may produce a spectrum of pro- and anti-inflammatory mediators.

24.4.6 Xenobiotic Metabolism in the Respiratory Tract

The respiratory tract is a portal of entry into the body and, as such, has cells with high xenobiotic-metabolizing activity at strategic locations. These locations are sites where inhaled materials commonly deposit—the nose, the respiratory or terminal bronchioles, and the alveoli. The enzymes include phase I enzymes that activate the substrates into reactive electrophiles and phase II enzymes that hydrolyze the electrophiles or conjugate them with compounds that allow excretion of the xenobiotic metabolites into the urine. The major enzymes discussed in this section will be cytochrome P450 enzymes, flavin adenine dinucleotide (FAD) monooxygenases, oxidation/reduction enzymes, epoxide hydrolases, GSH transferases, sulfotransferases, glucuronosyltransferases, and other conjugating or phase II enzymes and esterases. There are many isoforms of these enzymes, each with its own substrate specificities.

24.4.6a Nose

The mucosae lining the nose include those lined by squamous, transitional, respiratory, and olfactory epithelia. Activity of the xenobiotic-metabolizing enzymes is highest in the olfactory and respiratory mucosae, with the olfactory mucosa showing a greater activity than the respiratory mucosa for most xenobiotic-metabolizing enzymes. The olfactory mucosa of small laboratory rodents has cytochrome P450 activities, which approach the activities found in the liver on a per milligram tissue weight basis. Mammals have a large number of exocrine glands called Bowman's glands in the olfactory mucosa. These glands contain large amounts of smooth endoplasmic reticulum, known to be a major site for cytochrome P450s. Chemicals such as aflatoxin, bromobenzene, chloroform, dibromoethane, dichlorobenzamide, dichlorobenzonitrile, and various nitrosoamines are activated to electrophiles in the Bowman's glands, where they bind to macromolecules and cause lesions. The respiratory mucosa also has cytochrome P450 activities but at lower levels than found in the olfactory mucosa. Human respiratory mucosa has cytochrome P450 levels about half those in the rat or dog and one-tenth those in the rabbit. The isoforms of cytochrome P450 found in the nose vary between species as illustrated in Table 24.1. The isoform 2G1 is found only in the nose and may be related to the metabolism of steroids.

Flavin-containing monooxygenases catalyze the oxidation of nitrogen-, phosphorus- and sulfur-containing xenobiotics and are present in the olfactory and respiratory epithelium of the nose. Carboxylesterases catalyze the hydrolysis of carboxylesters, carboxylamides, and carboxylthioesters and are among the highest enzyme activities present in the nose. Aldehyde dehydrogenases catalyze the oxidation of toxic aldehydes to less toxic carboxylic acids. Both formaldehyde and acetaldehyde dehydrogenase activities have been detected in varying amounts in the different nasal epithelia. The toxicities of the two aldehydes toward the nasal epithelia are inversely related to the substrate-specific enzyme activity in the epithelia, that is, the higher the aldehyde dehydrogenase activity at a specific site, the lower the toxicity of the aldehyde at that site.

Epoxide hydrolases catalyze the hydrolysis of reactive toxic epoxides, which may be formed via the cytochrome P450 enzymes. Both the cytochrome P450 enzymes and the epoxide hydrolases are located in the microsomes of the endoplasmic reticulum of the epithelia cells. The hydrolysis of epoxides to glycols is a detoxication reaction. Unlike many other xenobiotic enzymes, the epoxide hydrolases are present in about equal amounts in the respiratory and olfactory epithelia in the

TABLE 24.1 Cytochrome P450 Isoforms Present in the Nasal Cavity.

P450 Isoform[a]	Typical Substrate	Species	Location in the Naval Cavity
1A1	Benzo(a)pyrene	Rat	Bowman's gland
1A2	Isosafrole	Rat	Olfactory
		Rabbit	Olfactory
2A	Coumarin	Rat	Olfactory, respiratory, Bowman's gland
		Rabbit	Olfactory, respiratory
		Human	Olfactory, respiratory, Bowman's gland
2B1	Pentoxyresorufin	Rat	Olfactory, respiratory
2B4	Benzphetamine	Rabbit	Olfactory, respiratory
2E1	Dimethylnitrosamine	Rat	Olfactory, respiratory
		Rabbit	Olfactory
2G1	Testosterone	Rat	Olfactory
		Rabbit	Olfactory
3A	Various steroids	Rat	Olfactory, respiratory
4B1	2-Aminofluorene	Rabbit	Olfactory, respiratory

[a] Cytochrome P450 nomenclature is used as described by Nelson et al. (1993). (Taken from Thornton-Manning and Dahl, *Encyclopedia of Human Biol.*, Vol. 6, p. 34, Academic Press, 1997.)

species studies. A mitochondrial enzyme found in nasal tissue is rhodanese, which metabolizes cyanide to the less toxic thiocyanate.

Phase II or conjugating enzymes are also found in the nose. Unlike the cytochrome P450s and the epoxide hydrolases, the phase II enzymes are usually found in the cytoplasm. Glucuronosyltransferases are present in the noses of rats and dogs but not humans. These enzymes catalyze the binding of glucuronic acid to hydroxyl groups, making the compound more soluble and excretable. It is also thought that this enzyme plays a role in removing odorant molecules from olfactory receptors to allow for reactivation of the receptor. Olfaction is a function that is more developed in dogs and rats than in humans. GSH S-transferases catalyze the binding of the nucleophilic tripeptide GSH to electrophiles, making them more water soluble and more easily excreted. These reactions usually lead to detoxication but may, in special cases, lead to formation of toxic compounds. The GSH S-transferases are present to a greater extent in olfactory than in respiratory epithelia of the nose and have been noted in the noses of rats, cows, and humans.

24.4.6b Airway

Once an inhaled xenobiotic passes the nose, the compound must traverse the conducting airways to reach the lung. The terminal portion of these conducting airways contains highly metabolically active cells called Clara cells. These cells contain a large amount of endoplasmic reticulum, the site of microsomes containing cytochrome P450 enzymes and epoxide hydrolases. Like the olfactory cells of the nose, the Clara cell is a frequent site for the toxicity of inhaled pollutants because of the cell's enzymatic ability to activate the compounds into electrophiles. Inhaled chemicals that are known to result in Clara cell toxicity include napthalene, 4-ipomeanol, 3-methylfuran, carbon tetrachloride, and 2-acetylaminofluorene. Enzymatic activities high in the Clara cells include cytochrome P450s, GSH transferase, GSH peroxidase, GSH reductase, NAPDH quinone oxidoreductase, and NADPH- and NADH cytochrome c reductase.

24.4.6c Alveolus

In the deep lung, the alveolar portion of the respiratory tract, the cells most active in xenobiotic metabolism are the alveolar type II cells (see Section 24.2 on anatomy of the respiratory tract). The major biochemical activity of these cells is in the synthesis of the surfactant of the lung (see below). Unlike the olfactory cells of the nose and the Clara cells of the bronchiolar region of the airways, the type II cells are not as sensitive to inhaled toxicants as their neighboring cells, the type I cells. In fact, the type II cells appear to proliferate and replace injured type I cells when the need arises. Enzymes associated with xenobiotic metabolism in the type II cells include cytochrome P450, NADH and NADPH cytochrome c reductase, NADPH quinone oxidoreductase, and enzymes related to GSH metabolism.

As discussed earlier, pulmonary alveolar macrophages are resident cells in the alveolar space and have high metabolic activity. The cells contain enzymes associated with xenobiotic and GSH metabolism, such as cytochrome P450s, GSH peroxidase, GSH reductase, NADPH cytochrome c reductase, and NADH cytochrome c reductase, but have either very little or no GSH-S-transferase or NADPH quinone oxidoreductase activities.

24.4.6d The Lung as a Route of Excretion

Although the respiratory tract is a portal of entry of inhaled xenobiotics, it is also a route of excretion of volatile gases and vapors. This information has been used to determine the vapors in the atmosphere to which people have been exposed. The chemical content of exhaled air from submariners coming off a 90-day cruise correctly reflects the air contaminants on the submarine. In toxicology, an increase in the concentration of an inhaled vapor in the exhaled air of an exposed animal can be used to determine the point at which the animal can no longer metabolize the vapor efficiently. Exhaled ethane and pentane, the products of lipid peroxidation, are used to monitor oxidative stress.

24.5 DEPOSITION AND CLEARANCE OF INHALED TOXICANTS: INFLUENCE ON TOXICITY

24.5.1 Particles

The size, shape, and density of inhaled particles influence the manner and site of deposition in the lung, which in turn influence the biologic effects of the particles. The aerodynamic diameter of a particle is the size parameter of greatest importance for deposition considerations and is equal to the diameter of a unit-density sphere having the same terminal-settling velocity as the particle in question. Four major processes are involved in deposition. Impaction is a process that occurs when an inhaled particle has too much momentum to change course with the bulk airflow and impacts against the airway surface. This process is the major manner of deposition of particles having an aerodynamic diameter greater than 0.5 microns, and it

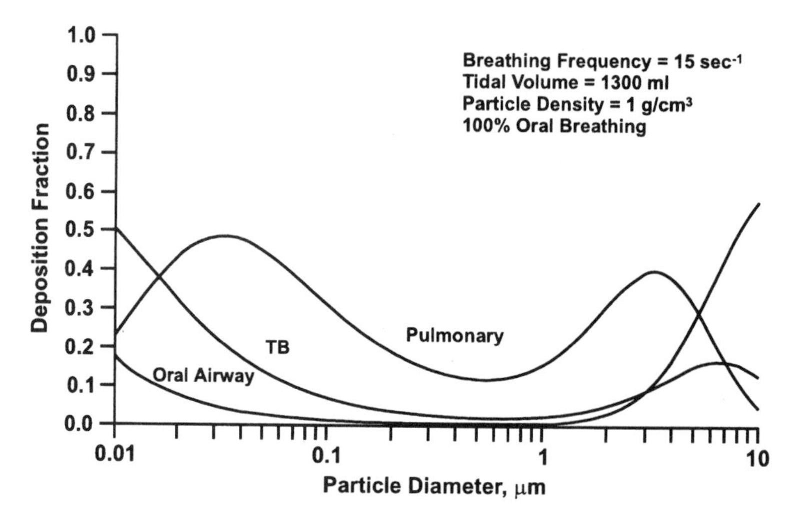

FIGURE 24.9 Effect of particle size on deposition in the respiratory tract. TB = tracheobronchial region.

Source: Based on model of the National Council on Radiation Protection and Measurements, Bethesda, MD, Report No. 125, 1997.

occurs mainly in the upper (oral, nasopharyngeal, or tracheobronchial) airways. Interception happens when the edge of an elongated particle (such as a fiber) contacts an airway wall; this form of deposition occurs when long, thin fibers intercept the bifurcation of airways. Sedimentation is a form of deposition due to gravity. Sedimentation is an important deposition mechanism for particles with an aerodynamic diameter greater than 0.5 microns and occurs in airways where the air velocity is relatively low as in the small bronchi and bronchioles. The final mechanism, diffusion, is important for submicron particles, usually less than 0.2 microns, in airways with almost no airflow, such as the terminal bronchioles and alveoli.

The influence of the aerodynamic diameter on the site of deposition of inhaled particles in the lung is shown in Figure 24.9.

24.5.2 Vapors and Gases

The mechanisms governing the deposition and uptake of inhaled gases and vapors are important determinants of their toxicity. The chemical properties of the compounds strongly influence where the agents will be deposited in the respiratory tract and the degree to which the compound will be taken up into the tissue and blood versus being re-exhaled. In addition, the physiological parameters of minute ventilation and cardiac output will influence the dose of an inhaled chemical received by a person or laboratory animal.

24.5.2a Chemical Characteristics

One of those chemical characteristics is the lipophilicity (lipid-seeking) or hydrophilicity (water-seeking) of the compound, which determines how it will partition between two body compartments or fluids. For example, a highly lipophilic

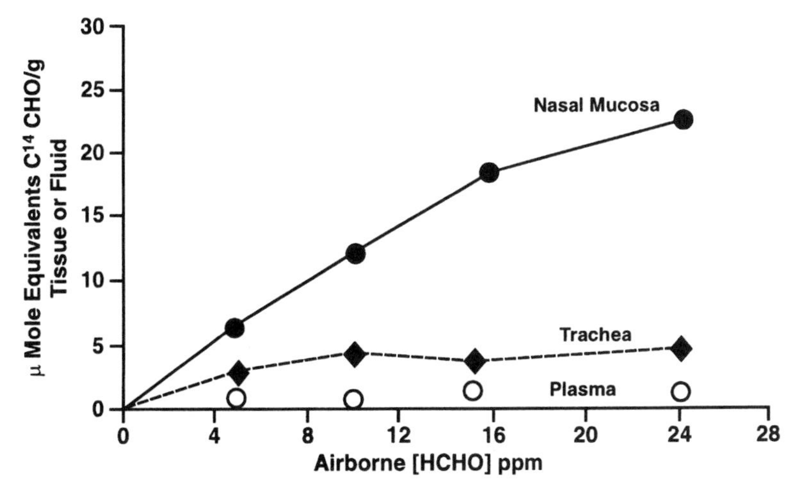

FIGURE 24.10 Absorption of a highly soluble and reactive chemical, formaldehyde, in the respiratory tract. Most of the chemical is taken up in the nose.

compound will tend to leave the blood and enter fat. A highly water-soluble compound will tend to leave the air in the respiratory tract and enter the mucus lining. The relative amount of a compound in two compartments of the body at equilibrium is called the partition coefficient for that compound in those two compartments.

A second chemical characteristic that strongly influences the deposition and clearance of an inhaled compound is its reactivity. If the chemical has a chemically reactive site (a site that is in a high-energy state, such as a formaldehyde or an epoxide moiety, and can easily bond with other chemicals), that chemical will tend to react with the first molecules it encounters in the body.

Gases or vapors that have a high mucus/air partition coefficient (high water solubility) and are highly reactive will deposit mainly in the upper respiratory tract, and only small fractions of what is inhaled will reach the lower respiratory tract or enter the blood. Toxicologists often refer to this phenomenon by saying the compound is "scrubbed" by the nose. Examples of such gases include formaldehyde, hydrochloric acid, sulfur dioxide, and butadiene diepoxide. Figure 24.10 illustrates the deposition of inhaled formaldehyde. Such compounds will exert most of their toxicity in the nasal region of the respiratory tract.

Gases and vapors that have relatively poor water solubility, but are highly reactive, deposit to a greater degree further down in the respiratory tract, in what is called the acinar region (see earlier Section 24.2 on the anatomy of the respiratory tract). Examples of these chemicals are nitrogen dioxide and ozone. These gases cause greater toxicity to the respiratory bronchioles in humans, or in animals that do not have respiratory bronchioles, they injure the terminal bronchioles.

Inhaled volatile organic compounds (VOCs), which are usually poorly soluble in water and have low reactivity, will reach the alveolar region of the lung. The uptake into the blood will depend on the blood/air partition coefficient and on yet another major factor: the metabolism of the compound. A compound that is rapidly metabolized by body tissues (usually mainly in the liver) will be rapidly removed from

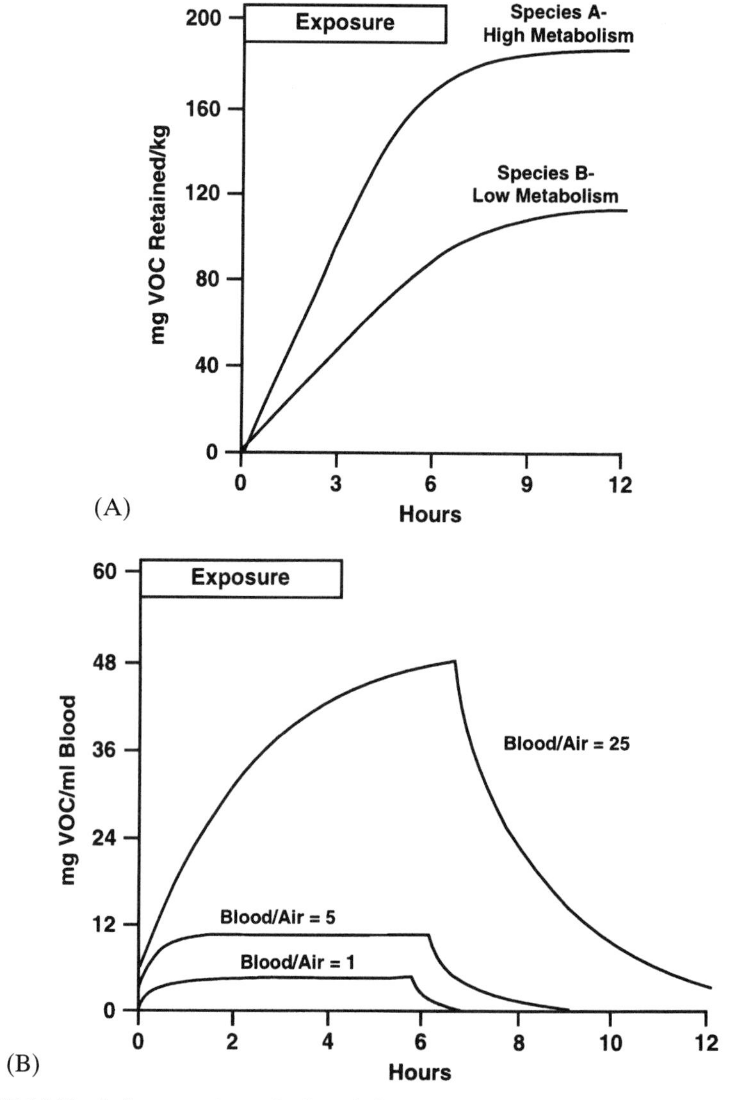

FIGURE 24.11 Influence of metabolism (A) and blood/air partition coefficient (B) on the uptake of inhaled volatile organic compounds (VOCs) in the respiratory tract.

blood, allowing more of the compound to partition into the blood. The influence of these two factors (blood/air partition coefficient and metabolism) on the uptake of a VOC is illustrated in Figure 24.11. The more soluble the compound is in blood and the more rapidly it is metabolized, the larger the uptake of the inhaled compound will be. For many compounds, there will be a limit to what the body can metabolize, that is, the rate of metabolism will not increase linearly with increasing dose. An example of this is shown in Table 24.2, in which mice and rats began to exhale unmetabolized benzene at the higher benzene exposure concentrations. It can also be noted that species differ in the ability to metabolize benzene with mice

TABLE 24.2 Total Metabolism of Inhaled Benzene (Percentage of Dose Retained at End of 6-Hour Exposure).

Exposure Concentration (ppm)	Retained Internal Dose (mg/kg)		Total Metabolites		Exhaled Benzene	
	Rats	Mice	Rats	Mice	Rats	Mice
11	3.3	7.5	96	92	4	3
130	24	60	95	99	5	1
925	116	152	52	86	48	15

(Data from Sabourin et al., *Toxicol. Appl. Pharmacol. 87:* 325–336, 1987.)

TABLE 24.3 Species Differences in Body Weights and Ventilation Rates.

Species	Body Weight (kg)	Minute Volume		
		L/min	L/min/kg	L/min/g of Lung
Mouse	0.03	0.030	1.0	0.15
Syrian hamster	0.12	0.076	0.63	0.095
Rat	0.25	0.13	0.53	0.072
Adult man—light activity	70	20	0.29	0.020

having a greater capacity than rats. These species differences in metabolism will influence the dose of both the inhaled parent compound and its metabolites delivered by inhalation.

24.5.2b Physiological Characteristics

The uptake (or internal dose) of an inhaled airborne compound depends on the amount of air drawn into the respiratory tract (ventilation) and on the amount of blood passing through the lung to pick up the compound and carry it to the rest of the body (cardiac output). Thus, a person who is exercising will receive a higher dose of an inhaled soluble compound than a sedentary individual inhaling the same atmosphere. If a compound is readily soluble in the blood, the rate-limiting factor for its uptake will be the rate of ventilation or minute volume. The greater the minute volume, the greater the dose delivered to the body. If the compound is poorly soluble in the blood, the rate of ventilation will not be rate limiting, because only a small fraction of what is inhaled will be taken up into the blood. The cardiac output then becomes an important factor in how much of the compound gets distributed to the body.

As is true with rates of metabolism, species differ in physiological parameters that influence the internal dose received via inhalation. Table 24.3 illustrates this for ventilation rates. In general, the smaller the species, the greater the proportional dose per kilogram body weight or per gram lung received by inhalation of the same atmosphere.

24.6 TOXICANT-INDUCED ACUTE RESPONSES OF THE RESPIRATORY TRACT

24.6.1 Irritant Responses

Some airborne pollutants cause an irritant response in the respiratory tract. Such irritant responses have been classified as sensory irritation, pulmonary irritation, bronchoconstriction, or respiratory irritation. Irritant responses are often detected by changes in pulmonary function, such as alterations in breathing patterns or tidal volumes.

A sensory irritant stimulates the trigeminal nerve endings in the upper respiratory tract evoking a burning sensation in the nose and leading to reduced respiration rates in some species. Some toxicology tests for the irritancy of chemicals are based on reduced respiration rates in mice. Other names for a sensory irritant include upper respiratory tract irritant, nasal trigeminal stimulant, chemical sense stimulant, and lachrymator. The last listed term refers to the fact that such substances irritate the eyes as well as the respiratory tract, leading to lacrimation. Examples of sensory irritants are acrolein, ammonia, formaldehyde, and sulfur dioxide.

A pulmonary irritant stimulates sensory receptors (C-fiber receptors) in the lung, causing dyspnea, increases in the respiration rate, and decreases in tidal volume, producing rapid, shallow breathing. A pulmonary irritant may also be called a lower respiratory irritant, a lung irritant, or a deep lung irritant. Examples of pulmonary irritants are phosgene, nitrogen dioxide, ozone, and sulfuric acid mist.

In addition to the irritant effects described above, some chemicals can induce bronchoconstriction. Bronchoconstrictors induce an increase in resistance to airflow in conducting airways by constriction of those airways. The mechanism can directly affect smooth muscles, stimulate nerve endings, or release histamine (see Section 24.6.2 below on immune-based acute responses). Examples of bronchoconstrictors include sulfur dioxide, ammonia, some particles, and allergens.

A respiratory irritant is a compound that can act as either a sensory irritant, a bronchoconstrictor, or a pulmonary irritant. Examples include chlorine gas, diepoxybutane, and ketene.

24.6.2 Immune-Based Responses

When an adaptive immune response occurs in an exaggerated form, it is called a hypersensitivity reaction, which can result in inflammation and tissue damage. There are four types of hypersensitivity reactions and three of them, type I, type III, and type IV hypersensitivity, are especially important in the lung. In type I (immediate) hypersensitivity, mast cells of a previously sensitized individual have antigen-specific immunoglobulin E (IgE) bound to their surface. When the antigen (allergen) is encountered, the IgE becomes cross-linked; this causes degranulation of mast cells and rapid release of inflammatory mediators such as histamine and synthesis of additional mediators such as prostaglandins and leukotrienes. These mediators are important to the recruitment of additional inflammatory cells and a late-phase (12–24 hour) or chronic response. In the respiratory tract, for example, mucosal mast

TABLE 24.4 Biological Agents as Etiological Factors in Type III Hypersensitivity.

Disease	Exposure	Antigen
Farmer's lung	Moldy hay or grain	*Micropolyspora faeni, Thermoactinomyces vulgaris*
Bagassosis	Stored sugar cane fiber (bagasse)	*T. saccharii* and possibly other organisms
Mushroom picker's disease	Moldy vegetable compost	*M. faeni, T. vulgaris*
Humidifier, air-conditioner, or heating system disease	Contaminated forced air system	Thermophilic actinomycetes and other organisms
Maple bark stripper's disease	Maple tree logs or bark	*Cryptostroma corticale*
Sequoiosis	Redwood sawdust	Graphium, Pullularia (*A. pullulans*), and other fungi
Paper mill worker's disease	Moldy wood pulp	Alternaria
Brewer's or malt worker's lung	Malt or barley dust	*Aspergillus clavatus, A. fumigatus*
Cheese washer's lung	Cheese mold	*P. casei*
Paprika slicer's disease	Moldy paprika pods	Mucor stolonifer
Pigeon breeder's disease	Pigeon serum and droppings	Avian proteins
Turkey handler's disease	Contact with turkeys	Turkey proteins
Bible printer's disease	Moldy typesetting water	Unknown
Ceptic or mummy disease	Cloth wrappings of mummies	Unknown

cells may have bound IgE molecules specific for pollen or animal dander. When the specific pollen or dander is inhaled, the mediator release results in symptoms such as mucosal edema, hypersecretion, and bronchoconstriction associated with rhinitis or asthma. Type III hypersensitivity, also known as immune complex disease, can occur in the lung with repeated inhalation of antigenic materials from mold, plants, or animals. Examples include Farmer's lung and pigeon fancier's lung (Table 24.4). The initial (sensitizing) inhalation of antigen induces specific circulating antibodies primarily of the immunoglobulin G (IgG) class. When antibody is inhaled again, immune complexes (antibody-antigen complexes) are formed in the alveoli. Complement (a complex system of proteolytic enzymes) is bound by the immune complexes, which leads to release of vasoactive amines, chemotactic factors, and cytokines resulting in inflammatory cell accumulation, tissue damage, and fibrosis. This disease is also known as extrinsic allergic alveolitis. Type IV (delayed) hypersensitivity reactions take more than 12 hours to develop and involve cell-mediated instead of antibody-mediated immune responses. T cells that have been specifically sensitized by a previous encounter with the antigen are responsible for recruiting other cells to the site of the delayed hypersensitivity reaction. In the lung, the most important variant of type IV hypersensitivity is granulomatous hypersensitivity. Granulomas are formed by the aggregation and proliferation of macrophages.

These macrophages contain intracellular microorganisms or sensitizing particles (e.g., beryllium) that the cell cannot destroy. The macrophage core of the granuloma is surrounded by a cuff of lymphocytes and often fibrosis. Occasionally, persistent immune complexes incite a type IV reaction, and type IV hypersensitivity may contribute to allergic alveolitis. Nonimmunologic granulomas, called foreign-body granulomas, also occur in the lung. Foreign-body granulomas form in response to particles such as talc and silica that macrophages cannot digest. Often, foreign-body granulomas can be distinguished from immune granulomas by the absence of lymphocytes.

24.6.3 Acute Inflammation

An acute inflammatory response as the result of inhaled toxicants can occur at any point along the respiratory tract. The site of the inflammation is determined by the deposition patterns of the inhaled materials. Inflammation may occur in the nose as a result of inhaled water-soluble and/or reactive irritant particles or gases and result in rhinitis (runny nose). Tobacco smoking and oxidant gases may produce a bronchitis or bronchiolitis in the airways. Infectious agents, agents inducing hypersensitivity, radiation, and toxic particles of a size that allows them to deposit in the alveoli, can induce alveolitis or pneumonia. The induction of acute inflammation in the respiratory tract is often used in toxicology studies to determine the pulmonary toxicity of inhaled or instilled noxious agents.

Early biochemical mediators of pulmonary inflammation are tumor necrosis factor-α (TNF-α) and interleukin-1 (IL-1), cytokines released from the resident macrophages that promote the adherence of circulating inflammatory cells to the endothelium. These cytokines stimulate the release of chemoattractant factors such as IL-8 (major chemoattractant for neutrophils in primates), macrophage inflammatory protein-2 (MIP-2; neutrophilic chemoattractant in rodents), IL-6, and macrophage chemoattractant protein-1 (MCP-1) that attract inflammatory cells into the alveoli.

A screening tool to evaluate the inflammatory potential of an inhaled agent is the analysis of bronchoalveolar lavage fluid from the exposed animals to quantitate the degree of the inflammatory response. Animals are exposed by inhalation or by intratracheal instillation to varying amounts of the agent of interest. Subsequently, the lungs of the animals are lavaged (washed) with physiological saline to sample the epithelial lining fluid of the respiratory tract. For large animals, such as dogs, this procedure can be done serially in the same animal using a fiberoptic bronchoscope to instill and remove the wash fluid. In smaller laboratory animals, the procedure is normally done in the excised lung as part of a serial sacrifice of animals at selected time points after or during the exposures. The lavage fluid is analyzed for cellular and biochemical indicators or toxicity and inflammation. In this manner both the degree of the inflammatory response and the time course of recovery from or intensification of the process can be followed. The most sensitive indicator of the inflammatory response in the lung is the influx of neutrophils. Protein content in the lavage fluid can be used to monitor increased permeability of the alveolar/capillary barrier. Lactate dehydrogenase is a cytoplasmic enzyme whose presence extracellularly indicates cell death. Beta-glucuronidase activity in the extracellular fluid is a measure of activated macrophages, and alkaline phosphatase is an enzyme

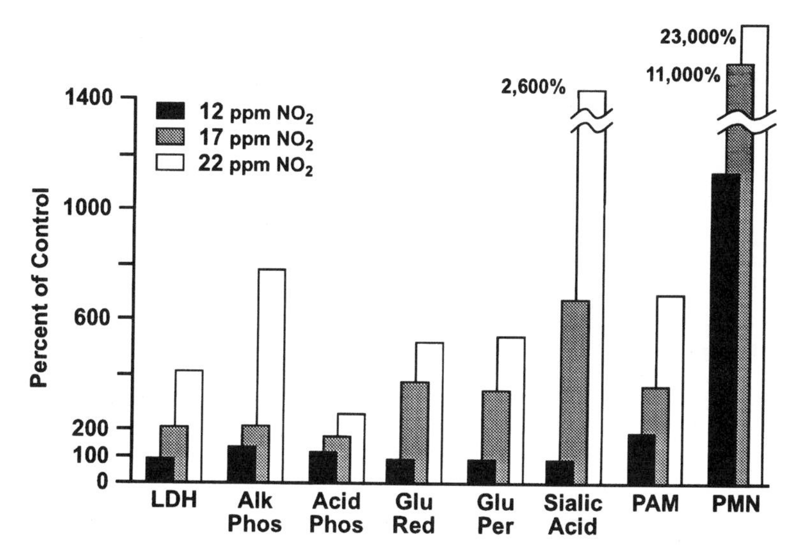

FIGURE 24.12 Analysis of bronchoalveolar lavage fluid from Syrian hamsters after 48 hours of exposure to 12, 17 or 22 ppm NO_2.
Values are means of samples from 6 animals. LDH = lactate dehydrogenase activity; alk phos = alkaline phosphatase activity; acid phos = acid phosphatase activity; glu red = glutathione reductase activity; glu per = glutathione peroxidase activity; PAM = number of pulmonary alveolar macrophages; PMN = number of polymorphonuclear leukocytes. Sialic acid was assayed as a measure of glycoproteins.

Source: From Henderson et al., *Chest* 80S: 12S–15S, 1981.

associated with type II cell secretions. Proinflammatory cytokines can be measured in the lavage fluid or, as is done more frequently, as secretions from cultured cells retrieved from the lavage fluid. The acute inflammatory response in lungs of hamsters exposed to nitrogen dioxide is shown in Figure 24.12. If the exposure to an inflammatory material continues, a chronic inflammatory response may occur as illustrated in the analysis of bronchoalveolar lavage fluid from rats chronically exposed to diesel engine exhaust (Fig. 24.13). In this case, the inflammatory response was maintained by the continued exposure and by the increasing load of diesel soot in the lung.

24.7 TOXICANT-INDUCED CHRONIC DISEASE OF THE RESPIRATORY TRACT: BIOCHEMICAL FACTORS

Once an inhaled or systemically delivered toxic agent damages the pulmonary endothelium or epithelium, edema and inflammation usually follow. Subsequently, several outcomes are possible. If the noxious agent is rapidly cleared, and the damage is not too severe, the lung can be repaired to a normal condition. For example, if a few endothelial cells are acutely injured or killed, vascular permeability and interstitial edema increase. But, if the basement membrane remains intact, the neighboring endothelial cells will proliferate and repair the defect, and the edema

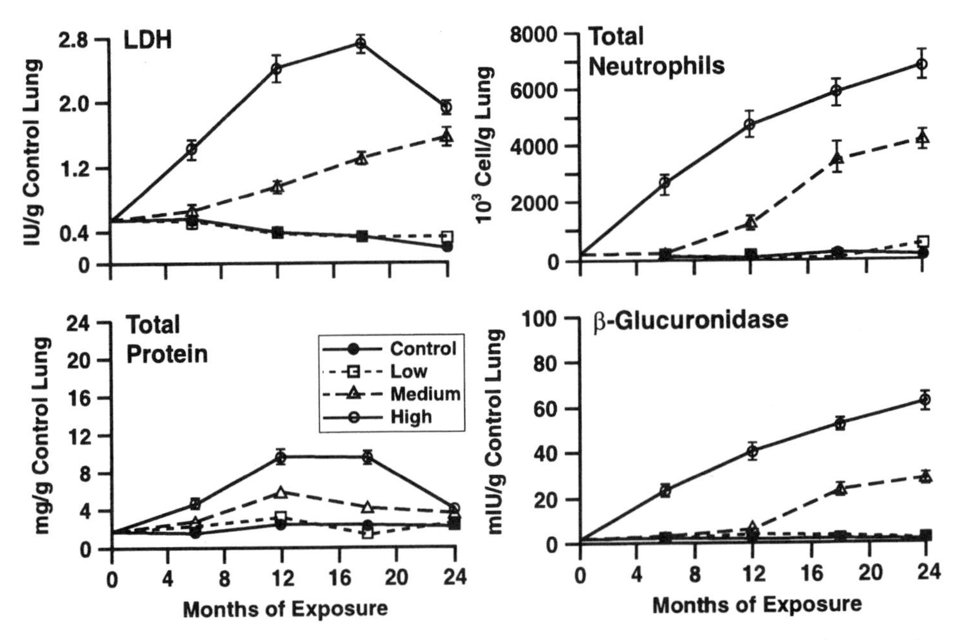

FIGURE 24.13 Analysis of bronchoalveolar lavage fluid from rats chronically (6 hours/day, 5 days/week for 24 months) exposed to 0, 0.35. 3.5 or 7.0 mg/m^3 of diluted diesel engine exhaust.

will rapidly resolve. Similarly, if epithelial cells lining alveoli are acutely injured, an exudative phase will be characterized by edema, fibrin, and inflammatory cells in the alveolus. But, if the damage is not severe and the basement membrane is intact, type II cells will rapidly proliferate, then differentiate into type I cells to repair the epithelium, and the exudate will resolve. Alternatively, if the toxic material remains in the lung, or if exposures are repeated, chronic inflammation and lung damage may ensue. For example, chronic edema or inflammation of the interstitium or persistent alveolar exudation will result in interstitial or intra-alveolar fibrosis, respectively. If the cause of the damage and inflammation persists in the upper airways, as in many cigarette smokers, a chronic condition characterized clinically by excess mucous production and cough, known as chronic bronchitis, may develop. If the cause of the damage in the alveolar region persists, and the balance between proteases and antiproteases is tilted in favor of proteases, the lung matrix and alveolar walls may be destroyed leading to emphysema. If electrophilic metabolites or direct-acting electrophiles have bound to critical parts of the DNA, neoplasia may result. In the following sections, the biochemical aspects of these chronic diseases of the respiratory tract will be discussed.

24.7.1 Chronic Inflammation and Fibrosis

Functionally, fibrosis is characterized by a small, stiff lung with poor compliance and difficulty in expanding to allow the inhalation of the needed amount of air. The condition is called a "restrictive" lung disease by pulmonary physiologists. The term *fibrosis* refers to an increase in collagen, an extracellular structural protein, or the presence of collagen in an abnormal site, such as the alveolar lumen. Marked fibrosis with distortion of normal architecture or cellular organization of a tissue is often called scarring. Fibrosis can be detected by microscopic examination of a tissue or by measuring an increase in hydroxyproline concentration biochemically. When severe epithelial damage results in delayed repair and persistent inflammation, intra-alveolar fibrosis may develop. For example, after oral administration of the insecticide paraquat, the compound is preferentially taken up by lung alveolar epithelial cells via a diaminelpolyamine transport system. There, it undergoes cyclic oxidation and reduction with generation of reactive oxygen species including superoxide anion, H_2O_2, and hydroxyl free radical. Consequently, there is peroxidation of membrane lipids and depletion of cellular NADPH causing epithelial cell death. The widespread death of epithelial cells results in a severe, persistent, fibrinous, and cellular exudate. The lack of an epithelial barrier and chemokines and cytokines from the inflammatory exudate causes fibroblasts to migrate into the fibrinous exudate and produce extracellular matrix and collagen, resulting in fibrosis. The inflammation may involve both the alveolar lumens and interstitium, resulting in fibrosis (scarring) that markedly distorts the parenchymal architecture.

Interstitial fibrosis results from chronic interstitial edema and inflammation. For example, right heart failure with pulmonary hypertension can lead to chronic pulmonary interstitial edema and interstitial fibrosis. Exposures to hyperoxia, smoke inhalation, and radiation can cause inflammation of the interstitium that involves the parenchyma fairly uniformly. This is called diffuse interstitial pneumonia, alveolitis, or diffuse alveolar damage. The interstitium is initially thickened by edema, fibrin, inflammatory cells, and fibroblast proliferation. The fibroblasts produce collagen. If the animal survives the early inflammatory period, the inflammation may largely resolve, but the collagen persists, thus, the end-stage is diffuse fibrosis. Both diffuse interstitial pneumonia and its sequellum, diffuse fibrosis, can interfere with gas diffusion across the blood–air barrier, resulting in hypoxemia.

Interstitial fibrosis can also be a consequence of repeated type III hypersensitivity responses due to repeated exposures to antigen, and it is a prominent component of type IV hypersensitivity diseases. For example, chronic beryllium disease in a sensitized person who has inhaled beryllium is manifest as multiple firm nodules in the lung. These nodules are interstitial granulomas, and a major component of each nodule is collagen. Immune-type granulomas with multifocal fibrosis also occur with type IV hypersensitivity to inhaled zirconium and in sarcoidosis, a type IV hypersensitivity disorder of unknown etiology.

Interstitial fibrosis is also a major component of several diseases, called pneumoconioses, due to inhalation of poorly soluble dusts. Table 24.5 gives examples of inhaled dusts and the associated pneumoconiotic diseases. The inhaled dusts accumulate in the interstitium and induce an inflammatory response (a nonimmune granulomatous inflammation), which may be low-grade or marked, depending on the burden and the toxicity of the dust. Dust particles of relatively low toxicity (e.g., coal dust) usually accumulate within the cytoplasm of interstitial macrophages, but dust particles of higher toxicity (e.g., silica or asbestos) may be phagocytosed by macrophages, cause injury or death of the macrophages, and be released extracellularly. These particles may be phagocytosed by other macrophages, thus continuing the macrophage activation and release of cytokines and chemokines. In the various pneumoconioses, interstitial fibrosis occurs multifocally as a result of the

TABLE 24.5 Mineral Dusts and Associated Pneumoconiotic Diseases.

Coal dust	Anthracosis (pigment accumulation without cellular reaction) Coal workers' pneumoconiosis (pigment and macrophage accumulation with little fibrosis or pulmonary dysfunction) Progressive massive fibrosis (extensive fibrosis and compromised lung function)
Silica	Silicosis
Asbestos	Asbestosis
Beryllium	Acute berylliosis and granulomatous inflammation in nonimmune individuals Chronic beryllium disease (immune-mediated disease)
Aluminum oxide (bauxite)	Shaver's disease
Barium sulfate	Baritosis
Iron oxide	Siderosis
Tin oxide	Stannosis

disruption of the interstitium by the particle accumulation and the granulomatous inflammation.

Macrophages are key to fibrogenic responses in the lung. Macrophages secrete chemical mediators called cytokines and chemokines, for example, TNF-α that recruit and activate additional inflammatory cells. They also produce transforming growth factor-β1 (TGF-β1), which stimulates fibroblast proliferation and synthesis of extracellular matrix, and MIP-2, which is an angiogenic CXC chemokine that enhances the vascularization of fibrotic areas. Macrophage produced IL-1 and fibronectin both stimulate fibroblast proliferation. Lymphokines from activated T lymphocytes are also important in fibrosis associated with immune responses and granulomas. These lymphokines affect macrophages (macrophage-activating factor, macrophage-migrating inhibitory factor, and monocyte-chemotactic factor) and fibroblasts (fibroblast-chemotactic factor, fibroblast-activating factor) to stimulate fibroblast proliferation and collagen synthesis indirectly and directly.

24.7.2 Chronic Obstructive Pulmonary Disease

Chronic obstructive pulmonary disease is defined as a disease state characterized by the presence of airflow obstruction due to chronic bronchitis, chronic bronchiolitis, or emphysema; the airflow obstruction is generally progressive, may be accompanied by airway hyperreactivity, and is primarily irreversible. Chronic bronchitis is defined as the presence of chronic productive cough for 3 months in each of 2 successive years in a person in whom other causes of chronic cough have been excluded. Chronic bronchiolitis, also known as small airways disease, is difficult to define clinically but may be detected by sophisticated tests of small airway function. Emphysema is defined anatomically by abnormal permanent enlargement of the airspaces distal to the terminal bronchioles, accompanied by destruction of their walls and without obvious fibrosis. Asthma is sometimes viewed as a form of COPD, because both asthma and COPD are identified by similar symptoms and functional abnormalities, with airway obstruction being the defining characteristic in both diseases. The airway obstruction in asthma must be reversible to establish a diagnosis, whereas COPD is defined as a syndrome characterized by abnormal expiratory flow that does not change markedly over a period of several months. Both diseases are now known to be caused by lung inflammation induced by different initiating factors, most likely environmental allergens, occupational-sensitizing agents, or viral respiratory infections in the case of asthma, and cigarette smoking in the case of COPD. Asthma is discussed in the next section.

The reduced airflow in COPD is caused by increased resistance to airflow due to inflammation, fibrosis, goblet cell metaplasia, and smooth muscle hypertrophy in small airways, as well as loss of elastic recoil due to inflammation and destruction of alveolar attachments to bronchioles. Functionally, persons with COPD have reduced FEV_1 values. A person with emphysema will have an overly compliant, hyperinflated lung with reduced ability to exhale air.

Chronic irritation by inhaled substances and microbial infections are key factors in chronic bronchitis and bronchiolitis. Cigarette smoking is the most important factor. People using biomass fuels indoors without proper ventilation are also more prone to have COPD. Environmental and occupational exposures to irritant gases, such as SO_2, and to poorly soluble particles are minor contributors to this disease. Infections do not initiate chronic bronchitis/bronchiolitis, but they play a role in maintaining chronic airway inflammation and trigger acute exacerbations of chronic bronchitis. Cigarette smoking has multiple roles in chronic bronchitis, bronchiolitis, and exacerbations of these diseases. Cigarette smoke causes increased mucous gland size (hypertrophy) and goblet cell number (hyperplasia) in large airways. Small airways of smokers show goblet cell metaplasia with mucous plugging, inflammatory infiltrates, and fibrosis of the airways. All of these changes contribute to airway narrowing and increased resistance to airflow. In addition, cigarette smoke interferes with ciliary action, damages the airway epithelium, and inhibits the ability of bronchial and alveolar leukocytes to clear bacteria, thus predisposing to infection and exacerbations of bronchitis.

Cigarette smoking is the major risk factor for the development of emphysema. Development of the disease is associated with chronic inflammation in smoker's lungs. The destruction of lung matrix, especially elastin, which leads to the loss of alveolar walls, is thought to be due to an imbalance between proteases released from inflammatory cells and anti-proteases of the lung that protect against proteases. This is referred to as the proteinase-antiproteinase hypothesis for the development of emphysema. Support for this hypothesis comes from (1) the fact that persons having a genetic deficiency of alpha1-antiproteinase (also known as alpha1-anti-trypsin, the major lung inhibitor of neutrophil elastase) are predisposed toward development of emphysema whether they smoke or not, and (2) animal models in which emphysema can be induced by intrapulmonary instillation of pancreatic or neutrophil elastase. Many proteinases may be present in inflamed lungs. Proteinases from neutrophils include the serine proteinases known as neutrophil elastase, cathepsin G, and proteinase 3; and the matrix metalloproteinase MMP-9. Alveolar macrophages can release metalloproteinases (e.g., MMP-2), a metalloelastase (MMP-12), several collagenases, and cysteine proteinases known as cathepsins. Macrophages can also internalize neutrophil elastase and store it for later release. In addition, several antiproteinases are in the lung, including alpha1-antiproteinase, which is derived from the liver; alpha2 macroglobulin, derived from liver; fibroblasts; macrophages; secretory leukoproteinase inhibitor (SLPI), produced by lung epithelial cells; monocyte/neutrophil elastase inhibitor, produced by monocytes and neutrophils; and members of the tissue inhibitors of metalloproteinases (TIMPs) family, which are generally co-produced in the same cells that produce MMPs. Regulation of the balance between proteinases and antiproteinases is not well understood, nor is it clear which inflammatory cell types and proteinases are most important in the pathogenesis of emphysema. Various experiments have provided evidence for the importance of neutrophil elastase, matrix metalloproteinases (especially MMP-9 and MMP-2, which have elastolytic activity), and macrophage metalloelastase (MMP-12). The exact mechanism by which cigarette smoke causes emphysema in some smokers is unknown. However, it is known that cigarette smoking inactivates alpha1-antiproteinase by oxidizing amino acids at its active site. In addition, cigarette smoke activates and recruits macrophages and neutrophils that secrete proteases. Furthermore, the neutrophil elastase inactivates inhibitors of the MMPs. such as the TIMPs, and some MMPs inactivate alpha1-antiproteinase. Thus, cigarette smoking is thought to lead to an imbalance of inflammatory cell-derived, active proteases in the lung.

24.7.3 Asthma

Asthma is considered to be an immunologically based large airway disease and is a clinical state of heightened reactivity of the tracheobronchial tree to numerous stimuli, manifest as variable airways obstruction. Characteristically, episodes of dyspnea and wheezing, which are symptomatic of airway obstruction, are features of the disorder. Some patients may cough, with or without production of sputum. The symptoms are the result of IgE-mediated sensitivity to antigen resulting in bronchospasm, bronchial wall edema, and hypersecretion by mucous glands. There is an increase in TH2 cytokines (IL-4, IL-5, IL-13), eosinophils, lymphocytes (mainly CD4 type), mast cells, and smooth muscle hyperplasia in the airways of asthmatics, and neutrophil numbers increase during an asthma attack.

Over the past few decades, the prevalence of asthma has increased and affects up to 10% of the population in developed countries. The reason for the increase is not clear. Childhood asthma has a strong genetic component; work-related asthma can be directly related to workplace exposures to sensitizing materials, particularly organic dusts (see Table 24.4). Recent studies indicate that immunological memory to inhaled environmental allergens is established at an early age (infancy to 5 years).

Although the airway obstruction seen in asthma is reversible, some patients with asthma have been shown, over time, to develop irreversible changes in lung function due to airway remodeling and repair associated with asthmatic attacks. Such structural changes include a thickened basement membrane as well as an increase in smooth muscle mass in the large airways, and an increase in the number of mucous glands and goblet cells.

24.7.4 Neoplasia

As a portal of entry for inhaled materials, the lung is exposed directly to carcinogens or their precursors. Lung cancer is the leading cause of death from cancer in

TABLE 24.6 Age-Adjusted Lung Cancer Incidence per 100,000.[a]

Year of Diagnosis	White		Black	
	Men	Women	Men	Women
1975	75.7	21.9	101.2	20.6
1980	82.0	28.3	131.2	34.0
1985	81.9	35.9	131.7	40.8
1990	78.6	41.5	116.0	45.3
1995	71.5	44.2	114.7	42.9

[a] U.S. Surveillance, Epidemiology and End Results (SEER).

men and women in the United States. The age-adjusted incidence of lung cancer per 100,000 men (1986–1990) was 83, second only to prostate cancer (108). In women the incidence for the same period was 39, second to breast cancer (108). The major risk factor for lung cancer is tobacco smoke. Occupational exposures to arsenic, asbestos, bis-(chloromethyl)ether, chloromethyl methylether, chromium, nickel, polycyclic aromatic hydrocarbons (PAHs), radon, and vinyl chloride are associated with increased incidence of lung cancer. The higher incidence of lung cancer in men versus women is based on the higher rate of smoking in men versus women. The recent increase in smoking among women has led to a doubling in lung cancer incidence in women, whereas the recent decline in smoking among men has led to a slight decline in lung cancer incidence in men over the past two decades (Table 24.6). Various studies have indicated some increased risk for lung cancer in nonsmoking wives of heavy smokers.

The major histological types of human lung tumor are squamous cell carcinoma (29%), adenocarcinoma (32%), small cell carcinoma (18%), and large cell carcinoma (9%). Clinically the distinction is often made between small cell lung cancer (SCLC) and non-small cell lung cancer (NSCLC), because the small cell cancer is much more aggressive and requires a different therapeutic regimen than the other types of lung cancer. The adenocarcinomas and large cell carcinomas tend to occur more in the peripheral lung, whereas the squamous cell carcinomas and the small cell carcinomas are more likely to occur in the central lung, in and around the bronchi. The most common phenotype of lung tumors associated with cigarette smoking has shifted in both males and females over the past two decades, with a decreasing percentage of the squamous type and an increasing percentage of adenocarcinomas. The reason for this is unknown, but it has been suggested that the greater use of filtered cigarettes has removed larger particles from the tobacco smoke, changing the overall deposition pattern for cigarette-derived particles to favor deposition of the remaining, smaller particles in the deep lung. There is also speculation that use of lower tar and nicotine-filtered cigarettes has led some smokers to inhale more deeply and to hold each bolus of inhaled smoke longer before exhalation. This maneuver would enhance deposition in the deep lung.

The major histological types of lung cancer found in rats and mice include adenomas, adenocarcinoma (also called bronchiolalveolar carcinoma), and squamous cell carcinoma. In the rat, one also finds a cystic keratinizing epithelioma, mainly in response to heavy burdens of particles in the lung. This lesion appears to be limited to rats and is rarely seen in other species. Missing from these commonly used rodent models of human carcinogenesis is the SCLC, which has only rarely been reported in rodents. In humans, more respiratory tract carcinomas are in the central lung, in or around bronchi, whereas in rodent models of pulmonary carcinogenesis, the major site of lung tumors is in the deep lung in the pulmonary parenchyma. The reason for these species differences may relate to species differences in deposition of inhaled material, differences in epithelial thickness and metabolic capability, and perhaps most importantly, the fact that the main human pulmonary carcinogen is tobacco smoke, which cigarette smokers intentionally inhale. Rodents in carcinogen bioassays are exposed to a wide variety of chemicals that have different deposition and metabolism patterns based on their chemical and physical characteristics. Some studies on dosimetry suggest that the human bronchi are susceptible to tumors from cigarette smoke because the thickness of the barrier between the airways and the blood is greater in the bronchi than in the alveoli. This barrier allows inhaled precursors to carcinogens (such as PAHs in tobacco smoke) to remain in the tissues long enough to be metabolized to reactive electrophiles and to form adducts with DNA before reaching the blood stream.

The molecular biology of lung cancer has become more delineated in recent years. (For a detailed discussion of the mechanisms of carcinogenesis and the progression of events from normal cell to malignant tumor, see Chapter 15 Carcinogenesis.) Table 24.7 lists the molecular changes associated with human lung cancer. The most consistent changes observed in all cancer types is the loss of portions of the short arm of chromosome 3 (3p) and loss of 9p. A candidate tumor-suppressor gene identified with the 3p region is the fragile histidine triad (FHIT) gene. The p16 tumor suppressor gene is localized to chromosome 9p21 and is inactivated in a high percentage of NSCLC primarily by aberrant hypermethylation of its promoter region. In contrast, the RB gene is inactivated in most SCLCs. Mutation in the p53 gene is seen in 95%, 65%, and 35% of SCLCs, squamous cell carcinomas, and adenocarcinomas, respectively. Other factors frequently altered in lung cancers include overexpression of epidermal growth factor receptor, cyclin D1, and c-myc.

The mouse has proved to be a good model for the study of the progression of lung cancer (Fig. 24.14). Initially, formation of DNA adducts with the administered carcinogen and methylation of DNA are observed. The formation of the DNA adducts can be detected in the metabolically active cells such as the type II and the Clara cell. Extensive methylation of the promoter region of genes interferes with the transcription of DNA to RNA. Increased activity of cytosine DNA methyltransferase, the enzyme responsible for methylation of cytosines, can be detected at all stages of the tumor development, from the hyperplastic epithelium to the benign adenomas to the neoplastic adenocarcinomas. As early as the hyperplastic stage, mutations in K-ras can be observed. At the stage of adenoma formation there is an enhanced expression of c-myc; eventually, with the loss of heterozygosity on mouse chromosome 4 where the p16 gene is located or loss of other tumor suppressor genes (p53, RB1), the tumor becomes an adenocarcinoma. The mutations observed in K-ras depend on the carcinogen, and some mutational spectra characteristic of individual carcinogens have been found. In tetranitromethane-induced tumors, there is

TABLE 24.7 Molecular Changes[a] Associated with Human Lung Cancer.

			Information on Specific NSCLC Type		
Molecular Change	Small Cell Lung Cancer	Non-Small Cell Lung Cancer (NSCLC)	Squamous Cell Carcinoma	Adenocarcinoma	Large-Cell Carcinoma
Oncogene expression					
Myc	18	8	—	—	—
K-ras	0	16	—		—
H-ras	0	2	—	[30]	—
N-ras	0	10	—		—
HER2/neu	0	—	—	30	—
Suppressor genes (mutation or loss)					
Retinoblastoma	60	10	—	—	—
p53	95	49	65	35	—
Chromosome 3p	~100	75	—	—	—
Chromosome 9p	75	60	—	—	—
Growth factor expression					
Bombesin/gastrin releasing peptide	31	6	—	—	—
Epidermal growth factor	0	—	86	50	55
Insulin-like growth factor	95	82	—	—	—

[a] Percent of total tumor type showing change.

Source: Modified from F. F. Hahn, Lung Carcinogenesis. In: *Carcinogenicity*, K. T. Kitchim, ed., Marcel Dekker, Inc., NY, 1998.

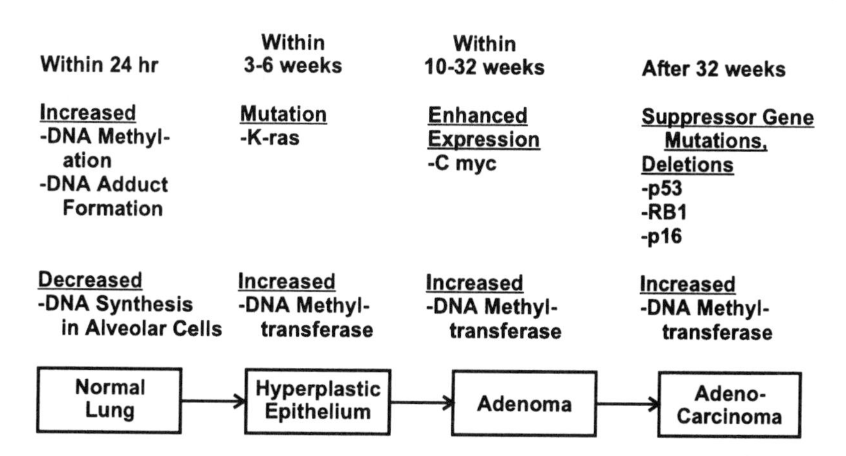

FIGURE 24.14 Sequence of molecular changes in mouse lung after carcinogen exposure.

Source: Malkison and Belinsky, Chapter 16 in *Lung Cancer*, Pass, H.I., Mitchell, J.B., Johnson, D.H. and Turris, A.T., eds., Lippincott-Raven, New York, 1996 as presented in Hahn, Chapter 19 in *Lung Carcinogenesis*, Kitchim, K.T., ed., Marcell Dekker, Inc., New York, 1998.

TABLE 24.8 Pulmonary Tumor Response of Laboratory Rodents to Inhalation of Known Human Pulmonary Carcinogens.[a]

	Human	Rat	Mouse
Chemicals and groups of chemicals			
Arsenic and arsenic compounds	+[b]	ND	ND
Asbestos	+	+	±
Beryllium compounds	+	+	±
Bis(chloromethyl)ether and chloromethyl methyl ether	+	+	+
Cadmium compounds	+	+	±
Chromium compounds, hexavalent	+	±	ND
Coal tars	+	+	+
Coal tar pitches	+	ND	ND
Mustard gas (sulfur mustard)	+	ND	±
Nickel and nickel compounds	+	+	±
Soots	+	+	ND
Talc containing asbestiform fibers	+	+	−
Vinyl chloride	+	+	+
Environmental agents and cultural risk factors			
Tobacco smoke	+	+	±
Radon and its decay products	+	+	−

[a] Not included are industrial processes known to be associated with human lung cancer.

[b] + = Positive, ND = no data, ± = limited, − = negative.

Source: From F. F. Hahn, Lung Carcinogenesis. In: *Carcinogenicity*, K. T. Kitchim, ed., Marcel Dekker, Inc., NY, 1998.

a change in codon 12 from GGT to GAT. In lung tumors produced by exposure of mice to 1,3-butadiene, GGC changes to CGC in codon 13.

Rat lung cancer appears to be a good model for investigation of the role of DNA methylation in p16 dysfunction and the progression of events leading to neoplasia. Promoter methylation has been shown to be an alternative to mutation in inactivating tumor suppressor genes in human lung cancer; this mechanism has also been demonstrated in rat lung cancer models.

The molecular changes observed in rodent lung tumors vary in frequency from those observed in human lung tumors. The most striking difference is in the mutation or loss of the tumor suppressor genes, RB1 and p53. These changes are common in human lung tumors but are not prominent in mouse lung tumors (<10%) and are only present in <10% of lung squamous cell carcinomas (and not other tumor types) in rats. K-ras mutations vary with the species and strain of rodent and the nature of the carcinogen. Ninety percent of the lung tumors in the highly susceptible strain A mouse have K-ras mutations, whereas lung tumors from less susceptible strains of mice have a lower frequency of such mutations. K-ras mutations in lung tumors in rats vary from 0% in rats exposed to 4-(N-methyl-N-nitrosamino)-1-(3-pyridyl)-butanone (commonly referred to as NNK) to 6% in rats exposed to diesel exhaust to 100% in rats exposed to tetranitromethane. On the other hand, rodent models have been excellent for detection of lung tumors associated with inactivation of suppressor genes by aberrant DNA methylation.

The use of rodent models to detect human carcinogens have value in that one would prefer to detect cancer-causing agents in laboratory studies and regulate the chemicals accordingly rather than to detect the carcinogenic potential of chemicals first in humans in epidemiology studies. However, the species differences noted above suggest some caution in interpretation of the animal data. The degree of concordance in pulmonary tumor response in exposed humans compared to exposed rodents (Table 24.8) suggests that the rodent studies can detect most human carcinogens. The rodent models of lung cancer are also valuable for mechanistic studies to determine factors influencing the progression of events leading to neoplasia.

SUGGESTED READING

Gardner, D.E., Crapo, J.D. and McClellan, R.O. (Eds), *Toxicology of the Lung*, 3rd Ed., Raven Press, New York, 1999.

McClellan, R.O. and Henderson, R.F. (Eds), *Concepts in Inhalation Toxicology*, 2nd Ed., Taylor & Francis, Washington, DC, 1995.

Pass, H.I., Mitchell, J.B., Johnson, D.H. and Turrisi, A.T. (Eds), *Lung Cancer, Principles and Practice*, Lippincott-Raven, New York, 1995.

West, J.B. *Respiratory Physiology—The Essentials*, 6th Ed., Lippincott, Williams and Wilkins, Philadelphia, 1999.

CHAPTER TWENTY-FIVE

Biochemical Mechanisms of Renal Toxicity

JOAN B. TARLOFF

25.1 INTRODUCTION

Nephrotoxicity can be a potentially serious complication of drug therapy or chemical exposure. Although in most instances, the mechanisms mediating nephrotoxicity are unclear, susceptibility of the kidney to toxic injury appears to be related, at least in part, to the complexities of renal anatomy and physiology.

The focus of this chapter is three fold: (1) to review components of renal physiology contributing to susceptibility to chemically induced nephrotoxicity; (2) to examine current methodologies for assessment of nephrotoxicity; and (3) to provide examples of a few specific nephrotoxicants, emphasizing mechanisms thought to contribute to the unique or selective susceptibility of specific nephron segments to these toxicants.

25.2 FUNDAMENTAL ASPECTS OF RENAL PHYSIOLOGY

25.2.1 Structural Organization of the Kidney

Upon gross examination, three major anatomical areas of the kidney are apparent: cortex, medulla, and papilla (Fig. 25.1). The cortex is the outermost portion of the kidney and contains proximal and distal tubules, glomeruli, and peritubular capillaries. Cortical blood flow is high relative to cortical volume and oxygen consumption; the cortex receives about 90% of total renal blood flow. A blood-borne toxicant will be delivered preferentially to the renal cortex and therefore have a greater potential to influence cortical, rather than medullary or papillary, functions.

The renal medulla is the middle portion of the kidney and consists of the loops of Henle, vasa recta, and collecting ducts. Medullary blood flow (about 6% of total renal blood flow) is considerably lower than cortical flow. However, by virtue of its countercurrent arrangement between tubular and vascular components, the medulla may be exposed to high concentrations of toxicants within tubular and interstitial structures.

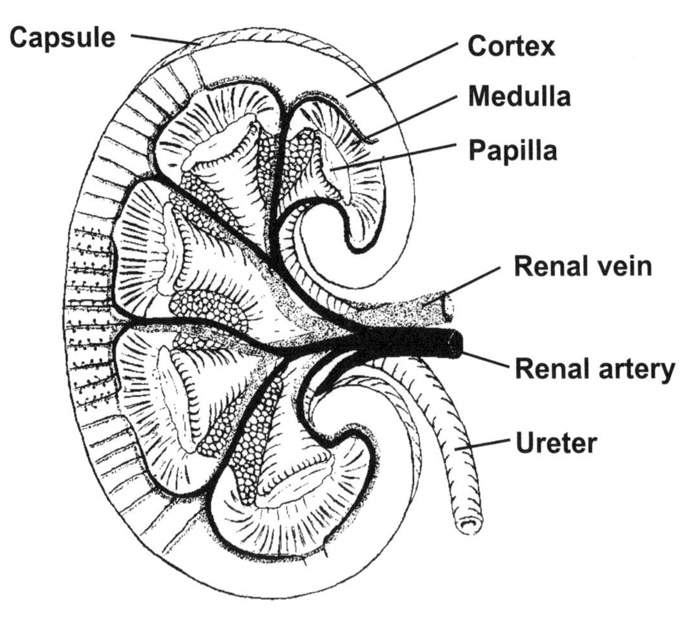

FIGURE 25.1 Sagittal section of a human kidney, showing major gross anatomical features and renal blood supply.

Source: Modified from: R.F. Pitts, *Physiology of the Kidney and Body Fluids*, 3rd Edition (1974), Year Book Medical Publishers Incorporated.

The papilla is the smallest anatomical portion of the kidney. Papillary tissue consists primarily of terminal portions of the collecting duct system and the vasa recta. Papillary blood flow is low relative to cortex and medulla; less than 1% of total renal blood flow reaches the papilla. However, tubular fluid is maximally concentrated and the volume of luminal fluid is maximally reduced within the papilla. Potential toxicants trapped in tubular lumens may attain extremely high concentrations within the papilla during the process of urinary concentration. High intraluminal concentrations of potential toxicants may result in diffusion of these chemicals into papillary tubular epithelial and/or interstitial cells, leading to cellular injury.

25.2.2 Nephron Structure and Function

The nephron is the functional unit of the kidney and consists of vascular and tubular elements (Fig. 25.2). Both elements of the nephron have multiple specific functions, any one or more of which may be influenced by toxicants.

25.2.3 Renal Vasculature and Glomerular Filtration

The renal vasculature serves to (1) deliver waste and other materials to the tubule for excretion, (2) return reabsorbed and synthesized materials to the systemic circulation, and (3) deliver oxygen and metabolic substrates to the nephron. Vascular components of the nephron include afferent and efferent arterioles, glomerular capillaries, peritubular capillary network, and the vasa recta (Fig. 25.2).

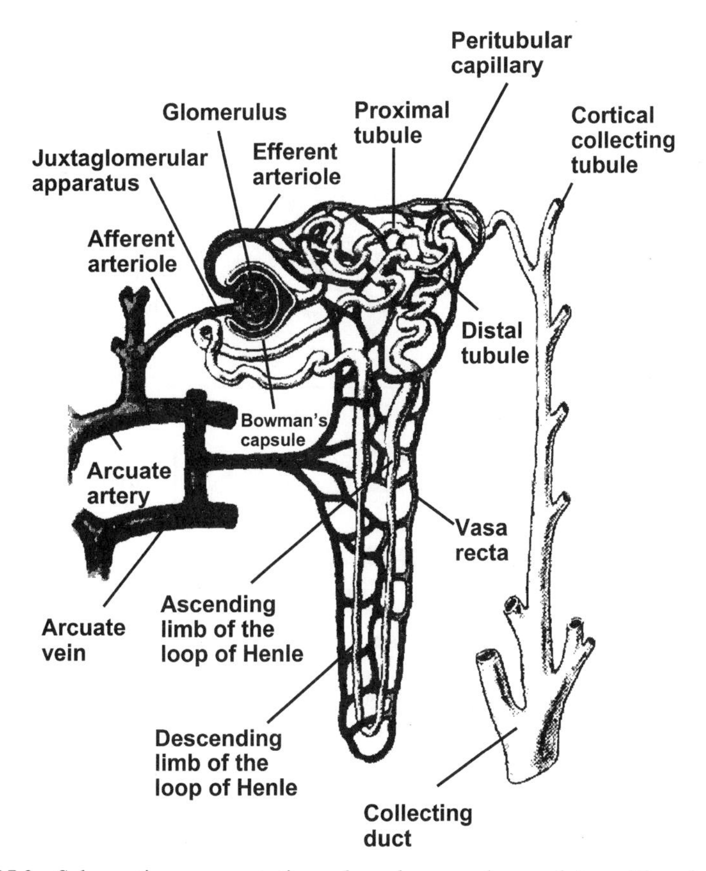

FIGURE 25.2 Schematic representation of nephron and vasculature. The glomerulus is positioned between afferent and efferent arterioles and the juxtaglomerular apparatus is the point of contact between the vascular pole and distal tubule of the nephron. A capillary network surrounds tubular structures.

Source: From A.C. Guyton and J.E. Hall, *Textbook of Medical Physiology*, 9th edition (1996), W.B. Saunders Company, Philadelphia. Reproduced with permission.

25.2.3a Arterioles

The glomerulus is unique in that it is the only capillary bed in the body positioned between two vasoactive arterioles (Fig. 25.3). The afferent and efferent arterioles control blood flow and hydrostatic pressure through the glomerulus. Afferent and efferent arterioles are vascular smooth muscle cells, innervated by sympathetic nerve fibers, and contract in response to nerve stimulation, endothelin, angiotensin II, antidiuretic hormone (ADH, vasopressin), and other stimuli. Additionally, the afferent and efferent arterioles form the vascular component of the juxtaglomerular apparatus, an important mechanism in the regulation of glomerular filtration rate.

25.2.3b Morphologic Basis for Glomerular Filtration

Urine formation begins at the glomerulus where an ultrafiltrate of plasma is formed. The process of filtration is governed by physical processes that determine fluid

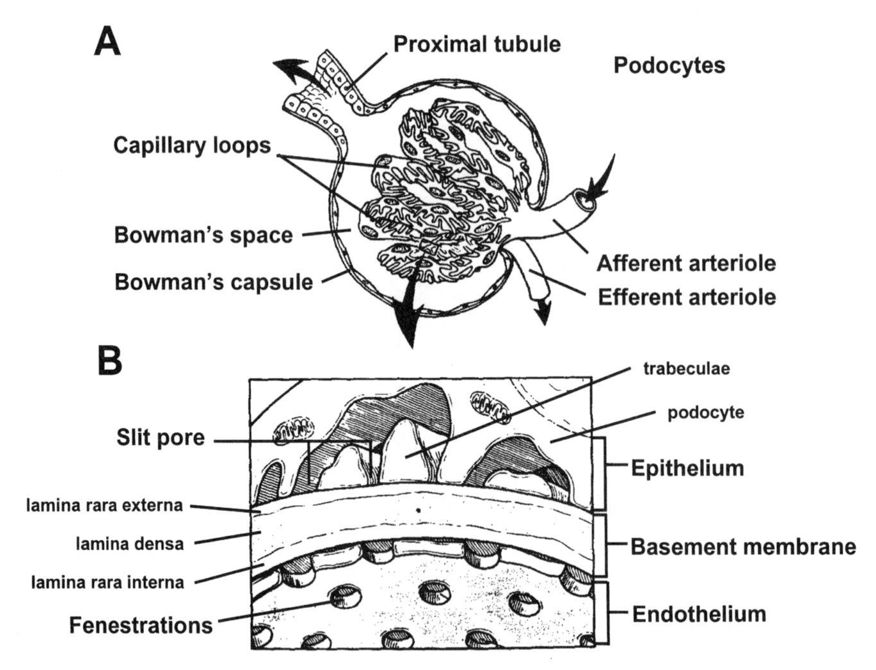

FIGURE 25.3 *A*. Glomerular capillaries are positioned between two arterioles that regulate hydrostatic pressure and blood flow through the capillaries, thereby controlling glomerular filtration rate. *B*. The glomerular basement membrane consists of three layers: the capillary endothelium, the capillary basement membrane, and podocytes from epithelial cells of Bowman's capsule. The capillary basement membrane also contains three layers: the lamina rara interna (blood side), lamina densa, and lamina rara externa (epithelial side).

Source: From A.C. Guyton and J.E. Hall, *Textbook of Medical Physiology*, 9th edition (1996), W.B. Saunders Company, Philadelphia. Reproduced with permission.

movement across capillary beds: transcapillary hydrostatic (hydraulic) pressure and colloid oncotic (osmotic) pressure. Glomerular filtration rate (GFR) can be represented mathematically as follows:

$$\text{GFR} = K_f(P_{GC} + \pi_{BS}) - (P_{BS} + \pi_{GC})$$

where P represents hydrostatic pressure and π indicates colloid oncotic pressure in the glomerular capillaries (GC) and Bowman's space (BS). GFR is determined primarily by the ultrafiltration coefficient (K_f) and by the difference between mean capillary pressure (ΔP_{GC}) and mean plasma oncotic pressure ($\Delta\pi_{GC}$). K_f is a measure of capillary permeability and is determined by the total surface area available for filtration and the hydraulic permeability of the capillary wall. Filtration is favored when transcapillary hydrostatic pressure exceeds plasma oncotic pressure. Plasma oncotic pressure does not remain constant along the length of the glomerular capillary, but increases as water is filtered from plasma. When plasma oncotic pressure is equal to or exceeds transcapillary hydrostatic pressure, glomerular filtration ceases.

Glomerular filtrate passes through three barriers before entering Bowman's space: (1) capillary endothelium; (2) capillary basement membrane; and (3) visceral epithelium (podocytes) (Fig. 25.3). These three filtration barriers contain anionic molecules, such as glycosaminoglycans and heparan sulfate, which are thought to restrict the filtration of anionic molecules. In general, for a given molecular size, filtration of anionic dextran sulfate is less than that of neutral dextran, suggesting that fixed negative charges of the glomerular capillary may retard or hinder the filtration of anions via electrostatic repulsion.

The glomerular capillary endothelium behaves as a porous membrane and forms an effective filtration barrier only to molecules larger than 50–70 Å. In comparison, hemoglobin and albumin in solution have molecular radii of approximately 32 and 36 Å, respectively. The glomerular basement membrane is a trilamellar structure consisting of the lamina rara interna (glomerular capillary side), lamina densa, and lamina rara externa (Bowman's space side), as shown in Figure 25.3. The glomerular basement membrane is an effective barrier for filtration: digestion, removal, or neutralization of fixed anionic sites of the glomerular basement membrane results in increased permeability to large, anionic molecules such as ferritin and albumin. The visceral epithelium is an unusual epithelium consisting of a cell body (podocyte) from which many trabeculae extend. These trabeculae, in turn, are in contact with the lamina rara externa of the glomerular basement membrane via many pedicles (foot processes). The distance between foot processes varies between 25 and 60 nm and forms the basis of the filtration slit (Fig. 25.3). Foot processes also greatly increase the capillary surface area available for filtration.

25.2.3c Mesangial Cells

The exact function(s) of mesangial cells is (are) unknown. These cells are thought to provide mechanical support for the loops of the glomerular capillary tuft. As modified smooth muscle cells, mesangial cells contain contractile proteins and receptors for substances such as angiotensin II, vasopressin, and catecholamines. Contraction of mesangial cells results in reduction of glomerular capillary surface area, thereby reducing glomerular filtration rate. Additionally, mesangial cells have phagocytic properties. Experimental studies utilizing injection of tracer materials (such as ferritin and colloidal carbon) have demonstrated uptake of these tracers by mesangial cells. In some forms of immune-complex glomerulonephritis, mesangial cells appear to play an important role in the sequestration and elimination of immune complexes deposited within the glomerular capillary tuft.

25.2.3d Peritubular Capillaries

Post-glomerular blood flows into peritubular capillaries or the vasa recta (Fig. 25.2). The peritubular capillaries perfuse the renal cortex and contain blood in which plasma colloid oncotic pressure is elevated due to concentration of plasma proteins during glomerular filtration. This elevated plasma oncotic pressure provides a driving force for fluid and electrolyte reabsorption from renal interstitium into the peritubular capillary network. In addition, the peritubular capillaries deliver oxygen and nutrients to tubular epithelial cells.

25.2.3e Vasa Recta

The arrangement of postglomerular capillary loops into the vasa recta (Fig. 25.2) is an efficient arrangement for delivery of blood-borne nutrients to medullary and papillary tubular structures. Additionally, relatively low blood flow in medulla and papilla ensures hypertonicity in these areas. Indeed, a common complication of vasodilator therapy is a reduction in medullary hypertonicity, resulting in impaired urinary concentrating ability.

25.2.3f Juxtaglomerular Apparatus and Renin Secretion

The juxtaglomerular apparatus is a specialized nephron portion consisting of both tubular and vascular elements. Within an individual nephron, cells of the afferent arteriole come into contact with the distal tubule at the juxtaglomerular apparatus. Cells of the distal tubule form the macula densa, the tubular component of the juxtaglomerular apparatus. Both afferent arteriolar and distal tubular cells contain renin. Renin is secreted by arteriolar cells in response to sympathetic nerve stimulation and/or decreased stretch, as in hypotension. Additionally, in response to some poorly understood signal at the macula densa (i.e., increased distal tubular flow rate, reduced Na^+/Cl^- flux at the distal tubule, and/or increased tubular fluid concentrations of Na^+/Cl^-), cells of the juxtaglomerular apparatus increase renin secretion, resulting in increased formation of angiotensin II. By preferentially constricting afferent arterioles, as well as mesangial cells, angiotensin II reduces single nephron blood flow and glomerular filtration rate.

25.2.4 Tubular Function and Formation of Urine

The renal tubule begins as a blind pouch surrounding the glomerulus. Renal tubules consist of multiple segments, shown schematically in Figure 25.4. These tubular elements selectively modify the composition of glomerular ultrafiltrate, enabling conservation of electrolytes and metabolic substrates while allowing elimination of waste products. For example, renal tubules reabsorb 98–99% of filtered electrolytes and water, and virtually 100% of filtered glucose and amino acids. Additionally, renal tubules participate in the reabsorption of bicarbonate and secretion of protons, thereby participating in acid-base balance.

25.2.4a Proximal Tubule

The initial tubular segment, the proximal tubule, reabsorbs about 70% of water filtered at the glomerulus. In other words, for each 100 ml of glomerular ultrafiltrate formed, only 30 ml will enter the loop of Henle (Fig. 25.5). The proximal tubule consists of the proximal convoluted (pars convoluta) and proximal straight (pars recta) segments (Fig. 25.4). Filtered HCO_3^-, proteins, and glucose are reabsorbed primarily by the proximal convoluted tubule, whereas organic ion secretion occurs primarily in the proximal straight tubule.

Water reabsorption is an isosmotic, passive process driven primarily by reabsorption of Na^+ occurring via Na^+-K^+ ATPase. the proximal tubule contains numerous active transport systems capable of driving concentrative, uphill transport of many metabolic substrates, including amino acids, glucose, and citric acid cycle intermediates. Proximal tubular transport systems also scavenge virtually all of the filtered proteins by specific endocytotic protein reabsorption mechanisms.

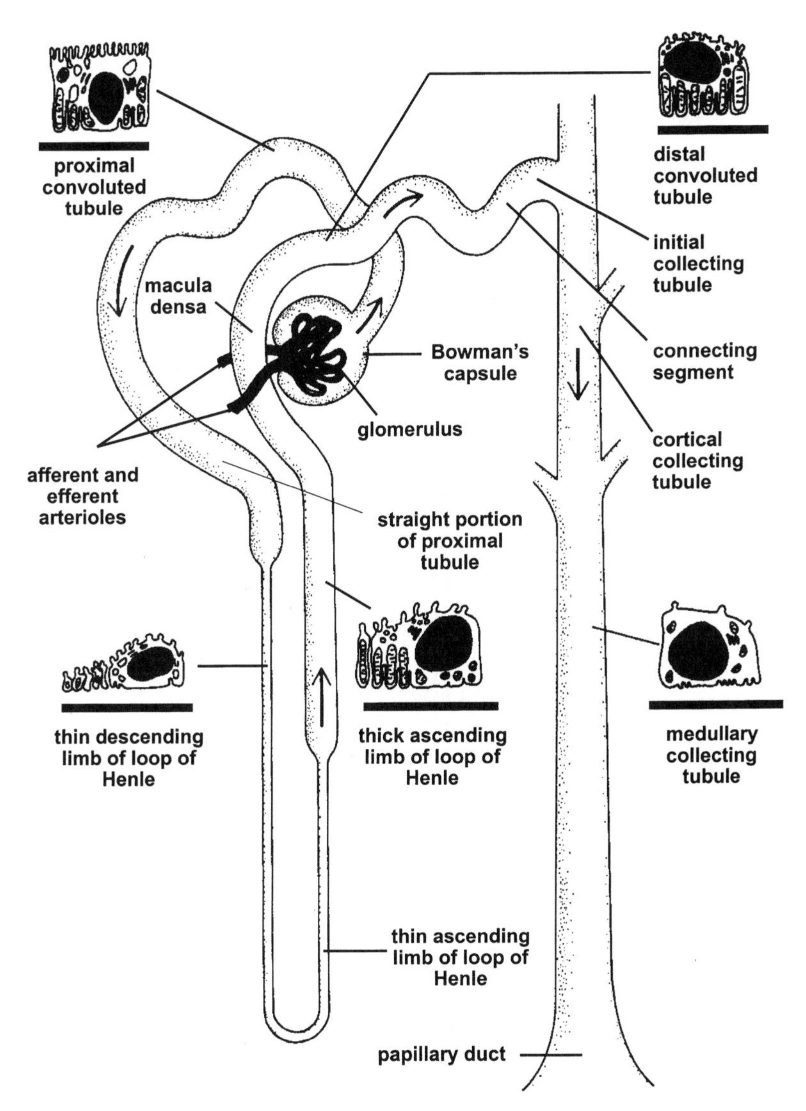

FIGURE 25.4 Schematic representation of a renal tubule. The epithelial cells have differing morphology and function along the length of the tubule.

Source: From A.J. Vander, *Renal Physiology*, 3rd edition (1985), McGraw-Hill Book Company, New York. Reproduced with permission.

In addition, proximal tubular brush border enzymes metabolize proteins and/or peptides to constituent dipeptides and amino acids. Proximal tubular transport systems mediate the reabsorption of sulfate, phosphate, calcium, magnesium, and other ions following glomerular filtration. An important excretory function of the proximal tubule is secretion of weak organic anions and cations by specialized transport systems that drive concentrative, uphill movement of these substances from postglomerular blood into proximal tubular cells followed by excretion into tubular fluid.

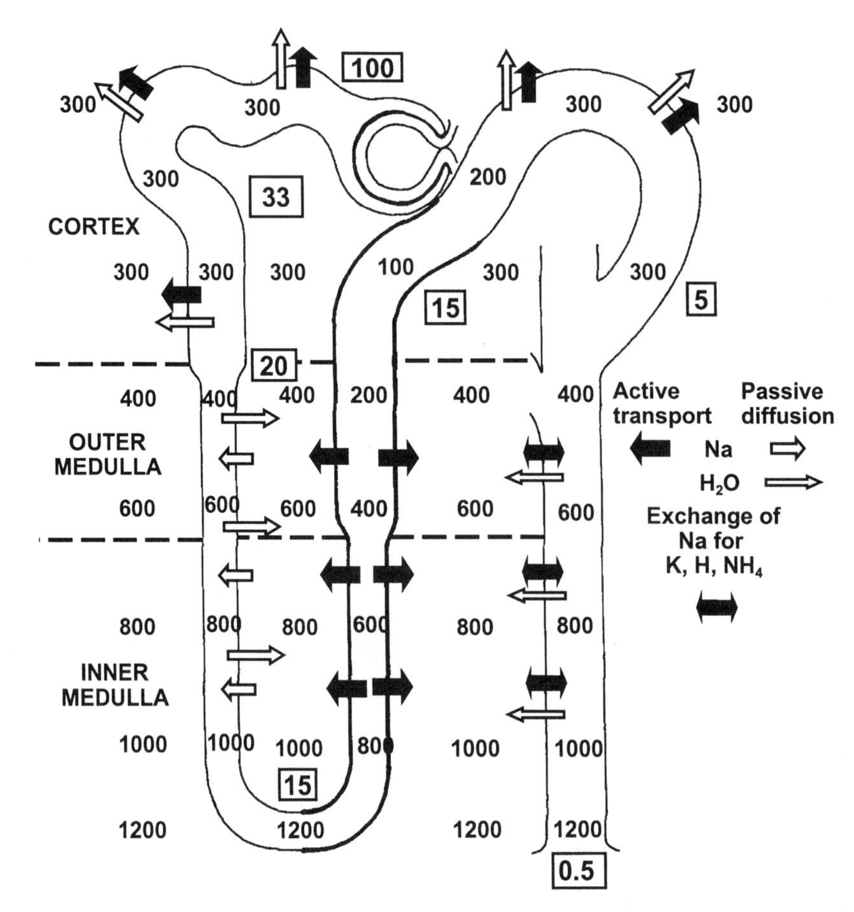

FIGURE 25.5 Schematic illustration of passive and active movement of electrolytes and water following glomerular filtration. Concentrations of tubular urine and peritubular fluid are given in milliosmoles per liter; large, boxed numerals are estimated volumes of glomerular filtrate (per 100 ml of plasma filtered) remaining within the tubule.

Source: From Pitts, *Physiology of the Kidney and Body Fluids*, 3rd Edition (1974), Year Book Medical Publishers, Chicago.

25.2.4b Loop of Henle and Countercurrent System

In addition to conservation of electrolytes and metabolic intermediates, mammalian nephrons are adapted for maximal conservation of water. The thin descending and ascending limbs of the loop of Henle and the thick ascending limb of the loop of Henle are critical to the processes mediating urinary concentration (Fig. 25.4). The descending and ascending limbs of the loop of Henle have differential permeabilities to water and electrolytes, establishing medullary hypertonicity necessary for urinary concentration. The loop system removes an additional 10–20% of fluid filtered at the glomerulus so that, of 100 ml of ultrafiltrate initially formed, only 10–20 ml are delivered to the distal tubule (Fig. 25.5).

Four factors contribute to the ability to concentrate urine: (1) Active reabsorption of Na^+, K^+, and Cl^- without water reabsorption by the thick ascending limb of the loop of Henle results in interstitial hypertonicity and hypoosmotic tubular fluid;

(2) Selective permeability to water, but not small electrolytes, in the descending thin limb of the loop of Henle allows passive reabsorption of water, facilitated by interstitial hypertonicity; (3) Relatively low medullary blood flow maintains medullary hypertonicity, allowing continued elaboration of concentrated urine; (4) In the presence of ADH, the distal tubule and collecting ducts are permeable to water so that water may diffuse out of the tubular lumen into the medullary and papillary interstitium. Because of the ability of the thick ascending limb of the loop of Henle to move solutes but not water into the medullary interstitium, the medullary and papillary interstitium are hyperosmotic and hypertonic compared to plasma and cortical interstitium (Fig. 25.5).

25.2.4c Distal Tubule

Fluid entering the distal tubule is hypoosmotic compared to blood plasma. The distal tubule further modifies tubular fluid by reabsorbing most of the remaining intraluminal Na^+ in loose exchange for K^+ and H^+. Both H^+ and K^+ secretion are driven by Na^+ reabsorption. In turn, Na^+ reabsorption is determined both by plasma Na^+ concentration and distal tubular flow rate. As flow rate is increased, either by inhibition of reabsorptive processes in the proximal tubule and/or loop of Henle or by enhanced glomerular filtration, the amount of Na^+ reabsorbed by the distal tubule is increased. Because a loose exchange of Na^+ for K^+ and H^+ occurs, increases in Na^+ reabsorption concomitantly increase K^+ and H^+ secretion. Consequences of increased K^+ and H^+ secretion include hypokalemia and metabolic alkalosis, frequent complications of diuretic therapy in which distal tubular flow rate is markedly increased and Na^+ reabsorption is increased. During transit through the distal tubule, tubular fluid volume is reduced an additional 20–30%; of the original 100 ml of glomerular filtrate, only about 5 ml enters the collecting system (Fig. 25.5).

25.2.4d Collecting Duct

The collecting tubule and duct performs final regulation and fine-tuning of urinary volume and composition. Active transport systems in the collecting tubule reabsorb Na^+ and secrete K^+ and H^+. Additionally, the combination of medullary and papillary hypertonicity generated by countercurrent multiplication and the action of ADH combine to enhance water permeability of the collecting duct system. Water permeability of the collecting tubule and duct is dependent on the presence of antidiuretic hormone, and agents that interfere with the secretion or action of antidiuretic hormone may impair urinary concentrating ability. Additionally, maximal urinary concentration depends upon medullary and papillary hypertonicity. Thus, agents that increase medullary blood flow may impair urinary concentrating ability by dissipating the medullary osmotic gradient. In the presence of antidiuretic hormone, intraluminal volume may be reduced to leave a scant 0.5 ml of 100 ml of original glomerular filtrate (Fig. 25.5).

25.3 FACTORS CONTRIBUTING TO NEPHROTOXICITY

Several factors contribute to the unique susceptibility of the kidney to toxicants (Table 25.1). First, renal blood flow is high relative to organ weight. For an organ constituting less than 1% of body weight, the kidneys receive about 25% of the

TABLE 25.1 Factors Influencing Susceptibility of the Kidney to Toxicants.

High renal blood flow
Concentration of chemicals in intraluminal fluid
Reabsorption and/or secretion of chemicals through tubular cells
Biotransformation of protoxicant to reactive intermediates

resting cardiac output. Thus, the kidneys will receive higher concentrations of toxicants (per gram of tissue) than poorly perfused tissue such as skeletal muscle, skin, and fat. Renal blood flow is unequally distributed, with cortex receiving a disproportionately high flow compared to medulla and papilla. Therefore, a blood-borne toxicant will be delivered preferentially to the renal cortex and thereby have a greater potential to influence cortical rather than medullary or papillary functions.

Second, the processes involved in forming concentrated urine also will serve to concentrate potential toxicants present in the glomerular filtrate. Reabsorptive processes along the nephron may raise the intraluminal concentration of a toxicant from 10 to 50 mM by the end of the proximal tubule, 66 mM at the hairpin turn of the loop of Henle, 200 mM at the end of the distal tubule, and as high as 2000 mM in the collecting duct. Progressive concentration of toxicants may result in intraluminal precipitation of poorly soluble compounds, causing acute renal failure secondary to mechanical obstruction. Additionally, high intraluminal concentrations of a toxicant may enable passive diffusion of toxicant from tubular lumen into epithelium. The potentially tremendous concentration gradient for passive diffusion between lumen and cell may drive even a relatively nondiffusible toxicant into tubular cells. Furthermore, the proximal tubular epithelium is "leaky"; that is, it is fairly permeable to solutes and water. In contrast, the distal tubule is a relatively "tight" epithelia and less permeable to solutes and water than the proximal tubule. Thus, high intraluminal concentrations of toxicants may result in fairly high concentrations in proximal tubular epithelium and selectively injure the proximal tubule.

Third, active transport processes within the proximal tubule may further raise the intracellular concentration of an actively transported toxicant. During active secretion and/or reabsorption, substrates generally accumulate in proximal tubular cells in much higher concentrations than present in either luminal fluid or peritubular blood.

Fourth, certain segments of the nephron have a capacity for metabolic bioactivation. For example, the proximal and distal tubules contain isozymes of the cytochrome P450 monooxygenase system that may mediate intrarenal bioactivation of several protoxicants. Additionally, prostaglandin synthetase activity in medullary and papillary interstitial cells may be involved in cooxidation of protoxicants, resulting in selective papillary injury.

25.4 ASSESSMENT OF NEPHROTOXICITY

Evaluation of the effects of a toxicant on renal function can be accomplished by several methods. The method used depends on the complexity of the question to be

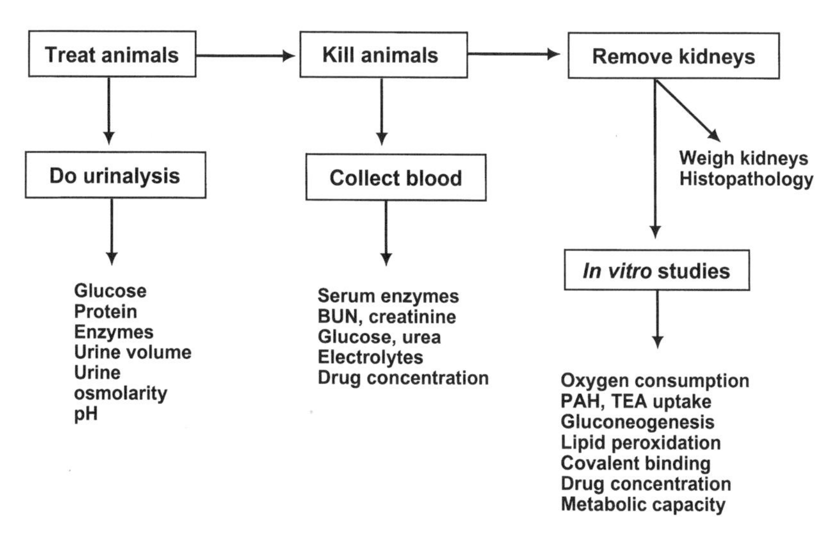

FIGURE 25.6 Representative design of experimental protocols to test renal function following drug or chemical administration.

answered. For example, unanesthetized or anesthetized animals may be used to determine if a chemical has an effect on kidney function. Investigations also may be conducted in vitro to examine specific biochemical or functional lesions. Finally, histopathologic techniques can provide a great deal of information about structural integrity.

25.4.1 In Vivo Methods

Initial evaluation of the effect of a chemical on renal function often is performed in intact, unanesthetized animals. An advantage of this type of study is that a great deal of information concerning overall renal function may be obtained relatively quickly using noninvasive methods. A major limitation of these noninvasive tests is lack of specificity: these tests provide no information concerning an intrarenal site of toxicity, but rather assess overall kidney function.

The standard battery of noninvasive tests includes measurement of urine volume and osmolality, urinary pH, and excretion of Na^+, K^+, glucose and protein (Fig. 25.6). An increase in urine volume and decrease in osmolality following toxicant administration may suggest an impaired ability to concentrate urine or interference with antidiuretic hormone secretion or action. Alternately, a toxicant may produce an abrupt decline in urine volume, possibly due to intraluminal obstruction by precipitation of the toxicant or cellular debris following tubular injury. Abnormalities in urine osmolality may indicate toxicant effects on renal medullary function. Alternations in the ability to acidify or alkalinize urine may suggest alterations in distal tubular function. Because both glucose and protein are reabsorbed almost completely by the proximal tubule, glucosuria and/or proteinuria may indicate proximal tubular dysfunction. However, glucosuria also may be secondary to toxicant-induced hyperglycemia so that measurement of serum glucose concentrations may

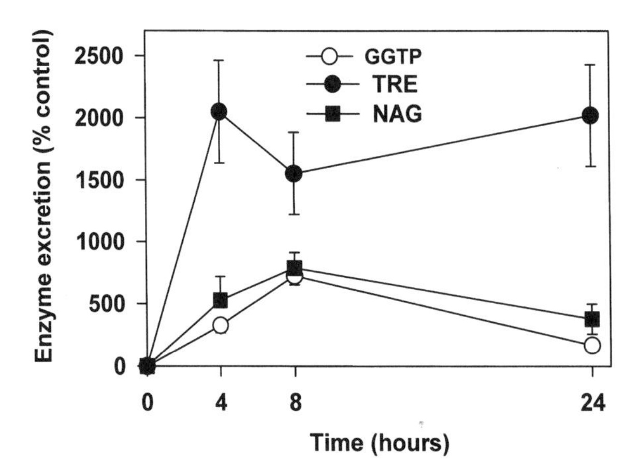

FIGURE 25.7 Urinary excretion of proximal tubular enzymes following *para*-aminophenol (PAP) administration to female Sprague-Dawley rats.
Rats received PAP (300 mg/kg ip) and urine was collected for 24 h following treatment. Urine was assayed for activities of γ-glutamyl transpeptidase (GGTP, ○), trehalase (TRE, ●), and *N*-acetyl-β-glucosaminidase (NAG, ■). Data are expressed as a percentage of enzyme excretion in rats treated with saline. Each data point represents the mean ± standard error of at least 4 determinations. GGTP and TRE are enzymes located primarily on the brush border of proximal tubule cells and NAG is an intracellular enzyme.

be warranted. Proteinuria may reflect either proximal tubular or glomerular damage. Excretion of high molecular weight proteins, such as albumin, is suggestive of glomerular injury whereas excretion of low molecular weight proteins, such as β_2-microglobulin, is more suggestive of proximal tubular injury. Serial blood samples may be obtained and blood urea nitrogen and plasma creatinine concentrations determined. Elevations in blood urea nitrogen and/or plasma creatinine concentrations often suggest decreased glomerular filtration rate. However, increased blood urea nitrogen and creatinine concentrations may be secondary to dehydration, hypovolemia, muscle injury, or protein catabolic states, and therefore may not necessarily reflect renal damage.

Attempts have been made to develop noninvasive tests that might provide more specific information about the site of injury. For example, excretion of enzymes in urine may reflect renal injury. Urinary excretion of enzymes of renal origin (enzymes that are specific to the kidney, such as maltase, γ-glutamyl transpeptidase, or trehalase) could indicate specific destruction of renal proximal tubules, whereas alkaline phosphatase in urine could arise from renal or prerenal (e.g., hepatic) damage (Fig. 25.7).

The determination of renal function in anesthetized animals provides specific information on the effects of chemicals on glomerular filtration rate and renal blood flow. In addition, the ability of the kidney to reabsorb or secrete electrolytes may be determined by fractional clearance of Na^+, K^+, HCO_3^-, Cl^-, etc. Fractional clearance involves comparison of electrolyte clearance to the clearance of a substance such as inulin, which is removed from plasma by glomerular filtration. Thus, fractional clearance takes glomerular filtration rate into account, allowing comparisons of electrolyte transport between treated and control animals even if renal hemodynamics have changed. Nephron function may be assessed by free water clearance, representing the ability of the kidney to remove almost all Na^+ from urine.

25.4.2 In Vitro Methods

A variety of in vitro techniques may be employed to help elucidate mechanisms of chemical-induced nephrotoxicity. Toxic effects of chemicals may be evaluated in vitro in tissue obtained from naive animals by adding the toxicant in vitro and/or in tissue obtained from animals treated in vivo with the proposed toxicant. These approaches may be used to distinguish between an effect on the kidney due to direct chemical insult and secondary effects, such as those due to extrarenal metabolites and/or alterations in pharmacokinetics or hemodynamics.

Renal cortical slices have been used extensively to evaluate the influence of nephrotoxicants on the transport of organic anions such as para-aminohippurate (PAH) and organic cations such as N-methylnicotinamide (NMN) and tetraethylammonium (TEA). Reabsorptive transport in slices may be evaluated using the nonmetabolized amino acid analog, α-aminoisobutyric acid, and the nonmetabolized sugar, α-methyl-D-glucose. The ability of renal cortical slices to produce ammonia and glucose in vitro can provide specific information about metabolic alterations produced by a toxicant. Renal cortical slices also have been used to investigate biochemical alterations following toxicant exposure, such as lipid peroxidation (reflected by malondialdehyde (MDA) production), glutathione (GSH) depletion, ATP depletion, and oxygen consumption. Subtle biochemical effects of toxicants measured in vitro may provide important information regarding mechanisms of cytotoxicity that are difficult to assess in intact animals.

Several laboratories have developed methods to physically separate and characterize the metabolic capacity of glomeruli, proximal tubules, and distal tubules. These techniques can offer insight into biochemical changes associated with site-specific nephrotoxicity. Additionally, micropuncture and microperfusion experiments have been utilized to help identify specific loci of action of nephrotoxicants.

25.4.3 Histopathology

Finally, histopathologic examination of tissue can reveal structural changes that have occurred in response to nephrotoxicants, often allowing identification of affected areas of the nephron. For instance, light microscopy can identify changes in renal morphology caused by chemicals, such as the presence of intraluminal protein casts, sloughed brush border, and crystals or stones in the kidney and urine. Many histochemical techniques are available to evaluate renal responses to toxicants (Fig. 25.8). Electron microscopy provides information concerning subcellular localization of tubular injury. Electron microscopy has been employed extensively in efforts to understand the changes in glomerular structure that might account for changes in permeability following nephrotoxicant exposure.

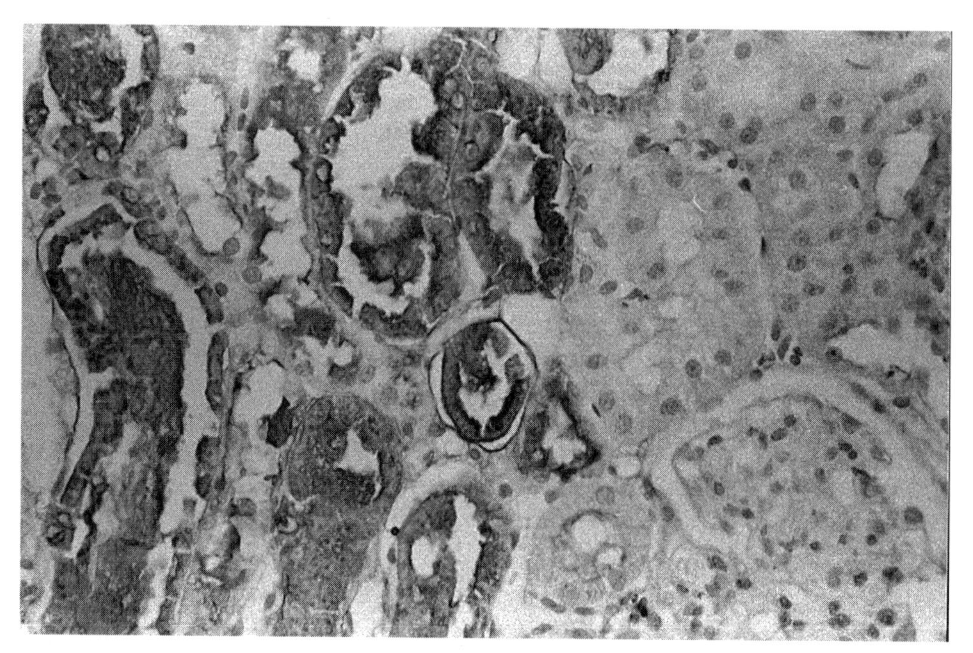

FIGURE 25.8 Kidney tissue from a rat treated with *para*-aminophenol (300 mg/kg ip). Tissue was stained with hemotoxylin and eosin and was subjected to terminal deoxynucleotidyl transferase-diamonibenzidine staining to reveal DNA strand breaks. Final magnification was 360.

25.4.4 Compensation for Renal Damage

Fortunately, the kidney has a remarkable ability to compensate for the loss of renal functional mass. Within a short time after unilateral nephrectomy, the remaining kidney hypertrophies such that overall renal function appears normal by standard clinical tests. Compensation becomes a problem when evaluating the effects of nephrotoxicants; specifically, changes in kidney function may not be detected until the ability of the kidney to compensate is exceeded. Then, within a short period of time, an animal might develop life-threatening renal failure.

25.5 SITE-SPECIFIC NEPHROTOXICITY

Many compounds have been implicated as nephrotoxicants (Table 25.2). Only rarely have specific receptors for specific nephrotoxicants been identified. Rather, in many cases it appears that toxicants exert multiple effects on intracellular systems. This is not to say, however, that there are not specific targets for certain nephrotoxicants in the kidney. For example, the proximal convoluted tubule seems to be more susceptible than other nephron segments to certain metals, such as chromium. The straight portion of the proximal tubule seems to be more susceptible to damage due to halogenated hydrocarbons (i.e., hexachlorobutadiene and dichlorovinyl-L-cysteine). Some agents, such as analgesic mixtures (usually aspirin, phenacetin, and caffeine) taken over long periods can produce a unique toxicity characterized by

TABLE 25.2 Segments of the Nephron Affected by Selected Toxicants.

Glomerulus	**Distal tubule/collecting duct**
Immune complexes	Lithium
Aminoglycoside antibiotics	Tetracyclines
Puromycin aminonucleoside	Amphotericin
Adriamycin	Fluoride
Penicillamine	Methoxyflurance
Proximal tubule	**Papilla**
Antibiotics	Aspirin
Cephalosporins	Phenacetin
Aminoglycosides	Acetaminophen
Antineoplastics	Nonsteroidal anti-inflammatory agents
Nitrosoureas	2-Bromoethylamine
Cisplatin and analogs	
Radiographic contrast agents	
Halogenated hydrocarbons	
Chlorotrifluoroethylene	
Hexafluropropene	
Hexachlorobutadiene	
Trichloroethylene	
Chloroform	
Carbon tetrachloride	
Maleic acid	
Cirtinin	
Metals	
Mercury	
Uranyl nitrate	
Cadmium	
Chromium	

renal medullary and papillary necrosis. Histological evaluation following intoxication with analgesic mixtures reveals damage to the ascending limbs of the loop of Henle. Likewise, fluoride ion and outdated tetracyclines produce damage in this area.

25.5.1 Glomerulus as a Site of Toxicity

Aminoglycoside antibiotics (Fig. 25.9), such as gentamicin, amikacin and netilmicin, are powerful drugs for the treatment of serious Gram-negative infections. However, about 10% of patients treated with aminoglycosides will develop moderate but significant declines in glomerular filtration rate and elevations in serum creatinine concentration. The therapeutic utility of aminoglycosides is limited by nephrotoxicity, ototoxicity and neuromuscular junction blockade. Aminoglycoside nephrotoxicity is characterized by proximal tubular necrosis, proteinuria, and a profound decline in glomerular filtration rate.

Gentamicin-induced declines in glomerular filtration rate could be due to alterations in one or more factors determining glomerular filtration rate (i.e., changes in K_f, P_{GC}, π_{GC}). Micropuncture experiments indicate that, following low dosages of

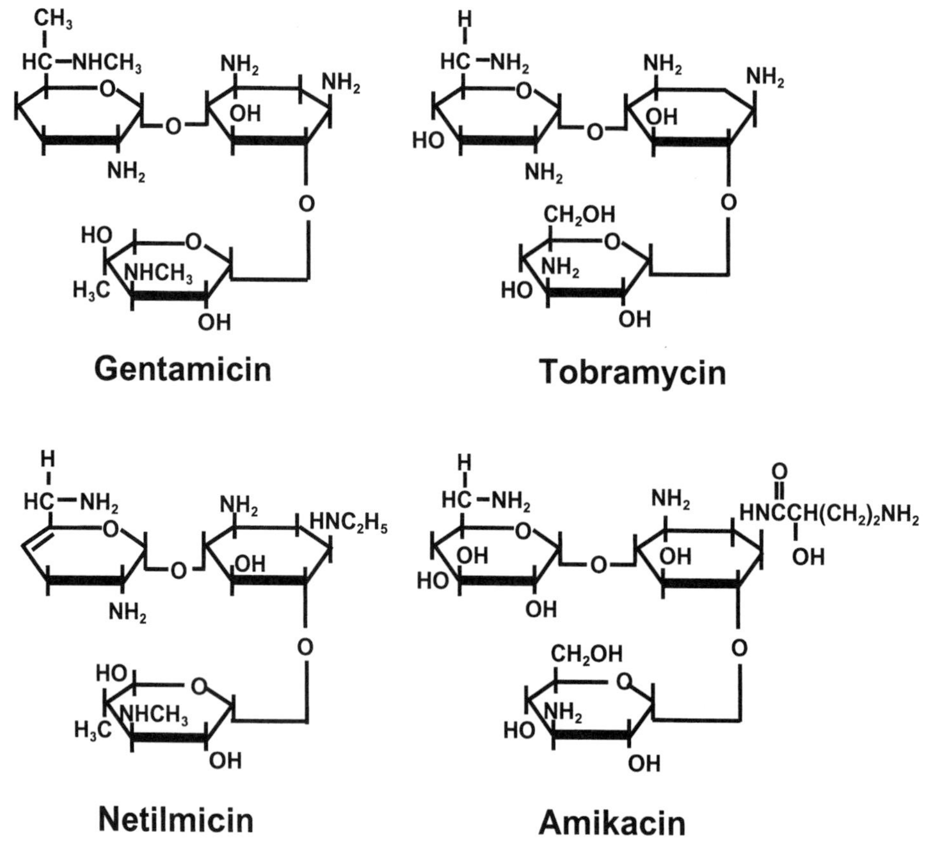

Gentamicin

Tobramycin

Netilmicin

Amikacin

FIGURE 25.9 Chemical structures of several aminoglycoside antibiotics.

gentamicin in rats, both single nephron glomerular filtration rate and whole kidney glomerular filtration rate are markedly reduced. However, glomerular plasma flow rate, and glomerular capillary hydrostatic and colloid oncotic pressures are unaltered by low-dose gentamicin treatment, suggesting that changes in renal hemodynamics do not contribute significantly to the gentamicin-induced decline in glomerular filtration rate. In contrast to the lack of effect on renal hemodynamics, gentamicin treatment produces a significant and marked reduction in capillary ultrafiltration coefficient (K_f) sufficient to account for the decline in single nephron glomerular filtration rate. Thus, gentamicin may affect the structural integrity of glomerular capillaries, resulting in a reduction in the surface area available for filtration. Ultrastructural examination of glomeruli following gentamicin treatment indicates a reduction in the size and number of endothelial fenestra, an effect which would reduce capillary ultrafiltration coefficient and hence, decrease glomerular filtration rate.

Aminoglycosides and polycationic compounds, and the observed ultrastructural alterations of glomerular capillaries following gentamicin treatment would be consistent with interactions between cationic aminoglycoside molecules and anionic sites on the glomerular capillaries. However, there is little correlation between net positive charge on various aminoglycoside molecules and capacity to reduce the size and number of endothelial fenestra. For example, netilmicin, gentamicin, and tobramycin all contain five ionizable groups. Based on electrostatic interactions alone, each compound would be expected to bind equally to glomerular anionic sites and have equal propensities to reduce capillary ultrafiltration coefficient and hence, glomerular filtration rate. Determinations of whole kidney glomerular filtration rate and electron microscopic examination of glomerular endothelial fenestra indicate the following rank order of toxicity: gentamicin > tobramycin > netilmicin. Thus, factors other than or in addition to electrostatic interactions may contribute to aminoglycoside-induced changes in glomerular structure and function.

A prominent component of aminoglycoside nephrotoxicity is acute proximal tubular necrosis. Interestingly, the rank order of toxicity for both tubular damage and reduction of glomerular filtration rate is similar: gentamicin > tobramycin > netilmicin, suggesting the possibility that reduced glomerular filtration rate may be secondary to tubular damage. Several laboratories have suggested that at least a portion of glomerular alterations produced by nephrotoxic aminoglycosides may be related to activation of the renin-angiotensin system. With extensive proximal tubular necrosis, normal reabsorptive processes may be impaired. Indeed, nephrotoxic aminoglycosides produce a nonoliguric renal failure, suggesting that tubular reabsorption of water and electrolytes may be reduced. As previously discussed, increased flow rate at the macula densa is a stimulus for secretion of renin and hence, formation of the vasoconstrictor, angiotensin II. In support of an involvement of angiotensin II in aminoglycoside-induced glomerular alterations, maneuvers designed to suppress angiotensin II formation (e.g., inhibition of renin secretion by sodium loading and converting enzyme inhibitor therapy) attenuate aminoglycoside-induced decrements in glomerular filtration rate. Thus, direct glomerular toxicity, as well as activation of angiotensin II, may contribute to aminoglycoside-induced alterations in glomerular filtration rate.

25.5.2 Proximal Tubule as a Site of Toxicity

The proximal tubule is the most frequently identified site of toxicant-induced renal dysfunction. The reasons for this enhanced susceptibility may relate, in part, to clearly identifiable proximal tubular functions (e.g., gluconeogenesis, ammoniagenesis, organic ion transport, glucose reabsorption) that can be assessed easily in vivo or in vitro following toxicant exposure. This, proximal tubular damage may be easier to detect than damage to the loop of Henle or distal tubule where functions are more integrated and less easily identified and assessed. Additionally, enhanced susceptibility of the proximal tubule relative to other nephron segments may be related to one or a combination of the following factors: (1) Increased intraluminal concentration of potential toxicants coupled with epithelial permeability to small, lipophilic organic compounds, may contribute to intracellular accumulation of potential toxicants due to passive diffusion; (2) Active transport functions, both secretory and reabsorptive, may result in high intracellular concentrations of toxicants; (3) Xenobiotic metabolism, mediated by cytochrome P450 monooxygenases in the proximal tubule, are capable of biotransforming protoxicants to reactive, toxic intermediates.

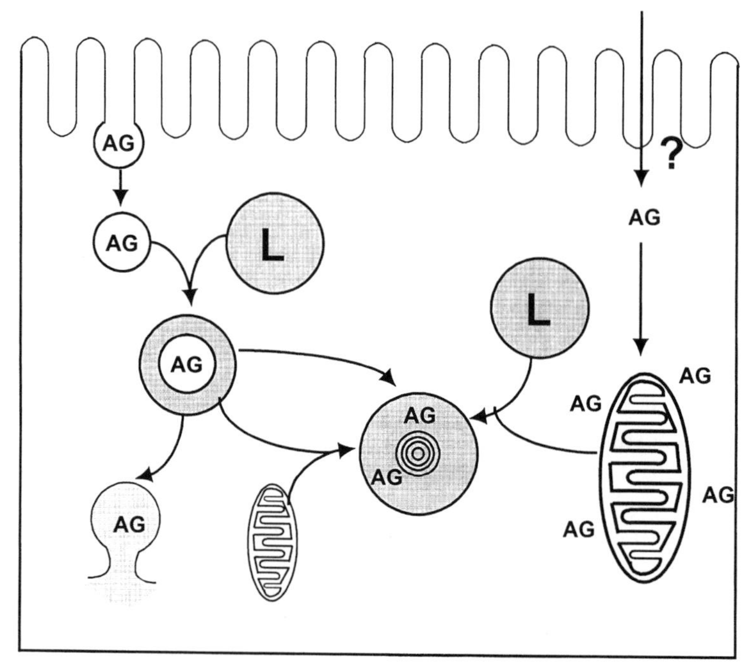

FIGURE 25.10 Postulated pathways of aminoglycoside-induced cellular injury. On the left, aminoglycoside (AG) enters the cell by pinocytosis and endocytosis, subsequently fusing with a primary lysosome (L). Aminoglycosides may interfere with normal lysosomal function, forming myeloid bodies (center). Additionally, aminoglycosides may destabilize lysosomes leading to release of intralysosomal enzymes (lower left). Intracellular aminoglycosides may produce direct injury to intracellular organelles such as mitochondria.

Source: From Kaloyanides and Pastoriza-Munoz, Kidney International *18*: 571–582 (1980), Springer-Verlag, New York. Reproduced with permission.

25.5.3 Aminoglycoside Nephrotoxicity: Role of Proximal Tubular Reabsorption

As discussed previously, aminoglycoside therapy is frequently limited by nephrotoxicity. Nonoliguric acute renal failure appears within 5–7 days after aminoglycoside therapy is initiated. Increased urine volume appears secondary to a concentrating defect and may precede increases in blood urea nitrogen and serum creatinine concentrations, delaying the recognition of acute renal failure. A small number of patients may have permanent deficits in renal function despite discontinuation of aminoglycoside therapy. Aminoglycoside-induced alterations in renal function include (1) reduced organic anion and cation secretion, (2) proteinuria, primarily involving low molecular weight proteins, (3) enzymuria, including excretion of alanylaminopeptidase and lysosomal enzymes (e.g., N-acetyl-β-glucosaminidase), and (4) decreased glomerular filtration rate. Morphologically, aminoglycoside nephrotoxicity is characterized by the presence of myeloid bodies, secondary lysosomes containing electron-dense lamellar membranous structures, and widespread proximal tubular necrosis (Fig. 25.10).

25.5.3a Renal Handling of Aminoglycosides

Aminoglycoside antibiotics are organic polycations and carry net positive charges (Fig. 25.9). These compounds have relatively low volumes of distribution and the primary route of elimination is by renal excretion. Gentamicin, a typical nephrotoxic aminoglycoside, is freely filtered at the glomerulus and appears to be reabsorbed via active transport processes at the proximal tubular brush border.

Intracellular accumulation of gentamicin appears to occur following binding to luminal membrane sites and incorporation of bound drug into apical vesicles (Fig. 25.10). Major binding sites are acidic phospholipids of the renal brush border membranes and initial binding is driven by anionic-cationic interactions. Small molecular weight proteins such as lysozyme appear to compete with gentamicin for intracellular uptake; small proteins both inhibit cortical accumulation of gentamicin and attenuate nephrotoxicity, suggesting a correlation between uptake and toxicity. However, cortical concentration and nephrotoxicity are not tightly correlated for all aminoglycosides. For example, using equivalent dosage regimens, gentamicin and netilmicin accumulate in renal cortical and medullary tissue to a similar extent, yet only gentamicin produces nephrotoxicity, suggesting that factors other than, or in addition to, cortical accumulation are important in gentamicin-induced nephrotoxicity.

Gentamicin nephrotoxicity is attenuated by treatment with other aminoglycosides or with organic polycations such as spermine, polyaspartate, and polylysine. Initially, competition for binding and intracellular uptake was the postulated mechanism of interaction between gentamicin and polycations. However, recent studies indicate that intracortical concentrations of gentamicin are not reduced in the presence of polycations, suggesting that mechanisms other than competitive uptake contribute to the amelioration in gentamicin nephrotoxicity.

25.5.3b Biochemical Mechanisms in Aminoglycoside Cytotoxicity

The sequence of biochemical events leading to gentamicin-induced proximal tubular dysfunction is unknown. Perhaps owing to its polycationic structure, gentamicin interferes with a number of intracellular proteins and macromolecules, producing a variety of biochemical effects (Table 25.3). Several mechanisms have been proposed to account for gentamicin cytotoxicity, including (1) lysosomal damage, (2) altered phospholipid metabolism, (3) inhibition of critical intracellular enzymes, (4) inhibition of mitochondrial respiration, (5) lipid peroxidation, and (6) misreading of mRNA.

Lysosomal alterations and the presence of myeloid bodies and cytosegresomes are characteristic of aminoglycoside nephrotoxicity. Aminoglycosides accumulate in lysosomes and may inhibit lysosomal enzymes, contributing to aminoglycoside-induced phospholipidosis. It is not entirely clear how lysosomal accumulation of aminoglycosides may contribute to aminoglycoside-induced toxicity. It is possible that accumulation of aminoglycosides within lysosomes may, due to mechanical or chemical injury, promote lysosomal labilization. With lysosomal rupture, lysosomal enzymes (i.e., acid hydrolases) may be released intracellularly and attack intracellular organelles leading to cell injury and necrosis. However, lysosomal alterations and myeloid bodies are found following therapy with other cationic, amphophilic compounds that do not produce nephrotoxicity. Similarly, embryonic rat fibroblasts incubated with gentamicin display lysosomal aberrations (including myelin

TABLE 25.3 Membrane Alterations in Aminoglycoside Nephrotoxicity.

Plasma membrane alterations

1. Phospholipid alterations

Aminoglycoside binding to membrane acidic phospholipids
Changes in content and metabolism of membrane phospholipids
Displacement of Ca^{2+} bound to phospholipids.

2. Transport defects

Inhibition of Na^+-K^+ ATPase and adenylate cyclase
Alterations in PAH transport
Renal wasting of K^+ and Mg^{2+}

Mitochondrial membrane alterations

1. In vivo gentamicin

Decreased state 3 and DNP-uncoupled respiration

2. In vitro gentamicin

Increased state 4, decreased state 3 respiration
Decreased DNP-uncoupled respiration
Alterations of Mg^{2+}-controlled, monovalent cation permeability pathway
Inhibition of Ca^{2+} transport

Lysosomal membrane alterations

1. Lysosomal instability in vivo and in vitro
2. Diminished phospholipase activity

bodies) with no loss of viability. Thus, lysosomal alterations and myeloid bodies alone are not sufficient to produce cytotoxicity. However, lysosomes may act as intracellular storage depots for gentamicin. When intralysosomal concentrations of gentamicin reach critically high values, lysosomes may rupture (Fig. 25.10), leading to very high intracellular gentamicin concentrations that may alter critical cellular functions.

Phospholipidosis in an early cellular alteration induced by gentamicin (Table 25.3), and is characterized by an increase in total phospholipid content with no change in relative amounts of individual phospholipids. Both in vivo and in vitro, gentamicin inhibits a variety of enzymes involved with phospholipid metabolism, including lysosomal phospholipases A and C and extralysosomal phosphatidylinositol-specific phospholipase C. However, cultured rat fibroblasts incubated with gentamicin display phospholipidosis with no loss in cell viability. The exact role of phospholipidosis in gentamicin-induced proximal tubular necrosis, therefore, is not clear.

Gentamicin inhibits several intracellular enzymes, including Na^+-K^+ ATPase, and alters intracellular functions including Na^+-dependent glucose transport and antidiuretic hormone stimulation of adenylate cyclase (Table 25.3). The significance of these gentamicin-induced alterations to cytotoxicity, however, is uncertain.

In vivo, gentamicin reduces mitochondrial state 3 but not state 4 respiration (Table 25.3), and reduces mitochondrial respiratory control ratio and kidney ATP concentrations. In vitro, gentamicin inhibits state 3 and stimulates state 4 respiration independent of ADP but does not completely release respiratory control or decrease ADP:O ratio (Table 25.3). Gentamicin in vitro displaces Mg^{2+} from inner mitochondrial membranes, alters Ca^{2+} fluxes across mitochondrial membranes by competing with Ca^{2+} for membrane binding and transport sites, and causes mitochondrial swelling (Table 25.3). Taken collectively, these data suggest that gentamicin-induced mitochondrial damage may contribute, in part, to the progression of gentamicin-induced proximal tubular injury.

Gentamicin nephrotoxicity is associated with lipid peroxidation and an increase in unsaturated fatty acids in renal cortex, indicative of oxidative stress. However, it is not clear whether lipid peroxidation is causally related to, or is secondary to, nephrotoxicity. Conflicting results have been reported concerning the effects of antioxidants on gentamicin-induced nephrotoxicity. For example, one laboratory has noted that, although gentamicin-induced lipid peroxidation can be prevented by antioxidant therapy (diphenyl-phenylenediamine, vitamin E), the induction of nephrotoxicity (reflected by blood urea nitrogen concentration, serum creatinine, or histopathology) is not altered. In contrast, another laboratory has reported that antioxidants (e.g., dimethylthiourea, deferoxamine, dimethyl sulfoxide, sodium benzoate) confer marked protection against gentamicin-induced nephrotoxicity (increases in blood urea nitrogen and serum creatinine concentrations, proximal tubular necrosis). Thus, the role of lipid peroxidation in gentamicin nephrotoxicity remains controversial.

The antimicrobial mechanism of aminoglycoside antibiotics involves binding to ribosomes, followed by misreading of mRNA and cessation or alteration of protein synthesis. A similar mechanism may mediate nephrotoxicity. However, gentamicin binding to ribosomes has not been investigated following gentamicin therapy, nor has fidelity of mRNA transcription or protein synthesis been determined. Thus, the mechanisms contributing to gentamicin-induced nephrotoxicity are unresolved. Rather than any single event mediating gentamicin nephrotoxicity, it is likely that a number of different mechanisms probably contribute to gentamicin-induced toxicity.

25.5.4 Cephalosporin Nephrotoxicity: Role of Proximal Tubular Secretion

Cephalosporins are broad-spectrum antibiotics similar in structure to penicillin. For several cephalosporins, therapy is limited by the development of nephrotoxicity. Cephaloridine-induced nephrotoxicity has been examined extensively in laboratory animals, and is characterized by an increase in blood urea nitrogen concentration within 24–48 hours, reductions in PAH and TEA transport, and inhibition of glucose production following treatment.

25.5.4a Renal Handling of Cephalosporins

Cephaloridine is zwitterionic and the principal route of elimination is by the kidneys. Cephaloridine clearance approximates inulin clearance, indicating absence of net secretion for cephaloridine. However, in rabbits, cortex/serum (C/S) ratios for cephaloridine are greater than unity, and higher than simultaneously determined C/S ratios for PAH and inulin, indicating that intracellular accumulation of cephaloridine occurs. Further, inhibitors of organic anion transport such as PAH,

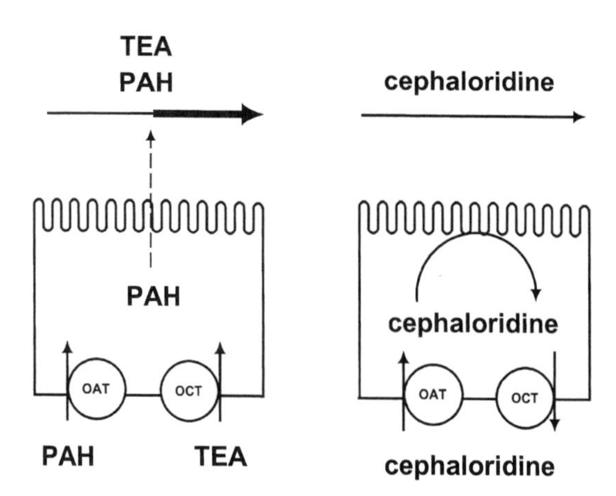

FIGURE 25.11 Schematic representation of proximal tubular transport and urinary excretion of *para*-aminophippurate (PAH), tetraethylammonium (TEA), and cephaloridine in rabbit kidney.
A. PAH and TEA are excreted following both filtration and active secretion by the proximal tubule. PAH is transported across the basolateral membrane by organic anion transporter(s) (OAT) and TEA is secreted by organic cation transporter(s) (OCT). Intracellular concentrations of PAH and TEA may become great enough to drive passive diffusion from intracellular fluid to tubular fluid. Alternately, anion and cation exchangers may facilitate movement across the luminal membrane. *B*. Cephaloridine is excreted primarily following filtration. Active cortical uptake of cephaloridine, inhibited by probenecid and PAH, indicates a secretory component for cephaloridine transport. However, diffusion of cephaloridine from proximal tubular cell to lumen is restricted, leading to high intracellular concentrations of cephaloridine. Some efflux of cephaloridine from proximal tubular cells appears to be mediated by organic cation transporter(s) since inhibitors of this transport system potentiate cephaloridine nephrotoxicity.

penicillin, and probenecid, inhibit both intracellular accumulation and nephrotoxicity of cephaloridine. Toxicity correlates with maturation of organic anion transport: neonatal rabbits have low rates of PAH transport compared to adults, accumulate less cephaloridine than do adults, and are not susceptible to cephaloridine nephrotoxicity. Stimulation of neonatal renal organic anion transport by pretreatment with penicillin increases cortical accumulation of cephaloridine and susceptibility to nephrotoxicity. Additionally, renal cortical cephaloridine concentrations are increased by inhibitors of organic cation transport (mepiperphenidol, TEA, cyanine 863) and cephaloridine nephrotoxicity is exacerbated by these inhibitors. Taken together, these data suggest that, owing to its zwitterionic charge, cephaloridine is actively accumulated into proximal tubular cells via the organic anion transport system (inhibited by probenecid, PAH) and that a portion of cephaloridine efflux occurs via the organic cation transport system (inhibited by mepiperphenidol, cyanine). Once cephaloridine is transported into proximal tubular cells, it diffuses across the luminal membrane into tubular fluid only to a limited extent (Fig. 25.11). Thus, active transport of cephaloridine into proximal tubular cells results in extremely high intracellular cephaloridine concentrations compared to other organs, which, in turn, contributes to selective nephrotoxicity.

However, cortical concentration does not appear to be the sole determinant of toxicity for other cephalosporin antibiotics, since several cephalosporins reach high cortical concentrations without producing nephrotoxicity. For example, cephaloglycin and cephaloridine both produce nephrotoxicity whereas cephalexin is not nephrotoxic. Cortical concentrations of cephaloridine, cephaloglycin and cephalexin are approximately equal initially (0–2 hours) following treatment with these antibiotics. While cortical cephaloridine concentration does not decline, cortical concentrations of cephaloglycin and cephalexin both decline in a similar fashion over 3 hours. Thus, neither cortical accumulation nor retention completely accounts for cephalosporin-induced nephrotoxicity, and suggest that some component of molecular structure also contributes to toxicity.

25.5.4b Biochemical Mechanism of Cephalosporin Cytotoxicity

Although the role of renal tubular transport in cephaloridine nephrotoxicity has been well defined, the exact molecular mechanisms mediating cephaloridine nephrotoxicity are less well understood. Several mechanisms have been postulated to mediate cephaloridine nephrotoxicity, including: (1) production of a highly reactive acylating metabolite(s) by cytochrome P450-dependent mixed function oxidases, (2) production of mitochondrial respiratory toxicity, and (3) production of lipid peroxidation.

Cobaltous chloride and piperonyl butoxide, inhibitors of cytochrome P450-dependent monooxygenase system, reduce cephaloridine nephrotoxicity in mice and rats, suggesting that a cytochrome P450-dependent reactive intermediate mediated cephaloridine nephrotoxicity. However, more recent studies fail to support the hypothesis of cytochrome P450-dependent metabolism for the following reasons: (1) Piperonyl butoxide protects against cephaloridine nephrotoxicity in rabbits, whereas cobaltous chloride does not protect. (2) In rabbits, piperonyl butoxide protection against cephaloridine nephrotoxicity may be related to reduction of cephaloridine uptake in renal cortex, rather than inhibitory effects on cytochrome P450. (3) Inducers of renal cytochrome P450 activity in rats (e.g., β-naphthoflavone, trans-stilbene oxide, polybrominated biphenyls) do not potentiate cephaloridine nephrotoxicity. (4) Phenobarbital, a renal cytochrome P450 inducer in rabbits, potentiates cephaloridine nephrotoxicity in rabbits but also increases renal cortical accumulation of cephaloridine. (5) A glutathione conjugate of cephaloridine has not been identified in kidneys of cephaloridine-treated animals.

Mitochondrial damage may at least partially mediate cephaloridine cytotoxicity. This hypothesis is supported by the following evidence: (1) Ultrastructural damage to renal mitochondria has been observed following cephaloridine administration. (2) Mitochondrial respiration is depressed following either in vivo or in vitro exposure of rabbit kidney tissue to cephaloridine and is characterized by marked inhibition of ADP-dependent respiration using succinate as substrate. (3) Functional impairment of renal mitochondria occurs as early as 1–2 hours following cephaloridine administration. (4) Nephrotoxic cephalosporins (cephaloridine, caphaloglycin) produce similar patterns of respiratory depression whereas non-nephrotoxic cephalosporins (e.g., cephalexin) do not alter mitochondrial function. Although the relatively early depression in mitochondrial respiration is suggestive of a pathogenic mechanism in cephaloridine nephrotoxicity, other biochemical changes following cephaloridine treatment have also been detected 1–2 hours following drug admin-

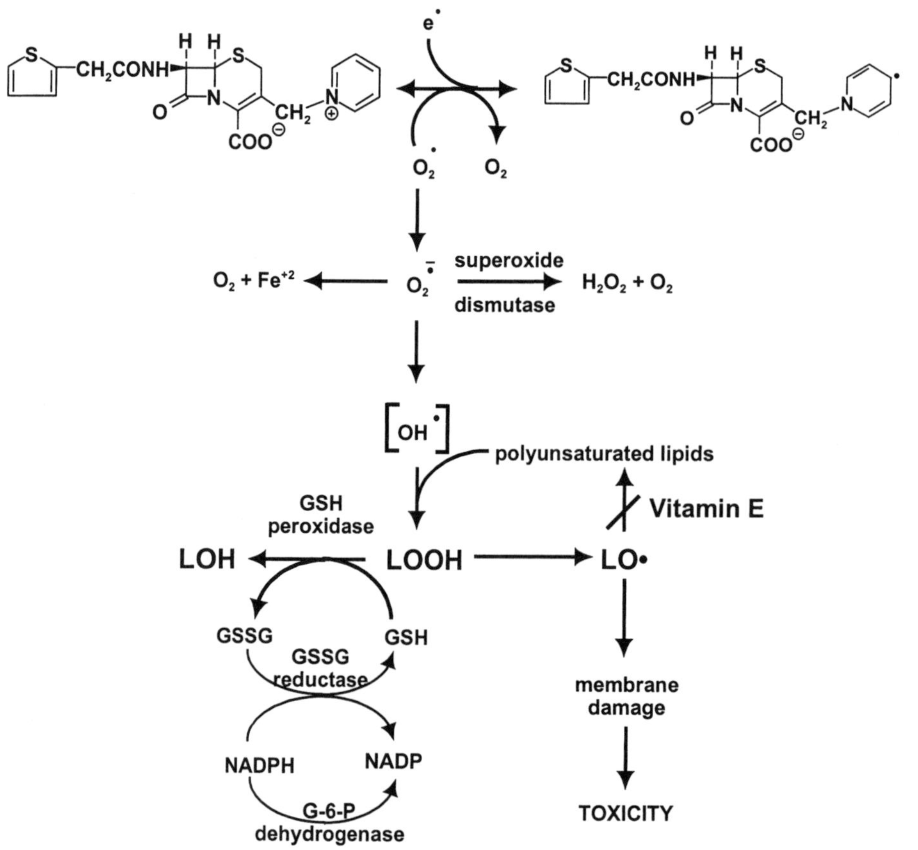

FIGURE 25.12 Proposed mechanism of cephalosporin-induced nephrotoxicity. LOOH = lipid hydroperoxide; LOH = lipid alcohol; LO· = lipid radical; G6P = glucose-6-phosphate.

istration, including lipid peroxidation (see below). Further studies are needed to document a cause-effect relationship between cephaloridine-induced mitochondrial dysfunction and proximal tubular necrosis.

Recently, lipid peroxidation has been proposed as a mediator of cephaloridine cytotoxicity (Fig. 25.12). Several lines of evidence support the involvement of peroxidative injury in cephaloridine nephrotoxicity: (1) Renal cortical concentrations of conjugated dienes, products of lipid peroxidation, are increased 1–2 hours following cephaloridine administration. (2) Rats fed vitamin E and/or selenium deficient diets are more susceptible to cephaloridine nephrotoxicity than rats fed vitamin E and/or selenium adequate diets. In addition, cephaloridine produces a dose-related depletion of renal cortical reduced glutathione, accompanied by an increase in renal cortical concentrations of oxidized glutathione. These observations are suggestive of oxidative stress and are consistent with the postulated role of lipid peroxidation in cephaloridine nephrotoxicity. Lipid peroxidation may be initiated by one electron reduction and redox cycling of the pyridinium ring of cephaloridine, resulting in the reduction of molecular oxygen to superoxide anion (Fig. 25.12).

In support of this mechanism, cephaloridine undergoes anaerobic reduction by isolated renal cortical microsomes with production of superoxide anion radicals and hydrogen peroxide. However, the postulated involvement of the pyridinium ring is not supported by the observation that cephalothin, which is structurally identical to cephaloridine with the exception that it lacks a pyridinium ring, also induces lipid peroxidation (as reflected by malondialdehyde production).

In vitro exposure of renal cortical slices to cephaloridine results in time- and concentration-dependent increases in lipid peroxidation, as reflected by malondialdehyde production. Further, the onset of cephaloridine-induced malondialdehyde production preceded cephaloridine-induced inhibition of organic ion accumulation, suggesting that cephaloridine-induced lipid peroxidation mediates the effects of cephaloridine on organic ion transport. Additionally, antioxidants (e.g., promethazine, N,N′-diphenyl-p-phenylenediamine) block the effects of cephaloridine on both lipid peroxidation and on organic ion transport, suggesting a cause-effect relationship between cephaloridine-induced lipid peroxidation and inhibition of organic ion transport.

Glutathione is known to play a critical role in the detoxification of lipid hydroperoxides by acting as a cosubstrate for glutathione peroxidase, which catalyzes the conversion of lipid hydroperoxides to lipid alcohols (Fig. 25.12). Incubation of renal cortical slices with cephaloridine depletes renal cortical glutathione content in a concentration- and time-dependent manner. Further, depletion of renal cortical glutathione content by cephaloridine precedes the earliest detectable increase in cephaloridine-induced lipid peroxidation, a phenomenon that is consistent with the role of glutathione in the detoxification of lipid hydroperoxides. However, antioxidant treatment does not protect against cephaloridine-induced glutathione depletion. The absence of antioxidant protection against cephaloridine-induced glutathione depletion may suggest that these antioxidants act subsequent to the formation of lipid hydroperoxides or that cephaloridine-induced glutathione depletion is not related to lipid peroxidation.

25.5.5 Chloroform Nephrotoxicity: Role of Metabolic Bioactivation

Chloroform is a nephrotoxicant that most likely undergoes metabolic bioactivation within the kidney. Chloroform ($CHCl_3$), a common organic solvent widely used in the chemical industry, produces hepatic and renal injury in humans and experimental animals. Renal necrosis due to chloroform is sex and species specific: for example, male mice exhibit primarily renal necrosis whereas female mice develop primarily hepatic necrosis following chloroform administration. Nephrotoxicity may be related, in part, to high intraluminal concentrations of chloroform attained following glomerular filtration and fluid reabsorption in the proximal tubular, coupled with relatively high permeability of the proximal tubule compared to other nephron segments.

Tissue injury by chloroform is probably not due to chloroform per se, but is mediated by a chloroform metabolite. The initial step leading to chloroform-induced tissue injury is believed to be the biotransformation of chloroform to a reactive intermediate, phosgene ($COCl_2$), by cytochrome P450-dependent monooxygenases (Fig. 25.13). Formation of phosgene may proceed through oxidative dechlorination involving oxidation of the C–H bond of chloroform, producing the trichloro-

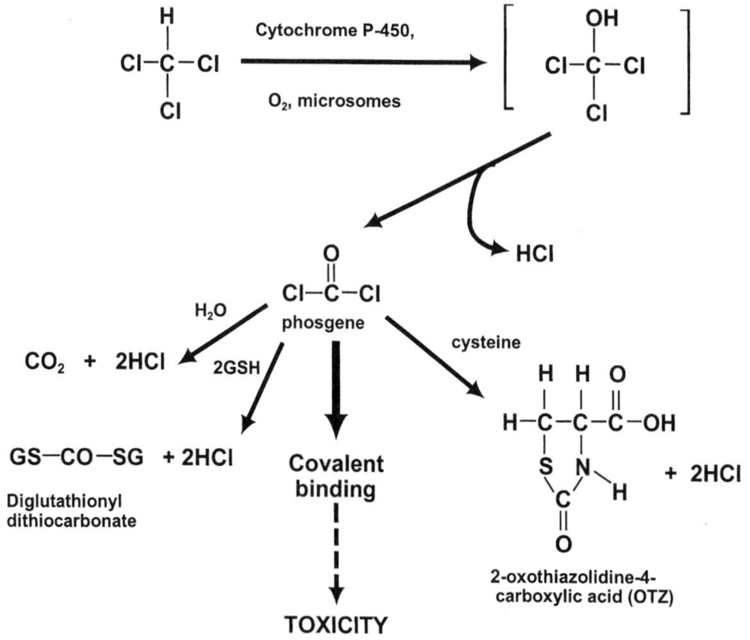

FIGURE 25.13 Proposed mechanism of chloroform biotransformation.
Chloroform undergoes cytochrome P450-catalyzed conversion to trichloromethanol (CCl_3–OH), which spontaneously decomposes to form phosgene. Phosgene is highly reactive and may be detoxified by reacting with sulfhydryl-containing chemicals (cysteine, glutathione [GSH]). Alternately, phosgene can react with sulfhydryl groups on protein, leading to covalent binding and possibly to toxicity.

methanol (CCl_3–OH) intermediate, a highly unstable species that would spontaneously dechlorinate to phosgene (Fig. 25.13). Phosgene is a highly reactive intermediate and may react with intracellular macromolecules to induce cell damage.

Kidneys have relatively low xenobiotic-metabolizing enzyme activities, and chemically induced nephrotoxicity has been assumed to be produced by toxic intermediates generated in the liver and transported to the kidney. If a single hepatic metabolite of chloroform produced both kidney and liver injury, species, strain, and sex differences in susceptibility to chloroform nephro- and hepatotoxicity should be similar. However, species, strain, and sex differences in susceptibility to chloroform nephrotoxicity are not consistent with those of chloroform hepatotoxicity. In addition, several modulators of tissue xenobiotic-metabolizing activities alter chloroform nephrotoxicity and hepatotoxicity differently. Because chloroform-induced kidney injury does not parallel liver damage, it is unlikely that hepatic metabolism of chloroform mediates renal toxicity. Furthermore, the highest concentration of cytochrome P-450 in the kidney is found in proximal tubular epithelial cells, consistent with preferential necrosis of the proximal tubule following chloroform treatment.

The concept that kidney injury is produced by a chloroform metabolite generated in the kidney has been demonstrated directly using in vitro techniques. In order to avoid hepatic metabolism of chloroform, renal cortical slices from naive animals were incubated with chloroform in vitro. Under these conditions, the only site of metabolism of chloroform is the kidney. In vitro exposure to chloroform produced toxicity in kidney slices from male, but not from female, mice. Furthermore, ^{14}C-labeled chloroform was metabolized to $^{14}CO_2$ and covalently bound radioactivity by male, but not female, renal cortical microsomes. In vitro metabolism of chloroform by male, but not female, renal slices is consistent with reduced susceptibility of female mice to in vivo chloroform nephrotoxicity. Metabolism required oxygen, an NADPH regenerating system, was dependent on incubation time, microsomal protein concentration, and substrate concentration and was inhibited by carbon monoxide. The negligible degree of chloroform metabolism and toxicity in female mice is consistent with lower renal cytochrome P450 concentrations and activities in female versus male mice.

Chloroform containing deuterium ($CDCl_3$) is metabolized in the liver to phosgene at approximately half the rate of $CHCl_3$ metabolism to $COCl_2$. $CDCl_3$ is also less hepatotoxic than chloroform. Because the C–D bond is stronger than the C–H bond, these data suggest that cleavage of the C–H bond is the rate-limiting step in the activation of chloroform. $CDCl_3$ is also less toxic to the kidney than chloroform. This deuterium isotope effect on chloroform-induced nephrotoxicity suggests that the kidney metabolizes chloroform in the same manner as the liver, that is, by oxidation to phosgene. Indeed, rabbit renal cortical microsomes incubated in media supplemented with L-cysteine metabolized ^{14}C-labeled chloroform to radioactive phosgene-cysteine 2-oxothiazolidine-4-carboxylic acid (Fig. 25.13). These in vitro data collectively support the hypothesis that mouse and rabbit kidneys biotransform chloroform to a metabolite (phosgene) that mediates nephrotoxicity.

25.5.6 Loop of Henle as a Site of Toxicity

The loop of Henle has not been identified as a primary target for toxicant-induced injury. This lack of susceptibility may be related, in part, to difficulties in identifying and localizing abnormalities in structure and function of this nephron segment. Segments within the loop of Henle (i.e., thin descending and ascending limbs) are composed of thin, poorly characterized epithelial cells and damage may be quite difficult to detect. Additionally, function of the loop structures is integrated closely with functions of other nephron segments, making it difficult to identify a specific effect in the loop of Henle. Functional abnormalities of the loop of Henle, such as those induced therapeutically by thiazide diuretics, manifest primarily as an impairment of urinary concentrating mechanisms and increased excretion of Na^+. Polyuria and hyposthenuria can occur following renal vasodilation, glomerular damage, distal and collecting duct damage, lack of antidiuretic hormone, or several other factors. Thus, identification of structural and functional abnormalities of the loop of Henle may hinder recognition of toxicant-induced damage in this nephron segment.

In contrast to ambiguities in vivo, specific damage to the thick ascending limb of the loop of Henle has been demonstrated in vitro using isolated perfused kidneys from rats. Within 15–30 minutes of in vitro perfusion, a marked concentrating defect, as well as progressive increase in Na^+ excretion, is observed in isolated perfused kidneys. Histologically, damage is localized to the thick ascending limb and is characterized by mitochondrial swelling, cytoplasmic disruption and progressive nuclear pyknosis. Selective damage to the medullary thick ascending limb in the isolated perfused kidney relates to the delicate balance between low O_2 tension and increased O_2 demand in this nephron segment. Medullary damage in the isolated perfused kidney can be attenuated by reducing tubular work and O_2 consumption, via diuretics (e.g., furosemide, ouabain) or by increasing O_2 supply by inclusion of oxygen carrier (e.g., hemoglobin, erythrocytes) in the perfusate. Attenuation of damage by diuretics or ouabain relates to the primary work of the medullary thick ascending limb. In this "diluting segment", oxygen consumption correlates strongly with Na^+ and Cl^- reabsorption. Ouabain and diuretics such as furosemide inhibit reabsorption of electrolytes, thereby reducing oxygen demand by the thick ascending limb. Alternatively, inclusion of oxygen carrier in perfusate increases oxygen delivery to the thick ascending limb, allowing supply and demand to match more closely. Thus, damage to the thick ascending limb in the isolated perfused rat kidney is related to relative hypoxia in this nephron segment coupled with high oxygen demand mandated by high rates of electrolyte reabsorption.

Theoretically, reductions in renal medullary perfusion in intact animals might produce selective damage to the medullary thick ascending limb. However, in vivo, the primary site for anoxic and/or hypoxic damage is the proximal straight tubule, rather than ascending thick limb. Therefore, the relevance of the selective injury in the in vitro isolated perfused kidney to the situation observed in vivo is questionable. Nevertheless, elucidation of the mechanism of thick ascending limb damage in the isolated perfused kidney has been important in defining the relationship between oxygen supply and demand to cell injury in medullary structures.

25.5.7 Distal Tubule as a Site of Injury

Distal tubular injury is difficult to detect compared with proximal tubular nephrotoxicants. Again, lack of highly specialized functions and poorly characterized tubular ultrastructure may contribute to this difficulty. Additionally, compared with the proximal tubule, the distal tubule is a "tight epithelium" with high electrical resistance. Therefore, the distal tubule may be intrinsically less permeable to intraluminal toxicants than the proximal tubule, despite significantly higher concentrations at the distal tubule compared with proximal. Thus, distal tubular cells may not be exposed to high intracellular concentrations of toxicants as are proximal tubular cells. Also, in contrast to the proximal tubule, the distal tubule has negligible mixed function oxidase activity so that conversion of protoxicants to reactive intermediates via this mechanism is unlikely in the distal tubule.

One compound that has been associated with distal tubular injury is amphotericin B, a polyene antifungal agent used in the treatment of systemic mycoses caused by opportunistic fungi. Clinical utility of amphotericin B is limited by its nephrotoxicity, characterized functionally by polyuria resistant to antidiuretic hormone administration, hyposthenuria, hypokalemia, and mild renal tubular acidosis.

Amphotericin B is highly lipophilic (Fig. 25.14) and interacts with membrane lipid sterols, such as cholesterol, to disrupt membrane permeability. Because amphotericin is freely filtered, it achieves high concentrations in distal tubular fluid and easily forms complexes with cholesterol and other lipids present in distal tubular luminal membranes. Amphotericin effectively transforms the "tight" distal tubular epithelium into an epithelium leaky to water, H^+ and K^+. Functional abnormalities

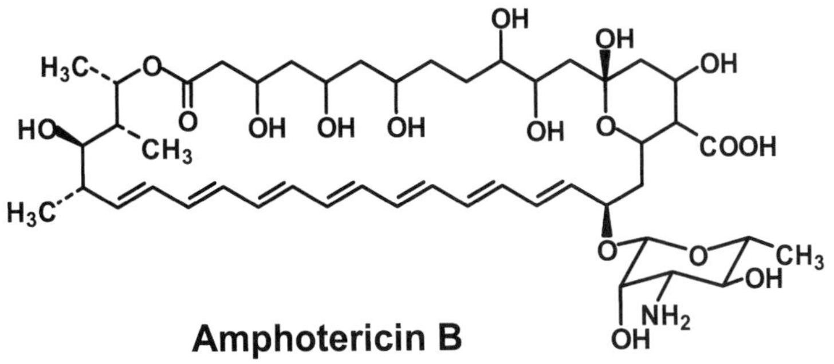

FIGURE 25.14 Structure of amphotericin B.

observed with amphotericin B are attenuated when the antifungal agent is administered as an emulsion formulation whereby amphotericin is incorporated into lipid micelles. Antifungal activity of emulsion-formulated amphotericin B is equivalent to the standard nonemulsion formulation, whereas polyuria and hyposthenuria are significantly reduced by emulsion formulation.

25.5.8 Collecting Tubules as a Site of Injury

Chronic consumption of large dosages of combination analgesics, typically phenacetin and/or caffeine-containing preparations, may be associated with renal papillary necrosis. This injury is described as an ischemic infarction of the inner medulla and papilla of the kidney. Renal function may be compromised modestly by a loss of concentrating ability or, in severe cases, anuria, sepsis, and rapid deterioration of renal function may occur. Morphologically, there is loss of renal papilla (containing terminal collecting ducts), medullary inflammation and interstitial fibrosis, and loss of renomedullary interstitial cells.

A variety of non-narcotic analgesics have been implicated in the etiology of renal papillary necrosis, including acetaminophen, aspirin, acetanilid, and nonsteroidal anti-inflammatory agents such as ibuprofen, phenylbutazone, and indomethacin. Although these agents are dissimilar structurally and chemically, they share a common mechanism of action, acting as analgesics by inhibiting prostaglandin synthesis. Specifically, these analgesics inhibit cyclooxygenase activity but not prostaglandin hydroperoxidase activity of prostaglandin H synthase complex. In the kidney, prostaglandin H synthase activity is distributed asymmetrically, with highest activity in renal medulla and lowest activity in renal cortex. Additionally, the prostaglandin hydroperoxidase component of prostaglandin H synthetase complex is capable of metabolizing xenobiotics, including acetaminophen and phenacetin, to reactive intermediates capable of covalent binding to cellular macromolecules. Thus, renal papilla may be injured selectively by non-narcotic analgesic agents due to the combination of high concentrations of potential toxicants present in tubular fluid and specialized enzymes capable of biotransforming protoxicants to active intermediates.

In laboratory animals, papillary necrosis due to non-narcotic analgesic has been extremely difficult to produce. However, papillary necrosis has been demonstrated following administration of 2-bromoethylamine; 2-bromoethylamine has been used to demonstrate the role of urinary concentrating mechanisms in the etiology of 2-bromoethylamine-induced papillary necrosis. Maneuvers that produce large volumes of dilute urine, such as lack of antidiuretic hormone (Brattleboro rats), diuretic therapy, or volume expansion with 5% glucose, prevent papillary necrosis due to 2-bromoethylamine. In contrast, maneuvers that restore urinary concentrating ability, such as exogenous antidiuretic hormone replacement in Brattleboro rats, render these animals susceptible to 2-bromoethylamine-induced papillary necrosis. Thus, papillary necrosis due to 2-bromoethylamine is critically dependent on the ability to produce concentrated urine.

Biochemical mechanisms responsible for papillary necrosis are not entirely clear. Several mechanisms have been proposed, including: (1) direct cellular injury, (2) reduction or redistribution of blood flow, (3) prostaglandin inhibition, and (4) free radical formation. A decrease in papillary blood flow would lead to ischemia of deeper portions of the kidney, ultimately leading to cellular necrosis. Renal prostaglandins and thromboxanes are proposed to play a role in control of intrarenal blood flow. Salicylates and nonsteroidal anti-inflammatory agents inhibit renal cyclooxygenase activity, preventing prostaglandin synthesis. As vasodilators, prostaglandins in medulla and papilla may have an important role in maintaining normal blood flow. However, it is not entirely clear that papillary and medullary blood flow are altered early in the development of papillary necrosis. Additionally, acetaminophen has been implicated in the etiology of papillary necrosis yet it is relatively ineffective as an inhibitor of prostaglandin synthesis, calling into question the relationship between inhibition of prostaglandin synthesis and papillary necrosis. Free radicals (generated from prostaglandin H synthetase-catalyzed cooxidation of xenobiotics) and/or covalent binding of activated metabolites to cellular macromolecules also may contribute to papillary necrosis. However, the relevance of these mechanisms to papillary necrosis is, as yet, unclear.

25.6 SUMMARY

Susceptibility of the kidney to chemically induced toxicity is related, at least in part, to several unique aspects of renal anatomy and physiology. By virtue of high renal blood flow, active transport processes for secretion and reabsorption, and progressive concentration of the glomerular filtrate following water removal during the formation of urine, renal tubular cells may be exposed to higher concentrations of potential toxicants than are cells in other organs. Additionally, intrarenal metabolism, via cytochrome P450 or prostaglandin H synthetase, may contribute to the generation of toxic metabolites within the kidney.

The precise biochemical mechanisms leading to irreversible cell injury and nephrotoxicity are not well defined. Many diverse biochemical activities occur within the kidney, and interference with one or more of these functions may lead to irreversible cell injury. Rather than any one single mechanism mediating chemically induced nephrotoxicity, it is likely that a toxicant alters a number of critical intracellular functions, ultimately leading to cytotoxicity and cellular necrosis.

SUGGESTED READING

Goldstein, R.S. and Schnellmann, R.G. Toxic responses of the kidney. In: Klaassen, C.D., Amdur, M.O. and Doull, J. (Eds), *Casarett and Doull's Toxicology: The Basic Science of Poisons.* McGraw-Hill, New York, 1996, pp 417–442.

Humes, H.D. and Weinberg, J. Toxic nephropathies. In: Brenner, B.M. and Rector, F.C., Jr. (Eds), *The Kidney*, 3rd ed. W.B. Saunders Company, Philadelphia, PA, 1986, p 1491.

Goldstein, R.S. (Ed). *Comprehensive Toxicology, Vol. 7: Renal Toxicology.* Pergamon Press, Elsevier Science Limited, Oxford, 1997.

CHAPTER TWENTY-SIX

Cardiovascular Toxicity

ALAN B. COMBS, KENNETH RAMOS, and DANIEL ACOSTA

26.1 INTRODUCTION

The cardiovascular system is comprised of the heart and a diverse vascular network, consisting of arteries, capillaries, and veins. The principal function of the cardiovascular system is to supply tissues and cells of the body with appropriate nutrients, respiratory gases, hormones, and metabolites, and to remove the waste products of tissue and cellular metabolism and foreign matter such as invading microorganisms. Because circulation of the blood through the cardiovascular system reaches every tissue of the body, it is responsible for maintaining the optimal internal homeostasis of the body, as well as critical regulation of body temperature and pH in the body fluids.

The introduction of drugs or xenobiotics into the body will by necessity cause them to come into contact with the cardiovascular system at times before there is contact with other organs. It is not surprising that some of these agents may have deleterious effects on the cardiovascular system, as well as on other highly vascularized organs. With the exception of major toxicology tests such as *Casarett and Doull's Toxicology*, most introductory toxicology books have little discussion of the potential toxic effects of xenobiotics on the cardiovascular system. The purpose of this chapter is to highlight chemical agents that have can significant toxic effects on the myocardium and vascular system, emphasizing subcellular sites of action and biochemical mechanisms of toxicity.

26.2 GENERAL PRINCIPLES IN CARDIOVASCULAR TOXICOLOGY

A wide spectrum of chemicals including therapeutic drugs, natural products, and anthropogenic chemicals cause structural and/or functional alterations in cardiovascular function. Toxicity can be expressed at the electrical, mechanical, metabolic, or genetic level. The concentration of the chemical in contact with target structures and the duration of exposure usually determines the severity of the toxic response. Other factors also may determine the severity of the toxic insult. These include: (1) the nature of the interaction of the chemicals with their molecular target,

(2) the extent to which the target is involved in cellular homeostasis, (3) genetic and environmental influences, and (4) the balance between cellular injury and repair mechanisms. The excitability of cardiovascular tissue represents a sensitive functional target of toxicity. In the case of myocardial cells, chemicals may cause arrhythmias as a result of alterations in impulse formation or conduction. Although a number of reactive chemicals can alter contractile function directly, mechanical alterations are often secondary to electrical disturbances in both cardiac and vascular cells. Most of the toxic responses of cardiovascular cells result from chemically induced metabolic alterations. Such disturbances range from alterations in enzymatic activity to discrete modulation of second messenger systems. Although genetic alterations in the cardiovascular system are just now starting to be studied in detail, several chemicals are known to interact with the genome of vascular smooth muscle cells to cause lesions of atherosclerotic etiology and/or tumors of epithelial or mesenchymal origin. If cellular repair mechanisms are not themselves the target of toxic insult, temporary derangements in cardiovascular function may be corrected.

The assessment of cardiovascular toxicity has traditionally focused on cellular alterations associated with acute chemical exposure. This is partly due to the high risk of lethality associated with acute cardiovascular dysfunction. More recently, chronic exposure to several toxic chemicals has been associated with the development of human cardiovascular diseases. The mechanisms that govern organ-specific cardiovascular toxicity of chemicals include: (1) selective alterations in cellular excitability, (2) unique biochemical pathways for the bioactivation of protoxicants, (3) erratic chemical detoxication, (4) preferential accumulation of the active toxicant, and (5) extension of organ-specific pharmacologic effects. These mechanisms are illustrated in subsequent sections where selected cardiovascular toxicants are discussed.

26.2.1 Physiologic Determinants

The coordinated sequence of electrical and mechanical events in muscle cells allows the circulation of oxygen and nutrients throughout the body and the removal of the waste products of cellular metabolism. Within the myocardium, a finite number of muscle cells is specialized for the conduction of electrical impulses. The degree of functional specificity is not absolute but rather susceptible to toxicant interference. In general, the level of vascular smooth muscle cell differentiation is less rigid than that of cardiac muscle cells. This is reflected by a reduced degree of structural organization and an enhanced synthetic capacity. An enhancement in the synthetic capacity of smooth muscle cells may result from toxicant-induced modulation of phenotype expression. Such alterations reduce the capacity of the cells for force development and thus, the maintenance of vascular tone. Fibroblastic and endothelial cells serve to provide structural support to the heart and blood vessels. For example they are thought to contribute to the regulation of parenchymal cell function. The ability of xenobiotics to disrupt the integrity of one or more of these cell systems would result in cardiovascular toxicity.

The conduction of electrical impulses in the myocardium is an orderly series of events coupled to contractile function. This relationship is often referred to as excitation-contraction coupling. Under normal conditions, electrical impulses generated spontaneously by cells in the sinoatrial node, the primary pacemaker of the heart, are conducted from the atria to the ventricles via the atrioventricular node and a specialized network of fibers known as the His-Purkinje system. The ability of cardiac muscle cells to generate electrical impulses spontaneously is referred to as automaticity. Intrinsic automaticity is due to enhanced permeability of the plasma membrane to sodium ions. The influx of sodium ions causes a partial depolarization of the cell that reaches the threshold potential to fire an action potential. Several cardiovascular toxicants enhance the automaticity of non-pacemaker cells and result in the formation of ectopic foci that disrupt this sequence of electrical events. The wave of depolarization initiated by nodal cells is propagated along the plasma membrane and enters the muscle cell through a tubular (T) system. Depolarization of cardiac cells is associated with sodium and calcium influx from the extracellular space through the sarcolemma via specialized ionic channels. In addition, calcium is released from intracellular stores to ultimately raise the concentration of free intracellular calcium from about 0.1 to 10 μM. The role of calcium in the contractile process is described in Section 26.2.3d. The efflux of potassium ions from cardiac cells into the extracellular space mediates the repolarization of the cells. The most common site for disturbances in conduction is the atrioventricular node followed by the His-Purkinje system. The ability of cardiovascular toxicants to interfere with the synchronicity of the excitation-contraction process often results in marked myocardial dysfunction.

The high degree of electrical activity characteristic of cardiac muscle cells is not observed in vascular smooth muscle cells. Although smooth muscle cells can potentially depolarize spontaneously, automaticity under normal physiologic conditions is rarely found in most animal species. Vascular cells rely on pharmacomechanical coupling for the initiation of muscle contraction. Surface receptors located on the plasma membrane must be activated by endogenous transmitters or by modulators under neurochemical control to signal the initiation of contractile events. In marked contrast to cardiac cells, initiation of muscle contraction in vascular cells is not associated with membrane depolarization, but rather with receptor-mediated activation of ionic currents. Although cardiac myocytes share with vascular smooth muscle cells the ability to respond to neurohumoral regulation, under normal conditions cardiac cells exhibit intrinsic automaticity.

Multiple mechanisms are involved in the regulation of calcium within cardiovascular cells. These include (a) membrane channels, (b) sarcolemmal pumps, (c) energy-dependent sequestration of calcium by intracellular structures, and (d) a number of calcium-binding proteins either free or associated with other proteins. The extent to which each of these mechanisms is involved in cardiovascular homeostasis can be altered by the presence of a toxic chemical. At the sarcolemmal level, calcium efflux is mediated by several energy-dependent mechanisms. Also, bidirectional exchange of calcium with sodium, potassium, hydronium, and possibly magnesium are thought to occur at the sarcolemmal level. The exact mechanisms by which these processes operate are not yet well understood. The sequestration of calcium by intracellular organelles such as mitochondria and sarcoplasmic reticulum also provides an important mechanism for calcium regulation. Although the biochemical events that mediate muscle relaxation are similar in both cardiac and vascular cells, the relative contributions of specific mechanisms to the regulation of cellular function are thought to be different.

26.2.2 Pathology versus Toxicology

In its broadest sense, pathology is the study of abnormal biology, whereas toxicology is the study of the adverse effects of xenobiotics on biological systems. Thus, both disciplines are concerned with the responses of biological systems to injurious stimuli. Whereas pathologists are concerned mainly with the study of lesions associated with specific disease processes, toxicologists center their attention on the processes by which biological systems respond to toxic agents. A common theme to both disciplines is a focus on the study of pathogenesis of a specific disease or the mechanism of toxicity of a particular chemical agent.

Often, advances in one discipline will have some bearing upon another. For example, the responses of the myocardium to ischemic injury initiated by atherosclerotic occlusion of coronary arteries have been investigated extensively by experimental pathologists. Cellular injury of the myocardium by ischemia has been shown to occur in very distinct phases, with prominent biochemical and ultrastructural changes of key subcellular organelles. When toxicologists explored the cardiotoxicity of catecholamines such as isoproterenol, it was observed that many of the toxicological responses of the heart to these agents were similar to changes produced by ischemic injury. Consequently, a chemical model of myocardial ischemic injury has been developed and has been instrumental in gaining a better understanding of the cell injury process. Although pathology and toxicology are separate and distinct disciplines, it is important to consider the contributions and perspective that each discipline brings to the other.

26.2.3 Subcellular Targets of Toxicity

Because the ultimate effects of toxicants on the cardiovascular system are reflections of cellular and subcellular alterations, the purpose of this section is to highlight those cellular organelles that may serve as targets for the toxic manifestations of chemical agents. A review of organelle responses to cardiovascular toxicants should provide valuable insight into the biochemical mechanisms that underlie the toxicity of the chemical agents. Some of the more important organelles to be reviewed include the sarcolemma, mitochondrion, sarcoplasmic reticulum, contractile apparatus, and lysosome.

26.2.3a Sarcolemma

Classically, the sarcolemma was considered to consist only of the trilaminar plasma membrane of cardiac and smooth muscle cells. However, in recent years, this concept has been expanded to include the external glycocalyx as part of the functional unit. The plasma membrane consists of a lipid bilayer with the hydrophobic ends of the lipids directed internally. The glycocalyx is a surface layer consisting of highly charged glycoproteins and glycolipids. Many different proteins penetrate into and through the sarcolemma. Some of these proteins are structural in nature and others are involved in active membrane processes. Points of toxicological vulnerability of the sarcolemma are the lipid membrane, various transmembrane enzymes and structures, including transport mechanisms, and receptors located on the surface. Some of these actions are general to many types of cells and not all represent specific actions of toxic agents.

Peroxidation of the sarcolemma and other lipid membranes would be expected to occur from exposure to any substance that causes oxidative stress. The widely used antineoplastic agent doxorubicin may cause a direct effect on the plasma membrane. This is indicated by work showing that doxorubicin can be toxic to cells in culture, even when tightly bound to a large agarose polymer that physically prevents the drug from entering into the cell. The ability of doxorubicin to cause membrane peroxidation is discussed in Section 26.3.2b.

Cardiovascular toxicants may act upon transmembrane ion channels. Verapamil and diltiazem produce negative inotropic effects because of their inhibition of calcium ion transport through the slow channels. This blockade of calcium channels becomes clinically dangerous in hearts exhibiting some previous degree of failure. On the other hand, enhancement of channel-mediated calcium influx may occur with A23187, Bay K 8644, and toxins such as the snake poison atrotoxin. Enhanced calcium influx may lead to intracellular calcium overload. The problems related to excess intracellular calcium are described later in this chapter.

Blockade of the sodium channels serves to explain many of the therapeutic and toxic actions of Class IA antiarrhythmic agents such as quinidine and procainamide. These drugs reduce automaticity and decrease conduction velocity. In overdose, they can disrupt cardiac conduction and lead to fatal ventricular fibrillation. Another dangerous arrhythmia caused by Class IA antiarrhythmic drugs is the potentially fatal torsades des pointes, which can be a precursor to ventricular fibrillation. The clinical use of the Class I antiarrhythmic drugs is undergoing extensive reevaluation because recent large studies have indicated that people on placebos have a greater life expectancy because of the proarrhythmogenic potential of these drugs. Sodium channel blocking actions and the consequent potential for cardiotoxicity are shared by many local anesthetics, by certain of the tricyclic antidepressant agents, and by natural products such as tetrodotoxin and saxitoxin. Additional details are presented in Section 26.2.4a.

The cardiac potassium channels are susceptible to toxicological action by several substances. The Class III antiarrhythmic drugs such as bretylium and amiodarone reduce channel-mediated potassium efflux, an effect that prolongs repolarization. In addition, the cardiotoxicity of barium and cesium ions may be related in part to potassium channel blockade.

The cardiotoxic glycosides also act upon structures within the sarcolemma. The natural sources of such glycosides are very diverse and include the digitalis, strophanthus, and oleander plants, and the skins of certain toads and salamanders. Their therapeutic and cardiotoxic actions result from a direct inhibition of the sarcolemmal sodium/potassium ATPase (sodium pump) and from simultaneously increased sympathotonia and vagotonia. Arrhythmias constitute the most dangerous consequences of digitalis intoxication. The arrhythmogenic consequences of digitalis include a decrease in the maximum diastolic potential and resting membrane potential, a decrease in the duration of the ventricular refractory period, an increase in phase 4 oscillatory afterpotentials, and a generally enhanced automaticity of the myocardium with the exception of the sinoatrial node. Digitalis-induced vagotonia decreases the automaticity of the sinoatrial node and the increase in sympathotonia helps to increase the automaticity of the rest of the heart. These effects are dose dependent, and the inevitable consequence of their progression is the formation of ectopic pacemakers and the generation of arrhythmias.

Another membrane transport mechanism that is vulnerable to toxic perturbation is the sodium/calcium exchange carrier of the sarcolemma. This exchanger is a facilitated diffusion process driven by the transmembrane sodium gradient. It helps to terminate contraction by transporting calcium out of the cell. Amiloride and certain of its derivatives are inhibitors of the exchanger. Doxorubicin also has been reported to inhibit this exchanger. The result of intoxication with substances inhibiting the sodium/calcium exchanger would be an eventual intracellular calcium overload.

Reaction of substances with receptors on the membrane surface may cause toxicity. The most prominent example of such toxicity results from the reaction of catecholamines with the beta-adrenergic receptors of the heart. Among other effects, the physiological responses of activation of the beta-receptors include modulation of the function of the voltage-dependent calcium channels. The toxicological effects of catecholamines upon the heart are described later in this chapter.

In the older literature, many substances such as cocaine, tricyclic antidepressants, and certain inhaled gases (cyclopropane and carbon tetrachloride, ethyl chloride, and other halogenated hydrocarbons) have been reported to be cardiotoxic because they enhance the sensitivity of the heart to catecholamines. The common mechanism behind the cardiotoxicity of all these agents appears to be blockade of reuptake of norepinephrine into adrenergic neurons. The result is a prolonged duration of norepinephrine at the postjunctional site and an increase in its activity. All of these compounds have the potential to trigger fatal arrhythmias, particularly when exogenous adrenergic drugs are present.

As described below, inhibition of glycolysis may disturb the formation of the ATP necessary for activity of sarcolemmal pumps. Problems related to excitation and propagation of the action potential can result.

26.2.3b Mitochondria

The primary function of mitochondria is to generate ATP through oxidative phosphorylation. One of the most prominent features of cardiac muscle is the large number of mitochondria present. This observation is consistent with the large energy requirement of cardiac myocytes relative to other cells. Mitochondria often align along the contractile fibers of muscle cells to facilitate the delivery of energy for contraction. Because the myocardium does not store energy as efficiently as other tissues, the maintenance of an adequate balance between energy supply and demand is essential for the preservation of organ function. In contrast, the energy requirement of vascular tissue is lower than that of cardiac cells. As a result, vascular cells preserve functional integrity for longer periods under lower oxygen tensions.

The capacity of mitochondria to exchange calcium provides a powerful intracellular calcium buffer system. A carrier-mediated process that is driven by the negative potential inside mitochondria regulates mitochondrial calcium uptake. This potential is generated by respiration-dependent or ATP-dependent ejection of protons across the inner mitochondrial membrane. Calcium uptake by mitochondria is inhibited by uncouplers of oxidative phosphorylation and enhanced in the presence of inorganic phosphate. Mitochondria utilize the same pool of energy for ATP synthesis and calcium uptake. If presented with both alternatives, isolated heart mitochondria will take up calcium before using energy to phosphorylate ADP. If intracellular calcium levels rise beyond physiologic levels, mitochondria act as an intracellular calcium sink. Although calcium efflux may occur by reversal of the uptake mechanism, the presence of a mitochondrial efflux pathway associated with sodium-induced calcium release is favored. Because mitochondrial calcium uptake occurs at the expense of ATP synthesis, calcium accumulation by mitochondria inhibits cellular energy production.

Irreversible myocardial injury is associated with swelling of the inner mitochondrial compartment, loss of matrix density, and the formation of electron-dense granules. A certain degree of dissociation among these alterations is apparent. Whereas exposure of isolated hearts to toxic concentrations of isoproterenol causes the appearance of electron-dense intramitochondrial granules, doxorubicin causes extensive mitochondrial swelling in the absence of such granules. In any case, the ability of these toxins to alter the structural integrity of mitochondria is consistent with their ability to disrupt intramitochondrial calcium regulation. Whether such alterations represent primary or secondary alterations during the cell injury process is not yet clear.

26.2.3c Enzymes of Glycolysis

During the last decade, there have been significant advances in the understanding of cellular energetics, one being the finding of the compartmentalization of ATP synthesis within the cell. The large amounts of ATP needed for contraction originate from oxidative mitochondrial metabolism. On the other hand, the ATP that is used in sarcolemmal membrane pump function comes from glycolysis, and not from the mitochondria. This division of labor in the synthesis of ATP has particular significance for cardiac toxicity. Substances that inhibit glycolysis may be very toxic to membrane pump activity. Consequences of this would include an inability of the sodium pump to maintain the resting membrane potential, a decreased rate of action potential propagation, and decreased ability to exclude calcium. The glycolysis intermediate fructose-1,6-bisphosphate had been shown in vitro to reduce the cardiac toxicity of doxorubicin and of emetine. In addition, emetine has been shown to be a potent inhibitor of cardiac phosphofructokinase, the rate-limiting enzyme in glycolytic throughput. Emetine in vitro decreases the action potential amplitude and slows ventricular conduction velocity, effects that might be expected from locally decreased availability of ATP for membrane pumps.

26.2.3d Sarcoplasmic Reticulum

The sarcoplasmic reticulum (SR) consists of an interconnected network of tubules surrounding the contractile elements of cardiac and smooth muscle cells. Many of the particulars of SR function have been learned from skeletal muscle, rather than from the heart. One of the primary functions of the SR appears to be control of the intracellular free calcium needed for contraction. The ability to trigger release of calcium from the SR upon influx of extracellular calcium and the ability to take calcium back up to allow relaxation are important elements in the contractile process (see below).

Toxicological processes affecting the SR may act to disrupt the handling of calcium. These toxicological processes are mostly nonspecific in nature and can accompany changes in other subcellular organelles. For example, compounds that cause oxidative stress may cause peroxidative injury to SR membranes and their calcium transport enzymes. Doxorubicin is an example of a cardiotoxin that may undergo futile redox cycling and cause oxidative stress. This process involves a one electron reduction of doxorubicin to the semiquinone, followed by its oxidation back to the quinone. The latter step is coupled to a one-electron reduction of molecular oxygen to the superoxide anion radical. The resulting SR membrane damage and disruption of calcium function is thought to be part of the etiology of cardiac damage. In vitro studies have shown that the redox cycling of doxorubicin may occur from action by mitochondrial cytochromes, or by cytochrome P450-mediated reduction within the SR itself.

26.2.3e Contractile Apparatus

The contractile machinery of cardiac muscle consists of four major proteins whose physical interaction results in shortening of the sarcomere. Actin and myosin are referred to as contractile proteins because they form the cross bridges that mediate muscle contraction. In cardiac muscle, two additional proteins known as troponin and tropomyosin modulate the interaction of the contractile proteins and thus are referred to as regulatory proteins. In vascular smooth muscle, the interaction of actin and myosin is regulated by calmodulin. Structural and functional differences in the regulation of contractile function in cardiac and vascular smooth muscle may render these tissues susceptible to selective toxic insult. For instance, the phosphorylation of myosin and the allosteric interaction of calcium with an unknown myofilament site are features unique to vascular smooth muscle cells. The contraction/relaxation cycles of both cardiac and vascular muscle cells are regulated by fluctuations in free intracellular calcium. The energy required to support contraction is generated by the hydrolysis of ATP by myosin ATPase. Thus, any derangement in the structural organization of the contractile apparatus, as well as in the rate of extent of ATP hydrolysis by myofibrils or calcium binding proteins, may result in severe cardiovascular dysfunction. The hypercontraction and agglutination of myofibrils induced by several cardiotoxic agents exemplify this.

In recent years a great deal of information has become available regarding the specific patterns of cytoplasmic matrix organization in a variety of cell types. These cell-specific patterns are due to a group of fibrous elements called the cytoskeleton. Toxic chemicals may alter the quantitative and qualitative profile of cytoskeletal elements. Because cytoskeletal components change as a function of the degree of cellular differentiation, prominent alterations may result from toxic insult. Apparent redistribution of cellular filamentous actin occurs during cell blebbing in injured endothelial cells. The fragmentation of cytoskeletal proteins upon oxidative injury to vascular endothelial cells has also been documented.

26.2.3f Lysosomes

The role of lysosomes in the cell injury process as initiated by specific disease states or by chemical agents has evolved through the years, ranging from one of primary importance to one of lesser influence in the mechanism of cardiovascular cell injury. Ever since de Duve and his colleagues first described lysosomes as intracellular organelles containing a large number of hydrolytic and degradative enzymes that were encased by a specialized semipermeable membrane system, investigators have implicated a major role for these lysosomal enzymes as initiating factors in the cell injury process. For instance, a "suicide bag hypothesis" was proposed for lysosomes in which a toxic agent or a specific disease state first damaged the lysosomal membranes as the initial step in the cell injury process. The injured lysosomes would then release their highly destructive enzymes (proteases, phospholipases, nucleases, etc.) into the cytosol, thus wreaking havoc on important intracellular structures and organelles whose membranes are composed of phospholipids, proteins, and other constituents subject to lysosomal enzyme hydrolysis. It was later established that many of these lysosomal changes were the result of the cell injury process, rather than the cause, because lysosomal effects occurred much later during cellular damage. Thus, lysosomes are thought to act as scavengers in the degradation of damaged cellular components. This interpretation does not eliminate the possibility that certain toxic agents may produce early effects on lysosomes that are the initial cause of the observed cell injury cascade of events. Such an example is silica that is taken up by alveolar macrophages as the initial event in pulmonary silicosis. The silica particles then lyse the lysosomes, resulting in release of the enzymes and subsequent cellular injury.

At present, there are few examples of cardiovascular toxic agents that have primary effects on lysosomes. However, doxorubicin has recently been shown to have some major effects on myocardial lysosomes. As will be discussed in Section 26.3.2b, doxorubicin's effectiveness as an anticancer agent is limited because of its severe cardiotoxicity that eventually results in clinical cardiomyopathy and heart failure. Singal and co-workers have suggested that doxorubicin induces lysosomal changes that occur concurrently with the formation of lipid peroxides in the heart. Thus, lysosomal alterations may be an early event in the myocardial cell injury process induced by doxorubicin treatment.

26.2.4 Biochemical Mechanisms of Toxicity

In many cases, disturbances of intracellular biochemical function have been shown to be responsible for toxic injury within the myocardial cell. The examples discussed in this section include changes leading to ionic perturbations, disturbances in energy production or utilization, alterations in membrane function, and disruption of intracellular defense mechanisms.

26.2.4a Ionic Perturbations

Calcium. The functional capacity of the myocardium and vascular smooth muscle is highly dependent on the ionic composition of intracellular and extracellular fluids, especially calcium and sodium. Calcium plays a critical role in the contractile process of both myocardial and vascular smooth muscle. It is also an important regulator or modulator of enzyme activities, transduction of hormonal information, release of secretory products, and other membrane-associated activities. Because calcium has many important functions in maintaining and regulating numerous cellular functions and processes, it is not surprising that an alteration in extracellular and cytosolic calcium concentrations may prove to be deleterious to the overall integrity of myocardial and vascular tissue.

A concept of chemically induced cell injury has been proposed by Farber and his colleagues that implicates a major role for the accumulation of intracellular calcium as a common final step in irreversible cell injury or cell death. The first step in toxic chemical cell injury is disruption of the plasma membrane that results in loss of membrane integrity and subsequent cytosolic accumulation of calcium through the "leaky" membranes. The increased free cytosolic calcium ions may then activate several types of degradative enzymes that may cause further breakdown of cellular membranes, including phospholipases and proteases. Furthermore, calcium may accumulate in mitochondria, causing uncoupling of oxidative phosphorylation and depletion of energy stores. Although calcium overload is a common feature of irreversibly damaged tissue, a caution should be added to the foregoing discussion in that it has not been definitively proven that the rise in cytosolic free calcium is the primary cause of cell death. It is possible that the increased cytosolic calcium is the result of some other lethal alteration in the cell.

The cytotoxic effects of calcium in the myocardium can be illustrated by the following examples. Zimmerman and co-workers first described the phenomenon known as the calcium paradox in which reperfusion of isolated rat hearts with calcium-containing buffers, after a short period of calcium-free perfusion, results in irreversible loss of electrical and mechanical activity, extensive ultrastructural damage, and rapid and massive loss of intracellular constituents. The consensus for explaining this phenomenon is that injury to intercalated disks, which have been weakened by prior perfusion of calcium-free medium, occur upon contracture of the myocardium after readmission of calcium.

Several prominent cardiotoxic agents have been identified as having an association with increased cytosolic calcium concentration and myocardial cell injury. These compounds include the catecholamines, doxorubicin, and certain halogenated anesthetic agents. Isoproterenol, a synthetic catecholamine, induces massive myocardial necrosis in high doses, and calcium overload is a common feature associated with this injury. Some investigators believe that oxidation of catecholamines can lead to the formation of highly cytotoxic free radicals, which are thought to be responsible for extensive membrane damage via lipid peroxidation and for subsequent calcium influx into the cells. However, other investigators suggest that the role of oxidation by-products of catecholamines in producing myocardial injury is not convincing because the levels of oxidative metabolites needed to produce injury are very high and physiologically difficult to reach in vivo. Other mechanisms that have been proposed for the development of myocardial injury by the various catecholamines include the role of ischemia and hypoxia, effects on coronary microcirculation, sarcolemmal permeability changes, and deficiencies in high-energy phosphates. More information on the catecholamines can be found in Section 26.3.2a.

A role for calcium has also been implicated in the cardiotoxicity of doxorubicin. Singal and co-workers have suggested that free radical metabolites of doxorubicin can react with oxygen to form superoxide anions and hydroxyl radicals. These in turn damage sarcolemmal membranes and cause an increased calcium influx and subsequent injurious effects on mitochondria and other cellular processes. Further discussion of the cardiotoxicity of doxorubicin may be found in Section 26.3.2b.

Whereas most of the cardiotoxic agents seem to produce their toxic effects on the myocardium by promoting calcium overload, there are compounds that have other effects on calcium metabolism and transport. The halogenated anesthetic agents, including halothane, are known to depress myocardial contractility. It is postulated that this depression of myocardial function by halothane and other similar agents rests on their ability to disrupt calcium exchange across cardiac sarcolemmal membranes, possibly by altering the lipid bilayer structure of the membranes.

Sodium. Depolarization of cardiac and vascular smooth muscle cells is due to the massive influx of sodium ions. The rapid rise in intracellular sodium is followed by changes in the permeability of the sarcolemma to calcium and potassium ions. The sodium current as well as other ionic currents are mediated by proteins that span the lipid bilayer and serve as transmembrane pores or channels. These ionic channels are operated by activation and inactivation gates that open and close to regulate the movement of ions down their electrochemical gradient. At rest, the plasma membrane is fully polarized and the activation gates of the sodium channel are closed to prevent the movement of ions through the channel. Under these conditions the channel is considered to be closed, but available for activation. Upon membrane depolarization to threshold potential levels, the activation gate open and allow the entry of sodium ions from the extracellular space into the cell. This transition of the sodium channel from a closed to an active state is susceptible to chemical interference. The opening of the channel is followed by rapid closure of the inactivation gates to preclude further sodium influx and thus shut off the sodium current. At this stage, the activation gates are open, but the channel is unavailable for further sodium entry.

The toxic effects of several cardiovascular toxins are due to selective interference with intracellular sodium homeostasis. The basis for regional and tissue selectivity, as well as ionic selectivity, may be due to structural differences in the sodium channel. In general, the state of excitability of the membrane and, thus, the kinetic state of the channel are sources of variability in toxicologic potential. Sodium influx stimulants such as the veratrum alkaloids and marine toxins cause depolarization of muscle cells and increase the contractility of cardiac and vascular muscle cells. The enhanced sodium influx induced by these agents may be due to direct interaction with channel proteins. The affinity of these agents for the sodium channel is modulated by the kinetic state of the channel. Veratrum alkaloids have greater affinity for open, conducting channels than for closed or inactivated channels. Positive cooperativity between veratrum alkaloids and the marine toxins has been described, suggesting that their channel binding sites are different. Alkaloids such as batrachotoxin, grayanotoxin, and aconitine increase resting sodium permeability to prolong the duration of the sodium current. Batrachotoxin is thought to enhance resting sodium permeability partly by direct interaction with the inactivation gates. This action results in an increase in the duration of the plateau phase of the action potential and a secondary increase in contractility. Grayanotoxin, a tetracyclic diterpene, enhances resting sodium permeability by the appearance of a secondary, slow, nonactivating sodium current. It has been hypothesized that grayanotoxin acts on a different population of sodium channels not normally involved in the fast sodium current.

In contrast to agents that interact with the sodium channel to increase sodium conductance, antiarrhythmic agents inhibit the sodium current in excitable cells. These agents are thought to interfere with the kinetic transitions of the sodium channel. Although these actions constitute the basis for their therapeutic efficacy, excessive sodium channel blockade results in prominent cardiovascular toxicity. Antiarrhythmic agents bind sodium channels at a faster rate when the channel is in the activated state. Ionic influx inhibitors such as tetrodotoxin, saxitoxin, and chiriquitoxin are present in marine animals including fish, octopus, and frogs. These agents block the sodium channel to prevent the fast sodium current. Interestingly, the potency of these agents as inhibitors of the sodium current is greater in nerves than cardiovascular tissue. Nonspecific effects of lipophilic agents may cause secondary inhibition of sodium conductance by modulation of plasma membrane fluidity.

Other Ions. Although major cardiotoxic ionic disturbances can involve calcium and sodium as described above, other ions occasionally may be involved in cardiotoxicity. Physiologic ions to be considered include potassium, manganese, and iron. Several nonphysiologic ions also are known to be cardiotoxic.

The action of agents on cardiac potassium channels was described in the section on the sarcolemma. Changes in plasma potassium levels have distinct effects on cardiovascular function. Hyperkalemia causes its most dangerous effects upon the heart and can lead to arrhythmias and ventricular asystole at concentrations of 8–9 mEq/l. Hypokalemia leads to acid-base imbalances, and chronic hypokalemia can lead to subendocardial cardiac necrosis. In addition, hypokalemia enhances the binding of the digitalis glycosides to the sodium pump. Potassium-depleting diuretics, therefore, can dangerously enhance the toxic effects of digitalis.

Cardiac toxicity has been implicated in acute iron poisoning, an action that may involve the generation of free radicals. Moreover, as described in the section on anthracyclines, the presence of iron appears to be necessary for the occurrence of anthracycline-mediated oxidative stress. Several ions can interfere with calcium channel function. Those ions that block the slow channels include manganese, nickel, cobalt, and lanthanum. Rather than blocking the channels, barium and strontium enhance positive charge influx through the calcium channels.

26.2.4b Energy Perturbations

Poisons that cause changes in cardiac energy availability act through direct or indirect disturbances of mitochondrial function. Recent data indicate that specific inhibition of glycolysis also may be associated with cardiotoxicity. The heart is particularly susceptible to metabolic perturbations because it has minimal energy storage capacity, and it must produce high-energy phosphates as needed, continually. The effects of toxicants that act upon the mitochondria were reviewed earlier in this chapter. The consequences of mitochondrial damage can include depletion of high-energy phosphates, reduction in intracellular oxidant defenses, and intracellular calcium overload. The anthracycline section of this review describes the potential inter-relationships between these factors. This section will briefly discuss various substances that can perturb myocardial energetics and cause cardiotoxicity.

Excessive cytosolic calcium may lead to mitochondrial accumulation of calcium and uncoupling of oxidative phosphorylation. Several investigators have proposed that the sequence of mitochondrial calcium accumulation and the consequent ATP depletion is the final common pathway for cellular death caused by many different toxicants. Mitochondrial metabolic inhibitors abound. Cyanide and sodium azide both inhibit the cytochrome complex known as cytochrome oxidase. The dangerous rodenticide sodium fluoroacetate (Compound 1080) and its metabolic precursor, fluoroacetamide, inhibit mitochondrial matrix aconitase, thereby, preventing the Krebs cycle conversion of citrate to isocitrate. In humans, fatalities occur with very low doses of these rodenticides, and the usual cause of death is ventricular fibrillation.

Potassium ionophores such as valinomycin and nigericin allow potassium to penetrate the mitochondrial membrane. This entry results in dissipation of the mitochondrial membrane potential and phosphorylation ceases. Classic uncouplers of oxidative phosphorylation such as dinitrophenol allow protons to penetrate back into the mitochondrial matrix. For this reason, dinitrophenol has been called a proton ionophore. According to the chemiosmotic hypothesis, phosphorylation is normally coupled to the controlled return of protons to the mitochondrial matrix. Dinitrophenol, therefore, results in uncoupling of oxidative phosphorylation. Such uncouplers may cause damage to high oxygen demand tissues including the heart and central nervous system. Doxorubicin may impair oxidative phosphorylation by inhibiting ubiquinone-dependent enzymes and by inactivating cytochrome *c* oxidase after binding to cardiolipin.

26.2.4c Membrane Perturbations

From the previous discussions, it is apparent that the toxicity of many cardiovascular toxicants may be directly related to specific membrane actions. Because these membrane actions have been discussed in the sections concerning subcellular organelles and ionic/energy perturbations and have been further illustrated with selected key cardiovascular agents in sections 26.3 and 26.4, the reader is referred to those sections for specific details.

26.2.4d Cellular Defense Perturbations

Because the myocardium and blood vessels are aerobic tissues, energy to support most electrical and mechanical functions is generated primarily through oxidative phosphorylation. Although the oxygen requirement of the myocardium is greater than that of vascular cells, the utilization of molecular oxygen in both tissues results in the formation of highly reactive oxygen-derived free radicals. Several cardiovascular toxins are converted to highly reactive intermediates or cause oxidative stress as a result of enhanced free radical formation. Under these conditions, endogenous mechanisms inactivate free radicals generated during cellular metabolism and protect cells from oxidant injury. The maintenance of cell structure and function in the presence of highly reactive oxidant species depends on the presence of an effective antioxidant defense system. These antioxidant defense mechanisms include glutathione/glutathione peroxidase system, superoxide dismutase, catalase, ascorbic acid, and alpha-tocopherol.

Superoxide anion is formed by the one-electron reduction of molecular oxygen during oxidative phosphorylation and during the enzymatic oxidation of organic substrates by several enzymes including NADPH-cytochrome c reductase and monoamine oxidase. The dismutation of superoxide to form hydrogen peroxide and oxygen is catalyzed by superoxide dismutase. Cytosolic and mitochondrial isozymes of superoxide dismutase have been identified in mammalian tissues. The cytosolic isozyme is a Cu-Zn containing protein, whereas the mitochondrial isozyme contains manganese. The activity of the mitochondrial isozyme is particularly high in organs with high respiration rates such as the myocardium. Known physiologic functions of glutathione include the maintenance of cellular membrane integrity and cytoskeletal organization, involvement in protein and DNA synthesis, and modulation of protein conformation and enzymic activity. In addition, various membrane functions including ion transport and hormonal translation depend upon a suitable thiol balance within cells. Glutathione plays a protective role in the detoxication of electrophilic substances and participates as a co-factor for the glutathione peroxidase by providing the reducing equivalents to metabolize oxygen intermediates and lipid hydroperoxides. The enzymatic detoxication of electrophilic chemicals is mediated by glutathione transferases. These enzymes comprise a family of proteins that promote the conjugation of glutathione with electrophiles and noncovalently bind a large number of compounds for transport across the cell membrane. A strong reducing potential allows glutathione to maintain the reduced state of the cellular sulfhydryl pool and to serve as a nonenzymatic scavenger of free radicals. Conjugation of glutathione with endogenous electrophiles, however, can result in acute depression of glutathione. Other important antioxidant mechanisms include catalase and the complimentary antioxidant activities of tocopherol and ascorbic acid. Catalase reduces hydrogen peroxide to water whereas ascorbate reduces the tocopherol radical formed upon tocopherol reduction of oxidant species.

As mentioned previously, several toxicants directly or indirectly overwhelm the endogenous antioxidant capacity of cardiovascular cells and result in extensive oxidant injury. Within this context, one of the perplexing features of cardiovascular tissue is the low antioxidant activity relative to other aerobic organs. This paradox may be accounted for by differences among various organ systems in the turnover rate of the antioxidant mechanisms described.

26.2.4e Disturbances in Genetic Control Mechanisms

Toxicant-induced disturbances in those genetically defined mechanisms that control metabolism and function of the heart have become of greater interest in recent years. Teratology, for example, is the study of toxicants that can derange the sequence of normal development. Drugs such as anthracyclines that bind to DNA can prevent transcription of needed messengers leading to many different adverse consequences.

Apoptosis. Apoptosis, programmed cell death, is a topic of intense recent interest for all subdivisions of biological sciences. Recently, there is recognition of the role that inappropriate apoptosis can play in toxicology. Inappropriate apoptosis might occur in two opposing scenarios, one in which apoptosis does not occur when it should, and alternatively, one in which it occurs when it should not.

The heart has not evaded interest in toxicological aspects of apoptosis. Facilitation of, or inhibition of, apoptosis is mediated through a very intricate cascade of control mechanisms. These mechanisms are only in the early stages of being defined, but such a complex scheme affords many points of vulnerability to toxic injury.

For example, apoptosis appears to be involved in the adverse remodeling that occurs with pressure overload of the heart, which can lead to congestive heart failure. It appears that the beneficial effects of angiotensin converting enzyme inhibitors may be mediated, at least in part, by decreasing apoptosis, perhaps by reducing the apoptotic effects of nitric oxide. A growing body of literature is exploring the cardiac apoptotic effects of ischemia, norepinephrine, cocaine, oxidized low-density lipoproteins, doxorubicin, and other anthracycline-related compounds. Many more reports undoubtedly will follow.

Stress Proteins. Metallothionein, heat shock protein (HSP), and other stress proteins are elaborated in response to several types of cellular stress. Though many of the specific functions of such stress proteins are not known, the assumption is that for the most part they are beneficial. Prevention of production of stress proteins might be expected to lead to very subtle toxic effects in the heart and other tissues where they occur.

Roles of Knockout Animals and Transgenic Animals. When the specific functions of a gene or protein are not known, one of the powerful investigative techniques of modern molecular biology is to produce animals that are lacking the factor in question (genetic knockout technique). This methodology can be used in cardiovascular toxicology to determine the protective or deleterious effects of specific factors. Remove the factor and see what happens. The creation of transgenic animals, ones containing genes they normally would not have, can also be used to mechanistically investigate the toxic effects of chemicals.

26.3 MYOCARDIUM

26.3.1 Classification of Cardiotoxic Xenobiotics

There is no clear-cut way of classifying the many cardiotoxic agents into simple categories. A brief listing of the drugs and chemicals that have major cardiotoxic effects includes such diverse classes of compounds as the tricyclic antidepressants, antibiotics, antineoplastic agents, alcohols, inhalation anesthetics, heavy metals, and other industrial and environmental chemicals. In addition, there are the many cardiovascular drugs used as therapeutic agents to treat specific cardiovascular problems, which if taken in overdose may adversely affect the heart. Examples of these drugs include digitalis, quinidine, beta-adrenergic-receptor agonists, beta-adrenergic-receptor antagonists, and certain antihypertensive agents. Therefore, rather than attempting to cover every cardiotoxic agent, this section will focus on selected key compounds for purposes of illustration: the catecholamines, alcohols, anthracycline analogs, and other miscellaneous chemicals.

26.3.2 Examples of Cardiotoxic Agents

26.3.2a Catecholamines

The principal catecholamines that have been extensively investigated for their cardiotoxic effects include the endogenous sympathetic amines, epinephrine and norepinephrine, and a synthetic derivative, isoproterenol.

The pharmacological actions of these compounds are mediated by their interactions with specific alpha- and beta-receptors found throughout the body. In general, epinephrine and norepinephrine activate both receptor subtypes, whereas isoproterenol is a relatively selective beta-receptor agonist. The cardiotoxic effects of these compounds represent an exaggeration of their pharmacological actions on beta-receptors found on the heart. Some of the more serious cardiac effects observed with toxic doses of the agents include arrhythmias, tachycardia, possible ventricular fibrillation, increased blood pressure, and myocardial damage and/or necrosis. Of the three compounds, isoproterenol is the only one capable of producing "infarct-like" myocardial necrosis. Because this cardiotoxic lesion produced by isoproterenol resembles human myocardial infarction, it has been investigated extensively and has served as a chemical model for better understanding the pathogenesis of myocardial infarction.

As alluded to in the previous sections, the mechanism of cardiotoxicity of the catecholamines as exemplified by isoproterenol is complex and multifactorial. The principal organelles of the myocardial cell have all been implicated as potential targets for the cardiotoxic effects of isoproterenol. The major membrane system most likely affected first by toxic doses of isoproterenol is the sarcolemma, which is responsible for maintenance of ionic homeostasis and overall cellular integrity. Because studies have shown that toxic levels of catecholamines compromise the permeability of the sarcolemma, the calcium overload theory has received the most support from investigators as playing a causative role in the pathogenesis of catecholamine-induced myocardial necrosis. As early as a few minutes after administration of a toxic dose of isoproterenol to experimental animals, calcium accumulation occurs in the heart. Although other hypotheses have been proposed to explain the cardiotoxicity of catecholamines, the calcium overload theory seems to explain more satisfactorily the pathogenesis of catecholamine-induced myocardial necrosis. From these experimental studies on catecholamines and cardiotoxicity, it is interesting to note that the calcium channel blockers were developed as experimental tools to study calcium fluxes into the heart. These blockers have now been established as valuable therapeutic agents in the treatment of many different types of cardiovascular disorders.

26.3.2b Anthracycline Antibiotics

The anthracycline antibiotics such as doxorubicin and daunorubicin are among the most effective cancer chemotherapeutic agents. The extensive use of the anthracyclines is hindered by their cardiotoxicity, which can be observed upon acute or chronic administration. Because cardiac damage is dose dependent and cumulative, the amount of drug that can be used in therapy is severely restricted. Histopathologically, the damage is characterized by destruction of muscle cells, loss of contractile elements in the remaining cells, and mitochondrial swelling.

There are several major themes proposed as mechanisms for anthracycline cardiotoxicity. These include oxidative stress, disturbances in myocardial energetics, disturbances in intracellular calcium homeostasis, and the release of cardiotoxic humoral mediators.

Oxidative Stress. It was reported in 1977 that the cardiac toxicity of doxorubicin in mice was associated with lipid peroxidation. Since then, many investigators have studied anthracycline-induced oxidative injury. Because the antioxidant defense capacity of the heart is reduced in comparison to other tissues, and because tocopherol and other antioxidants have been useful in protecting against experimental toxicity, the oxidative stress concept of doxorubicin cardiotoxicity was proposed. In this model, doxorubicin within the mitochondria or within the SR undergoes one electron reduction to its semiquinone, followed by redox cycling. Iron may be involved in this process. The redox cycling leads to the production of various reactive oxygen species that cause peroxidative damage to membrane lipids and injury to other cellular components. The heart appears to be particularly susceptible to oxidative damage because of its inherent low levels of antioxidant protection.

Evidence against the oxidative stress hypothesis comes from clinical trials that have been very disappointing using promising agents from animal studies. Agents tested include tocopherol, ubiquinone, and other antioxidants. On the other hand, the iron-chelating compound razoxone (ICRF-187) has been reported to reduce doxorubicin cardiotoxicity in several animal models and in human clinical trials. The efficacy of iron chelation provides evidence in favor of the iron-mediated, oxidative stress hypothesis.

Calcium Changes. Certain investigators have reported that increases in ventricular calcium content might be involved in doxorubicin-induced cardiotoxicity. Early in the toxic process, total cardiac calcium increases and mitochondrial calcium decreases. The action of doxorubicin in causing in vitro inhibition of the calcium pump of rat heart mitochondria is consistent with this finding. In addition, doxorubicin has been found to increase cardiac mitochondrial calcium accumulation in the rabbit, the rat, and the mouse. The reported accumulation and deposition of calcium within mitochondria appear to be a late toxic process.

Other effects on calcium function also have been reported. Small doses of doxorubicin appear to reduce calcium exchangeability in isolated guinea pig atria, while leaving the cardiac calcium content unchanged. Doxorubicin changes the calcium conductance in rat papillary muscle, it enhances slow channel-mediated calcium influx, and it may act to decrease the sarcolemmal Na/Ca exchange of the heart.

Energy Production. There is an extensive literature describing the effects of doxorubicin on myocardial energy production. Decreased mitochondrial respiratory control reflecting a decrease in the maximum oxidative capacity has been reported. Doxorubicin inhibits the succinate dehydrogenase-ubiquinone reductase, and the NADH oxidase-ubiquinone reductase systems of beef heart mitochondria. More recently, it has been shown that doxorubicin inhibits cytochrome *c* oxidase by covalently binding with and removing the mitochondrial membrane cardiolipin necessary for mitochondrial electron transport. By any or all of these actions, doxorubicin could act to decrease the ability of the heart to perform oxidative metabolism, especially under conditions of increased oxygen demand.

The cause and effect relationships between doxorubicin-induced oxidative stress, calcium changes, and decreased energy production have not been defined. It is not clear which of these processes, if any, is the primary or causal biochemical lesion. Theoretically, each has the potential to lead directly or indirectly to each of the other two. Definition of the relationships between these three cardiotoxic mechanisms remains one of the most intriguing challenges in cardiovascular toxicology.

Release of Cardiotoxic Humoral Mediators. The release of cardiotoxic humoral mediators is a recent hypothesis proposed for the cardiotoxic action of the anthracyclines. Although difficult to reconcile with the previously described mechanisms, the hypothesis is interesting and is proposed by established investigators in the field. Doxorubicin is extremely potent in causing release of histamine from mast cells; as potent as the standard mast cell-degranulating agent, 48/80. This histamine-releasing action is probably responsible for the acute cutaneous reactions, and possibly for the acute arrhythmias that sometimes occur following clinical doxorubicin administration. In addition, administration of H_1- and H_2-blockers reduces the toxicity of doxorubicin in several animal models.

Other Considerations. There have been many hundreds of papers dealing with the cardiotoxicity of doxorubicin and other anthracyclines. Because the cardiac effects of these compounds are so many and so varied (you will find them mentioned throughout this chapter), it may not be rational to hope to find simple mechanisms for the cardiotoxicity they cause. It is likely that much of the need to understand the underlying toxic mechanisms of anthracyclines may have been rendered moot by changes in current techniques of drug administration. In the past, intravenous administration of bolus doses of anthracyclines resulted in high initial concentrations reaching the heart. The pharmacokinetic adaptation resulting from this observation is to give the anthracycline by much slower infusion. The overall concentration at the tumor ends up the same after distribution throughout the body, but the heart receives much lower peak concentrations. This change in dosing routine has been very helpful in reducing the incidence of anthracycline clinical cardiotoxicity.

26.3.2c Emetine

The alkaloid emetine is the active ingredient of ipecac syrup. As usually used in the emergency treatment of oral poisoning by ingested substances, emetine does not cause cardiac problems. With chronic administration such as occurs with its use as an amebacide or with abuse by bulimics, emetine-induced damage accumulates as indicated by disturbances in the ECG and histopathological damage. Though the mechanism is not understood, much of the evidence indicates that emetine accumulation causes disturbances of myocardial energy production. In particular, recent data indicate that emetine acts to decrease glycolytic flux. This, in turn, will prevent synthesis of the ATP needed for sarcolemmal cation pump function. Other than the glycolysis studies, there has not been much recent interest in emetine-induced cardiac damage and current literature is scarce. The impetus for study is reduced by the recent trend to use much less ipecac in treating acute toxic ingestions and by the successful repression of bulimia by fluoxetine and the newer, mechanistically similar antidepressants.

26.3.2d Alcohols

The principal effects of alcohols on the heart are reflected by their acute and chronic exposures in animals. Acutely, alcohols depress myocardial contractile force as a function of their chain length. The longer the chain length, the more toxic is the compound. For example, methanol and ethanol produce a negative inotropic effect that is less than butanol and pentanol. Chronic exposure of animals to alcohols, especially ethanol, is thought to produce myocardial alterations that ultimately lead to myocardial failure and cardiomyopathy. However, for a long period of time, the role of alcohol in the pathogenesis of the syndrome was not well understood. This was because heart dysfunction was attributed to thiamine deficiency secondary to alcoholism (beriberi heart disease), and to other types of malnutrition, or to specific additives found in alcoholic beverages such as cobalt (beer drinker's cardiomyopathy). But more recent studies have clearly suggested that cardiac dysfunction in alcoholics can be dissociated from nutritional deficiencies and beriberi heart disease. In addition, the withdrawal of alcohol from patients with myocardial dysfunction may reverse many of the clinical symptoms and, thus, provides more evidence that alcohol is involved in the pathogenesis of the cardiomyopathy.

The classic gross clinical signs of alcoholic cardiomyopathy are hypertrophy, depressed mechanical activity, and abnormal ECGs. At the microscopic level, there is loss and disorganization of the myofibrils and sarcoplasmic reticulum. The number of mitochondria that are swollen with loss of cristae is decreased, whereas the numbers of lysosomes and fat vesicles, and amount of glycogen are increased. In some areas of damage, there are straight, thin, nonbranching tubules and the presence of hyaline, edema, vacuoles, granules, and abnormal nuclei.

Altered mechanical activity is a primary indicator of alcoholic cardiomyopathy. Alcohol, acutely and chronically, depresses mechanical activity in humans, dogs, and rats. Indices of altered mechanical activity include decreases in force-velocity relationship, ejection fraction, heart rate, peak-developed tension, and velocity of shortening; and increases in end-diastolic pressure, rate of left ventricular pressure rise, and pre-ejection period/left ventricular ejection time.

Both acute and chronic ethanol administrations affect the phospholipids and proteins of membranes. Initially, ethanol increases the fluidity of biological membranes, but upon chronic exposure to ethanol the lipid composition of the membranes changes and adapts by becoming more rigid (fluidity decreases), which makes the membranes more resistant to the disordering effects of ethanol. Some of these membrane effects may be reflected by alterations in calcium dynamics. Calcium binding and uptake by myocardial sarcoplasmic reticulum and mitochondria are decreased after exposure to ethanol. It is possible that excitation-contraction coupling and calcium dynamics are disturbed in the myocardium at the level of the sarcolemma, sarcoplasmic reticulum, mitochondria, and regulatory proteins after exposure to ethanol.

Although the myocardium has limited capacity to metabolize ethanol to acetaldehyde by alcohol dehydrogenase, there is evidence that suggests that livers of alcoholics may metabolize ethanol to acetaldehyde in sufficient amounts to produce high blood levels of the metabolite that then can reach the heart. Acetaldehyde has been reported to have a variety of deleterious effects on the myocardium including inhibition of protein synthesis, inhibition of calcium uptake by isolated sarcoplasmic reticulum, impairment of mitochondrial respiratory function, and interference with the association of actin and myosin in vitro. Thus, it is possible that both ethanol and acetaldehyde may be involved in the pathogenesis of alcoholic cardiomyopathy.

Alcohol may undergo a toxic interaction with cocaine, a frequently co-abused substance. One of the mechanisms suggested for this is the formation of a toxic condensation product, cocaethylene. Cocaethylene is very cardiotoxic, but whether it is actually formed in vivo has not been established.

Finally, there is the irony of the therapeutic benefit of alcohol in cardiovascular disease. There is controversy in the literature about whether it is the alcohol or other constituents in red wines that are beneficial. The preponderance of literature indicates that the benefit comes from the alcohol equivalent, regardless of the vehicle. It is very clear that alcohol is a double-edged sword in this regard. Small doses are helpful to the heart and vascular system, but as described, larger doses are very toxic. Because of this, alcohol use is an area where prudent clinicians hesitate to make recommendations.

26.3.2e Other Agents

Many of the heavy metals are toxic to the heart. Lead, for example, can inactivate the sulfhydryl-containing active sites of enzymes and thereby damage metabolically active organs such as the heart. An alternate cardiotoxic mechanism for lead is competition with calcium to perturb calcium cellular homeostasis. Cobalt and lanthanum also interfere with calcium function by blocking of the slow channels. Cadmium is associated with atherosclerosis and hypertension, and it can also be directly toxic to the myocardium.

Another area of active research interest is the field of cardiotoxic peptides. Fractions from many animal toxins have demonstrable cardiotoxic properties that for the most part remain to be defined chemically. Included in this category are stings from may different coelenterates, venoms from certain scorpions, and venoms from snakes such as the pit vipers, cobras, and mambas.

26.3.3 Assessment of Cardiotoxicity

The highest standard for judging the presence of cardiotoxicity is histopathology. Unless cardiac tissue biopsies are performed, however, histopathology is useful only after the fact. The ECG can provide intimate detail about what is going on in cardiac electrical activity and function, but it is technically difficult and may not be all that sensitive. For example, extensive histopathological lesions can be found in rats given emetine chronically but in whom the ECG changes are minimal.

The presence of proteins in the plasma that normally are found as intracellular constituents indicates extensive membrane damage to the cardiac cells. Classically, plasma lactate dehydrogenase (LDH) has been used as such a marker for cellular injury. The primary problems with this enzyme are that the analytical methods are not satisfactorily sensitive and that it is not cardiac specific. Recently, measurement of plasma troponin-T has been found to be very useful. It is very cardiac specific and the methodology is more sensitive than are the analyses for LDH and other plasma markers of cardiac injury that have been used. It has been used to measure cardiotoxicity of emetine and doxorubicin in rats. Cardiac troponin I has the potential to demonstrate cardiotoxicity in chronic in vivo studies as part of the routine hematology testing.

26.4 VASCULATURE

26.4.1 Classification of Vasculotoxic Xenobiotics

The classification of vasculotoxic chemicals would be simplified somewhat if the chemical identity, sources, uses, and/or biochemical actions of all known vascular toxins were available. Instead, the classes established to date have been described in general terms. The classification of compounds based on their primary mechanism of action is advantageous if knowledge of the biochemical basis of toxicity is accessible. Unfortunately, not enough information is yet available to permit accurate application of this approach. As in the previous sections, the chemicals to be discussed in this section have been selected to highlight compounds for which experimental and/or epidemiological data suggest prominent vascular toxicity.

26.4.2 Examples of Vasculotoxic Agents

26.4.2a Carbon Monoxide

Vascular injury has been observed upon exposure of several animal species to carbon monoxide. Short-term exposure is associated with damage to the intimal layer of blood vessels, whereas chronic exposure regimens result in the formation of atherosclerotic lesions. The incidence and severity of the lesions is enhanced if carbon monoxide is administered to animals consuming a lipid-enriched diet. Carbon monoxide-induced vascular injury is observed at concentrations of carbon monoxide to which humans may be exposed from environmental sources, The toxic effects of carbon monoxide have been attributed to severe tissue hypoxia. Carbon monoxide interacts reversibly with hemoglobin to form carboxyhemoglobin, which decreases the oxygen-carrying capacity of the blood and shifts the oxyhemoglobin saturation curve to the left. These actions cause functional anemia as a result of reduced oxygen delivery to tissues throughout the body.

Evidence is now available that suggests that carbon monoxide exerts toxic effects independent of those associated with carboxyhemoglobin formation. Elevated partial pressures of carbon monoxide are found in several tissues upon carbon monoxide intoxication. Thus, it has been proposed that the interaction of carbon monoxide with intracellular constituents may account for the occurrence of direct toxic effects. Carbon monoxide reacts in vivo and in vitro with a variety of metal-containing proteins including hemoglobin, myoglobin, and cytochrome *c* oxidase. These metalloproteins contain iron and/or copper centers that form metal-ligand complexes with carbon monoxide in competition with molecular oxygen.

Major environmental sources of carbon monoxide include automobile exhaust, tobacco smoke, and fossil fuels. Because these sources involve multiple exposures to a complex mixture of chemicals, extreme difficulty has been encountered in attempting to delineate the actions of carbon monoxide relative to the effects of other toxic chemicals. One must consider that sulfur oxides, nitrogen oxides, aldehydes, and hydrocarbons alone or in combination with carbon monoxide could be responsible for the angiotoxic effects of environmental pollution.

26.4.2b Heavy Metals

The contribution of heavy metals to the cardiovascular morbidity and mortality of industrialized nations has been one of the areas in cardiovascular toxicology that has received the most attention. Hard water elements (calcium and magnesium), food- and water-borne elements (selenium, chromium, copper, zinc, cadmium, lead, and mercury) and air-borne elements (vanadium and lead) are all potentially toxic to the cardiovascular system, as well as to other systems. The role of these metals in the development of cardiovascular diseases remains uncertain.

The toxic effects of metals have been attributed to nonspecific reactions with sulfhydryl, carboxyl, or phosphate groups or to the intercalation of the metal with cellular macromolecules. Some metals including cobalt, magnesium, manganese, nickel, cadmium, and lead act as physiological calcium antagonists. Recent evidence suggests that intracellular calcium-binding proteins are the targets of heavy metal toxicity. Thus, the toxic effects of heavy metals in the cardiovascular system may be due, at least in part, to inhibition of calcium-mediated cellular events.

Most of the experimental and epidemiological studies conducted to date have focused on cadmium and lead. Long-term exposure of laboratory animals to low levels of cadmium causes aortic atherosclerosis and/or hypertension in the absence of other toxic effects. Cadmium causes a variety of changes in blood vessel endothelium including vesiculation, vacuolization, widened intercellular junctions, and fragmentation. Cadmium also induces the stress protein metallothionein. Selenium, copper, and zinc inhibit where lead potentiates the hypertensive effects of cadmium. Pressor doses of cadmium fail to raise blood pressure if dissolved in the presence of calcium and magnesium. These observations have lead investigators to propose that calcium and magnesium interfere with the expression of cadmium toxicity. Subsequent studies have suggested that calcium, but not magnesium, prevents the toxic effects of cadmium. However, the role of calcium as a protective element is not yet clear. The calcium-antagonistic properties of magnesium may account for its ability to potentiate the vasculotoxic effects of lead and cadmium. Although evidence suggesting that cadmium and lead contribute to the development of several cardiovascular disorders continues to accumulate, numerous other studies have reported conflicting results. Such variability may be due to regional differences in water hardness that facilitate toxicologic interactions similar to those described previously.

26.4.2c Amines

Sympathomimetic amines exert prominent pharmacological effects mediated by alpha- and/or beta-receptors in vascular cells. Toxic concentrations of alpha-receptor agonists such as norepinephrine cause excessive vasoconstriction that results in prominent hemodynamic changes. In addition, administration of large doses of catecholamines causes atherosclerosis in several animal species. The development of atherosclerotic lesions is associated with extensive endothelial injury. These observations are consistent with the ability of catecholamines to stimulate the growth of vascular smooth muscle cells. The mitogenic effects of catecholamines seem to be mediated by alpha-receptors, as they are antagonized by phentolamine, an alpha-receptor antagonist. This mitogenic response may play an important role in the development of atherosclerosis associated with conditions in which elevated levels of catecholamines have been documented. It is of interest that isoproterenol, a synthetic catecholamine that preferentially stimulates beta-receptors, alters the morphologic expression of cultured rat aortic cells but does not promote cellular growth. The significance of these observations has not yet been clarified.

Unsaturated aliphatic amines are potent vascular toxins. Monoallylamine is more toxic than other unsaturated primary amines of higher molecular weight. Saturation of the double bond decreases vascular toxicity. Although the systemic toxicity of these compounds has yet to be fully elucidated, the outstanding pathologic feature observed upon chronic administration of allylamine is the development of atherosclerotic lesions. Administration of allylamine by a variety of routes in several animal species is associated with the development of vascular and myocardial lesions that closely resemble those found in atherosclerotic vessels and ischemic myocardium, respectively. Bioactivation of allylamine by an amine oxidase present in vascular tissue to form acrolein and hydrogen peroxide appears to be responsible for the toxicity of this compound. Acrolein is an extremely reactive aldehyde that disrupts the thiol balance of vascular cells. A large number of compounds react with acrolein under physiological conditions. The main reaction products result from addition at the terminal ethylenic carbon atom. It has been proposed that the biochemical and the toxic effects of acrolein are most likely caused by reaction with critical sulfhydryl groups. The adducts of acrolein with amino groups are considerably less stable and are formed more slowly than those obtained from sulfhydryl groups. Nevertheless, reactions of acrolein with amino groups are thought to be responsible, at least in part, for the formation of cross-links with proteins of the vascular wall. Such cross-links have been shown to occur with collagen, elastin, and connective tissue via the amino group of lysine and may result in the formation of atherosclerotic-like lesions. In addition, acrolein can be oxidized enzymatically to glycidaldehyde, a potent carcinogen. The results of in vitro metabolism studies of acrolein demonstrate that exposures to acrolein may entail simultaneous exposures to glycidaldehyde and acrylic acid, two toxic metabolites. Although acrolein lacks significant transforming potential, it is cytotoxic and a weak initiator following chemical promotion. Hydrogen peroxide generated during the deamination process may further disrupt cellular homeostasis and contribute to the overall toxic response. The effects of allylamine on the thiol balance of vascular cells disrupt the regulation of smooth muscle cell growth.

26.4.2d Polycyclic Aromatic Hydrocarbons

Epidemiologic and experimental evidence suggests that carcinogens may contribute to the development of atherosclerosis. Exposure of several rodent and avian species to benzo(a)pyrene or 7,12-dimethylbenz[a,h]anthracene causes atherosclerosis without altering serum cholesterol levels. The ability of these carcinogens to exert their vasculotoxic effects appears to depend upon their conversion to reactive metabolic intermediates. In this regard, the presence of aryl hydrocarbon hydroxylase has been correlated with the degree of susceptibility to atherosclerosis in avian species. The ability of chemical carcinogens to cause atherosclerosis is consistent with the proposal of Benditt and colleagues that the development of atherosclerotic lesions is the result of a mutagenic process initiated by viral or chemical insult. Lesions may be derived from the proliferation of a single cell and as such be considered a benign smooth muscle cell (SMC) tumor. This concept evolved from the observation that the SMC population in fibrous plaques from humans is mostly monotypic. Normal arterial SMC exhibit two isozymes of glucose 6-phosphate dehydrogenase. In contrast, only one of two isozymes is expressed in SMC of atherosclerotic tissue. The focal nature of early atherosclerotic lesions is compatible with the monoclonal theory of atherosclerosis.

26.4.2e Other Chemicals

Carbon Disulfide. Exposure of factory workers in rayon plants to carbon disulfide has been associated with a significant increase in mortality from atherosclerotic disease. Although the mechanism of toxicity is not known, effects on glucose and lipid metabolism or blood coagulation have been invoked. However, the possibility that direct toxic effects account for the atherogenic effect of carbon disulfide has not yet been ruled out.

Dinitrotoluene. Dinitrotoluene has been identified as a potent rodent carcinogen. Cancers of the liver, gall bladder, and kidney, as well as benign tumors of the connective tissues, have been documented. Retrospective human mortality studies in workers exposed daily to relatively low concentrations of dinitrotoluene have suggested that this compound causes atherosclerotic heart disease. Laboratory studies are needed to elucidate the biochemical basis of this response.

26.4.3 The Endothelium as a Target for Cardiotoxicity

The tonic control of the endothelium over its adjacent vascular smooth muscle has been known for a long time. A revolution in vascular biology occurred with the discovery that the endothelium derived relaxing factor is nitric oxide. Certain classic old observations became explained, for example, when it was demonstrated that the vasodilating, atropine-blockable properties of intravenous parasympathomimetics were mediated by the release or block of release of nitric oxide from the endothelium. Both the release of nitric oxide and the prevention of its release are amenable to pharmacologic and toxicologic manipulations. As described above, nitric oxide production can be associated with inappropriate apoptosis. Other observations will follow.

SUGGESTED READING

Artman, M., Olson, R.D., Boucek, R.J., Jr. and Boerth, R.C. Depression of contractility in isolated rabbit myocardium following exposure to iron: role of free radicals. *Toxicol. Appl. Pharmacol.* 72 (1984), 324.

Balazs, T., Hanig, J.P. and Herman, E.H. Toxic responses of the cardiovascular system. In: Klaassen, C.D., Amdur, M.O. and Doull, J. (Eds), *Casarett and Doull's Toxicology*, 3rd Ed., Macmillan Publishing Company, New York, 1986, pp 387–411.

Langer, G.A., Frank, J.S. and Philipson, K.D. Ultrastructure and calcium exchange of the sarcolemma, sarcoplasmic reticulum and mitochondria of the myocardium. *Pharmacol. Ther.* 16 (1982), 331.

Merin, R.G. Myocardial metabolism for the toxicologist. *Environ. Health Perspect.* 26 (1978), 169.

Rona, G. Catecholamine, cardiotoxicity. *J. Mol. Cell. Cardiol.* 17 (1965), 291.

INDEX